Calculus

Multivariable

Brian E. Blank
Steven G. Krantz
Washington University in St. Louis

Debut Edition

Brian E. Blank and Steven G. Krantz
Department of Mathematics
Washington University in St. Louis
St. Louis, MO 63130

Key College Publishing was founded in 1999 as a division of Key Curriculum Press® in cooperation with Springer-Verlag New York, Inc. We publish innovative texts and courseware for the undergraduate curriculum in mathematics and statistics as well as mathematics and statistics education. For more information, visit us at www.[illegible].

Published by Key College Publishing, an imprint of Key Curriculum Press.

Development Editor: Allyndreth Cassidy
Production Director: McKinley Williams
Production Coordinator: Ken Wischmeyer
Editorial Production Project Managers: Beth Masse and Laura Ryan
Project Manager: Eric Houts
Copyeditor: Tara Joffe
Proofreader: Andrea Fox
Indexer: Victoria Baker
Text Designer: Suzanne Montazer
Composition, Illustration: Interactive Composition Corporation
Cover Designer: Jensen Barnes
Cover Photo Credit: St. Louis Arch, Missouri, USA: GettyImages/Charles Thatcher
Printer: RR Donnelley

Editorial Director: Richard J. Bonacci
General Manager: Mike Simpson
Publisher: Steven Rasmussen

Library of Congress Cataloging-in-Publication Data

Blank, Brian E., 1953-
Calculus, multivariable / Brian E. Blank, Steven G. Krantz.
p. cm.
Includes index.
ISBN 1-931914-60-5 (pbk.)
1. Calculus. 2. Variables (Mathematics) I. Krantz, Steven G. (Steven George), 1951-II. Title.

QA303.2.B467 2005
515—dc22

2004050702

Printed in China
10 9 8 7 6 5 4 3 2 1   09 08 07 06 05

A pebble for Louis.
BEB

This book is for Hypatia, the sweetheart of my life.
SGK

Contents

Preface

Calculus is one of the milestones of human thought. Every well-educated person should be acquainted with the basic ideas of the subject. In today's technological world, in which more and more ideas are being quantified, knowledge of calculus has become essential to a broader cross section of the population.

Starting in the late 1980s, a vigorous discussion began about the approaches to and the methods of teaching calculus. This debut edition of *Calculus,* published in two volumes, *Single Variable* and *Multivariable,* offers the best in current calculus teaching. We have worked hard to properly assess, with realism and purpose, the calculus market as it actually exists and to address the needs of today's students, bringing together time-tested, as well as innovative, pedagogy and exposition.

The Changing Face of the Student

More than ever, today's students compose a highly heterogeneous group. Calculus students come from a wide variety of disciplines—some study the subject because it is required, others because it will widen their career options. Mathematics majors often go into law, medicine, genome research, the technology sector, and many other professions. As the teaching and learning of calculus is rethought, instructors must keep their students' backgrounds and futures in mind.

Instructors must also remember that an increasingly larger number of college and university students have already seen quite a bit of calculus in high school. Instructors must build on what students already know, refining their mathematical skills and expanding their conceptual horizons.

The goal is to empower students, enhance their critical thinking skills, and give them the intellectual equipment to proceed successfully in whatever major or discipline they ultimately choose to study. This text is intended to be a cornerstone of that process.

The Changing Role of the Textbook

Many resources are available to instructors and students today, from Web sites to interactive tutorials. The calculus textbook must be a tool that instructors can use to augment and bolster their lectures, classroom activities, and resources. It must enhance the classroom experience and speak compellingly to the students who are actively engaged in the class.

A calculus book must tell the truth. It must be carefully written in the accepted language of mathematics, but it also must be credible—for students and instructors alike—and it must be readable. The textbook should include useful and fascinating applications. It should acquaint students with the history of the subject and with a sense of what mathematics is all about. While teaching technique, it should also teach ideas. It should help students master basic methods and teach them how to discover and build their own concepts in a scientific subject. In today's world, it is particularly important that a calculus book illustrate ideas using modeling and numerical calculation. We have made every effort to ensure that this text is such a calculus book.

Calculus is designed to increase the student's role as an independent thinker, whether as a potential mathematician, scientist, or practitioner in another analytical field. The intent of this book is to make it natural for students to succeed in their calculus course as well as in future courses. We believe that a good calculus book is a crucial stepping-stone in the foundational education of a student in the twenty-first century.

Brian E. Blank
Steven G. Krantz

Features

Although *Calculus: Single Variable* and *Calculus: Multivariable* have many features that appeal to those teaching mathematics and science majors, it was written to offer a large segment of the calculus market a textbook that provides the necessary tools for students to succeed in their study and to appreciate the subject. Instructors teaching calculus to students with various backgrounds and educational goals will be able to shape their course to fit particular needs.

Students must understand concepts as well as calculations—they must be able to reason through word problems as well as complete drill problems. To strike this balance, the following features are included in the text.

Pedagogy

- The writing style is clear and readable. Students should not have to struggle with the exposition as they learn calculus.
- Motivation for important topics is crisp and clean, enabling students to get to key examples quickly and efficiently.
- All essential ideas are showcased with examples, offering a seamless link between concepts and applications.
- Concepts are reinforced by graphical interpretations. The large number of figures helps students visualize the concepts. The text also presents numerical examples when they will aid student's understanding.
- Material that is not required for subsequent sections is denoted by an asterisk (∗). Instructors can choose whether to include this material in their courses.
- At the end of each section, before the exercises, students can immediately reinforce the concepts learned by answering the Quick Quiz questions.

Exercises, Examples, and Applications

- Examples are carefully tied to the Problems for Practice exercises, allowing students to immediately practice and master the needed skills.
- Each example is presented as a problem with a clearly stated task. The authors have taken great care to explain the steps of the solutions to these problems. For example, when an equation is obtained by using a formula labeled by a number, that number is placed over the equals sign.
- Applications, both large and small, abound. Instructors may pick and choose those that best suit the course being taught. Applications to chemistry, biology, medicine, public policy, finance, economics, and other social sciences augment classic applications in physics and engineering.
- Each end-of-section exercise set comprises three types of exercises: Problems for Practice, Further Theory and Practice, and Calculator/Computer Exercises.
- The Problems for Practice develop essential computational skills. They are organized by type and include more exercises than will typically be assigned. Students can use the unassigned exercises for extra practice, as needed.
- The Further Theory and Practice exercises are mixed in nature. In general, these exercises cannot be done by following a worked example. Some exercises in this section extend the theory discussed in the text. Some Further Theory and Practice exercises fill in details of proofs. There are far more Further Theory and Practice exercises than will be assigned in a typical course. Instructors may choose the ones suitable for their courses.

Content of Multivariable

- Vectors and vector algebra, both in two and three dimensions, begin our study of multivariable calculus. Geometric operations, such as the dot and cross product, are both defined and illustrated. Chapter 11 provides the geometric background for the mathematical analysis that is to follow.
- The concept of distance is treated in detail. It is related to vector analysis in a natural way. The triangle inequality is analyzed and illustrated. The role of unit vectors is developed and conceptualized. The special unit vectors **i**, **j**, **k** are given detailed attention.
- Chapter 12 begins with a detailed study of the geometry and analysis of vector-valued functions. The difference between scalar-valued functions and vector-valued functions is precisely delineated. The role of derivatives is given special emphasis, and these are illustrated with the concepts of velocity and acceleration. Tangent lines play a special role in this discussion. An unusual illustration of the physics of baseball follows, with special emphasis on the sinking fastball.

- Arc length and reparametrization, with special attention to which integrals can actually be calculated, are discussed in detail. Curvature and osculating circles are given an especially careful treatment.
- A special section on the physics of motion, including central force fields and centripetal force, is capped off with a derivation of Kepler's Laws of planetary motion.
- In Chapter 13, functions of several variables are carefully defined and delineated from functions of one variable, and from vector-valued functions. Concepts are illustrated with mathematical and physical examples. Special attention is given to the derivative and the chain rule. Level sets are presented as a special tool for graphing. Focus is then turned to limits, continuity, and the derivative.
- The gradient is illustrated with (among other ideas) the concept of tangent plane. This thorough treatment is then followed by that of Taylor's theorem and numerical approximation. Applied maximum/minimum problems and Lagrange multipliers round out the discussion and illustrate the key concepts.
- Multiple integrals are presented with a geometric treatment in Chapter 14. Special emphasis is placed on the nature of the domain of integration, and the method of calculating the integral: what is the relationship between a double (or triple) integral and an integrated integral? The order of integration is seen to be a critical and valuable tool. The calculation of volume illustrates the ideas.
- Polar coordinates, cylindrical coordinates, and spherical coordinates are offered as important analytical tools. Integration in three or more variables delineate some of the scope of integration theory. There is an entire separate section on physical applications, including mass and moments.
- Green's theorem in two dimensions, Stokes's theorem on surfaces, and Gauss's theorem in three variables are described in detail and proved (in special cases) in Chapter 15. Most important, the physics of these theorems is emphasized. These theorems unify and illustrate all the geometry that has gone before. The book emphasizes this feature, and provides copious examples. There are many physical applications. It is stressed that these three key theorems are higher-dimensional versions of the fundamental theorem of calculus, and that insight is amply illustrated.

Content of Single Variable

- Transcendental functions (trigonometric, logarithmic, and exponential) receive a thorough, but accessible, treatment. Chapter 1 reviews trigonometric functions. Chapter 2 introduces exponential functions. The natural logarithm is introduced in Chapter 3. The order of these chapters is suitable for an "early transcendentals" course.
- Differential equations are treated throughout the text. Chapter 6 uses first order differential equations to provide a second look at the exponential and

logarithmic functions. Differential equations that involve the other transcendental functions are also discussed.

- Sequences, which are introduced in Chapter 1, are treated throughout the text. Chapter 2 defines and studies limits of sequences. This treatment takes advantage of students' prior familiarity with the topic. Although that knowledge, gained through the infinite decimal expansions that students learned before calculus, is informal, it makes for a good intuitive base. Sequences therefore help make the general concept of limit more concrete. The early and recurring discussion of sequences results in a more comfortable, expeditious treatment of infinite series. Mastering the notion of infinite sequence in advance better prepares students to understand the sophisticated idea on which the infinite series rests.

- When introduced, integration techniques (substitution, integration by parts, partial fractions, and the use of trigonometric identities) are discussed in considerable detail. These topics provide students with useful practice in the important mathematical techniques of substitution and algebraic manipulation. The development of partial fractions has been spread over two sections, allowing instructors greater flexibility in what they choose to cover.

- The chapter on applications of integration introduces and develops concepts from probability theory.

- The development of power series has been given its own chapter. Taylor polynomials are introduced before Taylor series.

Technology

- Computer modeling and numerical calculations are included both in the text and in the exercises. Screen shots from the popular computer algebra system (CAS), Maple™, are shown throughout so that students learn to recognize CAS interfaces.

- The use of a CAS or graphing calculator provides students with another avenue for exploring calculus.

- Each end-of-section exercise set concludes with several problems that are intended to be solved with CAS or a graphing calculator. We use Maple in our calculus courses, but the exercises are written so that any CAS or graphing calculator can be used.

- Technology complements and enhances traditional mathematical skills; it does not eliminate mathematical techniques.

- Throughout the text, the mathematical notation used is compatible with modern calculators and computer algebra systems. Rather than a bewildering array of brackets for algebraic groupings, parentheses are employed. The argument of every function appears within parentheses. For example, we write $\sin(x)$ and not $\sin x$.

Chapter Structure and Elements

- Each chapter starts with a preview of the topics that will be covered. This short initial discussion gives an overview and provides motivation for the chapter. Each chapter contains a section that summarizes the chapter's important formulas, theorems, definitions, and general concepts.

- Occasionally within the text, A Look Back/A Look Forward boxes remind students of concepts that were learned earlier in the text or offer previews of still-to-come material.

- Boxed features called Insights highlight further information about concepts. They occur both in margins and in the text; arrows indicate where they flow within the text. The Insights are remarks directed to the student. Sometimes an Insight clears up a point, sometimes it answers a question that arises frequently in lectures. The term *Insight* does not reflect any deep understanding of calculus that the authors claim to have; the remarks are simply the result of day-to-day teaching experience. Naturally, at the chalkboard, instructors will offer their own insights.

- Prior to exercise sets, students test their understanding with the Quick Quiz.

- Each chapter ends with a novel feature, *Genesis & Development*. These sections give history and perspective on key topics in the evolution of the subject. There are occasional references to these sections in the text, but by and large they are intended to be supplementary. Instructors can assign these sections as reading if they wish, but there is no core material that requires any of the *Genesis & Developments* to be covered.

Supplements

For the Instructor

Instructor Resources
Single Variable (1-931914-34-6) and *Multivariable* (1-931914-70-2)
Each volume of the *Instructor Resources* is structured similarly and is designed to provide you with a tool that guides your teaching of *Calculus*. Each volume is devoted to teaching suggestions, such as sample syllabi, lecture outlines, and course pacing. Sample midterms and finals are also included.

Instructor Solutions Manual
Single Variable (1-597570-21-4) and *Multivariable* (1-597570-22-2)
These volumes contain full solutions to all text exercises.

Test Generator (1-931914-73-7)
One comprehensive test generator CD-ROM is available for all *Calculus* chapters. You may build your own tests by choosing from a wide variety of both static and algorithmically generated problems. The problems are written in a style to match the prose of the text. Question types include multiple choice, true/false, and short answer. Tests can be administered with paper and pencil, or they can be delivered online through your school's local network.

You can also keep track of your courses with the software's built-in management. Tests administered over a network can be automatically graded and entered into a grade book. Each of your sections will be monitored separately.

Web site: www.keycollege.com/online
A portion of this course-specific Web site is devoted to instructors using *Calculus: Single Variable* and *Calculus: Multivariable*. If your school uses WebCT, you will find content that is portable into that system. Other resources include PowerPoint® presentations, extra worked examples, and modifiable sample syllabi. Each problem from the exercise sets has been keyed so that homework can be assigned and completed online. You will also be able to access all portions of the student Web site.

For the Student

Student Study and Solutions Companion
Single Variable (1-931914-71-0) and *Multivariable* (1-931914-72-9)
The *Student Study and Solutions Companions* are published in two volumes to correlate with each text volume. Students will find the same valuable resources in each *Companion*. Study hints and additional worked examples are provided for each chapter. Fully worked solutions are given for each text problem that has its answer at the end of the textbook. The *Single Variable* volume also includes algebra review. Both volumes have a section with common formulas and integral tables.

Web site: www.keycollege.com/online
To enter this site, students will use their unique access code, which is packaged with every new text. Once registered, students will be able to complete homework assignments online. They will also be able to access Maple keystrokes that correlate to the book's Computer/Calculator Exercises. Links are also provided to online calculus resources.

Access Code Package (1-931914-74-5)
Students who purchase used books will also need to purchase the Access Code Package to get their unique access code to register for www.keycollege.com/online. If the package is not available through the school's bookstore, students may call toll free 888-877-7240 to order the Access Code Package.

Acknowledgments

Over the years of its development, *Calculus: Single Variable* and *Calculus: Multivariable* have profited from the comments of our colleagues around the country. We are thankful for the reviews at all stages of development. We would particularly like to acknowledge the following people:

David Calvis, Baldwin-Wallace College
Gunnar Carlsson, Stanford University
Chi Keung Cheung, Boston College
Dennis DeTurck, University of Pennsylvania
Bruce Edwards, University of Florida
Saber Elaydi, Trinity University
David Ellis, San Francisco State University
Salvatrice Keating, Eastern Connecticut State University
Jerrold Marsden, California Institute of Technology
Jack Mealy, Austin College
Harold Parks, Oregon State University
Ronald Taylor, Berry College

From Brian E. Blank

I would like to thank William Moser, Kohur Gowrisankaran, Jeremy Hayhurst, and William Hoffman for their important contributions to the early conceptual periods of this project and Mike Simpson for his strategic insights at the critical marketing stage of the published text.

From Steven G. Krantz

In my career, I have benefited from my association with many editors and publishing professionals. Rich Jones was the first to encourage me to write a calculus book. He provided both direction and guidance and taught me how to write. Barbara Holland worked closely with me to develop the text and taught me to listen to and learn from reviewers. Jim Harrison provided me with encouragement, guidance, and my first computer. Seth Howell and Greg Knese carefully checked all of the mathematics for correctness. Richard Bonacci and Allyndreth Cassidy at Key College Publishing made it all come together and turned this nascent project into a finished book. Eric Houts, Beth Masse, and Laura Ryan, master production editors, lovingly shepherded the project through every stage of copy editing and composition. No project of this magnitude can be created without the combined efforts of many people. I am grateful to them all—both named and unnamed.

About the Authors

Brian E. Blank and Steven G. Krantz have a combined experience of more than 50 years teaching calculus. They are both award-winning teachers and highly respected writers. Their extensive experience in consulting for a variety of professions enables them to bring to this project diverse and motivational applications as well as realistic and practical uses of the computer.

Brian E. Blank received his B.Sc. degree from McGill University in 1975 and Ph.D. from Cornell University in 1980. He has taught calculus at The University of Texas, the University of Maryland, and Washington University in St. Louis.

Steven G. Krantz received his B.A. degree from the University of California at Santa Cruz in 1971 and his Ph.D. from Princeton University in 1974. He has been on the faculties of the University of California at Los Angeles, Pennsylvania State University, and Princeton University. He is former chair at Washington University in St. Louis. Krantz is holder of the Chauvenet Prize, the Beckenbach Book Award, and the Kemper Prize. He has written more than 120 research papers and more than 45 books, including *How to Teach Mathematics* and *Mathematical Apocrypha.*

11

Vectors

PREVIEW

In this chapter, we lay the geometric foundation for all of our coming work with functions of two or more variables. Those studies will require us to have a mathematical model of three-dimensional space. Just as we use a real number as the mathematical model of a point on a line and an ordered pair of real numbers as the mathematical model of a point in the plane, so we use an ordered triple (x, y, z) as the mathematical model of a point in three-dimensional space. Equations among these space variables describe curves and surfaces in space. In this chapter, we study the simplest such objects—lines and planes—in detail. As an aid to our investigations, we introduce a new concept, the *vector,* which will be a vital tool in every chapter to follow.

A vector may be thought of as an arrow that has a length and a direction but whose initial point is of no importance. If we reposition the arrow while preserving its direction, then it still represents the same vector. We learn how to add vectors and how to perform a number of other algebraic operations with them. The interplay between the geometry and algebra of vectors is the perfect device for understanding the relationship between the geometry of lines and planes and their Cartesian equations.

The vector construction is ideal for capturing many concepts in both algebra and geometry. But the use of vectors extends to other sciences as well. Many fundamental quantities in physics, such as force, velocity, and acceleration, are understood by means of their direction and magnitude. By using the length of a vector to represent magnitude, we can model these physical quantities by means of vectors.

11.1 Vectors in the Plane

For many purposes in calculus and physics, we need a concept that simultaneously contains the notions of direction and magnitude. For instance, force, velocity, and acceleration are all quantities that have both direction and magnitude. In this section, we develop a mathematical tool, the *vector,* for handling these concepts.

Definition A line segment $\overline{AB}$ between two points A and B is said to be a *directed line segment* if one endpoint is considered to be the initial point of the line segment and the other endpoint the terminal point. We denote the directed line segment with initial point A and terminal point B by $\overrightarrow{AB}$. The same pair of points determines a second directed line segment, $\overrightarrow{BA}$, which is said to be *opposite* in direction to $\overrightarrow{AB}$. See Figure 1.

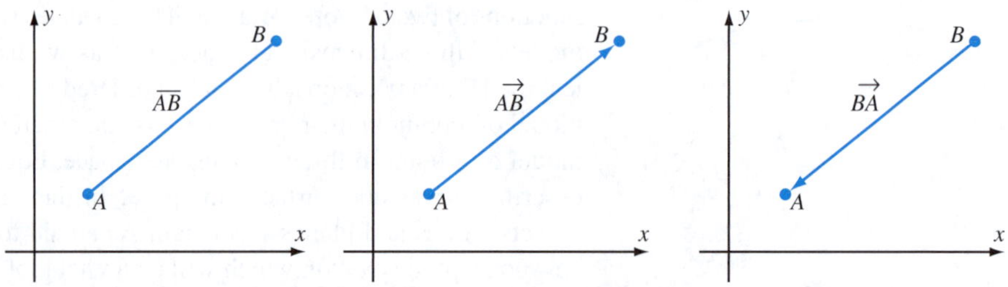

Figure 1
Line segment $\overline{AB}$, directed line segment $\overrightarrow{AB}$, and directed line segment $\overrightarrow{BA}$

The directed line segment $\overrightarrow{AB}$ may be thought of as a straight path from A to B. Its direction is indicated by an arrow, as in Figure 1. Algebraically, it is often convenient to represent a directed line segment by means of a parameterization, as the next example demonstrates.

Example 1 Suppose that $A = (5, 6)$ and $B = (9, 14)$. Find a parameterization $x = f(t)$, $y = g(t)$, $0 \le t \le 1$, of the directed line segment $\overrightarrow{AB}$ with $t = 0$ corresponding to the initial point and $t = 1$ the terminal point.

Solution The concept of parameterization was discussed in Section 1.5. We are to find functions f and g so that the point $(f(t), g(t))$ traverses the line segment from A to B as t increases from 0 to 1. From the initial point A to the terminal point B, the x-displacement is $9 - 5 = 4$ and the y-displacement is $14 - 6 = 8$. If we set $x = f(t) = 5 + 4t$ and $y = g(t) = 6 + 8t$, then $(f(0), g(0)) = (5, 6) = A$ and $(f(1), g(1)) = (5 + 4, 6 + 8) = B$. Since $t = (x - 5)/4$, we see that $y = 6 + 8(x - 5)/4$, or $y = 2x - 4$, which is the equation of a line. Thus, the equations $x = 5 + 4t$, $y = 6 + 8t$, $0 \le t \le 1$, parameterize the directed line segment $\overrightarrow{AB}$. ■

inSIGHT

The calculation of Example 1 may be carried out with any two points $A = (x_0, y_0)$ and $B = (x_1, y_1)$. The result is that the equations

$$x = x_0 + t(x_1 - x_0),\ \ y = y_0 + t(y_1 - y_0) \quad (0 \le t \le 1) \qquad \textbf{(11.1)}$$

parameterize the directed line segment $\overrightarrow{AB}$.

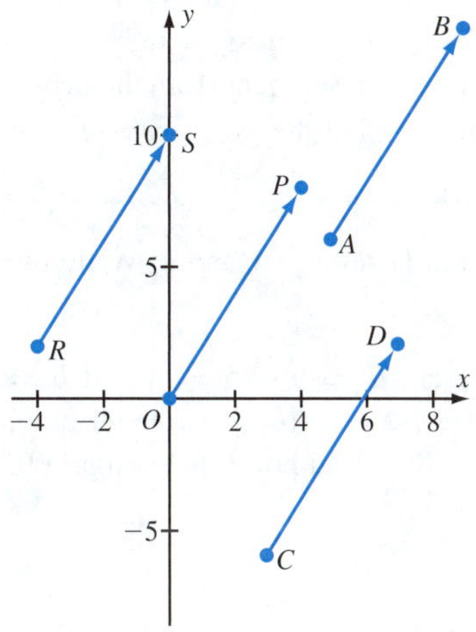

Figure 2
Directed line segments with the same length and direction as $\overrightarrow{AB}$

A directed line segment is determined by its initial point, direction, and length. Given a directed line segment $\overrightarrow{AB}$, it is important to be able to construct directed line segments with the same length and direction as $\overrightarrow{AB}$ but with other initial points. The next example shows how this is done.

Example 2 Let $A = (5, 6)$ and $B = (9, 14)$, as in Example 1. Let O denote the origin $(0, 0)$. If $C = (3, -6)$ and $S = (0, 10)$, then find points D, P, and R so that the directed line segments $\overrightarrow{OP}$, $\overrightarrow{CD}$, and $\overrightarrow{RS}$ have the same length and direction as $\overrightarrow{AB}$.

Solution The length and direction of $\overrightarrow{AB}$ are both determined by its x-displacement $9 - 5 = 4$ and y-displacement $14 - 6 = 8$. If we add these displacements to the coordinates of the initial points O and C, then we obtain the terminal points $P = (0 + 4, 0 + 8) = (4, 8)$ and $D = (3 + 4, -6 + 8) = (7, 2)$. If we subtract these displacements from the coordinates of the terminal point S, then we obtain the initial point $R = (0 - 4, 10 - 8) = (-4, 2)$. In this way, $\overrightarrow{OP}$, $\overrightarrow{CD}$, and $\overrightarrow{RS}$ have the same x- and y-increments as $\overrightarrow{AB}$ and therefore the same directions and length. See Figure 2. ■

Vectors

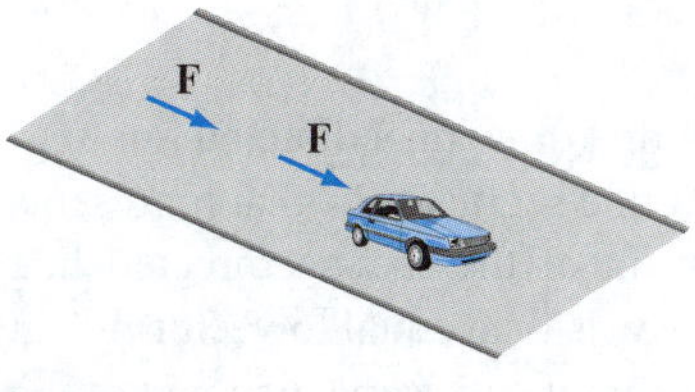

Figure 3

Imagine that you are exerting a force of constant magnitude to push a stalled automobile (Figure 3). If the street is straight, then the direction of the force **F** that you exert is also constant. An arrow may be used to represent the *direction* of the force **F**, as in Figure 3. We can use the length of the arrow to represent the *magnitude* of **F** and the initial point of the arrow to represent the point of application of **F**. As the car moves, the *position* of the arrow changes but the direction and length do not. We can therefore regard the force **F** as a directed line segment that (i) has a fixed direction, (ii) has a fixed length, and (iii) can be applied at any point. These considerations suggest creating a mathematical object that captures the direction and length of a directed line segment *but that disregards its initial point*. Such an object is called a *vector*.

Definition Let A and B be points in the plane. The collection of all directed line segments having the same length and direction as $\overrightarrow{AB}$ is said to be a *vector*. Every directed line segment in this collection is said to *represent* the vector.

To help distinguish between vectors and ordinary numbers, we often refer to real numbers as *scalars*. In textbooks, boldface type such as **v** is frequently used to denote a vector. When rendered by hand, the vector **v** is often denoted by $\overrightarrow{v}$. Informally, we can

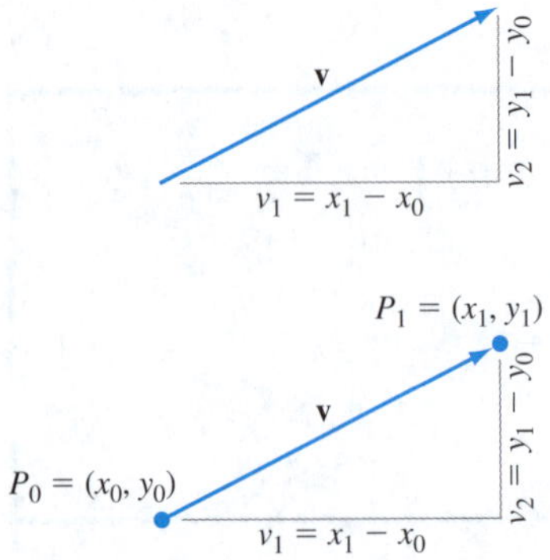

Figure 4
The components v_1 and v_2 of vector **v** are the same in each directed line segment representation.

think of a vector as a "floating" arrow with an initial point that can be chosen in any convenient way. Figure 2 (previous page) illustrates four representations of a *single vector*. In practice, we often refer to a vector by one of the directed line segments that represent it. This can lead to no confusion and is often very helpful. Thus, we may refer to the (one) vector that appears in Figure 2 as *vector* $\overrightarrow{AB}$, *vector* $\overrightarrow{PQ}$, *vector* $\overrightarrow{ON}$, *or vector* $\overrightarrow{RS}$. Even though these four directed line segments are all different, they all represent the same vector.

When a particular directed line segment $\overrightarrow{P_0P_1}$ is used to represent a vector **v** (as in Figure 4), the x-displacement $v_1 = x_1 - x_0$ and the y-displacement $v_2 = y_1 - y_0$ do not depend on the choice of initial point P_0. We can therefore use the notation $\langle v_1, v_2\rangle$ to *unambiguously* denote vector **v**. The quantities v_1 and v_2 are said to be the *components*, or *entries*, of the vector $\langle v_1, v_2\rangle$. Notice that if $P = (x, y)$ and $O = (0, 0)$, then vector $\overrightarrow{OP}$ is equal to $\langle x - 0, y - 0\rangle$, or $\langle x, y\rangle$. Thus, every *point* $P = (x, y)$ gives rise to the *vector* $\langle x, y\rangle$ that is represented by the directed line segment from the origin to P (Figure 5). The vector $\overrightarrow{OP} = \langle x, y\rangle$ is sometimes called the *position vector* of $P = (x, y)$.

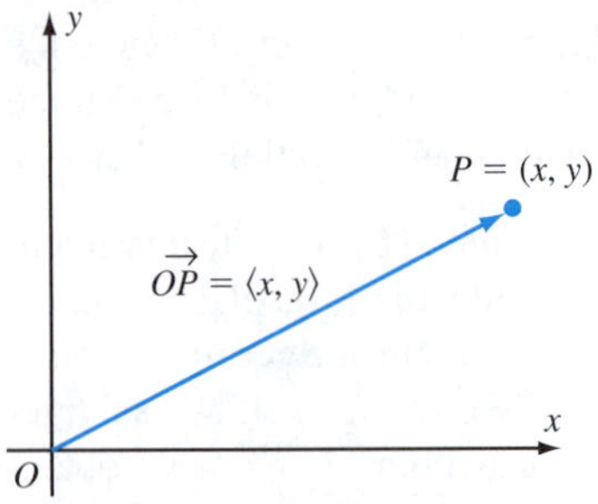

Figure 5
The position vector of point $P = (x, y)$

Example 3 Let $A = (5, 6)$ and $B = (9, 14)$, as in Examples 1 and 2. Write the vector $\overrightarrow{AB}$ in terms of its components.

Solution The components of vector $\overrightarrow{AB} = \langle a, b\rangle$ are $a = 9 - 5 = 4$ and $b = 14 - 6 = 8$. Therefore, $\overrightarrow{AB} = \langle 4, 8\rangle$. Notice that the *point* $P = (4, 8)$ is the terminal point of *vector* $\langle 4, 8\rangle$ when the origin O is chosen as the initial point. Refer again to Figure 2. ■

Vector Algebra

In this subsection, we introduce some algebraic operations that can be performed with vectors.

Definition

The *sum* $\mathbf{v} + \mathbf{w}$ of two vectors $\mathbf{v} = \langle v_1, v_2\rangle$ and $\mathbf{w} = \langle w_1, w_2\rangle$ is formed by adding the vectors componentwise:

$$\mathbf{v} + \mathbf{w} = \langle v_1, v_2\rangle + \langle w_1, w_2\rangle = \langle v_1 + w_1, v_2 + w_2\rangle.$$

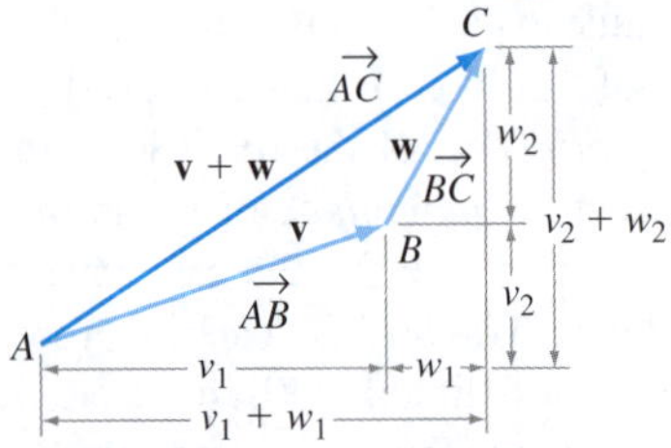

Figure 6

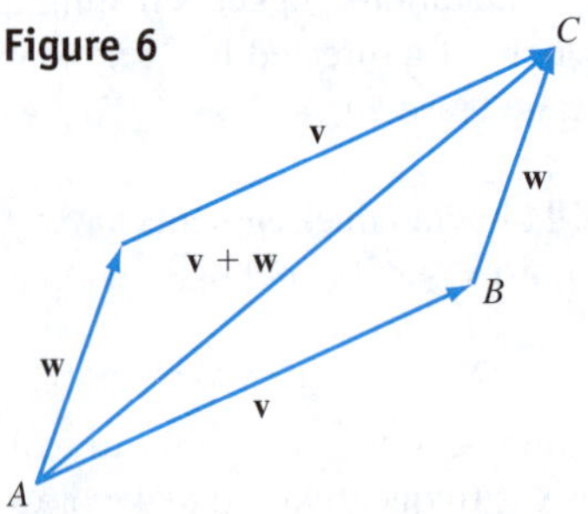

Figure 7

We can geometrically interpret vector addition by first drawing a directed line segment $\overrightarrow{AB}$ that represents **v** and then drawing a directed line segment $\overrightarrow{BC}$ that represents **w**. Notice (as shown in Figure 6) that the initial point of the second directed line segment is the terminal point of the first. The sum $\mathbf{v} + \mathbf{w}$ is then equal to vector $\overrightarrow{AC}$. In other words, $\overrightarrow{AB} + \overrightarrow{BC} = \overrightarrow{AC}$. Figure 7 shows that the sum $\mathbf{v} + \mathbf{w}$ is the diagonal of the parallelogram determined by **v** and **w**. It follows that $\mathbf{v} + \mathbf{w} = \mathbf{w} + \mathbf{v}$, as is also evident from the algebraic definition.

Example 4 Add the vectors $\mathbf{v} = \langle -3, 9\rangle$ and $\mathbf{w} = \langle 1, 8\rangle$.

Solution We have $\mathbf{v} + \mathbf{w} = \langle -3 + 1, 9 + 8\rangle = \langle -2, 17\rangle$. ■

Next we define the vector analogue of the scalar zero.

Definition

The *zero vector* $\vec{\mathbf{0}}$ is the vector with components that are both 0. That is, $\vec{\mathbf{0}} = \langle 0, 0\rangle$.

The zero vector is the identity for vector addition. This means that if $\mathbf{v}$ is any vector, then $\mathbf{v} + \vec{\mathbf{0}} = \mathbf{v}$.

Definition

If $\mathbf{v} = \langle v_1, v_2\rangle$ is a vector and λ is a real number, then we define the *scalar multiplication* of $\mathbf{v}$ by λ to be

$$\lambda\mathbf{v} = \langle \lambda v_1, \lambda v_2\rangle.$$

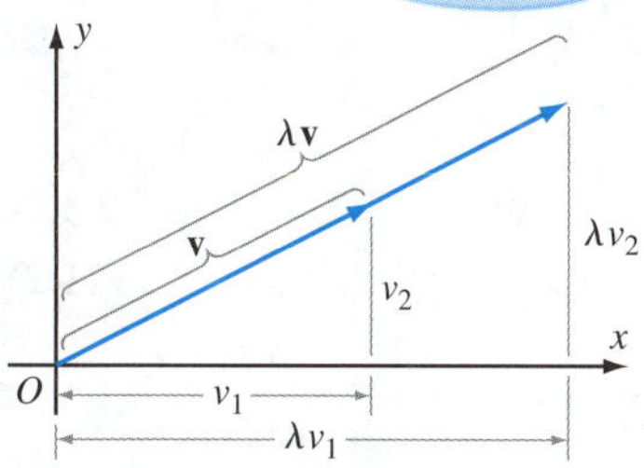

Figure 8a
$\lambda > 0$

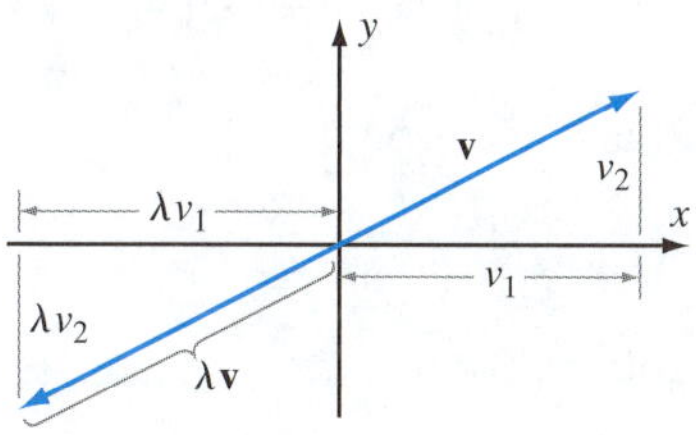

Figure 8b
$\lambda < 0$

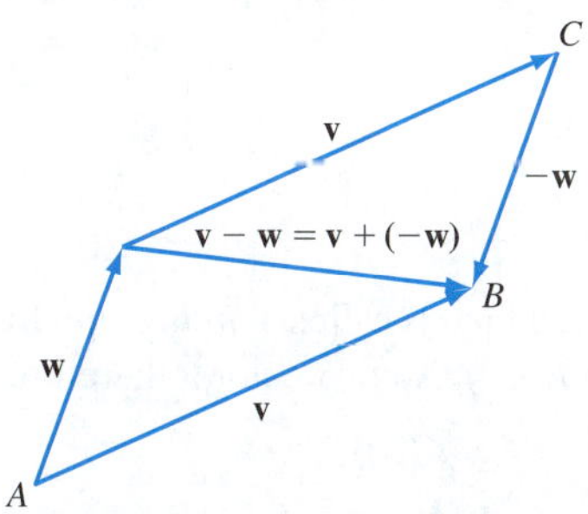

Figure 9

Geometrically, we think of scalar multiplication as producing a vector that is "parallel" to $\mathbf{v}$. We may use the idea of similar triangles to check this idea, as Figure 8 suggests. If $\lambda > 0$, then $\lambda\mathbf{v}$ and $\mathbf{v}$ have the same direction (see Figure 8a). If $\lambda < 0$, then $\lambda\mathbf{v}$ and $\mathbf{v}$ have opposite directions (see Figure 8b). These intuitive notions will be made more precise when a formal definition of *direction* is given below.

Example 5 If $\mathbf{v} = \langle 2, -1\rangle$, then calculate $2\mathbf{v}$ and $-3\mathbf{v}$.

Solution We have $2\mathbf{v} = \langle 2 \cdot 2, 2 \cdot (-1)\rangle = \langle 4, -2\rangle$ and $-3\mathbf{v} = \langle (-3) \cdot 2, (-3) \cdot (-1)\rangle = \langle -6, 3\rangle$. ■

Vector addition and scalar multiplication can be used together to define *vector subtraction*. If $\mathbf{v}$ and $\mathbf{w}$ are given vectors, then the expression $\mathbf{v} - \mathbf{w}$ is interpreted to mean $\mathbf{v} + ((-1)\mathbf{w})$. Refer to Figure 9, which features the parallelogram that is generated by $\mathbf{v}$ and $\mathbf{w}$. In Figure 7, this same parallelogram is used to visualize the sum $\mathbf{v} + \mathbf{w}$. Notice that $\mathbf{v} - \mathbf{w}$ is represented by the *other* (directed) diagonal of the parallelogram, as can be observed by focusing on the three sides of the upper right triangle. As the three sides of the lower left triangle of Figure 9 suggest, $\mathbf{v} - \mathbf{w}$ is the vector that we add to $\mathbf{w}$ to obtain $\mathbf{v}$.

A simple calculation shows that subtraction of vectors is performed componentwise, just like addition of vectors: if $\mathbf{v} = \langle v_1, v_2\rangle$ and $\mathbf{w} = \langle w_1, w_2\rangle$, then

$$\mathbf{v} - \mathbf{w} = \langle v_1, v_2\rangle + (-1)\,\langle w_1, w_2\rangle = \langle v_1, v_2\rangle + \langle -w_1, -w_2\rangle = \langle v_1 - w_1, v_2 - w_2\rangle.$$

Example 6 If $\mathbf{v} = \langle -6, 12\rangle$ and $\mathbf{w} = \langle 19, -7\rangle$, then calculate $\mathbf{v} - \mathbf{w}$.

Solution We have $\mathbf{v} - \mathbf{w} = \langle -6 - 19, 12 - (-7)\rangle = \langle -25, 19\rangle$. ■

The Length of a Vector

If $\mathbf{v} = \langle v_1, v_2\rangle$ is a vector, then its *length* is defined to be

$$\|\mathbf{v}\| = \sqrt{(v_1)^2 + (v_2)^2}.$$

This quantity is also called the *magnitude* of $\mathbf{v}$ or the *norm* of $\mathbf{v}$. If $\overrightarrow{AB}$ is a directed line segment that represents $\mathbf{v}$, then $\|\mathbf{v}\|$ is simply the distance between the points A

and B. The length of a nonzero vector is positive. The zero vector $\vec{\mathbf{0}}$ is the only vector that has length equal to 0.

Example 7 Let $O = (0, 0)$ denote the origin. If $P = (-9, 6)$ and $Q = (2, 1)$, then calculate $\|\overrightarrow{PQ}\|$, $\|\overrightarrow{OP}\|$, and $\|\overrightarrow{OQ}\|$.

Solution Since $\overrightarrow{PQ} = \langle 2 - (-9), 1 - 6\rangle = \langle 11, -5\rangle$, $\overrightarrow{OP} = \langle -9, 6\rangle$, and $\overrightarrow{OQ} = \langle 2, 1\rangle$, we calculate $\|\overrightarrow{PQ}\| = \sqrt{11^2 + (-5)^2} = \sqrt{146}$, $\|\overrightarrow{OP}\| = \sqrt{(-9)^2 + 6^2} = \sqrt{117}$, and $\|\overrightarrow{OQ}\| = \sqrt{2^2 + 1^2} = \sqrt{5}$. ■

Theorem 1 If $\mathbf{v}$ is a vector and λ is a number, then

$$\|\lambda\mathbf{v}\| = |\lambda|\|\mathbf{v}\|. \qquad \textbf{(11.2)}$$

In words: The length of $\lambda\mathbf{v}$ is $|\lambda|$ times the length of $\mathbf{v}$.

Proof Equation (11.2) is easily derived from the definitions of length and scalar multiplication:

$$\|\lambda\mathbf{v}\| = \|\langle\lambda v_1, \lambda v_2\rangle\| = \sqrt{(\lambda v_1)^2 + (\lambda v_2)^2} = \sqrt{\lambda^2}\sqrt{(v_1)^2 + (v_2)^2} = |\lambda|\|\mathbf{v}\|.$$ ■

Example 8 Verify equation (11.2) for $\mathbf{v} = \langle 4, 12\rangle$ and $\lambda = -5$.

Solution We calculate

$$|\lambda|\|\mathbf{v}\| = |-5|\sqrt{4^2 + 12^2} = 5\sqrt{160} = 20\sqrt{10}$$

and

$$\|\lambda\mathbf{v}\| = \|(-5)\langle 4, 12\rangle\| = \|\langle -20, -60\rangle\| = \sqrt{(-20)^2 + (-60)^2}$$
$$= 20\sqrt{(-1)^2 + (-3)^2} = 20\sqrt{10}.$$ ■

Unit Vectors and Directions

If a vector $\mathbf{u} = \langle u_1, u_2\rangle$ has length 1, that is, if $\|\mathbf{u}\| = 1$, then $\mathbf{u}$ is called a *unit vector.* In this case, the *point* (u_1, u_2) lies on the unit circle (Figure 10). A unit vector $\mathbf{u}$ therefore has the form

$$\mathbf{u} = \langle\cos(\alpha), \sin(\alpha)\rangle \qquad \textbf{(11.3)}$$

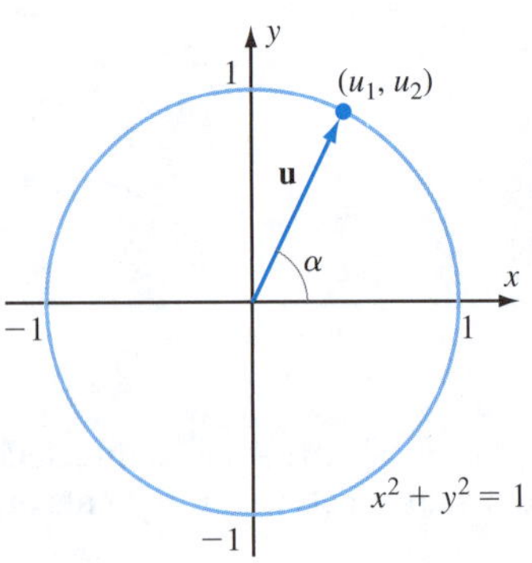

Figure 10
A unit vector $\mathbf{u} = \langle u_1, u_2\rangle$

for some angle α. Sometimes we refer to a unit vector $\mathbf{u}$ as a *direction vector* or simply a *direction*. For any nonzero vector $\mathbf{v} = \langle v_1, v_2\rangle$, the vector

$$\mathbf{u} = \frac{1}{\|\mathbf{v}\|}\mathbf{v} \qquad \textbf{(11.4)}$$

is a unit vector because, according to equation (11.2),

$$\|\mathbf{u}\| = \left\|\frac{1}{\|\mathbf{v}\|}\mathbf{v}\right\| = \left|\frac{1}{\|\mathbf{v}\|}\right|\|\mathbf{v}\| = \frac{1}{\|\mathbf{v}\|}\|\mathbf{v}\| = 1.$$

We refer to the direction vector **u** defined by equation (11.4) as the *direction of* **v** and write

$$\text{dir}(\mathbf{v}) = \frac{1}{\|\mathbf{v}\|}\mathbf{v} \qquad (\mathbf{v} \neq \vec{\mathbf{0}}).$$

By rearranging this equation, we see that every nonzero vector **v** can be expressed as the scalar multiplication of the direction vector of **v** by the magnitude of **v**:

$$\mathbf{v} = \|\mathbf{v}\|\text{dir}(\mathbf{v}). \tag{11.5}$$

We do not define the direction of the zero vector.

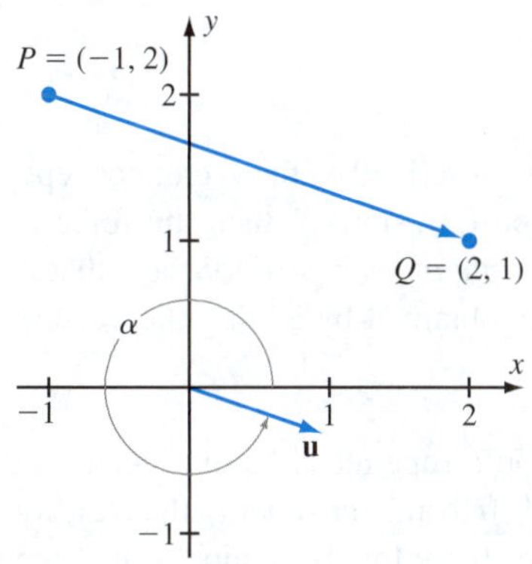

Figure 11

Example 9 Let $P = (-1, 2)$ and $Q = (2, 1)$. What is the direction of $\overrightarrow{PQ}$?

Solution We calculate $\overrightarrow{PQ} = \langle 2-(-1), 1-2\rangle = \langle 3, -1\rangle$ and $\|\overrightarrow{PQ}\| = \sqrt{3^2+(-1)^2} = \sqrt{10}$. The unit vector

$$\mathbf{u} = \frac{1}{\|\overrightarrow{PQ}\|}\overrightarrow{PQ} = \frac{1}{\sqrt{10}}\langle 3, -1\rangle = \left\langle \frac{3}{\sqrt{10}}, \frac{-1}{\sqrt{10}} \right\rangle$$

is the direction of $\overrightarrow{PQ}$. Figure 11 illustrates both $\overrightarrow{PQ}$ and its direction **u**. The angle α that $\mathbf{u} = \langle \cos(\alpha), \sin(\alpha)\rangle$ makes with the positive x-axis is the value of α between $3\pi/2$ and 2π such that $\cos(\alpha) = 3/\sqrt{10}$ and $\sin(\alpha) = -1/\sqrt{10}$. With the aid of a calculator, we find that $\alpha \approx 5.96$ radians. ■

Definition Let **v** and **w** be nonzero vectors. We say that **v** and **w** are *opposite in direction* if $\text{dir}(\mathbf{v}) = -\text{dir}(\mathbf{w})$. We say that **v** and **w** are *parallel* if either (i) **v** and **w** have the same direction or (ii) **v** and **w** are opposite in direction. Although the zero vector $\vec{\mathbf{0}}$ does not have a direction, it is conventional to say that $\vec{\mathbf{0}}$ is parallel to every vector.

Notice that nonzero vectors **v** and **w** are parallel if and only if $\text{dir}(\mathbf{v}) = \pm\text{dir}(\mathbf{w})$. Our next theorem provides a simple algebraic condition for recognizing parallel vectors.

Theorem 2 Vectors **v** and **w** are parallel if and only if $\mathbf{v} = \lambda\mathbf{w}$ or $\mathbf{w} = \lambda\mathbf{v}$ for some scalar λ.

Proof Suppose that **v** and **w** are parallel. If one of them, say **v**, is $\vec{\mathbf{0}}$ then $\mathbf{v} = \lambda\mathbf{w}$ for $\lambda = 0$. If neither is $\vec{\mathbf{0}}$ then, by definition,

$$\frac{1}{\|\mathbf{v}\|}\mathbf{v} = \pm\frac{1}{\|\mathbf{w}\|}\mathbf{w}.$$

It follows that $\mathbf{v} = \lambda\mathbf{w}$ for $\lambda = \pm\|\mathbf{v}\|/\|\mathbf{w}\|$. Conversely, if $\mathbf{v} = \lambda\mathbf{w}$, then $\|\mathbf{v}\| = |\lambda|\|\mathbf{w}\|$. If $\lambda = 0$ or $\|\mathbf{w}\| = 0$, then $\mathbf{v} = \vec{\mathbf{0}}$, which, by convention, is parallel to all vectors, **w** included. Otherwise, dir(**v**) and dir(**w**) are unambiguously defined and

$$\text{dir}(\mathbf{v}) = \frac{1}{\|\mathbf{v}\|}\mathbf{v} = \left(\frac{1}{|\lambda|\|\mathbf{w}\|}\right)\lambda\mathbf{w} = \left(\frac{\lambda}{|\lambda|}\right)\frac{1}{\|\mathbf{w}\|}\mathbf{w} = \left(\frac{\lambda}{|\lambda|}\right)\text{dir}(\mathbf{w}) = \pm\text{dir}(\mathbf{w}).$$

This equation tells us that **v** and **w** have either the same direction or opposite directions. By definition, it follows that **v** and **w** are parallel. ■

inSIGHT

> A closer examination of the proof of Theorem 2 reveals that two nonzero vectors have *the same direction* if and only if each is a positive scalar multiple of the other. They have *opposite direction* if and only if each is a negative scalar multiple of the other.

An Application to Physics

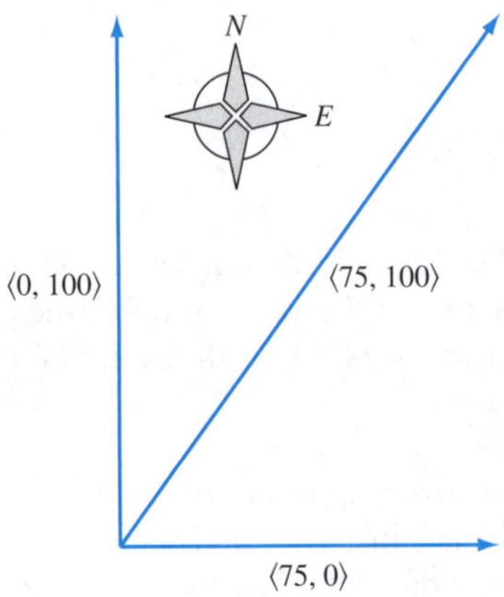

Figure 12

As discussed earlier, *force* is a quantity that is naturally described by the vector concept. If a force of magnitude F is applied in direction $\langle \cos(\alpha), \sin(\alpha) \rangle$, then the force is written as the vector $\langle F\cos(\alpha), F\sin(\alpha) \rangle$. The *principle of superposition* says that if two forces act on a body, then the *resultant force* is obtained by adding the vectors corresponding to the two given forces.

Example 10 If two workers are each pulling on a rope attached to a dead tree stump, one in the northerly direction with a force of 100 lb and the other in the easterly direction with a force of 75 lb, compute the resultant force that is applied to the tree stump.

Solution The force vector for the first worker is $\langle 0, 100 \rangle$, and that for the second worker is $\langle 75, 0 \rangle$. The resultant force is the sum of these, or $\langle 75, 100 \rangle$. The resultant force vector is shown in Figure 12. The associated directed line segment has magnitude (length) equal to $(75^2 + 100^2)^{1/2} = 125$. ■

The Special Unit Vectors i and j

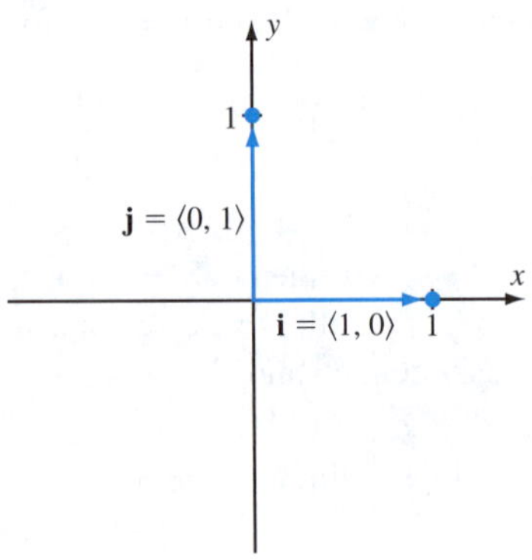

Figure 13

Let **i** denote the vector $\langle 1, 0 \rangle$ and **j** denote the vector $\langle 0, 1 \rangle$, as shown in Figure 13. If $\mathbf{v} = \langle a, b \rangle$ is any vector, then we may express **v** in terms of **i** and **j** as follows:

$$\mathbf{v} = \langle a, b \rangle = a\langle 1, 0 \rangle + b\langle 0, 1 \rangle = a\mathbf{i} + b\mathbf{j}.$$

Example 11 If $\mathbf{v} = \langle 3, -5 \rangle$ and $\mathbf{w} = \langle 2, 4 \rangle$, then express **v**, **w**, and $\mathbf{v} + \mathbf{w}$ in terms of **i** and **j**.

Solution We have $\mathbf{v} = 3\mathbf{i} - 5\mathbf{j}$ and $\mathbf{w} = 2\mathbf{i} + 4\mathbf{j}$. We can calculate

$$\mathbf{v} + \mathbf{w} = \langle 3, -5 \rangle + \langle 2, 4 \rangle = \langle 5, -1 \rangle = 5\mathbf{i} - \mathbf{j}.$$

Alternatively, we can express **v** and **w** in terms of **i** and **j**, as we have already done, and add these expressions:

$$\mathbf{v} + \mathbf{w} = (3\mathbf{i} - 5\mathbf{j}) + (2\mathbf{i} + 4\mathbf{j}) = (3 + 2)\,\mathbf{i} + (-5 + 4)\,\mathbf{j} = 5\mathbf{i} - \mathbf{j}.$$ ■

The Triangle Inequality

A glance back at $\triangle ABC$ in Figure 7 suggests, that if $\mathbf{v}$ and $\mathbf{w}$ are vectors, then the length of $\mathbf{v}+\mathbf{w}$ is never greater than the sum of the lengths of $\mathbf{v}$ and $\mathbf{w}$. In other words,

$$\|\mathbf{v}+\mathbf{w}\| \leq \|\mathbf{v}\| + \|\mathbf{w}\|.$$

In fact, this inequality, known as the *Triangle Inequality,* is merely a vector interpretation of the familiar theorem in Euclidean geometry that states that the sum of the lengths of any two sides of a triangle is greater than the length of the third side.

Example 12 Verify the Triangle Inequality for the vectors $\mathbf{v} = \langle -3, 4\rangle$ and $\mathbf{w} = \langle 8, 6\rangle$.

Solution After calculating $\|\mathbf{v}\| + \|\mathbf{w}\| = \sqrt{(-3)^2+4^2} + \sqrt{8^2+6^2} = 5+10 = 15$, we find that

$$\|\mathbf{v}+\mathbf{w}\| = \|\langle 5, 10\rangle\| = \sqrt{125} \approx 11.18 < 15 = \|\mathbf{v}\| + \|\mathbf{w}\|.$$ ■

We conclude this section by stating the commutative, associative, and distributive laws for scalar multiplication and vector addition. If $\mathbf{u}$, $\mathbf{v}$, $\mathbf{w}$ are vectors and λ, μ are scalars, then

a. $\mathbf{v}+\mathbf{w} = \mathbf{w}+\mathbf{v}$,
b. $\mathbf{u}+(\mathbf{v}+\mathbf{w}) = (\mathbf{u}+\mathbf{v})+\mathbf{w}$,
c. $\lambda(\mu\mathbf{v}) = (\lambda\mu)\mathbf{v}$,
d. $\lambda(\mathbf{v}+\mathbf{w}) = \lambda\mathbf{v}+\lambda\mathbf{w}$,
e. $(\lambda+\mu)\mathbf{v} = \lambda\mathbf{v}+\mu\mathbf{v}$.

quickquiz

1. What is a vector in the plane?
2. What does it mean for a directed line segment to "represent" a vector?
3. How do we add vectors geometrically? Algebraically?
4. How do we calculate the norm (length) of a vector?

EXERCISES

Problems for Practice

In Exercises 1–8, sketch $\overrightarrow{PQ}$, $\overrightarrow{QP}$, $\overrightarrow{OP}$, and $\overrightarrow{OQ}$. Write each vector in the form $\langle a, b\rangle$.

1. $P = (5, 8)$, $Q = (4, 4)$
2. $P = (1, 1)$, $Q = (3, 9)$
3. $P = (0, 0)$, $Q = (-5, 1)$
4. $P = (1, 1)$, $Q = (2, 2)$
5. $P = (-1, 8)$, $Q = (4, -3)$
6. $P = (0, 1)$, $Q = (0, 0)$
7. $P = (-5, 6)$, $Q = (1, 0)$
8. $P = (2, 2)$, $Q = (2, -2)$

In Exercises 9–12, calculate the lengths of $\overrightarrow{PQ}$ and $\overrightarrow{RS}$. Also, determine whether these vectors are parallel.

9. $P = (1, 2)$, $Q = (2, 0)$, $R = (3, 6)$, $S = (6, 0)$
10. $P = (1, 3)$, $Q = (3, 1)$, $R = (1, 2)$, $S = (6, 3)$
11. $P = (3, 1)$, $Q = (-6, 4)$, $R = (7, -2)$, $S = (5, 5)$
12. $P = (1, 1)$, $Q = (-7, 5)$, $R = (1, 0)$, $S = (17, -8)$

In Exercises 13–16, calculate the indicated vector given that $\mathbf{v} = \langle 4, 2\rangle$ and $\mathbf{w} = \langle 1, -3\rangle$. For each exercise, draw a sketch of **v**, **w**, and the vector you have calculated.

13. $2\mathbf{v} + \mathbf{w}$ **14.** $(1/2)\mathbf{v} + 3\mathbf{w}$
15. $\mathbf{v} - 2\mathbf{w}$ **16.** $-3\mathbf{v} + 2\mathbf{w}$

In Exercises 17–20, determine the following:

- **a.** The vector **v** that is represented by $\overrightarrow{PQ}$
- **b.** The length $\|\mathbf{v}\|$ of **v**
- **c.** The vector that is parallel to **v** and of the same length but pointing in the opposite direction
- **d.** A unit vector parallel to **v** and pointing in the same direction as **v**
- **e.** A unit vector parallel to **v** but pointing in the opposite direction

17. $P = (5, -7), Q = (0, 2)$
18. $P = (-7, 1), Q = (-1, 4)$
19. $P = (-3, 7), Q = (1, -9)$
20. $P = (3, 2), Q = (1, 1)$

In Exercises 21–24, vectors **v** and **w** are given. Express $\mathbf{v}, \mathbf{w}, -4\mathbf{v}, 3\mathbf{v} - 2\mathbf{w}$, and $4\mathbf{v} + 7\mathbf{j}$ in terms of the unit vectors **i** and **j**.

21. $\mathbf{v} = \langle -7, 2\rangle, \mathbf{w} = \langle -2, 9\rangle$
22. $\mathbf{v} = \langle 0, 3\rangle, \mathbf{w} = \langle -5, 0\rangle$
23. $\mathbf{v} = \langle 6, -2\rangle, \mathbf{w} = \langle 9, -3\rangle$
24. $\mathbf{v} = \langle 1/3, 1\rangle, \mathbf{w} = \langle -3/2, 1/2\rangle$

In Exercises 25–28, calculate the resultant force $\mathbf{F} = \mathbf{F}_1 + \mathbf{F}_2$. What is the magnitude of **F**?

25. $\mathbf{F}_1 = \langle 3, 0\rangle, \mathbf{F}_2 = \langle 0, 4\rangle$
26. $\mathbf{F}_1 = \langle 1, 2\rangle, \mathbf{F}_2 = \langle 2, 2\rangle$
27. $\mathbf{F}_1 = \mathbf{i} - 2\mathbf{j}, \mathbf{F}_2 = 2\mathbf{i} + \mathbf{j}$
28. $\mathbf{F}_1 = \mathbf{i} - \mathbf{j}, \mathbf{F}_2 = -5\mathbf{i} + 13\mathbf{j}$

In Exercises 29–32, write down the direction vector **u** that makes the given positive angle α with the positive x-axis.

29. $\pi/6$ **30.** $3\pi/4$
31. π **32.** $5\pi/3$

Further Theory and Practice

33. Mr. and Mrs. Woodman are pulling on ropes tied to a heavy wagon. Refer to Figure 14. If Mr. Woodman pulls with a strength of 100 lb, then how hard must Mrs. Woodman pull so that the wagon moves along the dotted line? What is the magnitude of the resultant force?

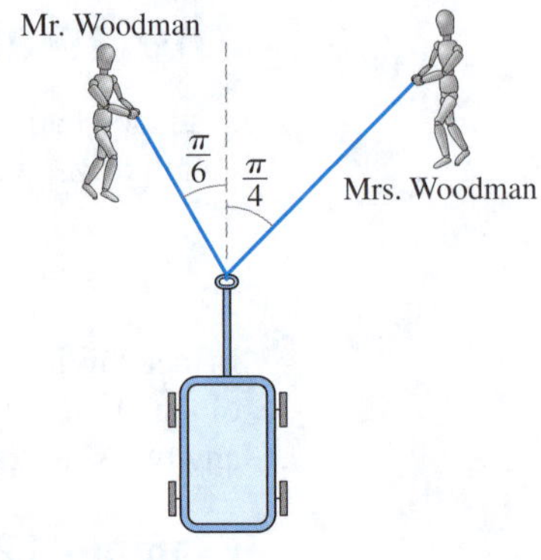

Figure 14

34. Three forces $\mathbf{F}_1$, $\mathbf{F}_2$, and $\mathbf{F}_3$ are applied to a point mass. Suppose that $\mathbf{F}_1 = 100\mathbf{j}$ newtons (N) and that $\mathbf{F}_2$ has magnitude 120 N applied in the direction $\langle 3/5, 4/5\rangle$. What must $\mathbf{F}_3$ be if the point mass is to remain at rest?

35. A boat maintains a straight course across a 500 m wide river at a rate of 50 m/min. The current pulls the boat down river at a rate of 20 m/min. When the boat docks on the other shore, how far down river will it have been carried?

36. A river is 1 mi wide and flows south with a current of 3 mi/h. What speed and heading should a motorboat adopt to cross the river in 10 min and reach a point on the opposite bank due east of its point of departure?

37. Bjarne, Leif, and Sammy are towing their vessel. The forces that they exert are directed along the tow lines, as indicated in Figure 15, which also provides the magnitudes of their forces. (Note that in Figure 15, the force vectors are not drawn to scale.) What is the resultant force?

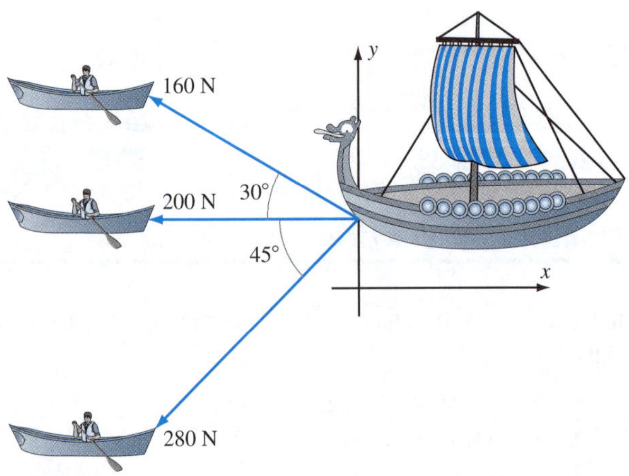

Figure 15

38. Let $P = (p_1, p_2)$ and $Q = (q_1, q_2)$ be distinct points in the plane. Let M be the midpoint of the segment joining P and Q. Use the relationship $\overrightarrow{OM} = \overrightarrow{OP} + (1/2)\overrightarrow{PQ}$ to determine the Cartesian coordinates of M.

39. Let $\mathbf{v}$ and $\mathbf{w}$ be two given nonzero vectors. Prove that there is a unique number λ such that $\mathbf{v} + \lambda\mathbf{w}$ is as short as possible. In other words, the function from $\mathbb{R}$ to $\mathbb{R}$ defined by $\lambda \mapsto \|\mathbf{v} + \lambda\mathbf{w}\|$ attains an absolute minimum value.

40. If $\mathbf{v} = \langle a, b\rangle$ is a vector, then verify that $\mathbf{w} = \langle -b, a\rangle$ is a vector that is perpendicular to $\mathbf{v}$. Give a formula for all possible vectors that are perpendicular to $\mathbf{v}$.

41. Let $\mathbf{v}$ and $\mathbf{w}$ be two nonzero vectors in the plane that are *not* parallel. Let $\mathbf{u}$ be any other vector. Prove that there are unique scalars λ and μ such that

$$\mathbf{u} = \lambda\mathbf{v} + \mu\mathbf{w}.$$

(If $\mathbf{v} = \mathbf{i}$ and $\mathbf{w} = \mathbf{j}$, then λ and μ are the entries of vector $\mathbf{u}$. In general, we can think of $\mathbf{v}$ and $\mathbf{w}$ generating a coordinate system with λ and μ the entries of vector $\mathbf{u}$ in this new coordinate system.)

Exercises 42 and 43 develop a formula for the distance of a point to a line.

42. Suppose that A and B are not 0. Consider the line ℓ whose Cartesian equation is $Ax + By + D = 0$. Suppose that $P_0 = (x_0, y_0)$ does not lie on ℓ. Show that $\mathbf{n} = \langle A, B\rangle$ is a vector that is perpendicular to ℓ. Let $Q_x = (x, -Ax/B - D/B)$ be a point on ℓ. For what value of x is $\overrightarrow{P_0Q_x}$ parallel to $\mathbf{n}$?

43. For the value of x determined in Exercise 42 show that

$$\|\overrightarrow{P_0Q_x}\| = \frac{|Ax_0 + By_0 + D|}{\sqrt{A^2 + B^2}}.$$

What does it mean to state that this is the distance of P_0 to ℓ?

44. Let $P = (p_1, p_2)$ and $Q = (q_1, q_2)$ be distinct points in the plane. Suppose that $0 < t < 1$. What are the Cartesian coordinates of the point P_t on line segment $\overline{PQ}$ whose distance from P is $t \cdot \|\overrightarrow{PQ}\|$? Use vectors to find an elegant solution to this problem.

45. Let m and b be constants. Describe all vectors that are parallel to the line $y = mx + b$.

Let A, B, and C denote any three distinct points that are not collinear. Let α, β, and γ denote the midpoints of line segments $\overline{BC}$, $\overline{AC}$, and $\overline{AB}$, respectively. The line segments $\overline{A\alpha}$, $\overline{B\beta}$, and $\overline{C\gamma}$ are called *medians* of $\triangle ABC$. Exercises 46–48 concern these medians and their common point of intersection, the *centroid* of $\triangle ABC$.

46. Prove that $\overrightarrow{A\alpha} + \overrightarrow{B\beta} + \overrightarrow{C\gamma} = \vec{\mathbf{0}}$.

47. Let M be point of intersection of $\overline{A\alpha}$ and $\overline{B\beta}$. Show that $\overrightarrow{AM} = (2/3)\overrightarrow{A\alpha}$ and $\overrightarrow{BM} = (2/3)\overrightarrow{B\beta}$. Symmetrically, the same relationship holds when the medians $\overline{B\beta}$ and $\overline{C\gamma}$ are selected. Deduce that all three medians intersect in one point. This point of intersection is called the *centroid* of $\triangle ABC$.

48. Express the coordinates of the centroid of $\triangle ABC$ in terms of the coordinates of the vertices A, B, and C.

Calculator/Computer Exercises

49. Plot $y = \exp(x)$ and $y = 2 - x^2$ for $-2 \le x \le 1$. Let P and Q be the points of intersection of the two curves in the second and first quadrants, respectively. What is vector $\overrightarrow{PQ}$?

50. Let $f(x) = x^2 + 2x + 2$. Let P_x be the point $(x, f(x))$. Plot $y = \|\overrightarrow{OP_x}\|$. What is the point on the graph of f that is closest to the origin?

51. For $-2 \le t \le 3$, let $Q(t)$ denote the point $(t, \exp(-t) + t^2)$. Let $f(t) = \|\overrightarrow{PQ(t)}\|$ where $P = (1, 4)$. Plot $f(t)$. At what value of t is $f(t)$ minimized?

52. Suppose that for each point $P_t = (t, t^2)$ on the curve $y = x^2$, a point $Q_t = (\xi(t), \eta(t))$ is found such that (i) $\xi(t) > t$, (ii) $\overrightarrow{P_tQ_t}$ is a unit vector, and (iii) $\overrightarrow{P_tQ_t}$ is tangent to the curve. Plot the curve with parametric equations $x = \xi(t)$, $y = \eta(t)$. What point on this parametric curve is farthest below the y-axis?

53. Suppose that for each point $P_t = \left(t, t^3\right)$ on the curve $y = x^3$, a point $Q_t = (\xi(t), \eta(t))$ is found such that (i) $\xi(t) > t$, (ii) $\overrightarrow{P_tQ_t}$ is a unit vector, and (iii) $\overrightarrow{P_tQ_t}$ is tangent to the curve. Plot the curve with parametric equations $x = \xi(t)$, $y = \eta(t)$. What are its x- and y-intercepts? To what values of t do these intercepts correspond?

11.2 Vectors in Three-Dimensional Space

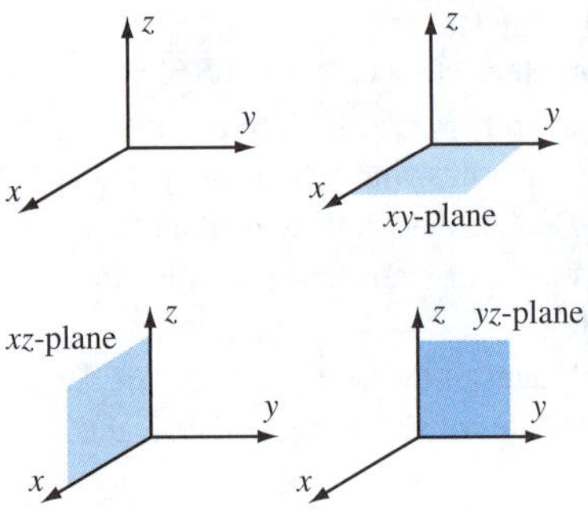

Figure 1

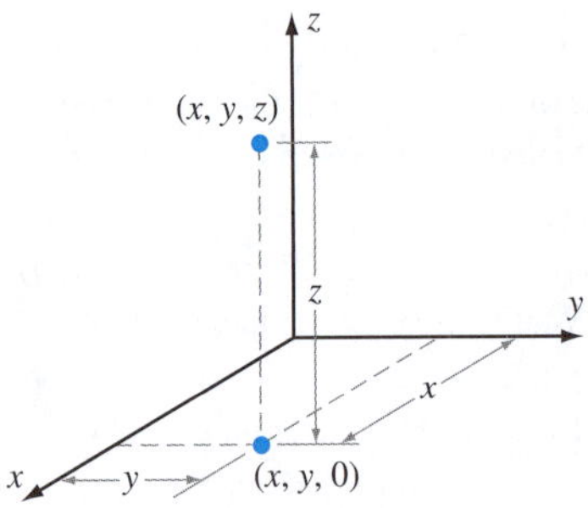

Figure 2

It is a familiar idea to locate points on the line using one coordinate and points in the plane using two coordinates. Now we turn to three-dimensional space: the locating of points will require three coordinates.

Refer to Figure 1, which exhibits three mutually perpendicular axes: the x-axis, the y-axis, and the z-axis. The yz-plane coincides with the plane of this page. The xy- and xz-planes emerge perpendicularly from this page. A point with spatial coordinates $(x, y, 0)$ lies in the xy-plane. The third coordinate z of (x, y, z) indicates the *height* above or below the xy-plane. Figure 2 exhibits a point (x, y, z) with $z > 0$: This point lies z units above the point $(x, y, 0)$ in the xy-plane. Notice how we use dotted lines in our picture to create a sense of perspective so that we can picture a three-dimensional concept on two-dimensional paper. The coordinates x, y, and z of the point $P = (x, y, z)$ are called the *Cartesian coordinates* of P. An equation among Cartesian coordinates is said to be a *Cartesian equation*. The terms *rectangular coordinates* and *rectangular equation* are sometimes used instead.

Just as the coordinate axes in the plane divide the plane into *quadrants,* so do the xy-plane, the xz-plane, and the yz-plane divide space into *octants*. The octant that contains points with all three coordinates positive is often called the *first octant*. It is sometimes helpful to think of the xy-plane as the floor of a room. Think of the origin $O = (0, 0, 0)$ as a corner of the floor and of the positive z-axis as the join of two walls (the two walls are the xz-plane and the yz-plane). Look at Figure 3, which depicts the first octant from this point of view. Let $P = (x_0, y_0, z_0)$ be a point with all three coordinates positive. The first two coordinates of P tell which position $P' = (x_0, y_0, 0)$ on the floor that P lies over. The last coordinate tells how high above the floor P is.

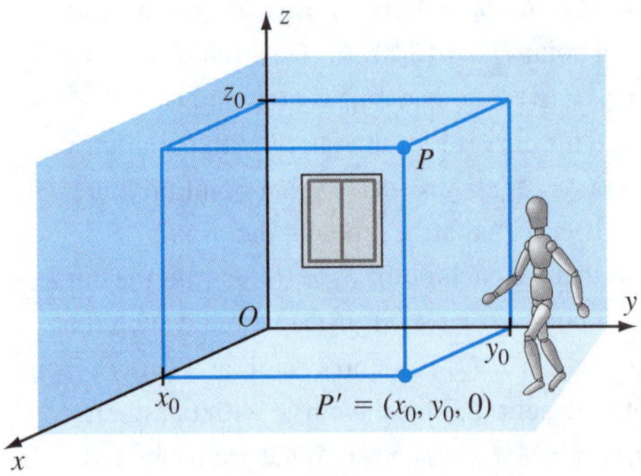

Figure 3
A point $P = (x_0, y_0, z_0)$ in the first octant and the projection of P to the xy-plane

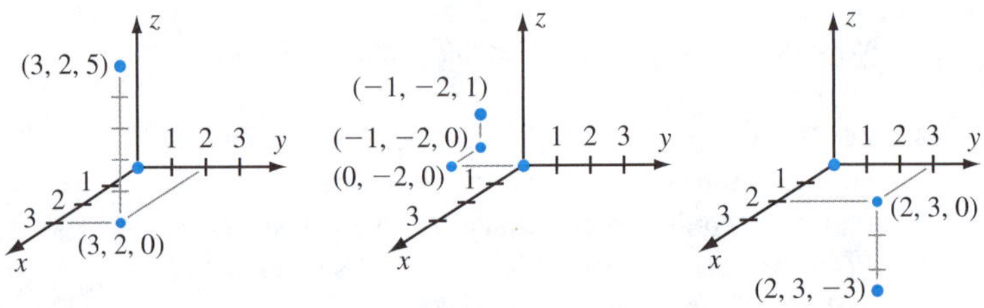

Figure 4

Example 1 Sketch the points $(3, 2, 5)$, $(-1, -2, 1)$, and $(2, 3, -3)$.

Solution These points are graphed in Figure 4. Some polygonal paths connecting these points to the origin are also plotted for perspective. For example, to reach the point $(-1, -2, 1)$ from $O = (0, 0, 0)$, we may go 2 units in the negative y-direction, arriving at the point $(0, -2, 0)$, then 1 unit in the negative x-direction, arriving at the point $(-1, -2, 0)$, and finally 1 unit in the positive z-direction, finishing at $(-1, -2, 1)$. ■

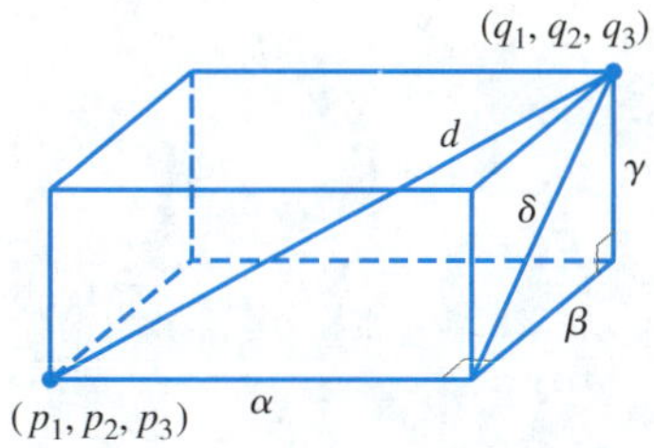

Figure 5

Distance

Two distinct points in space determine a box with sides parallel to the coordinate axes. See Figure 5. The length of a main diagonal of this box coincides with the *distance d* between these two points. We have labeled length, width, and height of the box as α, β, and γ, respectively. Figure 5 shows one main diagonal with length d as well as the length δ of a minor diagonal. Notice that, by the Pythagorean Theorem, $\delta^2 = \beta^2 + \gamma^2$ and $d^2 = \alpha^2 + \delta^2$. Substituting the first of these equations into the second yields $d^2 = \alpha^2 + \beta^2 + \gamma^2$. Translating this calculation into coordinates gives us a formula for the distance between two points in space.

Theorem 1 The distance between the points (p_1, p_2, p_3) and (q_1, q_2, q_3) is

$$\sqrt{(q_1 - p_1)^2 + (q_2 - p_2)^2 + (q_3 - p_3)^2}. \qquad \textbf{(11.6)}$$

Example 2 Calculate the distance between the points $(4, -9, 7)$ and $(2, 1, 4)$.

Solution According to Theorem 1, the required distance is

$$\sqrt{(2-4)^2 + (1-(-9))^2 + (4-7)^2} = \sqrt{(-2)^2 + (10)^2 + (-3)^2} = \sqrt{113}.$$ ■

If $P = (x_0, y_0, z_0)$ is a fixed point in space and if $r > 0$ is a fixed number, then the collection of all points with distance r from P is the *sphere* with center P and radius r. From Theorem 1, we see that the set of points that satisfy the equation

$$\sqrt{(x - x_0)^2 + (y - y_0)^2 + (z - z_0)^2} = r,$$

or, equivalently,

$$(x - x_0)^2 + (y - y_0)^2 + (z - z_0)^2 = r^2, \qquad \textbf{(11.7)}$$

is the sphere with center $P = (x_0, y_0, z_0)$ and radius r. Notice that the left side of equation (11.7) is quadratic in the three space variables x, y, and z. The coefficients of x^2, y^2, and z^2 are all 1. There are no mixed products xy, xz, or yz.

Example 3 Determine what set of points is described by the equation $x^2 + y^2 + z^2 - 6x + 4y + 6 = 0$.

Solution We complete the squares for x and y (because of the presence of the linear terms $-6x$ and $4y$): The equation $x^2 + y^2 + z^2 - 6x + 4y + 6 = 0$ is equivalent to $(x^2 - 6x) + (y^2 + 4y) + z^2 = -6$ or

$$\left(x^2 - 6x + \left(-\frac{6}{2}\right)^2\right) + \left(y^2 + 4y + \left(\frac{4}{2}\right)^2\right) + z^2 = -6 + \left(-\frac{6}{2}\right)^2 + \left(\frac{4}{2}\right)^2.$$

After simplification, we have $(x-3)^2+(y+2)^2+z^2=7$, or $(x-3)^2+(y-(-2))^2+(z-0)^2=7$. We see that the given equation describes a sphere with center $(3,-2,0)$ and radius $\sqrt{7}$. Refer to Figure 6.

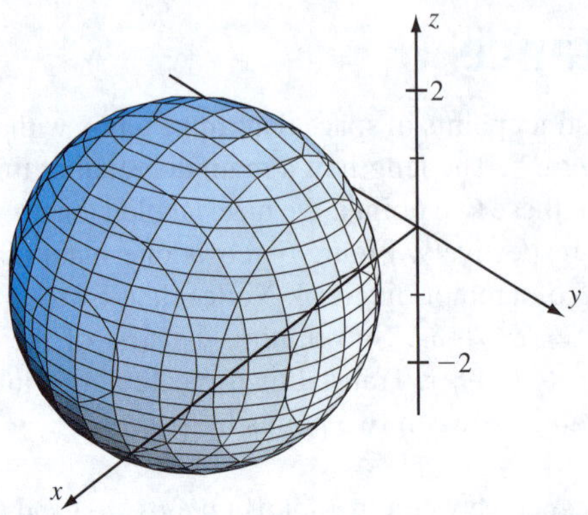

Figure 6
The sphere $(x-3)^2+(y-(-2))^2+(z-0)^2=7$

Definition The set of points inside the sphere $(x-x_0)^2+(y-y_0)^2+(z-z_0)^2=r^2$, that is, the set of points that satisfy

$$(x-x_0)^2+(y-y_0)^2+(z-z_0)^2<r^2,$$

is called the *open ball with center* (x_0, y_0, z_0) *and radius r*. The set of points that satisfy

$$(x-x_0)^2+(y-y_0)^2+(z-z_0)^2\le r^2$$

is the preceding open ball plus its boundary, the sphere. We call this set the *closed ball with center* (x_0, y_0, z_0) *and radius r*.

Example 4 Describe the set of points S that satisfy the equation $x^2+y^2+z^2+80+8y+16z<6x$.

Solution This set can be identified by rearranging the given inequality as $\left(x^2-6x\right)+\left(y^2+8y\right)+\left(z^2+16z\right)<-80$ and completing the squares in each parenthesis:

$$\left(x^2-6x+\left(\frac{6}{2}\right)^2\right)+\left(y^2+8y+\left(\frac{8}{2}\right)^2\right)+\left(z^2+16z+\left(\frac{16}{2}\right)^2\right)<-80+\left(\frac{6}{2}\right)^2+\left(\frac{8}{2}\right)^2+\left(\frac{16}{2}\right)^2.$$

This inequality simplifies to $(x-3)^2+(y+4)^2+(z+8)^2<9$ or $(x-3)^2+(y-(-4))^2+(z-(-8))^2<3^2$. From this last inequality, we see that the given set consists of those points with a distance from the point $(3,-4,-8)$ that is less than 3. In other words, S is the open ball with center $(3,-4,-8)$ and radius 3. ■

Vectors in Space

A *vector* in three-dimensional space is an ordered triple $\langle v_1, v_2, v_3\rangle$. This is the algebraic definition of vector and is the one that we use in calculations. However, there is a geometric interpretation as well. Like a vector in the plane, a vector in space can be represented by a directed line segment that specifies magnitude and direction. We say that a directed line segment $\overrightarrow{PQ}$ with initial point P and terminal point Q *represents* the vector $\mathbf{v} = \langle v_1, v_2, v_3\rangle$ if the points $P = (p_1, p_2, p_3)$ and $Q = (q_1, q_2, q_3)$ satisfy

$$v_1 = q_1 - p_1, \qquad v_2 = q_2 - p_2, \qquad \text{and} \qquad v_3 = q_3 - p_3. \tag{11.8}$$

If $P = (x, y, z)$ is a point in space, then the vector $\overrightarrow{OP} = \langle x, y, z\rangle$ is sometimes called the *position vector* of P.

Example 5 Let $P = (3, -1, 4)$, $Q = (2, 1, 0)$, $R = (-6, 9, 7)$, and $S = (-7, 11, 3)$. What vectors are represented by the directed line segments $\overrightarrow{PQ}$, $\overrightarrow{PR}$, and $\overrightarrow{QS}$? Do any of these directed line segments represent the same vector?

Solution We know the following:

The vector represented by $\overrightarrow{PQ}$ is $\langle 2-3, 1-(-1), 0-4\rangle = \langle -1, 2, -4\rangle$.
The vector represented by $\overrightarrow{PR}$ is $\langle -6-3, 9-(-1), 7-4\rangle = \langle -9, 10, 3\rangle$.
The vector represented by $\overrightarrow{QS}$ is $\langle -7-2, 11-1, 3-0\rangle = \langle -9, 10, 3\rangle$.

We see that $\overrightarrow{PQ}$ and $\overrightarrow{PR}$ represent different vectors. However, $\overrightarrow{PR}$ and $\overrightarrow{QS}$ represent the same vector, namely, $\langle -9, 10, 3\rangle$. ■

Vector Operations

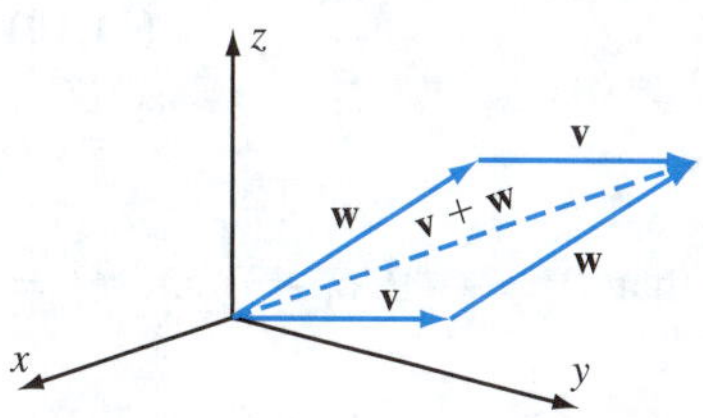

Figure 7

Addition. If $\mathbf{v} = \langle v_1, v_2, v_3\rangle$ and $\mathbf{w} = \langle w_1, w_2, w_3\rangle$, then we define

$$\mathbf{v} + \mathbf{w} = \langle v_1 + w_1, v_2 + w_2, v_3 + w_3\rangle.$$

As in the plane, the geometric interpretation of vector addition is obtained by using the Parallelogram Rule. The sum $\mathbf{v} + \mathbf{w}$, indicated by the dotted directed line segment in Figure 7, is the diagonal of the parallelogram determined by $\mathbf{v}$ and $\mathbf{w}$.

Example 6 Suppose that $\mathbf{v} = \langle 3, 0, 1\rangle$ and $\mathbf{w} = \langle 0, 4, 2\rangle$. Calculate $\mathbf{v} + \mathbf{w}$ and sketch the three vectors.

Solution We calculate $\mathbf{v} + \mathbf{w} = \langle 3+0, 0+4, 1+2\rangle = \langle 3, 4, 3\rangle$. The three vectors are drawn in Figure 8. ■

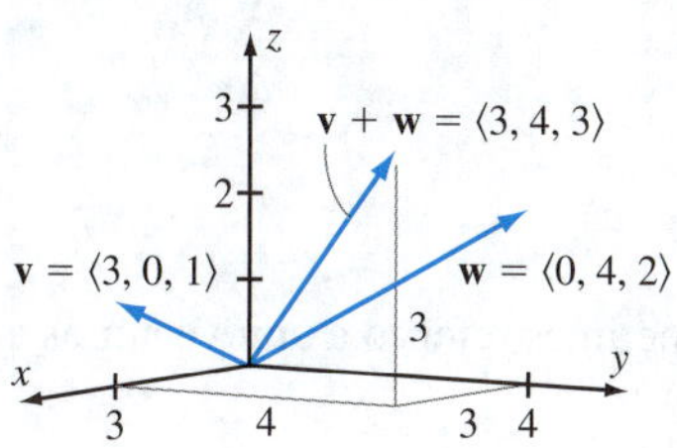

Figure 8

Scalar Multiplication. If $\mathbf{v} = \langle v_1, v_2, v_3\rangle$ is a vector and λ is a real number (that is, a *scalar*), then we define the *scalar multiplication* of $\mathbf{v}$ by λ to be

$$\lambda\mathbf{v} = \langle \lambda v_1, \lambda v_2, \lambda v_3\rangle.$$

Geometrically, we think of scalar multiplication as producing a vector that is parallel to $\mathbf{v}$ but with different length when $|\lambda| \neq 1$. An argument with similar triangles establishes that two nonzero vectors have *the same direction* if one is a *positive* scalar multiple of the other. They have *opposite directions* if one is a *negative* scalar multiple of the other. For now, these notions are intuitive generalizations of facts that we already know for vectors in the plane. They will be made more precise in the discussion that follows later in this section.

Example 7 Suppose that $\mathbf{v} = \langle 2, -1, 1\rangle$. Calculate $3\mathbf{v}$ and $-4\mathbf{v}$.

Solution We see that $3\mathbf{v} = \langle 6, -3, 3\rangle$ and $-4\mathbf{v} = \langle -8, 4, -4\rangle$. ■

Subtraction. If $\mathbf{v}$ and $\mathbf{w}$ are given vectors, then the expression $\mathbf{v} - \mathbf{w}$ is interpreted to mean $\mathbf{v} + ((-1)\mathbf{w})$. If $\mathbf{v} = \langle v_1, v_2, v_3\rangle$ and $\mathbf{w} = \langle w_1, w_2, w_3\rangle$, then

$$\mathbf{v} - \mathbf{w} = \mathbf{v} + (-1)\mathbf{w} = \langle v_1 - w_1, v_2 - w_2, v_3 - w_3\rangle.$$

Example 8 Suppose that $\mathbf{v} = \langle -6, 12, 5\rangle$ and $\mathbf{w} = \langle 19, -7, 4\rangle$. Calculate $\mathbf{v} - \mathbf{w}$ and $\mathbf{w} - 3\mathbf{v}$.

Solution We see that $\mathbf{v} - \mathbf{w} = \langle -6 - 19, 12 - (-7), 5 - 4\rangle = \langle -25, 19, 1\rangle$ and $\mathbf{w} - 3\mathbf{v} = \langle 19 - 3(-6), -7 - 3 \cdot 12, 4 - 3 \cdot 5\rangle = \langle 37, -43, -11\rangle$. ■

Vector Length

If $\mathbf{v} = \langle v_1, v_2, v_3\rangle$ is a vector, then its *length* or *magnitude* or *norm* is defined to be

$$\|\mathbf{v}\| = \sqrt{(v_1)^2 + (v_2)^2 + (v_3)^2}. \tag{11.9}$$

If $\overrightarrow{PQ}$ is a directed line segment that represents $\mathbf{v}$, then $\|\mathbf{v}\|$ is simply the distance between the points P and Q (as can be seen from formulas (11.6) and (11.8)). If $\mathbf{v}$ is a vector and λ is a scalar, then the magnitudes of vectors $\mathbf{v}$ and $\lambda\mathbf{v}$ are related by

$$\|\lambda\mathbf{v}\| = |\lambda|\|\mathbf{v}\|. \tag{11.10}$$

Thus, the length of $\lambda\mathbf{v}$ is $|\lambda|$ times the length of $\mathbf{v}$.

Example 9 Let O denote the origin. Suppose that $P = (-9, 6, -3)$ and $Q = (2, 1, 7)$. Calculate $\|\overrightarrow{PQ}\|$, $\|\overrightarrow{OP}\|$, and $\|\overrightarrow{OQ}\|$.

Solution Because $\overrightarrow{PQ} = \langle 11, -5, 10\rangle$, we have

$$\|\overrightarrow{PQ}\| = \sqrt{11^2 + (-5)^2 + 10^2} = \sqrt{246}.$$

Also, $\|\overrightarrow{OP}\| = \sqrt{(-9)^2 + 6^2 + (-3)^2} = \sqrt{126}$ and $\|\overrightarrow{OQ}\| = \sqrt{2^2 + 1^2 + 7^2} = \sqrt{54}$. ■

Unit Vectors and Directions

A vector with length 1 is called a *unit vector.* We sometimes refer to a unit vector as a *direction vector* or a *direction*.

Example 10 Is there a value of r for which $\mathbf{u} = \langle -1/3, 2/3, r\rangle$ is a unit vector? Is there a value of s for which $\mathbf{v} = \langle -4/5, 4/5, s\rangle$ is a unit vector?

Solution We set $\|\mathbf{u}\| = 1$ and attempt to solve for r. This gives us $(-1/3)^2 + (2/3)^2 + r^2 = 1$ or $r^2 = 1 - 1/9 - 4/9 = 4/9$. Thus, if $r = 2/3$ or $r = -2/3$ then $\mathbf{u}$ is a unit vector. On the other hand, $\|\mathbf{v}\| = \sqrt{(-4/5)^2 + (4/5)^2 + s^2} = \sqrt{32/25 + s^2} > \sqrt{1 + s^2}$. Because this inequality is strict, we see that $\|\mathbf{v}\| > 1$ for every value of s. Therefore, vector $\mathbf{v}$ cannot be a unit vector for any value of s. ■

If $\mathbf{v}$ is any nonzero vector, then we may form the vector

$$\text{dir}(\mathbf{v}) = \frac{1}{\|\mathbf{v}\|}\mathbf{v} \qquad (\mathbf{v} \neq 0),$$

which, according to equation (11.10), has length 1. We call this unit vector the *direction of* $\mathbf{v}$.

Definition Let $\mathbf{v}$ and $\mathbf{w}$ be nonzero vectors. We say that $\mathbf{v}$ and $\mathbf{w}$ are *opposite in direction* if $\text{dir}(\mathbf{v}) = -\text{dir}(\mathbf{w})$. We say that $\mathbf{v}$ and $\mathbf{w}$ are *parallel* if either (i) $\mathbf{v}$ and $\mathbf{w}$ have the same direction or (ii) $\mathbf{v}$ and $\mathbf{w}$ are opposite in direction. Although the zero vector $\vec{\mathbf{0}}$ does not have a direction, it is conventional to say that $\vec{\mathbf{0}}$ is parallel to every vector.

Notice that nonzero vectors $\mathbf{v}$ and $\mathbf{w}$ are parallel if and only if $\text{dir}(\mathbf{v}) = \pm\text{dir}(\mathbf{w})$. Our next theorem provides us with a simple algebraic condition for recognizing parallel vectors.

Theorem 2 Vectors $\mathbf{v}$ and $\mathbf{w}$ are parallel if and only if $\mathbf{v} = \lambda\mathbf{w}$ or $\mathbf{w} = \lambda\mathbf{v}$ for some scalar λ.

The proof of this theorem is the same as that of its planar analogue (Theorem 2 of Section 11.1).

Example 11 Suppose that $\mathbf{v} = \langle 4, 3, 1\rangle$ and $\mathbf{w} = \langle 2, b, c\rangle$. Are there values of b and c for which $\mathbf{v}$ and $\mathbf{w}$ are parallel?

Solution If $\mathbf{v} = \lambda\mathbf{w}$ for some scalar λ, then $\langle 4, 3, 1\rangle = \lambda\langle 2, b, c\rangle = \langle 2\lambda, \lambda b, \lambda c\rangle$. It follows that $4 = 2\lambda$ or $\lambda = 2$. Therefore, $3 = \lambda b = 2b$ and $1 = \lambda c = 2c$. We conclude that $\mathbf{v} = \langle 4, 3, 1\rangle$ and $\mathbf{w} = \langle 2, b, c\rangle$ are parallel if and only if $b = 3/2$ and $c = 1/2$. ■

The Special Unit Vectors i, j, and k

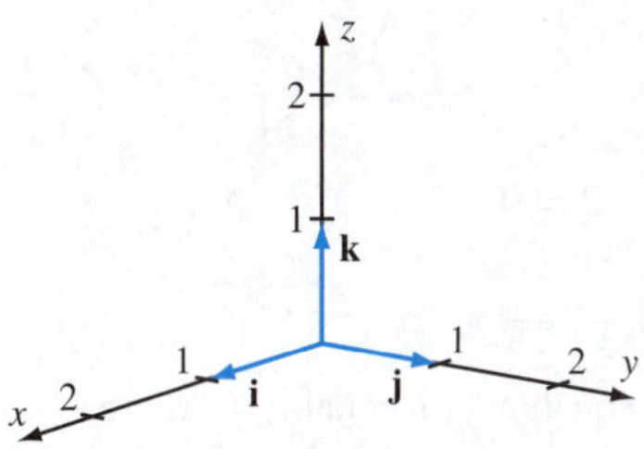

Figure 9

Let $\mathbf{i}$ denote the vector $\langle 1, 0, 0\rangle$, $\mathbf{j}$ the vector $\langle 0, 1, 0\rangle$, and $\mathbf{k}$ the vector $\langle 0, 0, 1\rangle$. These are the unit vectors that give the positive directions of the coordinate axes (see Figure 9). If $\mathbf{v} = \langle v_1, v_2, v_3\rangle$ is any vector, then we may express $\mathbf{v}$ in terms of $\mathbf{i}$, $\mathbf{j}$, and $\mathbf{k}$ as follows: $\mathbf{v} = v_1\mathbf{i} + v_2\mathbf{j} + v_3\mathbf{k}$.

Example 12 For what values of b and c are the vectors $\mathbf{v} = 9\mathbf{i} + b\mathbf{j} - \mathbf{k}$ and $\mathbf{w} = -3\mathbf{i} + 2\mathbf{j} + c\mathbf{k}$ opposite in direction?

Solution The vectors $\mathbf{v} = 9\mathbf{i} + b\mathbf{j} - \mathbf{k}$ and $\mathbf{w} = -3\mathbf{i} + 2\mathbf{j} + c\mathbf{k}$ are opposite in direction if and only if $\mathbf{v} = \lambda\mathbf{w}$ for some *negative* scalar λ. The vector equation $\mathbf{v} = \lambda\mathbf{w}$ is equivalent to the three simultaneous scalar equations $9 = -3\lambda$, $b = 2\lambda$, and $-1 = c\lambda$. The first scalar equation tells us that $\lambda = -3$, which indeed is negative. The second and third scalar equations give us $b = 2(-3) = -6$ and $c = -1/(-3) = 1/3$. Substituting these values for b and c in the formulas for $\mathbf{v}$ and $\mathbf{w}$, we obtain the vectors $9\mathbf{i} - 6\mathbf{j} - \mathbf{k}$ and $-3\mathbf{i} + 2\mathbf{j} + (1/3)\mathbf{k}$, which are opposite in direction. ■

Relations among Addition, Length, and Scalar Product

We conclude this section by summarizing the key properties of vectors in space. If **u**, **v**, **w** are vectors and λ, μ are scalars, then

a. $\mathbf{v} + \mathbf{w} = \mathbf{w} + \mathbf{v}$,
b. $\mathbf{u} + (\mathbf{v} + \mathbf{w}) = (\mathbf{u} + \mathbf{v}) + \mathbf{w}$,
c. $\lambda(\mu\mathbf{v}) = (\lambda\mu)\mathbf{v}$,
d. $\lambda(\mathbf{v} + \mathbf{w}) = \lambda\mathbf{v} + \lambda\mathbf{w}$,
e. $(\lambda + \mu)\mathbf{v} = \lambda\mathbf{v} + \mu\mathbf{v}$,
f. $\|\lambda\mathbf{v}\| = |\lambda|\,\|\mathbf{v}\|$,
g. $\|\mathbf{v} + \mathbf{w}\| \leq \|\mathbf{v}\| + \|\mathbf{w}\|$.

Equations a–f may be routinely verified. Inequality g is the Triangle Inequality. The proof in Section 11.1 for vectors in the plane applies to vectors in space as well.

quickquiz

1. How is a point located analytically in three-dimensional space?
2. How is distance calculated in three-dimensional space?
3. How does a directed line segment represent a vector in three-dimensional space?
4. What is the significance of the vectors **i**, **j**, **k**?

EXERCISES

Problems for Practice

In Exercises 1–4, graph the given point P. Also, graph the point that lies directly above or below P in the xy-plane.

1. $P = (3, 2, 1)$
2. $P = (2, 0, 5)$
3. $P = (-4, 1, -2)$
4. $P = (3, -1, 2)$

In Exercises 5–8, calculate the distance between the pair of points:

5. $(2, 0, -3), (8, 3, 6)$
6. $(-2, -3, -5), (4, 2, 0)$
7. $(4, 4, 1), (1, 1, 4)$
8. $(-4, 2, 1), (8, 0, 4)$

In Exercises 9–14, determine the center and radius of each sphere with the given Cartesian equation.

9. $x^2 + y^2 + z^2 + 2x + 4y + 6z = 8$
10. $3x^2 + 3y^2 + 3z^2 - 12x + 6y = 0$
11. $4x^2 + 4y^2 + 4z^2 + 4x + 4y + 4z = 2$
12. $-x^2 - y^2 - z^2 = 6x - 8y + 12z - 4$
13. $x^2 + y^2 + z^2 = x$
14. $-4x^2 - 4y^2 - 4z^2 = -2x - 2y - 2z - 6$

In Exercises 15–20, explain how to recognize that the given Cartesian equation is not the equation of a sphere.

15. $x^2 + y^2 + 2z^2 + 2x + 4z = 8$
16. $2x^2 - 2y^2 + 2z^2 = 1$
17. $x^2 + y^2 + z^2 + 4xy = 2$
18. $x^2 + y^2 + z^2 + 2z + 2 = 0$
19. $(x - y)^2 + z^2 = 1$
20. $x^2 + y^2 = 2x + 4y - 7 - z^2$

In Exercises 21–26, find an equation or inequality $P(x, y, z)$ such that the given set is described by $\{(x, y, z) : P(x, y, z)\}$. For example, the origin is the set $\{(x, y, z) : x^2 + y^2 + z^2 = 0\}$ and the xy-plane is the set $\{(x, y, z) : z = 0\}$.

21. The open ball with center $(3, -2, 6)$ and radius 4
22. The closed ball with center $(1, 2, 9)$ and radius 1

23. The set of points with distance to the point $(\pi, \pi, -\pi)$ of π

24. The set of points with a distance to the point $(1, -2, 0)$ that is not greater than 3

25. The set of points *outside* the closed ball with center $(2, 1, 0)$ and radius 2

26. The set of points *outside* the open ball with center $(-3, -1, -5)$ and radius 5

In Exercises 27–30, sketch the directed line segments $\overrightarrow{PQ}$, $\overrightarrow{OQ}$, and $\overrightarrow{OP}$. Express the vectors represented by these directed line segments in the form $\langle a, b, c\rangle$. Verify that the sum of vectors $\overrightarrow{OP}$ and $\overrightarrow{PQ}$ is equal to vector $\overrightarrow{OQ}$.

27. $P = (1, 4, -2)$, $Q = (2, 1, 6)$

28. $P = (1, 1, 1)$, $Q = (2, 3, 4)$

29. $P = (-1, 6, 1)$, $Q = (2, -1, 0)$

30. $P = (1, 0, 0)$, $Q = (0, 0, 3)$

In Exercises 31–34, calculate the lengths of vectors $\overrightarrow{PQ}$ and $\overrightarrow{RS}$. Also, determine if the two vectors are parallel.

31. $P = (1, 2, -1)$, $Q = (2, 0, -7)$, $R = (2, 6, 9)$, $S = (4, 2, -3)$

32. $P = (-1, 3, 6)$, $Q = (4, 1, 5)$, $R = (1, 1, 0)$, $S = (1, 3, 5)$

33. $P = (0, 1, 4)$, $Q = (-1, 3, 6)$, $R = (6, -3, 0)$, $S = (1, 1, 2)$

34. $P = (1, 1, 2)$, $Q = (-7, 5, 4)$, $R = (1, 0, 6)$, $S = (19, -12, 0)$

In Exercises 35–40, let $\mathbf{v} = \langle 7, 1, 3\rangle$ and $\mathbf{w} = \langle 4, -1, -7\rangle$. Calculate the specified vector.

35. $-6\mathbf{w}$

36. $2\mathbf{v} - 4\mathbf{w}$

37. $-5\mathbf{v} + \mathbf{w}$

38. $(1/2)\mathbf{w} - (1/3)\mathbf{v}$

39. $\mathbf{v} + 3\mathbf{k}$

40. $\mathbf{w} - 4\mathbf{i} + 7\mathbf{k}$

In Exercises 41–44 state the following:

a. The vector represented by $\overrightarrow{PQ}$

b. The vector that is parallel to $\overrightarrow{PQ}$ and that has the same length but points in the opposite direction

c. The length of $\overrightarrow{PQ}$

d. A unit vector pointing in the same direction as $\overrightarrow{PQ}$

41. $P = (1, 2, -1)$, $Q = (2, 0, -3)$

42. $P = (-2, 2, 3)$, $Q = (2, -2, 5)$

43. $P = (3, 1, 3)$, $Q = (-5, 0, -1)$

44. $P = (9, 8, -1)$, $Q = (6, 10, 5)$

In Exercises 45–50, let $\mathbf{v} = \langle 3, -4, 1\rangle$ and $\mathbf{w} = \langle -5, 2, 0\rangle$. Express the given vector in the form $a\mathbf{i} + b\mathbf{j} + c\mathbf{k}$.

45. $-5\mathbf{v}$

46. $4\mathbf{w}$

47. $\mathbf{w} + 2\mathbf{v}$

48. $2\mathbf{v} - \mathbf{w}$

49. $\mathbf{v} - 3\mathbf{i}$

50. $2\mathbf{w} + 4\mathbf{j}$

Further Theory and Practice

51. Let $P_0 = (x_0, y_0, z_0)$ be a point in space. Which point in the xy-plane is nearest to P_0? Which point in the yz-plane is nearest to P_0? How about the xz-plane? Give *reasons* for your answers.

52. Fix two distinct points Q_1 and Q_2 in space. Let S be the set of points $P = (x, y, z)$ in space such that the distance of P to Q_1 equals the distance of P to Q_2. What familiar geometric object is S? Explain why.

In Exercises 53–58, find an equation or inequality $P(x, y, z)$ such that the given set is described by $\{(x, y, z) : P(x, y, z)\}$. For example, the origin is the set $\{(x, y, z) : x^2 + y^2 + z^2 = 0\}$ and the xy-plane is the set $\{(x, y, z) : z = 0\}$.

53. The yz-coordinate plane

54. The plane that is parallel to the xy-plane and that passes through the point (π, π^2, π^3)

55. The open half-space that contains all points that lie on the same side of the xz-coordinate plane as the point $(1, 1, 1)$

56. The half-space that comprises the yz-plane as well as all points that are on the same side of the yz-plane as the point $(1, 1, 1)$

57. The set of points with a distance to the xy-coordinate plane that is greater than 5

58. The set of points that lie in one or more of the three coordinate planes (in other words, the union of the three coordinate planes)

59. Give a geometric description of the set of all vectors in space that have the form $a\mathbf{i} + b\mathbf{k}$ where a and b are real numbers.

60. Give a geometric description of the set of all vectors of the form

$$\langle \cos(\theta)\sin(\phi), \sin(\theta)\sin(\phi), \cos(\phi)\rangle \quad 0 \le \theta, \phi \le 2\pi.$$

61. Consider the vectors $\mathbf{u} = \langle 1, 3, 4 \rangle$, $\mathbf{v} = \langle -2, 1, 6 \rangle$, and $\mathbf{w} = \langle 0, 1, 5 \rangle$. If $\mathbf{m}$ is any other vector in space, then show that there is a unique triple of numbers a, b, c such that

$$\mathbf{m} = a\mathbf{u} + b\mathbf{v} + c\mathbf{w}.$$

62. A student walking at the rate of 4 ft/s crosses a 12 ft high pedestrian bridge. A car passes directly underneath traveling at constant speed 60 ft/s. How fast is the distance between the student and the car changing 2 s later?

63. What are the lengths of the diagonals of the parallelogram determined by the vectors $\langle 1, 1, 0 \rangle$ and $\langle 0, 1, 2 \rangle$?

64. Suppose that P, Q, and R are three noncollinear points. Show that there are three points A, B, and C, any one of which, together with P, Q, and R, forms a parallelogram. Show that

$$\overrightarrow{OA} + \overrightarrow{OB} + \overrightarrow{OC} = \overrightarrow{OP} + \overrightarrow{OQ} + \overrightarrow{OR}.$$

65. On which sphere do the four points $(5, 2, 3)$, $(1, 6, -1)$, $(3, -2, 5)$, and $(-1, 2, -3)$ lie?

66. For which vectors $\mathbf{v}$ and $\mathbf{w}$ does equality hold in the Triangle Inequality? That is, when is it true that

$$\|\mathbf{v} + \mathbf{w}\| = \|\mathbf{v}\| + \|\mathbf{w}\|?$$

67. Atomic particles may carry a positive charge (like a proton does) or a negative charge (like an electron does). A charged particle exerts a force on every other charged particle. Suppose that P is the location of a charge p and Q is the location of a charge q. Let r be the distance between P and Q. Coulomb's law states that the electric force exerted by the particle at P on the particle at Q is given by

$$\mathbf{F} = \frac{pq}{4\pi\epsilon_0 r^2}\mathbf{u}$$

where $\mathbf{u}$ is the unit vector in the direction of $\overrightarrow{PQ}$ and ϵ_0 is a positive constant, the numerical value of which depends on the units of measurement. Discuss the physical meaning of the sign of the coefficient of $\mathbf{u}$ in this equation. In particular, explain how the formula for force captures the principle "Like charges repel and opposite charges attract."

68. The magnitude of the gravitational attraction between two bodies with masses m and M is $GmMr^{-2}$ where r is the distance between the two masses and $G = 6.67 \times 10^{-11}$ N m^2 kg^{-2} is Newton's gravitational constant. The masses of the sun, Earth, and the moon are 1.99×10^{30} kg, 5.97×10^{24} kg, and 7.35×10^{22} kg, respectively. At the moment these three bodies are aligned as in Figure 10, which is not to scale, Earth is 3.84×10^5 km from the moon and 1.49×10^8 km from the sun. Calculate the vector that represents the resultant force on the moon.

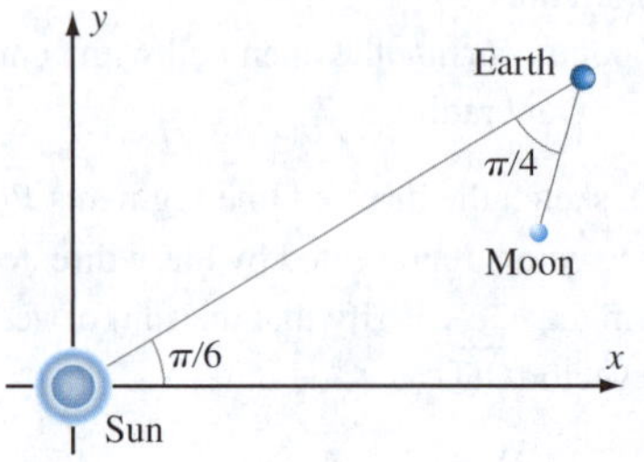

Figure 10

69. Describe the intersections of the surface $x^2 + y^2/9 + z^2/16 = 1$ with the coordinate planes. Sketch the surface.

70. Describe the intersections of the surface $z = x^2 + y^2/4$ with the coordinate planes and with the planes $z = h$ where h is a constant. Sketch the surface.

71. Describe the intersections of the surface $z = \sqrt{x^2 + y^2/4}$ with the coordinate planes and with the planes $z = h$ where h is a constant. Sketch the surface.

72. Suppose that $\mathbf{u}_1, \mathbf{u}_2, \ldots, \mathbf{u}_N$ are vectors and $\lambda_1, \lambda_2, \ldots, \lambda_N$ are scalars. The vector $\mathbf{v} = \lambda_1\mathbf{u}_1 + \lambda_2\mathbf{u}_2 + \ldots + \lambda_N\mathbf{u}_N$ is said to be a *linear combination* of the vectors $\mathbf{u}_1, \mathbf{u}_2, \ldots, \mathbf{u}_N$. If every vector in space is a linear combination of $\mathbf{u}_1, \mathbf{u}_2, \ldots, \mathbf{u}_N$, then we say that the vectors $\mathbf{u}_1, \mathbf{u}_2, \ldots, \mathbf{u}_N$ *span* or *generate* $\mathbb{R}^3$. The vectors $\mathbf{i}, \mathbf{j}, \mathbf{k}$, for example, span $\mathbb{R}^3$. Determine which of the following sets of three vectors also have this property.

a. $\{\mathbf{i}+\mathbf{j}, \mathbf{j}+\mathbf{k}, \mathbf{i}+\mathbf{k}\}$
b. $\{\mathbf{i}+\mathbf{j}+\mathbf{k}, \mathbf{i}+\mathbf{j}, \mathbf{i}-\mathbf{j}\}$
c. $\{\mathbf{i}+\mathbf{j}+\mathbf{k}, \mathbf{i}+\mathbf{j}, -\mathbf{i}-\mathbf{j}+2\mathbf{k}\}$
d. $\{\mathbf{i}+2\mathbf{j}+3\mathbf{k}, \mathbf{i}-2\mathbf{j}+3\mathbf{k}, \mathbf{i}-3\mathbf{k}\}$

Calculator/Computer Exercises

73. If $a = b = c > 0$, then the plot of $ax^2 + by^2 + cz^2 = 1$ is a sphere. Plot $ax^2 + by^2 + cz^2 = 1$ for several choices of positive a, b, and c that are not all equal. Describe the intersections of the resulting surfaces with the coordinate planes.

74. Plot $z = ax^2 + by^2$ for several choices of positive a and b. Describe the intersections of the resulting surfaces with the coordinate planes. Describe the intersections of the resulting surfaces with the planes $z = h$ for positive constants h.

75. Plot $ax^2 + by^2 - z^2 = 1$ for several choices of positive a and b. Describe the intersections of the resulting surfaces with the coordinate planes. Describe the intersections of the resulting surfaces with the planes $z = h$ for constants h.

76. At time $t \geq 0$, a particle's position P_t is the point $(t, t\cos(t), t\sin(t))$. Does the vector $\overrightarrow{P_t P_{t+1}}$ ever have length 2?

11.3 The Dot Product and Applications

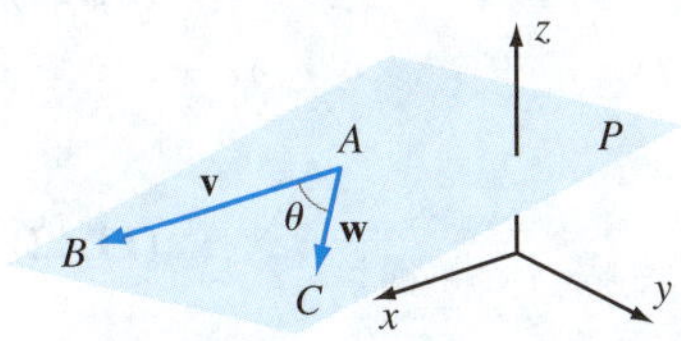

Figure 1

If two nonzero vectors **v** and **w** have the same direction, then the angle between them is 0. If the two vectors are in opposite directions, then the angle between them is π. In all other cases, we may use the two vectors to form a plane P by selecting directed line segment representatives $\overrightarrow{AB}$ and $\overrightarrow{AC}$ that have the same initial point. See Figure 1. The angle between **v** and **w** is understood to be the angle $\theta \in (0, \pi)$ between $\overrightarrow{AB}$ and $\overrightarrow{AC}$ in plane P. The angle does not depend on the initial point A that is selected. In this section, we study an algebraic operation that can be used to calculate the angle between two vectors.

Definition The *dot product* of two vectors $\mathbf{v} = \langle v_1, v_2, v_3 \rangle$ and $\mathbf{w} = \langle w_1, w_2, w_3 \rangle$ is denoted by $\mathbf{v} \cdot \mathbf{w}$ and defined by

$$\mathbf{v} \cdot \mathbf{w} = v_1 w_1 + v_2 w_2 + v_3 w_3.$$

inSIGHT

> When appropriate, we will continue to use the multiplication dot to signify the product of two scalars. For example, when we multiply the scalars c and $x - y$, the notation $c \cdot (x - y)$ clearly indicates a product, whereas $c(x - y)$ may be confused with functional evaluation. In this book we use bold type, such as **v**, or arrow notation, such as $\overrightarrow{PQ}$, to represent vectors. It will therefore always be easy to infer the meaning of the multiplication dot from its context.

Some texts refer to the dot product as the *inner product*. There is a third name for the dot product, one that requires some caution. Notice that the dot product of two vectors is a *scalar*, or real number. For that reason, $\mathbf{v} \cdot \mathbf{w}$ is often called the *scalar product* of **v** and **w**. This terminology should not be confused with the operation of *scalar multiplication* that has been considered in the preceding two sections. *Scalar multiplication* refers to the multiplication of one vector by a scalar. The result of scalar multiplication is a vector. By contrast the scalar product is an operation between two vectors. The result is a scalar. In this book, we always say "dot product" when we refer to $\mathbf{v} \cdot \mathbf{w}$.

Example 1 Let $\mathbf{u} = \langle 2, 3, -1 \rangle$, $\mathbf{v} = \langle 4, 6, -2 \rangle$, and $\mathbf{w} = \langle -2, -1, -7 \rangle$. Calculate $\mathbf{u} \cdot \mathbf{v}$, $\mathbf{u} \cdot \mathbf{w}$, and $\mathbf{v} \cdot \mathbf{w}$.

Solution We have

$$\begin{aligned} \mathbf{u} \cdot \mathbf{v} &= (2)(4) + (3)(6) + (-1)(-2) = 28 \\ \mathbf{u} \cdot \mathbf{w} &= (2)(-2) + (3)(-1) + (-1)(-7) = 0 \\ \mathbf{v} \cdot \mathbf{w} &= (4)(-2) + (6)(-1) + (-2)(-7) = 0. \end{aligned}$$

■

Algebraic Rules for the Dot Product

In the next theorem, we gather several simple rules for working with the dot product.

Theorem 1 Suppose that **u**, **v**, and **w** are vectors. The dot product satisfies the following elementary properties:

a. $\mathbf{u}\cdot(\mathbf{v}+\mathbf{w}) = \mathbf{u}\cdot\mathbf{v}+\mathbf{u}\cdot\mathbf{w}$;
b. $\mathbf{v}\cdot\mathbf{w} = \mathbf{w}\cdot\mathbf{v}$; and
c. If λ is a scalar, then $(\lambda\mathbf{v})\cdot\mathbf{w} = \mathbf{v}\cdot(\lambda\mathbf{w}) = \lambda(\mathbf{v}\cdot\mathbf{w})$.

A Geometric Formula for the Dot Product

Since $\mathbf{v}\cdot\mathbf{v} = (v_1)^2+(v_2)^2+(v_3)^2$, we can relate the dot product $\mathbf{v}\cdot\mathbf{v}$ to the length of **v** by rewriting equation (11.9) from Section 11.2:

$$\|\mathbf{v}\| = \sqrt{\mathbf{v}\cdot\mathbf{v}} \qquad \textbf{(11.11)}$$

or

$$\|\mathbf{v}\|^2 = \mathbf{v}\cdot\mathbf{v}. \qquad \textbf{(11.12)}$$

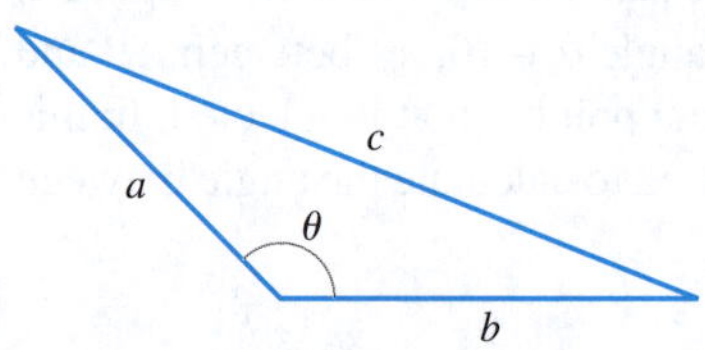

Figure 2
Law of Cosines:
$c^2 = a^2+b^2-2ab\cos(\theta)$

We use this formula to develop a method for calculating the angle between two vectors and for calculating projections. To that end, recall that the Law of Cosines says that if the sides of a triangle measure a, b, c and if θ is the angle between the first two sides, as in Figure 2, then

$$c^2 = a^2+b^2-2ab\cos(\theta). \qquad \textbf{(11.13)}$$

Let us apply the Law of Cosines to the triangle determined by **v**, **w**, and $\mathbf{v}-\mathbf{w}$ in Figure 3. It tells us that

$$\|\mathbf{v}-\mathbf{w}\|^2 = \|\mathbf{v}\|^2+\|\mathbf{w}\|^2-2\|\mathbf{v}\|\|\mathbf{w}\|\cos(\theta). \qquad \textbf{(11.14)}$$

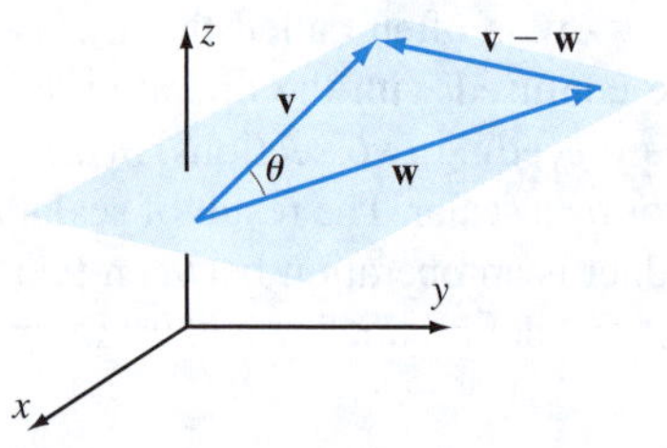

Figure 3

Using equation (11.12) and the rules of Theorem 1, we may expand the left side of equation (11.14) to obtain

$$\|\mathbf{v}-\mathbf{w}\|^2 = (\mathbf{v}-\mathbf{w})\cdot(\mathbf{v}-\mathbf{w}) = \mathbf{v}\cdot\mathbf{v}-\mathbf{v}\cdot\mathbf{w}-\mathbf{w}\cdot\mathbf{v}+\mathbf{w}\cdot\mathbf{w} = \|\mathbf{v}\|^2-2\mathbf{v}\cdot\mathbf{w}+\|\mathbf{w}\|^2.$$

Substituting this into equation (11.14) gives

$$\|\mathbf{v}\|^2-2\mathbf{v}\cdot\mathbf{w}+\|\mathbf{w}\|^2 = \|\mathbf{v}\|^2+\|\mathbf{w}\|^2-2\|\mathbf{v}\|\|\mathbf{w}\|\cos(\theta).$$

Cancelling like terms from either side results in $-2\mathbf{v}\cdot\mathbf{w} = -2\|\mathbf{v}\|\cdot\|\mathbf{w}\|\cdot\cos(\theta)$, or

$$\mathbf{v}\cdot\mathbf{w} = \|\mathbf{v}\|\|\mathbf{w}\|\cos(\theta). \qquad \textbf{(11.15)}$$

This result is summarized in our next theorem.

Theorem 2 If $\mathbf{v}$ and $\mathbf{w}$ are nonzero vectors, then the angle θ between $\mathbf{v}$ and $\mathbf{w}$ satisfies the equation

$$\cos(\theta) = \frac{\mathbf{v} \cdot \mathbf{w}}{\|\mathbf{v}\|\|\mathbf{w}\|}. \tag{11.16}$$

Example 2 Calculate the angle between the two vectors $\mathbf{v} = \langle 2, 2, 4 \rangle$ and $\mathbf{w} = \langle 2, -1, 1 \rangle$.

Solution If θ is the angle between $\mathbf{v}$ and $\mathbf{w}$, then Theorem 2 tells us that

$$\begin{aligned} \cos(\theta) &= \frac{\mathbf{v} \cdot \mathbf{w}}{\|\mathbf{v}\|\|\mathbf{w}\|} \\ &= \frac{\langle 2, 2, 4 \rangle \cdot \langle 2, -1, 1 \rangle}{\|\langle 2, 2, 4 \rangle\|\|\langle 2, -1, 1 \rangle\|} \\ &= \frac{2 \cdot 2 + 2 \cdot (-1) + 4 \cdot 1}{\sqrt{2^2 + 2^2 + 4^2} \cdot \sqrt{2^2 + (-1)^2 + 1^2}} \\ &= \frac{6}{\sqrt{24} \cdot \sqrt{6}} \\ &= \frac{1}{2}. \end{aligned}$$

Therefore, θ is $\pi/3$. ■

Equation (11.15) gives rise to an important inequality. Since $|\cos(\theta)| \leq 1$ for every angle θ we have

$$|\mathbf{v} \cdot \mathbf{w}| \leq \|\mathbf{v}\|\|\mathbf{w}\|. \tag{11.17}$$

This relationship is called the *Cauchy-Schwarz Inequality*.

Example 3 Verify that the two vectors $\mathbf{v} = \langle 2, 2, 4 \rangle$ and $\mathbf{w} = \langle 12, 13, 24 \rangle$ satisfy the Cauchy-Schwarz Inequality.

Solution We calculate $\mathbf{v} \cdot \mathbf{w} = (2)(12) + (2)(13) + (4)(24) = 146$ and

$$\|\mathbf{v}\|\|\mathbf{w}\| = \sqrt{2^2 + 2^2 + 4^2}\sqrt{12^2 + 13^2 + 24^2} = \left(2\sqrt{6}\right)\left(\sqrt{889}\right) = 146.068\ldots,$$

which is greater than $|\mathbf{v} \cdot \mathbf{w}|$, as the Cauchy-Schwarz Inequality asserts. ■

Definition Let θ be the angle between nonzero vectors $\mathbf{v}$ and $\mathbf{w}$. If $\theta = \pi/2$, then we say that the vectors $\mathbf{v}$ and $\mathbf{w}$ are *orthogonal* or *mutually perpendicular*. Although the zero vector $\vec{\mathbf{0}}$ has no direction, it is conventional to say that the zero vector $\vec{\mathbf{0}}$ is perpendicular to every vector.

Theorem 3 If we let $\mathbf{v}$ and $\mathbf{w}$ be any vectors, then

a. The vectors $\mathbf{v}$ and $\mathbf{w}$ are mutually perpendicular if and only if $\mathbf{v} \cdot \mathbf{w} = 0$.
b. The vectors $\mathbf{v}$ and $\mathbf{w}$ are parallel if and only if $|\mathbf{v} \cdot \mathbf{w}| = \|\mathbf{v}\|\|\mathbf{w}\|$.

Proof Statement a follows immediately from equation (11.15) because $\cos(\pi/2) = 0$. To prove statement b, recall that vectors $\mathbf{v}$ and $\mathbf{w}$ are parallel if and only if $\mathbf{v} = \lambda\mathbf{w}$ or $\mathbf{w} = \lambda\mathbf{v}$ for some scalar λ (Theorem 2, Section 11.2). Thus, statement b asserts that equality holds in the Cauchy-Schwarz Inequality—that is, $|\mathbf{v}\cdot\mathbf{w}| = \|\mathbf{v}\|\|\mathbf{w}\|$—if and only if $\mathbf{v} = \lambda\mathbf{w}$ or $\mathbf{w} = \lambda\mathbf{v}$ for some scalar λ. One direction of this equivalence reduces to a simple calculation: If, for example, we have $\mathbf{w} = \lambda\mathbf{v}$, then

$$|\mathbf{v}\cdot\mathbf{w}| = |\mathbf{v}\cdot(\lambda\mathbf{v})| = |\lambda||\mathbf{v}\cdot\mathbf{v}| = |\lambda|\|\mathbf{v}\|^2 = \|\mathbf{v}\|(|\lambda|\|\mathbf{v}\|) = \|\mathbf{v}\|\|\lambda\mathbf{v}\| = \|\mathbf{v}\|\|\mathbf{w}\|.$$

Exercise 60 outlines a proof of the converse. ■

inSIGHT

Let θ denote the angle between two nonzero vectors $\mathbf{v}$ and $\mathbf{w}$. The Cauchy-Schwarz Inequality tells us that the dot product $|\mathbf{v}\cdot\mathbf{w}|$ lies in the range $0 \le |\mathbf{v}\cdot\mathbf{w}| \le \|\mathbf{v}\|\|\mathbf{w}\|$. Theorem 3 tells us that each extreme value of $|\mathbf{v}\cdot\mathbf{w}|$ has a geometric interpretation: $|\mathbf{v}\cdot\mathbf{w}| = 0$ corresponds to perpendicular vectors and $|\mathbf{v}\cdot\mathbf{w}| = \|\mathbf{v}\|\|\mathbf{w}\|$ corresponds to parallel vectors. In the latter case, we can be more specific. If $\mathbf{v}\cdot\mathbf{w} = \|\mathbf{v}\|\|\mathbf{w}\|$, then $\cos(\theta) = 1$ by equation (11.15). In this case, $\theta = 0$ and $\mathbf{v}$ and $\mathbf{w}$ have the same direction. If $\mathbf{v}\cdot\mathbf{w} = -\|\mathbf{v}\|\,\|\mathbf{w}\|$, then $\cos(\theta) = -1$ by equation (11.15). In this case, $\theta = \pi$ and $\mathbf{v}$ and $\mathbf{w}$ are in opposite directions.

Example 4 Consider the vectors $\mathbf{u} = \langle 2, 3, -1\rangle$, $\mathbf{v} = \langle 4, 6, -2\rangle$, and $\mathbf{w} = \langle -2, -1, -7\rangle$ from Example 1. Are any of these vectors mutually perpendicular? Parallel?

Solution From the dot products $\mathbf{u}\cdot\mathbf{w} = 0$ and $\mathbf{v}\cdot\mathbf{w} = 0$ that we calculated in Example 1, we conclude that $\mathbf{u}$ and $\mathbf{v}$ *are* perpendicular to $\mathbf{w}$. Since $\mathbf{u}\cdot\mathbf{v} = 28 \ne 0$, we can tell that $\mathbf{u}$ is *not* perpendicular to $\mathbf{v}$. Indeed, the equation

$$\|\mathbf{u}\|\|\mathbf{v}\| = \sqrt{2^2+3^2+(-1)^2}\sqrt{4^2+6^2+(-2)^2} = \sqrt{14\cdot 56} = \sqrt{784} = 28 = |\mathbf{u}\cdot\mathbf{v}|$$

tells that $\mathbf{u}$ and $\mathbf{v}$ are parallel (by Theorem 3b). Since $\|\mathbf{u}\|\|\mathbf{v}\|$ actually equals $+\mathbf{u}\cdot\mathbf{v}$, we may further observe that $\mathbf{u}$ and $\mathbf{v}$ have the same direction. Noticing that $\mathbf{v} = 2\mathbf{u}$ is another way to reach this conclusion. ■

Projection

One of the most powerful constructions in geometry is the *projection*. Figure 4a (next page) shows two nonzero vectors $\mathbf{v}$ and $\mathbf{w}$, represented by directed line segments that share the same initial point. The projection $\mathbf{P}_{\mathbf{w}}(\mathbf{v})$ of $\mathbf{v}$ onto $\mathbf{w}$ is shown in Figure 4b. The main thing to notice in Figure 4c is that the projection of $\mathbf{v}$ onto $\mathbf{w}$ is parallel to $\mathbf{w}$ and determines, along with $\mathbf{v}$ and $\mathbf{Q}$, a right triangle. Figure 5 (next page) shows two other projections that will aid your geometric understanding of the concept. The next theorem provides us with an analytic expression for $\mathbf{P}_{\mathbf{w}}(\mathbf{v})$.

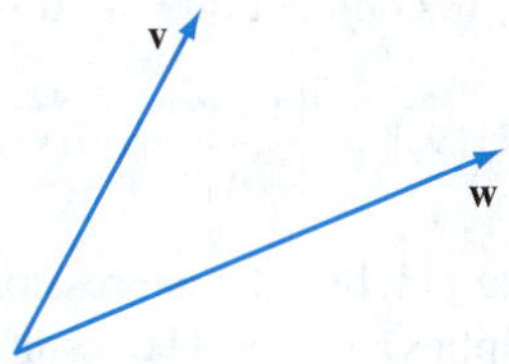

Figure 4a

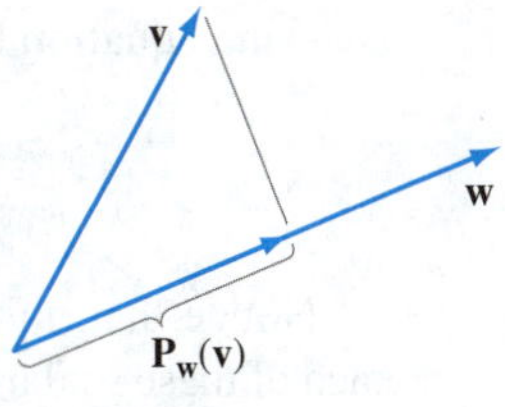

Figure 4b

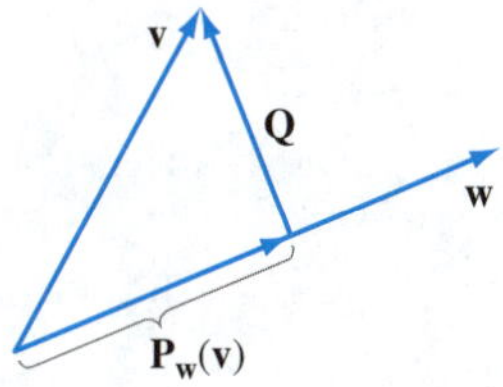

Figure 4c

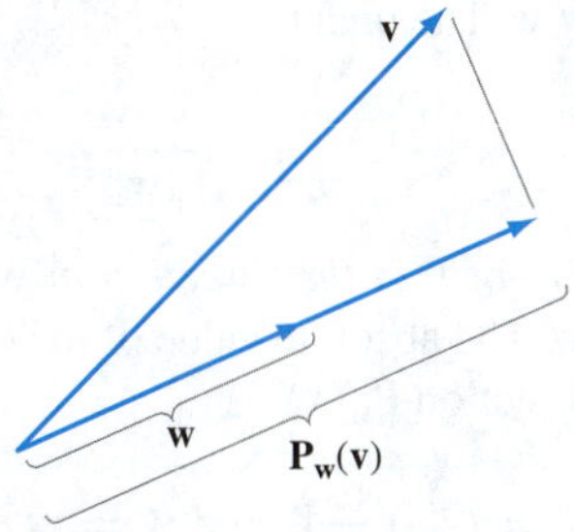

Figure 5a

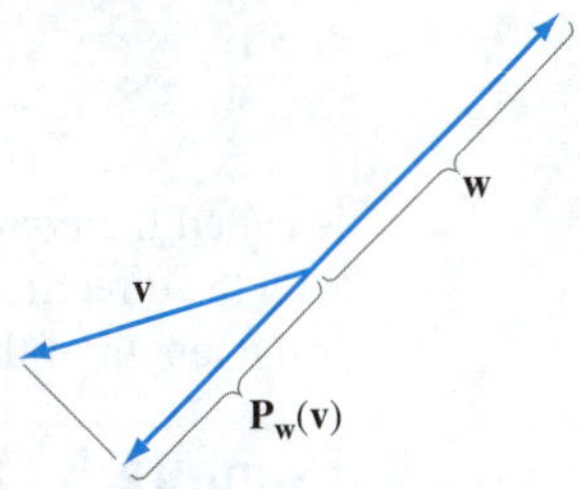

Figure 5b

Theorem 4 If $\mathbf{v}$ and $\mathbf{w}$ are nonzero vectors, then the projection of $\mathbf{v}$ onto $\mathbf{w}$ is given by

$$\mathbf{P_w(v)} = \left(\frac{\mathbf{v}\cdot\mathbf{w}}{\mathbf{w}\cdot\mathbf{w}}\right)\mathbf{w} = \left(\frac{\mathbf{v}\cdot\mathbf{w}}{\|\mathbf{w}\|}\right)\operatorname{dir}(\mathbf{w}) \tag{11.18}$$

where $\operatorname{dir}(\mathbf{w})$ is the direction $(1/\|\mathbf{w}\|)\,\mathbf{w}$ of $\mathbf{w}$. The length of $\mathbf{P_w(v)}$ is given by

$$\|\mathbf{P_w(v)}\| = \frac{|\mathbf{v}\cdot\mathbf{w}|}{\|\mathbf{w}\|}. \tag{11.19}$$

Proof Let $\mathbf{Q}$ denote the vector orthogonal to $\mathbf{w}$ that is depicted in Figure 4c. We have

$$\mathbf{P_w(v)} + \mathbf{Q} = \mathbf{v}. \tag{11.20}$$

Because $\mathbf{P_w(v)}$ is parallel to $\mathbf{w}$, we may write

$$\mathbf{P_w(v)} = c\mathbf{w} \tag{11.21}$$

for some scalar c. Substituting this expression for $\mathbf{P_w(v)}$ in equation (11.20) results in the equation $c\mathbf{w} + \mathbf{Q} = \mathbf{v}$. Taking the dot product of each side of this equation with $\mathbf{w}$ and using the distributive law for the dot product gives $c\mathbf{w}\cdot\mathbf{w} + \mathbf{Q}\cdot\mathbf{w} = \mathbf{v}\cdot\mathbf{w}$, or $(\mathbf{w}\cdot\mathbf{w})c + 0 = \mathbf{v}\cdot\mathbf{w}$. Thus, $c = (\mathbf{w}\cdot\mathbf{w})^{-1}\mathbf{v}\cdot\mathbf{w}$. With this value for c, equation (11.21) becomes

$$\mathbf{P_w(v)} = \left(\frac{\mathbf{v}\cdot\mathbf{w}}{\mathbf{w}\cdot\mathbf{w}}\right)\mathbf{w} = \left(\frac{\mathbf{v}\cdot\mathbf{w}}{\|\mathbf{w}\|^2}\right)\mathbf{w} = \left(\frac{\mathbf{v}\cdot\mathbf{w}}{\|\mathbf{w}\|}\right)\frac{1}{\|\mathbf{w}\|}\mathbf{w} = \left(\frac{\mathbf{v}\cdot\mathbf{w}}{\|\mathbf{w}\|}\right)\operatorname{dir}(\mathbf{w}).$$

By using equation (11.10) in conjunction with this formula for $\mathbf{P_w(v)}$, we see that

$$\|\mathbf{P_w(v)}\| = \left|\frac{\mathbf{v}\cdot\mathbf{w}}{\|\mathbf{w}\|}\right| \|\mathrm{dir}(\mathbf{w})\| = \frac{|\mathbf{v}\cdot\mathbf{w}|}{\|\mathbf{w}\|}.$$

Notice that, in formula (11.18), the expressions $(\frac{\mathbf{v}\cdot\mathbf{w}}{\mathbf{w}\cdot\mathbf{w}})$ and $(\frac{\mathbf{v}\cdot\mathbf{w}}{\|\mathbf{w}\|})$ are *scalars,* and each of these scalars multiplies a *vector.* The result of the scalar multiplication, namely, the projection of $\mathbf{v}$ onto $\mathbf{w}$, is a *vector*. Because equation (11.20) decomposes $\mathbf{v}$ as a sum of two mutually perpendicular vectors, $\mathbf{P_w(v)}$ is also called the *orthogonal projection of* $\mathbf{v}$ *in the direction of* $\mathbf{w}$. The quantity

$$\frac{\mathbf{v}\cdot\mathbf{w}}{\|\mathbf{w}\|}$$

is called the *component of* $\mathbf{v}$ *in the direction of* $\mathbf{w}$. It is equal to the dot product of $\mathbf{v}$ with the direction of $\mathbf{w}$. The absolute value of the component of $\mathbf{v}$ in the direction of $\mathbf{w}$ is the length of the projection $\mathbf{P_w(v)}$.

Example 5 Let $\mathbf{v} = \langle 2, 0, -4\rangle$ and $\mathbf{w} = \langle 1, -1, 2\rangle$. Calculate the projection of $\mathbf{v}$ onto $\mathbf{w}$, the projection of $\mathbf{w}$ onto $\mathbf{v}$, and calculate the lengths of these projections. Also, calculate the component of $\mathbf{v}$ in the direction of $\mathbf{w}$ and the component of $\mathbf{w}$ in the direction of $\mathbf{v}$.

Solution The projection of $\mathbf{v}$ onto $\mathbf{w}$ is

$$\begin{aligned}\mathbf{P_w(v)} &= \left(\frac{\mathbf{v}\cdot\mathbf{w}}{\|\mathbf{w}\|^2}\right)\mathbf{w}\\ &= \left(\frac{\langle 2, 0, -4\rangle\cdot\langle 1, -1, 2\rangle}{\|\langle 1, -1, 2\rangle\|^2}\right)\langle 1, -1, 2\rangle\\ &= \frac{-6}{6}\langle 1, -1, 2\rangle\\ &= \langle -1, 1, -2\rangle.\end{aligned}$$

On the other hand, the projection of $\mathbf{w}$ onto $\mathbf{v}$ is

$$\begin{aligned}\mathbf{P_v(w)} &= \left(\frac{\mathbf{w}\cdot\mathbf{v}}{\|\mathbf{v}\|^2}\right)\mathbf{v}\\ &= \left(\frac{\langle 1, -1, 2\rangle\cdot\langle 2, 0, -4\rangle}{\|\langle 2, 0, -4\rangle\|^2}\right)\langle 2, 0, -4\rangle\\ &= \frac{-6}{20}\langle 2, 0, -4\rangle\\ &= \left\langle -\frac{3}{5}, 0, \frac{6}{5}\right\rangle.\end{aligned}$$

As we could have anticipated from the geometry of projections, the two projections $\mathbf{P_w(v)}$ and $\mathbf{P_v(w)}$ are different. Notice that

$$\|\mathbf{P_w(v)}\| = \frac{|\mathbf{v}\cdot\mathbf{w}|}{\|\mathbf{w}\|} = \frac{6}{\sqrt{6}} = \sqrt{6}$$

and

$$\|\mathbf{P_v}(\mathbf{w})\| = \frac{|\mathbf{w} \cdot \mathbf{v}|}{\|\mathbf{v}\|} = \frac{6}{\sqrt{20}} = \frac{3}{\sqrt{5}}.$$

Finally, the component of $\mathbf{v}$ in the direction of $\mathbf{w}$ is $(\mathbf{v} \cdot \mathbf{w})/\|\mathbf{w}\| = -6/\sqrt{6} = -\sqrt{6}$ and the component of $\mathbf{w}$ in the direction of $\mathbf{v}$ is $(\mathbf{v} \cdot \mathbf{w})/\|\mathbf{v}\| = -6/\sqrt{20} = -3/\sqrt{5}$. ■

inSIGHT

Projection onto a unit vector is simple. If $\mathbf{u}$ is a unit vector and $\mathbf{v}$ is any other vector, then equations (11.18) and (11.19) simplify to

$$\mathbf{P_u}(\mathbf{v}) = (\mathbf{v} \cdot \mathbf{u})\mathbf{u} \qquad \text{and} \qquad \|\mathbf{P_u}(\mathbf{v})\| = |\mathbf{v} \cdot \mathbf{u}|.$$

Also, the component of $\mathbf{v}$ in the direction of $\mathbf{u}$ is simply the dot product $\mathbf{v} \cdot \mathbf{u}$.

Example 6 Let $\mathbf{v} = \langle \sqrt{2}, 8, -2 \rangle$ and $\mathbf{u} = \langle 1/\sqrt{2}, 1/2, -1/2 \rangle$. Calculate the projection of $\mathbf{v}$ onto $\mathbf{u}$. What is the component of $\mathbf{v}$ in the direction of $\mathbf{u}$?

Solution We calculate $\mathbf{v} \cdot \mathbf{u} = \langle \sqrt{2}, 8, -2 \rangle \cdot \langle 1/\sqrt{2}, 1/2, -1/2 \rangle = 1 + 4 + 1 = 6$. Notice that $\mathbf{u}$ is a unit vector. Therefore the component of $\mathbf{v}$ in the direction of $\mathbf{u}$ is $\mathbf{v} \cdot \mathbf{u} = 6$ and

$$\mathbf{P_u}(\mathbf{v}) = (\mathbf{v} \cdot \mathbf{u})\mathbf{u} = 6\left\langle \frac{1}{\sqrt{2}}, \frac{1}{2}, -\frac{1}{2} \right\rangle = \left\langle 3\sqrt{2}, 3, -3 \right\rangle.$$ ■

The Standard Basis Vectors

As noted in Section 11.2, it is often useful to express vectors in terms of the three unit vectors $\mathbf{i} = \langle 1, 0, 0 \rangle$, $\mathbf{j} = \langle 0, 1, 0 \rangle$, and $\mathbf{k} = \langle 0, 0, 1 \rangle$.

Example 7 Let $\mathbf{v} = \langle 2, -6, 12 \rangle$. Calculate $\mathbf{P_i}(\mathbf{v})$, $\mathbf{P_j}(\mathbf{v})$, and $\mathbf{P_k}(\mathbf{v})$.

Solution We have

$$\mathbf{P_i}(\mathbf{v}) = (\mathbf{v} \cdot \mathbf{i})\mathbf{i} = \big(\langle 2, -6, 12 \rangle \cdot \langle 1, 0, 0 \rangle\big)\mathbf{i} = 2\mathbf{i} = \langle 2, 0, 0 \rangle.$$

Similarly,

$$\mathbf{P_j}(\mathbf{v}) = -6\mathbf{j} = \langle 0, -6, 0 \rangle$$

and

$$\mathbf{P_k}(\mathbf{v}) = 12\mathbf{k} = \langle 0, 0, 12 \rangle.$$ ■

inSIGHT

We see from Example 7 that the projections of $\mathbf{v}$ on $\mathbf{i}$, $\mathbf{j}$, and $\mathbf{k}$ give the displacement of $\mathbf{v}$ in the x-, y-, and z-directions, respectively. In particular,

$$\mathbf{v} = \mathbf{P_i}(\mathbf{v}) + \mathbf{P_j}(\mathbf{v}) + \mathbf{P_k}(\mathbf{v}).$$

Direction Cosines and Direction Angles

Suppose that $\mathbf{u} = \langle u_1, u_2, u_3 \rangle = u_1\mathbf{i} + u_2\mathbf{j} + u_3\mathbf{k}$ is a unit vector. Since $u_1^2 + u_2^2 + u_3^2 = 1$, it follows that the numbers u_1, u_2, u_3 all lie between -1 and 1. Thus, there are unique numbers α, β, γ between 0 and π such that $u_1 = \cos(\alpha)$, $u_2 = \cos(\beta)$, and $u_3 = \cos(\gamma)$. As a result,

$$\mathbf{u} = \cos(\alpha)\mathbf{i} + \cos(\beta)\mathbf{j} + \cos(\gamma)\mathbf{k}.$$

The numbers α, β, and γ are called the *direction angles* for $\mathbf{u}$ and u_1, u_2, u_3 are called the *direction cosines* of $\mathbf{u}$. Notice that, by Theorem 2,

α is the angle that $\mathbf{u}$ makes with the positive x-axis,
β is the angle that $\mathbf{u}$ makes with the positive y-axis,
γ is the angle that $\mathbf{u}$ makes with the positive z-axis.

Refer to Figure 6.

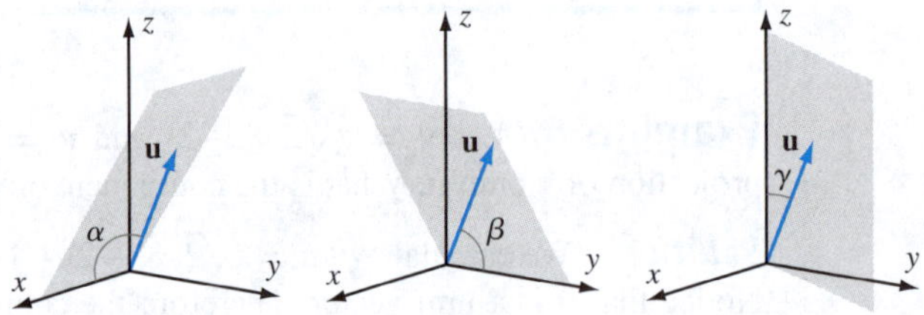

Figure 6
$\mathbf{u} = \cos(\alpha)\mathbf{i} + \cos(\beta)\mathbf{j} + \cos(\gamma)\mathbf{k}$

If $\mathbf{v}$ is *any* nonzero vector, then the direction angles and direction cosines for $\mathbf{v}$ are understood to be the direction angles and cosines for the unit vector $(1/\|\mathbf{v}\|)\mathbf{v}$. The direction angles for $\mathbf{v}$ are the angles that $\mathbf{v}$ makes with the positive x-, y-, and z-axes.

Example 8 Calculate the direction cosines and direction angles for the vector $\mathbf{v} = \langle 0, -3\sqrt{3}, 3 \rangle$.

Solution The associated unit vector is

$$\mathbf{u} = \frac{1}{\|\mathbf{v}\|}\mathbf{v} = \frac{1}{\sqrt{0^2 + \left(-3\sqrt{3}\right)^2 + 3^2}}\langle 0, -3\sqrt{3}, 3 \rangle = \frac{1}{6}\langle 0, -3\sqrt{3}, 3 \rangle = \left\langle 0, -\frac{\sqrt{3}}{2}, \frac{1}{2} \right\rangle.$$

The direction cosines for $\mathbf{v}$ are therefore $\cos(\alpha) = 0$, $\cos(\beta) = -\sqrt{3}/2$, and $\cos(\gamma) = 1/2$. It follows that $\alpha = \pi/2$, $\beta = 5\pi/6$, and $\gamma = \pi/3$. ■

Applications

If a constant force F is applied along the line of motion to move an object a distance d, then, as we learned in Chapter 8, the *work* performed is $W = Fd$. In many applications,

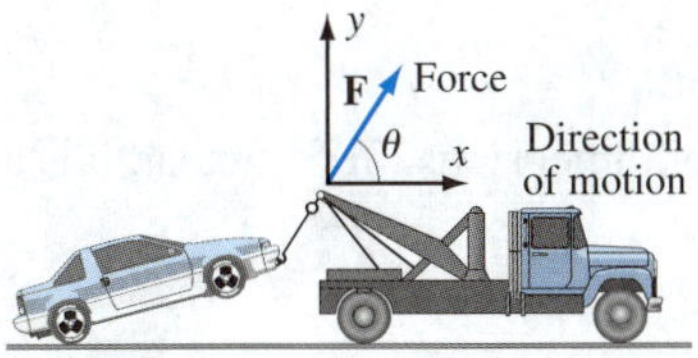

Figure 7

however, the force **F** is a vector that is not applied in the direction of motion. Imagine a truck towing a car (Figure 7). The force is applied in a direction that makes an angle θ with the direction of motion. Let **u** be a unit vector along the direction of motion. Let $g = \mathbf{F} \cdot \mathbf{u} = \|F\| \cos(\theta)$ be the component of **F** in the direction **u**. The work performed in moving the body in the figure a distance d is $W = gd$.

Example 9 A tow truck pulls a disabled vehicle a total of 20,000 ft. To keep the vehicle in motion, the truck must apply a constant force of 3000 lb. The hitch is set up so that the force is exerted at an angle of 30 deg with the horizontal. How much work is performed?

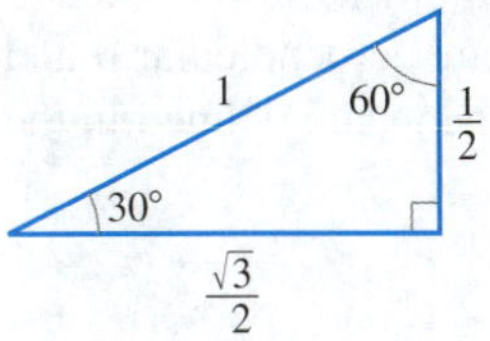

Figure 8a

Solution We use a 30°-60°-90° triangle, as shown in Figure 8a, to resolve the force vector **F** as a sum of two mutually perpendicular vectors, one of which is in the direction **i** of motion. We find that the direction of **F** is given by $\mathbf{v} = (\sqrt{3}/2)\mathbf{i} + (1/2)\mathbf{j}$. See Figure 8b. Since the force vector **F** has magnitude 3000 and direction **v**, we have

$$\mathbf{F} = \|\mathbf{F}\|\mathbf{v} = 3000\left(\frac{\sqrt{3}}{2}\mathbf{i} + \frac{1}{2}\mathbf{j}\right) = 1500\sqrt{3}\mathbf{i} + 1500\mathbf{j},$$

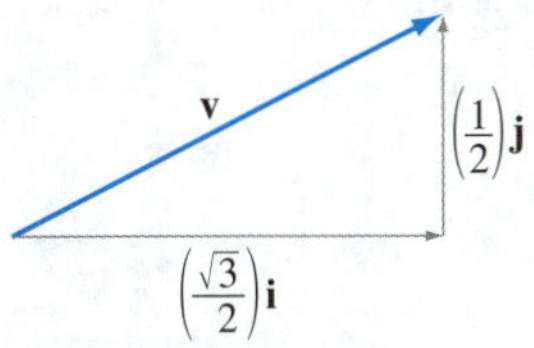

Figure 8b
$\mathbf{v} = (\sqrt{3}/2)\mathbf{i} + (1/2)\mathbf{j}$

as is indicated in Figure 8c. The component of vector **F** in the direction **i** of motion is $g = 1500\sqrt{3}$. As a result, the work performed is

$$W = gd = 1500\sqrt{3} \cdot 20000 = 3\sqrt{3} \cdot 10^7 \text{ ft-lb}.$$

Example 10 The force of the wind is given by the vector $\mathbf{F} = \langle 2, 1, 3\rangle$. In navigation, it is useful to resolve this force into its component in the direction of motion and its component perpendicular to the direction of motion. If $\langle 1, 1, -4\rangle$ represents the direction of motion, then perform this resolution.

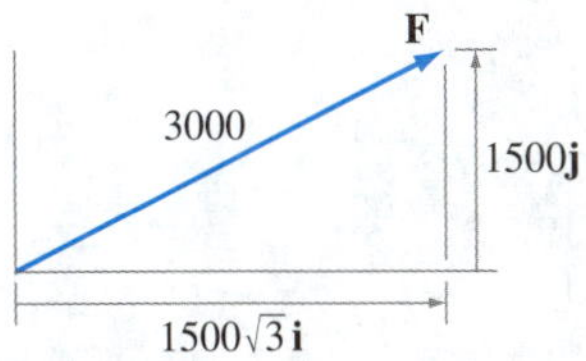

Figure 8c
$\mathbf{F} = 1500\sqrt{3}\mathbf{i} + 1500\mathbf{j}$

Solution The component of **F** in the direction $\mathbf{v} = \langle 1, 1, -4\rangle$ is just the projection $\mathbf{P}_\mathbf{v}(\mathbf{F})$. We calculate

$$\begin{aligned} \mathbf{P}_\mathbf{v}\mathbf{F} &= \left(\frac{\mathbf{F} \cdot \mathbf{v}}{\|\mathbf{v}\|^2}\right)\mathbf{v} \\ &= \frac{-9}{18}\langle 1, 1, -4\rangle \\ &= \left\langle -\frac{1}{2}, -\frac{1}{2}, 2\right\rangle. \end{aligned}$$

Then we write

$$\begin{aligned} \langle 2, 1, 3\rangle = \mathbf{F} &= \mathbf{P}_\mathbf{v}\mathbf{F} + (\mathbf{F} - \mathbf{P}_\mathbf{v}\mathbf{F}) \\ &= \left\langle -\frac{1}{2}, -\frac{1}{2}, 2\right\rangle + \left\langle \frac{5}{2}, \frac{3}{2}, 1\right\rangle. \end{aligned}$$

Notice that the two vectors into which we have decomposed **F** are mutually perpendicular. The first is a multiple of **v** and the second is perpendicular to **v**, which is the resolution that we desire.

A Final Remark

The dot product is a valid operation in any number of dimensions. In particular, it is a useful idea in the plane. We define

$$\langle v_1, v_2\rangle \cdot \langle w_1, w_2\rangle = v_1 w_1 + v_2 w_2.$$

By identifying the planar vectors $\mathbf{v} = \langle v_1, v_2\rangle$ and $\mathbf{w} = \langle w_1, w_2\rangle$ with the vectors $\langle v_1, v_2, 0\rangle$ and $\langle w_1, w_2, 0\rangle$ in space, we may see that all of the ideas in this section apply to vectors in the plane. In particular, two planar vectors are perpendicular if and only if their dot product is equal to 0. Projections and components are defined in the plane just as they were in this section for vectors in space.

quickquiz

1. Define the dot product of two vectors.
2. What does it signify when the dot product of two vectors equals 0?
3. How do we calculate the angle between two vectors?
4. How do we calculate the projection of the vector **v** onto the vector **w**?

EXERCISES

Problems for Practice

In Exercises 1–6, calculate the dot product of the given vectors.

1. $\langle 3, -2, 4\rangle, \langle 2, 1, 6\rangle$
2. $\langle -6, -3, -5\rangle, \langle -8, 1, 1\rangle$
3. $\langle 0, 4, 0\rangle, \langle 5, 1, 0\rangle$
4. $\langle 0, -2, 6\rangle, \langle -4, 2, 5\rangle$
5. $\langle 1, 4, 9\rangle, \langle 2, 3, -5\rangle$
6. $\langle -2, -10, 4\rangle, \langle -8, 0, 2\rangle$

In Exercises 7–12, calculate the angle between the given vectors.

7. $\langle 1, 1, 0\rangle, \langle 0, 1, 1\rangle$
8. $\langle 3, 1, 2\rangle, \langle -13, 13, \sqrt{126}\rangle$
9. $\langle 3, 0, 4\rangle, \langle 0, 8\sqrt{7}, -40\rangle$
10. $\langle 3, 4, 7\rangle, \langle 18, 24, 5\rangle$
11. $\langle 2, -1, 9\rangle, \langle -4, 1, 1\rangle$
12. $\langle 1, 1, 1\rangle, \langle -10, -12 - 3\sqrt{11}, -5\rangle$

In Exercises 13–16, determine whether the pair of vectors is perpendicular. Give a reason for your answer.

13. $\langle -3, 1, 5\rangle, \langle 4, -2, 3\rangle$
14. $\langle 0, -6, 7\rangle, \langle 8, 14, 12\rangle$
15. $\langle 2, -5, 8\rangle, \langle -2, 4, 3\rangle$
16. $\langle 1, 6, 5\rangle, \langle -8, 3, -2\rangle$

In Exercises 17–22, find $\mathbf{P_v}(\mathbf{w})$ and $\mathbf{P_w}(\mathbf{v})$. State the component of **w** in the direction of **v** and the component of **v** in the direction of **w**.

17. $\mathbf{v} = \langle -3, 7, 2\rangle, \mathbf{w} = \langle 1, 4, -4\rangle$
18. $\mathbf{v} = \langle 0, 0, 1\rangle, \mathbf{w} = \langle -9, 7, 5\rangle$
19. $\mathbf{v} = \langle 2, 4, 6\rangle, \mathbf{w} = \langle 4, 8, 12\rangle$
20. $\mathbf{v} = \langle 2, 1, 9\rangle, \mathbf{w} = \langle 1, 0, 1\rangle$
21. $\mathbf{v} = \langle 1, 0, 1\rangle, \mathbf{w} = \langle 0, 1, 0\rangle$
22. $\mathbf{v} = \langle 2, 4, 6\rangle, \mathbf{w} = \langle -2, -4, -6\rangle$

In Exercises 23–28, find the direction of the vector. Then find the direction angles for the vector.

23. $\langle 6, -2\sqrt{3}, 0\rangle$
24. $\langle 0, 3\sqrt{3}, -9\rangle$
25. $\langle 0, 1, 0\rangle$
26. $\langle 3\sqrt{2}, -3, 3\rangle$
27. $\langle -6, 0, 6\rangle$
28. $\langle 0, 1, 1\rangle$

In Exercises 29–32, calculate the dot product of the vectors and their lengths. Verify that the Cauchy-Schwarz Inequality holds for the pair.

29. $\langle 2, -4, 4\rangle, \langle -2, 1, -2\rangle$
30. $\langle 1, 2, 3\rangle, \langle 1, 3, 2\rangle$
31. $\langle 2, 0, 1\rangle, \langle 4, \sqrt{3}, 1\rangle$
32. $\langle 2, 1, -2\rangle, \langle 4, 1, -8\rangle$

In Exercises 33–36, calculate the projection of the vector **v** onto the direction **u**. Verify that $\mathbf{P_u}(\mathbf{v})$ and $\mathbf{v} - \mathbf{P_u}(\mathbf{v})$ are mutually perpendicular.

33. $\mathbf{v} = \langle 4, 1, -8\rangle$, $\mathbf{u} = \langle 2/3, 2/3, -1/3\rangle$

34. $\mathbf{v} = \langle 4, 2, -3\rangle$, $\mathbf{u} = \langle 3/5, 0, -4/5\rangle$

35. $\mathbf{v} = \langle \sqrt{12}, 0, \sqrt{48}\rangle$, $\mathbf{u} = \langle 1/\sqrt{3}, 1/\sqrt{3}, -1/\sqrt{3}\rangle$

36. $\mathbf{v} = \langle 3, 2, -1\rangle$, $\mathbf{u} = \langle 1/9, -4/9, 8/9\rangle$

In Exercises 37–40, find all values of s for which the two given vectors are mutually perpendicular.

37. $\langle 3, 2, -1\rangle$, $\langle s, 1, -4\rangle$

38. $\langle s, 5, -12\rangle$, $\langle 1, s, 2\rangle$

39. $\langle 3, 1, 1\rangle$, $\langle -7, s^2, 5\rangle$

40. $\langle s, -5, 1\rangle$, $\langle s, s, 6\rangle$

Further Theory and Practice

41. Find a unit vector **u** that is perpendicular to both $\mathbf{v} = \mathbf{i} + 2\mathbf{j} + \mathbf{k}$ and $\mathbf{w} = \mathbf{i} - \mathbf{j}$. Show that any other vector that is perpendicular to both **v** and **w** is parallel to **u**.

42. Show that if **v** is any vector, then

$$\mathbf{v} = (\mathbf{v} \cdot \mathbf{i})\mathbf{i} + (\mathbf{v} \cdot \mathbf{j})\mathbf{j} + (\mathbf{v} \cdot \mathbf{k})\mathbf{k}.$$

43. Prove that the vectors $\mathbf{a} = \langle \sqrt{3}/2, 1/(2\sqrt{2}), -1/(2\sqrt{2})\rangle$, $\mathbf{b} = \langle 0, 1/\sqrt{2}, 1/\sqrt{2}\rangle$, and $\mathbf{c} = \langle 1/2, -\sqrt{6}/4, \sqrt{6}/4\rangle$ are each perpendicular to the other two and are of unit length.

44. If **v** is any vector and if **a**, **b**, and **c** are the three vectors from Exercise 43, then prove that

$$\mathbf{v} = (\mathbf{v} \cdot \mathbf{a})\mathbf{a} + (\mathbf{v} \cdot \mathbf{b})\mathbf{b} + (\mathbf{v} \cdot \mathbf{c})\mathbf{c}.$$

45. Mrs. Woodman pulls a railroad car along a track using a rope that makes a 30 deg angle with the track (Figure 9). Calculate how much work is performed if she exerts a force of 200 lb and succeeds in pulling the car 1000 ft.

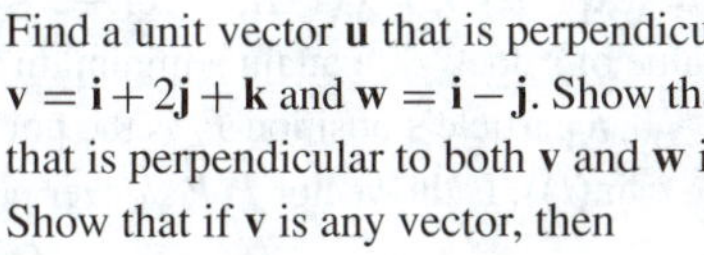

Figure 9

46. Show that the direction cosines for any vector **v** satisfy

$$\cos^2(\alpha) + \cos^2(\beta) + \cos^2(\gamma) = 1.$$

If **v** lies in the xy-plane, then to what familiar identity does this equation reduce?

47. Prove the Parallelogram Law:

$$\|\mathbf{v} + \mathbf{w}\|^2 + \|\mathbf{v} - \mathbf{w}\|^2 = 2\|\mathbf{v}\|^2 + \|\mathbf{w}\|^2.$$

Give a geometric interpretation for this equality.

48. Prove the Polarization Formula:

$$\|\mathbf{v} + \mathbf{w}\|^2 - \|\mathbf{v} - \mathbf{w}\|^2 = 4\mathbf{v} \cdot \mathbf{w}.$$

49. In each of the following, find a real number λ such that $\mathbf{P_u}(\mathbf{v}) = \lambda\mathbf{w}$.

a. $\mathbf{u} = \langle 3, 6, 9\rangle$, $\mathbf{v} = \langle 1, -2, 5\rangle$, $\mathbf{w} = \langle 1, 2, 3\rangle$

b. $\mathbf{u} = \langle -2, 0, 4\rangle$, $\mathbf{v} = \langle 1, -4, 6\rangle$, $\mathbf{w} = \langle -1, 0, 2\rangle$

c. $\mathbf{u} = \langle 2, 10, -6\rangle$, $\mathbf{v} = \langle -3, 4, 8\rangle$, $\mathbf{w} = \langle 1, 5, -3\rangle$

d. $\mathbf{u} = \langle 12, 0, -16\rangle$, $\mathbf{v} = \langle 1, 1, 3\rangle$, $\mathbf{w} = \langle 3, 0, -4\rangle$

50. Prove that for any two vectors **v** and **w**, the angle between $\mathbf{u} = \|\mathbf{w}\|\mathbf{v} + \|\mathbf{v}\|\mathbf{w}$ and **v** equals the angle between **u** and **w**.

51. Let $\overrightarrow{OP}$ be the position vector for an arbitrary point $P = (x, y, z)$ in space. Using vector methods, find an equation involving $\overrightarrow{OP}$, **j**, the dot product, and the norm to describe the cone with vertex at the origin, with an angle of 60 deg at the vertex, and with axis of symmetry the y-axis.

52. Let $\overrightarrow{OP}$ be the position vector for an arbitrary point $P = (x, y, z)$ in space. Using vector methods, find an equation involving $\overrightarrow{OP}$, **k**, the dot product, and the norm to describe the cone with vertex at the origin, with an angle of 45 deg at the vertex, and with axis of symmetry the z-axis.

53. Show that the points (2, 1, 4), (5, 3, 2), and (7, 4, 6) are the vertices of a right triangle.

54. Suppose that P, Q, and R are vertices of a cube with $\overline{PQ}$ the diagonal of the cube and $\overline{PR}$ the diagonal of a face of the cube. What is the angle between $\overrightarrow{PQ}$ and $\overrightarrow{PR}$?

55. Suppose that a and b are nonnegative. When the Cauchy-Schwarz Inequality is applied to $\mathbf{v} = \sqrt{a}\mathbf{i} + \sqrt{b}\mathbf{j}$ and $\mathbf{w} = \sqrt{b}\mathbf{i} + \sqrt{a}\mathbf{j}$, what inequality results?

56. Let v_1, v_2, and v_3 be any numbers. Find a suitable vector **w** such that the Cauchy-Schwarz Inequality applied to $\mathbf{v} = \langle v_1, v_2, v_3\rangle$ and **w** results in the inequality

$$(v_1 + v_2 + v_3)^2 \leq 3\left(v_1^2 + v_2^2 + v_3^2\right).$$

57. Use the Cauchy-Schwarz Inequality to prove the Triangle Inequality:

$$\|\mathbf{v} + \mathbf{w}\| \leq \|\mathbf{v}\| + \|\mathbf{w}\|.$$

58. Suppose that a, b, and c are positive numbers. The equation $x/a + y/b + z/c = 1$ represents a plane. What are the intercepts A, B, and C with the x-axis, y-axis, and z-axis, respectively? Calculate the cosines of the three angles of $\triangle ABC$. What arccosine identity can be deduced from these angles?

59. Verify the identity

$$\left(v_1^2 + v_2^2 + v_3^2\right)\left(w_1^2 + w_2^2 + w_3^2\right) - (v_1 w_1 + v_2 w_2 + v_3 w_3)^2 = (v_1 w_2 - w_1 v_2)^2 + (v_1 w_3 - w_1 v_3)^2 + (v_2 w_3 - w_2 v_3)^2.$$

Let $\mathbf{v} = \langle v_1, v_2, v_3 \rangle$ and $\mathbf{w} = \langle w_1, w_2, w_3 \rangle$. Deduce that

$$\|\mathbf{v}\|^2 \|\mathbf{w}\|^2 - (\mathbf{v} \cdot \mathbf{w})^2 = (v_1 w_2 - w_1 v_2)^2 + (v_1 w_3 - w_1 v_3)^2 + (v_2 w_3 - w_2 v_3)^2.$$

Prove the Cauchy-Schwarz Inequality from this identity.

60. Suppose that $\mathbf{v} = \langle v_1, v_2, v_3 \rangle$ and $\mathbf{w} = \langle w_1, w_2, w_3 \rangle$ are parallel. Use Theorem 3b to deduce that

$$\|\mathbf{v}\|^2 \|\mathbf{w}\|^2 - (\mathbf{v} \cdot \mathbf{w})^2 = 0.$$

Use the second identity from Exercise 59 to conclude that

$$(v_1 w_2 - w_1 v_2)^2 + (v_1 w_3 - w_1 v_3)^2 + (v_2 w_3 - w_2 v_3)^2 = 0,$$

which implies the following three equations

$$v_1 w_2 = w_1 v_2, \quad v_1 w_3 = w_1 v_3, \quad \text{and} \quad v_2 w_3 = w_2 v_3.$$

If one entry v_i of $\mathbf{v}$ is nonzero, then show that $\mathbf{w} = (w_i/v_i)\mathbf{v}$. Deduce that in any event, one of $\mathbf{v}$ and $\mathbf{w}$ is a scalar multiple of the other.

61. Let V denote the set of planar vectors. Let $T : V \to V$ be defined by

$$T(\langle a, b \rangle) = \langle a\cos(\theta) - b\sin(\theta), a\sin(\theta) + b\cos(\theta) \rangle.$$

Show that the angle between $\langle a, b \rangle$ and $T(\langle a, b \rangle)$ is θ. Also show that $\|T(\langle a, b \rangle)\| = \|\langle a, b \rangle\|$.

Calculator/Computer Exercises

62. Calculate the angles of the triangle with vertices $(1, 1, 2)$, $(2, 0, 3)$, and $(1, 2, -5)$. What is the sum of these numbers?

63. Calculate the angles of the triangle with vertices $(1, 1, 2)$, $(2, 0, 3)$, and $(1, 2, -5)$. What is the sum of these numbers?

64. For what value(s) of a are the vectors $\langle a^2, 2a, -3 \rangle$ and $\langle a, 3a, 1 \rangle$ perpendicular?

65. Plot $f(t) = \langle \exp(-t), t, 1 \rangle \cdot \langle 1, 2t, 3 \rangle$ for $-2 \le t \le 2$. At what value of t does $f(t)$ attain a minimum?

66. At time $t \ge 0$, a particle's position P_t is the point $(t, t\cos(t), t\sin(t))$. Is the vector $\overrightarrow{P_t P_{t+1}}$ ever parallel to $\langle 1, 2, 3 \rangle$?

11.4 The Cross Product and Triple Product

Imagine we are given two vectors $\mathbf{v}$ and $\mathbf{w}$ in space. It is frequently useful to be able to find a vector that is perpendicular to both of them. Of course, such a vector is not unique: If we have found one such vector $\mathbf{u}$, then $c\mathbf{u}$ is also perpendicular to $\mathbf{v}$ and $\mathbf{w}$ for any scalar c. We begin this section by presenting an algebraic method for finding one such vector $\mathbf{u}$.

Definition If $\mathbf{v} = \langle v_1, v_2, v_3 \rangle$ and $\mathbf{w} = \langle w_1, w_2, w_3 \rangle$, then we define their cross product to be

$$\mathbf{v} \times \mathbf{w} = (v_2 w_3 - w_2 v_3)\mathbf{i} - (v_1 w_3 - w_1 v_3)\mathbf{j} + (v_1 w_2 - w_1 v_2)\mathbf{k}. \qquad \textbf{(11.22)}$$

Let us first verify that the vector $\mathbf{v} \times \mathbf{w}$ does the job.

Theorem 1 If $\mathbf{v}$ and $\mathbf{w}$ are vectors, then $\mathbf{v} \times \mathbf{w}$ is perpendicular to both $\mathbf{v}$ and $\mathbf{w}$.

Proof Let $\mathbf{v} = \langle v_1, v_2, v_3 \rangle$ and $\mathbf{w} = \langle w_1, w_2, w_3 \rangle$. We calculate the dot product of $\mathbf{v}$ with $\mathbf{v} \times \mathbf{w}$:

$$\begin{aligned}\mathbf{v} \cdot (\mathbf{v} \times \mathbf{w}) &= \langle v_1, v_2, v_3 \rangle \cdot \langle (v_2 w_3 - w_2 v_3), -(v_1 w_3 - w_1 v_3), (v_1 w_2 - w_1 v_2) \rangle \\ &= v_1(v_2 w_3 - w_2 v_3) - v_2(v_1 w_3 - w_1 v_3) + v_3(v_1 w_2 - w_1 v_2) \\ &= 0.\end{aligned}$$

The verification that $\mathbf{w} \cdot (\mathbf{v} \times \mathbf{w}) = 0$ is similar. ■

Example 1 Let $\mathbf{v} = \langle 2, -1, 3 \rangle$ and $\mathbf{w} = \langle 5, 4, -6 \rangle$. Calculate $\mathbf{v} \times \mathbf{w}$. Verify that $\mathbf{v} \times \mathbf{w}$ is orthogonal to both $\mathbf{v}$ and $\mathbf{w}$.

Solution We calculate:

$$\begin{aligned}\mathbf{v} \times \mathbf{w} &= \langle 2, -1, 3 \rangle \times \langle 5, 4, -6 \rangle \\ &= ((-1) \cdot (-6) - 4 \cdot 3)\mathbf{i} - (2 \cdot (-6) - 5 \cdot 3)\mathbf{j} + (2 \cdot 4 - 5 \cdot (-1))\mathbf{k} \\ &= \langle -6, 27, 13 \rangle.\end{aligned}$$

Observe that $\mathbf{v} \times \mathbf{w}$ is perpendicular to both $\mathbf{v}$ and $\mathbf{w}$:

$$\mathbf{v} \cdot (\mathbf{v} \times \mathbf{w}) = \langle 2, -1, 3 \rangle \cdot \langle -6, 27, 13 \rangle = 2 \cdot (-6) + (-1) \cdot 27 + 3 \cdot 13 = 0$$

and

$$\mathbf{w} \cdot (\mathbf{v} \times \mathbf{w}) = \langle 5, 4, -6 \rangle \cdot \langle -6, 27, 13 \rangle = 5 \cdot (-6) + 4 \cdot 27 + (-6) \cdot 13 = 0.$$ ■

The Relationship between Cross Products and Determinants

There is a nice way to remember the formula for a cross product using the language of determinants. The determinant is a procedure for calculating a number based on the entries of a square array (or *matrix*) of numbers such as

$$\underbrace{\begin{bmatrix} a & b \\ c & d \end{bmatrix}}_{2 \times 2 \text{ array}} \quad \text{or} \quad \underbrace{\begin{bmatrix} \alpha & \beta & \gamma \\ a & b & c \\ d & e & f \end{bmatrix}}_{3 \times 3 \text{ array}}$$

By "square," we mean that the number of rows in the array equals the number of columns. In the general theory of determinants, the number of rows and columns may be any positive integer. For our purposes, we need only consider 2×2 and 3×3 arrays.

The determinant $\det(M)$ of a 2×2 array $M = \begin{bmatrix} a & b \\ c & d \end{bmatrix}$ is defined by

$$\det\left(\begin{bmatrix} a & b \\ c & d \end{bmatrix}\right) = ad - bc.$$

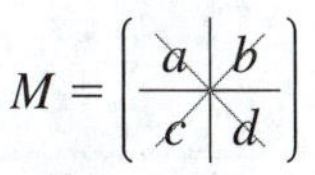

$\det(M) = ad - bc$

Figure 1

Notice that the determinant is the alternating sum of the product of diagonal entries indicated in Figure 1.

The determinant of a 3×3 array is defined by

$$\det\left(\begin{bmatrix} \alpha & \beta & \gamma \\ a & b & c \\ d & e & f \end{bmatrix}\right) = \alpha \det\left(\begin{bmatrix} b & c \\ e & f \end{bmatrix}\right) - \beta \det\left(\begin{bmatrix} a & c \\ d & f \end{bmatrix}\right) + \gamma \det\left(\begin{bmatrix} a & b \\ d & e \end{bmatrix}\right). \qquad \textbf{(11.23)}$$

This definition is said to be an *expansion along the first row* of the array. The right side of equation (11.23) is an alternating sum of three products, one for each entry in the first row. The factor of each entry is the determinant of the 2×2 array obtained by crossing out the row and column of the entry. Figure 2 illustrates this procedure. An equivalent scheme for calculating the determinant of a 3×3 array is often used. As Figure 3 shows, a copy of M is placed to the right of M. The products of the six indicated diagonals are then computed. The determinant is the sum of the products that arise from the diagonals that slope down to the right minus the sum of the products that arise from the diagonals that slope down to the left.

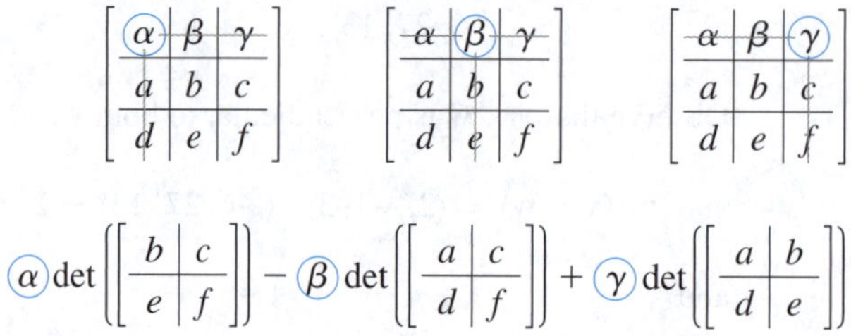

Figure 2

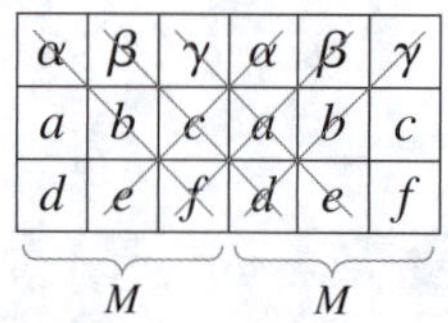

Figure 3

$\det(M) = \alpha bf + \beta cd + \gamma ae - \alpha ce - \beta af - \gamma bd$

If $\mathbf{v} = \langle v_1, v_2, v_3 \rangle$ and $\mathbf{w} = \langle w_1, w_2, w_3 \rangle$, then a simple calculation shows that

$$\mathbf{v} \times \mathbf{w} = \det\left(\begin{bmatrix} \mathbf{i} & \mathbf{j} & \mathbf{k} \\ v_1 & v_2 & v_3 \\ w_1 & w_2 & w_3 \end{bmatrix}\right). \qquad \textbf{(11.24)}$$

Here the determinant is expanded as if $\mathbf{i}, \mathbf{j}, \mathbf{k}$ were scalars. By doing so, we obtain

$$\begin{aligned} \mathbf{v} \times \mathbf{w} &= \det\left(\begin{bmatrix} v_2 & v_3 \\ w_2 & w_3 \end{bmatrix}\right)\mathbf{i} - \det\left(\begin{bmatrix} v_1 & v_3 \\ w_1 & w_3 \end{bmatrix}\right)\mathbf{j} + \det\left(\begin{bmatrix} v_1 & v_2 \\ w_1 & w_2 \end{bmatrix}\right)\mathbf{k} \\ &= (v_2 w_3 - w_2 v_3)\mathbf{i} - (v_1 w_3 - w_1 v_3)\mathbf{j} + (v_1 w_2 - w_1 v_2)\mathbf{k}, \end{aligned}$$

which is the formula for the cross product, as given by equation (11.22).

Example 2 Calculate the cross product of $\mathbf{v} = \langle 2, -1, 6 \rangle$ and $\mathbf{w} = \langle -3, 4, 1 \rangle$ using a determinant.

Solution We write

$$\begin{aligned} \mathbf{v} \times \mathbf{w} &= \det\left(\begin{bmatrix} \mathbf{i} & \mathbf{j} & \mathbf{k} \\ 2 & -1 & 6 \\ -3 & 4 & 1 \end{bmatrix}\right) \\ &= \mathbf{i}((-1) \cdot 1 - 4 \cdot 6) - \mathbf{j}(2 \cdot 1 - (-3) \cdot 6) + \mathbf{k}(2 \cdot 4 - (-3) \cdot (-1)) \\ &= -25\mathbf{i} - 20\mathbf{j} + 5\mathbf{k}. \end{aligned}$$

Thus, $\mathbf{v} \times \mathbf{w} = \langle -25, -20, 5 \rangle$. ■

inSIGHT

> Recall that the *dot product* $\mathbf{v} \cdot \mathbf{w}$ of two vectors is a *scalar;* however, the *cross product* of two vectors is *another vector*. Texts that call $\mathbf{v} \cdot \mathbf{w}$ the *scalar product* often call $\mathbf{v} \times \mathbf{w}$ the *vector product*. The present text does not use this alternative terminology.

If $\mathbf{u}$, $\mathbf{v}$, and $\mathbf{w}$ are vectors and λ and μ are scalars, then

$$\begin{aligned} \mathbf{u} \times (\mathbf{v} + \mathbf{w}) &= (\mathbf{u} \times \mathbf{v}) + (\mathbf{u} \times \mathbf{w}), \\ (\mathbf{v} + \mathbf{w}) \times \mathbf{u} &= (\mathbf{v} \times \mathbf{u}) + (\mathbf{w} \times \mathbf{u}), \\ \mathbf{v} \times \mathbf{w} &= -\mathbf{w} \times \mathbf{v}, \text{ and} \\ (\lambda \mathbf{v}) \times (\mu \mathbf{w}) &= (\lambda\mu)(\mathbf{v} \times \mathbf{w}). \end{aligned}$$

These algebraic properties for the cross product can be verified by routine calculation. The properties of distribution are as expected. However, pay particular attention to the law $\mathbf{v} \times \mathbf{w} = -\mathbf{w} \times \mathbf{v}$, which shows that the cross product is not a commutative operation. In fact, because of this property, the cross product is said to be *anticommutative*.

Example 3 Show that, if $\mathbf{u}$ and $\mathbf{v}$ are parallel vectors, then $\mathbf{u} \times \mathbf{v} = \vec{\mathbf{0}}$. In other words, the cross product of any two parallel vectors is the zero vector.

Solution The anticommutative property of the cross product states that $\mathbf{v} \times \mathbf{w} = -\mathbf{w} \times \mathbf{v}$ for any vectors $\mathbf{v}$ and $\mathbf{w}$. It follows that if $\mathbf{w} = \mathbf{v}$, then $\mathbf{v} \times \mathbf{v} = -\mathbf{v} \times \mathbf{v}$. By adding $\mathbf{v} \times \mathbf{v}$ to each side of this equation, we conclude that $2\mathbf{v} \times \mathbf{v} = \vec{\mathbf{0}}$, or $\mathbf{v} \times \mathbf{v} = \vec{\mathbf{0}}$. Thus, *the cross product of any vector with itself is the zero vector*. Next, suppose that $\mathbf{u}$ and $\mathbf{v}$ are parallel. According to Theorem 2 from Section 11.2, one of these vectors can be written as a scalar multiple of the other, say $\mathbf{u} = \lambda\mathbf{v}$. Then $\mathbf{u} \times \mathbf{v} = (\lambda\mathbf{v}) \times \mathbf{v} = \lambda(\mathbf{v} \times \mathbf{v}) = \lambda\vec{\mathbf{0}} = \vec{\mathbf{0}}$. ■

A Geometric Understanding of the Cross Product

Now we would like to develop a geometric way to think of the cross product. If we have a picture of the vectors $\mathbf{v}$ and $\mathbf{w}$, how can we visualize the vector $\mathbf{v} \times \mathbf{w}$?

Definition We say that a vector $\mathbf{n}$ is *normal* to the vectors $\mathbf{v}$ and $\mathbf{w}$ if $\mathbf{n}$ is perpendicular to both $\mathbf{v}$ and $\mathbf{w}$. If $\mathbf{n}$ is also a unit vector, then it is said to be a *unit normal vector* for $\mathbf{v}$ and $\mathbf{w}$.

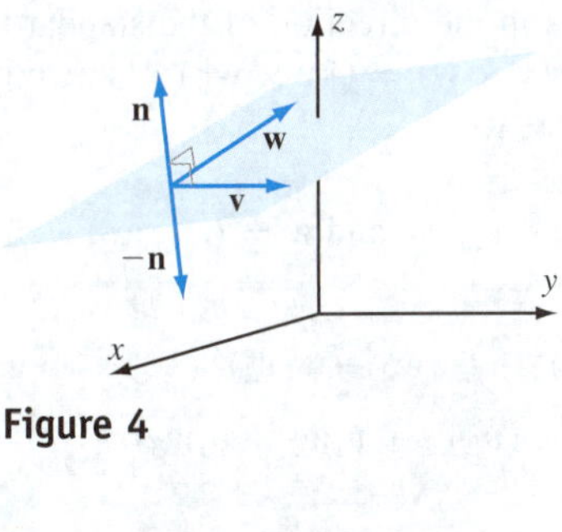

Figure 4

A pair of nonparallel vectors always has two unit normal vectors: If $\mathbf{n}$ is one of them, then $-\mathbf{n}$ is the other. See Figure 4. If we order the vectors $\mathbf{v}$ and $\mathbf{w}$, then we may declare one of the vectors $\mathbf{n}$ or $-\mathbf{n}$ to be the *standard unit normal vector* associated to the pair $\mathbf{v}$, $\mathbf{w}$ (in the given order). It is determined by the *Right Hand Rule*.

The Right Hand Rule for Finding the Direction of the Standard Unit Normal

Let **v** and **w** be two vectors that are not parallel to each other. Represent them by directed line segments with the origin as common initial point. Point the fingers of your right hand along **v**, then curl them toward **w**, as in Figure 5; the direction in which the thumb points during this process is the direction of the standard unit normal vector for the vectors **v**, **w**. Look again at Figure 4. The direction of vector **n** is the standard unit normal for the ordered pair **v**, **w**.

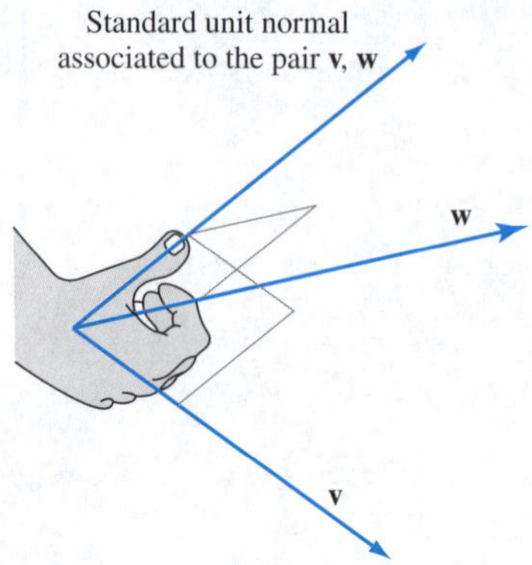

Figure 5

> **inSIGHT**
>
> According to the Right Hand Rule, the standard unit normal vector for the ordered pair **v**, **w** points in the opposite direction to the standard unit normal vector for the ordered pair **w**, **v**. Example 4 provides particular cases of this general fact.

Example 4 Find the standard unit normal for the pairs **i**, **j** and **j**, **k** and **k**, **i**. Find also the standard unit normal for the pairs **j**, **i** and **k**, **j** and **i**, **k**.

Solution If you curl the fingers of your right hand from **i** toward **j**, then your thumb will point in the direction of the positive z-axis. Thus, the standard unit normal for the pair **i**, **j** is **k**. Similar reasoning shows that the standard unit normal for **j**, **i** is $-\mathbf{k}$. We leave it as an exercise to check that the standard unit normal for the pair **j**, **k** is **i**, and the standard unit normal for the pair **k**, **i** is **j**. In addition, the standard unit normal for **k**, **j** is $-\mathbf{i}$, and the standard unit normal for **i**, **k** is $-\mathbf{j}$. ■

As we saw in Example 3, it may happen that $\mathbf{v} \times \mathbf{w} = \vec{\mathbf{0}}$. In all other cases, we may form the unit vector $\text{dir}(\mathbf{v} \times \mathbf{w}) = (\|\mathbf{v} \times \mathbf{w}\|)^{-1}(\mathbf{v} \times \mathbf{w})$. Because $\mathbf{v} \times \mathbf{w}$ is perpendicular to both **v** and **w**, the vector $\text{dir}(\mathbf{v} \times \mathbf{w})$ must be the standard unit normal for either **v**, **w** or **w**, **v**. Our next theorem tells us which. It also specifies the length of $\mathbf{v} \times \mathbf{w}$ and tells us when the equation $\mathbf{v} \times \mathbf{w} = \vec{\mathbf{0}}$ holds.

Theorem 2 If we let **v** and **w** be vectors, then the following hold.

a. $\|\mathbf{v} \times \mathbf{w}\|^2 = \|\mathbf{v}\|^2\|\mathbf{w}\|^2 - (\mathbf{v} \cdot \mathbf{w})^2$.
b. If **v** and **w** are nonzero, then $\|\mathbf{v} \times \mathbf{w}\| = \|\mathbf{v}\|\,\|\mathbf{w}\|\sin(\theta)$ where $\theta \in [0, \pi]$ denotes the angle between **v** and **w**.
c. **v** and **w** are parallel if and only if $\mathbf{v} \times \mathbf{w} = \vec{\mathbf{0}}$.
d. If **v** and **w** are not parallel, then $\mathbf{v} \times \mathbf{w}$ points in the direction of the standard unit normal for the pair **v**, **w**. In particular, $\text{dir}(\mathbf{v} \times \mathbf{w}) = (\|\mathbf{v} \times \mathbf{w}\|)^{-1}(\mathbf{v} \times \mathbf{w})$ is the standard unit normal vector for the pair **v**, **w**.

Proof Part a is an identity among the entries of $\mathbf{v} = \langle v_1, v_2, v_3\rangle$ and $\mathbf{w} = \langle w_1, w_2, w_3\rangle$. We have

$$\begin{aligned}\|\mathbf{v} \times \mathbf{w}\|^2 &= (v_2w_3 - w_2v_3)^2 + (v_1w_3 - w_1v_3)^2 + (v_1w_2 - w_1v_2)^2 \\ &= \left(v_1^2 + v_2^2 + v_3^2\right)\left(w_1^2 + w_2^2 + w_3^2\right) - (v_1w_1 + v_2w_2 + v_3w_3)^2 \\ &= \|\mathbf{v}\|^2\|\mathbf{w}\|^2 - (\mathbf{v} \cdot \mathbf{w})^2.\end{aligned}$$

Although the second equality in this chain is not obvious, it may be verified by multiplying everything out.

If neither $\mathbf{v}$ nor $\mathbf{w}$ is the zero vector, then the scalars $\|\mathbf{v}\|$ and $\|\mathbf{w}\|$ are nonzero. In this case, the identity of part a becomes

$$\begin{aligned}\|\mathbf{v}\times\mathbf{w}\|^2 &= \|\mathbf{v}\|^2\|\mathbf{w}\|^2 - (\mathbf{v}\cdot\mathbf{w})^2 = \|\mathbf{v}\|^2\|\mathbf{w}\|^2\left(1-\frac{(\mathbf{v}\cdot\mathbf{w})^2}{\|\mathbf{v}\|^2\|\mathbf{w}\|^2}\right)\\ &= \|\mathbf{v}\|^2\|\mathbf{w}\|^2\big(1-\cos^2(\theta)\big)\end{aligned}$$

where the last equality is obtained by using formula (11.16). It follows that $\|\mathbf{v}\times\mathbf{w}\|^2 = \|\mathbf{v}\|^2\|\mathbf{w}\|^2\sin^2(\theta)$ and, on taking the square root, $\|\mathbf{v}\times\mathbf{w}\| = \|\mathbf{v}\|\,\|\mathbf{w}\||\sin(\theta)|$. Since $0\le\theta\le\pi$, it follows that $\sin(\theta)\ge 0$. Therefore, $|\sin(\theta)| = \sin(\theta)$ and $\|\mathbf{v}\times\mathbf{w}\| = \|\mathbf{v}\|\|\mathbf{w}\|\sin(\theta)$, as asserted in part b.

Theorem 3 from Section 11.3 tells us that vectors $\mathbf{v}$ and $\mathbf{w}$ are parallel if and only if $|\mathbf{v}\cdot\mathbf{w}| = \|\mathbf{v}\|\|\mathbf{w}\|$. Using the identity of part a, we conclude that $\mathbf{v}$ and $\mathbf{w}$ are parallel if and only if $\|\mathbf{v}\times\mathbf{w}\| = 0$. Since $\vec{\mathbf{0}}$ is the only zero-length vector, assertion c follows.

For part d, we limit our verification to a few special cases. Notice that $\mathbf{i}\times\mathbf{j}=\mathbf{k}$, $\mathbf{j}\times\mathbf{k}=\mathbf{i}$, and $\mathbf{k}\times\mathbf{i}=\mathbf{j}$. Based on our observations in Example 4, we conclude that, for the basic vectors $\mathbf{i}$, $\mathbf{j}$, and $\mathbf{k}$, the operation of cross product produces the standard unit normal. Geometric reasoning can be used to show that the cross product $\mathbf{v}\times\mathbf{w}$ always points in the direction of the standard unit normal for the vectors $\mathbf{v}$, $\mathbf{w}$. ■

Example 5 What is the standard unit normal vector for the pair $\mathbf{v} = \langle 1,-3,2\rangle$, $\mathbf{w} = \langle 1,-1,4\rangle$?

Solution We calculate the vector

$$\begin{aligned}\mathbf{v}\times\mathbf{w} &= ((-3)\cdot 4-(-1)\cdot 2)\mathbf{i}-(1\cdot 4-1\cdot 2)\mathbf{j}+(1\cdot(-1)-1\cdot(-3))\mathbf{k}\\ &= -10\mathbf{i}-2\mathbf{j}+2\mathbf{j}\end{aligned}$$

and its length

$$\|\mathbf{v}\times\mathbf{w}\| = \sqrt{(-10)^2+(-2)^2+2^2} = \sqrt{108} = \sqrt{(36)(3)} = 6\sqrt{3}.$$

Theorem 2d tells us that

$$\operatorname{dir}(\mathbf{v}\times\mathbf{w}) = (\|\mathbf{v}\times\mathbf{w}\|)^{-1}(\mathbf{v}\times\mathbf{w}) = \frac{1}{6\sqrt{3}}(-10\mathbf{i}-2\mathbf{j}+2\mathbf{j}) = -\frac{5}{3\sqrt{3}}\mathbf{i}-\frac{1}{3\sqrt{3}}\mathbf{j}+\frac{1}{3\sqrt{3}}\mathbf{j}$$

is the vector we seek. ■

Example 6 Give an example to show that the cross product does not satisfy a cancellation property. In other words, the equality $\mathbf{v}\times\mathbf{w} = \mathbf{v}\times\mathbf{u}$ does not imply that $\mathbf{w}=\mathbf{u}$.

Solution A counterexample is given by the parallel vectors $\mathbf{v} = \langle 1,0,0\rangle$, $\mathbf{w} = \langle 2,0,0\rangle$, and $\mathbf{u} = \langle 3,0,0\rangle$. We have $\mathbf{v}\times\mathbf{w} = \vec{\mathbf{0}} = \mathbf{v}\times\mathbf{u}$, yet $\mathbf{w}\neq\mathbf{u}$. ■

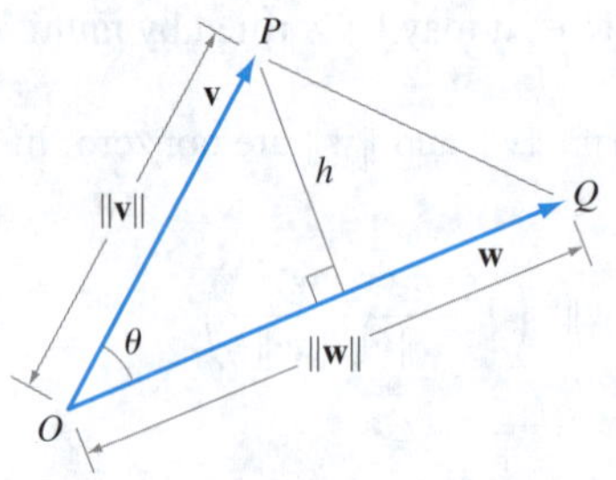

Figure 6
$h = \|\mathbf{v}\| \sin(\theta)$

Cross Products and the Calculation of Area

Now we learn a connection between cross products and areas of triangles and parallelograms. Look at Figure 6. If vectors $\mathbf{v} = \overrightarrow{OP}$ and $\mathbf{w} = \overrightarrow{OQ}$ are two nonparallel vectors with common initial point O, then we will speak of $\triangle OPQ$ as the triangle determined by the vectors $\mathbf{v}$ and $\mathbf{w}$. The altitude from vertex P to base $\overline{OQ}$ has length $\|\mathbf{v}\| \sin(\theta)$. The area of $\triangle OPQ$, half the product of its base and height, is therefore $(\|\mathbf{v}\| \sin(\theta)) \cdot \|\mathbf{w}\|/2$. This quantity is $\|\mathbf{v} \times \mathbf{w}\|/2$ by Theorem 2b. To summarize:

> The area of the triangle determined by the vectors $\mathbf{v}$ and $\mathbf{w}$ is $(1/2)\|\mathbf{v} \times \mathbf{w}\|$.

Example 7 Find the area of the triangle determined by the vectors $\mathbf{v} = \langle 0, 2, 1\rangle$ and $\mathbf{w} = \langle 3, 1, -1\rangle$.

Solution We calculate that

$$\frac{1}{2}\|\mathbf{v} \times \mathbf{w}\| = \frac{1}{2}\| -3\mathbf{i} + 3\mathbf{j} - 6\mathbf{k}\| = \frac{1}{2}\sqrt{54} = \frac{3}{2}\sqrt{6}.$$

■

Of course, the area of the parallelogram determined by the vectors $\mathbf{v}$ and $\mathbf{w}$ is twice the area of the triangle that they determine. In other words,

> The area of the parallelogram determined by the vectors $\mathbf{v}$ and $\mathbf{w}$ is $\|\mathbf{v} \times \mathbf{w}\|$.

Example 8 Calculate the area of the parallelogram determined by the vectors $\mathbf{v} = \langle -2, 1, 3\rangle$ and $\mathbf{w} = \langle 1, 0, 4\rangle$.

Solution The required area is $\|\mathbf{v} \times \mathbf{w}\| = \|4\mathbf{i} + 11\mathbf{j} - \mathbf{k}\| = \sqrt{138}$. ■

A Physical Application of the Cross Product

Experimental evidence teaches us that if a magnetic field $\mathbf{M}$ acts on a charged particle with charge s, and if the charged particle has velocity $\mathbf{v}$, then the resultant force $\mathbf{F}$ that is exerted is

$$\mathbf{F} = (s\mathbf{v}) \times \mathbf{M}.$$

Example 9 In the picture tube for an oscilloscope, a magnetic field is used to control the path of ions that transmit the image to the screen. If s is the charge of the particle, $\mathbf{v}$ is its velocity, and $\mathbf{M}$ is the magnetic field, then the force $\mathbf{F}$ exerted on the particle is $\mathbf{F} = (s\mathbf{v}) \times \mathbf{M}$. Suppose that the velocity vector for the ions is $\langle c, -c, 0\rangle$ where c is a physical constant. The magnetic field will have the form $\langle a, 1, 1\rangle$ where the value of a is varied to force the ions to go in different directions. If we want to exert a force on a positively charged ion in the direction $\langle c/2, c/2, c/2\rangle$, then how should we select a?

Solution We need to solve the equation $\lambda\langle c/2, c/2, c/2\rangle = s\langle c, -c, 0\rangle \times \langle a, 1, 1\rangle$. The constant λ on the left allows us to adjust for length. Calculating the cross product $\langle c, -c, 0\rangle \times \langle a, 1, 1\rangle$, we find that

$$\frac{\lambda}{2}\langle c, c, c\rangle = s\langle -c, -c, c + ca\rangle.$$

We choose $\lambda = -2s$ so that the first two entries on either side match up. For this vector equation to hold, the third entries $\lambda c/2$ and $s(c + ca)$ must also be equal. We must therefore choose a so that $(-2s)c/2 = s(c + ca)$. This forces us to set $a = -2$. ■

The Triple Scalar Product

We have just seen that the cross product is useful for finding the areas of triangles and parallelograms. We now develop this idea further by introducing a new type of operation, a product that involves three vectors.

Definition If $\mathbf{u}$, $\mathbf{v}$, and $\mathbf{w}$ are given vectors, then we define their *triple scalar product* to be the number $(\mathbf{u} \times \mathbf{v}) \cdot \mathbf{w}$.

Notice that the triple scalar product $(\mathbf{u} \times \mathbf{v}) \cdot \mathbf{w}$ of three vectors $\mathbf{u}$, $\mathbf{v}$, and $\mathbf{w}$ is a *scalar* because it is the dot product of the two vectors $\mathbf{u} \times \mathbf{v}$ and $\mathbf{w}$. The parentheses in the triple scalar product are not really necessary: The association $\mathbf{u} \times (\mathbf{v} \cdot \mathbf{w})$ makes no sense since the cross product of vector $\mathbf{u}$ with scalar $\mathbf{v} \cdot \mathbf{w}$ is not defined. However, the expression $\mathbf{u} \cdot (\mathbf{v} \times \mathbf{w})$ *is* defined, and by expanding both it and the triple scalar product $(\mathbf{u} \times \mathbf{v}) \cdot \mathbf{w}$ in terms of the entries of $\mathbf{u}$, $\mathbf{v}$, and $\mathbf{w}$, we find that

$$(\mathbf{u} \times \mathbf{v}) \cdot \mathbf{w} = \mathbf{u} \cdot (\mathbf{v} \times \mathbf{w}). \tag{11.25}$$

In other words, to compute the triple scalar product of $\mathbf{u}$, $\mathbf{v}$, and $\mathbf{w}$, we write down the vectors *in the given order,* insert the operations $\cdot$ and $\times$ in either order, and then associate in the only way possible—refer to Figure 7.

$$\underbrace{\mathbf{u}\ \mathbf{v}\ \mathbf{w}}_{\text{Three vectors, in order}} \quad\Rightarrow\quad \underbrace{\begin{matrix}\mathbf{u} \times \mathbf{v} \cdot \mathbf{w} \\ \text{or} \\ \mathbf{u} \cdot \mathbf{v} \times \mathbf{w}\end{matrix}}_{\text{Insert} \cdot \text{and} \times} \quad\Rightarrow\quad \underbrace{\begin{matrix}(\mathbf{u} \times \mathbf{v}) \cdot \mathbf{w} \\ \text{or} \\ \mathbf{u} \cdot (\mathbf{v} \times \mathbf{w})\end{matrix}}_{\text{Associate}}$$

Figure 7
The triple scalar product of $\mathbf{u}$, $\mathbf{v}$, and $\mathbf{w}$

Example 10 Calculate the triple scalar product of $\mathbf{u} = \langle 2, -1, 4\rangle$, $\mathbf{v} = \langle 7, 2, 3\rangle$, and $\mathbf{w} = \langle -1, 1, 2\rangle$ in two different ways.

Solution We may calculate the cross product as $\mathbf{u} \times \mathbf{v} = \langle -11, 22, 11\rangle$. Therefore

$$(\mathbf{u} \times \mathbf{v}) \cdot \mathbf{w} = \langle -11, 22, 11\rangle \cdot \langle -1, 1, 2\rangle = 11 + 22 + 22 = 55,$$

or,

$$\mathbf{u} \cdot (\mathbf{v} \times \mathbf{w}) = \langle 2, -1, 4\rangle \cdot \langle 1, -17, 9\rangle = 2 + 17 + 36 = 55.$$ ■

The next theorem shows that we can compute a triple scalar product without first calculating a cross product.

Theorem 3 The triple scalar product of $\mathbf{u} = \langle u_1, u_2, u_3\rangle$, $\mathbf{v} = \langle v_1, v_2, v_3\rangle$, and $\mathbf{w} = \langle w_1, w_2, w_3\rangle$ is given by the formula

$$(\mathbf{u} \times \mathbf{v}) \cdot \mathbf{w} = \det\left(\begin{bmatrix} u_1 & u_2 & u_3 \\ v_1 & v_2 & v_3 \\ w_1 & w_2 & w_3 \end{bmatrix}\right).$$

Proof Since $(\mathbf{u} \times \mathbf{v}) \cdot \mathbf{w} = \mathbf{u} \cdot (\mathbf{v} \times \mathbf{w})$ and

$$\mathbf{v} \times \mathbf{w} = \det\left(\begin{bmatrix} v_2 & v_3 \\ w_2 & w_3 \end{bmatrix}\right)\mathbf{i} - \det\left(\begin{bmatrix} v_1 & v_3 \\ w_1 & w_3 \end{bmatrix}\right)\mathbf{j} + \det\left(\begin{bmatrix} v_1 & v_2 \\ w_1 & w_2 \end{bmatrix}\right)\mathbf{k},$$

we have

$$\begin{aligned}(\mathbf{u} \times \mathbf{v}) \cdot \mathbf{w} &= u_1 \det\left(\begin{bmatrix} v_2 & v_3 \\ w_2 & w_3 \end{bmatrix}\right) - u_2 \det\left(\begin{bmatrix} v_1 & v_3 \\ w_1 & w_3 \end{bmatrix}\right) + u_3 \det\left(\begin{bmatrix} v_1 & v_2 \\ w_1 & w_2 \end{bmatrix}\right) \\ &= \det\left(\begin{bmatrix} u_1 & u_2 & u_3 \\ v_1 & v_2 & v_3 \\ w_1 & w_2 & w_3 \end{bmatrix}\right).\end{aligned}$$

■

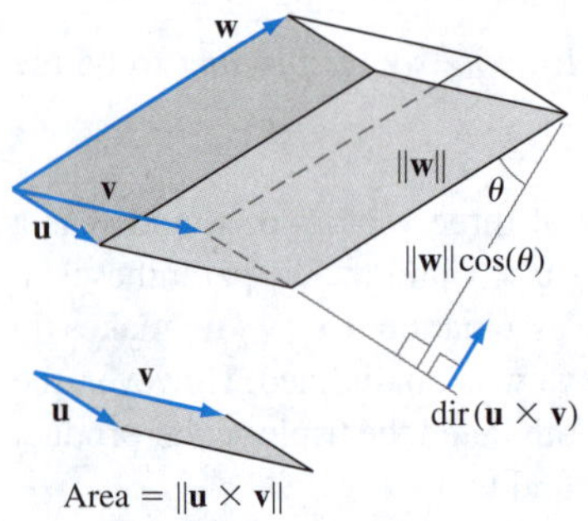

Figure 8

To understand the geometric significance of the triple scalar product, we look at Figure 8. The solid region determined by $\mathbf{u}$, $\mathbf{v}$, and $\mathbf{w}$ is called a parallelepiped. By the Right Hand Rule, $\mathbf{u} \times \mathbf{v}$ points upward in the figure and is perpendicular to the base of the parallelepiped. Moreover, $\|\mathbf{u} \times \mathbf{v}\|$ is the area of the base. The height of the parallelepiped is $\|\mathbf{w}\| \cos(\theta)$ where θ is the angle between $\mathbf{w}$ and $\mathbf{u} \times \mathbf{v}$ (as shown in Figure 8). We may always take θ to be between 0 and $\pi/2$. We calculate the volume of the parallelepiped as follows:

$$\begin{aligned}\text{volume of parallelepiped} &= \text{area of base} \cdot \text{height} \\ &= \|\mathbf{u} \times \mathbf{v}\| \cdot \|\mathbf{w}\| \cos\theta \\ &= |(\mathbf{u} \times \mathbf{v}) \cdot \mathbf{w}|.\end{aligned}$$

Since the volume of the parallelepiped is obviously independent of the order in which we take the vectors $\mathbf{u}$, $\mathbf{v}$, and $\mathbf{w}$, we deduce that the absolute value of the triple product, taken in any order, gives the same answer. We record the results of this investigation as the following theorem.

Theorem 4 Let $\mathbf{u}$, $\mathbf{v}$, $\mathbf{w}$ be vectors. The volume of the parallelepiped determined by these vectors is given by any of the expressions

$$|(\mathbf{u} \times \mathbf{v}) \cdot \mathbf{w}| = |(\mathbf{u} \times \mathbf{w}) \cdot \mathbf{v}| = |(\mathbf{v} \times \mathbf{w}) \cdot \mathbf{u}| = \left|\det\left(\begin{bmatrix} u_1 & u_2 & u_3 \\ v_1 & v_2 & v_3 \\ w_1 & w_2 & w_3 \end{bmatrix}\right)\right|.$$

If $\mathbf{u} \times \mathbf{v}$ and $\mathbf{w}$ point to the same side of the plane determined by $\mathbf{u}$ and $\mathbf{v}$ (the vectors form a right-hand system), then $(\mathbf{u} \times \mathbf{v}) \cdot \mathbf{w}$ is already positive and equals the volume of the parallelepiped.

Example 11 Use the determinant to calculate the volume of the parallelepiped determined by the vectors $\langle -3, 2, 5\rangle$, $\langle 1, 0, 3\rangle$, and $\langle 3, -1, -2\rangle$.

Solution The required volume is

$$\left|\det\left(\begin{bmatrix} -3 & 2 & 5 \\ 1 & 0 & 3 \\ 3 & -1 & -2 \end{bmatrix}\right)\right| = \left|-3\det\left(\begin{bmatrix} 0 & 3 \\ -1 & -2 \end{bmatrix}\right) - 2\det\left(\begin{bmatrix} 1 & 3 \\ 3 & -2 \end{bmatrix}\right)\right.$$
$$\left. + 5\det\left(\begin{bmatrix} 1 & 0 \\ 3 & -1 \end{bmatrix}\right)\right|$$
$$= |-3(3) - 2(-11) + 5(-1)| = 8.$$

Vectors are said to be *coplanar* when they lie in the same plane. Our next theorem gives us a simple way to tell when three vectors are coplanar.

Theorem 5 Three vectors $\mathbf{u} = \langle u_1, u_2, u_3\rangle$, $\mathbf{v} = \langle v_1, v_2, v_3\rangle$, and $\mathbf{w} = \langle w_1, w_2, w_3\rangle$ are coplanar if and only if $\mathbf{u}\cdot(\mathbf{v}\times\mathbf{w}) = 0$. Equivalently, they are coplanar if and only if $\det\left(\begin{bmatrix} u_1 & u_2 & u_3 \\ v_1 & v_2 & v_3 \\ w_1 & w_2 & w_3 \end{bmatrix}\right) = 0.$

Proof According to Theorem 4, this determinant is zero if and only if the parallelepiped determined by the three vectors has zero volume. But this means that the three vectors are coplanar.

Example 12 Show that $\mathbf{u} = \langle 3, 1, 1\rangle$, $\mathbf{v} = \langle 1, 2, 0\rangle$, and $\mathbf{w} = \langle 1, -3, 1\rangle$ are coplanar.

Solution We calculate

$$\det\left(\begin{bmatrix} 3 & 1 & 1 \\ 1 & 2 & 0 \\ 1 & -3 & 1 \end{bmatrix}\right) = 3\det\left(\begin{bmatrix} 2 & 0 \\ -3 & 1 \end{bmatrix}\right) - 1\cdot\det\left(\begin{bmatrix} 1 & 0 \\ 1 & 1 \end{bmatrix}\right) + 1\cdot\det\left(\begin{bmatrix} 1 & 2 \\ 1 & -3 \end{bmatrix}\right)$$
$$= 6 - 1 - 5 = 0.$$

quickquiz

1. Give the geometric description of the cross product of two vectors $\mathbf{v}$ and $\mathbf{w}$.
2. What is the determinant formula for the cross product of two vectors $\mathbf{v} = \langle v_1, v_2, v_3\rangle$ and $\mathbf{w} = \langle w_1, w_2, w_3\rangle$?
3. Find the standard unit normal vector for the pair $\langle 3, -1, 2\rangle$, $\langle 2, 1, 1\rangle$.
4. True or false: $\mathbf{u}\cdot\mathbf{v}\times\mathbf{w} = \mathbf{u}\times\mathbf{v}\cdot\mathbf{w}$.

EXERCISES

Problems for Practice

In Exercises 1–8, use equation (11.22) or (11.24) to compute $\mathbf{v} \times \mathbf{w}$ and $\mathbf{w} \times \mathbf{v}$. Verify the anticommutative property $\mathbf{v} \times \mathbf{w} = -\mathbf{w} \times \mathbf{v}$. Also calculate $\mathbf{v} \cdot (\mathbf{v} \times \mathbf{w})$ and $\mathbf{w} \cdot (\mathbf{v} \times \mathbf{w})$. Verify that $\mathbf{v}$ and $\mathbf{w}$ are perpendicular to $\mathbf{v} \times \mathbf{w}$.

1. $\mathbf{v} = \langle 3, 1, -6 \rangle$, $\mathbf{w} = \langle -2, 4, 4 \rangle$
2. $\mathbf{v} = \langle 0, 1, 1 \rangle$, $\mathbf{w} = \langle 1, 1, 0 \rangle$
3. $\mathbf{v} = \langle 4, -5, 8 \rangle$, $\mathbf{w} = \langle 2, 1, 0 \rangle$
4. $\mathbf{v} = \langle 2, 0, -2 \rangle$, $\mathbf{w} = \langle 2, 0, 2 \rangle$
5. $\mathbf{v} = \langle 2, -3, 5 \rangle$, $\mathbf{w} = \langle -6, 9, -15 \rangle$
6. $\mathbf{v} = \langle -9, 4, 6 \rangle$, $\mathbf{w} = \langle 0, 0, 1 \rangle$
7. $\mathbf{v} = \langle 1, 6, 8 \rangle$, $\mathbf{w} = \langle 1, 2, 8 \rangle$
8. $\mathbf{v} = \langle 2, 3, 3 \rangle$, $\mathbf{w} = \langle -2, -2, 3 \rangle$

In Exercises 9–14, find the standard unit normal vector associated to the given pair of vectors.

9. $\langle 2, 1, 3 \rangle$, $\langle -3, -2, 0 \rangle$
10. $\langle 0, 1, 1 \rangle$, $\langle 1, 1, 0 \rangle$
11. $\langle 4, -1, 1 \rangle$, $\langle 6, 4, 7 \rangle$
12. $\langle -4, -1, -6 \rangle$, $\langle 9, 2, 1 \rangle$
13. $\langle 1, 2, 2 \rangle$, $\langle 2, 2, 1 \rangle$
14. $\langle 0, 1/2, 1 \rangle$, $\langle 1/2, 0, 1 \rangle$

In Exercises 15–18, find the area of the triangle determined by the two vectors.

15. $\langle 2, 1, 2 \rangle$, $\langle 3, 2, 7 \rangle$
16. $\langle -5, -5, -1 \rangle$, $\langle 1, 3, 3 \rangle$
17. $\langle -3, 4, 1 \rangle$, $\langle -5, 0, 6 \rangle$
18. $\langle 2, 2, 5 \rangle$, $\langle -2, 11, -13 \rangle$

In Exercises 19–22, calculate the area of the parallelogram determined by the two vectors.

19. $\langle 4, 6, 1 \rangle$, $\langle 0, -5, 3 \rangle$
20. $\langle -6, 9, 2 \rangle$, $\langle 18, 4, -12 \rangle$
21. $\langle 0, 0, 2 \rangle$, $\langle 1, -3, 5 \rangle$
22. $\langle 1, 3, 7 \rangle$, $\langle -7, -3, -1 \rangle$

In Exercises 23–26, calculate $\mathbf{u} \times (\mathbf{v} \times \mathbf{w})$ and $(\mathbf{u} \times \mathbf{v}) \times \mathbf{w}$. Are the two vectors equal?

23. $\mathbf{u} = \langle -4, 1, 0 \rangle$, $\mathbf{v} = \langle 7, -1, -3 \rangle$, $\mathbf{w} = \langle 8, -5, 3 \rangle$
24. $\mathbf{u} = \langle 0, 4, -3 \rangle$, $\mathbf{v} = \langle 9, 0, -6 \rangle$, $\mathbf{w} = \langle -6, -3, -2 \rangle$
25. $\mathbf{u} = \langle 1, 1, 2 \rangle$, $\mathbf{v} = \langle 2, 1, 2 \rangle$, $\mathbf{w} = \langle -3, 4, -3 \rangle$
26. $\mathbf{u} = \langle -4, -3, -2 \rangle$, $\mathbf{v} = \langle -6, 1, -5 \rangle$, $\mathbf{w} = \langle 2, 1, 0 \rangle$

In Exercises 27–30, calculate $\mathbf{u} \times \mathbf{v}$ and $\mathbf{v} \times \mathbf{w}$. Verify that

$$(\mathbf{u} \times \mathbf{v}) \cdot \mathbf{w} = \mathbf{u} \cdot (\mathbf{v} \times \mathbf{w}).$$

27. $\mathbf{u} = \langle 3, 0, 1 \rangle$, $\mathbf{v} = \langle 2, -1, -3 \rangle$, $\mathbf{w} = \langle -1, -3, 2 \rangle$
28. $\mathbf{u} = \langle 0, 4, -3 \rangle$, $\mathbf{v} = \langle 2, 0, -6 \rangle$, $\mathbf{w} = \langle -3, -1, -2 \rangle$
29. $\mathbf{u} = \langle 1, 1, -2 \rangle$, $\mathbf{v} = \langle 2, 1, -2 \rangle$, $\mathbf{w} = \langle 3, 2, 3 \rangle$
30. $\mathbf{u} = \langle 4, 3, -2 \rangle$, $\mathbf{v} = \langle 2, 1, -5 \rangle$, $\mathbf{w} = \langle 2, 1, 0 \rangle$

In Exercises 31–34, use a determinant to calculate the triple scalar product $(\mathbf{u} \times \mathbf{v}) \cdot \mathbf{w}$.

31. $\mathbf{u} = \langle 1, -2, 4 \rangle$, $\mathbf{v} = \langle 2, 0, 1 \rangle$, $\mathbf{w} = \langle 3, 1, 1 \rangle$
32. $\mathbf{u} = \langle 2, 1, 2 \rangle$, $\mathbf{v} = \langle 2, 2, 3 \rangle$, $\mathbf{w} = \langle 3, 3, -5 \rangle$
33. $\mathbf{u} = \langle 1, 1, 1 \rangle$, $\mathbf{v} = \langle 3, 0, 2 \rangle$, $\mathbf{w} = \langle -2, -3, 3 \rangle$
34. $\mathbf{u} = \langle -5, -1, 0 \rangle$, $\mathbf{v} = \langle 1, 0, 3 \rangle$, $\mathbf{w} = \langle 0, 2, 2 \rangle$

In Exercises 35–38, use the triple scalar product to verify that the three vectors are coplanar.

35. $\mathbf{u} = \langle 1, -2, 4 \rangle$, $\mathbf{v} = \langle 2, 0, 1 \rangle$, $\mathbf{w} = \langle 5, -2, 6 \rangle$
36. $\mathbf{u} = \langle 2, 1, 2 \rangle$, $\mathbf{v} = \langle 2, 2, 3 \rangle$, $\mathbf{w} = \langle 6, 4, 7 \rangle$
37. $\mathbf{u} = \langle 1, 1, 1 \rangle$, $\mathbf{v} = \langle 3, 0, 2 \rangle$, $\mathbf{w} = \langle 8, 2, 6 \rangle$
38. $\mathbf{u} = \langle -5, -1, 0 \rangle$, $\mathbf{v} = \langle 1, 0, 3 \rangle$, $\mathbf{w} = \langle -6, -1, -3 \rangle$

Further Theory and Practice

In Exercises 39–42, find scalars s and t for which $\mathbf{u} \times (\mathbf{v} \times \mathbf{w}) = s\mathbf{v} + t\mathbf{w}$.

39. $\mathbf{u} = \langle 1, -2, 4 \rangle$, $\mathbf{v} = \langle 2, 0, 1 \rangle$, $\mathbf{w} = \langle 5, -3, 2 \rangle$
40. $\mathbf{u} = \langle 2, 1, 2 \rangle$, $\mathbf{v} = \langle 2, 2, 3 \rangle$, $\mathbf{w} = \langle 1, 1, 1 \rangle$
41. $\mathbf{u} = \langle 1, 1, -1 \rangle$, $\mathbf{v} = \langle 3, 0, 2 \rangle$, $\mathbf{w} = \langle 4, 2, 1 \rangle$
42. $\mathbf{u} = \langle 5, -1, 0 \rangle$, $\mathbf{v} = \langle 1, 0, 3 \rangle$, $\mathbf{w} = \langle -2, -1, 7 \rangle$

43. Suppose that $\mathbf{u}$, $\mathbf{v}$, and $\mathbf{w}$ are spatial vectors. Prove that

$$\mathbf{u} \times (\mathbf{v} + \mathbf{w}) = (\mathbf{u} \times \mathbf{v}) + (\mathbf{u} \times \mathbf{w}).$$

44. Suppose that $\mathbf{u}$, $\mathbf{v}$, and $\mathbf{w}$ are spatial vectors. Use the result of Exercise 43 to deduce that

$$(\mathbf{v} + \mathbf{w}) \times \mathbf{u} = (\mathbf{v} \times \mathbf{u}) + (\mathbf{w} \times \mathbf{u}).$$

45. Suppose that $\mathbf{v}$ and $\mathbf{w}$ are spatial vectors and that λ and μ are scalars. Prove that

$$(\lambda\mathbf{v}) \times (\mu\mathbf{w}) = (\lambda\mu)(\mathbf{v} \times \mathbf{w}).$$

46. Suppose that $\mathbf{v}$ and $\mathbf{w}$ are spatial vectors. Prove that

$$(\mathbf{u} \times \mathbf{v}) \times \mathbf{w} = \mathbf{w} \times (\mathbf{v} \times \mathbf{u}).$$

47. Suppose that $\mathbf{u}$, $\mathbf{v}$, and $\mathbf{w}$ are spatial vectors. Prove that

$$\mathbf{u} \times (\mathbf{v} \times \mathbf{w}) = (\mathbf{u} \cdot \mathbf{w})\mathbf{v} - (\mathbf{u} \cdot \mathbf{v})\mathbf{w}.$$

Deduce that

$$(\mathbf{u} \times \mathbf{v}) \times \mathbf{w} = (\mathbf{w} \cdot \mathbf{u})\mathbf{v} - (\mathbf{w} \cdot \mathbf{v})\mathbf{u}.$$

Notice that the cross product is *not* associative in general.

The cross product can be interpreted physically in terms of *moment of force*. Namely, let $\mathbf{F}$ be a force (vector) applied at a point Q in space and let P be another point in space. The *moment* or *torque* $\boldsymbol{\tau}$ of the force $\mathbf{F}$ at the point Q about the point P is defined to be

$$\boldsymbol{\tau} = \overrightarrow{PQ} \times \mathbf{F}.$$

See Figure 9. Notice that $\boldsymbol{\tau}$ is a vector. If a certain body pivots about the point P and extends in length to the point Q and if the force is applied at Q, then $\|\boldsymbol{\tau}\|$ measures the tendency of the force $\mathbf{F}$ to make the body rotate about P. The direction of $\boldsymbol{\tau}$ gives the axis of rotation. A bolt at point P will be driven by a wrench in the direction of $\boldsymbol{\tau}$ (see Figure 9). Exercises 48–50 concern torque.

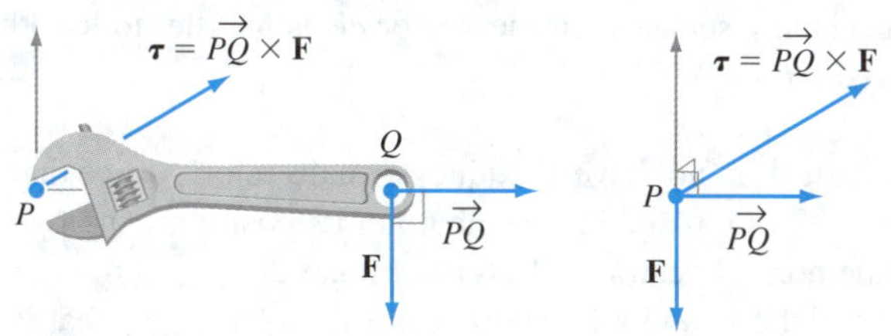

Figure 9

48. A diver who weighs 120 lb stands on the end of a diving board. The diving board is 10 ft long. The diving board is rigid; it is *not* a spring board. It extends in an upward direction from the edge of the pool at a 30 degree angle with the horizontal. What is the moment of force, due to the weight of the diver, at the end of the board where it is attached?

49. An 8 in. wrench is used to drive a bolt at point P. A force $\mathbf{F}$ is applied to the end of the handle (point Q). If $\|\mathbf{F}\| = 60$ lb and the angle between $\mathbf{F}$ and $\overrightarrow{PQ}$ is 90° then what is the magnitude (in foot-pounds) of the torque that is produced?

50. Repeat Exercise 49 but suppose that the angle of application is 120°.

51. Suppose that ℓ is the line through distinct points P and Q. Suppose that R is a point not on ℓ. The distance $d(R, \ell)$ of R to ℓ is defined to be the minimum value of $|\overline{RT}|$ as T varies over all points of ℓ. It may be shown that there is a unique point S on ℓ such that $\overrightarrow{RS}$ and $\overrightarrow{PQ}$ are mutually perpendicular. Show that $d(R, \ell) = \|\overrightarrow{RS}\|$. Use the cross product to calculate $d(R, \ell)$ in terms of the given points P, Q, and R.

52. The *Gram determinant* of two spatial vectors $\mathbf{v}$ and $\mathbf{w}$ is defined to be

$$G = \det\left(\begin{bmatrix} \mathbf{v} \cdot \mathbf{v} & \mathbf{v} \cdot \mathbf{w} \\ \mathbf{w} \cdot \mathbf{v} & \mathbf{w} \cdot \mathbf{w} \end{bmatrix}\right).$$

Show that $\mathbf{v}$ and $\mathbf{w}$ are parallel if and only if $G = 0$.

53. Prove the Lagrange Identity:

$$(\mathbf{v} \times \mathbf{w}) \cdot (\mathbf{p} \times \mathbf{q}) = (\mathbf{v} \cdot \mathbf{p})(\mathbf{w} \cdot \mathbf{q}) - (\mathbf{v} \cdot \mathbf{q})(\mathbf{w} \cdot \mathbf{p}).$$

54. Prove the Jacobi Identity:

$$(\mathbf{u} \times \mathbf{v}) \times \mathbf{w} + (\mathbf{v} \times \mathbf{w}) \times \mathbf{u} + (\mathbf{w} \times \mathbf{u}) \times \mathbf{v} = \vec{\mathbf{0}}.$$

Calculator/Computer Exercises

55. Let $P = (x, y, z)$, $P_0 = (1, 2, 3)$, $\mathbf{v} = \langle 2, 1, -4 \rangle$, and $\mathbf{w} = \langle 5, -2, 1 \rangle$. Write the equation $\overrightarrow{P_0P} \cdot (\mathbf{v} \times \mathbf{w}) = 0$ as a Cartesian equation in x, y, and z. Plot the solution set S of the equation. What geometric figure results? What is the relationship of this figure to the vector $\mathbf{v} \times \mathbf{w}$?

56. Consider the curve that is parameterized by $t \mapsto (5 + \cos(t), 3 + \sin(t), 5), 0 \le t \le 2\pi$. Find the point P on the curve that is closest to the origin. Find a vector $\mathbf{v}$ that is tangent to the curve at P. Calculate $\|\mathbf{v} \times \overrightarrow{OP}\|$ and $\|\mathbf{v}\| \cdot \|\overrightarrow{OP}\|$. What significance for $\mathbf{v}$ and $\overrightarrow{OP}$ do these calculations have?

57. Suppose that $\mathbf{v}_t = \langle 1 - t^2, t, 4 - t^2 \rangle$ and $\mathbf{w} = \langle 1, 2, 3 \rangle$. Plot

$$f(t) = \|\mathbf{v}_t \times \mathbf{w}\| - \mathbf{v}_t \cdot \mathbf{w}$$

for $-3 \le t \le 3$. What is the minimum value of f?

58. Suppose that $\mathbf{v}_t = \langle t, t^2, 1 - t^2 \rangle$ and $\mathbf{w}_t = \langle 1 - t^2, t, t^2 \rangle$. Plot $f(t) = \|\mathbf{v}_t \times \mathbf{w}_t\|^2$ for $-1 \le t \le 1$. What are the local and global minima of f?

11.5 Lines and Planes in Space

When we learned to graph with Cartesian coordinates in the two-dimensional plane, we found that the set of points satisfying one linear equation is a line, whereas the set of points satisfying two linear equations is (usually) a point. The philosophy here is that each equation "removes a degree of freedom" or takes away a dimension.

The same philosophy works in three dimensions: The set of points in space satisfying one linear equation will be a plane; the set of points in space satisfying two linear equations will (usually) be a line. Notice that we had to add the word "usually" because sometimes two equations have no solution or have too many solutions.

Cartesian Equations of Planes in Space

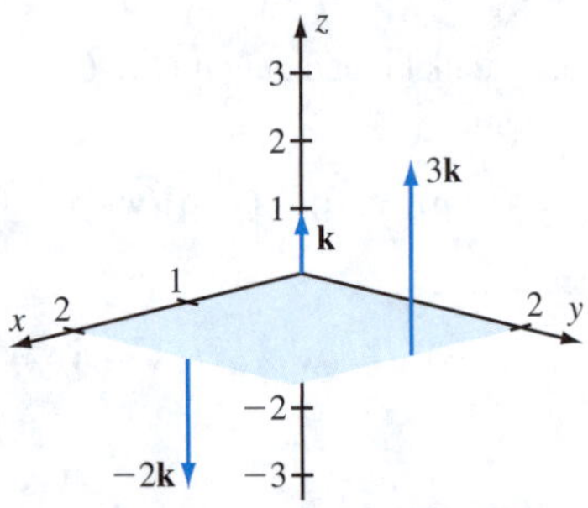

Figure 1

We learned to specify a line in the plane by giving two pieces of information, such as a point it passes through and the direction or slope. We want to use the same idea when specifying a plane in space. However, we specify the "direction" of a plane in a rather indirect way. Namely, we give a vector that is perpendicular to the plane. By this, we mean that the vector is perpendicular to *every* vector that lies in the plane. Such a vector is called a *normal vector* for the plane.

Example 1 Give several examples of normal vectors for the xy-plane.

Solution The vector $\mathbf{k}$ is a normal vector for the xy-plane. So are the vectors $3\mathbf{k}$ and $-2\mathbf{k}$ (Figure 1). Indeed, if c is any scalar, then the vector $c\mathbf{k}$ is parallel to $\mathbf{k}$ and therefore perpendicular to the xy-plane. ■

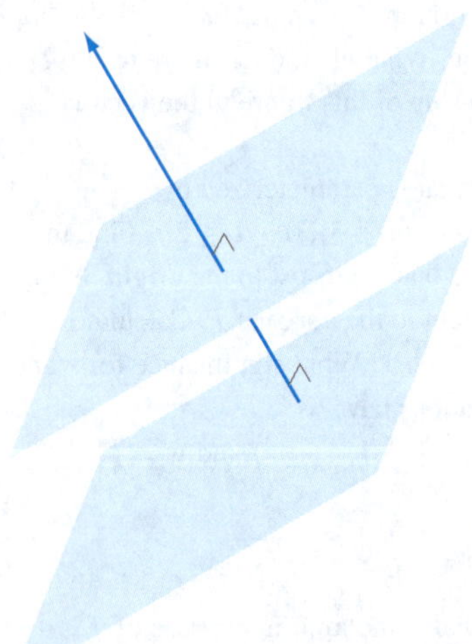

Figure 2
Parallel planes have the same normal vectors.

As Example 1 illustrates, a plane does not have a unique normal vector. Notice that a normal vector determines the "tilt" of a plane in space but not its position. Any two parallel planes will have the same normal vector(s). Refer to Figure 2.

Figure 3 (next page) shows a plane V and a nonzero vector $\mathbf{n} = \langle A, B, C\rangle$ that is normal to V. Figure 3 also shows a fixed point $P_0 = (x_0, y_0, z_0)$ on the plane. Now let $P = (x, y, z)$ be *any point* on the plane. The key geometric fact that we need is that the vector $\overrightarrow{P_0P}$ lies in the plane. This is true precisely when the vector $\overrightarrow{P_0P}$ is perpendicular to $\mathbf{n}$. In other words, P lies on the given plane V if and only if $\mathbf{n} \cdot \overrightarrow{P_0P} = 0$. In coordinates, the vector $\overrightarrow{P_0P}$ is given by $\langle x - x_0, y - y_0, z - z_0\rangle$ and the equation becomes $\langle A, B, C\rangle \cdot \langle x - x_0, y - y_0, z - z_0\rangle = 0$, or

$$A(x - x_0) + B(y - y_0) + C(z - z_0) = 0. \tag{11.26}$$

Notice that the expression $D = Ax_0 + By_0 + Cz_0$ is a constant. Using this constant, we may rewrite equation (11.26) as

$$Ax + By + Cz = D. \tag{11.27}$$

We summarize these observations in the following theorem.

Theorem 1 Suppose that V is a plane for which $\mathbf{n} = \langle A, B, C\rangle$ is a normal vector. If $P_0 = (x_0, y_0, z_0)$ is any point on V, then $A(x - x_0) + B(y - y_0) + C(z - z_0) = 0$ is a Cartesian equation for V. This equation may be written in the form $Ax + By + Cz = D$ where $D = Ax_0 + By_0 + Cz_0 = \mathbf{n} \cdot \overrightarrow{OP_0}$.

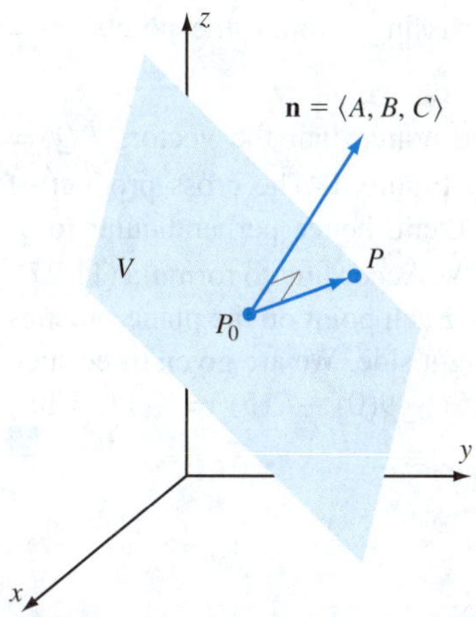

Figure 3
$\mathbf{n} \cdot \overrightarrow{P_0P} = 0$

Example 2 Determine a Cartesian equation for the plane that has normal vector $\mathbf{n} = \langle -2, 7, 3 \rangle$ and passes through the point $P_0 = (-5, -2, 1)$.

Solution Equation (11.27) tells us that the required equation has the form $-2x + 7y + 3z = D$ where $D = -2x_0 + 7y_0 + 3z_0$ for any point (x_0, y_0, z_0) on the plane. We are given one such point, P_0. It follows that $D = -2(-5) + 7(-2) + 3(1) = -1$, and, as a consequence, $-2x + 7y + 3z = -1$ is a Cartesian equation for the plane. Therefore, $-2x + 7y + 3z = -1$ is a Cartesian equation for the plane. ■

inSIGHT

If the three variables x, y, z have no equations or restrictions imposed on them, then they have three "degrees of freedom" and generate three-dimensional space. In general, each new equation imposed on x, y, z removes one degree of freedom. We therefore expect the solution set of *one* Cartesian equation to be two-dimensional. The same thing holds for curved shapes. As we have seen, a sphere can be described by one Cartesian equation $(x - x_0)^2 + (y - y_0)^2 + (z - z_0)^2 = r^2$ and that reflects its two-dimensional nature.

The chain of reasoning that we have used to determine a Cartesian equation for a given plane can be reversed. That is, if we are given either equation (11.26) or equation (11.27), and if at least one of the coefficients A, B, C is nonzero, then we can be sure that the solution set of the equation is a plane that is perpendicular to $\mathbf{n} = \langle A, B, C \rangle$. We state this observation as a theorem.

Theorem 2 Suppose that at least one of the coefficients A, B, C is nonzero. The solution set of the equation $A(x - x_0) + B(y - y_0) + C(z - z_0) = 0$ is the plane that has $\langle A, B, C \rangle$ as a normal vector and that passes through the point (x_0, y_0, z_0). The solution set of the equation $Ax + By + Cz = D$ is a plane that has $\langle A, B, C \rangle$ as a normal vector.

Example 3 Find a vector that is normal to, and two points that lie on, the plane V with Cartesian equation $z = 3y - 2x$.

Solution By writing the given equation in the standard form $2x - 3y + z = 0$, we may conclude that the vector $\langle 2, -3, 1 \rangle$ is a normal vector to V. To find a point $P_0 = (x_0, y_0, z_0)$ on V, we may choose any two of the entries of P_0 in an arbitrary way and use the equation to solve for the third entry. For example, if we arbitrarily set $x_0 = 7$ and $y_0 = 5$, then we obtain $z_0 = 3y_0 - 2x_0 = (3)(5) - (2)(7) = 1$. Thus, the point $(7, 5, 1)$ lies on V. Similarly, if we (arbitrarily) set $y_0 = 3$ and $z_0 = 1$, then we have $2x_0 - 3(3) + (1) = 0$, or $x_0 = 4$. Therefore the point $(4, 3, 1)$ is also on V. ■

It is intuitively clear that any three points that are not on the same straight line determine a plane. Imagine that the three points are three fingertips and that you balance your notebook on the fingertips—that is the plane we seek. To use equation (11.27), we must be able to obtain a normal vector from the three points. The next example shows how this is done.

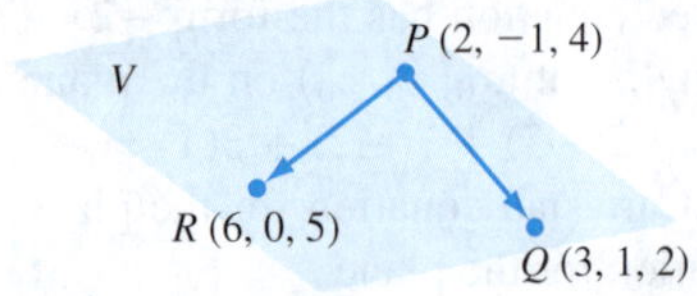

Figure 4

Example 4 Find an equation for the plane V passing through the points $P = (2, -1, 4)$, $Q = (3, 1, 2)$, and $R = (6, 0, 5)$.

Solution To determine a normal vector to V, we notice that the vectors $\overrightarrow{PQ} = \langle 1, 2, -2\rangle$ and $\overrightarrow{PR} = \langle 4, 1, 1\rangle$ lie in the plane. See Figure 4. The cross product of these two vectors will be perpendicular to both of them, hence perpendicular to V. Thus, a normal vector will be $\overrightarrow{PQ} \times \overrightarrow{PR} = 4\mathbf{i} - 9\mathbf{j} - 7\mathbf{k}$. According to formula (11.27), the desired equation has the form $4x - 9y - 7z = D$. Each point on the plane satisfies this equation, yielding the same value for D on the right side. We are given three such points. By choosing one, say R, we obtain $D = 4(6) - 9(0) - 7(5) = -11$. Thus, $4x - 9y - 7z = -11$ is a Cartesian equation for V. ■

inSIGHT

There is a lot of choice in working the type of problem illustrated by Example 4. We started by choosing two pairs of points: P, Q and P, R. We might have chosen P, Q and Q, R (or P, R and Q, R) equally well. We used each pair of points to determine a vector in the plane, but we could just as well have chosen the opposite vectors. For example, in taking the cross product to find a normal vector, we could have used $\overrightarrow{QP}$ instead of $\overrightarrow{PQ}$. You should try these other choices for yourself to see that an equivalent equation results.

We define the *angle θ between two planes* to be the angle between their normals. Note the ambiguity here: We could declare the angle to be θ or $2\pi - \theta$. We always select the angle θ such that $0 \le \theta \le \pi$.

Example 5 Find the angle between the planes $x - y - z = 7$ and $-x + y - 3z = 6$.

Solution As we have noticed, $\langle 1, -1, -1\rangle$ is a normal vector for the first plane and $\langle -1, 1, -3\rangle$ is a normal vector for the second plane. The angle θ between the two given planes satisfies

$$\begin{aligned}\cos(\theta) &= \frac{\langle 1, -1, -1\rangle \cdot \langle -1, 1, -3\rangle}{\|\langle 1, -1, -1\rangle\|\|\langle -1, 1, -3\rangle\|}\\ &= \frac{1}{\sqrt{3}\sqrt{11}}\\ &= \frac{1}{\sqrt{33}}.\end{aligned}$$

Thus, using a calculator, we see that the angle between the two planes is about 1.4 radians. ■

Parametric Equations of Planes in Space

Suppose that $\mathbf{u} = \langle u_1, u_2, u_3\rangle$ and $\mathbf{v} = \langle v_1, v_2, v_3\rangle$ are nonparallel vectors that lie in a plane V. Let $P_0 = (x_0, y_0, z_0)$ be any fixed point that lies on V. We can think of P_0 as an "origin" in the plane. Any point $P = (x, y, z)$ on the plane can be reached from P_0 by adding a scalar multiple of $\mathbf{u}$ and then a scalar multiple of $\mathbf{v}$, as illustrated in

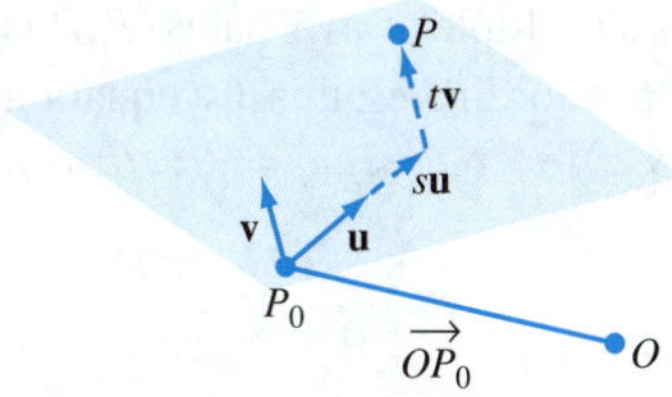

Figure 5
$\overrightarrow{OP} = \overrightarrow{OP_0} + s\mathbf{u} + t\mathbf{v}$

Figure 5. The position vector $\overrightarrow{OP}$ can then be obtained by vector addition as

$$\overrightarrow{OP} = \overrightarrow{OP_0} + s\mathbf{u} + t\mathbf{v}.$$

If we write out this vector equation in coordinates, then we get three scalar parametric equations for the plane:

$$x = x_0 + su_1 + tv_1,$$
$$y = y_0 + su_2 + tv_2,$$
$$z = z_0 + su_3 + tv_3.$$

The parameters s and t may be thought of as coordinates in the plane V: Each point of V is determined by specifying the values of s and t.

inSIGHT

The cross product is useful for producing a vector that is orthogonal to *two* given vectors. Obtaining two vectors orthogonal to *one* given vector $\mathbf{v} = \langle A, B, C\rangle$ is easier. To do that, we may replace one component of $\mathbf{v}$ by 0, interchange the other two, and change the sign of one of them. Thus, $\langle 0, C, -B\rangle$, $\langle C, 0, -A\rangle$, and $\langle B, -A, 0\rangle$ are all orthogonal to $\mathbf{v}$.

Example 6 Find parametric equations for the plane V with Cartesian equation $3x - y + 2z = 7$.

Solution We must find one point $P_0 = (x_0, y_0, z_0)$ and two nonparallel vectors $\mathbf{u}$ and $\mathbf{v}$ in V. To find P_0, we may select any values for x_0 and y_0 and use the Cartesian equation $3x - y + 2z = 7$ to solve for z_0. For example, $P_0 = (1, 0, 2)$ will do. The vectors $\mathbf{u}$ and $\mathbf{v}$ can be any nonparallel vectors that are perpendicular to the normal vector $\mathbf{n} = \langle 3, -1, 2\rangle$. For example, we may take $\mathbf{u} = \langle 0, 2, 1\rangle$ and $\mathbf{v} = \langle 2, 0, -3\rangle$, as you can verify by calculating $\mathbf{u}\cdot\mathbf{n} = 0$ and $\mathbf{v}\cdot\mathbf{n} = 0$. With these choices, the parametric equations become $x = 1 + 0s + 2t$, $y = 0 + 2s + 0t$, $z = 2 + s - 3t$. We may (and should) verify these parametric equations by substituting them into the Cartesian equation and checking that it holds for all s and t:

$$3(1 + 2t) - (2s) + 2(2 + s - 3t) \equiv 7 \quad \text{Identically for all } s \text{ and } t$$ ■

The following theorem summarizes everything that we have learned about planes.

Theorem 3 The equation $Ax + By + Cz = D$ describes a plane that has $\langle A, B, C\rangle$ as a normal vector. Conversely, any plane V that is perpendicular to the vector $\langle A, B, C\rangle$ has the form $Ax + By + Cz = D$ for some constant D. If $P_0 = (x_0, y_0, z_0)$ is a point on V, then the Cartesian equation of V may be written as $A(x - x_0) + B(y - y_0) + C(z - z_0) = 0$. If $\mathbf{u} = \langle u_1, u_2, u_3\rangle$ and $\mathbf{v} = \langle v_1, v_2, v_3\rangle$ are any two nonparallel vectors that are perpendicular to $\langle A, B, C\rangle$, then

$$x = x_0 + su_1 + tv_1,$$
$$y = y_0 + su_2 + tv_2,$$
$$z = z_0 + su_3 + tv_3$$

are parametric equations for V.

Parametric Equations of Lines in Space

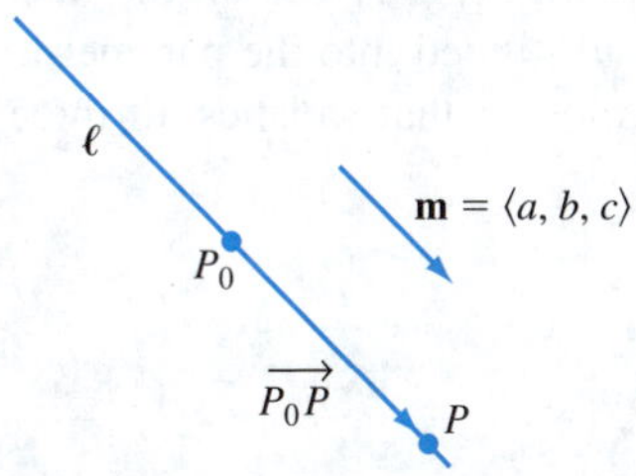

Figure 6

A line ℓ in space can be described by a point that it passes through and a vector that is parallel to it. Look at Figure 6. Let $P_0 = (x_0, y_0, z_0)$ be the point and $\mathbf{m} = \langle a, b, c\rangle$ the vector parallel to ℓ. An arbitrary point $P = (x, y, z)$ lies on ℓ if and only if the vector

$\overrightarrow{P_0P}$ is parallel to **m**. This happens if and only if $\overrightarrow{P_0P}$ is a multiple of **m**. That is, $\overrightarrow{P_0P}$ is parallel to **m** if and only if $\overrightarrow{P_0P} = t\mathbf{m}$ for some scalar t. In coordinates, this last equation becomes $\langle x - x_0, y - y_0, z - z_0\rangle = t\langle a, b, c\rangle$, or $\langle x, y, z\rangle - \langle x_0, y_0, z_0\rangle = \langle ta, tb, tc\rangle$. By adding vector $\langle x_0, y_0, z_0\rangle$ to each side, we obtain

$$\langle x, y, z\rangle = \langle x_0 + ta, y_0 + tb, z_0 + tc\rangle.$$

Notice that t is a scalar and will play the role of a parameter. After we match up coordinates, the parametric equations for a line in space become

$$\begin{aligned} x &= x_0 + ta, \\ y &= y_0 + tb, \\ z &= z_0 + tc. \end{aligned}$$

Let us summarize our observations with a theorem.

Theorem 4 ***The Parametric Form for a Line in Space*** The line in space that passes through the point $P_0 = (x_0, y_0, z_0)$ and that is parallel to the vector $\mathbf{m} = \langle a, b, c\rangle$ has equation

$$\overrightarrow{P_0P} = t\mathbf{m}$$

where $P = (x, y, z)$ is a variable point on the line. In coordinates, the equation may be written as three parametric equations:

$$\begin{aligned} x &= x_0 + ta, \\ y &= y_0 + tb, \\ z &= z_0 + tc. \end{aligned}$$

Example 7 Give three points that lie on the line ℓ that has parametric equations

$$\begin{aligned} x &= -5 + 3t, \\ y &= 7 - 8t, \\ z &= 1 + 2t. \end{aligned}$$

Does the point $(1, 2, 3)$ lie on ℓ? How about the point $(-11, 23, -3)$? Give parametric equations for the line through the origin that is parallel to ℓ.

Solution Each value assigned to t will generate a point on ℓ. For instance, $t = 0$ gives $x = -5$, $y = 7$, $z = 1$, or the point $(-5, 7, 1)$. The value $t = 3$ gives $x = 4$, $y = -17$, $z = 7$, or the point $(4, -17, 7)$. The value $t = -1$ gives $x = -8$, $y = 15$, $z = -1$, or the point $(-8, 15, -1)$. To determine whether the point $(1, 2, 3)$ lies on the line, we need to determine whether some value of t, when substituted into the parametric equations, yields this point. Thus, we need to find a *single* t that satisfies all three equations

$$\begin{aligned} 3t - 5 &= 1, \\ -8t + 7 &= 2, \\ 2t + 1 &= 3 \end{aligned}$$

simultaneously. The first equation gives $t = 2$, the second gives $t = 5/8$. We need look no further: *There is no single value of t that satisfies all three equations.* Thus, the point $(1, 2, 3)$ does not lie on the line.

We have better luck with the point $(-11, 23, -3)$. We endeavor to find a (single) t by solving

$$\begin{aligned} 3t - 5 &= -11, \\ -8t + 7 &= 23, \\ 2t + 1 &= -3. \end{aligned}$$

The first equation gives $t = -2$, and so does the second, and so does the third. Thus, substituting $t = -2$ into our parametric equations yields the point $(-11, 23, -3)$, and this point lies on the line. Finally, the parametric equations tell us that the vector $\langle 3, -8, 2 \rangle$ is parallel to ℓ. It follows that the line parameterized by $x = 0 + 2t$, $y = 0 - 8t$, $z = 0 + 2t$ is parallel to ℓ and passes through the origin. ■

Cartesian Equations of Lines in Space

The Cartesian equations of a line are obtained by eliminating the parameter from a parameterization. For example, suppose that a line ℓ is given by the equations $x = x_0 + ta$, $y = y_0 + tb$, $z = z_0 + tc$ with a, b, and c nonzero. (For the cases with one or two of a, b, c being zero, see Exercise 90.) After solving for the parameter t in each of these equations, we have three expressions, $t = (x - x_0)/a$, $t = (y - y_0)/b$, $t = (z - z_0)/c$, for t, and we may equate them:

$$\frac{x - x_0}{a} = \frac{y - y_0}{b} = \frac{z - z_0}{c}. \tag{11.28}$$

These are the Cartesian equations for ℓ. Notice that a line in space is described by two Cartesian equations: $(x - x_0)/a = (y - y_0)/b$ and $(y - y_0)/b = (z - z_0)/c$. A third equation, $(x - x_0)/a = (z - z_0)/c$, can be formed from these two, but it would be redundant. The two equations of line (11.28) are said to be the *symmetric form* of ℓ.

Example 8 Line ℓ from Example 7 is parameterized by $x = -5 + 3t$, $y = 7 - 8t$, $z = 1 + 2t$. Find Cartesian equations for ℓ. Is the point $(1, -9, 5)$ on ℓ? How about the point $(4, -1, 7)$?

Solution We solve for the parameter t in each of the parametric equations: $t = (x+5)/3$, $t = (y-7)/(-8)$, and $t = (z-1)/2$. We now equate these three expressions for t to obtain the symmetric form of ℓ:

$$\frac{x + 5}{3} = \frac{y - 7}{(-8)} = \frac{z - 1}{2}.$$

A point $P = (x, y, z)$ lies on ℓ if and only if all three of these quantities are equal. For example, $(1, -9, 5)$ lies on the line because all of these quantities become 2 when we substitute $x = 1$, $y = -9$, $z = 5$. However, $(4, -1, 7)$ is not on the line because when we substitute $x = 4$, $y = -1$, $z = 5$, the first quantity becomes 3 and the second becomes 1. ■

Just as two points in the plane determine a straight line, so do two points in space. The next example shows how to find the symmetric form of a line that passes through two given points.

Example 9 Let $P = (2, -3, 5)$ and $Q = (-6, 1, 12)$. Write Cartesian equations for the line ℓ passing through P and Q.

Solution The symmetric form of a line requires a point on the line and a vector $\mathbf{m}$ that is parallel to the line. The piece of information we are initially missing is $\mathbf{m}$, but the vector $\overrightarrow{PQ} = \langle -6-2, 1-(-3), 12-5\rangle = \langle -8, 4, 7\rangle$ will certainly do the job. We can use either P or Q as our point. If we choose P, then the symmetric form for ℓ is

$$\frac{x-2}{(-8)} = \frac{y+3}{4} = \frac{z-5}{7}.$$

■

Example 10 Convert the symmetric equations

$$\frac{x+1}{3} = \frac{y}{2} = \frac{2z-3}{6}$$

to parametric form.

Solution We set each of these three equal quantities equal to t. We obtain

$$\frac{x+1}{3} = t, \qquad \frac{y}{2} = t, \qquad \frac{2z-3}{6} = t,$$

or, equivalently,

$$x = 3t - 1, \qquad y = 2t, \qquad z = 3t + \frac{3}{2}.$$

These are parametric equations for the given line. ■

Example 11 Find symmetric equations for the line perpendicular to the plane $3x - 7y + 4z = 2$ and passing through the point $(1, 4, 5)$.

Solution Notice that the vector $\mathbf{n} = \langle 3, -7, 4\rangle$ is normal to the plane. So the line we seek will be parallel to $\mathbf{n}$. Because the line passes through $(1, 4, 5)$, its parametric equations will be

$$\begin{aligned} x &= 1 + 3t, \\ y &= 4 - 7t, \\ z &= 5 + 4t. \end{aligned}$$

The symmetric equations for the line are

$$\frac{x-1}{3} = \frac{y-4}{-7} = \frac{z-5}{4}.$$

■

insight

As we have discussed, a line is described by *two* linear equations. Each of these equations is, taken by itself, the equation of a plane. Therefore, the symmetric equations of a line permit us to see the line as the intersection of two planes. The Cartesian equations in Example 11 may be rewritten as

$$\frac{x-1}{3} = \frac{y-4}{-7} \qquad \text{and} \qquad \frac{y-4}{-7} = \frac{z-5}{4}.$$

Rewriting these equations as

$$7x + 3y = 19 \qquad \text{and} \qquad 4y + 7z = 51,$$

we see that our line is *the set of points that lie in the intersection of two planes*. These ideas are illustrated in Figure 7.

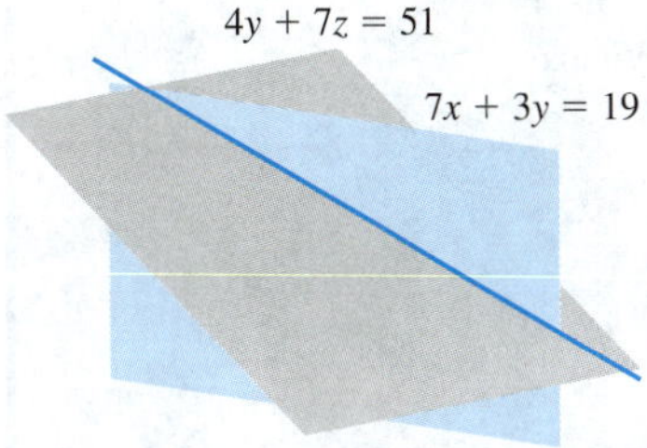

Figure 7
The line $\frac{x-1}{3} = \frac{y-4}{-7} = \frac{z-5}{4}$ as an intersection of two planes.

Example 12 Find parametric equations for the line of intersection of the two planes

$$x - 2y + z = 4 \qquad \text{and} \qquad 2x + y - z = 3.$$

Solution To do so, we set $x = t$ and substitute this into the two planar equations:

$$-2y + z = 4 - t \qquad \text{and} \qquad y - z = 3 - 2t.$$

We may solve these equations simultaneously for y and z in terms of t:

$$\begin{aligned} y &= 3t - 7 \\ z &= 5t - 10. \end{aligned}$$

Parametric equations for the line of intersection are therefore

$$\begin{aligned} x &= t, \\ y &= 3t - 7, \\ z &= 5t - 10. \end{aligned}$$

The symmetric equations are $x = (y + 7)/3 = (z + 10)/5$. ■

inSIGHT

We could have done the last example by setting either $y = t$ or $z = t$. This would have resulted in *different parameterizations for the same line*. For clarity, let us set $y = s$ (instead of t) and see what happens. We obtain $x + z = 4 + 2s$ and $2x - z = 3 - s$. Solving these simultaneously for x and z in terms of s yields

$$x = \frac{s}{3} + \frac{7}{3}$$
$$z = \frac{5}{3}s + \frac{5}{3}.$$

This leads to parametric equations

$$x = \frac{s}{3} + \frac{7}{3},$$
$$y = s,$$
$$z = \frac{5}{3}s + \frac{5}{3}.$$

These three equations describe the very same line as in Example 12, but they look quite different from the parametric equations that we found there. What is the relationship between these two parameterizations? The answer is that the change of variable $s = 3t - 7$ transforms one parametrization to the other.

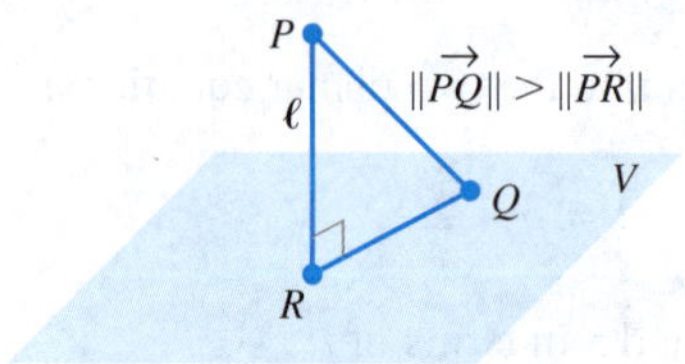

Figure 8
R is the point on V that is closest to P.

Suppose that a plane V and a point P that is not on V are given. Let R be the intersection of V with the line ℓ through P that is normal to V. The Pythagorean Theorem provides us with one way to see that R is the point on V that is closest to P. See Figure 8. The length $\|\overline{PR}\|$ of line segment $\overline{PR}$ is said to be the *distance* of the point P to the plane V.

Example 13 Find the distance between the point $P = (5, 9, 3)$ and the plane V with Cartesian equation $2x + 4y - z = 1$.

Solution Let ℓ be the line through P that is normal to V. Let R be the point at which ℓ passes through V. The distance we seek is $\|\overrightarrow{PR}\|$. To calculate this number, we select a point Q at random on the plane: for instance, substituting $x = 0$ and $y = 0$ in the equation for the plane gives $z = 1$. So $Q = (0, 0, -1)$ lies in the plane. The vector $\overrightarrow{PQ} = \langle -5, -9, -4 \rangle$ connects P to the plane, but it is not the shortest such vector. The shortest vector connecting P to the plane will be the *projection* of $\overrightarrow{PQ}$ onto the normal. The geometry is clear from Figure 8. The length of this shortest vector is the distance from P to the plane. A normal to the plane is $\mathbf{n} = \langle 2, 4, -1 \rangle$. The length of the projection of $\overrightarrow{PQ}$ onto $\mathbf{n}$ is

$$\left| \frac{\overrightarrow{PQ} \cdot \mathbf{n}}{\|\mathbf{n}\|} \right| = \frac{|(-5)(2) + (-9)(4) + (-4)(-1)|}{\sqrt{21}} = \frac{42}{\sqrt{21}} = 2\sqrt{21}.$$

Thus, the distance of the point P to the plane is $2\sqrt{21}$. ■

inSIGHT

Example 13 demonstrates how powerful the concept of projection can be. By calculating the projection, we are able to find the length $\|\overrightarrow{PR}\|$ without first finding the point R. In the following example, we learn how to find the point of intersection of a line and a plane, a procedure that provides a second method for finding the distance between a point and a plane.

Example 14 Find parametric equations for the line ℓ that passes through $P = (5, 9, 3)$ and that is parallel to the vector $\mathbf{n} = \langle 2, 4, -1\rangle$. Find the point R at which ℓ passes through the plane V that has Cartesian equation $2x + 4y - z = 1$.

Solution Line ℓ has parameterization $x = 5 + 2t$, $y = 9 + 4t$, $z = 3 - t$. We substitute the parametric equations for x, y, and z into the equation of the plane. This gives $2(5 + 2t) + 4(9 + 4t) - (3 - t) = 1$. Solving for t, we obtain $t = -2$. The coordinates of R are therefore $x = 5 + 2(-2) = 1$, $y = 9 + 4(-2) = 1$, and $z = 3 - (-2) = 5$. The point $R = (1, 1, 5)$ is the point of intersection that we seek. (The point P and plane V are the same as in Example 13. Because line ℓ is normal to V, the distance of P to R, $\sqrt{(5-1)^2 + (9-1)^2 + (3-5)^2} = 2\sqrt{21}$, is the distance of P to V.) ■

Example 15 Determine whether the lines parameterized by

$$x = 3t - 7,\quad y = -2t + 5,\quad z = t + 1$$

and

$$x = -3t + 2,\quad y = t + 1,\quad z = 2t - 2$$

intersect.

Solution First of all, let us note that we cannot expect that the point of intersection, if there is one, will correspond to the same value of t for each line. (Imagine two cars crossing each other's path at an intersection. When all goes well, the two cars pass through the same point but *at different times*.) We overcome this complication by using a different parameter, say s, for the first line. Now we equate the expressions for x, y, and z. We have

$$\begin{aligned} 3s - 7 &= -3t + 2, \\ -2s + 5 &= t + 1, \\ s + 1 &= 2t - 2. \end{aligned}$$

Notice that there are three equations in only two unknowns. Solving the first two equations simultaneously gives $s = 1$ and $t = 2$. There will be a point of intersection if and only if these values for s and t also satisfy the third equation. That is the case in this example. The value $s = 1$ corresponds to the point $(-4, 3, 2)$ on the first line; the value $t = 2$ corresponds to the same point on the second line. This is the only point of intersection. ■

inSIGHT

When we determine the (line of) intersection of *two planes,* we solve two equations in three unknowns. Such a system generally has a solution (corresponding to the fact that two nonparallel planes in space intersect). However, when we determine the (point of) intersection of *two lines,* we solve three equations in two unknowns. Such a system generally does *not* have a solution (corresponding to the fact that two lines in space generally do not intersect). Much of the best mathematics arises from the elegant fashion in which algebra reinforces geometry. The situation of intersecting planes and intersecting lines is a simple instance of this principle.

Triple Vector Product

In Section 11.4, we learned how to define a triple product of vectors that results in a scalar. It is reasonable for us to ask, "Is there a triple product of vectors that results in a vector?" An affirmative answer is contained in the next definition.

Definition If $\mathbf{u}$, $\mathbf{v}$, and $\mathbf{w}$ are given vectors, then we define their *triple vector product* to be the vector $\mathbf{u} \times (\mathbf{v} \times \mathbf{w})$.

inSIGHT

Although the triple scalar product $\mathbf{u} \cdot \mathbf{v} \times \mathbf{w}$ does not require parentheses, the triple vector product *does*. The expression $\mathbf{u} \times \mathbf{v} \times \mathbf{w}$ is ambiguous because its meaning depends on how the terms are associated: In general, $(\mathbf{u} \times \mathbf{v}) \times \mathbf{w} \neq \mathbf{u} \times (\mathbf{v} \times \mathbf{w})$. For example, $(\mathbf{i} \times \mathbf{i}) \times \mathbf{j} = \vec{0} \times \mathbf{j} = \vec{0}$, whereas $\mathbf{i} \times (\mathbf{i} \times \mathbf{j}) = \mathbf{i} \times \mathbf{k} = -\mathbf{j}$.

If $\mathbf{v}$ and $\mathbf{w}$ are not parallel vectors, then they determine a plane V. Notice that V is the plane of vectors perpendicular to $\mathbf{v} \times \mathbf{w}$. Since the triple vector product $\mathbf{u} \times (\mathbf{v} \times \mathbf{w})$ is perpendicular to $\mathbf{v} \times \mathbf{w}$, it must lie in the plane V. In our work on the parametric equations of planes, we observed that every vector in V can be written as $s\mathbf{v} + t\mathbf{w}$ for some scalars s and t. The consequence is that there are scalars s and t such that $\mathbf{u} \times (\mathbf{v} \times \mathbf{w}) = s\mathbf{v} + t\mathbf{w}$. It is not difficult to verify that $s = \mathbf{u} \cdot \mathbf{w}$ and $t = -\mathbf{u} \cdot \mathbf{v}$ (Exercise 88). In summary, we have the formula

$$\mathbf{u} \times (\mathbf{v} \times \mathbf{w}) = (\mathbf{u} \cdot \mathbf{w})\mathbf{v} - (\mathbf{u} \cdot \mathbf{v})\mathbf{w}. \tag{11.29}$$

When we derive Kepler's laws of planetary motion in Chapter 12, we will find equation (11.29) useful.

Example 16 Let $\mathbf{u} = \langle 7, -1, 2\rangle$, $\mathbf{v} = \langle 11, 4, 3\rangle$, and $\mathbf{w} = \langle 5, 0, 2\rangle$. Calculate $\mathbf{u} \times (\mathbf{v} \times \mathbf{w})$ and verify equation (11.29).

Solution We have

$$\mathbf{v} \times \mathbf{w} = \det\left(\begin{bmatrix} \mathbf{i} & \mathbf{j} & \mathbf{k} \\ 11 & 4 & 3 \\ 5 & 0 & 2 \end{bmatrix}\right) = 8\mathbf{i} - 7\mathbf{j} - 20\mathbf{k}$$

and

$$\mathbf{u} \times (\mathbf{v} \times \mathbf{w}) = \det\left(\begin{bmatrix} \mathbf{i} & \mathbf{j} & \mathbf{k} \\ 7 & -1 & 2 \\ 8 & -7 & -20 \end{bmatrix}\right) = 34\mathbf{i} + 156\mathbf{j} - 41\mathbf{k}.$$

Also, $\mathbf{u} \cdot \mathbf{w} = (7)(5) + (-1)(0) + (2)(2) = 39$ and $\mathbf{u} \cdot \mathbf{v} = (7)(11) + (-1)(4) + (2)(3) = 79$. Therefore, $(\mathbf{u} \cdot \mathbf{w})\mathbf{v} - (\mathbf{u} \cdot \mathbf{v})\mathbf{w} = 39(11\mathbf{i} + 4\mathbf{j} + 3\mathbf{k}) - 79(5\mathbf{i} + 0\mathbf{j} + 2\mathbf{k}) = (429 - 395)\mathbf{i} + (156 - 0)\mathbf{j} + (117 - 158)\mathbf{k}, = 34\mathbf{i} + 156\mathbf{j} - 41\mathbf{k}$, which agrees with our direct calculation of $\mathbf{u} \times (\mathbf{v} \times \mathbf{w})$. ■

quickquiz

1. What is the Cartesian equation for the plane through the point $(0, 0, -3)$ and perpendicular to the line parameterized by $x = 3t$, $y = -2 + t$, $z = t/5$?
2. What are the parametric equations for the line through the point $(0, -3, 2)$ and parallel to the line whose symmetric equations are $x/2 = 2y = 1 - z$?
3. True or false: $\mathbf{u} \times (\mathbf{v} \times \mathbf{w}) = (\mathbf{u} \times \mathbf{v}) \times \mathbf{w}$.
4. How can we tell if three vectors are coplanar?

EXERCISES

Problems for Practice

In Exercises 1–4, a Cartesian equation that describes a plane V is given. Using only the given equation, describe all normals to V. Then verify that the given points P, Q, and R are on V. Calculate $\mathbf{N} = \overrightarrow{PQ} \times \overrightarrow{PR}$, and verify that $\mathbf{N}$ is normal to V.

1. $x - 3y + 4z = 2$; $P = (2, 0, 0)$, $Q = (1, 1, 1)$, $R = (5, 1, 0)$
2. $z = 5 - 7x + 11y$; $P = (0, 0, 5)$, $Q = (1, 0, -2)$, $R = (2, 1, 2)$
3. $3x - 4z = 1$; $P = (3, 0, 2)$, $Q = (7, 1, 5)$, $R = (-5, 2, -4)$
4. $x = 2y$; $P = (0, 0, 0)$, $Q = (-4, -2, 3)$, $R = (2, 1, 1)$

In Exercises 5–8, describe all normals to the plane that is parameterized by the equations.

5. $x = 3 + 2s + t$, $y = -1 + 2s - t$, $z = 2 + s - 3t$
6. $x = 1 + s + t$, $y = 2 + 2s + 2t$, $z = 3 + 2s + 3t$
7. $x = 2s$, $y = 1 - t$, $z = s - 2t$
8. $x = 2s + t$, $y = 2s$, $z = -3t$

In Exercises 9–12, find the Cartesian equation of the plane with the normal vector $\mathbf{n}$ and passing through the point P.

9. $\mathbf{n} = \langle -3, 7, 9 \rangle$, $P = (1, 4, 6)$
10. $\mathbf{n} = \langle 9, -3, 7 \rangle$, $P = (6, 0, -9)$
11. $\mathbf{n} = \langle 2, 2, 5 \rangle$, $P = (-5, 9, 2)$
12. $\mathbf{n} = \langle 7, 1, 1 \rangle$, $P = (2, 0, 0)$

In Exercises 13–16, find the Cartesian equation of the plane determined by the three points.

13. $(0, 1, 3)$, $(3, -9, 5)$, $(4, 1, 6)$
14. $(1, 1, 8)$, $(0, 1, 2)$, $(1, 9, 5)$
15. $(2, -6, 8)$, $(1, -9, -4)$, $(-5, 1, 4)$
16. $(6, 1, 3)$, $(8, 2, 5)$, $(-5, 4, 6)$

In Exercises 17–20, find parametric equations for the line passing through the point P_0 and parallel to the vector $\mathbf{m}$.

17. $P_0 = (2, 1, 9)$, $\mathbf{m} = \langle -3, 1, 7 \rangle$
18. $P_0 = (7, -11, 3)$, $\mathbf{m} = \langle 4, 6, -8 \rangle$
19. $P_0 = (0, 1, 0)$, $\mathbf{m} = \langle 2, -1, 1 \rangle$
20. $P_0 = (2, 2, -5)$, $\mathbf{m} = \langle 1, -1, -3 \rangle$

In Exercises 21–24, find symmetric equations for the line that passes through the given point and that is parallel to the given vector.

21. $(2, 1, 5)$, $\langle -5, 8, 2\rangle$
22. $(-4, -2, -1)$, $\langle -6, -3, -7\rangle$
23. $(0, 1, 0)$, $\langle 1, 1, 9\rangle$
24. $(-5, -3, -9)$, $\langle 9, 3, -1\rangle$

In Exercises 25–28, find parametric equations for the line perpendicular to the given plane and passing through the given point.

25. $x - 3y + 7z = 6$, $(7, -4, 6)$
26. $3x + 5y + z = 9$, $(2, 3, 8)$
27. $x - y + z/3 = 0$, $(-1, -1, -8)$
28. $x/2 + y + 2z = 1$, $(0, 2, 1)$

In Exercises 29–32, find symmetric equations for the line that passes through the two given points.

29. $(1, 0, 1)$, $(2, 1, 0)$
30. $(1, 1, -1)$, $(-1, -1, 1)$
31. $(1, 2, 1)$, $(3, 1, 0)$
32. $(2, 2, 1)$, $(0, 0, 0)$

In Exercises 33–36, find, in parametric form, the line of intersection of the two given planes.

33. $x - 3y + 5z = 2$, $2x + 6y + 3z = 4$
34. $x - 3y = 4$, $y + 8z = 6$
35. $4x + 6y - z = 1$, $4x + z = 0$
36. $x + 2y - z = 1$, $2x - y - z = 2$

In Exercises 37–40, find symmetric equations for the line of intersection of the two given planes.

37. $x - y + z = 2$, $x + y + 3z = 6$
38. $x - 3y = 1$, $y + 8z = 1$
39. $2x + y - z = 2$, $4x - z = 3$
40. $x + 2y - z = 0$, $2x - y - z = 1$

In Exercises 41–44, find the point of intersection of the given plane and the given line.

41. $x - 3y + 5z = 0$; $x = 2t + 6$, $y = 6t + 4$, $z = -t - 3$
42. $2x + 3y - 5z = 16$; $x = t - 4$, $y = -3t + 2$, $z = 4t + 1$
43. $x + y = 4$; $x = -t$, $y = 2t + 6$, $z = -2t + 3$
44. $x + y - 3z = 5$; $x = 4 - t$, $y = 2t + 6$, $z = 2t + 5$

In Exercises 45–48, find the point of intersection of the given plane and the given line.

45. $x - 3y + 5z = 19$, $x = y + 1 = z/2$
46. $2x + 3y - 5z = 0$, $(x - 1)/2 = (y + 1)/4 = 2z - 5$
47. $x + y = 4$, $(x + 1)/2 = y + 1 = 3z + 2$
48. $x + y - 2z = 5$, $(x - 2)/2 = (y + 1)/3 = z - 1$

In Exercises 49–52, find the cosine of the angle between the two given planes.

49. $4x - 3z = 4$, $2x + y + 2z = 2$
50. $2x - 2y + z = 1$, $x - 2y + 2z = 5$
51. $-5x - 3y - 4z = 0$, $x + y + z = 1$
52. $8z - x - 4y = 1$, $z - x - y = 5$

In Exercises 53–56, use the method from Example 13 to find the distance between the given point and the given plane.

53. $(1, 0, 1)$, $2x + y + z = 9$
54. $(-2, 1, 1)$, $2x - 2y + z = 13$
55. $(1, 1, 1)$, $2x + 2y + z = 14$
56. $(0, 0, 0)$, $x - 6y + 4z = 3$

In Exercises 57–60, determine whether the given point lies on the given line.

57. $(-7, 6, 2)$, $x = 3t - 1$, $y = -t + 4$, $z = 2t + 6$
58. $(-4, 7, 8)$, $x = -3t + 5$, $y = t + 4$, $z = 5t - 7$
59. $(1, 5, 1)$, $x = t$, $y = 5$, $z = -3t + 1$
60. $(21, 4, -4)$, $x = 6t - 3$, $y = t$, $z = -t$

In Exercises 61–64, determine whether the pairs of lines intersect. If they do, find the point(s) of intersection.

61. $x = 3t - 5$, $y = -4t - 5$, $z = t + 1$; $x = t + 1$, $y = t - 6$, $z = t + 5$
62. $x = -t + 6$, $y = t + 3$, $z = 4t - 6$; $x = 2t - 4$, $y = t - 5$, $z = -3t + 4$
63. $x = t - 1$, $y = -2t + 14$, $z = 2t + 3$; $x = 3t + 7$, $y = -t + 4$, $z = -2t + 11$
64. $x = t + 4$, $y = -2t + 1$, $z = -3t - 3$; $x = t - 2$, $y = t + 7$, $z = 2t + 6$

In Exercises 65–68, a point Q and a plane $\mathcal{V}$ are given. Compute the distance of Q to $\mathcal{V}$ as follows: First, find the line through Q that is perpendicular to $\mathcal{V}$; next, find the point R of intersection of this line and $\mathcal{V}$; finally, calculate the distance from Q to R to obtain the distance between Q and $\mathcal{V}$.

65. $Q = (8, -2, 4)$, $x - y + z = 4$
66. $Q = (-6, -3, 13)$, $2x + y - 3z = 2$
67. $Q = (2, -3, 6)$, $2x - 5y + 7z = 4$
68. $Q = (-1, 5, 3)$, $-3x + y - 4z = 2$

In Exercises 69–72, verify equation (11.29) for the triple vector product.

69. $\mathbf{u} = \langle 1, -2, 4\rangle$, $\mathbf{v} = \langle 2, 0, 1\rangle$, $\mathbf{w} = \langle 3, 1, 1\rangle$
70. $\mathbf{u} = \langle 2, 1, 2\rangle$, $\mathbf{v} = \langle 2, 2, 3\rangle$, $\mathbf{w} = \langle 3, 3, -5\rangle$
71. $\mathbf{u} = \langle 1, 1, 1\rangle$, $\mathbf{v} = \langle 3, 0, 2\rangle$, $\mathbf{w} = \langle -2, -3, 3\rangle$
72. $\mathbf{u} = \langle -5, -1, 0\rangle$, $\mathbf{v} = \langle 1, 0, 3\rangle$, $\mathbf{w} = \langle 0, 2, 2\rangle$

Further Theory and Practice

In Exercises 73–76, find the Cartesian equation of the plane that is parallel to the given plane and that passes through the point $(1, 2, 3)$.

73. $3x + 4y + z = 0$

74. $z = 2x$

75. $(x - 1) + 2(y - 2) + 3(z - 3) = 1$

76. $z = 2x - 3y$

Sometimes it is helpful to sketch a plane by drawing its intercepts with the three coordinate axes and then drawing the resulting triangle. In Exercises 77–80, use this method to sketch that part of the given plane that lies in the first octant.

77. $2x + 3y + z = 6$

78. $x + 4y + 6z = 12$

79. $3x + 5y + 15z = 15$

80. $2x + 3y + 9z = 18$

81. If the x- and y-intercepts of a line in the xy-plane are the nonzero numbers a and b, respectively, then the Cartesian equation of the line is $x/a + y/b = 1$. State and prove an analogous Cartesian equation for a plane in space.

82. State and prove a simple analytic characterization of the Cartesian equations of planes that pass through the origin.

83. Determine whether the following pairs of planes are parallel.

a. $x - 2y + 4z = 7,\ 2x - 4y + 8z = 5$

b. $x + y + 5z = 1,\ 3x + 3y + 15z = -3$

c. $4x - 6y + 2z = 1,\ x - y + 5z = 8$

d. $x + 4y - 6z = -5,\ -2x - 8y + 12z = 1$

e. $x + y + z = 0,\ 2x + 2y + 2z = 9$

84. The planes with Cartesian equations $Ax + By + Cz = D_1$ and $Ax + By + Cz = D_2$ are parallel. Explain why. Find a formula for the distance between these two planes.

85. Suppose you are given parametric equations of a line ℓ and the Cartesian equation of a plane V. How can you tell if ℓ lies in V? If ℓ does not lie in V, how can you tell if ℓ and V intersect (without attempting to solve the equations simultaneously)? In parts a–d, apply your answer to the given line and plane. If the line and plane do intersect, find the point of intersection.

a. $x - 5y + 2z = 8;$
$x = -3t + 4,\ y = -t - 6,\ z = -t + 2$

b. $3x - 7y + 2z = -22;$
$x = 3t - 6,\ y = t + 2,\ z = -t + 5$

c. $x + y - 2z = 10;\ x = 4t,\ y = t + 6,\ z = -3t + 9$

d. $-3x - 2y + 6z = 1;$
$x = 2t + 1,\ y = 3t - 4,\ z = 2t + 8$

86. Prove that the distance of the point (x_0, y_0, z_0) to the plane $Ax + By + Cz = D$ is given by

$$\frac{|Ax_0 + By_0 + Cz_0 - D|}{\sqrt{A^2 + B^2 + C^2}}.$$

87. Suppose that six distinct points $V_1, V_2, V_3, V_4, L_1, L_2$ are given. Suppose that the first three are not collinear so that they determine a plane. Suppose that V_4 does not lie in that plane. The four points are then the vertices of a solid tetrahedron. How can you tell if the line through L_1 and L_2 passes through this tetrahedron?

88. Verify equation (11.29) for the triple vector product $\mathbf{u} \times (\mathbf{v} \times \mathbf{w})$. Show that equation (11.29) is valid even when $\mathbf{v}$ and $\mathbf{w}$ are parallel.

89. Let s and t be scalars. Show that $\mathbf{u}$, $\mathbf{v}$, and $s\mathbf{u} + t\mathbf{v}$ are coplanar.

90. Let ℓ be a line with direction $\mathbf{m} = \langle a, b, c\rangle$ that has one or two zero entries. Show that ℓ has the Cartesian equations

$$\frac{x - x_0}{a} = \frac{y - y_0}{b},\ z = z_0 \text{ when } \mathbf{m} = \langle a, b, 0\rangle,$$
$$\frac{x - x_0}{a} = \frac{z - z_0}{c},\ y = y_0 \text{ when } \mathbf{m} = \langle a, 0, c\rangle,$$
$$\frac{y - y_0}{b} = \frac{z - z_0}{c},\ x = x_0 \text{ when } \mathbf{m} = \langle 0, b, c\rangle$$

and

$$y = y_0,\ z = z_0 \text{ when } \mathbf{m} = \langle a, 0, 0\rangle,$$
$$x = x_0,\ z = z_0 \text{ when } \mathbf{m} = \langle 0, b, 0\rangle,$$
$$x = x_0,\ y = y_0 \text{ when } \mathbf{m} = \langle 0, 0, c\rangle.$$

Calculator/Computer Exercises

91. Plot the sphere with center $(2, 1, 2)$ and radius 3. Verify that $P_0 = (3, 3, 4)$ is a point on this sphere. Add to your plot the plane that passes through P_0 and that is tangent to the sphere.

92. Plot the plane $2x + y + z = 5$. Add to this plot a sphere that is tangent to this plane at the point $(1, 2, 1)$ and that has radius 1.

93. The line ℓ that is parallel to $\langle -1, -2, 3\rangle$ and passes through the point $(1, 1, 0)$ intercepts the solution set S of the Cartesian equation $z = x^4 + y^4$ in two points. Find them.

Summary of Key Topics

Points in Space (Section 11.2)

A point in space is located with three coordinates: (x, y, z). The distance between two points (a_1, b_1, c_1) and (a_2, b_2, c_2) is

$$\sqrt{(a_1 - a_2)^2 + (b_1 - b_2)^2 + (c_1 - c_2)^2}.$$

It follows that the equation of a sphere with center (a, b, c) and radius $r > 0$ is

$$(x - a)^2 + (y - b)^2 + (z - c)^2 = r^2.$$

The sets

$$\{(x, y, z) : (x - a)^2 + (y - b)^2 + (z - c)^2 < r^2\}$$

and

$$\{(x, y, z) : (x - a)^2 + (y - b)^2 + (z - c)^2 \leq r^2\}$$

are called, respectively, the *open* and *closed balls* with center (a, b, c) and radius r.

Vectors (Sections 11.1, 11.2)

A vector in two dimensions is an ordered pair $\mathbf{v} = \langle v_1, v_2\rangle$. A vector in three dimensions is an ordered triple $\mathbf{w} = \langle w_1, w_2, w_3\rangle$. If $P = (p_1, p_2)$ and $Q = (q_1, q_2)$ are points in the plane and if $\mathbf{v} = \langle v_1, v_2\rangle$ where $v_1 = q_1 - p_1$ and $v_2 = q_2 - p_2$, then the directed line segment $\overrightarrow{PQ}$ represents the vector $\mathbf{v}$. A similar definition is used in three dimensions. The most important attributes of a vector are its direction and its magnitude.

If $\mathbf{v} = \langle v_1, v_2, v_3\rangle$ and $\mathbf{w} = \langle w_1, w_2, w_3\rangle$ are vectors, then

$$\mathbf{v} + \mathbf{w} = \langle v_1 + w_1, v_2 + w_2, v_3 + w_3\rangle.$$

Geometrically, two vectors are added by following the displacement of the first by the displacement of the second.

If $\mathbf{v} = \langle v_1, v_2, v_3\rangle$ is a vector, then its *length* is

$$\|\mathbf{v}\| = \sqrt{(v_1)^2 + (v_2)^2 + (v_3)^2}.$$

The terms *magnitude* and *norm* are also used for this quantity. If α is a real number, then $\alpha\mathbf{v}$ is defined to be the vector $\langle \alpha v_1, \alpha v_2, \alpha v_3\rangle$. If α is positive, then $\alpha\mathbf{v}$ points in the same direction as $\mathbf{v}$. If α is negative, then $\alpha\mathbf{v}$ points in the direction opposite to $\mathbf{v}$. In both cases, the length of $\|\alpha\mathbf{v}\|$ is obtained by dilating the length of $\mathbf{v}$ by a factor of $|\alpha|$:

$$\|\alpha\mathbf{v}\| = |\alpha|\,\|\mathbf{v}\|.$$

The Triangle Inequality tells us that the length of a sum of vectors is less than the sum of the lengths of the summands:

$$\|\mathbf{v}+\mathbf{w}\| \leq \|\mathbf{v}\| + \|\mathbf{w}\|.$$

Dot Product (Section 11.3)

If $\mathbf{v} = \langle v_1, v_2, v_3\rangle$ and $\mathbf{w} = \langle w_1, w_2, w_3\rangle$ are vectors, then their dot product is

$$\mathbf{v}\cdot\mathbf{w} = v_1w_1 + v_2w_2 + v_3w_3.$$

The dot product is commutative and distributes over addition. It commutes with scalar multiplication. The length of a vector is related to the dot product by the equation

$$\mathbf{v}\cdot\mathbf{v} = \|\mathbf{v}\|^2.$$

If $\mathbf{v}$ and $\mathbf{w}$ are nonzero vectors, then

$$\cos(\theta) = \frac{\mathbf{v}\cdot\mathbf{w}}{\|\mathbf{v}\|\|\mathbf{w}\|}$$

where θ is the angle between the two vectors, $0 \leq \theta \leq \pi$. It follows that

Two vectors are mutually perpendicular if and only if their dot product is zero.

They are parallel if and only if $\cos(\theta) = \pm 1$. This means that two vectors $\mathbf{v}$ and $\mathbf{w}$ are parallel if and only if $\|\mathbf{v}\|\|\mathbf{w}\| = |\mathbf{v}\cdot\mathbf{w}|$. An equivalent condition is that one is a scalar multiple of the other:

Two vectors $\mathbf{v}$ and $\mathbf{w}$ are parallel if and only if there is a scalar λ such that $\mathbf{v} = \lambda\mathbf{w}$ or $\mathbf{w} = \lambda\mathbf{v}$.

Projection (Section 11.3)

The projection of a vector $\mathbf{v}$ onto a vector $\mathbf{w}$ is

$$\mathbf{P}_{\mathbf{w}}(\mathbf{v}) = \left(\frac{\mathbf{v}\cdot\mathbf{w}}{\|\mathbf{w}\|^2}\right)\mathbf{w}.$$

The component of $\mathbf{v}$ in the direction of $\mathbf{w}$ is $\mathbf{v}\cdot\mathbf{w}/\|\mathbf{w}\|$. The length of the projection is $|\mathbf{v}\cdot\mathbf{w}|/\|\mathbf{w}\|$.

Direction Vectors (Sections 11.1, 11.2, 11.3)

If $\mathbf{v}$ is any nonzero vector, then the vector $(1/\|\mathbf{v}\|)\mathbf{v}$ is a unit vector pointing in the same direction. A unit vector is called a direction. The unit vectors

$$\begin{aligned}\mathbf{i} &= \langle 1, 0, 0\rangle,\\ \mathbf{j} &= \langle 0, 1, 0\rangle, \text{ and}\\ \mathbf{k} &= \langle 0, 0, 1\rangle\end{aligned}$$

are called the standard basis vectors. Any vector $\mathbf{u} = \langle u_1, u_2, u_3\rangle$ can be written uniquely in the form

$$\mathbf{u} = u_1\mathbf{i} + u_2\mathbf{j} + u_3\mathbf{k}.$$

If $\mathbf{u}$ is a unit vector, then we can write

$$\mathbf{u} = \cos(\alpha)\mathbf{i} + \cos(\beta)\mathbf{j} + \cos(\gamma)\mathbf{k}.$$

The coefficients of $\mathbf{i}$, $\mathbf{j}$, and $\mathbf{k}$ are called the direction cosines for $\mathbf{u}$, and the angles α, β, and γ are called the direction angles.

Cross Product (Section 11.4)

The cross product of two vectors $\mathbf{v} = \langle v_1, v_2, v_3\rangle$ and $\mathbf{w} = \langle w_1, w_2, w_3\rangle$ is defined to be the vector

$$\mathbf{v} \times \mathbf{w} = (v_2w_3 - w_2v_3)\mathbf{i} - (v_1w_3 - w_1v_3)\mathbf{j} + (v_1w_2 - w_1v_2)\mathbf{k}.$$

The geometric interpretation of cross product is that

$$\mathbf{v} \times \mathbf{w} = \|\mathbf{v}\|\,\|\mathbf{w}\|\sin(\theta)\mathbf{n}$$

where θ is the angle between the two vectors and $\mathbf{n}$ is the standard unit normal to $\mathbf{v}$ and $\mathbf{w}$ determined by the Right Hand Rule. We may use determinant notation to express the cross product as

$$\mathbf{v} \times \mathbf{w} = \det\left(\begin{bmatrix} \mathbf{i} & \mathbf{j} & \mathbf{k} \\ v_1 & v_2 & v_3 \\ w_1 & w_2 & w_3 \end{bmatrix}\right).$$

The cross product is anticommutative: $\mathbf{v} \times \mathbf{w} = -\mathbf{w} \times \mathbf{v}$. It distributes over addition.

The area of the triangle determined by the two vectors $\mathbf{v}$ and $\mathbf{w}$ is $\|\mathbf{v} \times \mathbf{w}\|/2$.

Lines and Planes (Section 11.5)

The plane with normal vector $\mathbf{n} = \langle A, B, C\rangle$ and passing through the point $P_0 = (x_0, y_0, z_0)$ has equation

$$\overrightarrow{P_0P} \cdot \mathbf{n} = 0,$$

or

$$A(x - x_0) + B(y - y_0) + C(z - z_0).$$

Here $P = (x, y, z)$ is a variable point on the plane. A line in space that passes through $P_0 = (x_0, y_0, z_0)$ and that is parallel to a vector $\mathbf{m} = \langle a, b, c\rangle$ is given parametrically by

$$\begin{aligned} x &= x_0 + ta \\ y &= y_0 + tb \\ z &= z_0 + tc. \end{aligned}$$

By solving each of these equations for t and equating, we can express the line—in terms of x, y, and z only—as the intersection of two planes. This is called the symmetric form for a line. The Cartesian equations that result are

$$\frac{x - x_0}{a} = \frac{y - y_0}{b} = \frac{z - z_0}{c}.$$

Triple Scalar Product (Section 11.4)

If $\mathbf{u}$, $\mathbf{v}$, and $\mathbf{w}$ are vectors, then the volume of the parallelepiped that they determine is given by the triple product $|(\mathbf{u} \times \mathbf{v}) \cdot \mathbf{w}|$. The product may be taken in any order. The vectors $\mathbf{u}$, $\mathbf{v}$, and $\mathbf{w}$ are coplanar if and only if $(\mathbf{u} \times \mathbf{v}) \cdot \mathbf{w} = 0$.

genesis & DEVELOPMENT

The roots of vector calculus can be traced back to 1799. In that year, both Carl Friedrich Gauss and a Norwegian surveyor, Caspar Wessel (1745–1818), "identified" the complex numbers $\mathbb{C}$ with the real plane $\mathbb{R}^2$. They did this by means of the one-to-one correspondence

$$(x, y) \mapsto x + y\sqrt{-1}$$

between $\mathbb{R}^2$ and $\mathbb{C}$. By associating each point of the plane with a complex number, they were able to transfer the geometry of the plane to the field of complex numbers. This geometric interpretation of complex numbers gave rise to a new mathematical subject, the calculus of complex-valued functions of a complex variable.

In fact, the identification of $\mathbb{R}^2$ with $\mathbb{C}$ is a two-way street: It also allows us to transfer the algebraic structure of the complex number system to the plane. For example, the multiplication of complex numbers,

$$\left(x + y\sqrt{-1}\right)\left(x' + y'\sqrt{-1}\right) = (xx' - yy') + (xy' + yx')\sqrt{-1},$$

can be used to define a product of points in the plane:

$$(x, y) \odot (x', y') = (xx' - yy', xy' + yx').$$

The search for an analogous algebraic structure that could be imposed on the points of three-dimensional space led to the creation of vector calculus in 1843. As so often happens, the breakthrough was attained more or less simultaneously by two mathematicians, Hermann Grassmann and Sir William Rowan Hamilton, working independently.

Hermann Günther Grassmann and Vector Calculus

Hermann Günther Grassmann (1809–1877) was born and raised in Stettin, a major city in the Pomeranian region of Germany. Since the end of World War II, when part of Pomerania was ceded to Poland, Stettin has been known by its Polish name, Szczecin. In Grassmann's time, Stettin was a coveted port that opened to the Baltic Sea. It was not, however, an important center of learning. Grassmann acquired his education in Berlin, where he studied philology and theology. He received no mathematical training at the university level. After returning to Stettin, Grassmann became a high school teacher and father.

The demands of rearing 11 children notwithstanding, Grassmann maintained his interest in languages and also carried out a program of original research in mathematics and physics. He first published his work on vector calculus in a book that appeared in 1844. Almost no notice was taken of it during his lifetime. After selling only a few copies, his publisher rendered the remaining inventory of 600 copies into waste paper. Grassmann's hopes for landing a university position were dashed. Spurned in mathematics, he returned to the study of languages, a field in which he finally earned a measure of respect for his toil. At the time of his death, and for some time thereafter, his Sanskrit-to-German translation of the *Rig-Veda* looked to be his primary legacy. In fact, Grassmann's mathematical work did not attract any serious attention until nearly 50 years after his death.

One of Grassmann's most original ideas was the creation of an anticommutative algebra of vectors that is now called "exterior algebra." In the 20th century, this algebra played a prominent role in algebra, geometry, and analysis. In the course of developing his theory of exterior algebra, Grassmann discovered not only the cross product but also the dot product, the vector norm, orthogonal projection, and some sophisticated concepts of modern mathematics, such as the "Hodge star operator." Frustrated by the apathy with which his research was greeted, Grassmann did not waver in his belief of its importance. As he expressed it in the foreword to his translation of *Rig-Veda*,

> *I remain completely confident that the labor I have expended on the science presented here and which has demanded a significant part of my life as well as the most strenuous application of my powers, will not be lost. . . . I know and feel obliged to state (though I run the risk of seeming arrogant) that even if this work should again remain unused for another seventeen years or even longer, without entering into the actual development of science, still that time will come when it will be brought forth from the dust of oblivion and when ideas now dormant will bring forth fruit.*

Sir William Rowan Hamilton and Vector Calculus

In stark contrast to Grassmann, Sir William Rowan Hamilton was the best-known scientist of his day. Even before his talents in mathematics and physics became evident, Hamilton attracted attention for his precocious ability to learn languages. By five years of age, he was already proficient in Latin, Greek, and Hebrew. As a teenager, he was versed in the rudiments of 13 languages. It should be noted that all this knowledge was the result of self-study under the tutelage of an uncle. Hamilton received his only formal schooling when he attended university.

Hamilton's interest in mathematics was kindled by the reading of a Latin copy of *Euclid* when he was 10 years old. When he was 16, Hamilton detected a hitherto unsuspected error in Laplace's treatise on celestial mechanics. He published original mathematical research one year later and made his name in physics over the next 10 years by applying the *calculus of variations* to discover fundamental principles of mechanics and optics. The Nobel laureate Erwin Schrödinger summarized Hamilton's influence as follows:

> *The central conception of all modern physics is the "Hamiltonian." If you wish to apply modern theory to any particular problem, you must start with putting the problem "in Hamiltonian form."*

On the basis of his theoretical investigations, Hamilton predicted the phenomenon of conical refraction. In the words of a prominent scientist of the time, this phenomenon was "unheard of and without analogy." The experimental verification of conical refraction created a sensation and brought fame to Hamilton beyond scientific circles.

Hamilton's introduction of vector calculus was a by-product of his creation of the *quaternions* $\mathcal{Q}$, an algebraic system that may be identified with $\mathbb{R}^4$. A typical quaternion is of the form

$$t + x\mathbf{i} + y\mathbf{j} + z\mathbf{k}.$$

Quaternions of the form $x\mathbf{i}+y\mathbf{j}+z\mathbf{k}$ may be identified with points of three-dimensional space $\mathbb{R}^3$ by the one-to-one correspondence $(x, y, z) \to x\mathbf{i}+y\mathbf{j}+z\mathbf{k}$. At first, Hamilton referred to these space quaternions as *triplets*.

Addition and subtraction of quaternions was no problem. Finding a suitable multiplication proved difficult. The stumbling block was Hamilton's initial reluctance to abandon the commutative law: $\mathbf{pq} = \mathbf{qp}$ $(\mathbf{p}, \mathbf{q} \in \mathcal{Q})$. The answer came to him on October 16, 1843. Shortly before his death in 1865, Hamilton described the circumstances in a letter to his son Archibald:

> *Every morning in the early part of [October 1843], on my coming down from breakfast your (then) little brother William Edwin, and yourself, used to ask me "Well, Papa, can you multiply triplets?" Whereto I was always obliged to reply, with a sad shake of the head: "No. I can only add and subtract them."*
>
> *But on the sixteenth day of the same month ... I was walking ... along the Royal Canal .... An electric current seemed to close; and a spark flashed forth, the herald ... of many long years to come of definitely directed thought and work .... Nor could I resist the impulse to cut with a knife on a stone of Brougham Bridge ... the fundamental formula with the symbols* **i**, **j**, **k**; *namely*
>
> $$\mathbf{i}^2 = \mathbf{j}^2 = \mathbf{k}^2 = \mathbf{ijk} = -1,$$
>
> *which contains the Solution of the problem, but of course, as an inscription, has long since mouldered away.**

Hamilton retained the *associative* law of multiplication: $(\mathbf{pq})\mathbf{r} = \mathbf{p}(\mathbf{qr})(\mathbf{p}, \mathbf{q}, \mathbf{r} \in \mathcal{Q})$; he even coined the term. Therefore, no bracketing is necessary in the relation $\mathbf{ijk} = -1$ that Hamilton carved in stone. But Hamilton recognized that he would have to give up the commutative property of multiplication. To see how to multiply two quaternions, imagine the face of a clock with **i**, **j**, and **k** positioned at 12, 4, and 8 o'clock, respectively (Figure 1). A clockwise product of any two adjacent terms results in the third term, a counterclockwise product results in the negative of the remaining term. Thus, $\mathbf{ij} = \mathbf{k} = -\mathbf{ji}$. Two general quaternions are then multiplied by the ordinary rules of arithmetic except that the order of factors must be preserved. For example,

$$\begin{aligned}(2 + 3\mathbf{j})(5\mathbf{i} + 6\mathbf{j} + 7\mathbf{k}) &= 10\mathbf{i} + 12\mathbf{j} + 14\mathbf{k} + 15\mathbf{ji} + 18\mathbf{j}^2 + 21\mathbf{jk} \\ &= -18 + (10 + 21)\mathbf{i} + 12\mathbf{j} + (14 - 15)\mathbf{k} \\ &= -18 + 31\mathbf{i} + 12\mathbf{j} - \mathbf{k}.\end{aligned}$$

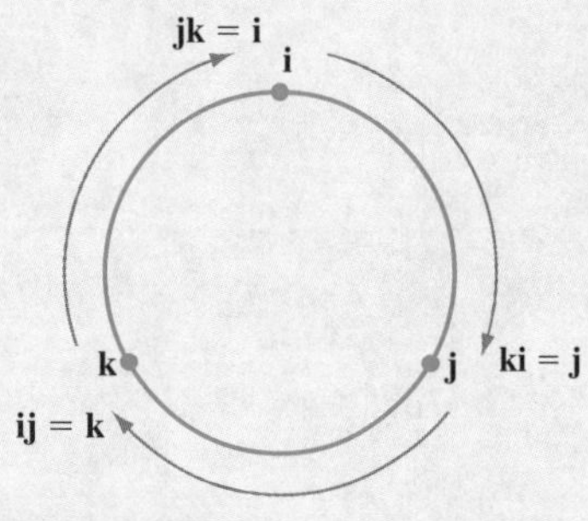

Figure 1

*Since 1954 a plaque on Brougham Bridge has commemorated the vanished engraving.

By 1846, Hamilton was calling the triplet $x\mathbf{i} + y\mathbf{j} + z\mathbf{k}$ a *vector*. This terminology was derived from *radius vector,* a term that was already used in analytic geometry. At the same time, he wrote that "the algebraically real part [namely, the quantity t in the quaternion $t + x\mathbf{i} + y\mathbf{j} + z\mathbf{k}$] may receive... all values contained on the one scale of progression of number from negative to positive infinity; we shall call it therefore the *scalar part*." In his notation, he wrote the quaternion $\mathbf{q} = t + x\mathbf{i} + y\mathbf{j} + z\mathbf{k}$ as $-S \cdot \mathbf{q} + \mathbf{V} \cdot \mathbf{q}$ with $-S \cdot \mathbf{q}$, the scalar component of $\mathbf{q}$, equal to t and $\mathbf{V} \cdot \mathbf{q}$, the vector component of $\mathbf{q}$, equal to $x\mathbf{i} + y\mathbf{j} + z\mathbf{k}$.

Hamilton also introduced the *scalar product* (that is, the dot product) and the *vector product* (that is, the cross product). These products arise simultaneously when the scalar and vector parts of a quaternionic product of two *vectors* are separated. In other words, if $\mathbf{u}$ and $\mathbf{v}$ are vectors (quaternions $\mathbf{u}$ and $\mathbf{v}$ with $S \cdot \mathbf{u} = S \cdot \mathbf{v} = 0$), then $\mathbf{uv} = -S \cdot \mathbf{uv} + \mathbf{V} \cdot \mathbf{uv}$ with $S \cdot \mathbf{uv} = \mathbf{u} \cdot \mathbf{v}$ and $\mathbf{V} \cdot \mathbf{uv} = \mathbf{u} \times \mathbf{v}$.

Despite his professional success, Hamilton's personal life was a shambles. At the age of 20, Hamilton was rejected in love by Catherine Disney, the sister of a friend. He never really recovered from this unrequited first love. A second courtship six years later led to a second rejection. Hamilton finally proposed to a woman who was too timid to reject him but who was also too timid for very much else. Throughout his marriage, he remained in love with Catherine Disney but was torn by guilt over it. Depression and drink became frequent problems. In 1848, Catherine initiated a correspondence with Hamilton, but her conscience was not up to it. She soon confessed to her husband and attempted suicide. Hamilton responded to the stress by drinking more.

By 1853, Hamilton had collected his work on quaternions and vectors into a book that he titled *Lectures on Quaternions.* It ran to more than 800 pages. The famous astronomer, Sir John Herschel, complained that it would "take any man... half a lifetime to digest." In the United States, the mathematician Thomas Hill, who later become president of Harvard, reviewed the book in extremely favorable terms. He concluded that "in quaternions... there is as much real promise of benefit to mankind as in any event of [Queen] Victoria's reign." Hill lamented that "Perhaps not fifty men on this side of the Atlantic have seen [Hamilton's book], certainly not five have read it." His review was somewhat compromised when he admitted that he himself was not actually among that handful.

After Catherine's death in 1853, Hamilton's life was marked by grief, reclusiveness, concentrated work at irregular intervals, and a struggle with alcoholism. As one biographer put it, "When thirsty, he would visit the locker, and the one blemish in the man's personal character is that these latter visits were sometimes paid too often." It was during this last phase of his work, when he was organizing his final thoughts on vectors and quaternions, that Hamilton discovered the theorem on square arrays (or matrices) that is fundamental to matrix algebra.

Although the theory of quaternions never assumed the central importance that Hamilton expected, his influence on mathematics has been as enduring as his influence on physics. The study of noncommutative algebraic systems that he initiated was taken up by first-rate mathematicians such as Arthur Cayley (1821–1895) and William Kingdon Clifford (1845–1879). Their work gave rise to the subject of *noncommutative algebra*. To this day, noncommutative algebra continues to be an important branch of mathematical research.

12

Vector-Valued Functions

PREVIEW

Until now, all of the functions $x \mapsto f(x)$ that we have studied have been scalar valued: We evaluate f at a real number x, and we obtain a real number $f(x)$ as the result. In this chapter, we study the calculus of *vector-valued* functions $t \mapsto \mathbf{r}(t)$. Here we evaluate a function $\mathbf{r}$ at a real number t, and we obtain a *vector* $\mathbf{r}(t)$ as the result. By writing out the vector $\mathbf{r}(t)$ in terms of its components $\langle x(t), y(t), z(t) \rangle$, we see that we can identify the value of $\mathbf{r}(t)$ with the point $(x(t), y(t), z(t))$ in space. As t varies in the domain of $\mathbf{r}$, the values $\mathbf{r}(t)$ trace out a curve in three-dimensional space. By developing the calculus of vector-valued functions, we will be able to analyze such space curves. For example, if we differentiate each component of $\mathbf{r}(t)$, then we obtain a vector that is tangent to the space curve. That much is analogous to the scalar-valued theory that we already know. But the geometry of curves in three-dimensional space is very rich, and there is much new material in this chapter.

Early in the 16th century, Johannes Kepler deduced the three laws of planetary motion that now bear his name. These laws were among the first precise quantitative laws known to science. What was lacking, however, was any theory that provided a reason for those laws. Half a century later, Isaac Newton demonstrated the power of the calculus, then in its infancy, by using it to derive Kepler's laws from basic physical principles. The derivation of Kepler's laws is still a highlight of any calculus course; it is with these fundamental ideas that we conclude the chapter.

12.1 Vector-Valued Functions—Limits, Derivatives, and Continuity

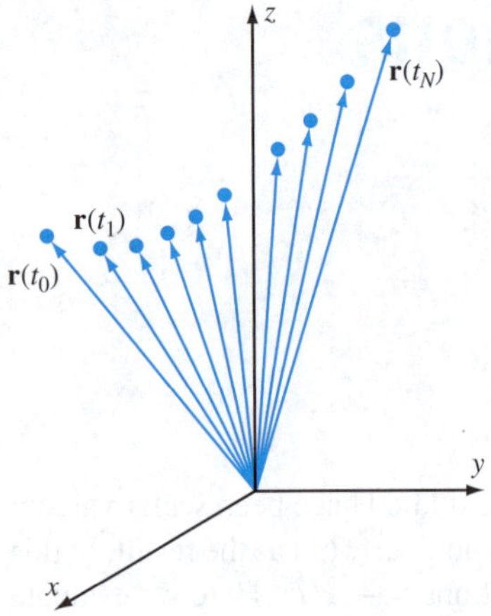

Figure 1a

A function whose range consists of vectors is called a *vector-valued function.* For example, the formula $\mathbf{r}(t) = \cos(t)\mathbf{i} + \sin(t)\mathbf{j} + t\mathbf{k}$ defines a vector-valued function $\mathbf{r}$ that depends on a single variable t. Such functions are the focus of our study in this chapter. One reason for our interest in this matter is that we can use a vector-valued function $\mathbf{r}$ to describe a curve $\mathcal{C}$ in space. To do so, we draw $\mathbf{r}(t)$ as a directed line segment with initial point at the origin. Figure 1a shows such directed line segments for several values, $t_0, t_1, \ldots, t_N$, in the domain of $\mathbf{r}$. The curve $\mathcal{C}$ is the collection of terminal points of the vectors $\mathbf{r}(t)$ as t runs through all values of the domain of $\mathbf{r}$, as Figure 1b illustrates. We say that the curve $\mathcal{C}$ is *parameterized* by $\mathbf{r}$, and we refer to the vector-valued function $\mathbf{r}$ as a *parametric curve.*

Example 1 Describe geometrically the curve that is parameterized by the vector-valued function $\mathbf{r}(t) = (5 - t)\mathbf{i} + (1 + 2t)\mathbf{j} - 3t\mathbf{k}$.

Solution From Section 11.5, we know that the equations $x = 5 - t$, $y = 1 + 2t$, $z = -3t$ are parametric equations of the line L that (i) is parallel to the vector $\langle -1, 2, -3 \rangle$ and (ii) passes through the point $(5, 1, 0)$. This straight line is therefore the curve described by $\mathbf{r}$. ■

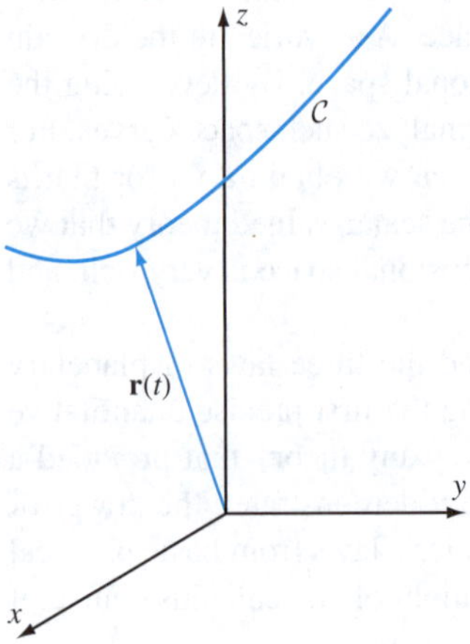

Figure 1b

Suppose that a curve $\mathcal{C}$ is parameterized by $t \mapsto \mathbf{r}(t)$. If we think of the variable t as time, then we can imagine a particle tracing the curve so that its position at time t is the terminal point of $\mathbf{r}(t)$. With this interpretation of the curve, we say that $\mathbf{r}(t)$ is the *position vector* of the particle at time t. If we write the position vector as $\mathbf{r}(t) = x(t)\mathbf{i} + y(t)\mathbf{j} + z(t)\mathbf{k}$ or, equivalently, $\mathbf{r}(t) = \langle x(t), y(t), z(t) \rangle$, then $(x(t), y(t), z(t))$ is the position of the particle at time t and $x = x(t)$, $y = y(t)$, $z = z(t)$ are parametric equations for the motion. In this context, we often call $\mathcal{C}$ a *trajectory.*

Example 2 The position of a particle moving through space is given by $\mathbf{r}(t) = \cos(t)\mathbf{i} + \sin(t)\mathbf{j} + t\mathbf{k}$. What is the position of the body at $t = 0$, $t = \pi/2$, $t = \pi$, $t = 3\pi/2$, and $t = 2\pi$? Describe the curve $\mathcal{C}$ along which the particle moves.

Solution At time 0, the body has position vector

$$\mathbf{r}(0) = \cos(0)\mathbf{i} + \sin(0)\mathbf{j} + 0\mathbf{k} = \mathbf{i} = \langle 1, 0, 0 \rangle.$$

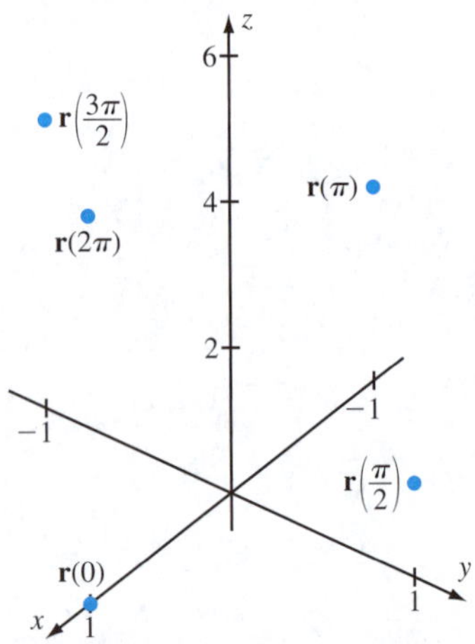

Figure 2

This means that at time $t = 0$, the particle is at the point $(1, 0, 0)$ in space. At time $t = \pi/2$, the particle has position vector

$$\mathbf{r}(0) = \cos\left(\frac{\pi}{2}\right)\mathbf{i} + \sin\left(\frac{\pi}{2}\right)\mathbf{j} + \frac{\pi}{2}\mathbf{k} = \left\langle 0, 1, \frac{\pi}{2} \right\rangle.$$

Thus, at $t = \pi/2$, the particle is at the point $(0, 1, \pi/2)$ in space. Similarly, we find that the particle is at the points $(-1, 0, \pi)$, $(0, -1, 3\pi/2)$, and $(1, 0, 2\pi)$ at times $t = \pi$, $t = 3\pi/2$, and $t = 2\pi$, respectively. These five points are plotted in Figure 2; they do not, however, give a very good idea of the particle's path!

To better understand the trajectory, we observe that the values $x = \cos(t)$, $y = \sin(t)$, $z = t$ satisfy the equation $x^2 + y^2 = 1$. Now, the set $\{(x, y, 0) : x^2 + y^2 = 1\}$

is a circle in the xy-plane of xyz-space, as shown in Figure 3. The coordinates of every point above or below that circle also satisfy the equation $x^2 + y^2 = 1$, since the equation imposes no requirement on the z-coordinate. Therefore, in xyz-space, the graph of $x^2 + y^2 = 1$ is a cylinder, as can also be seen in Figure 3. Because curve $\mathcal{C}$ lies on this cylinder, we can picture how $\mathcal{C}$ passes through the five plotted points. See Figure 4 for the arc of $\mathcal{C}$ that joins the five plotted points and Figure 5 for the arc of $\mathcal{C}$ that is obtained by plotting $\mathbf{r}(t)$ for values of t between -3π and 3π. We call $\mathcal{C}$ a *helix*.

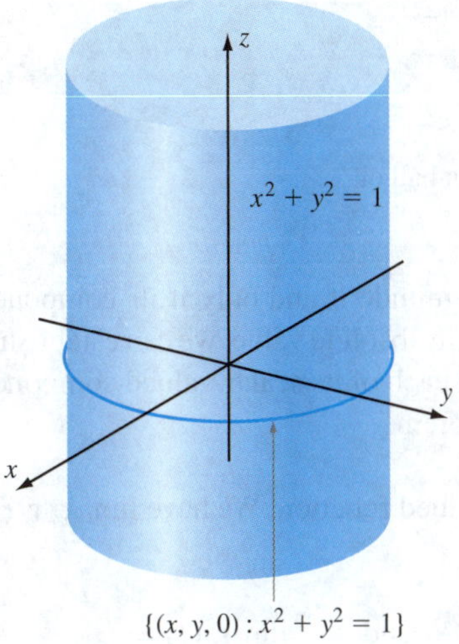

Figure 3

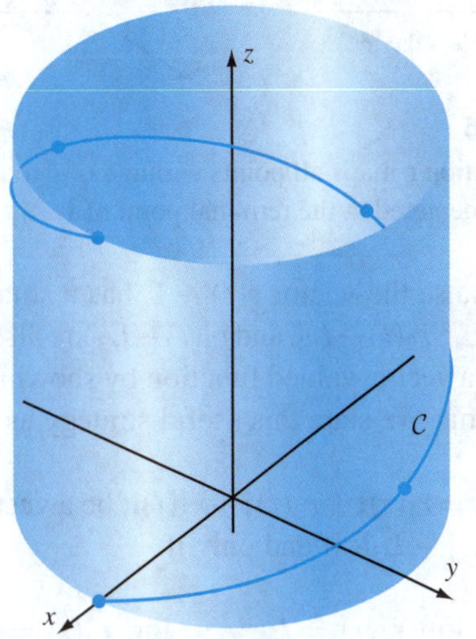

Figure 4

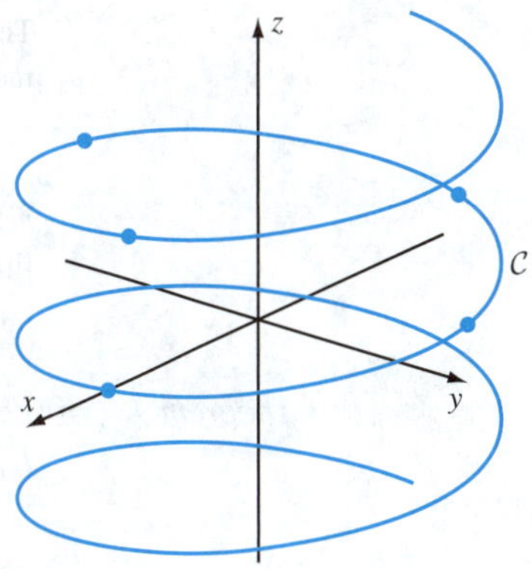

Figure 5

When studying a particle moving through space, we want to calculate its velocity and its acceleration. Therefore, we need to calculate derivatives. We begin our study by defining limits of vector-valued functions.

Limits of Vector-Valued Functions

If $\mathbf{L}$ is a vector and $\mathbf{r}$ is a vector-valued function, then we say that the limit of $\mathbf{r}(t)$ is $\mathbf{L}$ as t tends to c if we can make $\mathbf{r}(t)$ be arbitrarily close to $\mathbf{L}$ by taking t sufficiently close to c. The following definition states this idea precisely.

Definition We say that $\mathbf{r}(t) = r_1(t)\mathbf{i} + r_2(t)\mathbf{j} + r_3(t)\mathbf{k}$ *converges* to the vector $\mathbf{L} = \langle L_1, L_2, L_3 \rangle$ as t tends to c if, for any $\epsilon > 0$, there is a $\delta > 0$ such that

$$0 < |t - c| < \delta \quad \text{implies} \quad \|\mathbf{r}(t) - \mathbf{L}\| < \epsilon.$$

If $\mathbf{r}(t)$ converges to $\mathbf{L}$ as t tends to c, then we write

$$\lim_{t \to c} \mathbf{r}(t) = \mathbf{L}$$

and we say that $\mathbf{L}$ is the *limit* of $\mathbf{r}(t)$ as t tends to c.

Is there anything new in this definition? It looks just like the rigorous definition of limit in Section 2.2. The difference is that, because our function is vector valued, we measure distance in the range by using the norm $\| \ \|$ instead of the absolute value $| \ |$. Figure 6 illustrates the geometry behind the definition.

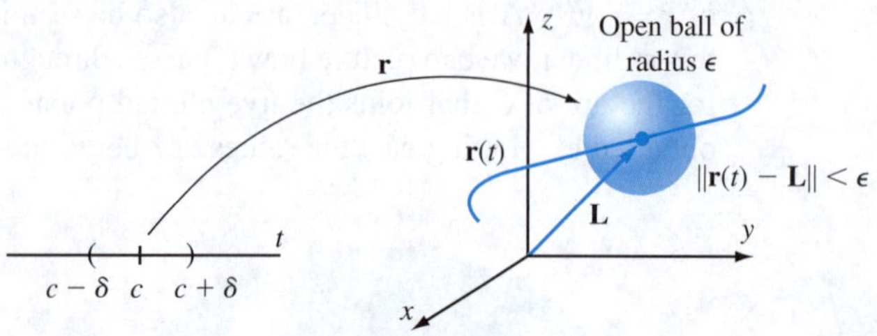

Figure 6
The function **r** maps all points within δ of c to an open ball of radius ϵ centered at the terminal point of **L**.

Because the vector $\mathbf{r}(t) - \mathbf{L}$ has a small magnitude if and only if its components $r_1(t)-L_1$, $r_2(t)-L_2$, and $r_3(t)-L_3$ are all small in absolute value, we can establish the limit of a vector-valued function by showing that each of its scalar-valued components has a limit. We state this useful strategy as a theorem.

Theorem 1 Let $\mathbf{r}(t) = r_1(t)\mathbf{i} + r_2(t)\mathbf{j} + r_3(t)\mathbf{k}$ be a vector-valued function. We have $\lim_{t\to c} \mathbf{r}(t) = L_1\mathbf{i} + L_2\mathbf{j} + L_3\mathbf{k}$ if and only if

$$\lim_{t\to c} \mathbf{r}_1(t) = L_1, \qquad \lim_{t\to c} \mathbf{r}_2(t) = L_2, \qquad \text{and} \qquad \lim_{t\to c} \mathbf{r}_3(t) = L_3.$$

We mention in passing that if the limit of a vector-valued function exists, then it is unique. This assertion follows from reasoning similar to that used in Chapter 2 to see that the limit of a scalar-valued function is unique when it exists. Now we can legitimately calculate limits in the most convenient way, and the resulting answer will have the right physical significance.

Example 3 Let $\mathbf{r}(t) = (t^2 - 4t)\mathbf{i} + (t^2 - 9)/(t-3)\mathbf{j} + \sin(\pi t)/(t-3)\mathbf{k}$. What is $\lim_{t\to 3} \mathbf{r}(t)$?

Solution We calculate that

$$\lim_{t\to 3}(t^2 - 4t) = -3 \qquad \text{and} \qquad \lim_{t\to 3} \frac{t^2-9}{t-3} = \lim_{t\to 3}(t+3) = 6.$$

To calculate the limit of the third component, notice that $\lim_{t\to 3} \sin(\pi t) = \sin(3\pi) = 0$ and $\lim_{t\to 3}(t-3) = 0$. To resolve the indeterminate form $\frac{0}{0}$, we may use l'Hôpital's Rule:

$$\lim_{t\to 3} \frac{\sin(\pi t)}{(t-3)} = \lim_{t\to 3} \frac{\frac{d}{dt}\sin(\pi t)}{\frac{d}{dt}(t-3)} = \lim_{t\to 3} \frac{\pi\cos(\pi t)}{1} = \pi\cos(3\pi) = -\pi.$$

Thus, according to Theorem 1, we have $\lim_{t\to 3} \mathbf{r}(t) = -3\mathbf{i} + 6\mathbf{j} - \pi\mathbf{k}$. ■

The next theorem collects a number of results about calculating limits of vector-valued functions.

Theorem 2 Let f and g be vector-valued functions, and let ϕ be a scalar-valued function. Let c be a real number. Assume that $\lim_{t\to c}\mathbf{f}(t)$, $\lim_{t\to c}\mathbf{g}(t)$, and $\lim_{t\to c}\phi(t)$ exist. Then

a. $\lim_{t\to c}(\mathbf{f}+\mathbf{g})(t) = \lim_{t\to c}\mathbf{f}(t) + \lim_{t\to c}\mathbf{g}(t)$;
b. $\lim_{t\to c}(\mathbf{f}-\mathbf{g})(t) = \lim_{t\to c}\mathbf{f}(t) - \lim_{t\to c}\mathbf{g}(t)$;
c. $\lim_{t\to c}(\mathbf{f}\cdot\mathbf{g})(t) = (\lim_{t\to c}\mathbf{f}(t))\cdot(\lim_{t\to c}\mathbf{g}(t))$;
d. $\lim_{t\to c}(\lambda\mathbf{f}(t)) = \lambda\lim_{t\to c}\mathbf{f}(t)$, for any constant λ;
e. $\lim_{t\to c}(\mathbf{f}\times\mathbf{g})(t) = (\lim_{t\to c}\mathbf{f}(t))\times(\lim_{t\to c}\mathbf{g}(t))$; and
f. $\lim_{t\to c}(\phi(t)\mathbf{f}(t)) = (\lim_{t\to c}\phi(t))(\lim_{t\to c}\mathbf{f}(t))$.

inSIGHT

These formulas are similar to those in Theorem 2 from Section 2.4. You should notice three things, however. First, the "$\mathbf{f}\cdot\mathbf{g}$" in part c signifies a dot product: It *must*, because $\mathbf{f}(t)$ is a vector and $\mathbf{g}(t)$ is a vector. Second, the "$\mathbf{f}\times\mathbf{g}$" in part e represents the cross product of **f** and **g**. Third, there is no result about the limit of a quotient, simply because we may not take the quotient of two vectors.

Example 4 Calculate $\lim_{t\to 4}((t^2\mathbf{i}-3t\mathbf{j})\cdot(4\mathbf{i}-\sqrt{t}\mathbf{j}+5t\mathbf{k}))$.

Solution We calculate

$$\lim_{t\to 4}\left((t^2\mathbf{i}-3t\mathbf{j}+0\mathbf{k})\cdot(4\mathbf{i}-\sqrt{t}\mathbf{j}+5t\mathbf{k})\right) = \lim_{t\to 4}\left(4t^2+3t\sqrt{t}+0\right) = 64+24 = 88.$$

According to Theorem 2c, we will obtain the same answer if we first compute the limits of the terms that appear in the dot product:

$$\begin{aligned}\lim_{t\to 4}(t^2\mathbf{i}-3t\mathbf{j})\cdot\lim_{t\to 4}\left(4\mathbf{i}-\sqrt{t}\mathbf{j}+5t\mathbf{k}\right) &= (16\mathbf{i}-12\mathbf{j}+0\mathbf{k})\cdot(4\mathbf{i}-2\mathbf{j}+20\mathbf{k})\\ &= (16)(4)+(-12)(-2)+(0)(20) = 88.\end{aligned}$$

■

Continuity

The definition of continuity for vector-valued functions is the same as the definition in Chapter 2 for functions with scalar values. The essence of the definition is that the actual value of the function *at c* agrees with the value approached as $t\to c$. A more precise definition follows.

Definition We say that a vector-valued function **r** is *continuous* at a value c of its domain if

$$\lim_{t\to c}\mathbf{r}(t) = \mathbf{r}(c).$$

If **r** is not continuous at a value c in its domain, then we say that it is *discontinuous* there.

Because of Theorem 1, we find that we can test for continuity componentwise.

Theorem 3 If c belongs to the domain of a vector-valued function **r**, then **r** is continuous at c if and only if each component function of **r** is continuous at c.

Example 5 Discuss the continuity properties of the vector-valued function **r** that is defined on the set $\mathcal{D} = \{t\in\mathbb{R} : t\neq 2\}$ by means of the formula $\mathbf{r}(t) = \cos(t)\mathbf{i}+\ln(|t-2|)\mathbf{j}+t^3\mathbf{k}$.

Solution The functions $t\mapsto\cos(t)$ and $t\mapsto t^3$ are defined for every real t. Both of these functions are continuous everywhere. The function $t\mapsto\ln(|t-2|)$ is defined and continuous at all real $t\neq 2$. From these observations, we use Theorem 3 to conclude that the function **r** is continuous at all points in its domain $\mathcal{D}$. ■

inSIGHT

We do not say that $\mathbf{r}(t)$ is discontinuous at $t = 2$, because $\mathbf{r}$ is undefined at 2. We only discuss continuity and discontinuity at points of a function's domain.

The next theorem collects a number of results that are often convenient for establishing continuity.

Theorem 4 Let λ be a constant. Suppose that ϕ is a scalar-valued function and that $\mathbf{f}$ and $\mathbf{g}$ are vector-valued functions. Suppose that these three functions are continuous at a common value c of their domains. Then

a. $\mathbf{f} + \mathbf{g}$ and $\mathbf{f} - \mathbf{g}$ are continuous at c,
b. $\mathbf{f} \cdot \mathbf{g}$ is continuous at c,
c. $\lambda \mathbf{f}$ is continuous at c,
d. $\mathbf{f} \times \mathbf{g}$ is continuous at c, and
e. $\phi \mathbf{f}$ is continuous at c.

inSIGHT

The facts about continuous functions in Theorem 4 are similar to those in Section 2.3 about scalar-valued continuous functions. Two things should be noted. First, there is no statement about quotients of continuous vector-valued functions because we do not have a way to form the quotient of two vectors. Second, there are three statements about products: one for the dot product, one for the cross product, and one for scalar multiplication. Just as in Section 2.3, the proofs of these rules are immediate applications of the rules for limits.

Example 6 Discuss the continuity properties of the functions $\mathbf{f}(t) = |t|\mathbf{i} - 3\mathbf{j} + t^3\mathbf{k}$ and $\mathbf{g}(t) = (1/(t+1))\mathbf{i} - \sec(t)\mathbf{k}$ at the value $t = 0$. Next, discuss the continuity properties of $\mathbf{f} \cdot \mathbf{g}$ and $\mathbf{f} \times \mathbf{g}$ at $t = 0$.

Solution Both $\mathbf{f}$ and $\mathbf{g}$ are continuous at $t = 0$ because all of their component functions are. Therefore, $(\mathbf{f} \cdot \mathbf{g})(t)$ and $(\mathbf{f} \times \mathbf{g})(t)$ are continuous at $t = 0$ by Theorem 4b and 4d. ■

inSIGHT

Of course we can always calculate

$$(\mathbf{f} \cdot \mathbf{g})(t) = \frac{|t|}{t+1} - t^3 \sec(t) \quad \text{and}$$

$$(\mathbf{f} \times \mathbf{g})(t) = 3\sec(t)\mathbf{i} + \left(|t|\sec(t) + \frac{t^3}{t+1}\right)\mathbf{j} + \frac{3}{t+1}\mathbf{k}.$$

The first of these, $\mathbf{f} \cdot \mathbf{g}$, is a scalar-valued function that we can show to be continuous at $t = 0$ using the methods from Chapter 2. We can use Theorem 3 to show the continuity of the second function $\mathbf{f} \times \mathbf{g}$ at $t = 0$, since each of its components is continuous at $t = 0$.

Derivatives of Vector-Valued Functions

Now that we understand limits, it is a simple matter to define derivatives.

Definition Suppose that $\mathbf{r}$ is a vector-valued function defined on an open interval that contains c. If the limit

$$\lim_{\Delta t \to 0} \frac{1}{\Delta t}(\mathbf{r}(c + \Delta t) - \mathbf{r}(c))$$

exists, then we call this limit the *derivative* of the function $\mathbf{r}$ at the point c, and we denote this quantity by

$$\mathbf{r}'(c) \qquad \text{or} \qquad \left.\frac{d\mathbf{r}}{dt}\right|_{t=c}.$$

The notation $\dot{\mathbf{r}}(c)$ is also used, especially when t represents time. The process of calculating $\mathbf{r}'$ is called *differentiation* of $\mathbf{r}$. If the derivative $\mathbf{r}'(c)$ exists, then $\mathbf{r}$ is said to be *differentiable* at c.

inSIGHT

Notice how similar the definition of $\mathbf{r}'(c)$ is to the definition of the derivative given in Section 3.2. The difference, of course, is that we are now considering vector-valued functions; therefore, our limit process is a bit different. The quantity $\mathbf{r}(c + \Delta t) - \mathbf{r}(c)$ is a vector, and the product $[\mathbf{r}(c + \Delta t) - \mathbf{r}(c)]/\Delta t$ signifies the operation of scalar multiplication. The result of this operation is a vector. Thus, the derivative $\mathbf{r}'(c)$ is a vector.

By Theorem 1, we may perform the process of calculating a limit by calculating the limit in each component separately. It follows that we may differentiate a vector-valued function by differentiating each component separately, *provided that all of the components are differentiable*.

Theorem 5 A vector-valued function $\mathbf{r}(t) = r_1(t)\mathbf{i} + r_2(t)\mathbf{j} + r_3(t)\mathbf{k}$ is differentiable at $t = c$ if and only if each component function of $\mathbf{r}$ is differentiable at c. In this case,

$$\mathbf{r}'(c) = r_1'(c)\mathbf{i} + r_2'(c)\mathbf{j} + r_3'(c)\mathbf{k}.$$

Example 7 Let $\mathbf{r}(t) = \exp(2t)\mathbf{i} + |t|\mathbf{j} - \cos(t)\mathbf{k}$. For what values of t is $\mathbf{r}$ differentiable? What is $\mathbf{r}'(t)$ at these values?

Solution We calculate $\mathbf{r}'(t)$ by differentiating each component:

$$\mathbf{r}'(t) = \frac{d}{dt}(\exp(2t))\mathbf{i} + \frac{d}{dt}(|t|)\mathbf{j} - \left(\frac{d}{dt}\cos(t)\right)\mathbf{k} = 2\exp(2t)\mathbf{i} + \frac{d}{dt}(|t|)\mathbf{j} + \sin(t)\mathbf{k}.$$

We see that $\mathbf{r}(t)$ is *not* differentiable at $t = 0$ because the second component, $|t|$, is not. For $t > 0$, however, we have $|t| = t$ and $\frac{d}{dt}(|t|) = 1$. It follows that $\mathbf{r}(t)$ is differentiable for $t > 0$ and $\mathbf{r}'(t) = 2\exp(2t)\mathbf{i} + \mathbf{j} + \sin(t)\mathbf{k}$ for positive values of t. For $t < 0$, we have $|t| = -t$ and $\frac{d}{dt}(|t|) = -1$. It follows that $\mathbf{r}(t)$ is also differentiable for $t < 0$ and $\mathbf{r}'(t) = 2\exp(2t)\mathbf{i} - \mathbf{j} + \sin(t)\mathbf{k}$ for negative values of t. ■

The derivative of a vector-valued function is another vector-valued function. Therefore we can differentiate the derivative function to create a second derivative, and so on. Just as for real-valued functions, the second derivative of $\mathbf{r}(t)$ is denoted by

$$\mathbf{r}''(t) \qquad \text{or} \qquad \frac{d^2\mathbf{r}}{dt^2}.$$

The notation $\ddot{\mathbf{r}}(t)$ is also used, especially when t represents time.

Example 8 Let $\mathbf{r}(t) = (3/t - t)\mathbf{i} + \cos(\pi t)\mathbf{j} + \exp(-t)\mathbf{k}$. Calculate $\mathbf{r}'$ and $\mathbf{r}''$. What is $\mathbf{r}''(1)$?

Solution We have $\mathbf{r}'(t) = (-3/t^2 - 1)\mathbf{i} - \pi\sin(\pi t)\mathbf{j} - \exp(-t)\mathbf{k}$ and $\mathbf{r}''(t) = (6/t^3)\mathbf{i} - \pi^2\cos(\pi t)\mathbf{j} + \exp(-t)\,\mathbf{k}$. In particular, $\mathbf{r}''(1) = 6\mathbf{i} + \pi^2\mathbf{j} + (1/e)\,\mathbf{k}$. ■

The next theorem gathers together several useful differentiation rules.

Theorem 6 Let $\mathbf{f}$ and $\mathbf{g}$ be vector-valued functions. If $\mathbf{f}'(c)$ and $\mathbf{g}'(c)$ both exist, then

a. $(\mathbf{f}+\mathbf{g})'(c)$ exists and

$$(\mathbf{f}+\mathbf{g})'(c) = \mathbf{f}'(c) + \mathbf{g}'(c);$$

b. $(\mathbf{f}-\mathbf{g})'(c)$ exists and

$$(\mathbf{f}-\mathbf{g})'(c) = \mathbf{f}'(c) - \mathbf{g}'(c);$$

c. $(\mathbf{f}\cdot\mathbf{g})'(c)$ exists and

$$(\mathbf{f}\cdot\mathbf{g})'(c) = \mathbf{f}'(c)\cdot\mathbf{g}(c) + \mathbf{f}(c)\cdot\mathbf{g}'(c);$$

d. If λ is a constant, then $(\lambda\mathbf{f})'(c)$ exists and

$$(\lambda\mathbf{f})'(c) = \lambda(\mathbf{f}'(c));$$

e. $(\mathbf{f}\times\mathbf{g})'(c)$ exists and

$$(\mathbf{f}\times\mathbf{g})'(c) = \mathbf{f}'(c)\times\mathbf{g}(c) + \mathbf{f}(c)\times\mathbf{g}'(c).$$

f. If $\phi(t)$ is a scalar-valued function and $\phi'(c)$ exists, then $(\phi\mathbf{f})'(c)$ exists and

$$(\phi\mathbf{f})'(c) = \phi'(c)\mathbf{f}(c) + \phi(c)\mathbf{f}'(c);$$

g. If ψ is a scalar-valued differentiable function on an open interval I that contains the point a, if $\psi(a) = c$, and if the composition $\mathbf{f}\circ\psi$ makes sense on I, then $\mathbf{f}\circ\psi$ is differentiable at a and

$$(\mathbf{f}\circ\psi)'(a) = \psi'(a)\mathbf{f}'(\psi(a)).$$

inSIGHT

The differentiation rules in Theorem 6 are similar to those for functions with scalar values (Chapter 3). However, two points should be noted. First, there is no quotient rule because we do not have a way to form the quotient of two vectors. Second, we have three product rules. Rule c is a product rule for the dot product. Rule e is a product rule for the cross product. Rule f is a product rule for scalar multiplication. Care should be taken to distinguish among these three rules.

Take particular note that it is important that the order of **f** and **g** on the right side of formula e match the order of **f** and **g** on the left side. (Remember that the cross product is an anticommutative operation: Interchanging the operands of a cross product changes the sign of the result.) Notice also the form of the Chain Rule stated in part g. On the right side of the formula, $\psi'(a)$ is a scalar and $\mathbf{f}'(\psi(a))$ is a vector. Thus, the right side of the formula is the scalar multiplication of a vector.

Example 9 Let $\mathbf{f}(t) = \cos(t)\mathbf{j} - \ln(t)\mathbf{k}$ and $\mathbf{g}(t) = t^2\mathbf{i} - (1/t^2)\mathbf{j} + t\mathbf{k}$. Calculate $(\mathbf{f} \cdot \mathbf{g})'(t)$.

Solution We apply the formula from Theorem 6c to obtain

$$
\begin{aligned}
(\mathbf{f} \cdot \mathbf{g})'(t) &= \mathbf{f}'(t) \cdot \mathbf{g}(t) + \mathbf{f}(t) \cdot \mathbf{g}'(t) \\
&= \left(-\sin(t)\mathbf{j} - \left(\frac{1}{t}\right)\mathbf{k}\right) \cdot \left(t^2\mathbf{i} - \left(\frac{1}{t^2}\right)\mathbf{j} + t\mathbf{k}\right) \\
&\quad + (\cos(t)\mathbf{j} - \ln(t)\mathbf{k}) \cdot \left(2t\mathbf{i} + \left(\frac{2}{t^3}\right)\mathbf{j} + \mathbf{k}\right) \\
&= \frac{\sin(t)}{t^2} - 1 + \cos(t)\left(\frac{2}{t^3}\right) - \ln(t).
\end{aligned}
$$

■

inSIGHT

We are not obliged to use Theorem 6 in Example 9. As an alternative, we can first calculate $\mathbf{f} \cdot \mathbf{g}$ and then differentiate:

$$(\mathbf{f} \cdot \mathbf{g})'(t) = \frac{d}{dt}\left(0 - \frac{\cos(t)}{t^2} - t\ln(t)\right) = \frac{\sin(t)}{t^2} - 1 + \cos(t)\left(\frac{2}{t^3}\right) - \ln(t).$$

Example 10 Define $\mathbf{f}(t) = \cos(t)\mathbf{i} - \sin(t)\mathbf{j}$, $\mathbf{g}(t) = t^3\mathbf{j} + t^{-1}\mathbf{k}$, $\phi(t) = t^2$, and $\lambda = 5$. Calculate $(\phi\mathbf{f} + \lambda\mathbf{g})'(t)$.

Solution We use several of our differentiation rules:

$$\begin{aligned}(\phi\mathbf{f}+\lambda\mathbf{g})'(t) &= (\phi\mathbf{f})'(t)+(\lambda\mathbf{g})'(t) && \text{Theorem 6a}\\ &= \phi'(t)\mathbf{f}(t)+\phi(t)\mathbf{f}'(t)+\lambda(\mathbf{g}'(t)) && \text{Theorem 6f, 6d}\\ &= 2t(\cos(t)\mathbf{i}-\sin(t)\mathbf{j})+t^2(-\sin(t)\mathbf{i}-\cos(t)\mathbf{j})+5(3t^2\mathbf{j}-t^{-2}\mathbf{k})\\ &= (2t\cos(t)-t^2\sin(t))\mathbf{i}+(-2t\sin(t)-t^2\cos(t)+15t^2)\mathbf{j}-5t^{-2}\mathbf{k}.\end{aligned}$$

As an alternative, we can first calculate $(\phi\mathbf{f}+\lambda\mathbf{g})(t) = t^2\cos(t)\mathbf{i}+(-t^2\sin(t)+5t^3)\mathbf{j}+5t^{-1}\mathbf{k}$. We can then directly differentiate this expression. As an exercise, verify that this second method results in the same formula for $(\phi\mathbf{f}+\lambda\mathbf{g})'(t)$. ■

Example 11 Let $\mathbf{f}(t)=\tan(t)\mathbf{i}-\cos(t)\mathbf{k}$ and $\mathbf{g}(t)=\sin(t)\mathbf{j}$. Calculate $(\mathbf{f}\times\mathbf{g})'(t)$.

Solution By Theorem 6e,

$$\begin{aligned}(\mathbf{f}\times\mathbf{g})'(t) &= \mathbf{f}'(t)\times\mathbf{g}(t)+\mathbf{f}(t)\times\mathbf{g}'(t)\\ &= (\sec^2(t)\mathbf{i}+\sin(t)\mathbf{k})\times\sin(t)\mathbf{j}+(\tan(t)\mathbf{i}-\cos(t)\mathbf{k})\times\cos(t)\mathbf{j}\\ &= (-\sin^2(t)\mathbf{i}+\sec^2(t)\sin(t)\mathbf{k})+(\cos^2(t)\mathbf{i}+\sin(t)\mathbf{k})\\ &= (\cos^2(t)-\sin^2(t))\mathbf{i}+(\sin(t)+\sec^2(t)\sin(t))\mathbf{k}.\end{aligned}$$

Verify this answer by first calculating the cross product $\mathbf{f}\times\mathbf{g}$ and then differentiating. ■

We next turn to a statement relating differentiability and continuity of a vector-valued function. The proof illustrates how we often analyze a vector-valued function by reducing the investigation to its scalar-valued components.

Theorem 7 If $\mathbf{f}$ is a vector-valued function that is differentiable at c, then $\mathbf{f}$ is continuous at c.

Proof If $\mathbf{f}(t)=f_1(t)\mathbf{i}+f_2(t)\mathbf{j}+f_3(t)\mathbf{k}$ is differentiable at c, then, by Theorem 5, the scalar-valued expressions $f_1(t)$, $f_2(t)$, and $f_3(t)$ are all differentiable at c. Each of the functions f_1, f_2, f_3 is therefore continuous at c by Theorem 1 from Section 3.2. Theorem 3 of this section now tells us that $\mathbf{f}$ is continuous at c. ■

Antidifferentiation

If $\mathbf{f}(t)=f_1(t)\mathbf{i}+f_2(t)\mathbf{j}+f_3(t)\mathbf{k}$ is continuous, then we may consider the antiderivative

$$\begin{aligned}\mathbf{F}(t) &= \int \mathbf{f}(t)\,dt\\ &= \left(\int f_1(t)\,dt\right)\mathbf{i}+\left(\int f_2(t)\,dt\right)\mathbf{j}+\left(\int f_3(t)\,dt\right)\mathbf{k}+\mathbf{C}.\end{aligned}$$

Here it should be clearly understood that each of the antidifferentiations in the $\mathbf{i}$, $\mathbf{j}$, and $\mathbf{k}$ components will have a constant of integration. Therefore the constant of integration $\mathbf{C}$ is a vector of the form $\mathbf{C}=C_1\mathbf{i}+C_2\mathbf{j}+C_3\mathbf{k}$.

Example 12 Find the antiderivative $\mathbf{F}(t)$ of $\mathbf{f}(t)=3t^2\mathbf{i}-4t\mathbf{j}+8\mathbf{k}$ that satisfies the equation $\mathbf{F}(1)=2\mathbf{i}-3\mathbf{j}+2\mathbf{k}$.

Solution We have

$$\mathbf{F}(t) = \int \mathbf{f}(t)\,dt$$
$$= t^3\mathbf{i} - 2t^2\mathbf{j} + 8t\mathbf{k} + \mathbf{C}.$$

Now, the additional condition that has been given tells us that $2\mathbf{i} - 3\mathbf{j} + 2\mathbf{k} = \mathbf{F}(1) = \mathbf{i} - 2\mathbf{j} + 8\mathbf{k} + \mathbf{C}$. Solving for $\mathbf{C}$ gives $\mathbf{C} = \mathbf{i} - \mathbf{j} - 6\mathbf{k}$. Therefore, $\mathbf{F}(t) = (t^3 + 1)\mathbf{i} - (2t^2 + 1)\mathbf{j} + (8t - 6)\mathbf{k}$. ■

quickquiz

1. How is a vector-valued function different from the functions studied in earlier chapters?
2. How do we calculate the derivative of a vector-valued function?
3. State three analogues of the Product Rule that involve the differentiation of one or more vector-valued functions.
4. State a Chain Rule for the differentiation of vector-valued functions.

EXERCISES

Problems for Practice

In Exercises 1–6, describe and sketch the curve defined by the given vector-valued function.

1. $\mathbf{r}(t) = t\mathbf{i} + t^2\mathbf{j}$
2. $\mathbf{r}(t) = t^2\mathbf{i} + t^2\mathbf{j}$
3. $\mathbf{r}(t) = t\mathbf{i} + t^2\mathbf{j} + 2\mathbf{k}$
4. $\mathbf{r}(t) = t^2\mathbf{i} + t^2\mathbf{j} + 3\mathbf{k}$
5. $\mathbf{r}(t) = \mathbf{i} + \cos(t)\mathbf{j} + \sin(t)\mathbf{k}$
6. $\mathbf{r}(t) = t\mathbf{i} + \cos(t)\mathbf{j} + \sin(t)\mathbf{k}$

In Exercises 7–10, the position vector $\mathbf{r}(t)$ of a particle is given. Describe the particle's trajectory.

7. $\mathbf{r}(t) = t^2\mathbf{i} + t^2\mathbf{j} + \mathbf{k}, -1 \le t \le 2$
8. $\mathbf{r}(t) = t^2\mathbf{i} + t^4\mathbf{j}, -1 \le t \le 2$
9. $\mathbf{r}(t) = t\mathbf{i} + t\mathbf{j} + \sin(t)\mathbf{k}, 0 \le t \le 2\pi$
10. $\mathbf{r}(t) = \cos(t^2)\mathbf{i} + \sin(t^2)\mathbf{j} + t\mathbf{k}, \sqrt{2\pi} \le t \le 2\sqrt{\pi}$

In Exercises 11–14, a vector-valued function $\mathbf{r}$ and a point c are given. Calculate the limit of $\mathbf{r}(t)$ as $t \to c$.

11. $\mathbf{r}(t) = \cos(\pi t)\mathbf{i} + 2^t\mathbf{j}, c = -1$
12. $\mathbf{r}(t) = \ln(1 + t^2)\mathbf{i} + \mathbf{j}, c = 0$
13. $\mathbf{r}(t) = \langle t^2, t^t, e^{2\ln(t-1)}\rangle, c = 2$
14. $\mathbf{r}(t) = ((t-1)/(t^2-1))\mathbf{i} + ((t+1)/(t^2+1))\mathbf{j}, c = 1$

In Exercises 15–20, calculate $\mathbf{r}'(t)$.

15. $\mathbf{r}(t) = t^{1/2}\mathbf{i} - 3\mathbf{j} + t^2\mathbf{k}$
16. $\mathbf{r}(t) = t^{3/2}\mathbf{i} - t^2\mathbf{j} - 5\sqrt{t}\mathbf{k}$
17. $\mathbf{r}(t) = \tan(t)\mathbf{i} - \cot(t)\mathbf{j} + \sec(t)\mathbf{k}$
18. $\mathbf{r}(t) = (1/t)\mathbf{i} - t\mathbf{j} + \ln(t)\mathbf{k}$
19. $\mathbf{r}(t) = t\mathbf{i} + (t+5)^{-1}\mathbf{j} + (t-7)^{-2}\mathbf{k}$
20. $\mathbf{r}(t) = \left\langle \arctan(\arcsin(t)), \dfrac{\arcsin(t)}{\sqrt{1-t^2}}, \dfrac{\arctan(t)}{1+t^2} \right\rangle$

In Exercises 21–26, calculate $\mathbf{r}''(t)$.

21. $\mathbf{r}(t) = t^2\mathbf{i} - t^{3/2}\mathbf{j} + (t+1)^{5/2}\mathbf{k}$
22. $\mathbf{r}(t) = (t^2+1)^{-1}\mathbf{i} - \sec(t)\mathbf{k}$
23. $\mathbf{r}(t) = e^{-t}\mathbf{i} - \ln(t^2 + t^4)\mathbf{j} + \mathbf{k}$
24. $\mathbf{r}(t) = \sec(t)\mathbf{j} - \cot(t)\mathbf{k}$
25. $\mathbf{r}(t) = t^{-1/3}\mathbf{i} - (1+t^2)^{-1/2}\mathbf{j} + t^{1/4}\mathbf{k}$
26. $\mathbf{r}(t) = \arctan(t)\mathbf{i} - t\sin(t)\mathbf{j} + \arcsin(t)\mathbf{k}$

In Exercises 27–30, calculate the indefinite integral.

27. $\int (t^2\mathbf{i} - t^{-1/2}\mathbf{j} + \cos(2t)\mathbf{k})\,dt$
28. $\int (e^{3t}\mathbf{i} - e^{-2t}\mathbf{j} + te^t\mathbf{k})\,dt$
29. $\int (\ln(t)\mathbf{i} - \cos^2(t)\mathbf{j} + \tan(t)\mathbf{k})\,dt$
30. $\int (\sqrt{t}\mathbf{i} - t^{-1/3}\mathbf{j} + \sec^2(t)\mathbf{k})\,dt$

In Exercises 31–34, find the antiderivative $\mathbf{F}(t)$ for $\mathbf{f}(t)$ that satisfies the additional condition that is given.

31. $\mathbf{f}(t) = 2t\mathbf{i} - 3t^2\mathbf{j} + 4t^3\mathbf{k}$
    $\mathbf{F}(2) = 5\mathbf{i} - 2\mathbf{j} + \mathbf{k}$

32. $\mathbf{f}(t) = \sqrt{t}\mathbf{i} - t^2\mathbf{j} + \mathbf{k}$
$\mathbf{F}(2) = 5\mathbf{i} - 2\mathbf{j}$

33. $\mathbf{f}(t) = \cos(t)\mathbf{i} - \sin(2t)\mathbf{j} + \sec(t)\tan(t)\mathbf{k}$
$\mathbf{F}(0) = 3\mathbf{i} + (3/2)\mathbf{j} - 2\mathbf{k}$

34. $\mathbf{f}(t) = \ln(t)\mathbf{j}$
$\mathbf{F}(e) = -2\mathbf{i} + \mathbf{j} - 4\mathbf{k}$

Suppose that $\mathbf{f}(t) = e^t\mathbf{i} - \cos(t)\mathbf{j} + \ln(1+t^2)\mathbf{k}$, $\mathbf{g}(t) = t^3\mathbf{i} - t\mathbf{k}$, $\phi(t) = 1 + 5t$, and $\lambda = 3$. In Exercises 35–48, calculate the derivative of the given function.

35. $\mathbf{r}(t) = \mathbf{f}(\lambda t)$

36. $\mathbf{r}(t) = \phi(t)^{\lambda}\mathbf{f}(t)$

37. $\mathbf{r}(t) = \mathbf{g}(\phi(t))$

38. $\mathbf{r}(t) = \lambda\mathbf{f}(t) - \phi(t)\mathbf{g}(t)$

39. $\mathbf{r}(t) = \lambda\mathbf{f}(t) + \phi(t)\mathbf{g}(t)$

40. $\mathbf{r}(t) = \mathbf{g}(\lambda t) - \lambda\mathbf{g}(t)$

41. $\mathbf{r}(t) = \mathbf{g}(1/t^3)$

42. $\psi(t) = \mathbf{f}(t) \cdot \mathbf{g}(t)$

43. $\psi(t) = \|\mathbf{g}(t)\|^2$

44. $\mathbf{r}(t) = \mathbf{f}(t) \times \mathbf{f}(t)$

45. $\psi(t) = \phi(\mathbf{f}(t) \cdot \mathbf{g}(t))$

46. $\mathbf{r}(t) = t\mathbf{j} \times \mathbf{g}(t)$

47. $\mathbf{r}(t) = (\phi(t)\mathbf{i} + \lambda\mathbf{k}) \times \mathbf{g}(t)$

48. $\mathbf{r}(t) = \mathbf{f}(t) \times \mathbf{g}(t)$

Further Theory and Practice

Suppose that **f** and **g** are the functions of Exercises 35–48. In Exercises 49–52, calculate the derivative of the given function.

49. $\mathbf{r}(t) = (\mathbf{i} \cdot \mathbf{f}(t))(\mathbf{i} \times \mathbf{g}(t))$

50. $\mathbf{r}(t) = ((\phi(t)\mathbf{j} + \lambda\mathbf{k}) \cdot \mathbf{g}(t))\mathbf{f}(t)$

51. $\psi(t) = \mathbf{f}(t) \cdot (\mathbf{f}(t) \times \mathbf{g}(t))$

52. $\mathbf{r}(t) = (\phi(t)\mathbf{j} \times \mathbf{g}(t)) \times \mathbf{g}(t) - \phi(t)\mathbf{j} \times (\mathbf{g}(t) \times \mathbf{g}(t))$

In Exercises 53–56, calculate the given limit.

53. $\lim_{t\to 0}(|t|^t\mathbf{i} + \cos^{1/t}(t)\mathbf{j} + (1-t)^t\mathbf{k})$

54. $\lim_{t\to 0}\left(\dfrac{\sin(t)}{t}\mathbf{i} + \dfrac{t}{\cos(t)}\mathbf{j} + \dfrac{\tan(t)}{t}\mathbf{k}\right)$

55. $\lim_{t\to \pi}\left\langle \dfrac{\sin(t)}{t-\pi}, \dfrac{\ln(t^2)}{\cos(t)}, |\sec(t)| \right\rangle$

56. $\lim_{t\to 0}((1+t)^{1/t}\mathbf{i} + (1+2t)^{1/t}\mathbf{j} + (1-t)^{1/t}\mathbf{k})$

Determine where each function in Exercises 57–60 is continuous.

57. $\mathbf{r}(t) = \phi(t)\mathbf{i} - 7t^2\mathbf{j}$ where

$$\phi(t) = \begin{cases} 1 & \text{if } t < 5 \\ -1 & \text{if } t \geq 5 \end{cases}$$

58. $\mathbf{r}(t) = \tan(t)\mathbf{i} - t^3\mathbf{j} + \phi(t)\mathbf{k}$ where

$$\phi(t) = \begin{cases} \sin(1/t) & \text{if } t \neq 0 \\ 1 & \text{if } t = 0 \end{cases}$$

59. $\mathbf{r}(t) = \phi(t)\mathbf{i} + \ln(1+t^2)\mathbf{j} + |t|\mathbf{k}$ where

$$\phi(t) = \begin{cases} \sin(t)/t & \text{if } t > 0 \\ \cos(t) & \text{if } t \leq 0 \end{cases}$$

60. $\mathbf{r}(t) = \mathbf{i} + \phi(t)\mathbf{j} + (1/(1+|t|))\mathbf{k}$ where

$$\phi(t) = \begin{cases} t^3 - 3t - 1 & \text{if } t < 2 \\ 3t - 5 & \text{if } t \geq 2 \end{cases}$$

61. If $\mathbf{r}(t) = p(t)\mathbf{i} + q(t)\mathbf{j} + s(t)\mathbf{k}$ where p, q, and s are polynomials, then prove that there exists a positive integer N such that the Nth derivative of $\mathbf{r}$ is identically the zero vector. What is the least N that will suffice?

62. Verify that the function $\mathbf{r}(t) = \cos(t)\mathbf{i} - \sin(t)\mathbf{k}$ satisfies the differential equation

$$\frac{d^2}{dt}\mathbf{r}(t) + \mathbf{r}(t) = \vec{\mathbf{0}}.$$

63. Verify that the curve described by $\mathbf{r}(t) = \cos^2(t)\mathbf{i} + \cos(t)\sin(t)\mathbf{j} + \sin(t)\mathbf{k}$ lies on the sphere $x^2 + y^2 + z^2 = 1$.

64. Let $\mathcal{S}$ be the surface that consists of all points (x, y, z) that satisfy the equation $x^2 + y^2 = z^2$. What are the intersections of $\mathcal{S}$ with horizontal planes $z = h$? What is the intersection of $\mathcal{S}$ with the yz-plane $x = 0$? Describe $\mathcal{S}$. Describe the space curve $\mathcal{C}$ defined by $\mathbf{r}(t) = t\cos(t)\mathbf{i} + t\sin(t)\mathbf{j} + t\mathbf{k}$ and verify that $\mathcal{C}$ lies on $\mathcal{S}$.

65. Give an example of a vector-valued function that is continuous at a point c but that is not differentiable at c.

66. Formulate a definition of right-hand limit and a definition of left-hand limit for vector-valued functions. Prove that these limits may be calculated componentwise.

67. If $\mathbf{r}(t)$ is a vector-valued function, then prove that $\lim_{t\to c}\mathbf{r}(t) = \mathbf{L}$ can hold for, at most, one value of $\mathbf{L}$. *Hint:* Use the analogous theorem for scalar-valued functions.

68. If $\mathbf{r}(t)$ is a vector-valued function satisfying $\lim_{t\to c}\mathbf{r}(t) = \mathbf{L}$, then prove that $\lim_{t\to c}\|\mathbf{r}(t)\| = \|\mathbf{L}\|$. *Hint:* The Triangle Inequality is relevant here. Prove that the converse is false.

69. Prove that if $t \mapsto \mathbf{r}(t)$ is a vector-valued function that is continuous at $t = c$, then $t \mapsto \|\mathbf{r}(t)\|$ is a scalar-valued function that is continuous at $t = c$. Prove that the converse is false.

70. Prove that if $t \mapsto \mathbf{r}(t)$ is a vector-valued function that is differentiable at $t = c$, then $t \mapsto \|\mathbf{r}(t)\|$ is a scalar-valued function that is differentiable at $t = c$ *provided that* $\mathbf{r}(c) \neq \vec{\mathbf{0}}$. Prove that the converse is false.

71. Let $\mathbf{f}$ be a vector-valued function. Suppose that

$$\|\mathbf{f}(s) - \mathbf{f}(t)\| \leq |s - t|$$

for all s and t. Prove that $\mathbf{f}$ is continuous.

Calculator/Computer Exercises

In Exercises 72–79, plot the planar curve associated with the given vector-valued function $\mathbf{r}(t) = r_1(t)\mathbf{i} + r_2(t)\mathbf{j}$. In each case, choose a point $P_0 = \mathbf{r}(t_0)$ at which $r_1'(t_0) \neq 0$. To your plot, add the straight line through P_0 with slope given by $r_2'(t_0)/r_1'(t_0)$. Use the Chain Rule to explain the relationship of this quantity to $\frac{dy}{dx}$ at P_0.

72. $\mathbf{r}(t) = (5 + 2\cos(t))\mathbf{i} + (3 + \sin(t))\mathbf{j},\ 0 \leq t \leq 2\pi$

73. $\mathbf{r}(t) = (t^2 + t)\mathbf{i} + (t^2 - t)\mathbf{j},\ -2 \leq t \leq 2$

74. $\mathbf{r}(t) = \tan(t)\mathbf{i} + \sec(t)\mathbf{j},\ -\pi/2 < t < \pi/2$

75. $\mathbf{r}(t) = 3t/(1 + t^3)\mathbf{i} + 3t^2/(1 + t^3)\mathbf{j},\ -\infty < t < \infty$ (The Folium of Descartes)

76. $\mathbf{r}(t) = (e^t + e^{-t})\mathbf{i} + (e^t - e^{-t})\mathbf{j},\ -\infty < t < \infty$

77. $\mathbf{r}(t) = (t^3 - t)\mathbf{i} + (t^2 - 1)\mathbf{j},\ -2 \leq t \leq 2$

78. $\mathbf{r}(t) = 2t/(1 + t^2)\mathbf{i} + (1 - t^2)/(1 + t^2)\mathbf{j},\ -\infty < t < \infty$

79. $\mathbf{r}(t) = (t - \sin(t))\mathbf{i} + (1 - \cos(t))\mathbf{j},\ 0 \leq t \leq 4\pi$ (Two arches of the cycloid)

In Exercises 80 and 81 plot the space curve associated with the given vector-valued function $\mathbf{r}(t) = r_1(t)\mathbf{i} + r_2(t)\mathbf{j} + r_3(t)\mathbf{k}$.

80. $\mathbf{r}(t) = \cos^2(t)\mathbf{i} + \cos(t)\sin(t)\mathbf{j} + \sin(t)\mathbf{k},\ 0 \leq t \leq 2\pi$

81. $\mathbf{r}(t) = (\cos(t)/t)\mathbf{i} + (\sin(t)/t)\mathbf{j} + (1/t)\mathbf{k}$, $4\pi \leq t \leq 12\pi$

12.2 Velocity and Acceleration

Imagine a particle traveling through space with position vector $\mathbf{r}(t) = r_1(t)\mathbf{i} + r_2(t)\mathbf{j} + r_3(t)\mathbf{k}$. In Figure 1, we see that the vector $\mathbf{r}(t + \Delta t) - \mathbf{r}(t)$ is the displacement of the particle's position from $\mathbf{r}(t)$ to $\mathbf{r}(t + \Delta t)$: The difference vector represents how far and in what direction the particle moves from time t to time $t + \Delta t$. Therefore,

$$\frac{1}{\Delta t}(\mathbf{r}(t + \Delta t) - \mathbf{r}(t))$$

represents the average change in position over the time interval $[t, t+\Delta t]$ or the *average velocity* of the body over the time interval $[t, t + \Delta t]$. Following the line of reasoning used in Chapter 3, we define the *instantaneous velocity* of the body at time t to be

$$\mathbf{v}(t) = \mathbf{r}'(t) = \lim_{\Delta t \to 0} \frac{1}{\Delta t}(\mathbf{r}(t + \Delta t) - \mathbf{r}(t)),$$

provided that this limit exists.

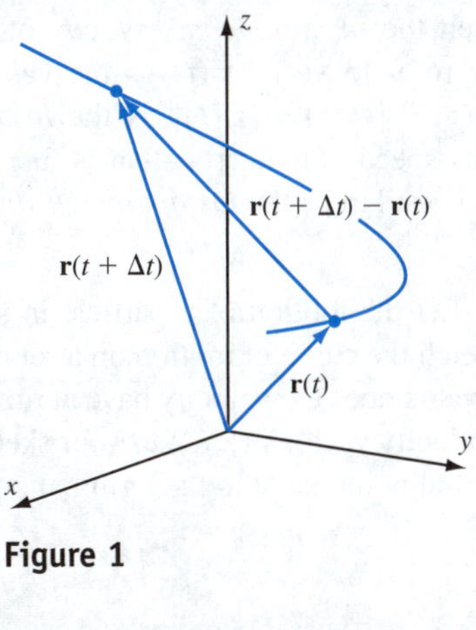

Figure 1

inSIGHT

Our definition of instantaneous velocity is similar to the one we adopted for scalar-valued functions in Section 3.1. Notice that the expression that defines average velocity and that appears in the formula for instantaneous velocity does *not* contain the quotient of two vectors; it is actually the scalar multiplication of the vector $\mathbf{r}(t+\Delta t)-\mathbf{r}(t)$ by the scalar $1/\Delta t$. Thus average velocity is a vector, and instantaneous velocity is a vector. This is an important point to keep in mind.

How could we have been so clumsy in Chapter 3 to have thought that velocity was a number, when it now turns out to be a vector? Notice that, in space, a vector has three components. In the plane, a vector has two components. But in one dimension, a vector has just one component; in other words, in one dimension, a vector is just a number. Therefore, in one dimension, there is no need to use vector language. Now that we are working in space, there is definitely a need to do so.

The advantage of treating instantaneous velocity as a vector is that the direction of the vector is the instantaneous direction of motion of the moving body. If we wish to ignore direction and merely talk about magnitude, then we can proceed as follows: $\|\mathbf{r}(t+\Delta t)-\mathbf{r}(t)\|$ is the *magnitude* of the displacement vector over the time interval $[t, t+\Delta t]$. Magnitude represents distance traveled, without regard to direction traveled. Therefore

$$\left\|\frac{1}{\Delta t}(\mathbf{r}(t+\Delta t)-\mathbf{r}(t))\right\|$$

represents average rate of change of distance traveled over the time interval $[t, t+\Delta t]$. This number is always nonnegative. We define the *instantaneous speed* to be

$$v(t)=\lim_{\Delta t\to 0}\left\|\frac{1}{\Delta t}(\mathbf{r}(t+\Delta t)-\mathbf{r}(t))\right\|,$$

provided that this limit exists. Notice that $v(t)=\|\mathbf{v}(t)\|=\|\mathbf{r}'(t)\|$. We now summarize what we have learned.

Definition Suppose that a particle moving through space has position vector $\mathbf{r}(t)$. If $\mathbf{r}$ is differentiable at t, then the *instantaneous velocity* of the particle at time t is the vector $\mathbf{v}(t)=\mathbf{r}'(t)$. We refer to $\mathbf{v}(t)=\mathbf{r}'(t)$ as the *velocity vector* for short. The nonnegative number $v(t)=\|\mathbf{v}(t)\|=\|\mathbf{r}'(t)\|$ is the *instantaneous speed* of the particle. If the instantaneous speed is nonzero—that is, if $\mathbf{r}'(t)\neq\vec{\mathbf{0}}$—then the direction vector $(1/\|\mathbf{r}'(t)\|)\mathbf{r}'(t)$ is said to be the *instantaneous direction of motion* of the particle.

Example 1 Let the motion of a particle in space be given by $\mathbf{r}(t)=\cos(t)\mathbf{i}+\sin(t)\mathbf{j}+t\mathbf{k}$. Sketch the curve of motion on a set of axes. Calculate the velocity vector for any t. What value does the velocity have at time $t=\pi/2$? What is the speed at this time? Add the velocity vector $\mathbf{v}(\pi/2)$ to your sketch, representing it by a directed line segment with initial point equal to the terminal point of $\mathbf{r}(t)$.

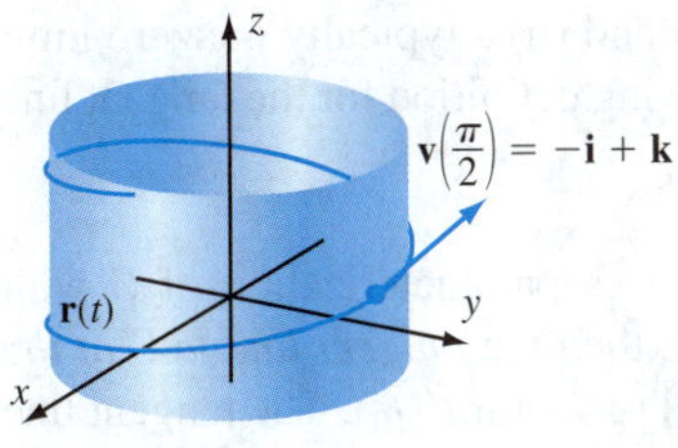

Figure 2

Solution We have studied the curve of motion—a helix—in Example 2 from Section 12.1. Recall that the $\mathbf{i}$ and $\mathbf{j}$ components represent motion around a circle and the $\mathbf{k}$ component represents a simple vertical motion. The velocity vector for any time t is $\mathbf{v}(t) = \mathbf{r}'(t) = -\sin(t)\mathbf{i} + \cos(t)\mathbf{j} + \mathbf{k}$. We therefore have $\mathbf{v}(\pi/2) = -\mathbf{i} + \mathbf{k}$. The speed of the particle at time $t = \pi/2$ is $v(t) = \|\mathbf{v}(\pi/2)\| = \sqrt{(-1)^2 + 0^2 + 1^2} = \sqrt{2}$. The curve of motion and this velocity vector are sketched in Figure 2. ■

inSIGHT

> Figure 2 suggests that the velocity vector $\mathbf{r}'(\pi/2)$ is tangent to the curve of motion at the point $\mathbf{r}(\pi/2)$. Imagine that the curve is a roller-coaster track and that the particle is a roller-coaster car. Suppose that the car loses its grip on the track and hurtles into space. If we ignore air resistance and the effect of gravity, then the car will continue to move in the direction it was moving at the instant it left the track; that direction is the direction of the instantaneous velocity vector. These ideas are developed in the remainder of this section and also in Section 12.5.

The Tangent Line to a Curve in Space

Let $\mathcal{C}$ be a space curve that is parameterized by $t \mapsto \mathbf{r}(t)$. Suppose $P_0 = \mathbf{r}(t_0)$ is a point on $\mathcal{C}$ at which $\mathbf{r}'(t_0)$ exists. If Δt is sufficiently small, then the secant line through the two points $\mathbf{r}(t_0)$ and $\mathbf{r}(t_0 + \Delta t)$ can be used as an approximation to the (as yet undefined) tangent line to $\mathcal{C}$ at P_0. Figure 3 illustrates this idea. As you look at Figure 3a, 3b, and 3c, notice that the approximation improves as Δt decreases to 0. The secant line in Figure 3c is very close to our intuitive conception of the tangent line, which is shown in Figure 3d.

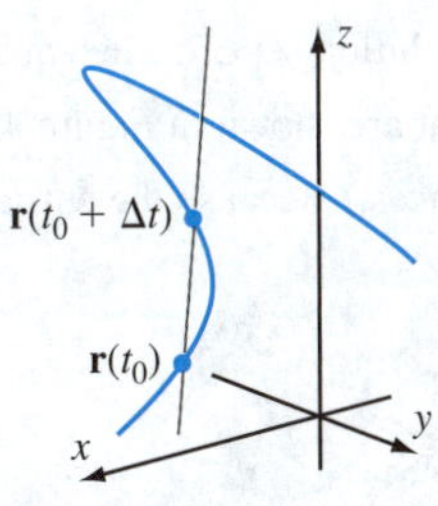

Figure 3a

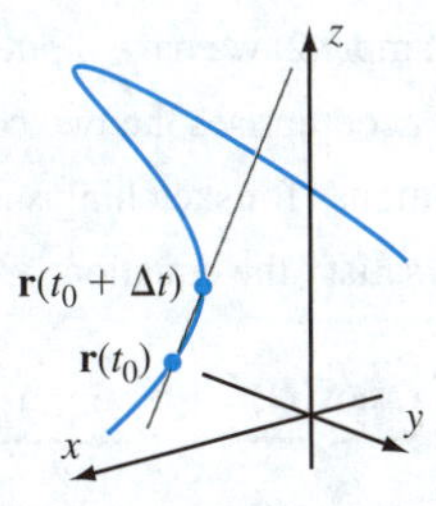

Figure 3b

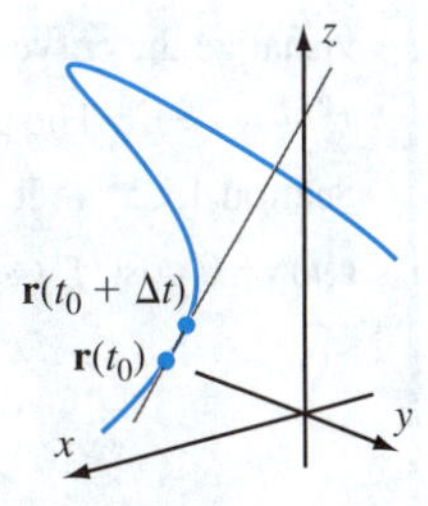

Figure 3c

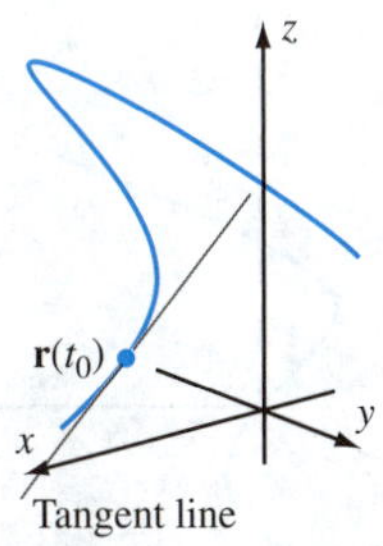

Figure 3d

Notice that the displacement vector $\mathbf{r}(t_0 + \Delta t) - \mathbf{r}(t_0)$ captures the direction of the secant line through the points $\mathbf{r}(t_0)$ and $\mathbf{r}(t_0 + \Delta t)$. We would like to let Δt tend to 0, but when we do so, the displacement vector $\mathbf{r}(t_0 + \Delta t) - \mathbf{r}(t_0)$ approaches $\vec{\mathbf{0}}$, and its direction becomes undefined. However, if Δt is positive, then the average velocity

$$\frac{1}{\Delta t}\left(\mathbf{r}(t_0 + \Delta t) - \mathbf{r}(t_0)\right)$$

has both (i) the same direction as $\mathbf{r}(t_0 + \Delta t) - \mathbf{r}(t_0)$ and (ii) a typically nonzero limit $\mathbf{r}'(t_0)$ as Δt tends to 0. These ideas suggest the following definition for the tangent line to $\mathcal{C}$ at the point P_0.

Definition Let P_0 be a point on a space curve $\mathcal{C}$. Suppose $\mathbf{r}$ is a parameterization of $\mathcal{C}$ with $P_0 = \mathbf{r}(t_0)$. If $\mathbf{r}'(t_0)$ exists and is not the zero vector, then the *tangent line to $\mathcal{C}$ at the point $P_0 = \mathbf{r}(t_0)$* is the line through P_0 that is parallel to vector $\mathbf{r}'(t_0)$. The tangent line is parameterized by $u \mapsto \mathbf{r}(t_0) + u\mathbf{r}'(t_0)$.

inSIGHT

Why do we introduce a new parameter u to describe the tangent line? Notice that the curve described by $t \mapsto \mathbf{r}(t)$ and its tangent line $u \mapsto \mathbf{r}(t_0) + u\mathbf{r}'(t_0)$ pass through the point $P_0 = \mathbf{r}(t_0)$ at the values $t = t_0$ and $u = 0$, respectively. It is therefore prudent to use different parameters for $\mathcal{C}$ and its tangent line.

Example 2 Consider the curve $\mathcal{C}$ defined by the vector-valued function $\mathbf{r}(t) = \langle t\cos(t),\ t\sin(t),\ t\rangle$. What are parametric equations for the tangent line to $\mathcal{C}$ at the point $P_0 = (0, \pi/2, \pi/2)$?

Solution By looking at the third entry of $\mathbf{r}(t)$, we see that P_0 corresponds to $t = \pi/2$. We calculate

$$\mathbf{r}'(t) = \left\langle \frac{d}{dt}(t\cos(t)),\ \frac{d}{dt}(t\sin(t)),\ \frac{d}{dt}t \right\rangle = \langle \cos(t) - t\sin(t),\ \sin(t) + t\cos(t),\ 1\rangle.$$

Therefore, $\mathbf{r}'(\pi/2) = \langle -\pi/2, 1, 1\rangle$. The tangent line to $\mathcal{C}$ at $\mathbf{r}(\pi/2) = \langle 0, \pi/2, \pi/2\rangle$ is therefore parameterized by

$$u \mapsto \mathbf{r}\left(\frac{\pi}{2}\right) + u\mathbf{r}'\left(\frac{\pi}{2}\right) = \left\langle 0, \frac{\pi}{2}, \frac{\pi}{2}\right\rangle + u\left\langle -\frac{\pi}{2}, 1, 1\right\rangle = \left\langle -\frac{\pi}{2}u, \frac{\pi}{2} + u, \frac{\pi}{2} + u\right\rangle.$$

The parametric equations for the tangent line are

$$x = -\frac{\pi}{2}u, \qquad y = \frac{\pi}{2} + u, \qquad z = \frac{\pi}{2} + u.$$

inSIGHT

A plot of $\mathcal{C}$ is shown in Figure 4a. A static, two-dimensional image of a space curve often does not impart a clear understanding of how the curve twists in space. To help visualize the curve $\mathcal{C}$ of Example 2, we have plotted the solution set $\mathcal{S}$ of the equation $x^2 + y^2 = z^2$. The graph of $\mathcal{S}$ comprises the two cones that are shown in Figure 4b. (In Section 13.2, we learn techniques for sketching such surfaces.) Because the entries of $\mathbf{r}(t) = \langle t\cos(t), t\sin(t), t\rangle$ satisfy the equation

$$\underbrace{(t\cos(t))^2}_{x^2} + \underbrace{(t\sin(t))^2}_{y^2} = \underbrace{t^2}_{z^2},$$

we see that $\mathcal{C}$ lies on $\mathcal{S}$. Superimposing the plots of $\mathcal{C}$ and $\mathcal{S}$ in Figure 4c reveals that $\mathcal{C}$ winds around $\mathcal{S}$. Finally, the tangent line that we calculated in Example 2 has been added in Figure 4d.

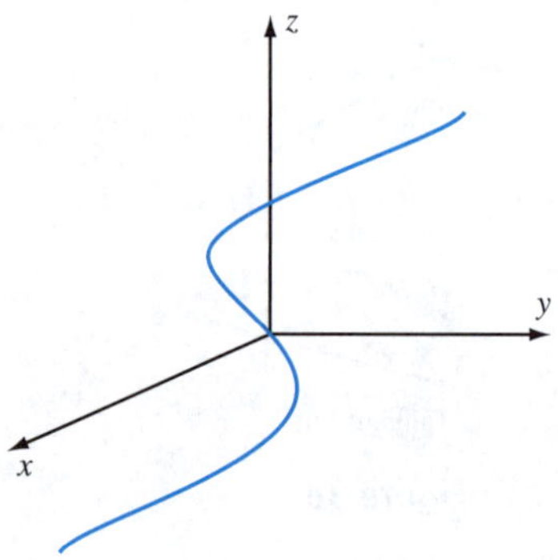

$\mathbf{r}(t) = \langle t\cos(t), t\sin(t), t\rangle$

Figure 4a

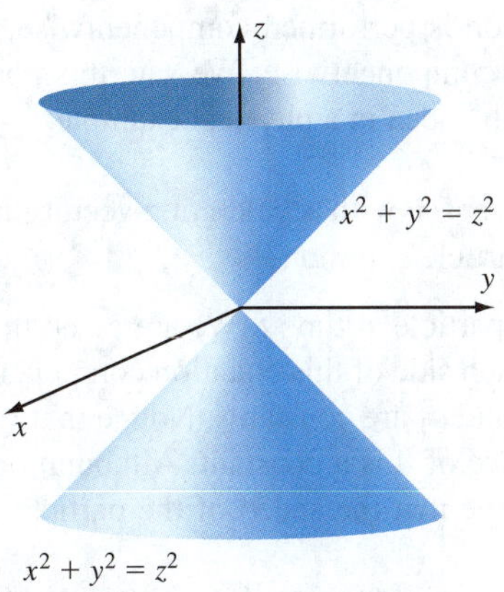

Figure 4b

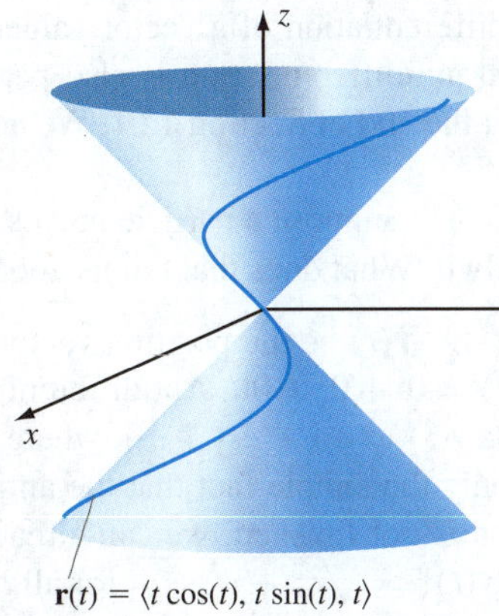

Figure 4c

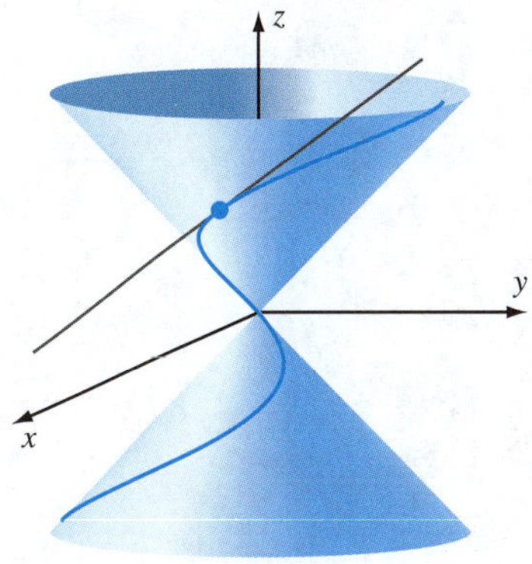

Figure 4d

Acceleration

As we might expect, acceleration is obtained by differentiating velocity. Acceleration is again represented by a vector.

Definition

If $\mathbf{r}(t)$ is the position vector of a body moving through space with velocity $\mathbf{v}(t) = \mathbf{r}'(t)$, then the *instantaneous acceleration* of the body at time t is $\mathbf{a}(t) = \mathbf{v}'(t)$, provided that this derivative exists. Equivalently, we may define $\mathbf{a}(t) = \mathbf{r}''(t)$, provided this second derivative exists.

inSIGHT

Although the velocity vector points in the direction of motion of the moving body, the acceleration vector usually does *not*. In fact, Newton's Second Law tells us that force $\mathbf{F}$ is equal to mass m times acceleration $\mathbf{a}$; that is,

$$\mathbf{F} = m\mathbf{a}.$$

Note that mass is a positive scalar and force and acceleration are vectors. Thus Newton's Second Law tells us that the acceleration vector points in the same direction as the force being applied to the body. The next example strikingly illustrates this idea.

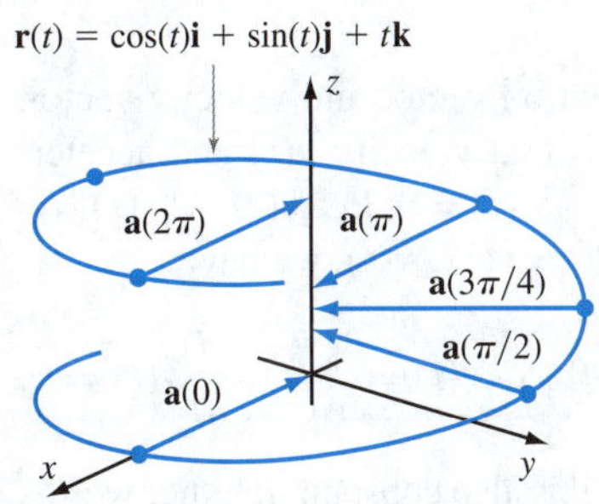

Figure 5

Example 3 Let $\mathbf{r}(t) = \cos(t)\mathbf{i} + \sin(t)\mathbf{j} + t\mathbf{k}$, as in Example 1. Calculate the acceleration $\mathbf{a}(t)$. Evaluate the acceleration at $t = 0, \pi/2, 3\pi/4, \pi$, and 2π.

Solution We have $\mathbf{v}(t) = -\sin(t)\mathbf{i} + \cos(t)\mathbf{j} + \mathbf{k}$ and $\mathbf{a}(t) = -\cos(t)\mathbf{i} - \sin(t)\mathbf{j}$. In particular, $\mathbf{a}(0) = -\mathbf{i}$, $\mathbf{a}(\pi/2) = -\mathbf{j}$, $\mathbf{a}(3\pi/4) = (+1/\sqrt{2})\mathbf{i} + (-1/\sqrt{2})\mathbf{j}$, $\mathbf{a}(\pi) = \mathbf{i}$, and $\mathbf{a}(2\pi) = -\mathbf{i}$. These acceleration vectors are exhibited in Figure 5. In each instance, the acceleration vector points horizontally and *into* the helix, which is also the direction of the force required to hold the particle on the spiral path. ■

Since differentiation of a vector-valued function is performed componentwise, it follows that antidifferentiation is also performed componentwise. We saw this phenomenon at the end of Section 12.1. We now use this idea in a physical example.

Example 4 Suppose a particle moves through space with acceleration vector that is identically $\vec{\mathbf{0}}$. What does that tell us about the particle's motion?

Solution Let $\mathbf{r}(t)$ be the position vector of the particle at time t. We are given that $\mathbf{v}'(t) = \mathbf{a}(t) = 0\mathbf{i} + 0\mathbf{j} + 0\mathbf{k}$. Antidifferentiating each side of this equation componentwise yields $\mathbf{v}(t) = c_1\mathbf{i} + c_2\mathbf{j} + c_3\mathbf{k}$ where c_1, c_2, and c_3 are constants. Notice that we are using only the simple fact that the antiderivative of 0 is a constant. Although our computation is not finished, we can already observe that the speed of the particle is constant: $\|\mathbf{v}(t)\| = \sqrt{c_1^2 + c_2^2 + c_3^2}$ for all t.

Next, we work with the equations $\mathbf{r}'(t) = \mathbf{v}(t)$ and $\mathbf{v}(t) = c_1\mathbf{i} + c_2\mathbf{j} + c_3\mathbf{k}$. We equate to obtain $\mathbf{r}'(t) = c_1\mathbf{i} + c_2\mathbf{j} + c_3\mathbf{k}$. Again, we antidifferentiate componentwise, this time obtaining

$$\mathbf{r}(t) = (c_1 t + d_1)\mathbf{i} + (c_2 t + d_2)\mathbf{j} + (c_3 t + d_3)\mathbf{k}.$$

Here we used the fact that the most general antiderivative of the constant c_1 is $c_1 t + d_1$ and so on. Of course, d_1, d_2, and d_3 are new constants of integration. Our calculation is complete. We have learned that $\mathbf{r}(t) = \langle d_1, d_2, d_3\rangle + t\langle c_1, c_2, c_3\rangle$, which is the parameterization of a line. In summary, we have shown that if a particle moves through space with an acceleration vector that is identically $\vec{\mathbf{0}}$, then the particle travels with constant speed along a line. ■

inSIGHT

Example 4 is essentially Newton's First Law: A body moves at constant speed along a straight line unless a force acts upon it. In particular, Example 4 tells us that whenever a body travels along a path that is not a straight line, there must be some acceleration (and hence some force) present.

Example 5 Show that if a body moves through space in such a way that its acceleration vector is always perpendicular to its velocity vector, then the body is traveling at constant speed.

Solution Let $\mathbf{r}(t)$ and $\mathbf{v}(t) = \mathbf{r}'(t)$ denote the position vector and velocity vector, respectively, of the body at time t. Our hypothesis is that velocity $\mathbf{v}(t)$ and acceleration $\mathbf{v}'(t)$ are perpendicular. We therefore have $\mathbf{v}(t) \cdot \mathbf{v}'(t) = 0$. Let $v(t) = \|\mathbf{v}(t)\| = \sqrt{\mathbf{v}(t) \cdot \mathbf{v}(t)}$ denote the speed of the body. Since $v(t)^2 = \mathbf{v}(t) \cdot \mathbf{v}(t)$, we have

$$\frac{d}{dt}(v(t)^2) = \frac{d}{dt}(\mathbf{v}(t) \cdot \mathbf{v}(t)) = \mathbf{v}(t) \cdot \frac{d}{dt}\mathbf{v}(t) + \left(\frac{d}{dt}\mathbf{v}(t)\right) \cdot \mathbf{v}(t) = 2\mathbf{v}(t) \cdot \frac{d}{dt}\mathbf{v}(t) = 0.$$

We conclude that $v(t)^2$ is constant. It follows that $|\mathbf{v}(t)|$ is also constant. In other words, the body travels with constant speed. ■

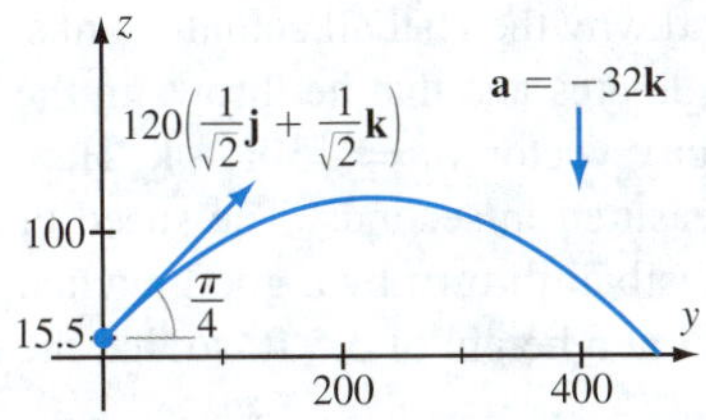

Figure 6

Example 6 An arrow is shot into the air from a height of 15.5 ft at a speed of 120 ft/s. The initial angle of the arrow from the horizontal is $\pi/4$ (see Figure 6). Assuming that the acceleration due to gravity is 32 ft/s^2, determine how far the arrow travels horizontally.

Solution The initial position vector is $0\mathbf{i} + 0\mathbf{j} + (31/2)\mathbf{k}$. The motion takes place entirely in the yz-plane (there is no side-to-side motion). The initial direction of the velocity is $(1/\sqrt{2})\mathbf{j} + (1/\sqrt{2})\mathbf{k}$. Writing the initial velocity as the scalar multiplication of its magnitude and direction vector, we have $120((1/\sqrt{2})\mathbf{j} + (1/\sqrt{2})\mathbf{k}) = 0\mathbf{i} + 60\sqrt{2}\mathbf{j} + 60\sqrt{2}\mathbf{k}$. The only acceleration in the problem is due to gravity: $\mathbf{a} = -32\mathbf{k}$.

We begin with the acceleration and work backward: $\mathbf{a}(t) = -32\mathbf{k} = 0\mathbf{i} + 0\mathbf{j} - 32\mathbf{k}$. Antidifferentiating in each component separately (don't forget the constants of integration!) gives $\mathbf{v}(t) = c_1\mathbf{i} + c_2\mathbf{j} + (-32t + c_3)\mathbf{k}$. We now have

$$0\mathbf{i} + 60\sqrt{2}\mathbf{j} + 60\sqrt{2}\mathbf{k} = \mathbf{v}(0) = c_1\mathbf{i} + c_2\mathbf{j} + (-32 \cdot 0 + c_3)\mathbf{k}.$$

This means that $c_1 = 0$, $c_2 = 60\sqrt{2}$, and $c_3 = 60\sqrt{2}$. We conclude that $\mathbf{v}(t) = 60\sqrt{2}\mathbf{j} + (-32t + 60\sqrt{2})\mathbf{k}$. Antidifferentiating once again gives

$$\mathbf{r}(t) = d_1\mathbf{i} + \left(60\sqrt{2}t + d_2\right)\mathbf{j} + \left(-16t^2 + 60\sqrt{2}t + d_3\right)\mathbf{k}.$$

But then

$$0\mathbf{i} + 0\mathbf{j} + \left(\frac{31}{2}\right)\mathbf{k} = \mathbf{r}(0) = d_1\mathbf{i} + d_2\mathbf{j} + d_3\mathbf{k}.$$

We conclude that $d_1 = 0$, $d_2 = 0$, and $d_3 = 31/2$. To summarize, the trajectory of the arrow is given by

$$\mathbf{r}(t) = 60\sqrt{2}t\mathbf{j} + \left(-16t^2 + 60\sqrt{2}t + \frac{31}{2}\right)\mathbf{k}.$$

The motion of the arrow ceases when the arrow hits the ground, which is when the $\mathbf{k}$ component of $\mathbf{r}$ is 0. We solve the quadratic equation $-16t^2 + 60\sqrt{2}t + 31/2 = 0$ to obtain

$$t = \frac{-60\sqrt{2} \pm \sqrt{(-60\sqrt{2})^2 - 4(-16)(31/2)}}{-32} = \frac{-60\sqrt{2} \pm \sqrt{8192}}{-32} = \frac{-60\sqrt{2} \pm 64\sqrt{2}}{-32}.$$

The negative solution $(-60\sqrt{2} + 64\sqrt{2})/(-32) = -\sqrt{2}/8$ has no physical relevance. For $t = (-60\sqrt{2} - 64\sqrt{2})/(-32) = 31\sqrt{2}/8$, the $\mathbf{j}$ component of $\mathbf{r}$ is $(60\sqrt{2})(31\sqrt{2}/8) = 465$. Thus the horizontal distance traveled by the arrow is 465 ft. ■

The Physics of Baseball

Many experts consider the fastball to be the most important pitch in baseball. If the pitcher imparts a certain backspin to the ball when it is released, then the ball can appear to abruptly sink shortly before it reaches the plate. The ball can sink by as much as several inches. Since the sweet spot of a standard baseball bat has diameter 2.75 in., and since the batter must get his "fix" on the ball during the first 30 ft of the ball's 60 ft journey to the plate, the "sinking fastball" can certainly fool a batter.

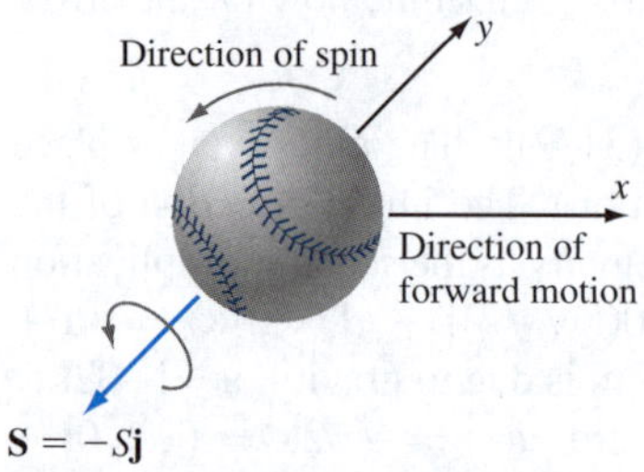

Figure 7

Let us use the ideas we have learned to understand why the fastball actually sinks. Suppose that the pitcher stands at the origin of coordinates and that he throws in the direction of the positive x-axis with an initial velocity vector $\mathbf{v}_0 = 130\mathbf{i} - \mathbf{k}$. Here the units are distance measured in feet and time measured in seconds. The speed of 130 ft/s is about 88.6 mi/h, a plausible speed for a fastball thrown by a good pitcher. Assume, in addition, that the pitcher releases the ball at a height of 5.5 ft, so that the initial position is $\mathbf{r}_0 = 5.5\mathbf{k}$.

The interesting part of this analysis is the acceleration. Of course there is a component of acceleration that is due to gravity; this component contributes $-32\mathbf{k}$. Since the ball is given an initial backward spin, there is also a component called *spin acceleration*. Use the Right Hand Rule to see that the so-called spin vector $\mathbf{S}$ for the ball is pointing in the direction of the negative y-axis (Figure 7). We write $\mathbf{S} = -S\mathbf{j}$.

According to the theory of aerodynamics, the spin causes a difference in air pressure on the sides of the ball and results in a contribution of spin acceleration that is given by

$$\mathbf{a}^s = c\mathbf{S} \times \mathbf{v}_0$$

where c is a physical constant that is determined through experimentation. Let us calculate this cross product:

$$\mathbf{S} \times \mathbf{v}_0 = \det\left(\begin{bmatrix} \mathbf{i} & \mathbf{j} & \mathbf{k} \\ 0 & -S & 0 \\ v_1 & v_2 & v_3 \end{bmatrix}\right) = (-Sv_3)\mathbf{i} + (Sv_1)\mathbf{k}.$$

Noting that $v_1 = 130$ and $v_3 = -1$, we have thus determined that $\mathbf{a}^s = c(S\mathbf{i} + 130S\mathbf{k})$ and

$$\mathbf{a} = \mathbf{a}^s - 32\mathbf{k} = c(S\mathbf{i} + 130S\mathbf{k}) - 32\mathbf{k} = cS\mathbf{i} + (130cS - 32)\mathbf{k}.$$

Typically, the amount of spin placed on the ball is about $S = 2.5$ revolutions per second. Experimental evidence suggests that $c = 0.08$ is plausible. Thus, we arrive at $\mathbf{a} = 0.2\mathbf{i} - 6\mathbf{k}$. As usual, we integrate this last equation to obtain $\mathbf{v} = 0.2t\mathbf{i} - 6t\mathbf{k} + \mathbf{C}$. The value of $\mathbf{C}$ is determined by setting $\mathbf{v}(0) = \mathbf{v}_0$. We find that $\mathbf{C} = 130\mathbf{i} - \mathbf{k}$. Hence $\mathbf{v} = (130 + 0.2t)\mathbf{i} + (-1 - 6t)\mathbf{k}$. Integrating a second time yields $\mathbf{r}(t) = (130t + 0.1t^2)\mathbf{i} + (-t - 3t^2)\mathbf{k} + \mathbf{D}$. The value of the constant vector $\mathbf{D}$ is determined by setting $\mathbf{r}(0) = \mathbf{r}_0$. We find that $\mathbf{D} = 5.5\mathbf{k}$. The final result is then $\mathbf{r}(t) = (130t + 0.1t^2)\mathbf{i} + (5.5 - t - 3t^2)\mathbf{k}$.

What is the meaning of this formula for the motion of the ball? It is subtle. The distance of the pitcher's mound from home plate is 60 ft. To find out when the ball crosses the plate, we solve $130t + 0.1t^2 = 60$ to obtain $t \approx 0.46$. Suppose that the batter gets his fix on the ball at the halfway point. Solving $130t + 0.1t^2 = 30$, we obtain $t \approx 0.23$ s. At that time, the height of the ball is $5.5 - 0.23 - 3 \cdot (0.23)^2$. Thus the batter sees that half way through the ball's course, it drops $0.23 + 3 \cdot (0.23)^2 = 0.3887$ ft. He would expect it to drop a similar amount on the second half of its journey. However, at time $t = 0.46$, the ball has height $(5.5 - 0.46 - 3 \cdot (0.46)^2)$. The ball has dropped a total of $0.46 + 3 \cdot (0.46)^2 = 1.0948$ ft *instead* of the $0.3887 + 0.3887 = 0.7774$ ft that the batter expected. So, in effect, the ball has "sunk" $1.0948 - 0.7774 = 0.3174$ ft, or 3.81 in. Refer to Figure 8 to see the path of the pitched ball.

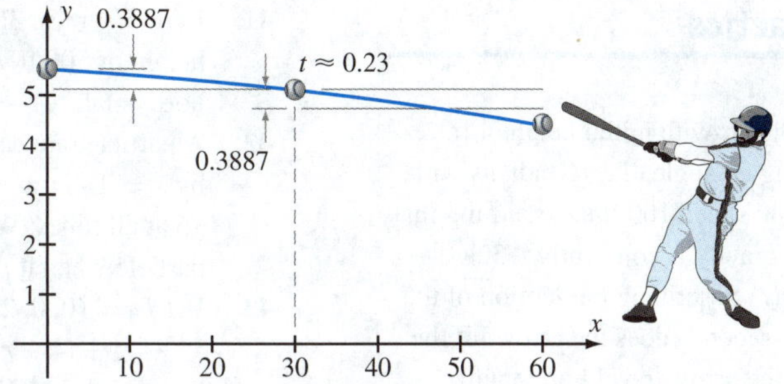

Figure 8
The expected drop after the halfway point is 0.3887 ft.

quickquiz

1. Why is instantaneous velocity a vector?
2. What physical significance do the attributes of the velocity vector have?
3. Why is acceleration a vector?
4. What is the physical significance of the attributes of the acceleration vector?

EXERCISES

Problems for Practice

For each position function **r** in Exercises 1–8, calculate the corresponding velocity **v**, speed v, and acceleration **a**.

1. $\mathbf{r}(t) = t\mathbf{i} + t^2\mathbf{j} + t^3\mathbf{k}$
2. $\mathbf{r}(t) = (2 - t^2)\mathbf{i} - 5t^4\mathbf{k}$
3. $\mathbf{r}(t) = (1 - t^2)\mathbf{i} + 2t\mathbf{j} + (1/(1 + t^2))\mathbf{k}$
4. $\mathbf{r}(t) = e^t\mathbf{i} + e^{-t}\mathbf{j} - e^{-2t}\mathbf{k}$
5. $\mathbf{r}(t) = t\mathbf{i} - 5\mathbf{j} + e^t\mathbf{k}$
6. $\mathbf{r}(t) = \ln(|t|)\mathbf{i} + t^{-1}\mathbf{j} - t^{-2}\mathbf{k}$
7. $\mathbf{r}(t) = -\cos(t^2)\mathbf{i} + \sin(t^2)\mathbf{j} + t^3\mathbf{k}$
8. $\mathbf{r}(t) = t^{1/2}\mathbf{i} + t^{1/3}\mathbf{j} + t^{1/4}\mathbf{k}$

In Exercises 9–12, a body traveling through space near Earth's surface has constant acceleration due to gravity given by $-32\mathbf{k}$. The initial velocity $\mathbf{v}_0 = \mathbf{v}(0)$ and initial position $\mathbf{r}_0 = \mathbf{r}(0)$ are given in each problem. Using antidifferentiation, compute a formula for $\mathbf{r}(t)$.

9. $\mathbf{v}_0 = 3\mathbf{i} - 2\mathbf{j} + \mathbf{k}$, $\mathbf{r}_0 = 2\mathbf{i} - 5\mathbf{k}$
10. $\mathbf{v}_0 = 4\mathbf{i} - 7\mathbf{j}$, $\mathbf{r}_0 = 2\mathbf{j} + 8\mathbf{k}$
11. $\mathbf{v}_0 = \mathbf{i} + \mathbf{j} + \mathbf{k}$, $\mathbf{r}_0 = 3\mathbf{i} - 2\mathbf{j} - \mathbf{k}$
12. $\mathbf{v}_0 = \mathbf{k}$, $\mathbf{r}_0 = \mathbf{j}$

In Exercises 13–22, determine parametric equations for the tangent line to the curve described by the function **r** at the point P.

13. $\mathbf{r}(t) = t\mathbf{i} + t^2\mathbf{j} + t^3\mathbf{k}$, $P = (1, 1, 1)$
14. $\mathbf{r}(t) = (2 - t^2)\mathbf{i} - (1 - t^2)\mathbf{j} - 5t\mathbf{k}$, $P = (1, 0, -5)$
15. $\mathbf{r}(t) = e^t\mathbf{i} + e^{-t}\mathbf{j} - e^{-2t}\mathbf{k}$, $P = (1, 1, -1)$
16. $\mathbf{r}(t) = \ln(1 + t)\mathbf{i} + 1/(1 + t)\mathbf{j} - 1/(1 - t)^2\mathbf{k}$, $P = (0, 1, -1)$
17. $\mathbf{r}(t) = t\mathbf{i} - \mathbf{j} + e^{t/2}\mathbf{k}$, $P = (2, -1, e)$
18. $\mathbf{r}(t) = -\cos(t)\mathbf{i} + \sin(t)\mathbf{j} + t\mathbf{k}$, $P = (1, 0, \pi)$
19. $\mathbf{r}(t) = t\mathbf{i} - 5\mathbf{j} + (5/t)\mathbf{k}$, $P = (5, -5, 1)$
20. $\mathbf{r}(t) = (5 - t^2)\mathbf{i} + 2t\mathbf{j} + (5/(1 + t^2))\mathbf{k}$, $P = (1, -4, 1)$
21. $\mathbf{r}(t) = (3 + t)/(1 + t^2)\mathbf{i} + t^2\mathbf{j} + \sin(\pi t)\mathbf{k}$, $P = (1, 4, 0)$
22. $\mathbf{r}(t) = e^t\mathbf{i} + te^t\mathbf{j} + t^2e^t\mathbf{k}$, $P = (1, 0, 0)$

In Exercises 23–32, determine symmetric equations for the tangent line to the curve described by the function **r** at the point P.

23. $\mathbf{r}(t) = t\mathbf{i} - t^2\mathbf{j} + t^3\mathbf{k}$, $P = (-2, -4, -8)$
24. $\mathbf{r}(t) = t\mathbf{i} - 5\mathbf{j} + e^t\mathbf{k}$, $P = (2, -5, e^2)$
25. $\mathbf{r}(t) = t\mathbf{i} - 5\mathbf{j} + (5/t)\mathbf{k}$, $P = (5, -5, 1)$
26. $\mathbf{r}(t) = (7 - t^2)\mathbf{i} + 2\mathbf{j} + (t + 1/t)\mathbf{k}$, $P = (6, 2, -2)$
27. $\mathbf{r}(t) = (2/t)\mathbf{i} + 2t\mathbf{j} + \mathbf{k}$, $P = (1, 4, 1)$
28. $\mathbf{r}(t) = 4\arctan(t)\mathbf{i} - \cos(\pi t)\mathbf{j} + (1/t)\mathbf{k}$, $P = (\pi, 4, 1)$
29. $\mathbf{r}(t) = -\cos(t^2)\mathbf{i} + \sin(t^2)\mathbf{j} + t\mathbf{k}$, $P = (1, 0, \sqrt{\pi})$
30. $\mathbf{r}(t) = e^t\mathbf{i} + e^{2t}\mathbf{j} + e^{3t}\mathbf{k}$, $P = (1, 1, 1)$
31. $\mathbf{r}(t) = (5 - t^2)\mathbf{i} + 2t\mathbf{j} + (5/(1 + t^2))\mathbf{k}$, $P = (1, -4, 1)$
32. $\mathbf{r}(t) = \ln(1 + t^2)\mathbf{i} + \arcsin(t)\mathbf{j} + \sec(t)\mathbf{k}$, $P = (0, 0, 1)$

Further Theory and Practice

33. An arrow is shot into the air with initial height 4 ft, initial trajectory forming an angle of $\pi/6$ radians with the horizontal, and initial speed 100 ft/s. Assuming that the acceleration due to gravity is constantly $-32\mathbf{k}$, compute the function $\mathbf{r}(t)$ describing the motion of the arrow. After how many seconds does the arrow hit the ground? How far does the arrow travel horizontally? What is the greatest height that the arrow achieves?

34. A rock is thrown from a 500 ft cliff, with a downward trajectory forming an initial angle of $\pi/6$ radians with the horizontal. The initial speed is 50 ft/s. The acceleration due to gravity is constantly $-32\mathbf{k}$. Derive a formula for the position $\mathbf{r}(t)$ of the rock. When does the rock hit the ground? How far does it travel horizontally?

35. A roller coaster is speeding along a track; its motion describes the path

$$\mathbf{r}(t) = -5t^2\mathbf{i} + t^3\mathbf{j} + (4t + 30)\mathbf{k}.$$

At time $t = 2$, the roller coaster leaves the track. Ignoring the effects of gravity and drag, determine where the roller coaster will be at time $t = 5$.

36. An arrow is shot into the air from an initial height of 6 ft and with trajectory making initial angle of $\pi/3$ radians with the horizontal. If it were a calm day, the force exerted by the bow would have launched the arrow with initial speed 120 ft/s. However, a sudden gust of wind reduces the *horizontal* component of the arrow's initial velocity by 20 ft/s (the wind blows horizontally and dies out after its instantaneous gusting). The acceleration due to gravity is constantly $-32\mathbf{k}$. What is the velocity of the arrow when it strikes the ground? How far does the arrow travel horizontally before it strikes the ground?

37. A body moving through space has constant acceleration vector and zero initial velocity. What can you say about the path traveled by the body? (Give a geometric description.)

38. A body in motion through space has the property that its velocity vector always equals its acceleration vector. What can you say about the motion? Can you write a formula for it?

39. A body travels through space in such a way that its position vector is always perpendicular to its velocity vector. Show that this implies that the motion is on the surface of a sphere.

40. Redo Exercise 35, but this time take into account the effect of gravity.

41. Redo Exercise 36 if the arrow is shot downward from a height of 500 ft with an initial angle of 60 deg from the horizontal.

42. A particle moves along the curve that is parameterized by $x = 1 + t$, $y = 2t^2$, $z = 6t + t^2$. Its vertical speed is 16 at all times. What is the velocity vector of the particle when it passes through the point $(2, 2, 7)$?

43. Let $P_0 = (0, 0, 2r)$ denote the north pole of the sphere $x^2 + y^2 + (z - r)^2 = r^2$. Let P_1 be any other point on the sphere. See Figure 9. Suppose that a rigid straight line joins P_0 to P_1 and that a particle initially at rest is allowed to slide down the line from P_0 to P_1 under a force equal to $-\mathrm{mg}\mathbf{k}$. Show that the elapsed time does not depend on the choice of the point P_1.

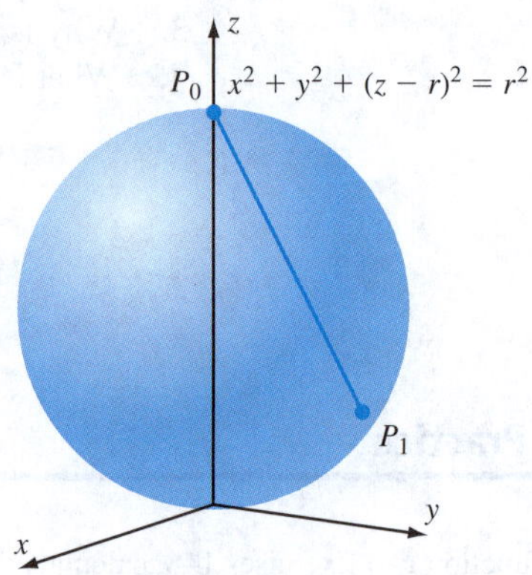

Figure 9

44. Let $\mathbf{r}_0$ and $\mathbf{v}_0$ be constant vectors. Let ω be a positive constant. Show that

$$\mathbf{r}(t) = \cos(\omega t)\mathbf{r}_0 + \frac{\sin(\omega t)}{\omega}\mathbf{v}_0$$

satisfies the initial value problem

$$\frac{d^2}{dt^2}\mathbf{r}(t) + \omega^2\mathbf{r}(t) = \vec{\mathbf{0}},\ \mathbf{r}(0) = \mathbf{r}_0,\ \mathbf{r}'(0) = \mathbf{v}_0.$$

45. Let $\mathbf{r}(t)$ denote the position at time t of a body moving through space. Suppose that $\mathbf{r}(t)$ is parallel to $\mathbf{a}(t) = \mathbf{r}''(t)$ for all t. Prove that $\mathbf{r}(t) \times \mathbf{r}'(t)$ is a constant vector.

46. To each point $P_t = (\cos(t), \sin(t), t)$ with $t \geq 0$, we associate another point Q_t as follows: $\overrightarrow{P_tQ_t}$ is tangent to the helix traced by P_t, $\|\overrightarrow{P_tQ_t}\| = \sqrt{2}t$, and $\overrightarrow{P_tQ_t} \cdot \mathbf{k} \geq 0$. Parameterize the curve traced by Q.

47. A particle moves in such a way that the angle between its velocity and acceleration vectors is always acute. Show that the speed of the particle is increasing. If the angle is obtuse, then show that the speed of the particle is decreasing.

48. Let $\mathbf{r}(t) = (t^3 - t)\mathbf{i} + (t^2 - 1)\mathbf{j}$. Plot the points $\mathbf{r}(-2)$, $\mathbf{r}(-1)$, $\mathbf{r}(-1/2)$, $\mathbf{r}(0)$, $\mathbf{r}(1/2)$, $\mathbf{r}(1)$, $\mathbf{r}(2)$. Sketch the curve. Describe the tangent lines to the curve at the point $(0, 0)$.

49. Let $\mathcal{C}$ be the curve with parameterization $\mathbf{r}(t) = \langle t, t^2, t^3 \rangle$ for $t > 0$. Find a point P_0 on $\mathcal{C}$ for which the tangent line to $\mathcal{C}$ at P_0 passes through the point $(0, -1/2, -1/\sqrt{2})$.

50. Show that the curves $t \mapsto \langle t^3 + t^2 - 3t + 1, t^2, 3t - 1 \rangle$ and $t \mapsto \langle 3t + 1, 2t, t^2 + t - 1 \rangle$ intersect at two points. At each of these points, what is the angle between the tangent lines to the curves?

Calculator/Computer Exercises

In Exercises 51–58, plot the planar curve associated with the given vector-valued function $\mathbf{r}(t) = r_1(t)\mathbf{i} + r_2(t)\mathbf{j}$. Calculate $\mathbf{r}'(t_0)$ at the given value of t_0. Add to your plot the tangent line to the curve at $\mathbf{r}(t_0)$.

51. $\mathbf{r}(t) = (5 + 2\cos(t))\mathbf{i} + (3 + \sin(t))\mathbf{j}$, $0 \le t \le 2\pi$, $t_0 = \pi/3$

52. $\mathbf{r}(t) = (t^2 + t)\mathbf{i} + (t^2 - t)\mathbf{j}$, $-2 \le t \le 2$, $t_0 = 1$

53. $\mathbf{r}(t) = \tan(t)\mathbf{i} + \sec(t)\mathbf{j}$, $-\pi/3 \le t \le \pi/3$, $t_0 = \pi/4$

54. $\mathbf{r}(t) = 3t/(1 + t^3)\mathbf{i} + 3t^2/(1 + t^3)\mathbf{j}$, $-0.6 \le t \le 15$, $t_0 = 1/3$

55. $\mathbf{r}(t) = (e^t + e^{-t})\mathbf{i} + (e^t - e^{-t})\mathbf{j}$, $-5 \le t \le 5$, $t_0 = 2$

56. $\mathbf{r}(t) = (t^3 - t)\mathbf{i} + (t^2 - 1)\mathbf{j}$, $-2 \le t \le 2$, $t_0 = 1/2$

57. $\mathbf{r}(t) = 2t/(1 + t^2)\mathbf{i} + (1 - t^2)/(1 + t^2)\mathbf{j}$, $-20 \le t \le 20$, $t_0 = 1/2$

58. $\mathbf{r}(t) = (t - \sin(t))\mathbf{i} + (1 - \cos(t))\mathbf{j}$, $0 \le t \le 2\pi$, $t_0 = \pi/6$

59. Plot the planar curve $\mathbf{r}(t) = 3t/(1 + t^3)\mathbf{i} + 3t^2/(1 + t^3)\mathbf{j}$, $t > -1/2$. It is an arc of the Folium of Descartes. Find all points at which the curve has a tangent line parallel to the line $y = x$. Find all points at which the curve has a horizontal tangent line. Find all points at which the curve has a vertical tangent line.

60. Plot $\mathbf{r}(t) = \cos(3t)\cos(t)\mathbf{i} + \sin(2t)\sin(t)\mathbf{j}$ for $0 < t < \pi/2$. Find the point P_0 where the curve crosses itself, and add the two tangents at P_0 to your plot.

61. Plot $\mathbf{r}(t) = \langle t^2 - 1, t^4 - 1, (t - 1)\ln(2 + t) \rangle$ for $-6/5 \le t \le 6/5$. To your plot, add the tangent lines at the point $(0, 0, 0)$.

62. The curve $\mathbf{r}(t) = \langle t\cos(t), t\sin(t), t \rangle$, $0 \le t \le \pi/6$ has one tangent line that intersects the vertical line $x = 1$, $y = 1$. What is the intersection point? Plot the curve, its tangent line, and the vertical line.

12.3 Tangent Vectors and Arc Length

In Section 12.2, we studied the derivative $\mathbf{r}'$ of a vector-valued function $\mathbf{r}$. Our focus was on the velocity and acceleration of the motion that $\mathbf{r}$ defines. We also learned how to use the derivative to obtain tangent lines to the graph of $\mathbf{r}$. In this section, we investigate the properties of tangents in greater depth. We also learn an application to arc length.

Unit Tangent Vectors

To effectively study the geometry of a space curve $\mathcal{C}$, we must use a parameterization $\mathbf{r}$ that provides us with information about the tangent lines of $\mathcal{C}$. This forces us to eliminate certain parameterizations from our discussions. As an example of what we must avoid, consider the vector-valued function $\mathbf{r}(t) = \langle \cos(t^3), \sin(t^3), 0 \rangle$ that parameterizes the circle $\mathcal{C} = \{(x, y, 0) : x^2 + y^2 = 1\}$. Notice that the derivative $\mathbf{r}'(t) = \langle -3t^2\sin(t^3), 3t^2\cos(t^3), 0 \rangle$ vanishes when $t = 0$. Because $\mathbf{r}'(0) = \vec{\mathbf{0}}$, we obtain no information about the direction of the tangent line to $\mathcal{C}$ at the point $\mathbf{r}(0)$. Our next definition excludes such inconvenient parameterizations.

Definition Suppose that $\mathbf{r}(t) = r_1(t)\mathbf{i} + r_2(t)\mathbf{j} + r_3(t)\mathbf{k}$ is continuous for $a \le t \le b$. We say that $\mathbf{r}$ is a *smooth* parameterization of the curve it defines if

a. the scalar-valued functions r_1, r_2, r_3 are all twice continuously differentiable on (a, b), and
b. $\mathbf{r}'(t) \neq \vec{\mathbf{0}}$ for every t in (a, b).

If we can divide the interval $[a, b]$ into finitely many subintervals $[a, x_1]$, $[x_1, x_2], \ldots, [x_{N-1}, b]$ such that the restriction of $\mathbf{r}$ to each subinterval is smooth, then we say that $\mathbf{r}$ is *piecewise smooth.*

Until Chapter 15, we will restrict our attention to smooth parameterizations. A moving particle with position vector $\mathbf{r}$ that is a smooth parameterization of its path always has positive speed, a differentiable velocity vector, and a continuous acceleration vector. For such a parameterization $\mathbf{r}$ of a curve $\mathcal{C}$, we define the *unit tangent vector* to $\mathcal{C}$ at the point $P = \mathbf{r}(t)$ to be the vector

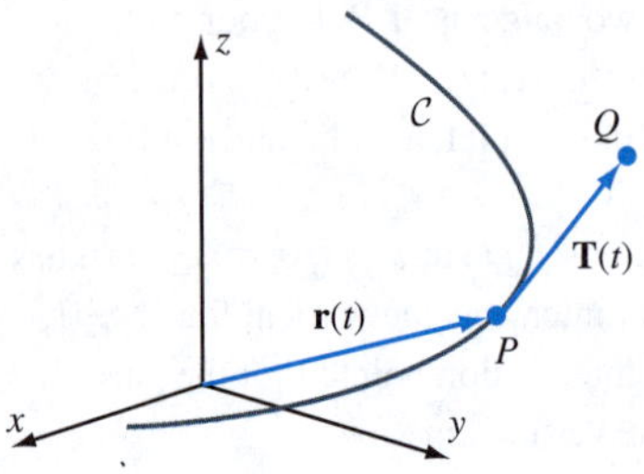

Figure 1

$$\mathbf{T}(t) = \left(\frac{1}{\|\mathbf{r}'(t)\|}\right)\mathbf{r}'(t) \qquad \textbf{(12.1)}$$

As its name implies, $\mathbf{T}(t)$ is a unit vector. Furthermore, if $\mathbf{T}(t)$ is represented by a directed line segment $\overrightarrow{PQ}$ with initial point P, then $\overrightarrow{PQ}$ is tangent to $\mathcal{C}$ at P. See Figure 1. Notice that a particle with position vector $\mathbf{r}(t)$ traces $\mathcal{C}$ in the direction indicated by $\mathbf{T}(t)$. Finally, observe that by rewriting equation (12.1), we may express the velocity vector $\mathbf{r}'(t)$ as the scalar multiplication of the unit tangent vector by the speed:

$$\mathbf{r}'(t) = \|\mathbf{r}'(t)\|\mathbf{T}(t) = v(t)\mathbf{T}(t). \qquad \textbf{(12.2)}$$

Example 1 Let $\mathbf{r}(t) = e^t\mathbf{i} + e^{-t}\mathbf{j} + \sqrt{2}t\mathbf{k}$. Calculate the unit tangent vector $\mathbf{T}(t)$. What are the unit tangent vectors at $P_0 = \mathbf{r}(0)$ and $P_1 = \mathbf{r}(\ln(2))$?

Solution We calculate that $\mathbf{r}'(t) = e^t\mathbf{i} - e^{-t}\mathbf{j} + \sqrt{2}\mathbf{k}$ is a tangent vector to the curve defined by $\mathbf{r}$ at the point $\mathbf{r}(t)$. The length of this tangent vector is

$$\|\mathbf{r}'(t)\| = \sqrt{(e^t)^2 + (-e^{-t})^2 + \left(\sqrt{2}\right)^2} = \sqrt{e^{2t} + e^{-2t} + 2} = \sqrt{(e^t + e^{-t})^2} = e^t + e^{-t}.$$

The *unit* tangent vector for any t is therefore

$$\mathbf{T}(t) = \frac{1}{e^t + e^{-t}}\left(e^t\mathbf{i} - e^{-t}\mathbf{j} + \sqrt{2}\mathbf{k}\right) = \frac{e^t}{e^t + e^{-t}}\mathbf{i} - \frac{e^{-t}}{e^t + e^{-t}}\mathbf{j} + \frac{\sqrt{2}}{e^t + e^{-t}}\mathbf{k}.$$

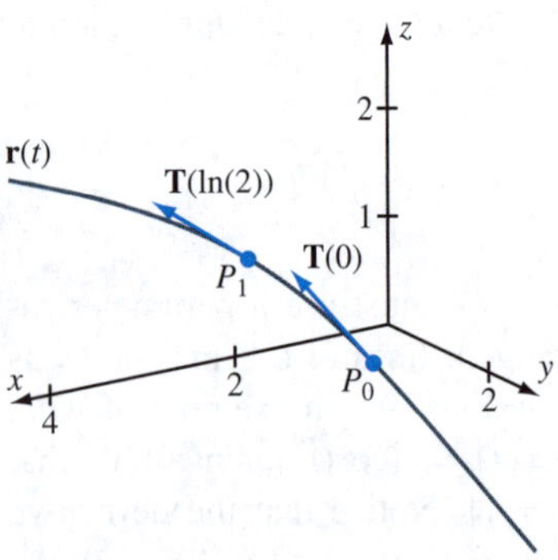

Figure 2

In particular, we have $\mathbf{T}(0) = (1/2)\mathbf{i} - (1/2)\mathbf{j} + (\sqrt{2}/2)\mathbf{k}$ and

$$\mathbf{T}(\ln(2)) = \frac{2}{2 + 1/2}\mathbf{i} - \frac{1/2}{2 + 1/2}\mathbf{j} + \frac{\sqrt{2}}{2 + 1/2}\mathbf{k} = \frac{4}{5}\mathbf{i} - \frac{1}{5}\mathbf{j} + \frac{2\sqrt{2}}{5}\mathbf{k}.$$

These two unit tangent vectors are plotted in Figure 2. ■

Example 2 Let $\mathbf{r}(t) = -t\mathbf{i} + t^2\mathbf{j} + (t^3 - 5t)\mathbf{k}$. The point $P_0 = (-1, 1, -4)$ corresponds to $t = 1$ and lies on this curve. Using the unit tangent vector at this point, write symmetric equations for the tangent line to the curve at P_0.

Solution The velocity vector at $t = 1$ is

$$\mathbf{r}'(1) = (-\mathbf{i} + 2t\mathbf{j} + (3t^2 - 5)\mathbf{k})\big|_{t=1} = -\mathbf{i} + 2\mathbf{j} - 2\mathbf{k},$$

and the speed is $\|\mathbf{r}'(1)\| = \sqrt{(-1)^2 + 2^2 + (-2)^2} = 3$. The unit tangent vector is therefore

$$\mathbf{T}(1) = \frac{1}{\|\mathbf{r}'(1)\|}\mathbf{r}'(1) = \frac{1}{3}(-\mathbf{i} + 2\mathbf{j} - 2\mathbf{k}) = -\frac{1}{3}\mathbf{i} + \frac{2}{3}\mathbf{j} + \left(-\frac{2}{3}\right)\mathbf{k}.$$

The tangent line we seek is

$$\frac{x - (-1)}{-1/3} = \frac{y - 1}{2/3} = \frac{z - (-4)}{-2/3}.$$

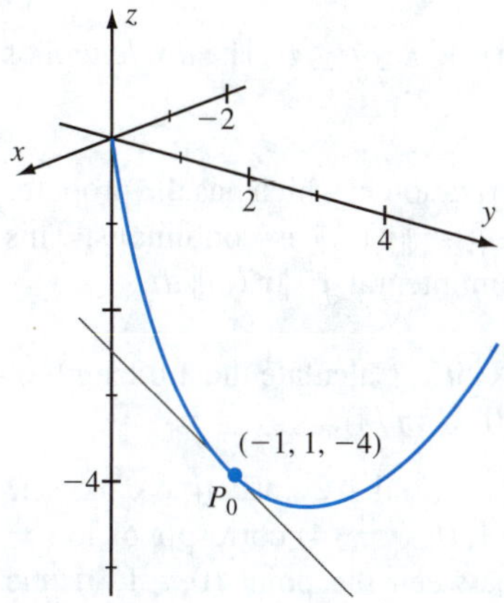

Figure 3

Figure 3 shows the plot of $\mathbf{r}(t)$ for $0 \le t \le 2$. The tangent line at P_0 has also been plotted. ■

Arc Length

If a curve $\mathcal{C}$ is defined by a smooth parameterization $\mathbf{r}(t)$ for $a \le t \le b$, then we may ask what the *length* of $\mathcal{C}$ is from $\mathbf{r}(a)$ to $\mathbf{r}(b)$. Here is an intuitive way to think about length: Lay a piece of string along the curve from $\mathbf{r}(a)$ to $\mathbf{r}(b)$. Mark the string where it touches $\mathbf{r}(a)$ and $\mathbf{r}(b)$, straighten out the string, and measure the distance between the marks. That is the length. To obtain a method for actually calculating length from the function $\mathbf{r}$, we follow the scheme that we used in Section 8.2 to compute the lengths of planar curves.

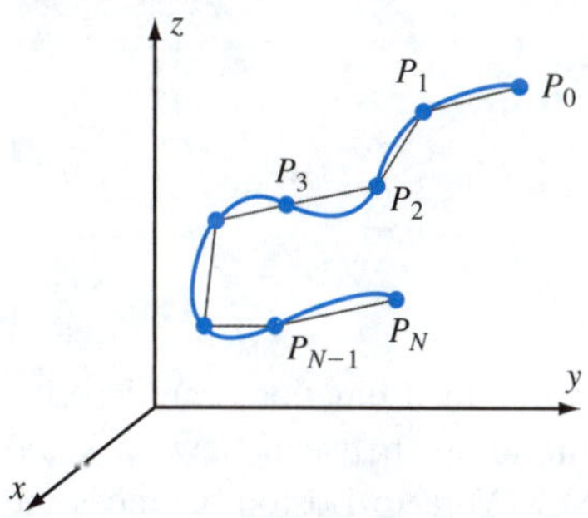

Figure 4

We divide the interval $[a, b]$ into small subintervals by a uniform partition

$$a = t_0 < t_1 < \cdots < t_N = b.$$

Let $\Delta t = (b - a)/N$ be the common width of the subintervals of this partition. Set $P_j = \mathbf{r}(t_j) = (x_j, y_j, z_j)$. The basic idea is that the length of the curve $\mathbf{r}$ is approximated by the sum of the lengths of the vectors $\overrightarrow{P_0P_1}, \overrightarrow{P_1P_2}, \overrightarrow{P_2P_3}, \ldots, \overrightarrow{P_{N-1}P_N}$. See Figure 4. Thus, the length of the curve is approximately given by

$$S_N = \sum_{j=1}^{N} \|\overrightarrow{P_{j-1}P_j}\| = \sum_{j=1}^{N} \sqrt{(x_j - x_{j-1})^2 + (y_j - y_{j-1})^2 + (z_j - z_{j-1})^2}.$$

The Mean Value Theorem tells us that

$$\begin{aligned} x_j - x_{j-1} &= r_1(t_j) - r_1(t_{j-1}) = r_1'(\rho_j)\Delta t, \\ y_j - y_{j-1} &= r_2(t_j) - r_2(t_{j-1}) = r_2'(\sigma_j)\Delta t, \\ z_j - z_{j-1} &= r_3(t_j) - r_3(t_{j-1}) = r_3'(\tau_j)\Delta t \end{aligned}$$

for some numbers ρ_j, σ_j, τ_j in the subinterval (t_{j-1}, t_j). We can rewrite our formula for S_N as

$$S_N = \sum_{j=1}^{N} \sqrt{r_1'(\rho_j)^2 + r_2'(\sigma_j)^2 + r_3'(\tau_j)^2}\,\Delta t.$$

On the one hand, this sum approximates the length of the curve. On the other hand, S_N approaches the Riemann integral $\int_a^b \|\mathbf{r}'(t)\|\, dt$ as N tends to infinity. We are led to the following definition.

Definition Let $\mathcal{C}$ be a curve with smooth parameterization $t \mapsto \mathbf{r}(t)$, $a \le t \le b$. The *arc length* of $\mathcal{C}$ is $\int_a^b \|\mathbf{r}'(t)\|\, dt$.

By restricting our attention to a smooth parameterization $\mathbf{r}$, which has the property that $\mathbf{r}'$ is continuous, we ensure that the function $t \mapsto \|\mathbf{r}'(t)\|$ is continuous. This property, in turn, assures the existence of the Riemann integral $\int_a^b \|\mathbf{r}'(t)\|\, dt$.

Example 3 Let $\mathbf{r}(t) = \sin(2t)\,\mathbf{i} - \cos(2t)\,\mathbf{j} + \sqrt{5}t\mathbf{k}$. Calculate the arc length of that portion of the curve between $(0, -1, 0)$ and $(1, 0, \sqrt{5}\pi/4)$.

Solution We differentiate $\mathbf{r}(t)$, obtaining $\mathbf{r}'(t) = 2\cos(2t)\mathbf{i} + 2\sin(2t)\mathbf{j} + \sqrt{5}\mathbf{k}$. The point $(0, -1, 0)$ corresponds to $t = 0$, and the point $(1, 0, \sqrt{5}\pi/4)$ corresponds to $t = \pi/4$. The arc length of the portion of the curve between the point $(0, -1, 0)$ and $(1, 0, \sqrt{5}\pi/4)$ is therefore

$$\begin{aligned}\int_0^{\pi/4} \|\mathbf{r}'(t)\|\, dt &= \int_0^{\pi/4} \sqrt{(2\cos(2t))^2 + (2\sin(2t))^2\mathbf{j} + \left(\sqrt{5}\right)^2}\, dt \\ &= \int_0^{\pi/4} \sqrt{4(\cos^2(2t) + \sin^2(2t)) + 5\, dt} \\ &= \int_0^{\pi/4} \sqrt{9}\, dt \\ &= 3\pi/4.\end{aligned}$$

Reparameterization

It is important to remember that many different vector-valued functions can describe the same curve in space. Imagine, for example, a railroad line between New York and Boston. Each time a train travels this track from New York to Boston, it traces the same curve $\mathcal{C}$. Yet, if the train travels at different speeds on different trips, then different parameterizations of $\mathcal{C}$ result.

To make this idea concrete, let us consider a specific example. The vector-valued functions

$$\mathbf{r}(t) = \cos(t)\mathbf{i} + \sin(t)\mathbf{j} + t\mathbf{k}, \quad 0 \le t \le 2\pi,$$

and

$$\mathbf{p}(s) = \cos(2s)\mathbf{i} + \sin(2s)\mathbf{j} + 2s\mathbf{k}, \quad 0 \le s \le \pi,$$

describe the same curve in space. To understand why, notice that as s runs through the interval $[0, \pi]$, the expression $t = 2s$ runs through the interval $[0, 2\pi]$. Furthermore, for every s in $[0, \pi]$, we have

$$\mathbf{p}(s) = \mathbf{p}\left(\frac{t}{2}\right) = \cos\left(2\frac{t}{2}\right)\mathbf{i} + \sin\left(2\frac{t}{2}\right)\mathbf{j} + 2\frac{t}{2}\mathbf{k} = \cos(t)\mathbf{i} + \sin(t)\mathbf{j} + t\mathbf{k} = \mathbf{r}(t),$$

which shows that $\mathbf{p}$ and $\mathbf{r}$ describe the same points in space (albeit at different values of their domains). The functions $\mathbf{r}$ and $\mathbf{p}$ are simply different parameterizations of the same curve. Notice that the increasing (and therefore invertible) function $\psi : [0, \pi] \to [0, 2\pi]$ defined by $\psi(s) = 2s = t$ affects the way we pass from one parameterization to the other: Given a value of s in $[0, \pi]$, we have $\mathbf{p}(s) = \mathbf{r}(\psi(s))$, or given a value of t in $[0, 2\pi]$, we have $\mathbf{r}(t) = \mathbf{p}(\psi^{-1}(t))$.

The process of changing the way that we parameterize a curve—that is, choosing a new vector-valued function to describe the same curve—is called *reparameterization.* If a curve $\mathcal{C}$ is parameterized by $\mathbf{r}(t)$ for $a \le t \le b$, then it is convenient to think of a reparameterization $\mathbf{p}$ as composition of $\mathbf{r}$ with a continuously differentiable increasing function $\psi : [c, d] \to [a, b]$. Thus

$$\mathbf{p}(s) = (\mathbf{r} \circ \psi)(s) = \mathbf{r}(\psi(s)), \quad c \le s \le d,$$

is a reparameterization of the curve $\mathcal{C}$, and $\mathbf{p}$ has domain $[c, d]$.

Example 4 Suppose that $\mathbf{r}$ is a smooth parameterization of a curve $\mathcal{C}$ with $P = \mathbf{r}(t)$. Suppse also that $\mathbf{p} = \mathbf{r} \circ \psi$ is a reparameterization with $\psi(s) = t$. Show that the tangent line to $\mathcal{C}$ at the point $\mathbf{p}(s) = \mathbf{r}(t) = P$ will be the same, whichever parameterization we use.

Solution The parameterization $\mathbf{r}$ yields $\mathbf{r}'(t)$ as a tangent vector to $\mathcal{C}$ at P. The reparameterization $\mathbf{p}$ yields $\mathbf{p}'(s)$ as a tangent vector to $\mathcal{C}$ at P. We relate the two vectors by means of Theorem 6g from Section 12.1:

$$\mathbf{p}'(s) = (\mathbf{r} \circ \psi)'(s) = \psi'(s)\mathbf{r}'(\psi(s)) = \psi'(s)\mathbf{r}'(t).$$

This chain of equalities tells us that the vectors $\mathbf{p}'(s)$ and $\mathbf{r}'(t)$ are parallel. It follows that the tangent line to $\mathcal{C}$ at the point $\mathbf{p}(s) = \mathbf{r}(t) = P$ will be the same, whichever parameterization we use. ■

The next theorem shows that the arc length of a curve does not depend on the chosen parameterization of the curve.

Theorem 1 Suppose that $\mathcal{C}$ is a curve with smooth parameterization $t \mapsto \mathbf{r}(t)$, $a \le t \le b$. Let $\psi : [c, d] \to [a, b]$ be a continuously differentiable increasing function, and let $\mathbf{p} = \mathbf{r} \circ \psi$ be a reparameterization of $\mathcal{C}$. The arc length of $\mathcal{C}$ when computed using the reparameterization $\mathbf{p}$ will equal the arc length of $\mathcal{C}$ when computed using the parameterization $\mathbf{r}$.

Proof We compute the arc length of $\mathcal{C}$ with our calculation based on the reparameterization $\mathbf{p}$. As in Example 4, we use Theorem 6g from Section 12.1 to obtain $\mathbf{p}'(s) = (\mathbf{r} \circ \psi)'(s) = \psi'(s)\mathbf{r}'(\psi(s))$. Thus

$$\int_c^d \|\mathbf{p}'(s)\|\, ds = \int_c^d \|\psi'(s)\mathbf{r}'(\psi(s))\|\, ds = \int_c^d \psi'(s)\|\mathbf{r}'(\psi(s))\|\, ds.$$

Next, we make the substitution $t = \psi(s), dt = \psi'(s)\, ds$. The last integral then becomes $\int_a^b \|\mathbf{r}'(t)\|\, dt$, which is the arc length of $\mathcal{C}$ as determined by the parameterization $\mathbf{r}$. ■

Parameterizing a Curve by Arc Length

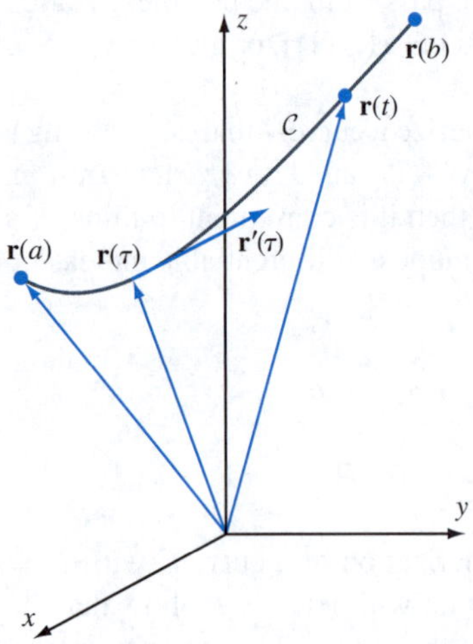

Figure 5
$\sigma(t) = \int_a^t \|\mathbf{r}'(\tau)\|\,d\tau$ is the arc length of C from $\mathbf{r}(a)$ to $\mathbf{r}(t)$.

Suppose that the vector-valued function $\mathbf{r}$ defined on the interval $[a, b]$ is a smooth parameterization of C. Let $L = \int_a^b \|\mathbf{r}'(\tau)\|\,d\tau$ denote the (total) arc length of C. The *arc length function* σ of $\mathbf{r}$ is the increasing function from $[a, b]$ to $[0, L]$ that is defined by

$$\sigma(t) = \int_a^t \|\mathbf{r}'(\tau)\|\,d\tau, \qquad a \le t \le b. \tag{12.3}$$

See Figure 5. When we apply the Fundamental Theorem of Calculus to equation (12.3), we obtain

$$\sigma'(t) = \|\mathbf{r}'(t)\|. \tag{12.4}$$

If we think of a particle with position vector $\mathbf{r}(t)$ tracing the curve, then equation (12.4) has a satisfying interpretation: *The instantaneous rate of change of distance along the curve with respect to time is equal to the speed of the particle.*

For some types of problems, it is convenient to reparameterize C by arc length. We use the arc length function σ of $\mathbf{r}$ to accomplish this task. As a consequence of our assumption that C is smooth, we have $\mathbf{r}'(t) \ne \vec{\mathbf{0}}$ for all t. In view of equation (12.4), we deduce that $\sigma'(t) > 0$. This inequality implies that σ is a strictly increasing function. Therefore, σ has an inverse function σ^{-1}, which is also increasing. We let $\mathbf{p}$ be the reparameterization of C defined by $\mathbf{p}(s) = \mathbf{r}(\sigma^{-1}(s))$ for $0 \le s \le L$. Using Theorem 1, with $\psi = \sigma^{-1}$ and $[c, d] = [0, s]$, we see that the arc length along C between $\mathbf{p}(0)$ and $\mathbf{p}(s)$ is

$$\int_0^s \|\mathbf{p}'(\tau)\|\,d\tau = \int_a^{\sigma^{-1}(s)} \|\mathbf{r}'(u)\|\,du = \sigma(\sigma^{-1}(s)) = s. \tag{12.5}$$

This means that the point $\mathbf{p}(s)$ is situated s units of arc length along C from its initial point—note the equality of the parameter s of $\mathbf{p}(s)$ and the distance. This property of the reparameterization $\mathbf{p}$, which is clearly very special, is the basis for the following definition.

Definition Let L be the length of a curve C. A parameterization $\mathbf{p}(s)$, $0 \le s \le L$, of C is called the *arc length parameterization of* C if the arc length between $\mathbf{p}(0)$ and $\mathbf{p}(s)$ is equal to s for every s in the interval $[0, L]$. We say that $\mathbf{p}$ *parameterizes* C *with respect to arc length.*

Theorem 2 If we let $s \mapsto \mathbf{r}(s)$, $0 \le s \le L$, be a smooth parameterization of a curve C, then $\mathbf{r}$ is the arc length parameterization of C if and only if $\|\mathbf{r}'(s)\| = 1$ for all s. In this case, the velocity vector $\mathbf{r}'(s)$ is equal to the unit tangent vector $\mathbf{T}(s)$ for every s:

$$\mathbf{T}(s) = \mathbf{r}'(s). \tag{12.6}$$

Proof Suppose that $s \mapsto \mathbf{r}(s)$, $0 \le s \le L$, is the arc length parameterization of C. By definition, we have $\int_0^s \|\mathbf{r}'(\tau)\|\,d\tau = s$. If we differentiate each side of this equation with respect to s, then according to the Fundamental Theorem of Calculus, we have

$$\|\mathbf{r}'(s)\| = \frac{d}{ds}\int_0^s \|\mathbf{r}'(\tau)\|\,d\tau = \frac{d}{ds}s = 1.$$

Conversely, if $\|\mathbf{r}'(s)\| = 1$ for all s, then the arc length along $\mathcal{C}$ between $\mathbf{r}(0)$ and $\mathbf{r}(s)$ is $\int_0^s \|\mathbf{r}'(\tau)\|\,d\tau = \int_0^s 1\,d\tau = s$, which is precisely the requirement for $\mathbf{r}$ to be the parametrization of $\mathcal{C}$ with respect to arc length.

Finally, if $\mathbf{r}$ is an arc length parameterization, then formula (12.6) is just a simplification of equation (12.2), taking into account $\|\mathbf{r}'(s)\| = 1$. ■

inSIGHT

Suppose we observe the path $\mathcal{C}$ of a moving particle, beginning at time $s = 0$. Theorem 2 tells us that the particle's position vector is the arc length parameterization of $\mathcal{C}$ if and only if the particle moves with constant speed 1.

Example 5 Are $\mathbf{r}(t) = \cos(2t)\mathbf{i} - \sin(2t)\mathbf{k}$, $0 \le t \le \pi$ and $\mathbf{p}(u) = \cos(u)\mathbf{i} - \sin(u)\mathbf{k}$, $0 \le u \le 2\pi$ arc length parameterizations of the curves that they define?

Solution The function $\mathbf{r}$ is *not* an arc length parameterization, because $\mathbf{r}'(t) = -2\sin(2t)\mathbf{i} - 2\cos(2t)\mathbf{k}$ and $\|\mathbf{r}'(t)\| = 2 \neq 1$. However, the function $\mathbf{p}(u) = \cos(u)\mathbf{i} - \sin(u)\mathbf{k}$ *is* an arc length parameterization, because $\mathbf{p}'(u) = -\sin(u)\mathbf{i} - \cos(u)\mathbf{k}$ so that $\|\mathbf{p}'(u)\| = 1$ for all u. Notice that the functions $\mathbf{r}$ and $\mathbf{p}$ describe the same set of points in space. Indeed, we obtain $\mathbf{p}$ from $\mathbf{r}$ by composing with the increasing function $\psi(u) = u/2$: $\mathbf{p}(u) = (\mathbf{r} \circ \psi)(u) = \mathbf{r}(\psi(u)) = \mathbf{r}(u/2)$.

Example 6 Reparameterize the curve $\mathbf{r}(t) = \langle \cos(t), \sin(t), t\rangle$, $0 \le t \le 2\pi$ so that it is parameterized with respect to arc length.

Solution We calculate the quantities $\mathbf{r}'(t) = \langle -\sin(t), \cos(t), 1\rangle$ and $\|\mathbf{r}'(t)\| = \sqrt{\sin^2(t) + \cos^2(t) + 1^2} = \sqrt{2}$. Therefore the arc length function of $\mathbf{r}$ is given by $s = \sigma(t) = \int_0^t \|\mathbf{r}'(\tau)\|\,d\tau = \int_0^t \sqrt{2}\,d\tau = \sqrt{2}t$. Then $t = s/\sqrt{2} = \sigma^{-1}(s)$ and

$$\mathbf{p}(s) = \mathbf{r}(\sigma^{-1}(s)) = \mathbf{r}\left(\frac{s}{\sqrt{2}}\right) = \left\langle \cos\left(\frac{s}{\sqrt{2}}\right), \sin\left(\frac{s}{\sqrt{2}}\right), \frac{s}{\sqrt{2}}\right\rangle, \quad 0 \le s \le 2\sqrt{2}\pi,$$

is the required parameterization by arc length. ■

Unit Normal Vectors

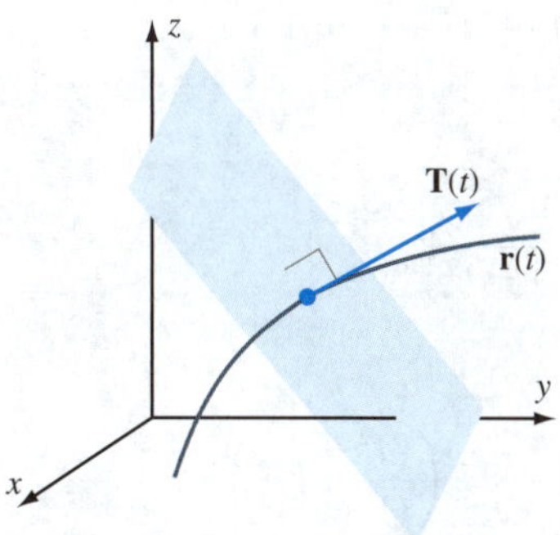

Figure 6

Suppose $t \mapsto \mathbf{r}(t)$ is a smooth parameterization of a curve $\mathcal{C}$. Fix a point $P = \mathbf{r}(t)$ on $\mathcal{C}$. Figure 6 shows the plane that passes through point P and that is perpendicular to the tangent vector to $\mathcal{C}$ at P. It is called the *normal plane* of $\mathcal{C}$ at P. Although every vector in the normal plane is perpendicular to $\mathbf{T}(t)$, we will see that two of these normal vectors are of particular interest. Our starting point is the identity $\|\mathbf{T}(t)\|^2 = 1$, which we may write as $\mathbf{T}(t)\cdot\mathbf{T}(t) = 1$. If we differentiate both sides of this last equation with respect to t, then we obtain $\mathbf{T}'(t)\cdot\mathbf{T}(t) + \mathbf{T}(t)\cdot\mathbf{T}'(t) = 0$, or $2\mathbf{T}(t)\cdot\mathbf{T}'(t) = 0$. The resulting equation,

$$\mathbf{T}(t)\cdot\mathbf{T}'(t) = 0, \tag{12.7}$$

tells us that the vector $\mathbf{T}'(t)$ *is perpendicular to* the unit tangent vector $\mathbf{T}(t)$. Therefore $\mathbf{T}'(t)$ lies in the normal plane of $\mathcal{C}$ at P. Although the vector $\mathbf{T}(t)$ has unit length, its derivative $\mathbf{T}'(t)$ generally does not have this property. If $\mathbf{T}'(t) \neq \vec{\mathbf{0}}$, then we obtain a unit vector by scalar multiplying $\mathbf{T}'(t)$ by $1/\|\mathbf{T}'(t)\|$.

Definition If $t \mapsto \mathbf{r}(t)$ is a smooth parameterization of a curve $\mathcal{C}$ with $\mathbf{T}'(t) \neq \vec{\mathbf{0}}$, then the vector

$$\mathbf{N}(t) = \frac{1}{\|\mathbf{T}'(t)\|}\mathbf{T}'(t) \tag{12.8}$$

is called the *principal unit normal* to $\mathcal{C}$ at $\mathbf{r}(t)$.

inSIGHT

If $s \mapsto \mathbf{r}(s)$ is the arc length parameterization of $\mathcal{C}$, then we can use equation (12.8) to write the principal unit normal as

$$\mathbf{N}(s) = \frac{1}{\|\mathbf{r}''(s)\|}\mathbf{r}''(s). \tag{12.9}$$

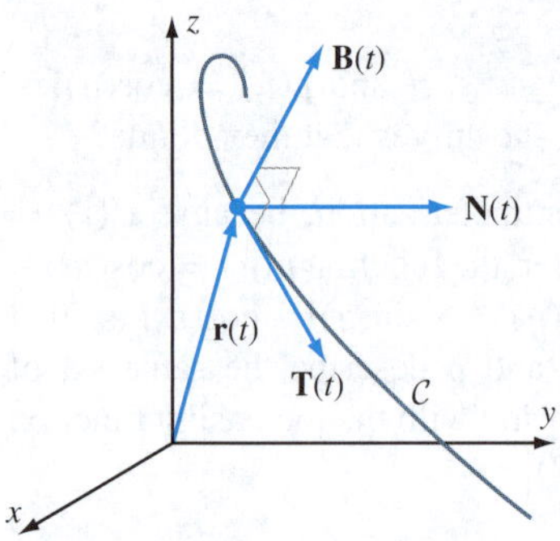

Figure 7
$\mathbf{B}(t) = \mathbf{T}(t) \times \mathbf{N}(t)$

inSIGHT

Imagine a particle with position vector $\mathbf{r}(s)$ tracing a curve $\mathcal{C}$ with unit speed. Formula (12.9) tells us that $\mathbf{N}(s)$ is the direction of the particle's acceleration vector, $\mathbf{r}''(s)$. Therefore, $\mathbf{N}(s)$ *points to the side to which* $\mathcal{C}$ *curves.* (Refer again to Figure 7.) This observation will be important in Section 12.4.

The cross product

$$\mathbf{B}(t) = \mathbf{T}(t) \times \mathbf{N}(t) \tag{12.10}$$

of the unit tangent vector and the principal unit normal vector is called the *binormal* vector to $\mathcal{C}$ at the point $\mathbf{r}(t)$. Our work in Section 11.4 shows that the binormal vector is a unit vector perpendicular to both $\mathbf{T}(t)$ and $\mathbf{N}(t)$. See Figure 7.

Example 7 Let $\mathcal{C}$ be the curve defined by $\mathbf{r}(t) = t\mathbf{i} + (t^3/3)\mathbf{j} + (1 - t^2)/\sqrt{2}\mathbf{k}$. Calculate the principal unit normal and the binormal vector of $\mathcal{C}$ at the point $P = (1, 1/3, 0)$.

Solution Observe that $\mathbf{r}'(t) = \mathbf{i} + t^2\mathbf{j} - \sqrt{2}t\mathbf{k}$ and

$$\|\mathbf{r}'(t)\| = \sqrt{1^2 + (t^2)^2 + \left(-\sqrt{2}t\right)^2} = \sqrt{1 + 2t^2 + t^4} = 1 + t^2.$$

It follows that

$$\mathbf{T}(t) = \frac{1}{1+t^2}\mathbf{i} + \frac{t^2}{1+t^2}\mathbf{j} - \frac{\sqrt{2}t}{1+t^2}\mathbf{k}$$

and

$$\mathbf{T}'(t) = \left(-\frac{2t}{(1+t^2)^2}\right)\mathbf{i} + \left(\frac{2t}{(1+t^2)^2}\right)\mathbf{j} - \sqrt{2}\left(\frac{t^2-1}{(1+t^2)^2}\right)\mathbf{k}.$$

The point P corresponds to $t = 1$; for this value of t, we have $\mathbf{T}(1) = (1/2)\mathbf{i} + (1/2)\mathbf{j} - (\sqrt{2}/2)\mathbf{k}$ and $\mathbf{T}'(1) = (-1/2)\mathbf{i} + (1/2)\mathbf{j} - 0\mathbf{k} = (1/2)(-\mathbf{i} + \mathbf{j})$. The required principal unit normal vector $\mathbf{N}(1)$ is the direction of $\mathbf{T}'(1)$, which is the direction of the vector $-\mathbf{i} + \mathbf{j}$ and which is quickly calculated to be

$$\mathbf{N}(1) = -\frac{1}{\sqrt{2}}\mathbf{i} + \frac{1}{\sqrt{2}}\mathbf{j}.$$

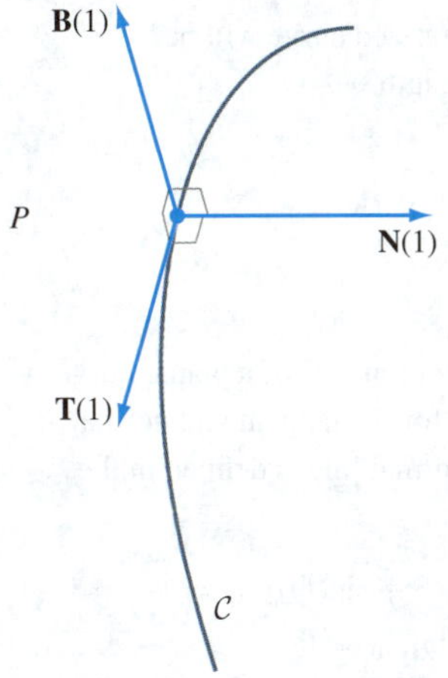

Figure 8
$\mathbf{T}(1)$, $\mathbf{N}(1)$, and $\mathbf{B}(1)$ are mutually perpendicular.

Therefore

$$\mathbf{B}(1) = \mathbf{T}(1) \times \mathbf{N}(1) = \left(\frac{1}{2}\mathbf{i} + \frac{1}{2}\mathbf{j} - \frac{\sqrt{2}}{2}\mathbf{k}\right) \times \left(-\frac{1}{\sqrt{2}}\mathbf{i} + \frac{1}{\sqrt{2}}\mathbf{j}\right)$$

$$= \det\left(\begin{bmatrix} \mathbf{i} & \mathbf{j} & \mathbf{k} \\ \frac{1}{2} & \frac{1}{2} & -\frac{\sqrt{2}}{2} \\ -\frac{1}{\sqrt{2}} & \frac{1}{\sqrt{2}} & 0 \end{bmatrix}\right) = \frac{1}{2}\mathbf{i} + \frac{1}{2}\mathbf{j} + \frac{\sqrt{2}}{2}\mathbf{k}.$$

In Figure 8, the vectors $\mathbf{T}(1)$, $\mathbf{N}(1)$, and $\mathbf{B}(1)$ have been added to a plot of $\mathbf{r}(t)$ for $0 \le t \le 2$. ■

inSIGHT

Like the vectors **i**, **j**, and **k**, the vectors $\mathbf{T}(t)$, $\mathbf{N}(t)$, and $\mathbf{B}(t)$ are mutually perpendicular and have length 1. Like **i**, **j**, and **k**, the vectors $\mathbf{T}(t)$, $\mathbf{N}(t)$, and $\mathbf{B}(t)$ form a *right-handed system*. Unlike **i**, **j**, and **k**, the vectors $\mathbf{T}(t)$, $\mathbf{N}(t)$, and $\mathbf{B}(t)$ are not fixed directions: They change as $\mathbf{r}(t)$ traverses the curve $\mathcal{C}$. The triple $(\mathbf{T}, \mathbf{N}, \mathbf{B})$ is said to be a *moving frame* and is particularly useful for studying the geometry of space curves and the trajectories of moving bodies.

quickquiz

1. How is the unit tangent vector to a curve $\mathbf{r}(t)$ calculated?
2. What is the formula for calculating arc length of a curve $\mathbf{r}$ parametrized over the interval $[a, b]$?
3. Explain what it means to reparameterize a curve.
4. How do we reparameterize a curve with respect to arc length?

EXERCISES

Problems for Practice

In Exercises 1–8, a parameterization $\mathbf{r}(t)$ of a curve is given. Calculate the tangent vector $\mathbf{r}'(t)$ at $\mathbf{r}(t)$, the unit tangent vector $\mathbf{T}(t)$ at $\mathbf{r}(t)$, and the tangent line at the given point P_0.

1. $\mathbf{r}(t) = t\mathbf{i} - t^2\mathbf{j} + t^3\mathbf{k}$, $P_0 = (1, -1, 1)$
2. $\mathbf{r}(t) = -\mathbf{j} + t\mathbf{k}$, $P_0 = (0, -1, 5)$
3. $\mathbf{r}(t) = t\mathbf{i} - t^{1/2}\mathbf{k}$, $P_0 = (4, 0, -2)$
4. $\mathbf{r}(t) = e^t\mathbf{i} + e^{2t}\mathbf{j} + e^{3t}\mathbf{k}$, $P_0 = (3, 9, 27)$
5. $\mathbf{r}(t) = (e^t + t^2)\mathbf{i} + 3\mathbf{j}$, $P_0 = (1, 3, 0)$
6. $\mathbf{r}(t) = \ln(t)\mathbf{i} - t \cdot \ln(t)\mathbf{j} + t\mathbf{k}$, $P_0 = (1, -e, e)$
7. $\mathbf{r}(t) = \cos(t)\mathbf{i} + \sin(t)\mathbf{j} + t\mathbf{k}$, $P_0 = (0, 1, \pi/2)$
8. $\mathbf{r}(t) = 2\ln(\cos(t))\mathbf{i} + 2\ln(\sin(t))\mathbf{j} + 4t\mathbf{k}$, $P_0 = (-\ln(2), -\ln(2), \pi)$

In Exercises 9–14, calculate the arc length of the given curve.

9. $\mathbf{r}(t) = \sin(t^2)\mathbf{i} - \cos(t^2)\mathbf{j} + t^2\mathbf{k}$, $0 \le t \le \sqrt{\pi}$
10. $\mathbf{r}(t) = e^{4t}\mathbf{i} + 8t\mathbf{j} + 2e^{-4t}\mathbf{k}$, $0 \le t \le \ln(2)$
11. $\mathbf{r}(t) = \cos^3(t)\mathbf{i} + 3\mathbf{j} - \sin^3(t)\mathbf{k}$, $\pi \le t \le 3\pi$
12. $\mathbf{r}(t) = \sin(3t)\mathbf{i} - \cos(3t)\mathbf{j} + t^{3/2}\mathbf{k}$, $\pi/2 \le t \le \pi$
13. $\mathbf{r}(t) = (1+t)^{3/2}\mathbf{i} + (1-t)^{3/2}\mathbf{j} + t^{3/2}\mathbf{k}$, $0 \le t \le 1$
14. $\mathbf{r}(t) = \sin(t)\mathbf{i} + \cos(t)\mathbf{j} + \ln(\cos(t))\mathbf{k}$, $0 \le t \le \pi/4$

In Exercises 15–18, determine whether the given function is an arc length parameterization of the curve it defines.

15. $\mathbf{r}(t) = t\mathbf{i} - (t^2/2)\mathbf{j} + (t^3/3)\mathbf{k}$
16. $\mathbf{r}(t) = \cos(t/\sqrt{2})\mathbf{i} + \sin(t/\sqrt{2})\mathbf{j} + (t/\sqrt{2})\mathbf{k}$
17. $\mathbf{r}(t) = (3\cos(t)/5)\mathbf{i} - (4\cos(t)/5)\mathbf{j} + \sin(t)\mathbf{k}$
18. $\mathbf{r}(t) = \cos^2(t)\mathbf{i} + \cos(t)\sin(t)\mathbf{j} + \sin(t)\mathbf{k}$

In Exercises 19–22, a parameterization $\mathbf{r}(t)$ of a curve is given. Calculate the arc length between the two specified points.

19. $\mathbf{r}(t) = t^2\mathbf{i} - \ln(t)\mathbf{j} + 2t\mathbf{k}$, between the points $(1, 0, 2)$ and $(e^2, -1, 2e)$

20. $\mathbf{r}(t) = (\sin(t) - t\cos(t))\mathbf{i} + t^2\mathbf{j} - (t\sin(t) + \cos(t))\mathbf{k}$, between the points corresponding to the values $t = 0$ and $t = 2\pi$

21. $\mathbf{r}(t) = e^t\mathbf{i} + e^{-t}\mathbf{j} + \sqrt{2}t\mathbf{k}$, between the points $(1, 1, 0)$ and $(e, 1/e, \sqrt{2})$

22. $\mathbf{r}(t) = \cos(t)\mathbf{i} + \sin(t)\mathbf{j} - t^{3/2}\mathbf{k}$, between the points $(1, 0, 0)$ and $(0, 1, -(\pi/2)^{3/2})$

In Exercises 23–30, a parameterization of a curve C and a point P_0 are specified. Give symmetric equations for the line through P_0 with direction vector equal to the principal unit normal vector to C at P_0. This line is called the *normal line* to C at P_0.

23. $\mathbf{r}(t) = t\mathbf{i} - t^2\mathbf{j} + t^3\mathbf{k}$, $P_0 = (1, -1, 1)$

24. $\mathbf{r}(t) = t^2\mathbf{i} + 2t^2\mathbf{j} + 3t^3\mathbf{k}$, $P_0 = (1, 2, -3)$

25. $\mathbf{r}(t) = \cos(t)\mathbf{i} + \sin(t)\mathbf{j} + t\mathbf{k}$, $P_0 = (0, 1, \pi/2)$

26. $\mathbf{r}(t) = t\mathbf{i} + (2/t)\mathbf{j} + t^2\mathbf{k}$, $P_0 = (2, 1, 4)$

27. $\mathbf{r}(t) = e^t\mathbf{i} + e^{-t}\mathbf{j} - t^2\mathbf{k}$, $P_0 = (1, 1, 0)$

28. $\mathbf{r}(t) = e^t\mathbf{i} + te^t\mathbf{j} + (t + e^t)\mathbf{k}$, $P_0 = (1, 0, 1)$

29. $\mathbf{r}(t) = \sqrt{t}\mathbf{i} + (4/t)\mathbf{j} + (4/\sqrt{t})\mathbf{k}$, $P_0 = (2, 1, 2)$

30. $\mathbf{r}(t) = \ln(t)\mathbf{i} + \arctan(t)\mathbf{j} + (1/t)\mathbf{k}$, $P_0 = (0, \pi/4, 1)$

In Exercises 31–38, a parameterization $\mathbf{r}(t)$ of a curve is given. Find the Cartesian equation for the normal plane to the curve at the given point P_0.

31. $\mathbf{r}(t) = t\mathbf{i} - t^2\mathbf{j} + t^3\mathbf{k}$, $P_0 = (1, -1, 1)$

32. $\mathbf{r}(t) = t^2\mathbf{i} + 2t^2\mathbf{j} + 3t^3\mathbf{k}$, $P_0 = (1, 2, -3)$

33. $\mathbf{r}(t) = t\mathbf{i} + (2/t)\mathbf{j} + t^2\mathbf{k}$, $P_0 = (2, 1, 4)$

34. $\mathbf{r}(t) = \cos(t)\mathbf{i} + \sin(t)\mathbf{j} + t\mathbf{k}$, $P_0 = (0, 1, \pi/2)$

35. $\mathbf{r}(t) = e^t\mathbf{i} + e^{-t}\mathbf{j} - t^2\mathbf{k}$, $P_0 = (1, 1, 0)$

36. $\mathbf{r}(t) = e^t\mathbf{i} + e^{2t}\mathbf{j} + e^{3t}\mathbf{k}$, $P_0 = (3, 9, 27)$

37. $\mathbf{r}(t) = \sqrt{t}\mathbf{i} + (4/t)\mathbf{j} + (t^2 - 15)\mathbf{k}$, $P_0 = (2, 1, 1)$

38. $\mathbf{r}(t) = \ln(t)\mathbf{i} + \arctan(t)\mathbf{j} + (1/t)\mathbf{k}$, $P_0 = (0, \pi/4, 1)$

Further Theory and Practice

In Exercises 39–42, a curve is parameterized by the given function $\mathbf{r}(t)$ for t in the interval $[0, 1]$. Calculate the arc length function

$$\sigma(t) = \int_0^t \|\mathbf{r}'(\tau)\|\, d\tau$$

explicitly. Calculate the inverse of σ by setting $s = \sigma(t)$ and calculating t as a function of s: $t = \sigma^{-1}(s)$. Next, substitute $t = \sigma^{-1}(s)$ into the given formula for $\mathbf{r}(t)$ to obtain $\mathbf{p}(s) = \mathbf{r}(\sigma^{-1}(s))$. The reparameterized curve will be parameterized according to arc length.

39. $\mathbf{r}(t) = t\mathbf{i} + t\mathbf{j} + t^{3/2}\mathbf{k}$

40. $\mathbf{r}(t) = \cosh(t)\mathbf{i} - \sinh(t)\mathbf{j} + t\mathbf{k}$

41. $\mathbf{r}(t) = e^t\mathbf{i} + e^{-t}\mathbf{j} + \sqrt{2}t\mathbf{k}$

42. $\mathbf{r}(t) = \cos^3(t)\mathbf{i} + \sin^3(t)\mathbf{k}$

In Exercises 43–50, a parameterization $\mathbf{r}(t)$ of a planar curve is given. Find symmetric equations for the tangent and normal lines to the curve at $\mathbf{r}(t_0)$. (The normal line is defined in the instructions to Exercises 23–30.)

43. $\mathbf{r}(t) = (5 + 2\cos(t))\mathbf{i} + (3 + \sin(t))\mathbf{j}$, $t_0 = \dfrac{\pi}{3}$

44. $\mathbf{r}(t) = (t^2 + t)\mathbf{i} + (t^2 - t)\mathbf{j}$, $t_0 = 1$

45. $\mathbf{r}(t) = \tan(t)\mathbf{i} + \sec(t)\mathbf{j}$, $t_0 = \dfrac{\pi}{4}$

46. $\mathbf{r}(t) = \dfrac{3t}{1 + t^3}\mathbf{i} + \dfrac{3t^2}{1 + t^3}\mathbf{j}$, $t_0 = 1$

47. $\mathbf{r}(t) = (e^t + e^{-t})\mathbf{i} + (e^t - e^{-t})\mathbf{j}$, $t_0 = 0$

48. $\mathbf{r}(t) = (t^3 - t)\mathbf{i} + (t^2 - 1)\mathbf{j}$, $t_0 = \dfrac{1}{2}$

49. $\mathbf{r}(t) = \dfrac{2t}{1 + t^2}\mathbf{i} + \dfrac{1 - t^2}{1 + t^2}\mathbf{j}$, $t_0 = \dfrac{1}{2}$

50. $\mathbf{r}(t) = (t - \sin(t))\mathbf{i} + (1 - \cos(t))\mathbf{j}$, $t_0 = \dfrac{\pi}{2}$

In Exercises 51–56, a parameterization $\mathbf{r}(t)$ of a space curve is given. Calculate the moving frame $(\mathbf{T}, \mathbf{N}, \mathbf{B})$ at $t(0)$.

51. $\mathbf{r}(t) = \cos(t)\mathbf{i} + \sin(2t)\mathbf{j} + t\mathbf{k}$, $t_0 = \pi/4$

52. $\mathbf{r}(t) = (t^2 + t)\mathbf{i} + (t^2 - t)\mathbf{j} + t^2\mathbf{k}$, $t_0 = 1$

53. $\mathbf{r}(t) = \tan(t)\mathbf{i} + \sec(t)\mathbf{j} + t\mathbf{k}$, $t_0 = \pi/4$

54. $\mathbf{r}(t) = t\mathbf{i} + t^2\mathbf{j} + t^3\mathbf{k}$, $t_0 = 1$

55. $\mathbf{r}(t) = e^t\mathbf{i} + e^{-t}\mathbf{j} + (e^t + e^{-t})\mathbf{k}$, $t_0 = \ln(2)$

56. $\mathbf{r}(t) = (t^3 - t)\mathbf{i} + (t^2 - 1)\mathbf{j} + t^2\mathbf{k}$, $t_0 = 1$

57. Given any positive number N, describe a way to define a piecewise smooth curve parameterized by $\mathbf{r}(t) = x(t)\mathbf{i} + y(t)\mathbf{j}$ that has length greater than N and components that satisfy $0 \le x(t), y(t) \le 1$.

58. Show that the tangent lines to the helix parameterized by $\mathbf{r}(t) = \langle\cos(t), \sin(t), t/2\rangle$ make a constant angle with $\mathbf{k}$. Show that the normal lines intersect the z-axis.

Calculator/Computer Exercises

In Exercises 59–62, use Simpson's Rule to approximate the arc length of the curve defined by the given vector-valued function $\mathbf{r}$ over the given interval to an accuracy of two decimal places (that is, with an error no greater than 0.005).

59. $\mathbf{r}(t) = t^2\mathbf{i} - t^4\mathbf{j} + t^{-1}\mathbf{k}$, $1 \le t \le 4$

60. $\mathbf{r}(t) = t\mathbf{i} - e^{-t}\mathbf{j} + t^5\mathbf{k}$, $0.4 \le t \le 0.9$

61. $\mathbf{r}(t) = \cos(t)\mathbf{i} + \sin(2t)\mathbf{j} + \sqrt{t}\mathbf{k}$, $1 \le t \le 4$

62. $\mathbf{r}(t) = \ln(t)\mathbf{i} + (2/t)\mathbf{j} + t^{3/2}\mathbf{k}$, $1 \le t \le 2$

12.4 Curvature

In this section, we learn what the second derivative $\mathbf{r}''$ of a vector-valued function $\mathbf{r}$ can tell us about the geometry of the curve $\mathcal{C}$ defined by $\mathbf{r}$. To be specific, we will see that $\mathbf{r}''$ can be used to measure the way $\mathcal{C}$ deviates from a straight line. The geometric tool for this study is the concept of *curvature*.

What does it mean to say that a circle is curved but a line is not? One answer is that as we move along a circle, the unit tangent vector changes direction; whereas if we move along a line, the unit tangent vector does not change direction. We define the curvature to be the magnitude of the instantaneous rate of change of direction with respect to arc length.

Definition Suppose that $s \mapsto \mathbf{r}(s)$ is a smooth arc length parameterization of a curve $\mathcal{C}$. The quantity

$$\kappa(s) = \left\| \frac{d\mathbf{T}(s)}{ds} \right\| = \left\| \frac{d^2\mathbf{r}(s)}{ds^2} \right\| \tag{12.11}$$

is called the *curvature* of $\mathcal{C}$ at the point $\mathbf{r}(s)$.

inSIGHT

The definition of curvature involves differentiation with respect to arc length. That choice is influenced by geometric considerations, but there is a subtle motivation as well. As we saw in Section 12.3, there are always many ways to parameterize a curve. The concept of curvature should depend only on the geometry of a curve and not on the choice of parameterization. By referring to the arc length parameterization, we avoid potential ambiguities that might arise from different choices of parameterization. Theorem 1 from Section 12.3 shows that any two parameterizations of $\mathcal{C}$ will give rise to the same arc length reparameterization. See also Theorem 1 on page 98.

Example 1 Let $\mathbf{r}(s) = \rho\cos(s/\rho)\mathbf{i} + \rho\sin(s/\rho)\mathbf{j}$ for some positive constant ρ. Calculate the curvature at each value of s.

Solution The curve $\mathcal{C}$ described by $\mathbf{r}$ is a circle in the xy-plane, $z = 0$. The radius of $\mathcal{C}$ is ρ, and the center of $\mathcal{C}$ is $(0, 0, 0)$. Notice that $\mathbf{r}'(s) = (-\sin(s/\rho))\mathbf{i} + (\cos(s/\rho))\mathbf{j}$, so that $\|\mathbf{r}'(s)\| = 1$ for all s. We deduce that $\mathbf{r}(s)$ is an arc length parameterization by Theorem 2 from Section 12.3. Finally,

$$\mathbf{r}''(s) = \left(\frac{-\cos(s/\rho)}{\rho}\right)\mathbf{i} + \left(\frac{-\sin(s/\rho)}{\rho}\right)\mathbf{j}$$

and therefore

$$\kappa(s) = \|\mathbf{r}''(s)\| = \sqrt{\left(\frac{-\cos(s/\rho)}{\rho}\right)^2 + \left(\frac{-\sin(s/\rho)}{\rho}\right)^2} = \frac{1}{\rho}\sqrt{\cos^2\left(\frac{s}{\rho}\right) + \sin^2\left(\frac{s}{\rho}\right)} = \frac{1}{\rho}. \quad \blacksquare$$

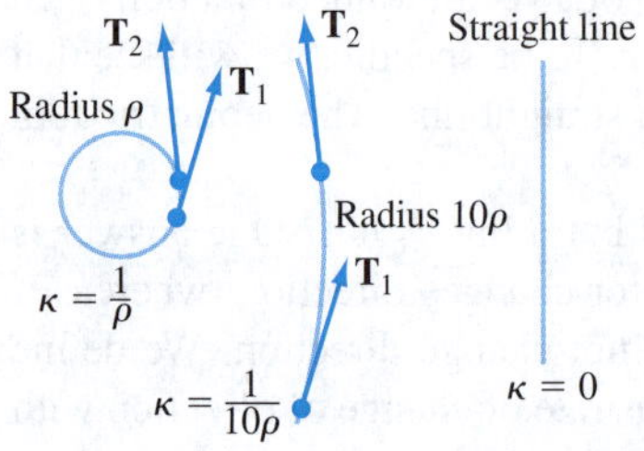

Figure 1

inSIGHT

Example 1 shows that our notion of curvature is a sensible one, because it says that the curvature of a circle of radius ρ is $1/\rho$ at all points of the circle. In particular, the curvature of a circle of large radius is small (the vector $\mathbf{T}$ changes slowly with respect to arc length), whereas the curvature of a circle of small radius is large (the vector $\mathbf{T}$ changes rapidly with respect to arc length). See Figure 1, in which arcs of circles with radii ρ and 10ρ are plotted. The same pair of tangent vectors, T_1 and T_2, appears on each circle. The change of direction shown on the small circle is the same as that shown on the large circle, yet it is attained with a much smaller change of arc length. The rate of change of direction with respect to arc length is therefore greater on the smaller circle.

Calculating Curvature without Reparameterizing

As we have discussed, there are good reasons for defining curvature by means of formula (12.11). However, this approach does have the drawback of requiring that we use the arc length parameterization of the curve, which is often tedious or difficult to calculate. It is therefore desirable to have an alternative expression for curvature, one that does not refer to arc length.

Theorem 1 If $\mathbf{r}$ is a smooth parameterization of a curve $\mathcal{C}$, then the curvature $\kappa_{\mathbf{r}}(t)$ of $\mathcal{C}$ at the point $\mathbf{r}(t)$ is given by

$$\kappa_{\mathbf{r}}(t) = \frac{\|\mathbf{r}'(t) \times \mathbf{r}''(t)\|}{\|\mathbf{r}'(t)\|^3}. \tag{12.12}$$

This quantity may also be expressed as

$$\kappa_{\mathbf{r}}(t) = \frac{\sqrt{(\|\mathbf{r}'(t)\| \cdot \|\mathbf{r}''(t)\|)^2 - (\mathbf{r}'(t) \cdot \mathbf{r}''(t))^2}}{\|\mathbf{r}'(t)\|^3}. \tag{12.13}$$

Proof We will derive the first of these formulas for $\kappa_{\mathbf{r}}(t)$; the second is left as Exercise 55. Let $\sigma(t) = \int_0^t \|\mathbf{r}'(\tau)\|\, d\tau$ denote the arc length function for the parameterization $\mathbf{r}$. Then, by equation (12.4), we may write the speed $\|\mathbf{r}'(t)\|$ as $d\sigma(t)/dt$. With this expression for the speed, equation (12.2) becomes

$$\mathbf{r}'(t) = \frac{d\sigma(t)}{dt}\mathbf{T}(t). \tag{12.14}$$

When we differentiate each side of equation (12.14) with respect to time, we obtain

$$\mathbf{r}''(t) = \frac{d^2\sigma(t)}{dt^2}\mathbf{T}(t) + \frac{d\sigma(t)}{dt}\mathbf{T}'(t). \tag{12.15}$$

Next, we form the cross products of the corresponding sides of equations (12.14) and (12.15), noting that $\mathbf{T}(t) \times \mathbf{T}(t) = \vec{\mathbf{0}}$:

$$\mathbf{r}'(t) \times \mathbf{r}''(t) = \left(\frac{d\sigma(t)}{dt}\right)^2 \mathbf{T}(t) \times \mathbf{T}'(t). \tag{12.16}$$

Recall that $\mathbf{T}(t)$ and $\mathbf{T}'(t)$ are perpendicular, by equation (12.7). Therefore $\|\mathbf{T}(t) \times \mathbf{T}'(t)\| = \|\mathbf{T}(t)\| \|\mathbf{T}'(t)\| = \|\mathbf{T}'(t)\|$ by Theorem 2a from Section 11.4. Comparing the magnitudes of each side of equation (12.16), we deduce that

$$\|\mathbf{r}'(t) \times \mathbf{r}''(t)\| = \left(\frac{d\sigma(t)}{dt}\right)^2 \|\mathbf{T}'(t)\|,$$

or

$$\|\mathbf{T}'(t)\| = \frac{\|\mathbf{r}'(t) \times \mathbf{r}''(t)\|}{\|\mathbf{r}'(t)\|^2}. \tag{12.17}$$

inSIGHT

> Each formula for curvature has advantages and disadvantages. Equation (12.11) is simple, natural, and easy to remember, but it is often useless for computation. Formula (12.12) appears more complicated, but it is usually the easier formula to work with because it does not require reparameterization. However, formula (12.12) does not have any obvious connection with the concept of curvature and certainly seems to depend on the choice of parameterization. Formula (12.13) shares these drawbacks and has the further disadvantage of being less memorable. On the other hand, formula (12.13) does not require a cross-product calculation.

Now let $s = \sigma(t)$ be the arc length of $\mathcal{C}$ between the points $\mathbf{r}(a)$ and $\mathbf{r}(t)$. Our work in Section 12.3 shows that the arc length reparameterization is $\mathbf{p}(s) = \mathbf{r}(\sigma^{-1}(s)) = \mathbf{r}(t)$. We have

$$\mathbf{p}'(s) = \mathbf{T}(t), \tag{12.18}$$

because each side is the unit tangent at the same point on the curve. Also,

$$\mathbf{T}'(t) = \frac{d}{dt}\mathbf{T}(t) = \frac{ds}{dt}\frac{d}{ds}\mathbf{T}(t) \overset{(12.18)}{=} \|\mathbf{r}'(t)\| \frac{d}{ds}\mathbf{p}'(s). \tag{12.19}$$

Finally, the curvature at the point in question is, by definition (12.11),

$$\kappa(s) = \left\|\frac{d\mathbf{T}(s)}{ds}\right\| = \left\|\frac{d}{ds}\mathbf{p}'(s)\right\| \overset{(12.19)}{=} \frac{\|\mathbf{T}'(t)\|}{\|\mathbf{r}'(t)\|} \overset{(12.17)}{=} \frac{\|\mathbf{r}'(t) \times \mathbf{r}''(t)\|}{\|\mathbf{r}'(t)\|^3},$$

which is $\kappa_{\mathbf{r}}(t)$. ∎

Example 2 Let $\mathcal{C}$ be the helix parameterized by $\mathbf{r}(t) = \cos(t)\mathbf{i} + \sin(t)\mathbf{j} + t\mathbf{k}$ for $0 \le t \le 4\pi$. Calculate the arc length function for $\mathbf{r}$. Use it to find the arc length reparameterization $\mathbf{p}(s)$ of $\mathcal{C}$. Use formula (12.11) to show that the curvature of the helix has the same value at each point. Verify that formulas (12.12) and (12.13) both yield the same value for the curvature.

Solution Observe that $\mathbf{r}'(t) = -\sin(t)\mathbf{i} + \cos(t)\mathbf{j} + \mathbf{k}$ and $\|\mathbf{r}'(t)\| = \sqrt{(-\sin(t))^2 + (\cos(t))^2 + 1^2} = \sqrt{2}$. It follows that the arc length function σ for the curve is $\sigma(t) = \int_0^t \|\mathbf{r}'(\tau)\|\, d\tau = \int_0^t \sqrt{2}\, d\tau = \sqrt{2}t$. Therefore, if $\sigma(t) = s$, then we have $\sqrt{2}t = s$, or $t = s/\sqrt{2}$. As we learned in Section 12.3, this implies that

$$\mathbf{p}(s) = \mathbf{r}(\sigma^{-1}(s)) = \mathbf{r}\left(\frac{s}{\sqrt{2}}\right) = \cos\left(\frac{s}{\sqrt{2}}\right)\mathbf{i} + \sin\left(\frac{s}{\sqrt{2}}\right)\mathbf{j} + \frac{s}{\sqrt{2}}\mathbf{k}$$

is the arc length parameterization of $\mathcal{C}$. We have

$$\mathbf{p}'(s) = -\frac{1}{\sqrt{2}}\sin\left(\frac{s}{\sqrt{2}}\right)\mathbf{i} + \frac{1}{\sqrt{2}}\cos\left(\frac{s}{\sqrt{2}}\right)\mathbf{j} + \frac{1}{\sqrt{2}}\mathbf{k}$$

and

$$\mathbf{p}''(s) = -\frac{1}{2}\cos\left(\frac{s}{\sqrt{2}}\right)\mathbf{i} - \frac{1}{2}\sin\left(\frac{s}{\sqrt{2}}\right)\mathbf{j}.$$

Therefore

$$\begin{aligned}\kappa(s) = \|\mathbf{p}''(s)\| &= \sqrt{\left(-\frac{1}{2}\cos\left(\frac{s}{\sqrt{2}}\right)\right)^2 + \left(-\frac{1}{2}\sin\left(\frac{s}{\sqrt{2}}\right)\right)^2} \\ &= \frac{1}{2}\sqrt{\cos^2\left(\frac{s}{\sqrt{2}}\right) + \sin^2\left(\left(\frac{s}{\sqrt{2}}\right)\right)} = \frac{1}{2}.\end{aligned}$$

Let us now use Theorem 1 to calculate curvature without using the arc length reparameterization. We have

$$\begin{aligned}\mathbf{r}'(t) \times \mathbf{r}''(t) &= (-\sin(t)\mathbf{i} + \cos(t)\mathbf{j} + \mathbf{k}) \times (-\cos(t)\mathbf{i} - \sin(t)\mathbf{j}) \\ &= \det\left(\begin{bmatrix} \mathbf{i} & \mathbf{j} & \mathbf{k} \\ -\sin(t) & \cos(t) & 1 \\ -\cos(t) & -\sin(t) & 0 \end{bmatrix}\right),\end{aligned}$$

which gives us $\mathbf{r}'(t) \times \mathbf{r}''(t) = \sin(t)\mathbf{i} - \cos(t)\mathbf{j} + \mathbf{k}$. Therefore we have $\|\mathbf{r}'(t) \times \mathbf{r}''(t)\| = \sqrt{\sin^2(t) + (-\cos(t))^2 + 1^2} = \sqrt{2}$. Formula (12.12) then tells us that curvature at $\mathbf{r}(t)$ is equal to

$$\frac{\|\mathbf{r}'(t) \times \mathbf{r}''(t)\|}{\|\mathbf{r}'(t)\|^3} = \frac{\sqrt{2}}{\left(\sqrt{2}\right)^3} = \frac{1}{2},$$

which agrees with our first calculation. Using formula (12.13), we find that curvature is given by

$$\begin{aligned}&\frac{\sqrt{(\|\mathbf{r}'(t)\| \cdot \|\mathbf{r}''(t)\|)^2 - (\mathbf{r}'(t) \cdot \mathbf{r}''(t))^2}}{\|\mathbf{r}'(t)\|^3} \\ &= \frac{\sqrt{\left(\sqrt{2}\cdot 1\right)^2 - ((-\sin(t)\mathbf{i} + \cos(t)\mathbf{j} + \mathbf{k}) \cdot (-\cos(t)\mathbf{i} - \sin(t)\mathbf{j}))^2}}{\left(\sqrt{2}\right)^3} = \frac{\sqrt{2 - 0^2}}{2\sqrt{2}},\end{aligned}$$

which agrees with the previously obtained value. ■

In Example 2, the curvature calculation is quite feasible whether or not we use Theorem 1. The next example better illustrates the power of Theorem 1.

Example 3 Let $\mathcal{C}$ be the curve parameterized by $\mathbf{r}(t) = e^t\mathbf{i} + e^{-t}\mathbf{j} + t\mathbf{k}$. Find the curvature $\kappa_{\mathbf{r}}(t)$ at point $\mathbf{r}(t)$.

Solution We calculate $\mathbf{r}'(t) = e^t\mathbf{i} - e^{-t}\mathbf{j} + \mathbf{k}$. Notice that, because the arc length function for $\mathbf{r}$ is

$$\sigma(t) = \int_0^t \|\mathbf{r}'(\tau)\|\,d\tau = \int_0^t \|e^\tau\mathbf{i} - e^{-\tau}\mathbf{j} + \mathbf{k}\|\,d\tau = \int_0^t (e^{2\tau} + e^{-2\tau} + 1)^{1/2}\,d\tau,$$

it is not practical to find the arc length parameterization of $\mathcal{C}$. We appeal to Theorem 1 instead. Since $\mathbf{r}''(t) = e^t\mathbf{i} + e^{-t}\mathbf{j}$, it follows that

$$\mathbf{r}'(t) \times \mathbf{r}''(t) = (e^t\mathbf{i} - e^{-t}\mathbf{j} + \mathbf{k}) \times (e^t\mathbf{i} + e^{-t}\mathbf{j}) = \det\left(\begin{bmatrix} \mathbf{i} & \mathbf{j} & \mathbf{k} \\ e^t & -e^{-t} & 1 \\ e^t & e^{-t} & 0 \end{bmatrix}\right) = -e^{-t}\mathbf{i} + e^t\mathbf{j} + 2\mathbf{k}.$$

Therefore the curvature is

$$\kappa_{\mathbf{r}}(t) = \frac{\|\mathbf{r}'(t) \times \mathbf{r}''(t)\|}{\|\mathbf{r}'(t)\|^3} = \frac{\|-e^{-t}\mathbf{i} + e^t\mathbf{j} + 2\mathbf{k}\|}{\|e^t\mathbf{i} - e^{-t}\mathbf{j} + \mathbf{k}\|^3} = \frac{\sqrt{e^{2t} + e^{-2t} + 4}}{(e^{2t} + e^{-2t} + 1)^{3/2}}.$$ ■

The Osculating Circle

Let $s \mapsto \mathbf{r}(s)$ be a smooth arc length parameterization of a curve $\mathcal{C}$. Recall that formula (12.9) expresses the principal unit normal vector to $\mathcal{C}$ at the point $\mathbf{r}(s)$ as

$$\mathbf{N}(s) = \frac{1}{\|\mathbf{r}''(s)\|}\mathbf{r}''(s) = \left(\left\|\frac{d^2\mathbf{r}(s)}{ds^2}\right\|^{-1}\right)\frac{d^2\mathbf{r}(s)}{ds^2}.$$

We can use curvature formula (12.11) to rewrite this last equation in the form

$$\frac{d^2\mathbf{r}(s)}{ds^2} = \kappa(s)\mathbf{N}(s). \tag{12.20}$$

Equation (12.20) states that the acceleration vector of a particle with position vector $\mathbf{r}(s)$ has the principal unit normal $\mathbf{N}(s)$ as its direction and the curvature $\kappa(s)$ for magnitude.

If $\mathcal{C}$ were a circle, then the circle would have radius $\rho(s) = 1/\kappa(s)$ at the point $P = \mathbf{r}(s)$, as we know from Example 1. Because $\mathcal{C}$ is curving like a circle of radius $\rho(s)$ at the point P, we should be able to approximate $\mathcal{C}$ by a circle of radius $\rho(s)$ with center at a distance $\rho(s)$ from the base point P. If $\mathcal{C}$ were actually a circle, then the direction of acceleration, $\mathbf{N}(s)$, would point to the center of the circle. Therefore the vector from P to the center of our approximating circle should have direction $\mathbf{N}(s)$. Finally, if $\mathcal{C}$ were a circle, then it would lie in a plane, and that plane would contain the circle's velocity and acceleration vectors. Therefore our approximating circle should lie in the plane determined by the vectors $\mathbf{T}(s)$ and $\mathbf{N}(s)$. We summarize all of these considerations with a precise definition of the approximating circle.

Definition Let $\mathbf{r}$ be a smooth arc length parameterization of a space curve $\mathcal{C}$. Let $P = \mathbf{r}(s)$ be a point on $\mathcal{C}$ at which $\kappa(s) > 0$. The *osculating circle* (or the *circle of curvature*) of $\mathcal{C}$ at P is the unique circle satisfying the following conditions:

a. The circle has radius $\rho(s) = 1/\kappa(s)$. This number is called the *radius of curvature* of $\mathcal{C}$ at P.
b. The circle has center $\mathbf{r}(s) + \rho(s)\mathbf{N}(s)$. This point is called the *center of curvature* of $\mathcal{C}$ at P.
c. The circle lies in the plane determined by the vectors $\mathbf{N}(s)$ and $\mathbf{T}(s)$. This plane is called the *osculating plane* of $\mathcal{C}$ at $\mathbf{r}(s)$.

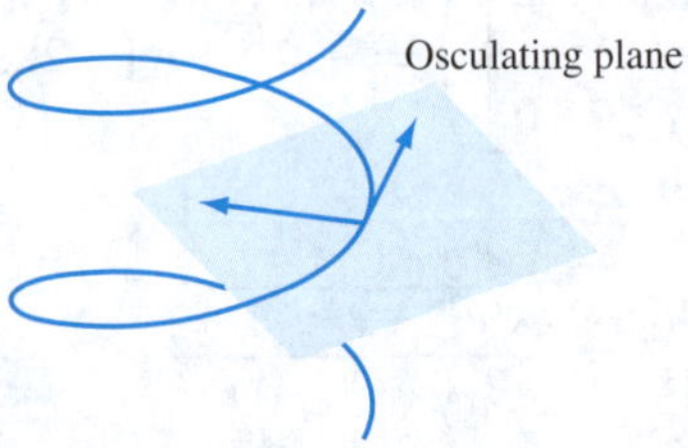

Figure 2a

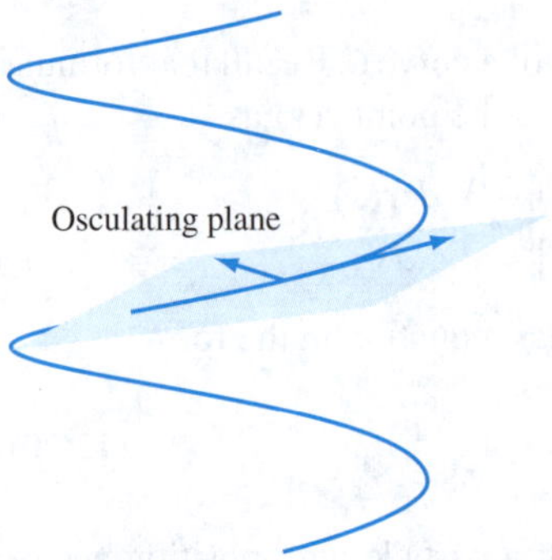

Figure 2b

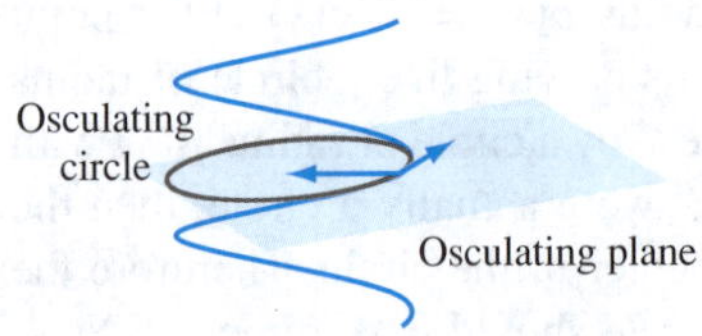

Figure 3

inSIGHT

> By construction, the osculating circle has the same principal unit normal vector and tangent vector as the curve $\mathcal{C}$ at the point of contact $\mathbf{r}(s)$. It therefore has the same binormal vector as well. The osculating circle and the curve $\mathcal{C}$ just touch at the point of contact P, since they share the same tangent line. That is the reason for the name: "Osculating" is a synonym for "kissing." An osculating plane for a helix is shown in Figure 2a. The unit tangent vector and the principal unit normal vector are also graphed. A different view of the same curve and osculating plane appears in Figure 2b, which shows how well the osculating plane "fits" the curve near the point of contact. The osculating circle is added in Figure 3.

Although we have used the arc length parameterization to simplify the discussion, it should be noted that all of the ingredients needed for the osculating plane and osculating circle can be calculated without reparameterization. The next example illustrates this point.

Example 4 Let $\mathcal{C}$ be the curve parameterized by $\mathbf{r}(t) = t^2\mathbf{i} + t\mathbf{j} + t\mathbf{k}$. Find the curvature $\kappa_{\mathbf{r}}(t)$ at the point $\mathbf{r}(t)$. Also, determine the radius of curvature, principal unit normal, and center of the osculating circle at the point $\mathbf{r}(t)$. What is the Cartesian equation of the osculating plane at $\mathbf{r}(t)$?

Solution We calculate $\mathbf{r}'(t) = 2t\mathbf{i} + \mathbf{j} + \mathbf{k}$ and $\mathbf{r}''(t) = 2\mathbf{i}$. It follows that

$$\mathbf{r}'(t) \times \mathbf{r}''(t) = (2t\mathbf{i} + \mathbf{j} + \mathbf{k}) \times (2\mathbf{i}) = \det\left(\begin{bmatrix} \mathbf{i} & \mathbf{j} & \mathbf{k} \\ 2t & 1 & 1 \\ 2 & 0 & 0 \end{bmatrix}\right) = 2\mathbf{j} - 2\mathbf{k}.$$

Therefore the curvature at the point $\mathbf{r}(t)$ is

$$\kappa_{\mathbf{r}}(t) = \frac{\|\mathbf{r}'(t) \times \mathbf{r}''(t)\|}{\|\mathbf{r}'(t)\|^3} = \frac{\|2\mathbf{j} - 2\mathbf{k}\|}{\|2t\mathbf{i} + \mathbf{j} + \mathbf{k}\|^3} = \frac{2\sqrt{2}}{(2 + 4t^2)^{3/2}} = \frac{1}{(1 + 2t^2)^{3/2}}.$$

In particular, the radius of curvature at $\mathbf{r}(t)$ is $(1 + 2t^2)^{3/2}$. The unit tangent vector at $\mathbf{r}(t)$ is

$$\mathbf{T}(t) = \frac{1}{\|\mathbf{r}'(t)\|}\mathbf{r}'(t) = \frac{1}{\sqrt{2 + 4t^2}}(2t\mathbf{i} + \mathbf{j} + \mathbf{k}) = \frac{2t}{\sqrt{2 + 4t^2}}\mathbf{i} + \frac{1}{\sqrt{2 + 4t^2}}\mathbf{j} + \frac{1}{\sqrt{2 + 4t^2}}\mathbf{k}.$$

We calculate

$$\mathbf{T}'(t) = \frac{4}{(2 + 4t^2)^{3/2}}\mathbf{i} - \frac{4t}{(2 + 4t^2)^{3/2}}\mathbf{j} - \frac{4t}{(2 + 4t^2)^{3/2}}\mathbf{k} = \frac{4}{(2 + 4t^2)^{3/2}}(\mathbf{i} - t\mathbf{j} - t\mathbf{k}).$$

The direction of $\mathbf{T}'(t)$ is the principal unit normal vector. This direction is the same as the direction of $\mathbf{i} - t\mathbf{j} - t\mathbf{k}$. It follows that

$$\mathbf{N}(t) = \frac{1}{\sqrt{1 + 2t^2}}\mathbf{i} - \frac{t}{\sqrt{1 + 2t^2}}\mathbf{j} - \frac{t}{\sqrt{1 + 2t^2}}\mathbf{k}$$

is the principal unit normal vector at $\mathbf{r}(t)$. The center of curvature is therefore

$$\begin{aligned}\mathbf{r}(t)+\rho(t)\mathbf{N}(t) &= (t^2\mathbf{i}+t\mathbf{j}+t\mathbf{k})+(1+2t^2)^{3/2}\left(\frac{1}{\sqrt{1+2t^2}}\mathbf{i}-\frac{t}{\sqrt{1+2t^2}}\mathbf{j}-\frac{t}{\sqrt{1+2t^2}}\mathbf{k}\right)\\ &= (t^2\mathbf{i}+t\mathbf{j}+t\mathbf{k})+((1+2t^2)\mathbf{i}-t(1+2t^2)\mathbf{j}-t(1+2t^2)\mathbf{k})\\ &= (1+3t^2)\mathbf{i}-2t^3\mathbf{j}-2t^3\mathbf{k}.\end{aligned}$$

The binormal vector

$$\mathbf{B}(t)=\mathbf{T}(t)\times\mathbf{N}(t)=\det\left(\begin{bmatrix}\mathbf{i} & \mathbf{j} & \mathbf{k}\\ \dfrac{2t}{\sqrt{2+4t^2}} & \dfrac{1}{\sqrt{2+4t^2}} & \dfrac{1}{\sqrt{2+4t^2}}\\ \dfrac{1}{\sqrt{1+2t^2}} & -\dfrac{t}{\sqrt{1+2t^2}} & -\dfrac{t}{\sqrt{1+2t^2}}\end{bmatrix}\right)=\frac{\sqrt{2}}{2}(\mathbf{j}-\mathbf{k})$$

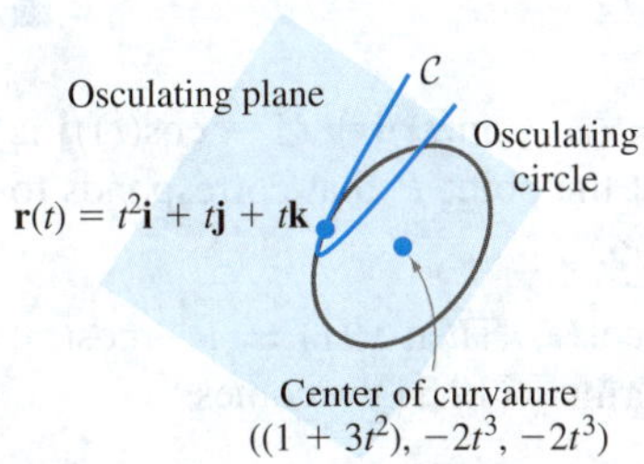

Figure 4

is perpendicular to the osculating plane. Because the osculating plane passes through the point $\mathbf{r}(t)=t^2\mathbf{i}+t\mathbf{j}+t\mathbf{k}$, its equation is

$$0(x-t^2)+\frac{\sqrt{2}}{2}(y-t)+\left(-\frac{\sqrt{2}}{2}\right)(z-t)=0,$$

which simplifies to $y=z$. The curve, its osculating plane, and osculating circle are shown in Figure 4. ■

inSIGHT

In Example 4, the entire curve $\mathbf{r}(t)=t^2\mathbf{i}+t\mathbf{j}+t\mathbf{k}$ is contained in the plane $y=z$. It is therefore not surprising that the plane containing the curve turns out to be the osculating plane at each point.

Planar Curves

The theory that we have developed for space curves can also be applied to curves in the xy-plane. Such a curve $\mathcal{C}$ might be the graph $y=f(x)$ of a function f or it might be given parametrically by functions $x(t)$ and $y(t)$: $\mathcal{C}=\{(x,y): x=x(t),\ y=y(t),\ a\le t\le b\}$. In fact, the graph $y=f(x)$ of a function can be realized parametrically by means of the equations $x=x(t)$, $y=y(t)$ where $x(t)=t$ and $y(t)=f(t)$. We can imagine the xy-plane to be the plane $z=0$ in xyz-space. The planar curve can then be realized by the vector-valued function $\mathbf{r}(t)=x(t)\mathbf{i}+y(t)\mathbf{j}+0\mathbf{k}$, and our formula for the curvature of space curves applies. The next theorem makes this explicit.

Theorem 2 Suppose that $x(t)$ and $y(t)$ are twice differentiable functions that define a planar curve $\mathcal{C}$ by means of the parametric equations $x=x(t)$, $y=y(t)$. Then, at any point $P=((x(t),y(t))$ for which the velocity vector $\langle x'(t),y'(t)\rangle$ is not $\vec{\mathbf{0}}$, the curvature of $\mathcal{C}$ is given by

$$\kappa(t)=\frac{|x'(t)y''(t)-y'(t)x''(t)|}{\big((x'(t))^2+(y'(t))^2\big)^{3/2}}. \tag{12.21}$$

In particular, the curvature of the graph of $y = f(x)$ at the point $(x, f(x))$ is

$$\kappa(x) = \frac{|f''(x)|}{\left(1 + (f'(x))^2\right)^{3/2}}. \tag{12.22}$$

Proof With the xy-plane realized as the plane $z = 0$ in xyz-space and with $\mathcal{C}$ parameterized by the equation $\mathbf{r}(t) = x(t)\mathbf{i} + y(t)\mathbf{j} + 0\mathbf{k}$, we have

$$\mathbf{r}'(t) \times \mathbf{r}''(t) = \det\left(\begin{bmatrix} \mathbf{i} & \mathbf{j} & \mathbf{k} \\ x'(t) & y'(t) & 0 \\ x''(t) & y''(t) & 0 \end{bmatrix}\right) = (x'(t)y''(t) - y'(t)x''(t))\mathbf{k}.$$

It follows that $\|\mathbf{r}'(t)\| = ((x'(t))^2 + (y'(t))^2)^{1/2}$ and $\|\mathbf{r}'(t) \times \mathbf{r}''(t)\| = |x'(t)y''(t) - y'(t)x''(t)|$. Formula (12.21) results when these values are substituted into equation (12.12). If $x(t) = t$ and $y(t) = f(t)$, then $x'(t) = 1$, $x''(t) = 0$, $y'(t) = f'(t)$, $y''(t) = f''(t)$, and formula (12.21) reduces to (12.22). ■

Example 5 The planar curve defined by $\mathbf{r}(t) = (t - \sin(t))\mathbf{i} + (1 - \cos(t))\mathbf{j}$ is called a *cycloid*. Calculate the cycloid's curvature at the point P that corresponds to $t = \pi/3$. Repeat for $t = 3\pi$.

Solution Let $x(t) = t - \sin(t)$ and $y(t) = 1 - \cos(t)$. Then $x'(t) = 1 - \cos(t)$, $x''(t) = \sin(t)$ and $y'(t) = \sin(t)$, $y''(t) = \cos(t)$. Formula (12.21) becomes

$$\kappa(t) = \frac{|x'(t)y''(t) - y'(t)x''(t)|}{\left((x'(t))^2 + (y'(t))^2\right)^{3/2}} = \frac{|(1 - \cos(t))\cos(t) - \sin(t)\sin(t)|}{\left((1 - \cos(t))^2 + (\sin(t))^2\right)^{3/2}},$$

which simplifies to

$$\kappa(t) = \frac{|\cos(t) - \cos^2(t) - \sin^2(t)|}{\left(1 - 2\cos(t) + \cos^2 t + \sin^2(t)^2\right)^{3/2}} = \frac{|\cos(t) - 1|}{2^{3/2}(1 - \cos(t))^{3/2}} = \frac{\sqrt{2}}{4\sqrt{1 - \cos(t)}}.$$

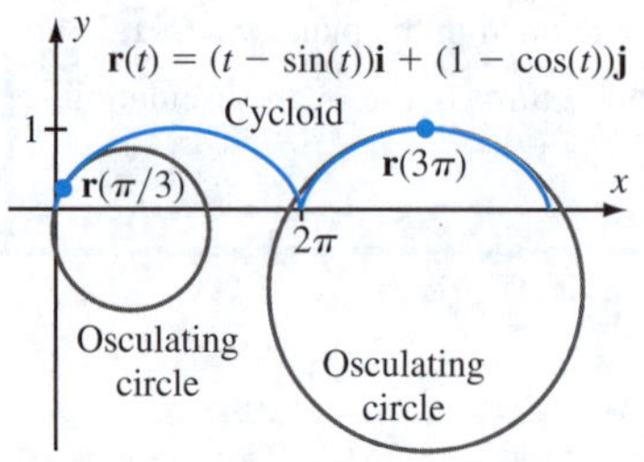

Figure 5

When $t = \pi/3$, the curvature is $\kappa(\pi/3) = \sqrt{2}/(4\sqrt{1 - \cos(\pi/3)}) = 1/2$. Similarly, we find that $\kappa(3\pi) = \sqrt{2}/(4\sqrt{1 - \cos(3\pi)}) = \sqrt{2}/(4\sqrt{1 - (-1)}) = 1/4$. Our calculations tell us that the osculating circles at $\mathbf{r}(\pi/3)$ and $\mathbf{r}(3\pi)$ have radii 2 and 4, respectively. The cycloid is plotted for $0 \le t \le 4\pi$ in Figure 5, which also shows the osculating circles at the points corresponding to $t = \pi/3$ and $t = 3\pi$. ■

Example 6 Find the circle of curvature of the curve $y = \sin(2x)$ at the point $P = (\pi/4, 1)$.

Solution We set $f(x) = \sin(2x)$ and notice that $f'(\pi/4) = 2\cos(\pi/2) = 0$ and $f''(\pi/4) = -4\sin(\pi/2) = -4$. Then, according to formula (12.22), we have

$$\kappa\left(\frac{\pi}{4}\right) = \frac{4}{(1 + 0^2)^{3/2}} = 4.$$

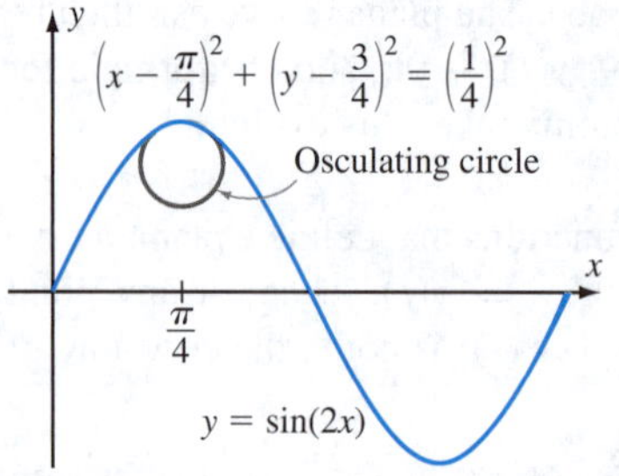

Figure 6

The radius of curvature is therefore 1/4. Because $f'(\pi/4) = 0$, the graph of $y = \sin(2x)$ has a horizontal tangent. The principal unit normal therefore points vertically. Because the graph of $y = \sin(2x)$ is concave down at P and because the principal unit normal points in the direction to which the curve bends, we conclude that the principal unit normal points downward. The center of curvature is the point $(\pi/4, 1 - 1/4)$ and the circle of curvature has Cartesian equation $(x - \pi/4)^2 + (y - 3/4)^2 = (1/4)^2$. The graph of f, together with this osculating circle, is shown in Figure 6. ■

A LOOK BACK

In Section 4.4, we were able to define and calculate the curvature of curves in the plane without appealing to the more sophisticated tools that we have brought to bear on space curves. It should be noted that formula (12.22) is exactly the same as formula (4.4).

quickquiz

1. How is curvature defined? (Answer with a descriptive sentence rather than a formula.)
2. What formula involving $\mathbf{r}'(t)$ and $\mathbf{r}''(t)$ can be used for calculating the curvature of a curve, even if the parameterization $\mathbf{r}$ is not by arc length?
3. If P is a point on a curve $\mathcal{C}$, then what vector is perpendicular to the osculating plane of $\mathcal{C}$ at P?
4. If P is a point on a curve $\mathcal{C}$, then what is the radius of the osculating circle of $\mathcal{C}$ at P?

EXERCISES

Problems for Practice

In Exercises 1–6, an arc length parameterization of a curve is given. Verify this assertion and then calculate the following:

a. $\mathbf{T}(s)$

b. $\mathbf{N}(s)$

c. $\kappa(s)$

1. $\mathbf{r}(s) = \sin(s)\mathbf{i} - \cos(s)\mathbf{j} + 3\mathbf{k}$

2. $\mathbf{r}(s) = (3 - \cos(s))\mathbf{i} - (6 + \sin(s))\mathbf{j}$

3. $\mathbf{r}(s) = (\cos(s/\sqrt{2}))\mathbf{i} - (\sin(s/\sqrt{2}))\mathbf{j} + (s/\sqrt{2})\mathbf{k}$

4. $\mathbf{r}(s) = (\sin(s/4))\mathbf{i} + (\cos(s/4))\mathbf{j} - (\sqrt{15}/4)s\mathbf{k}$

5. $\mathbf{r}(s) = (3\sin(s)/5)\mathbf{i} + \cos(s)\mathbf{j} + (4\sin(s)/5)\mathbf{k}$

6. $\mathbf{r}(s) = (2\sqrt{2}/3)s\mathbf{i} - (\sin(s)/3)\mathbf{j} + (\cos(s)/3)\,\mathbf{k}$

7. For the arc length parameterization given in Exercise 3, calculate the radius of curvature $\rho(s)$ and the center of the osculating circle at each point of the curve.

8. For the arc length parameterization given in Exercise 4, calculate the radius of curvature $\rho(s)$ and the center of the osculating circle at each point of the curve.

9. For the arc length parameterization given in Exercise 5, calculate the radius of curvature $\rho(s)$ and the center of the osculating circle at each point of the curve.

In Exercises 10–15, a parameterization of a planar curve is given. Calculate the curvature at each point of the curve.

10. $\mathbf{r}(t) = \cos^2(t^2)\mathbf{i} + \sin^2(t^2)\mathbf{j}$

11. $\mathbf{r}(t) = e^t\mathbf{i} - e^{2t}\mathbf{j}$

12. $\mathbf{r}(t) = t^2\mathbf{i} - t^3\mathbf{j}$

13. $\mathbf{r}(t) = \ln(|t|)\mathbf{i} - (1/t)\mathbf{j}$

14. $\mathbf{r}(t) = t^{1/2}\mathbf{i} + t^{3/2}\mathbf{j}$

15. $\mathbf{r}(t) = \tan(t)\mathbf{i} + \sec(t)\mathbf{j}$

For each function f in Exercises 16–21, calculate the curvature at every point on the graph (in the xy-plane) of the equation $y = f(x)$.

16. $f(x) = x^3$

17. $f(x) = x^{1/2}$

18. $f(x) = \ln(|x|)$

19. $f(x) = \ln(|\sec(x)|)$

20. $f(x) = (e^x + e^{-x})$

21. $f(x) = x^2/6 + 1/2x$

In Exercises 22–27, a parameterization of a space curve is given. At each point of the curve, calculate the radius of curvature and the center of curvature.

22. $\mathbf{r}(t) = t\mathbf{i} + t\mathbf{j} + t^2\mathbf{k}$

23. $\mathbf{r}(t) = t^2\mathbf{i} + t^2\mathbf{j} + (1 + 2t)\mathbf{k}$

24. $\mathbf{r}(t) = \sqrt{t}\mathbf{i} + (2 - t)\mathbf{j} + 3t$

25. $\mathbf{r}(t) = e^t\mathbf{i} + t\mathbf{j} + e^t\mathbf{k}$

26. $\mathbf{r}(t) = \ln(t)\mathbf{i} + t\mathbf{j} + 2t\mathbf{k}$

27. $\mathbf{r}(t) = t\mathbf{i} + \ln(\sec(t))\mathbf{j}$

In Exercises 28–32, a parameterization of a space curve is given. Find the Cartesian equation of its osculating plane at point P.

28. $\mathbf{r}(t) = (2 - t^2)\mathbf{i} + 2t^3\mathbf{j} + t^2\mathbf{k}$, $P = (1, 2, 1)$

29. $\mathbf{r}(t) = e^t\mathbf{i} + t\mathbf{j} + e^{2t}\mathbf{k}$, $P = (1, 0, 1)$

30. $\mathbf{r}(t) = \sqrt{t}\mathbf{i} + (4/t)\mathbf{j} + (t^2/4)\mathbf{k}$, $P = (2, 1, 4)$

31. $\mathbf{r}(t) = (t^2 + t)\mathbf{i} + (t^2 - t)\mathbf{j} + t^2\mathbf{k}$, $P = (2, 6, 4)$

32. $\mathbf{r}(t) = \sin(t)\mathbf{i} + \cos(2t)\mathbf{j} + 6t\mathbf{k}$, $P = (1/2, 1/2, \pi)$

Further Theory and Practice

Given a curve C, the locus of its centers of curvature is called the *evolute* of C. In Exercises 33–40, find the evolute of the given plane curve.

33. $\mathbf{r}(t) = (2 - t^2)\mathbf{i} + 2t^3\mathbf{j}$

34. $\mathbf{r}(t) = t\mathbf{i} + \cos(t)\mathbf{j}$

35. $\mathbf{r}(t) = t\mathbf{i} + e^t\mathbf{j}$

36. $\mathbf{r}(t) = 2\cos(t)\mathbf{i} + \sin(t)\mathbf{j}$

37. $\mathbf{r}(t) = (t - \sin(t))\mathbf{i} + (1 - \cos(t))\mathbf{j}$

38. $\mathbf{r}(t) = t\cos(t)\mathbf{i} + t\sin(t)\mathbf{j}$

39. $\mathbf{r}(t) = (1 + \sin(t))\cos(t)\mathbf{i} + (1 + \sin(t))\sin(t)\mathbf{j}$

40. $\mathbf{r}(t) = e^t\cos(t)\mathbf{i} + e^t\sin(t)\mathbf{j}$

In Exercises 41–44, find the point at which the graph $y = f(x)$ of the given function f has the greatest curvature.

41. $f(x) = x^2$

42. $f(x) = e^x$

43. $f(x) = \ln(2x^2)$

44. $f(x) = x^4/16 + 1/(2x^2)$, $x > 0$

In Exercises 45–48, find a parameterization for the curve that is the graph of the given equation. Find the curvature at the specified point.

45. $x^2/a^2 + y^2/b^2 = 1$, $(a/\sqrt{2}, b/\sqrt{2})$

46. $x^3 + y^3 = 4xy$, $(2, 2)$ *Hint:* Choose parameter $t = y/x$, divide the cubic equation by x^3, and find x in terms of t.

47. $x^{2/3} + y^{2/3} = 1$, $(1/8, 3\sqrt{3}/8)$ *Hint:* Take advantage of the identity $\cos^2(t) + \sin^2(t) = 1$.

48. $x(x + y)^2 = 9 + (x + y)^2$, $(2, 1)$ *Hint:* Try the parameter $t = x + y$, and solve for x in terms of t.

49. Prove that a curve with curvature zero at every point is a line.

Exercises 50–53 treat the concept of *torsion*.

50. Suppose that $\mathbf{r}(s)$ is an arc length parameterization of a curve. Prove that there is a scalar-valued function $s \mapsto \tau(s)$ such that the unit tangent vector $\mathbf{T}(s)$, the principal unit normal $\mathbf{N}(s)$, and the binormal $\mathbf{B}(s)$ satisfy the following equations:

$$\begin{aligned}\mathbf{T}'(s) &= \kappa(s)\mathbf{N}(s)\\ \mathbf{N}'(s) &= -\kappa(s)\mathbf{T}(s) + \tau(s)\mathbf{B}(s)\\ \mathbf{B}'(s) &= -\tau(s)\mathbf{N}(s).\end{aligned}$$

The function $s \mapsto \tau(s)$ is called the *torsion* of the curve. The three equations are collectively known as the *Frenet Formulas*.

51. Show that the torsion is given by the formula

$$\tau(s) = \frac{(\mathbf{r}'(t) \times \mathbf{r}''(t)) \cdot \mathbf{r}'''(t)}{\|\mathbf{r}'(t) \times \mathbf{r}''(t)\|^2}.$$

52. Compute the torsion of the helix $\mathbf{r}(s) = \sin(s/\sqrt{2})\mathbf{i} + \cos(s/\sqrt{2})\mathbf{j} + (s/\sqrt{2})\mathbf{k}$.

53. The torsion of a curve measures the extent to which the curve is twisting out of the osculating plane. In particular, prove that a curve with zero torsion lies in a plane. Conversely, if a curve is planar, it has zero torsion.

54. Let $p(x)$ be a polynomial of degree two or greater. Prove that there is a point on the plane curve $y = p(x)$ where the curvature is greatest.

55. Use Theorem 2a from Section 11.4 to prove formula (12.13):

$$\kappa_{\mathbf{r}}(t) = \frac{\sqrt{(\|\mathbf{r}'(t)\| \cdot \|\mathbf{r}''(t)\|)^2 - (\mathbf{r}'(t) \cdot \mathbf{r}''(t))^2}}{\|\mathbf{r}'(t)\|^3}.$$

56. Prove that

$$\kappa_{\mathbf{r}}(t) = \frac{\big\|\|\mathbf{r}'(t)\|^2\mathbf{r}''(t) - (\mathbf{r}'(t) \cdot \mathbf{r}''(t))\mathbf{r}'(t)\big\|}{\|\mathbf{r}'(t)\|^4}.$$

Calculator/Computer Exercises

In Exercises 57–62, a parameterization of a plane curve C is given. Plot the locus of the centers of curvature of C. The resulting plane curve is called the evolute of C.

57. $\mathbf{r}(t) = (2 - t^2)\mathbf{i} + 2t^3\mathbf{j}$

58. $\mathbf{r}(t) = t\mathbf{i} + \cos(t)\mathbf{j}$

59. $\mathbf{r}(t) = t\mathbf{i} + e^t\mathbf{j}$

60. $\mathbf{r}(t) = 2\cos(t)\mathbf{i} + \sin(t)\mathbf{j}$

61. $\mathbf{r}(t) = (t - \sin(t))\mathbf{i} + (1 - \cos(t))\mathbf{j}$

62. $\mathbf{r}(t) = t\cos(t)\mathbf{i} + t\sin(t)\mathbf{j}$

In Exercises 63–66, a parameterization of a space curve C is given. Plot the locus of the centers of curvature of C. The resulting space curve is called the evolute of C.

63. $\mathbf{r}(t) = \cos(t)\mathbf{i} + \sin(t)\mathbf{j} + t\mathbf{k}$

64. $\mathbf{r}(t) = t\mathbf{i} + t^2\mathbf{j} + t^3\mathbf{k}$

65. $\mathbf{r}(t) = \cos(t)\mathbf{i} + \sin(t)\mathbf{j} + t^2\mathbf{k}$

66. $\mathbf{r}(t) = (1 + t^2)\cos(t)\mathbf{i} + (1 + t^2)\sin(t)\mathbf{j} + t^2\mathbf{k}$

12.5 Applications of Vector-Valued Functions to Motion

Suppose that $\mathbf{r}$ is a smooth parameterization of a space curve $\mathcal{C}$. Then the second derivative $\mathbf{r}''$ has a physical interpretation: It is the acceleration of a moving particle that has $\mathbf{r}$ as its position vector. There is also a geometric significance to $\mathbf{r}''$ that shows up in the formula for curvature. We begin this section by establishing a connection between the two roles $\mathbf{r}''$ plays. Let us try to anticipate the relationship. When we go around a sharp curve in a car or a roller coaster, we feel a strong force in the outward direction. By Newton's Second Law, strong force means strong acceleration. So there is a correlation between large curvature and the acceleration vector having a large normal component.

To make these considerations precise, we let $\mathbf{T}(t)$ and $\mathbf{N}(t)$ denote the unit tangent and principal unit normal vectors at the point $\mathbf{r}(t)$. The velocity and acceleration vectors are given by $\mathbf{v}(t) = \mathbf{r}'(t)$ and $\mathbf{a}(t) = \mathbf{v}'(t) = \mathbf{r}''(t)$, respectively. The magnitude of the velocity vector is the speed, $v(t) = \| \mathbf{r}'(t) \|$. Let $\sigma(t) = \int_0^t \| \mathbf{r}'(\tau) \| \, d\tau$ be the arc length function of $\mathbf{r}$. Then $\frac{d\sigma}{dt}$ is another way to express the speed.

Next, we recall equation (12.15),

$$\mathbf{r}''(t) = \frac{d^2\sigma(t)}{dt^2}\mathbf{T}(t) + \frac{d\sigma(t)}{dt}\mathbf{T}'(t),$$

which decomposes the acceleration vector into components in mutually perpendicular directions in the osculating plane. We now rewrite this last equation as $\mathbf{a}(t) = (\frac{dv}{dt}(t))\mathbf{T}(t) + v(t)\mathbf{T}'(t)$, or

$$\mathbf{a}(t) = \frac{dv(t)}{dt}\mathbf{T}(t) + v(t)\| \mathbf{T}'(t) \|\mathbf{N}(t). \tag{12.23}$$

Formula (12.17) tells us that $\| \mathbf{T}'(t) \| = \| \mathbf{r}'(t) \times \mathbf{r}''(t) \| / \| \mathbf{r}'(t) \|^2$ and therefore

$$\frac{\| \mathbf{T}'(t) \|}{\| \mathbf{r}'(t) \|} = \frac{\| \mathbf{r}'(t) \times \mathbf{r}''(t) \|}{\| \mathbf{r}'(t) \|^3} = \kappa_{\mathbf{r}}(t).$$

We may rewrite this last equation as $\| \mathbf{T}'(t) \| = v(t)\kappa_{\mathbf{r}}(t)$ and substitute it into equation (12.23) to obtain:

$$\mathbf{a}(t) = \frac{dv(t)}{dt}\mathbf{T}(t) + v(t)^2\kappa_{\mathbf{r}}(t)\mathbf{N}(t). \tag{12.24}$$

Let us summarize these observations with a theorem.

Theorem 1 Suppose that $t \mapsto \mathbf{r}(t)$ is a smooth parameterization of a space curve $\mathcal{C}$. Let $v(t) = \| \mathbf{r}'(t) \|$ denote the speed of a particle moving along the curve with position vector $\mathbf{r}(t)$. The acceleration vector $\mathbf{a}(t) = \mathbf{r}''(t)$ can be decomposed as the sum of two vectors, one having the direction of $\mathbf{T}(t)$, the unit tangent to $\mathcal{C}$ at $\mathbf{r}(t)$, and the other having the direction of $\mathbf{N}(t)$, the principal unit normal to $\mathcal{C}$ at $\mathbf{r}(t)$. The decomposition has the form

$$\mathbf{a}(t) = a_T(t)\mathbf{T}(t) + a_N(t)\mathbf{N}(t), \tag{12.25}$$

where

$$a_T(t) = \frac{dv(t)}{dt} \qquad \text{and} \qquad a_N(t) = v(t)^2\kappa_{\mathbf{r}}(t).$$

The scalar $a_T(t)$ is called the *tangential component of acceleration,* and the *positive* scalar $a_N(t)$ is called the *normal component of acceleration*. We say that equation (12.25) expresses the acceleration vector $\mathbf{a}(t)$ as a *linear combination* of the vectors $\mathbf{T}(t)$ and $\mathbf{N}(t)$. This remarkable formula contains a great deal of information: The component of force in the direction of $\mathbf{N}(t)$ that is required to hold a moving body on its curved trajectory is called the *centripetal force*. By Newton's Second Law, force is mass times acceleration. Thus, centripetal force is the mass of the body times $a_N(t)\mathbf{N}(t)$, a vector called the *centripetal* or *normal acceleration*. This vector's magnitude $a_N(t)$ depends on the speed and the curvature (but not on the change of speed). The vector $a_T(t)\mathbf{T}(t)$ is called the *tangential acceleration,* and its magnitude $|a_T|$ depends on the rate of change of speed with respect to time. If the path of the moving body becomes more curved, then the normal component of the acceleration increases in magnitude, hence the normal component of force increases. If the speed increases, then the normal component of acceleration also increases, and the normal component of force increases.

When we calculate centripetal acceleration, the following theorem is often useful.

Theorem 2 The tangential and normal components of acceleration satisfy

a. $a_T(t) = \mathbf{a}(t) \cdot \mathbf{T}(t)$,
b. $a_N(t) = \mathbf{a}(t) \cdot \mathbf{N}(t)$, and
c. $\| \mathbf{a}(t) \|^2 = (a_T(t))^2 + (a_N(t))^2$.

Proof Each of the three formulas is obtained by calculating a dot product and using the equations $\mathbf{T}(t) \cdot \mathbf{T}(t) = 1$, $\mathbf{N}(t) \cdot \mathbf{N}(t) = 1$, and $\mathbf{T}(t) \cdot \mathbf{N}(t) = 0$. Thus, part a is obtained by taking the dot product of each side of equation (12.25) with $\mathbf{T}(t)$. The result is $\mathbf{a}(t) \cdot \mathbf{T}(t) = a_T(t)\mathbf{T}(t) \cdot \mathbf{T}(t) + a_N(t)\mathbf{N}(t) \cdot \mathbf{T}(t) = a_T(t)$, which is part a. Part b is obtained in a similar way: We take the dot product of each side of equation (12.25) with $\mathbf{N}(t)$. Finally,

$$\begin{aligned}
\|\mathbf{a}(t)\|^2 &= \mathbf{a}(t) \cdot \mathbf{a}(t) \\
&= \big(a_T(t)\mathbf{T}(t) + a_N(t)\mathbf{N}(t)\big) \cdot \big(a_T(t)\mathbf{T}(t) + a_N(t)\mathbf{N}(t)\big) \\
&= (a_T(t))^2\mathbf{T}(t) \cdot \mathbf{T}(t) + 2a_T(t)a_N(t)\mathbf{T}(t) \cdot \mathbf{N}(t) + (a_N(t))^2\mathbf{N}(t) \cdot \mathbf{N}(t) \\
&= (a_T(t))^2 + (a_N(t))^2.
\end{aligned}$$

■

Example 1 Let $\mathbf{r}(t) = \sin(t)\mathbf{i} - \cos(t)\mathbf{j} - (t^2/2)\mathbf{k}$. Calculate $a_T(t)$ and $a_N(t)$. Express $\mathbf{a}(t)$ as a linear combination of $\mathbf{T}(t)$ and $\mathbf{N}(t)$.

Solution We have $\mathbf{r}'(t) = \cos(t)\mathbf{i} + \sin(t)\mathbf{j} - t\mathbf{k}$ and $\mathbf{a}(t) = \mathbf{r}''(t) = -\sin(t)\mathbf{i} + \cos(t)\mathbf{j} - \mathbf{k}$. Therefore

$$v(t) = \| \mathbf{r}'(t) \| = \sqrt{\cos^2(t) + \sin^2(t) + t^2} = \sqrt{1 + t^2}.$$

It follows that

$$a_T(t) = \frac{dv(t)}{dt} = \frac{d}{dt}\sqrt{1 + t^2} = \frac{t}{\sqrt{1 + t^2}}.$$

Also, $\|\mathbf{a}(t)\|^2 = \|-\sin(t)\mathbf{i} + \cos(t)\mathbf{j} - \mathbf{k}\|^2 = (-\sin(t))^2 + (\cos(t))^2 + (-1)^2 = 2.$ According to Theorem 2,

$$a_N(t) = \sqrt{\|\mathbf{a}(t)\|^2 - (a_T(t))^2} = \sqrt{2 - \frac{t^2}{1+t^2}} = \sqrt{\frac{2+t^2}{1+t^2}}.$$

In conclusion, equation (12.25) gives us

$$\mathbf{a}(t) = \frac{t}{\sqrt{1+t^2}}\mathbf{T}(t) + \sqrt{\frac{2+t^2}{1+t^2}}\mathbf{N}(t). \tag{12.26}$$

■

inSIGHT

Equation (12.25) can be used in conjunction with Theorem 2b as an alternative means to calculate the principal unit normal vector $\mathbf{N}(t)$. For instance, in Example 1, we can use equation (12.26) to solve for $\mathbf{N}(t)$:

$$\mathbf{N}(t) = \sqrt{\frac{1+t^2}{2+t^2}}\left(\mathbf{a}(t) - \frac{t}{\sqrt{1+t^2}}\mathbf{T}(t)\right).$$

We can then substitute the values for vectors $\mathbf{a}(t)$ and $\mathbf{T}(t)$ into this formula to obtain $\mathbf{N}(t)$.

Central Force Fields

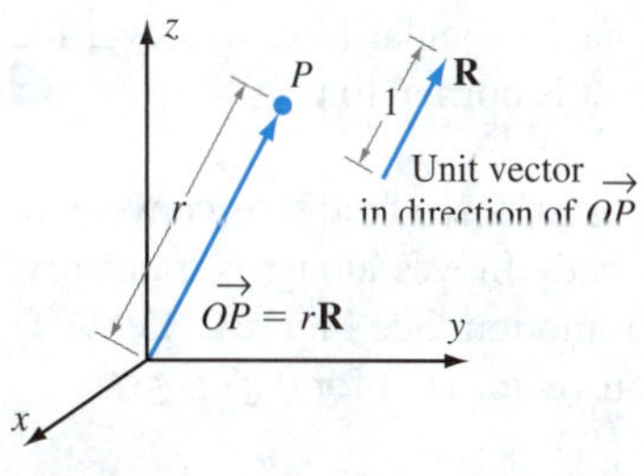

Figure 1
A central force field $\mathbf{F}(P) = f(r)\mathbf{R}$ acts in the direction of the position vector $\overrightarrow{OP}$ if $f(r) > 0$ and in the direction opposite to the position vector if $f(r) < 0$.

Let O be a fixed point in space. We will use it as the origin of our coordinate axes. We say that a force $\mathbf{F}$ is a *central force field* if, at each point $P \neq 0$, there is a scalar-valued function f of a real variable such that the force at P is given by $\mathbf{F}(P) = f(r)\mathbf{R}$ where $\mathbf{R}$ is the direction of $\overrightarrow{OP}$ and $r = \|\overrightarrow{OP}\|$ is the distance of O to P. See Figure 1. If $f(r) > 0$, then the force is directed away from O. If $f(r) < 0$, then the force is directed toward O and is said to be *attractive*. As an example of an attractive central force field, consider the sun as a point mass that is fixed at a point O in space. If a planet is at point P, then, according to Newton's Universal Law of Gravitation, the gravitational force exerted by the sun on the planet is equal to

$$\mathbf{F}(P) = -\frac{GMm}{r^2}\mathbf{R} \tag{12.27}$$

where G is a universal constant (that is the same for all planets), M is the mass of the sun, m is the mass of the planet, r is the distance of the planet to the sun, and $\mathbf{R}$ is the direction vector of $\overrightarrow{OP}$. By setting $f(r) = -GMm/r^2$, we see that the force of gravity is an attractive central force field.

Example 2 Suppose that $a > b > 0$. A force $\mathbf{F}$ acts on a particle of mass m in such a way that the particle moves in the xy-plane with position described by $\mathbf{r}(t) = a\cos(t)\mathbf{i} + b\sin(t)\mathbf{j}$. Show that $\mathbf{F}$ is a central force field.

Solution According to Newton's Second Law of Motion, $\mathbf{F} = m\mathbf{a}(t)$; therefore

$$\mathbf{F} = m\mathbf{r}''(t) = m(-a\cos(t)\mathbf{i} - b\sin(t)\mathbf{j}) = -m\mathbf{r}(t) = -m\|\mathbf{r}(t)\|\mathbf{R}.$$

This formula for $\mathbf{F}$ shows that it is a central force field directed toward the origin. ■

A key fact about central force fields is that they always give rise to trajectories that lie in a plane.

Theorem 3 If a particle moving in space is subject only to a central force field, then the particle's trajectory lies in a plane.

Proof Let $\mathbf{r}(t)$ be the position vector of the particle, and let m be its mass. The force on the particle can be written as $\mathbf{F} = f(r)\mathbf{R}$ where $r = \|\mathbf{r}(t)\|$ and $\mathbf{R} = (1/r)\mathbf{r}(t)$. Notice that $\mathbf{r}(t) \times \mathbf{F} = \vec{\mathbf{0}}$, since $\mathbf{r}(t)$ and $\mathbf{R}$ have the same direction. Newton's Second Law of Motion, force equals mass times acceleration, allows us to conclude that $\mathbf{r}(t) \times (m\mathbf{r}''(t)) = \vec{\mathbf{0}}$, or $\mathbf{r}(t) \times \mathbf{r}''(t) = \vec{\mathbf{0}}$. We can write this as

$$\mathbf{r}(t) \times \frac{d}{dt}\mathbf{r}'(t) = \vec{\mathbf{0}}.$$

From this equation, we deduce that

$$\frac{d}{dt}(\mathbf{r}(t) \times \mathbf{r}'(t)) = \mathbf{r}'(t) \times \mathbf{r}'(t) + \mathbf{r}(t) \times \frac{d}{dt}\mathbf{r}'(t) = \vec{\mathbf{0}} + \vec{\mathbf{0}} = \vec{\mathbf{0}}.$$

It follows that

$$\mathbf{r}(t) \times \mathbf{r}'(t) = \mathbf{c} \qquad \textbf{(12.28)}$$

for some constant vector $\mathbf{c}$. This tells us that $\mathbf{r}(t)$ is perpendicular to $\mathbf{c}$, whatever the value of t may be. In other words, $\mathbf{r}(t)$ is in a plane that is normal to $\mathbf{c}$. ■

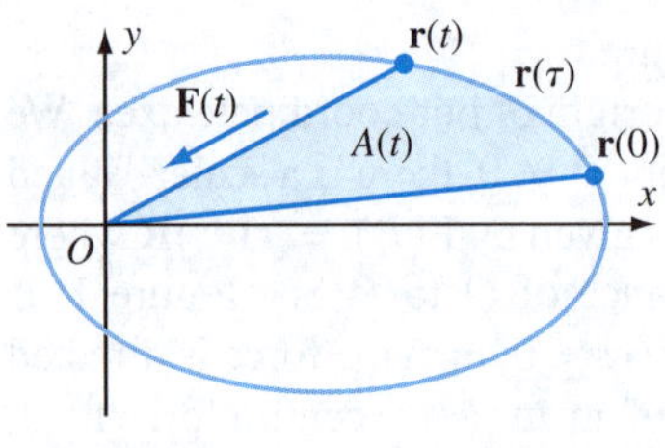

Figure 2

When we study the trajectory of a body under a central force field, it is convenient to assume that the motion is in the xy-plane. As the body moves along its trajectory, its position vector sweeps out a region in the plane of motion. See Figure 2. Let $A(t)$ denote the area of the region swept out by the position vector $\mathbf{r}(\tau)$ for $0 \le \tau \le t$.

Theorem 4 If a moving particle is subject only to a central force field, then the particle's position vector sweeps out a region with area $A(t)$ that has a constant rate of change with respect to t.

Proof Let Δt be an increment of time, $\Delta\mathbf{r}(t)$ the corresponding increment of position, and ΔA the increment of area swept out (Figure 3). We see that ΔA is approximately equal to half the area of the parallelogram determined by the vectors $\mathbf{r}(t)$ and $\mathbf{r}(t+\Delta t) = \mathbf{r}(t) + \Delta\mathbf{r}(t)$. The area of this parallelogram is $\| \mathbf{r}(t) \times \mathbf{r}(t + \Delta t)\| = \| \mathbf{r}(t) \times (\mathbf{r}(t) + \Delta\mathbf{r}(t)) \| = \| \mathbf{r}(t) \times \Delta\mathbf{r}(t)\|$. Thus

$$\frac{\Delta A}{\Delta t} \approx \frac{1}{2}\frac{\| \mathbf{r}(t) \times \Delta\mathbf{r}(t) \|}{\Delta t} = \frac{1}{2}\left\| \mathbf{r}(t) \times \frac{\Delta\mathbf{r}(t)}{\Delta t} \right\|.$$

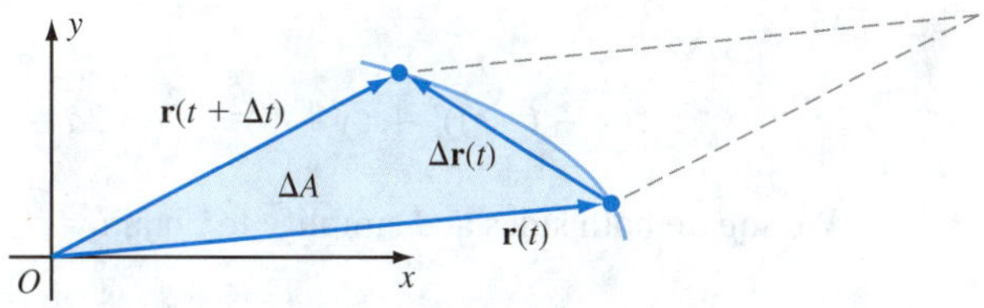

Figure 3

Letting $\Delta t \to 0$ gives

$$A'(t) = \lim_{\Delta t \to 0} \frac{\Delta A}{\Delta t} = \frac{1}{2}\left\| \mathbf{r}(t) \times \lim_{\Delta t \to 0} \frac{\Delta \mathbf{r}(t)}{\Delta t} \right\| = \frac{1}{2}\|\mathbf{r}(t) \times \mathbf{r}'(t)\| \overset{(12.28)}{=} \frac{1}{2}\|\mathbf{c}\|.$$

We conclude that area $A(t)$ is swept out at a constant rate. ■

Ellipses

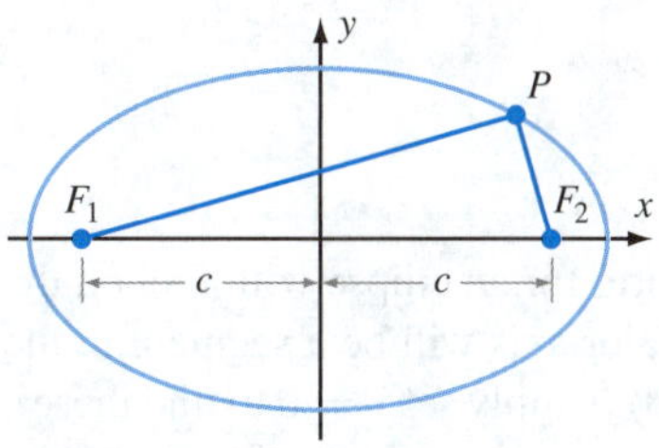

Figure 4
$|\overline{PF}_1| + |\overline{PF}_2| = 2a$

For the remainder of our work in this section, we will need an in-depth understanding of one particular planar curve, the ellipse. Fix two distinct points F_1 and F_2 in the xy-plane. Let c denote half of the distance between the two points. Suppose that a is a fixed positive constant greater than c. The set of points $P = (x, y)$ with the property that the distance from P to F_1 plus the distance from P to F_2 equals $2a$ is called an *ellipse*. See Figure 4.

Each of the points F_1 and F_2 is called a *focus* of the ellipse. Together they are called the *foci* (plural of "focus") of the ellipse. The midpoint of line segment $\overline{F_1 F_2}$ is called the *center* of the ellipse. The chord of the ellipse passing through the two foci is called the *major axis* of the ellipse. The chord perpendicular to the major axis and passing through the center is called the *minor axis* of the ellipse. These features are shown in Figure 5.

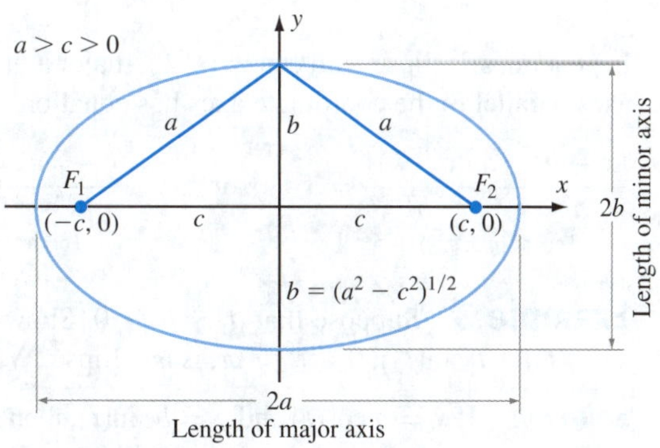

Figure 5

Let us derive the formula for an ellipse with major axis along the x-axis and center at the origin. Let c be a positive constant. We place the foci of the ellipse at points $F_1 = (-c, 0)$ and $F_2 = (c, 0)$, symmetrically situated on the x-axis. Let a be a number that exceeds c. A point (x, y) lies on the ellipse provided that

$$\underbrace{\left((x-(-c))^2 + (y-0)^2\right)^{1/2}}_{\text{Distance to } F_1} + \underbrace{\left((x-c)^2 + (y-0)^2\right)^{1/2}}_{\text{Distance to } F_2} = 2a,$$

or

$$\left((x-(-c))^2+(y-0)^2\right)^{1/2}=2a-\left((x-c)^2+(y-0)^2\right)^{1/2}.$$

We square both sides and simplify to obtain

$$a^2-cx=a(x^2-2cx+c^2+y^2)^{1/2}.$$

Squaring both sides again and simplifying gives

$$(a^2-c^2)x^2+a^2y^2=a^2(a^2-c^2).$$

Since $a>c>0$, it follows that each of the coefficients appearing in this equation is positive. Let us simplify matters by setting

$$b=(a^2-c^2)^{1/2}. \tag{12.29}$$

We then have $b^2x^2+a^2y^2=a^2b^2$ or, equivalently,

$$\frac{x^2}{a^2}+\frac{y^2}{b^2}=1$$

with $a>b>0$. This last equation is the standard form for an ellipse with foci on the x-axis and center at the origin. By definition, the major axis will be a segment in the x-axis. Since the x-intercepts of the ellipse are $(\pm a,0)$ (simply set $y=0$ to find these), we see that the major axis is the segment connecting $(-a,0)$ to $(a,0)$. Similarly, the minor axis is the segment connecting $(0,-b)$ to $(0,b)$. Refer to Figure 5.

If we were to repeat the preceding calculation with foci $(0,-c)$ and $(0,c)$ and $b=\sqrt{a^2-c^2}$, then we would obtain the equation

$$\frac{x^2}{b^2}+\frac{y^2}{a^2}=1.$$

In general, an ellipse with center (h,k), major axis length $2a$, minor axis length $2b$, and axes parallel to the coordinate axes has equation

$$\underbrace{\frac{(x-h)^2}{a^2}+\frac{(y-k)^2}{b^2}=1}_{\text{Foci at }(h\pm c,\,k)}\quad\text{or}\quad\underbrace{\frac{(x-h)^2}{b^2}+\frac{(y-k)^2}{a^2}=1}_{\text{Foci at }(h,\,k\pm c)}\quad(b^2+c^2=a^2). \tag{12.30}$$

Example 3 Suppose that $a>b>0$. Show that the curve described by $\mathbf{r}(t)=a\cos(t)\mathbf{i}+b\sin(t)\mathbf{j},\ 0\le t\le 2\pi$, is an ellipse. Where are the foci located?

Solution If $x=a\cos(t)$ and $y=b\sin(t)$, then $x^2/a^2+y^2/b^2=\cos^2(t)+\sin^2(t)=1$. Therefore, $\mathbf{r}(t)$ describes an ellipse. Although the half-distance c between the two foci does not explicitly appear in the equation $x^2/a^2+y^2/b^2=1$, we can use the formula $b=(a^2-c^2)^{1/2}$ to determine that $c=(a^2-b^2)^{1/2}$. Since $a>b$, the major axis of the ellipse lies along the x-axis. The foci are therefore located at $(-\sqrt{a^2-b^2},0)$ and $(\sqrt{a^2-b^2},0)$. ■

Definition Let c denote the half-distance between the foci of an ellipse. Let a denote half the length of the major axis of the ellipse. The quantity $e=c/a$ is called the *eccentricity* of the ellipse.

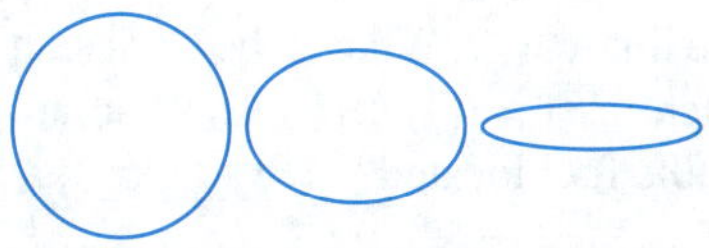

Eccentricity 0.1 Eccentricity 0.75 Eccentricity 0.98

Figure 6

It is traditional to use the letter e for the eccentricity of an ellipse. In this context, the letter e has nothing to do with the base of the natural logarithm. Notice that the eccentricity e of an ellipse is a nonnegative number that is less than 1. The nearer that e is to 0, the more the ellipse will look like a circle. The closer that e is to 1, the more the ellipse will look like a line segment. See Figure 6.

Example 4 If the length of the major axis of an ellipse is double the length of its minor axis, then what is the eccentricity of the ellipse?

Solution We suppose that $2a$ is the length of the major axis and $2b$ is the length of the minor axis. We are given that $b = a/2$. Therefore,

$$e = \frac{c}{a} = \frac{\sqrt{a^2 - b^2}}{a} = \frac{\sqrt{a^2 - (a/2)^2}}{a} = \sqrt{1 - \frac{1}{4}} = \frac{\sqrt{3}}{2} \approx 0.866. \quad \blacksquare$$

Until now, we have described an ellipse by means of its two foci and the length $2a$ of its major axis. However, it is often convenient to have an alternative characterization of an ellipse.

Theorem 5 If we let $\mathcal{C}$ be a curve in the plane, then $\mathcal{C}$ is an ellipse if and only if there is a real number e with $0 < e < 1$, a point F, and a line $\mathcal{D}$ such that $\mathcal{C}$ is the locus of all points P that satisfy $|\overline{PF}| = e \cdot |P\mathcal{D}|$. Here $|\overline{PF}|$ denotes the distance between the points P and F and $|P\mathcal{D}|$ the distance of P to the line $\mathcal{D}$. In other words, $|P\mathcal{D}|$ is the length of a perpendicular dropped from P to $\mathcal{D}$. If $\mathcal{C}$ is the ellipse defined by the equation $|\overline{PF}| = e \cdot |P\mathcal{D}|$, then

a. The eccentricity of $\mathcal{C}$ is e.
b. $\mathcal{C}$ lies on one side of $\mathcal{D}$.
c. F is a focus of $\mathcal{C}$ (the focus closest to $\mathcal{D}$).
d. The major axis of $\mathcal{C}$ is perpendicular to $\mathcal{D}$.

Proof Rather than give a general proof, we sketch the key algebraic procedure for establishing this theorem. Consider the ellipse $x^2/a^2 + y^2/b^2 = 1$ with foci $(-c, 0)$ and $F = (c, 0)$. The distance $|\overline{PF}|$ between F and a point $P = (x, y)$ on the ellipse is given by

$$|\overline{PF}| = \sqrt{(x-c)^2 + y^2} = \sqrt{(x-c)^2 + b^2\left(1 - \frac{x^2}{a^2}\right)}$$

$$= \sqrt{(x-c)^2 + (a^2 - c^2)\left(1 - \frac{x^2}{a^2}\right)} \overset{\text{Verify!}}{=} \sqrt{\frac{(a^2 - xc)^2}{a^2}} = \frac{|a^2 - xc|}{a}.$$

If we take the vertical line $x = a^2/c$ to be $\mathcal{D}$, then we have

$$|\overline{PF}| = \frac{c}{2} \cdot \left|\frac{2^2}{c} - x\right| = e\left|\frac{a^2}{c} - x\right| = e|P\mathcal{D}|.$$

We may reverse the steps of this algebraic argument to show that the locus of points P such that $|\overline{PF}| = e|P\mathcal{D}|$ is an ellipse. $\blacksquare$

Definition If an ellipse is defined by the equation $|\overline{PF}| = e|P\mathcal{D}|$, then the line $\mathcal{D}$ is called a *directrix* for the ellipse.

Example 5 Suppose that $F = (0, 0)$ and $\mathcal{D}$ is the line $x = 2$. What is the Cartesian equation for the locus of points $P = (x, y)$ for which $|\overline{PF}| = (1/2)|P\mathcal{D}|$? What are the lengths of the major and minor axes? Where are the foci located?

Solution If $P = (x, y)$ and $F = (0, 0)$, then $|\overline{PF}| = \sqrt{x^2 + y^2}$. The distance of the point (x, y) from the line $x = 2$ is $|2 - x|$. The equation $|\overline{PF}| = (1/2)|P\mathcal{D}|$ becomes

$$\sqrt{x^2 + y^2} = \frac{1}{2}|2 - x|.$$

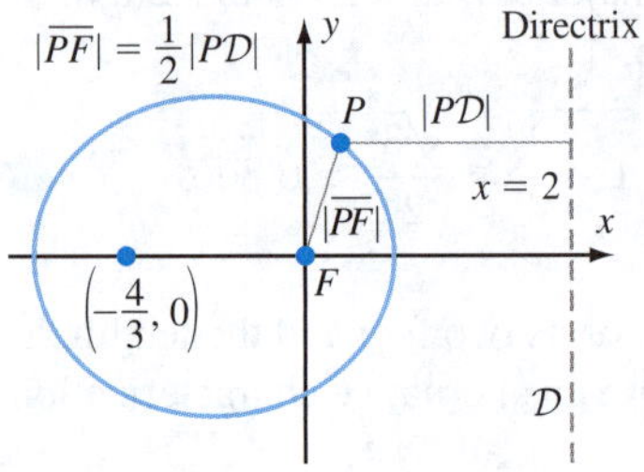

Figure 7

Squaring each side results in the equation $x^2 + y^2 = (2-x)^2/4$. If we expand the right side of this last equation and bring the terms involving x to the left side, then we obtain $3x^2/4 + x + y^2 = 1$, or $x^2 + 4x/3 + 4y^2/3 = 4/3$. We now complete the square on the left side: $(x^2 + 4x/3 + (2/3)^2) + 4y^2/3 = 4/3 + (2/3)^2$, or $(x + 2/3)^2 + 4y^2/3 = 16/9$. Finally, writing this last equation in standard form (12.30), we obtain

$$\frac{(x + 2/3)^2}{(4/3)^2} + \frac{y^2}{\left(2/\sqrt{3}\right)^2} = 1.$$

This equation tells us that the center of the ellipse is at $(-2/3, 0)$, the length of the major axis is $2a = 2(4/3) = 8/3$, and the length of the minor axis is $2b = 2(2/\sqrt{3}) = 4/\sqrt{3}$. One focus is $F = (0, 0)$. The distance c between focus $F = (0, 0)$ and center $(-2/3, 0)$ is $2/3$. The second focus is also situated on the major axis a distance c from the center; it is therefore the point $(-4/3, 0)$. See Figure 7. ■

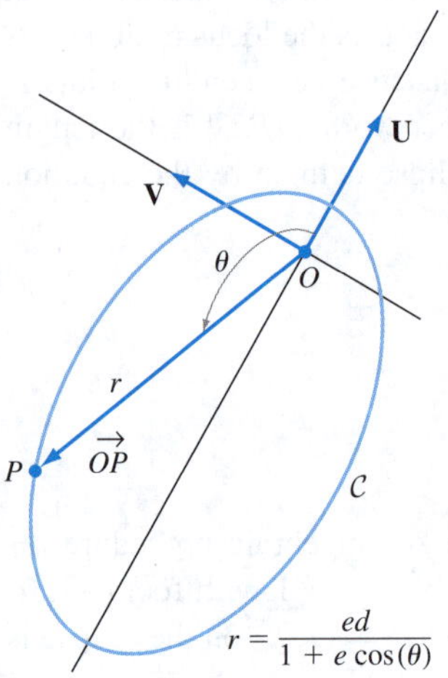

Figure 8a

inSIGHT

> In Example 5, there are several other ways to calculate the value of c. For instance, we may use the formula $c = \sqrt{a^2 - b^2}$ to obtain $c = \sqrt{16/9 - 4/3} = 2/3$. Alternatively, the equation $|\overline{PF}| = (1/2)|P\mathcal{D}|$ tells us that the eccentricity e equals $1/2$. It follows that $c = e \cdot a = (1/2)(4/3) = 2/3$.

Example 6 Let e and d be positive constants with $e < 1$. Suppose that **U** and **V** are mutually perpendicular direction vectors in space. Represent **U** and **V** by directed line segments that have the origin O as their common initial point. These directed line segments determine a unique plane $\mathcal{P}$. Figure 8a shows a curve $\mathcal{C}$ that lies in $\mathcal{P}$. Suppose that for every point P on $\mathcal{C}$, the distance r of P to the origin O and the angle θ that $\overrightarrow{OP}$ makes with **U** are related by the equation

$$r = \frac{ed}{1 + e\cos(\theta)}.$$

Show that $\mathcal{C}$ is an ellipse with one focus at O and eccentricity e.

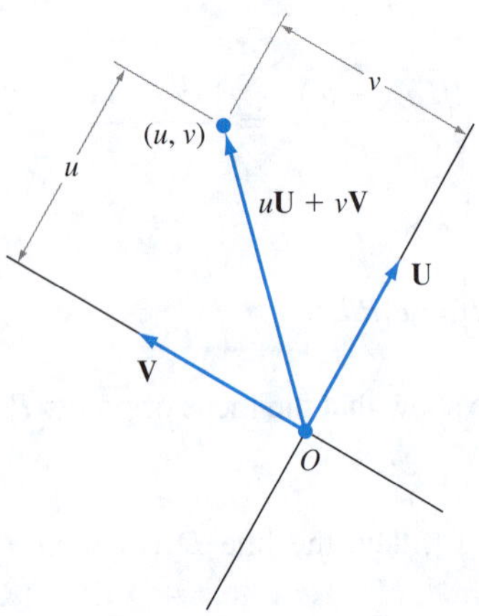

Figure 8b

Solution Let us write (u, v) as the coordinates in $\mathcal{P}$ of the point with position vector $u\mathbf{U} + v\mathbf{V}$. See Figure 8b. In these coordinates, elementary geometry shows that $r = \sqrt{u^2 + v^2}$ and $u = r\cos(\theta)$. On cross multiplication, the given equation becomes $\sqrt{u^2 + v^2} + eu = ed$. After isolating the radical and squaring, we obtain

$$u^2 + v^2 = (ed - eu)^2 = e^2d^2 - 2e^2du + e^2u^2,$$

or $(1-e^2)u^2+2e^2du+v^2=e^2d^2$. Division by the nonzero coefficient of u^2 results in the equation

$$u^2+\frac{2e^2d}{1-e^2}u+\frac{v^2}{1-e^2}=\frac{e^2d^2}{1-e^2}.$$

We complete the square by adding the square of half of the coefficient of u to both sides of the equation:

$$\left(u^2+\frac{2e^2d}{1-e^2}u+\left(\frac{e^2d}{1-e^2}\right)^2\right)+\frac{v^2}{1-e^2}=\frac{e^2d^2}{1-e^2}+\left(\frac{e^2d}{1-e^2}\right)^2.$$

This equation simplifies to

$$\left(u+\frac{e^2d}{1-e^2}\right)^2+\frac{v^2}{1-e^2}=\frac{e^2d^2}{(1-e^2)^2},$$

which we may write in standard form as

$$\frac{(u+(e^2d/1-e^2))^2}{(ed/(1-e^2))^2}+\frac{v^2}{\left(ed/\sqrt{1-e^2}\right)^2}=1.$$

Since $0<1-e^2<1$, we deduce that $1-e^2<\sqrt{1-e^2}$. From this inequality, it follows that $ed/(1-e^2)>ed/\sqrt{1-e^2}$. Thus the major axis has direction $\mathbf{U}$. The distance between focus and center is

$$\sqrt{\left(\frac{ed}{1-e^2}\right)^2-\left(\frac{ed}{\sqrt{1-e^2}}\right)^2}=\frac{e^2d}{1-e^2}.$$

Consequently, the eccentricity of the ellipse is the quotient of this distance $e^2d/(1-e^2)$ by half the length of the major axis, $ed/(1-e^2)$. The result is simply e. ■

Applications to Planetary Motion

In ancient times, it was believed that the planets traveled in circular orbits about Earth. The 16th-century astronomer Copernicus (1473–1543) was the first to give scientific arguments that the planets orbit the sun, but he still believed that the paths of motion were circles. Early in the 17th-century, Johannes Kepler (1571–1630), by studying many detailed planetary observations of Tycho Brahe (1546–1601), was able to devise the following three simple and elegant laws of planetary motion.

Kepler's Laws of Planetary Motion

I. The orbit of each planet is an ellipse, with the sun at one focus.
II. The line segment from the center of the sun to the center of an orbiting planet sweeps out area at a constant rate.
III. The square of the period of a planet's revolution is proportional to the cube of the length of the major axis of its elliptical orbit, with the same constant of proportionality for any planet.

Kepler's First Law is that each planet travels in an elliptical path. It turns out that the eccentricities of the ellipses that arise in the orbits of the planets are very small; the orbits are nearly circles, but they are definitely *not* circles. That is the importance of Kepler's First Law.

Kepler's Second Law tells us that when the planet is at its aphelion (farthest from the sun), it is traveling relatively slowly; whereas at its perihelion (nearest point to the sun), it is traveling relatively rapidly (see Figure 9).

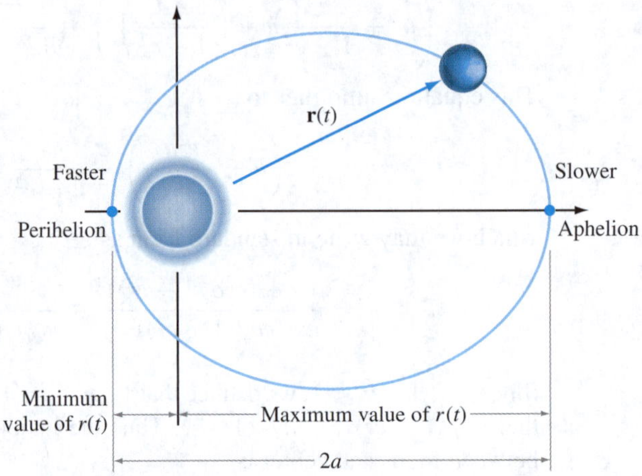

Figure 9

Kepler's Third Law allows us to calculate the length of a year on any given planet from knowledge of the shape of the planet's orbit.

Let us see how to derive Kepler's laws from Newton's inverse square law of gravitational attraction (12.27). To keep matters as simple as possible, we will assume that "our" solar system contains a fixed sun at the origin and only one planet. (The problem of analyzing interactions of gravity among three or more bodies is incredibly complicated and is still not thoroughly understood.)

We denote the position of the planet at time t by $\mathbf{r}(t)$. We can write this position vector as $\mathbf{r}(t) = r(t)\mathbf{R}(t)$ where $\mathbf{R}(t)$ is a unit vector pointing in the same direction as $\mathbf{r}(t)$ and $r(t)$ is a positive scalar representing the length of $\mathbf{r}(t)$. Since the gravitation force is a central force field, we know from Theorem 3 that the planet's motion is contained in a plane. To be specific, equation (12.28) tells us that there is a vector $\mathbf{c}$ such that $\mathbf{r}(t) \times \mathbf{r}'(t) = \mathbf{c}$.

If $\mathbf{F}$ is force, m is the mass of the planet, and $\mathbf{a}(t) = \mathbf{r}''(t)$ is its acceleration, then Newton's Second Law of Motion says that $\mathbf{F} = m\mathbf{r}''(t)$. We can also use Newton's Law of Gravitation, formula (12.27), to express the force. By equating the two expressions for the force, we conclude that $m\mathbf{r}''(t) = -(GMm/r(t)^2)\mathbf{R}(t)$ or, equivalently,

$$\mathbf{r}''(t) = -\frac{GM}{r(t)^2}\mathbf{R}(t).$$

Therefore

$$\mathbf{r}''(t) \times \mathbf{c} = -\frac{GM}{r(t)^2}\mathbf{R}(t) \times \mathbf{c} \overset{(12.27)}{=} -\frac{GM}{r(t)^2}\mathbf{R}(t) \times (\mathbf{r}(t) \times \mathbf{r}'(t)) = -\frac{GM}{r(t)}\mathbf{R}(t) \times (\mathbf{R}(t) \times \mathbf{r}'(t)).$$

Now

$$\mathbf{r}'(t) = \frac{d}{dt}(r(t)\mathbf{R}(t)) = r'(t)\mathbf{R}(t) + r(t)\mathbf{r}'(t).$$

Therefore

$$\mathbf{r}''(t) \times \mathbf{c} = -\frac{GM}{r}\mathbf{R}(t) \times (\mathbf{R}(t) \times (r'(t)\mathbf{R}(t) + r(t)\mathbf{r}'(t))) = -GM\,\mathbf{R}(t) \times (\mathbf{R}(t) \times \mathbf{r}'(t)). \tag{12.31}$$

According to formula (11.29), the vector triple product in equation (12.31) may be written as

$$\mathbf{R}(t) \times (\mathbf{R}(t) \times \mathbf{R}'(t)) = (\mathbf{R}(t) \cdot \mathbf{R}'(t))\mathbf{R}(t) - (\mathbf{R}(t) \cdot \mathbf{R}(t))\mathbf{R}'(t).$$

Because $\mathbf{R}(t) \cdot \mathbf{R}'(t) = 0$ and $\mathbf{R}(t) \cdot \mathbf{R}(t) = 1$, we have the simplification $\mathbf{R}(t) \times (\mathbf{R}(t) \times \mathbf{R}'(t)) = -\mathbf{R}'(t)$. If we substitute this result into equation (12.31), then we obtain

$$\mathbf{r}''(t) \times \mathbf{c} = GM\,\mathbf{R}'(t).$$

By integrating each side of this last equation, we see that

$$\mathbf{r}'(t) \times \mathbf{c} = GM\,\mathbf{R}(t) + \mathbf{h}$$

for some (constant) vector $\mathbf{h}$. Taking the dot product of each side with $\mathbf{r}(t)$ results in

$$\mathbf{r}(t) \cdot (\mathbf{r}'(t) \times \mathbf{c}) = GM\,\mathbf{r}(t) \cdot \mathbf{R}(t) + \mathbf{r}(t) \cdot \mathbf{h} = GM\,r(t) + \mathbf{r}(t) \cdot \mathbf{h} = GM\,r(t) + r(t)\|\mathbf{h}\|\cos(\theta(t))$$

where $\theta(t)$ is the angle between $\mathbf{r}(t)$ and $\mathbf{h}$. After we set $e = \|\,\mathbf{h}\,\|/GM$, our equation becomes

$$\mathbf{r}(t) \cdot (\mathbf{r}'(t) \times \mathbf{c}) = GMr(t)(1 + e\cos(\theta(t))).$$

According to the triple product formula (11.25), however,

$$\mathbf{r}(t) \cdot (\mathbf{r}'(t) \times \mathbf{c}) = (\mathbf{r}(t) \times \mathbf{r}'(t)) \cdot \mathbf{c} = \mathbf{c} \cdot \mathbf{c} = \|\,\mathbf{c}\,\|^2.$$

Thus

$$\|\,\mathbf{c}\,\|^2 = GMr(t)(1 + e\cos(\theta(t))).$$

Observe that the left side of this equation is positive. We conclude that $e < 1$ and

$$r(t) = \frac{\|\,\mathbf{c}\,\|^2/GM}{1 + e\cos(\theta(t))}.$$

Example 6 shows that this equation describes an ellipse, completing the derivation of Kepler's First Law.

Kepler's Second Law about the area function $A(t)$ is a special case of Theorem 4.

The proof of Kepler's Third Law begins with an observation from Figure 9. The length $2a$ of the major axis of the elliptical orbit is equal to the maximum value of $r(t)$

plus the minimum value of $r(t)$. From the equation for the ellipse, we see that these values occur when $\cos(\theta(t)) = -1$ and $\cos(\theta(t)) = 1$, respectively. Thus

$$2a = \frac{\|\mathbf{c}\|^2/GM}{1-e} + \frac{\|\mathbf{c}\|^2/GM}{1+e} = \frac{2\|\mathbf{c}\|^2}{GM(1-e^2)}.$$

We conclude that

$$\|\mathbf{c}\| = \sqrt{aGM(1-e^2)}.$$

Recall from the proof of Theorem 4 the area function A satisfies $A'(t) = \|\mathbf{c}\|/2$. By antidifferentiating, we find that $A(t) = \|\mathbf{c}\|t/2$. (The constant of integration is 0 because $A(0) = 0$.) It follows that

$$A(t) = \frac{t}{2}\sqrt{aGM(1-e^2)}.$$

Let T be the time it takes to sweep out one orbit. In other words, T is the year for the particular planet. Because the area inside an ellipse with major axis $2a$ and eccentricity e is $\pi a^2\sqrt{1-e^2}$ (as outlined in Exercise 40), we have

$$\pi a^2\sqrt{1-e^2} = A(T) = \frac{T}{2}\sqrt{aGM(1-e^2)}.$$

Solving for T, we obtain

$$T = \frac{2\pi}{\sqrt{GM}}a^{3/2},$$

or

$$\frac{T^2}{a^3} = \frac{4\pi^2}{GM}.$$

This is Kepler's Third Law.

quickquiz

1. How does the acceleration vector of a curve decompose into normal and tangential components?
2. What is centripetal force?
3. What do the laws of Kepler tell us about the rate at which an orbiting body sweeps out area?
4. How are the period of a planet's revolution and the major axis of its orbit related?

EXERCISES

Problems for Practice

In Exercises 1–8, the position vector $\mathbf{r}$ of a particle is given. Calculate $\mathbf{r}'(t)$, $\mathbf{r}''(t)$, $\mathbf{T}(t)$, $\mathbf{N}(t)$, $\kappa_{\mathbf{r}}(t)$, and the tangential and normal components, a_T and a_N, of acceleration for the planar motion.

1. $\mathbf{r}(t) = t\mathbf{i} - t^2\mathbf{k}$
2. $\mathbf{r}(t) = t^2\mathbf{i} - t^3\mathbf{j}$
3. $\mathbf{r}(t) = (t - t^2)\mathbf{i} + (t + t^2)\mathbf{j}$
4. $\mathbf{r}(t) = \ln(t)\mathbf{i} + t^{1/2}\mathbf{j}$
5. $\mathbf{r}(t) = (t - \sin(t))\mathbf{i} + (1 - \cos(t))\mathbf{j}$
6. $\mathbf{r}(t) = 5\cos(3t + \pi/6)\mathbf{i} + 5\sin(3t + \pi/6)\mathbf{j}$
7. $\mathbf{r}(t) = t^{3/2}\mathbf{i} + t^{1/2}\mathbf{j}$
8. $\mathbf{r}(t) = \cos^3(t)\mathbf{i} + \sin^3(t)\mathbf{j}$

In Exercises 9–16, the position vector $\mathbf{r}$ of a particle is given. Calculate $\mathbf{r}'(t)$, $\mathbf{r}''(t)$, $\mathbf{T}(t)$, $\mathbf{N}(t)$, $\kappa_{\mathbf{r}}(t)$, and the tangential and normal components, a_T and a_N, of acceleration for the motion in space.

9. $\mathbf{r}(t) = t\mathbf{i} + t^2\mathbf{j} + t^3\mathbf{k}$
10. $\mathbf{r}(t) = (t - t^2)\mathbf{i} + (t + t^2)\mathbf{j} + t^2\mathbf{k}$
11. $\mathbf{r}(t) = (1 + t)^{3/2}\mathbf{i} + (1 - t)^{3/2}\mathbf{j} + t\mathbf{k}$
12. $\mathbf{r}(t) = t^{1/2}\mathbf{i} + t^{-1/2}\mathbf{j} + t^{-1}\mathbf{k}$
13. $\mathbf{r}(t) = e^t\mathbf{i} + e^{-t}\mathbf{j} + \sqrt{2}t\mathbf{k}$
14. $\mathbf{r}(t) = e^t\mathbf{i} + te^t\mathbf{j} + 2e^t\mathbf{k}$
15. $\mathbf{r}(t) = \cos(t)\mathbf{i} + \sin(t)\mathbf{j} + t^2\mathbf{k}$
16. $\mathbf{r}(t) = \ln(|t|)\mathbf{i} + t^{-1}\mathbf{j} + t\mathbf{k}$

In Exercises 17–20, the position vector $\mathbf{r}$ of a particle is given. Calculate $\mathbf{a}(t) = \mathbf{r}''(t)$. Calculate the tangential and normal components, a_T and a_N, of the decomposition $\mathbf{a}(t) = a_T\mathbf{T}(t) + a_N\mathbf{N}(t)$ *without calculating* $\mathbf{T}(t)$ and $\mathbf{N}(t)$.

17. $\mathbf{r}(t) = (t + 2)\mathbf{i} + t^2\mathbf{j} + 2t\mathbf{k}$
18. $\mathbf{r}(t) = (t^2 + 1)\mathbf{i} + (t^2 - 1)\mathbf{j} + t^2\mathbf{k}$
19. $\mathbf{r}(t) = e^t\mathbf{i} + e^{-t}\mathbf{j} + t\mathbf{k}$
20. $\mathbf{r}(t) = \cos(t)\mathbf{i} + \sin(t)\mathbf{j} + \ln(\cos(t))\mathbf{k}$

In Exercises 21–30, write the Cartesian equation of the ellipse that is described.

21. center (0, 0), major axis 6, minor axis 4, foci on y-axis
22. foci (0, 3) and (0, −3), major axis 10
23. foci (4, 0) and (−4, 0), minor axis 3
24. foci (5, 0) and (−5, 0), eccentricity 5/13
25. foci (1, 14) and (1, −10), eccentricity 12/13
26. focus (2, 4), center (2, −2), minor axis 8
27. center (1, 2), focus (1, 10), eccentricity 0.8
28. center (1, 2), focus (−5, 2), minor axis 8
29. directrix $x = 6$, focus nearest the directrix (2, 0), eccentricity 1/2
30. directrix $y = 4$, focus nearest the directrix (1, 1), eccentricity 3/5

Further Theory and Practice

31. A particle moves at constant speed with position vector $\mathbf{r}$. If $\mathbf{r}'' \neq \vec{\mathbf{0}}$, then what is the relationship of $\mathbf{r}''$ to $\mathbf{r}'$?
32. A particle moves with position vector $\mathbf{r}$. If $\mathbf{r}' \cdot \mathbf{r}'' = 0$, then what can be said about the speed $\|\mathbf{r}'\|$?
33. Prove Huygens's law of centripetal force for uniform circular motion: If a body of mass m moves with constant speed v along a circle of radius r, then the centripetal force $\mathbf{F}$ acting on the body has magnitude $|\mathbf{F}| = mv^2/r$.
34. A car weighing 3000 lb races at a constant speed of 100 mi/h around a circular track with radius 1 mi. Calculate the centripetal force, in pounds, needed to hold the car on the track.
35. An artificial satellite orbits Earth in an elliptical orbit with major axis 1850 km. The orbit of Earth's moon has semimajor axis 384,400 km and period 27.322 days. Calculate the time for one orbit of the artificial satellite.
36. It is known that the eccentricity of Earth's orbit about the sun is about 0.0167 and the semiminor axis of the elliptical orbit is about 92,943,235 mi. Calculate the semimajor axis. Of course, it is also known that the length of time for one orbit is one Earth year (by definition). Using Kepler's Third Law and this other information, calculate GM where M is the sun's mass.
37. Earth's mass is known to be 5.976×10^{27} and the gravitational constant G equals 6.637×10^{-8} cm^3/(g · s^2). If an artificial satellite orbits Earth, with each orbit taking 15 h, then what will be the length of the major axis of the elliptical orbit?
38. The sun's mass is 2×10^{33} g and the gravitational constant is given in Exercise 37. Use Kepler's Third Law to calculate the length, in kilometers, of the semimajor axis of Earth's orbit around the sun.

39. An artificial satellite orbits Earth. It is known that at a certain moment in its orbit, the satellite's distance from Earth's center is 4310 mi and its speed is 900 mi/h. Moreover, astronomical observations determine that at that moment, the angle between the position vector and the velocity vector for this satellite is $\pi/3$. Give a numerical value for $\frac{dA}{dt}$.

40. Show that the area inside an ellipse with major axis $2a$ and minor axis $2b$ is πab. Show that this is equal to $\pi a^2(1-e^2)^{1/2}$ where e is the eccentricity of the ellipse.

Earth's semimajor axis is 199,597,887 km and its year is 365.256 days. Use this information as needed in Exercises 41–44.

41. The eccentricity of Mercury's orbit is 0.2056. Its year is 87.97 Earth days. Calculate mercury's perihelion distance to the sun.

42. Pluto's orbit of the sun lasts 248.08 Earth years. Its perihelion distance to the sun is 4,436,824,613 km. Determine the eccentricity of the orbit.

43. The eccentricity of Jupiter's orbit is 0.04839. At its farthest, Jupiter is 816,041,455 km from the sun. Determine the period of Jupiter's orbit.

44. Saturn's nearest and farthest distances to the sun are 1,349,467,375 km and 1,503,983,449 km, respectively. Determine the duration of its orbit.

In Exercises 45–50, an alternative derivation of elliptic planetary motion is outlined. This approach may be used once it is shown that the motion is planar, which we assume in these exercises. We also suppose that Kepler's Second Law has been proved. Imagine that the sun is at the origin of the plane of motion and that the x-axis passes through the aphelion of the planet, as in Figure 10. Let $r = r(t)$ denote the distance of the planet to the sun. Let θ be the angle between the planet's position vector and the x-axis. This angle is called the *true anomaly* in astronomy. As we will see in Chapter 14, Kepler's Second Law tells us that $r^2\dot{\theta} = C$ where C is a constant and $\dot{\theta}$ is the derivative of the true anomaly with respect to time. Let M and m denote the masses of the sun and planet, respectively, and let G denote the gravitational constant.

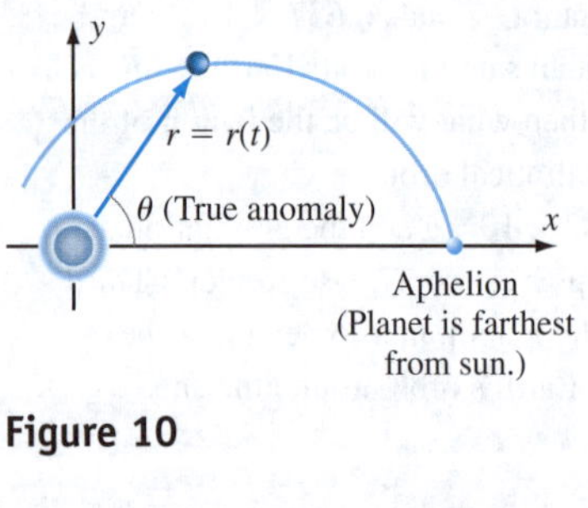

Figure 10

45. After resolving gravitational acceleration into the x- and y-directions, use the Chain Rule and the equation $r^2\dot{\theta} = C$ to show that

$$\frac{d\dot{x}}{d\theta} = -\frac{GM}{C}\cos(\theta)$$

and

$$\frac{d\dot{y}}{d\theta} = -\frac{GM}{C}\sin(\theta).$$

46. The hodograph of a motion $t \mapsto \langle x(t),\ y(t)\rangle$ is the parameterized velocity curve $t \mapsto \langle \dot{x}(t),\ \dot{y}(t)\rangle$. Integrate the equations of Exercise 45 to show that

$$\dot{x} = -\frac{GM}{C}\sin(\theta) + a$$

and

$$\dot{y} = \frac{GM}{C}\cos(\theta) + b$$

for some constants a and b. Deduce that the hodograph of planetary motion is a circle.

47. Substitute $x = r\cos(\theta)$ and $y = r\sin(\theta)$ into the equations of Exercise 46. Eliminate $\dot{r}$ from the resulting equations to obtain

$$r\cdot\dot{\theta} = \frac{GM}{C} - a\sin(\theta) + b\cos(\theta).$$

Deduce that

$$r = \frac{1}{GM/C^2 - (a/C)\sin(\theta) + (b/C)\cos(\theta)}.$$

48. Use the Addition Formula for the cosine to show that

$$r = \frac{ed}{1 + e\cos(\theta)}$$

for some constants e and d.

49. Show that

$$\frac{d\dot{x}}{dt} = -\frac{GM}{r^3}x$$

and

$$\frac{d\dot{y}}{dt} = -\frac{GM}{r^3}y.$$

Then show that

$$\frac{d}{dt}(\dot{x})^2 = -\frac{GM}{r^3}\frac{d}{dt}x^2$$

and

$$\frac{d}{dt}(\dot{y})^2 = -\frac{GM}{r^3}\frac{d}{dt}y^2.$$

50. Use the equations of Exercise 49 to show that there is a constant E such that

$$\frac{1}{2}m(\dot{x}^2 + \dot{y}^2) = \frac{GMm}{r} + E.$$

In Chapter 15, we will learn how to interpret this as a conservation of total energy principle.

51. Suppose that the acceleration $\mathbf{a}(t)$ of a smooth trajectory is differentiable. Show that if $\| \mathbf{a}(t_0)\| = a_N(t_0) = A$, with $A \neq 0$, then the graphs of $t \mapsto \| \mathbf{a}(t)\|$ and $t \mapsto a_N(t)$ are tangent to each other at the point (t_0, A).

Calculator/Computer Exercises

In Exercises 52–57, the trajectory $t \mapsto \mathbf{r}(t)$ of a moving particle is given. Plot $t \mapsto \| \mathbf{a}(t)\|$, $t \mapsto a_T(t)$, and $t \mapsto a_N(t)$ in the specified viewing window W before answering the questions about the acceleration.

52. $\mathbf{r}(t) = t^4\mathbf{i} + (2+t)\mathbf{j} + t^2\mathbf{k}$, $W = [0, 1] \times [0, 12.2]$
At what point on the curve defined by $\mathbf{r}$ are the normal and tangential components of acceleration equal? At what point does the normal component of acceleration have a local minimum?

53. $\mathbf{r}(t) = t^3\mathbf{i} + \exp(t)\mathbf{j} + t^2\mathbf{k}$, $W = [-1, 1] \times [-1, 5]$
At what point on the curve defined by $\mathbf{r}$ are the normal and tangential components of acceleration equal? At what point do the graphs of normal acceleration and magnitude of acceleration touch? What condition must be satisfied at such a point of contact?

54. $\mathbf{r}(t) = 1/(1+t^2)\mathbf{i} + t/(1+t^2)\mathbf{j} + t^3\mathbf{k}$,
$W = [-1, 1] \times [-6, 6]$
Over what time intervals is the speed decreasing? At what point do the graphs of normal acceleration and magnitude of acceleration touch? What condition must be satisfied at such a point of contact?

55. $\mathbf{r}(t) = \cos(t)\mathbf{i} + \sin(2t)\mathbf{j} + (\cos(3t) + \sin(5t))\mathbf{k}$,
$W = [1/4, 1] \times [-30, 35]$
On what interval is the particle slowing down? On what interval are both components of acceleration decreasing?

56. $\mathbf{r}(t) = (t + \cos(t))\mathbf{i} + (t - \sin(t))\mathbf{j} + t\mathbf{k}$,
$W = [0, \pi] \times [-3/4, 1]$
On what interval is tangential acceleration greater than normal acceleration? Find the points at which $|a_T|$ has a local extremum. Show that a_N has a local extremum at each of these points. Explain why this behavior happens for $\mathbf{r}$ but cannot be expected to happen for other trajectories.

57. $\mathbf{r}(t) = \exp(\sin(t))\mathbf{i} + \exp(\cos(t))\mathbf{j} + \ln(2 + \cos(t))\mathbf{k}$,
$W = [0, 2\pi] \times [-7/4, 3]$
Where does the absolute maximum of normal acceleration occur? Where does the absolute maximum value A of the magnitude of acceleration occur? (The quantities a_T, a_N, and A are required for Exercise 58.)

58. Let $a_T(t)$, $a_N(t)$, and A be as in Exercise 57. Plot the planar curve $t \mapsto a_T(t)\mathbf{i} + a_N(t)\mathbf{j}$, $0 < t < 2\pi$. To your plot, add the semicircle of radius A that is centered at the origin and lies in the upper half plane. Explain the relationship between the two plotted curves.

59. Calculate the tangential component $a_T(t)$ and normal component $a_N(t)$ of acceleration for the space curve $\mathbf{r}(t) = 1/(1+t^2)\mathbf{i} + t/(1+t^2)\mathbf{j} + t^3\mathbf{k}$ (the curve of Exercise 54). Plot the planar curve $t \mapsto a_T(t)\mathbf{i} + a_N(t)\mathbf{j}$ in the viewing window $[-0.2, 0.2] \times [1.9, 2.7]$. Explain how the loop that you see pertains to the particle's movement. Add vertical tangent lines to the plot and explain how they are obtained.

60. Calculate the tangential component $a_T(t)$ and normal component $a_N(t)$ of acceleration for the space curve $\mathbf{r}(t) = t^2\mathbf{i} + 1/(1+t^2)\mathbf{j} + (t+t^2)\mathbf{k}$. Plot the planar curve $t \mapsto a_T(t)\mathbf{i} + a_N(t)\mathbf{j}$, $3/8 < t < 9/8$. Find the two values t_0 and t_1 for which $a_T(t_0) = a_T(t_1)$ and $a_N(t_0) = a_N(t_1)$.

Summary of Key Topics

Vector-Valued Functions (Section 12.1)

A vector-valued function (in space) of a real variable has the form $\mathbf{r}(t) = r_1(t)\mathbf{i} + r_2(t)\mathbf{j} + r_3(t)\mathbf{k}$. For each t, we think of $\mathbf{r}(t)$ as the position vector of a point in space. We represent $\mathbf{r}(t)$ by a directed line segment with the origin as its initial point. As t runs through the domain of $\mathbf{r}$, the endpoints of the directed line segments representing

$\mathbf{r}(t)$ form a curve $\mathcal{C}$ in space. When t represents time, we can think of the curve $\mathcal{C}$ as the path of a particle moving through space. We say that $\mathbf{r}$ is a *parameterization* of $\mathcal{C}$.

If $\mathbf{L}$ is a vector, then we say that

$$\lim_{t \to c} \mathbf{r}(t) = \mathbf{L}$$

if for any $\epsilon > 0$ there is a $\delta > 0$ such that

$$0 < |t - c| < \delta \quad \text{implies} \quad \|\mathbf{r}(t) - \mathbf{L}\| < \epsilon.$$

We say that $\mathbf{r}$ is continuous at $t = c$ if c is in the domain of $\mathbf{r}$ and

$$\lim_{t \to c} \mathbf{r}(t) = \mathbf{r}(c).$$

We define the derivative of $\mathbf{r}$ at $t = c$ to be

$$\mathbf{r}'(t) = \lim_{\Delta t \to 0} \frac{1}{\Delta t}(\mathbf{r}(c + \Delta t) - \mathbf{r}(c)),$$

provided that the limit exists.

A vector-valued function is continuous at c if and only if all component functions r_j are continuous at c. It is differentiable at c if and only if all component functions r_j are differentiable at c.

Properties of Limits (Section 12.1)

Limits of vector-valued functions satisfy the following familiar properties: If $\mathbf{f}(t)$ and $\mathbf{g}(t)$ are vector-valued functions and if $\lim_{t \to c} \mathbf{f}$ and $\lim_{t \to c} \mathbf{g}$ exist, then

$$\begin{aligned}
&\lim_{t \to c}(\mathbf{f} \pm \mathbf{g})(t) = \lim_{t \to c} \mathbf{f}(t) \pm \lim_{t \to c} \mathbf{g}(t),\\
&\lim_{t \to c}(\mathbf{f} \cdot \mathbf{g})(t) = (\lim_{t \to c} \mathbf{f}(t)) \cdot (\lim_{t \to c} \mathbf{g}(t)),\\
&\lim_{t \to c}(\lambda \mathbf{f}(t)) = \lambda(\lim_{t \to c} \mathbf{f}(t)) \text{ for any constant } \lambda,\\
&\lim_{t \to c}(\mathbf{f} \times \mathbf{g})(t) = \lim_{t \to c} \mathbf{f}(t) \times \lim_{t \to c} \mathbf{g}(t)
\end{aligned}$$

It follows that the set of continuous functions is closed under addition and under multiplication by scalars.

If $\mathbf{f}$ and $\mathbf{g}$ are vector-valued functions that are continuous at c, then the scalar-valued function $\mathbf{f} \cdot \mathbf{g}$ and the vector-valued function $\mathbf{f} \times \mathbf{g}$ will also be continuous at c. If $\mathbf{f}$ and $\mathbf{g}$ are differentiable at c, then $\mathbf{f} \cdot \mathbf{g}$ and $\mathbf{f} \times \mathbf{g}$ will also be differentiable at c. We have the product rules

$$(\mathbf{f} \cdot \mathbf{g})'(t) = \mathbf{f}'(t) \cdot \mathbf{g}(t) + \mathbf{f}(t) \cdot \mathbf{g}'(t)$$

and

$$(\mathbf{f} \times \mathbf{g})'(t) = (\mathbf{f}'(t) \times \mathbf{g}(t)) + (\mathbf{f}(t) \times \mathbf{g}'(t)).$$

There is also a sum-difference rule for differentiation and a rule for multiplication by scalars. If ϕ is a scalar-valued differentiable function, then $\phi\mathbf{f}$ makes sense, is differentiable, and

$$(\phi \mathbf{f})'(t) = \phi'(t)\mathbf{f}(t) + \phi(t)\mathbf{f}'(t).$$

Finally, if ψ is a differentiable scalar-valued function and if $\mathbf{r} \circ \psi$ makes sense, then $\mathbf{r} \circ \psi$ is differentiable and

$$(\mathbf{r} \circ \psi)'(t) = \psi'(t)\mathbf{r}'(\psi(t)).$$

Velocity and Acceleration (Section 12.2)

If $\mathbf{r}$ is a differentiable, vector-valued function of a real variable and if we think of $\mathbf{r}$ as describing the motion of a body through space, then $\mathbf{r}'(t)$ is *velocity* at time t and $\mathbf{r}''(t)$ is *acceleration* at time t. Speed is a scalar-valued quantity and is given by

$$v(t) = \| \mathbf{r}'(t) \|.$$

Tangent Vectors and Arc Length (Section 12.3)

If $\mathbf{r}'(t)$ and $\mathbf{r}''(t)$ exist, if $\mathbf{r}''(t)$ is continuous, and if $\mathbf{r}'(t)$ is never the zero vector, then we say that $\mathbf{r}$ is a *smooth parameterization* of the curve it defines.

If $\mathbf{r}$ is a smooth parameterization of a curve $\mathcal{C}$, then the vector $\mathbf{r}'(t)$ can be interpreted as the tangent vector to $\mathcal{C}$ at the point $\mathbf{r}(t)$. The *unit tangent vector* is then

$$\mathbf{T}(t) = \frac{1}{\|\mathbf{r}'(t)\|}\mathbf{r}'(t).$$

The tangent line to $\mathcal{C}$ at the point $\mathbf{r}(t)$ is the line passing through the point $\mathbf{r}(t)$ with direction $\mathbf{T}(t)$.

If $t \mapsto \mathbf{r}(t), a \leq t \leq b$ is a smooth parameterization of a curve $\mathcal{C}$, then the arc length of $\mathcal{C}$ is

$$\int_a^b \| \mathbf{r}'(\tau) \| \, d\tau.$$

The arc length of $\mathcal{C}$ does not depend on the choice of parameterization.

A vector function $s \mapsto \mathbf{r}(s), 0 \leq s \leq L$ is said to *parameterize* a curve $\mathcal{C}$ *with respect to arc length* if $\mathbf{r}$ is a parameterization of $\mathcal{C}$ such that the arc length from $\mathbf{r}(0)$ to $\mathbf{r}(s)$ is s units of length. A smooth parameterization $t \mapsto \mathbf{r}(s), 0 \leq s \leq L$ of $\mathcal{C}$ is the arc length parameterization of $\mathcal{C}$ if and only if $\| \mathbf{r}'(s) \| = 1$ for all s. Otherwise, the reparameterization $\mathbf{p}(s) = \mathbf{r}(\sigma^{-1}(s))$ where

$$\sigma(t) = \int_a^t \| \mathbf{r}'(\tau) \| \, d\tau$$

is the arc length parameterization.

Curvature (Section 12.4)

If $\mathcal{C}$ is a curve that is parameterized according to arc length by $\mathbf{r}$, then the *curvature* of $\mathcal{C}$ at $\mathbf{r}(s)$ is defined to be

$$\kappa(s) = \|\mathbf{T}'(s)\| = \|\mathbf{r}''(s)\|.$$

The vector $\mathbf{r}''(s)$ is always perpendicular to $\mathbf{T}(s)$. We define the principal unit normal to $\mathbf{r}$ at t to be

$$\mathbf{N}(s) = \frac{1}{\|\mathbf{r}''(s)\|}\mathbf{r}''(s).$$

We have

$$\frac{d^2\mathbf{r}}{ds^2} = \kappa(s)\mathbf{N}(s).$$

The *radius of curvature* at $\mathbf{r}(s)$ is defined to be $\rho(s) = 1/\kappa(s)$. The *osculating plane* at $\mathbf{r}(s)$ is the plane through $\mathbf{r}(s)$ that contains the vectors $\mathbf{T}$ and $\mathbf{N}$. The *osculating circle* (or *circle of curvature*) for the curve at the point $\mathbf{r}(s)$ is the unique circle with center $\mathbf{r}(s) + \rho(s)\mathbf{N}(s)$ and radius $\rho(s)$ and that lies in the osculating plane.

If $\mathbf{r}$ is *any* smooth parameterization of a curve, then the curvature at $\mathbf{r}(t)$ is given by

$$\kappa_{\mathbf{r}}(t) = \frac{\| \mathbf{r}'(t) \times \mathbf{r}''(t) \|}{\|\mathbf{r}'(t)\|^3}.$$

For a planar curve that is parameterized by $\mathbf{r}(t) = x(t)\mathbf{i} + y(t)\mathbf{j}$, the curvature at $\mathbf{r}(t)$ is given by

$$\kappa_{\mathbf{r}}(t) = \frac{|x'(t)y''(t) - y'(t)x''(t)|}{(x'(t)^2 + y'(t)^2)^{3/2}}.$$

If the planar curve is the graph of the equation $y = f(x)$, then the curvature at $(x, f(x))$ is

$$\kappa(x) = \frac{|f''(x)|}{(1 + (f'(x))^2)^{3/2}}.$$

Tangential and Normal Components of Acceleration (Section 12.5)

If $\mathbf{r}(t)$ represents a motion through space, then the acceleration vector $\mathbf{r}''(t)$ can be decomposed as

$$\mathbf{r}''(t) = \left(\frac{d}{dt}v(t)\right)\mathbf{T}(t) + v(t)^2\kappa_{\mathbf{r}}(t)\mathbf{N}(t),$$

where $v(t) = \| \mathbf{r}'(t)\|$ is the speed of the motion.

Kepler's Three Laws of Planetary Motion (Section 12.5)

A planet travels in an elliptical orbit with the sun at one focus. The rate at which area is swept out by the ray from the sun to the moving planet is constant. The square of the period of revolution is proportional to the cube of the major axis of the elliptical orbit, with constant of proportionality independent of the particular planet.

genesis & DEVELOPMENT

Tycho Brahe

On the night of November 11, 1572, Tycho Brahe, a young Danish nobleman with a hobbyist's interest in astronomy, cast his eyes toward the constellation Cassiopeia. To his astonishment, he sighted a new star, one that was much brighter than any other. Tycho was well aware that the appearance of a new star in the firmament, a *nova* as he called it, was extraordinary. He characterized his discovery as "the greatest wonder that has ever shown itself in the whole of nature since the beginning of the world." At the very least, Tycho understood that the spectacle he witnessed contradicted the ancient astronomy of Aristotle.

We now know that Tycho observed a cataclysmic explosion called a type I supernova. This celestial event is the final evolutionary stage of a white dwarf star. It occurs when the core of the star collapses, releasing an enormous quantity of energy that blows the star to bits. In our galaxy, such supernovae occur very infrequently—only three have ever been observed. The first was recorded in 1054 CE by astronomers in China, Korea, and Japan. Tycho observed the second.

Tycho Brahe was born in his family's ancestral seat, Knutstorp Castle, in 1546. At the age of two, he was abducted by a paternal uncle and childless aunt, who brought him up. Tycho studied at the University of Copenhagen and at several renowned universities in the Germanic territories. It was at Basel that he learned the new theories of Copernicus, which he did not entirely accept. While he was a student in Rostok, Tycho engaged in a quarrel that escalated into a sword fight. During the battle, Tycho received a blow that left a diagonal slash across his forehead and hacked off a substantial chunk of his nose. (A nasal prosthetic can be seen in several subsequent portraits.)

After his return to the kingdom of Denmark, Tycho's social standing afforded him the leisure and wealth to pursue his interest in astronomy. In time, the king of Denmark, Frederick II, offered Tycho a choice of several fiefdoms. Tycho demurred. He wrote to a friend: "I am displeased with society here. Among people of my own class I waste much time." While staying in the castle made famous by Shakespeare's Hamlet, Frederick conceived a way of retaining Denmark's leading scholar. From his window, Frederick spotted the island of Hven. He granted the island to Tycho and awarded him a pension so that Tycho could build and operate an observatory there. Tycho took possession of the island in 1576. For the next 21 years, Tycho watched and recorded the planets.

Tycho's fortune took a turn for the worse when his royal patron died. Frederick's son and successor, Christian IV, was not inclined to continue the pension that his father had granted Tycho. Nor would he agree to honor his father's pledge to allow Tycho's children to inherit Hven. (Because Tycho had married a commoner, his children did not have automatic rights of inheritance.) Tycho abandoned Hven for Copenhagen, where he started over. When Christian had Tycho dismantle his new observatory because it obstructed the view from the royal palace, Tycho departed Denmark for good.

Johannes Kepler

When Tycho began his exile in 1597, Johannes Kepler was a 26-year-old teacher in Graz. His teaching duties were light enough that he had time to ponder the mechanics of the solar system. He published his first book on the planets, the *Mysterium Cosmographicum,* in 1596. Though this work is more flight of fancy than science, it reveals Kepler's early interest in the three mathematical questions that he would eventually answer with his laws of planetary motion.

By nature, Kepler was more a mathematician than a stargazer. He did not have the instruments of Tycho, and his eyesight was poor. Tycho, to whom Kepler sent a copy of his book, commented that the observations of Copernicus on which Kepler relied were not sufficiently accurate to be the basis for the conclusions Kepler reached. Moreover, Tycho did not think highly of Kepler's inclination to theorize. Tycho chided Kepler that the force behind the motion of the planets could only be established "a posteriori, after the motions have been definitely established, and not a priori as you would do." Kepler was frustrated: "I did not wish to be discouraged, but to be taught." He was aware that to make progress he needed Tycho's planetary data, data that Tycho did not share with other scholars. The problem was how to gain access. Referring to Tycho's wealth of data, Kepler wrote: "Like most rich men Tycho does not know how to make proper use of his riches. Therefore one must take pains to wring his treasures from him."

As it happened, fate brought the two men together. At the same time that royal hostility was driving Tycho from his native Denmark, religious persecution was driving Kepler from his post in Graz. Tycho made his way to Prague, the capital of the Holy Roman Empire, in 1599. Kepler arrived shortly afterward. Although Tycho maintained a tight grip on his secrets at first, he eventually allowed Kepler to work on the orbit of Mars. Its eccentricity had caused great difficulties for all the circular motion theories that Ptolemy, Copernicus, and Tycho had put forth.

In 1601, Kepler became Tycho's salaried assistant. Only a few days later, Tycho attended a banquet where, according to Kepler's account, he "drank a little overgenerously and experienced pressure on his bladder. He felt less concern for the state of his health than for etiquette," which required guests to remain seated at the dinner table. On the basis of this document, Tycho's death has traditionally been attributed to a burst bladder. Other reports, originating soon after Tycho's death, raised the suspicion of heavy metal poisoning. The modern verdict is that Tycho died of kidney failure brought on by enlargement of the prostate.

Tycho was laid to rest in the famous Tyn Church of Prague. During the religious turmoil of the 1620s, many graves were desecrated. According to legend, Tycho's corpse was removed from its burial site during one of those desecrations. In 1901, the 300th anniversary of his interment, Tycho's crypt was refurbished. That restoration provided an opportunity to investigate the contents of the crypt. It was opened and the grave-robbing story debunked: The male skeleton found there had a cranial wound that was consistent with the injury Tycho suffered during his sword fight. In the late 20th century, samples of Tycho's hair, stored since 1901, were analyzed. The results indicate a very high level of mercury present. There is no evidence, however, of criminal poisoning. It seems likely that Tycho ingested the mercury as a remedy for his urinary difficulties.

Kepler's Laws

Kepler succeeded Tycho in the position of Imperial Mathematician. By compensating Tycho's heirs, the emperor, Rudolf, was able to ensure that Tycho's observations were placed at Kepler's disposal. Those measurements were all Kepler had to work with—there were no established physical theories, such as gravitation, to guide him; no advanced mathematical tools, such as analytic geometry and calculus, to aid in calculation; no tools, such as logarithms, to ease the burden of numerical computation. After four years of intense mathematical labor, Kepler discovered the first two of the planetary laws that now carry his name. He published them in his *Astronomia Nova* of 1609.

For the most part, Kepler left remarkably candid accounts of the steps that precipitated his discoveries. In the case of his third law, however, Kepler was unusually silent. He did record the date, March 8, 1618, when the idea first came to him. His initial efforts to confirm his theory were not successful, but on May 15, 1618, he was able to write: "A fresh assault overcame the darkness of my reason . . . I feel carried away and possessed by an unutterable rapture over the divine spectacle of the heavenly harmony . . . I write a book for the present time, or for posterity. It is all the same to me. It may wait a hundred years for its readers, as God has also waited six thousand years for an onlooker."

Such moments of elation were brief. As one scientist has remarked, Kepler's standard biography has passages that can bring its readers to tears. Poverty, religious persecution, war, smallpox, typhus, plague—Kepler's life was an unrelenting struggle filled with hardship and sorrow. In his 58th year, he had a premonition of death and composed his own epitaph:

> *I used to measure the heavens, now I measure the shadows of the Earth.*
> *Although my mind was heaven-bound, the shadow of my body lies here.*

A few months after penning this distich, Kepler took ill and died. The peace that eluded Kepler in life eluded his mortal remains as well. Scarcely a few years passed before war ravaged the churchyard of St. Peter's, Regensburg, obliterating every trace of Kepler's burial site.

Sir Isaac Newton

The idea of a gravitational force originated in the early 16th century. By the last half of the 17th century, the foremost question of natural philosophy had become, *How could Kepler's laws of planetary motion be derived from a theory of gravitation?* The first step was to deduce the form of the gravity law. This was a problem with which Kepler wrestled unsuccessfully for 30 years. Ironically, his third law became one of the two keys that together were used to unlock the secrets of gravity. Christiaan Huygens (1629–1695) provided the second when, in 1659, he published the law of centrifugal force for uniform circular motion.

If a body of mass m moves with constant speed v along a circle of radius r, then its *centrifugal* force $\mathbf{F}$ is directed away from the center of the circle with magnitude

$$|\mathbf{F}| = \frac{mv^2}{r}.$$

To prevent a planet from flying out of its orbit, the sun must exert a gravitational force $-\mathbf{F}$ that exactly counterbalances the planet's centrifugal force. Now imagine planetary motion in an elliptic orbit with semimajor axis r and nearly 0 eccentricity—an orbit that, for all practical purposes, is a circle of radius r. According to Kepler's Third Law, there is a constant k such that the period T satisfies

$$T^2 = kr^3.$$

On the other hand, a body traveling around a circle of radius r with constant speed v completes one orbit in time

$$T = \frac{2\pi r}{v} \qquad \left(\text{time} = \frac{\text{distance}}{\text{rate}}\right).$$

By substituting this value of T in Kepler's Third Law, we see that $4\pi^2 r^2/v^2 = kr^3$, or $v^2 = 4\pi^2/(kr)$. If we now substitute this formula for v into Huygens's formula for centrifugal force, we find that

$$|\mathbf{F}| = \frac{mv^2}{r} = \frac{4\pi^2 m}{k} \cdot \frac{1}{r^2}.$$

Although the fall of an apple caused Newton to reflect on the nature of gravity, it was by these considerations that he deduced, in 1666, the law of gravity. Independently, Sir Edmond Halley, Sir Christopher Wren, and Robert Hooke came to the same conclusion several years later.

The next step was to show that the inverse square law *implied* Kepler's laws. During a visit to Cambridge in 1684, Halley asked Newton if he knew the curve that an inverse square law would entail. Without hesitation, Newton answered that it would be an ellipse. Three months later, he sent Halley a short manuscript that derived the three laws of Kepler from the inverse square law of gravitation. For the next 15 months, Newton went into seclusion, devoting himself to setting out a general science of dynamics that included the universal law of gravitation as well as Kepler's laws. The result was Newton's *Principia,* the most important scientific treatise ever written.

Publication was always an anxious process for Newton. In 1672, he was forced to defend a paper on optics against several criticisms, most notably from Hooke, who managed to question both Newton's conclusions and Newton's priority! Newton found the business tiresome. Shortly thereafter, he declined to publish a more complete treatment of optics, remarking that with further use of the press, "I shall not enjoy my former serene liberty." Many years later, when Newton looked back on the ensuing period of silence, he explained, "I began for the sake of a quiet life to decline correspondencies by Letters about Mathematical & Philosophical matters finding [that they] tend to disputes and controversies."

The *Principia* interrupted Newton's quiet life, embroiling him in a new dispute with his old nemesis, Hooke. As the *Principia* neared completion, Hooke began to stir things up. Halley, who was overseeing the publication of the *Principia,* communicated the problem to Newton: "Mr Hook has some pretensions upon the invention of ye rule of the decrease of Gravity ... He sais you had the notion from him ... Mr Hook seems to expect you should make some mention of him in the preface." Newton was aggravated: "Philosophy is such an impertinently litigious Lady that a man had as good be engaged in Law suits as have to do with her. I found it so formerly and now I no sooner come near her again but she gives me warning." Newton responded to the baseless charge of plagiarism by deleting some references to Hooke that were already in the manuscript. In private correspondence, he referred to his antagonist as an "ignoramus."

After completing the *Principia,* Newton rapidly lost interest in mathematical research. A number of documents leave no doubt that he suffered a complete mental breakdown in 1693. The cause remains uncertain, but mercury poisoning (resulting from carefree handling of substances in chemical experiments) is a plausible candidate. Samples of Newton's hair, taken from four preserved locks, were analyzed in 1979. The mercury levels, in parts per million, were found to be 7.2, 43, 54, and 197, compared with the normal level of 5 parts per million. Elevated levels of antimony, arsenic, gold, and lead were also present. Naturally, a conclusive diagnosis is not possible 300 years after the illness. What is not in question is that, in the remaining 34 years of his life, Newton's scientific activity was largely confined to polishing the exposition of earlier work. He retired from academic life in 1696, serving at the mint until his death in 1727.

In 1820, the apple tree that set Newton to think about the nature of gravity succumbed to disease and was felled. Scions were taken, and the line lives on, both in England and the United States. The fruit is pear shaped and, it is said, without flavor.

Many unexpected details concerning Newton's life came to light in the 20th century. Newton's interests in alchemy, theology, and biblical chronology had long been known. However, the *extent* of those interests remained hidden until 1936, when a large portion of his estate was put up for auction. Of the volumes in Newton's personal library, 3% and 7% pertained to physics and mathematics, respectively, whereas 8% and 27.5% concerned alchemy and theology. Of Newton's personal papers, about one million words were devoted to scientific subjects compared with two million words on alchemy, theology, and chronology. These revelations led the great economist John Maynard Keynes (1883–1946) to say of Newton, "Not the first of the age of Reason. He was the last of the magicians."

13

Functions of Several Variables

PREVIEW

Many of the quantities that we study in mathematics and other fields depend on two or more variables. For example, the output of a factory depends on the amount of capital allocated to labor and the amount allocated to equipment. The air pressure at a point in the atmosphere depends on the altitude of the point and on the temperature at the point. The height above sea level of a point on Earth's surface depends on the longitude and latitude of the point. Functions of two or more variables are, indeed, commonplace. To analyze these functions, we need to develop appropriate tools of calculus. That is the purpose of this chapter. We will ask and answer the same types of questions that we posed for functions of one variable.

We begin by learning how to plot a function f of two variables: Its graph is a surface in three-dimensional space. At a point at which the function is well-behaved, its graph will have a tangent plane, which is composed of tangent lines. The key mathematical concept for obtaining the tangent planes to the graph of f is the *partial derivative*.

If f is a function of variables x and y, then an increment Δx in x or an increment Δy in y will generally cause a change in the value of f. The numerator of the quotient

$$\frac{f(x_0 + \Delta x, y_0) - f(x_0, y_0)}{\Delta x}$$

represents the change of f as x varies from x_0 to $x_0 + \Delta x$, while the other variable y is held constant at y_0. The denominator Δx represents the change in x. The quotient itself, therefore, represents the *average* rate of change of f as its argument varies from x_0 to $x_0 + \Delta x$, with y being held constant. The limit

$$\frac{\partial f}{\partial x}(x_0, y_0) = \lim_{\Delta x \to 0} \frac{f(x_0 + \Delta x, y_0) - f(x_0, y_0)}{\Delta x}$$

represents the *instantaneous rate of change* of f at (x_0, y_0), with y being held constant. This quantity is called the *partial derivative of f with respect to x*. Similarly, if we allow y to vary while holding x fixed at x_0, then we obtain the *partial derivative of f with respect to y*:

$$\frac{\partial f}{\partial y}(x_0, y_0) = \lim_{\Delta y \to 0} \frac{f(x_0, y_0 + \Delta y) - f(x_0, y_0)}{\Delta y}.$$

As in the one-variable case, the derivative is essential for understanding and answering many of the questions that arise in the analysis of functions of many variables. As an application, we will learn how to use the partial derivatives of f to identify and classify the local maxima and minima of f.

13.1 Functions of Several Variables

The first eleven chapters of this book focus on scalar-valued functions of one variable. When we evaluate such a function f at an element x of its domain, the result, $f(x)$, is a real number. Chapter 12 is concerned with vector-valued functions of one variable. We evaluate such a function $\mathbf{r}$ at an element t of its domain, and the result, $\mathbf{r}(t)$, is a vector in the plane or in space. Notice that both types of functions receive only one input value. In this chapter, we learn about scalar-valued functions that are evaluated at two or more input values.

Definition Suppose that $\mathcal{D}$ is a set of ordered pairs of real numbers and that R is a set of real numbers. We say that f is a *scalar-valued function of two variables* with *domain* $\mathcal{D}$ and *range* R if for every ordered pair (x, y) in $\mathcal{D}$, there is associated a unique real number in R, which we denote by $f(x, y)$. The number $f(x, y)$ is said to be the *image* of the point (x, y) under f. We also say that f is *evaluated* at (x, y) or that f *maps* (x, y) to the value $f(x, y)$. The notation $(x, y) \mapsto f(x, y)$ is often used, with the arrowlike symbol being read as "maps to." A *function of three variables* is similarly defined, the only difference being that it has a set of ordered triples as its domain. The notation $(x, y, z) \mapsto F(x, y, z)$ is used for a function F of three variables.

Example 1 Express the surface area A and volume V of a rectangular box as functions of the side lengths.

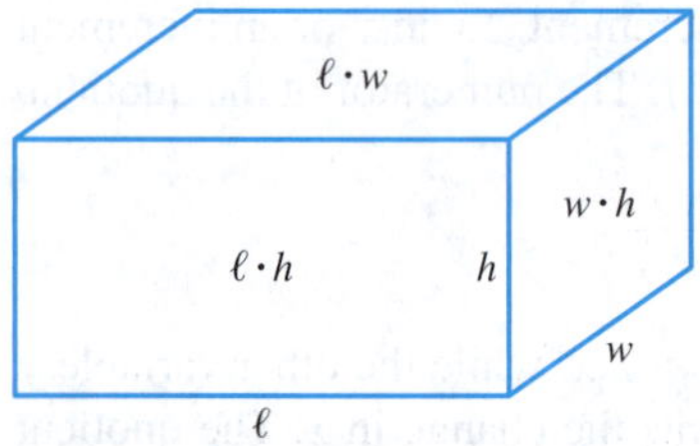

Figure 1

Solution Suppose that the side lengths of the box are ℓ, w, and h. See Figure 1. Then the top and bottom have area $\ell \cdot w$, the left and right sides have area $w \cdot h$, and the front and back have area $\ell \cdot h$. Thus, the surface area is $A(\ell, w, h) = 2\ell \cdot w + 2w \cdot h + 2\ell \cdot h$. The volume is, of course, $V(\ell, w, h) = \ell \cdot w \cdot h$. Each function has for its domain the set of ordered triples (ℓ, w, h), with all three entries positive. ■

It is often convenient to interpret an expression to be a function. When this is done, the domain of the function is taken to be the largest set of points for which the expression makes sense and evaluates to a real number. For example, the expression $\sqrt{25 - (x^2 + y^2)}$ defines the function $(x, y) \mapsto \sqrt{25 - (x^2 + y^2)}$ whose domain consists of all ordered pairs (x, y) with $x^2 + y^2 \le 25$. We employ the same convention when we use an equation such as $f(x, y) = \sqrt{25 - (x^2 + y^2)}$ to define a function without explicitly stating its domain.

inSIGHT

As in the one-variable case, it is convenient to think of a function as an input-output machine (Figure 2). The domain is thought of as the set of input values. Evaluation is the process of getting a unique output value from an input value that is fed into the machine. For example, if $f(x, y) = \sqrt{25 - (x^2 + y^2)}$, then the input-output machine is the function $(x, y) \mapsto \sqrt{25 - (x^2 + y^2)}$. For every input value (x, y) in the domain $\{(x, y) : x^2 + y^2 \leq 25\}$, the unique output is $\sqrt{25 - (x^2 + y^2)}$. For ease of reference, this function has been given the name f. The output value, $\sqrt{25 - (x^2 + y^2)}$, is then compactly denoted by $f(x, y)$.

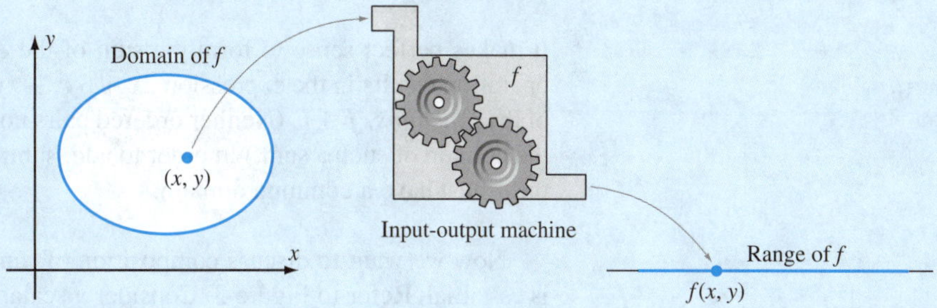

Figure 2

inSIGHT

Observe the distinction between functions g and h. When we define $g(x, y) = a/(x^2 + y^2)$, we regard a as a parameter of g, not a variable. Although a may have any value, its value does not change when we consider $g(x, y)$. Indeed, the *notation* $g(x, y)$ signals to us that a is understood to be constant. On the other hand, the notation $h(x, y, z) = z/(x^2 + y^2)$ tells us that the numerator z of $h(x, y, z)$ is a variable. Even though the values $g(x, y)$ and $h(x, y, a)$ are the same, the functions g and h are quite different.

Example 2 Let a be any constant. Discuss the domains of the functions $f(x, y) = x^2 + y^2$, $g(x, y) = a/(x^2 + y^2)$, and $h(x, y, z) = z/(x^2 + y^2)$.

Solution The expression $x^2 + y^2$ is defined for all values of x and y. Therefore, the domain of f is the set of all pairs (x, y) of real numbers. Whatever the value of a, the expression $a/(x^2 + y^2)$ is defined if x and y are not both 0 and undefined if $x = y = 0$. Therefore, the domain of g is the set of all ordered pairs (x, y) other than $(0, 0)$. For the same reason, the domain of h is the set of all ordered triples (x, y, z) for which x and y are not both 0. The set that is excluded from the domain of h is the z-axis in xyz-space. In other words, the domain of h is the set of all points not on the z-axis. ■

Combining Functions

Much of the elementary material that we learned about functions in Chapter 1 still applies here. For instance, we can add, subtract, multiply, and divide functions of two variables to form new functions:

1. $(\lambda f)(x, y) = \lambda \cdot f(x, y)$ (λ a constant);
2. $(f + g)(x, y) = f(x, y) + g(x, y)$;
3. $(f - g)(x, y) = f(x, y) - g(x, y)$;
4. $(f \cdot g)(x, y) = f(x, y)g(x, y)$;
5. $(f/g)(x, y) = f(x, y)/g(x, y)$, provided $g(x, y) \neq 0$.

There are similar formulas for functions of three variables.

Example 3 Let $f(x, y) = 2x + 3y^2$, $g(x, y) = 5 + x^3 \cdot y$, and $h(x, y, z) = xyz^2$. Compute $(f + g)(1, 2)$, $(f \cdot g)(1, 2)$, and $(f/g)(1, 2)$. Does the expression $f(x, y) + h(x, y, z)$ make sense? What about the function $f + h$?

Solution We have $f(1, 2) = 2(1) + 3(2)^2 = 14$ and $g(1, 2) = 5 + (1)^3 \cdot 2 = 7$. Therefore,

$$(f + 3g)(1, 2) = f(1, 2) + 3g(1, 2) = 14 + 3(7) = 35,$$
$$(f \cdot g)(1, 2) = f(1, 2)g(1, 2) = (14)(7) = 98,$$

and

$$\left(\frac{f}{g}\right)(1, 2) = \frac{f(1, 2)}{g(1, 2)} = \frac{14}{7} = 2.$$

It makes perfect sense to form the sum of the *expressions* $f(x, y) + h(x, y, z)$. This operation results in the expression $2x + 3y^2 + xyz^2$. However, we cannot form the sum of the *functions*, $f + h$. (Neither ordered pairs nor ordered triples would be suitable for the domain of such a sum.) In order to add, subtract, multiply, or divide two functions, they must have a common domain. ∎

Now we want to discuss composition of functions. For this, a schematic diagram is essential. Refer to Figure 3. Consider a scalar-valued function $(x, y) \mapsto f(x, y)$ of two variables. If g is a function of one variable, and if the values of f lie in the domain of g, then we may consider $g(f(x, y))$. We write

$$(g \circ f)(x, y) = g(f(x, y)),$$

and we call this the *composition* of g with f.

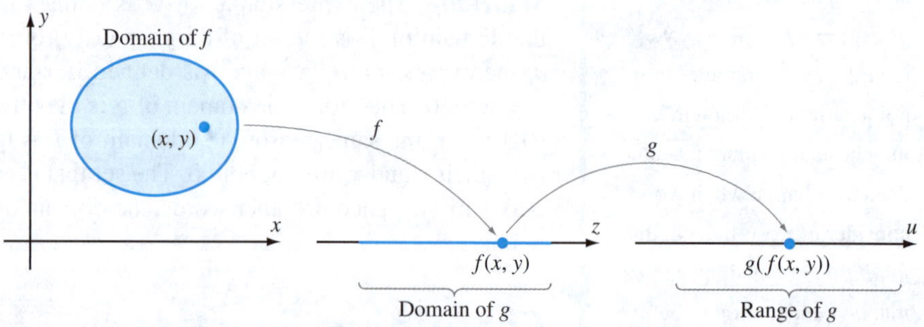

Figure 3
Schematic diagram of $(g \circ f)(x, y) = g(f(x, y))$

Example 4 Let g be defined on the set of all nonnegative numbers by the formula $g(u) = \sqrt{u}$. Suppose that f is defined on the set $\mathcal{D} = \{(x, y) : x^2 + y^2 \leq 25\}$ by the formula $f(x, y) = 25 - (x^2 + y^2)$. Show that the composition $g \circ f$ is defined on $\mathcal{D}$, and calculate $(g \circ f)(2, \sqrt{5})$. If $h(u) = 1/\sqrt{9 - u}$, then what is the largest subset of $\mathcal{D}$ on which $h \circ f$ is defined?

Solution The composition $(g \circ f)(x, y) = g(f(x, y))$ is defined precisely when the values $f(x, y)$ are in the domain of g. In other words, $(g \circ f)(x, y)$ is defined, provided $f(x, y) \geq 0$. This inequality holds if and only if $x^2 + y^2 \leq 25$. Because the coordinates of every ordered pair (x, y) in the domain $\mathcal{D}$ of f satisfies this inequality, the composition $g \circ f$ can be formed. In particular, for (x, y) in $\mathcal{D}$, we have

$$(g \circ f)(x, y) = g(f(x, y)) = \sqrt{f(x, y)} = \sqrt{25 - (x^2 + y^2)}.$$

Since $2^2 + (\sqrt{5})^2 = 9$ is not greater than 25, the point $(2, \sqrt{5})$ is in the domain of f and

$$(g \circ f)(2, \sqrt{5}) = \sqrt{25 - \left(2^2 + (\sqrt{5})^2\right)} = \sqrt{25 - 9} = 4.$$

For

$$(h \circ f)(x, y) = \frac{1}{\sqrt{9 - f(x, y)}} = \frac{1}{\sqrt{9 - (25 - (x^2 + y^2))}} = \frac{1}{\sqrt{(x^2 + y^2) - 16}}$$

to be defined, we must have $x^2 + y^2 > 16$. Therefore, the largest subset of $\mathcal{D}$ on which $h \circ f$ is defined is $\{(x, y) : 16 < x^2 + y^2 \leq 25\}$. This set is sketched in Figure 4. ■

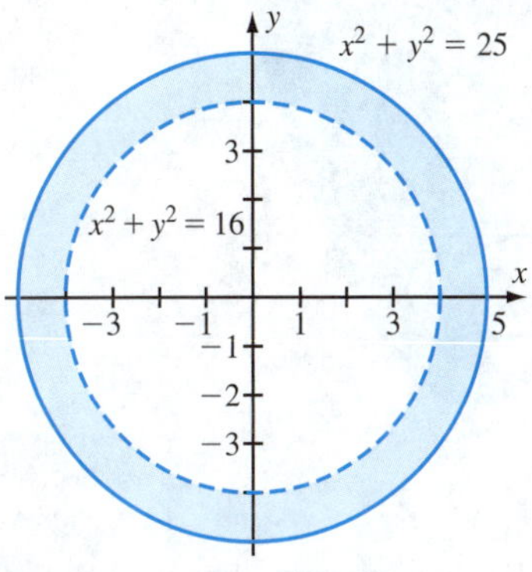

Figure 4
The set $\{(x, y) : 16 < x^2 + y^2 \leq 25\}$

Graphing Functions of Several Variables

When we graph a scalar-valued function of one variable, we use the two-dimensional plane: One dimension is for the domain of the function, and the other is for the range. Now we will learn to graph scalar-valued functions of two variables. This will be done in three-dimensional space, because we need two dimensions for the domain of the function and one dimension for the range. To be specific, the graph of the function $(x, y) \mapsto f(x, y)$ consists of all points (x, y, z) such that (i) (x, y) is in the domain of f, and (ii) $z = f(x, y)$. In other words, we plot a point on the graph of f by locating an element (x, y) of the domain of f in the xy-plane and then, in space, going up or down $f(x, y)$ units.

Some caution is in order here. In spite of all the sophisticated techniques that we know for graphing functions of one variable, we are often tempted to simply plot points and connect them with a curve. Sometimes we can get away with this, because curves are often fairly simple. But now we are working in three dimensions. Our graph will be a *surface,* and it is virtually impossible to plot some points by hand and "connect" them with a surface.

In fact, we need a new idea to graph functions of two variables. The idea is that a surface in three-dimensional space can be described by its two-dimensional slices. For instance, suppose that every horizontal slice of a surface is a circle of radius 1 with center lying on the z-axis (Figure 5). The only thing that this surface could be is a cylinder (Figure 6). On the other hand, suppose that every horizontal slice of a surface is a circle with center lying on the z-axis but the radius of the circle increases with the height. You may have a mental picture of the object in Figure 7a (next page) or perhaps the one in Figure 7b (there are other possibilities as well). We need to understand how to use horizontal slices effectively.

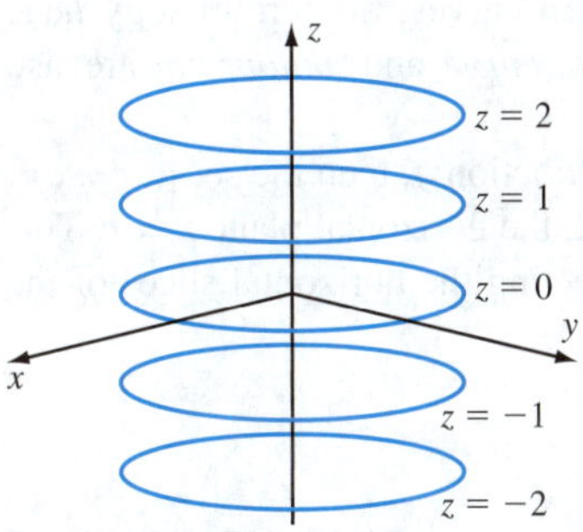

Figure 5
Horizontal slices of a surface $z = f(x, y)$

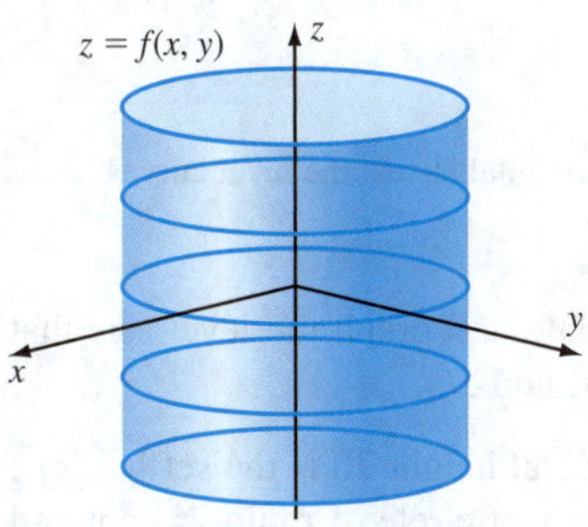

Figure 6

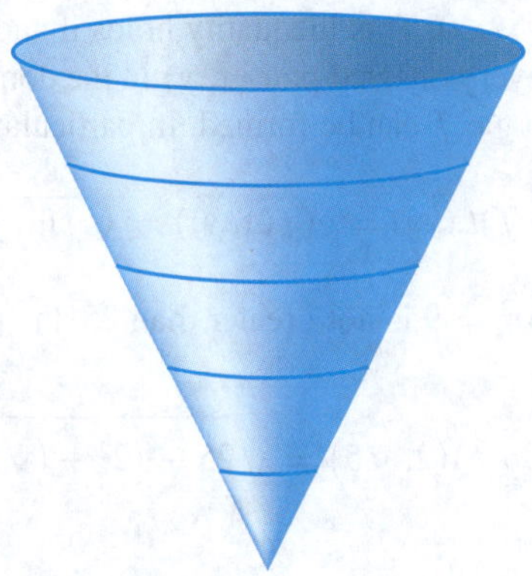

Figure 7a

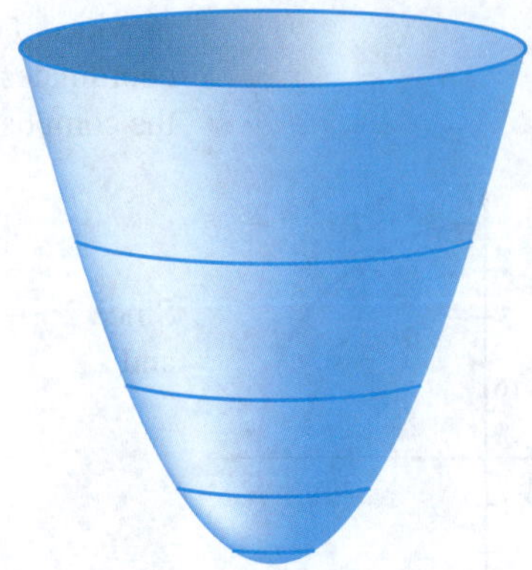

Figure 7b

Definition Let $(x, y) \mapsto f(x, y)$ be a function of two variables. If c is a constant, then we call the set $L_c = \{(x, y) : f(x, y) = c\}$ a *level set* of f. Notice that a level set of a function of x and y is a subset of the xy-plane.

Level sets of functions of two variables are often curves; the terminology *level curve* is used in such cases. The terms *contour, contour curve,* and *contour line* are also used.

If $L_c = \{(x, y) : f(x, y) = c\}$ is a level set of a function f, then the set $\{(x, y, c) : (x, y) \in L_c\}$ is the intersection of the graph of f with the horizontal plane $z = c$. This means that we obtain the level curves of f by projecting the horizontal slices of the graph of f into the xy-plane. See Figure 8.

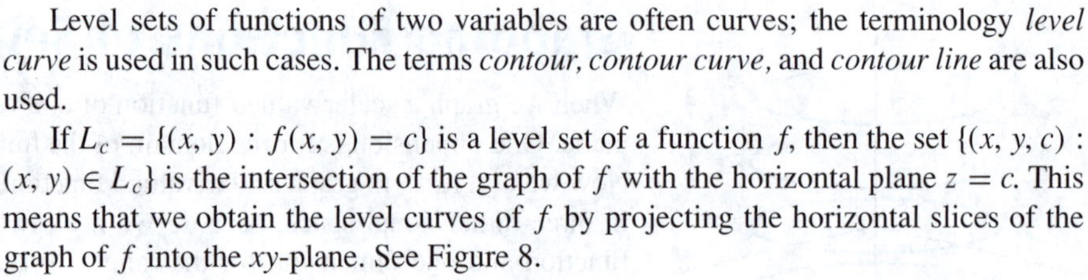

Figure 8
Graph of $z = f(x, y)$ with six horizontal slices; the six horizontal slices; the level curves corresponding to the slices

Example 5 Let $f(x, y) = x^2 + y^2 + 4$. Calculate and graph the level sets that correspond to horizontal slices at heights 20, 13, 5, 4, and 2.

Solution The level set that corresponds to the slice at height 20 is the set $\{(x, y) : x^2 + y^2 + 4 = 20\} = \{(x, y) : x^2 + y^2 = 16\}$. This is a circle of radius 4 centered at the origin of the xy-plane. The level set that corresponds to the slice at height 13 is $\{(x, y) : x^2 + y^2 + 4 = 13\} = \{(x, y) : x^2 + y^2 = 9\}$. This set is the circle of radius 3 centered at the origin of the xy-plane. Similarly, we find that the level set that corresponds to the slice at height 5 is the circle of radius 1 centered at the origin of the xy-plane. The level set that corresponds to the slice at height 4 is $\{(x, y) : x^2 + y^2 + 4 = 4\} = \{(x, y) : x^2 + y^2 = 0\} = \{(0, 0)\}$. This level set is just a single point. The graphs of these level sets appear in Figure 9. The level set that corresponds to the slice at height 2 is

$$\{(x, y) : x^2 + y^2 + 4 = 2\} = \{(x, y) : x^2 + y^2 = -2\} = \emptyset \quad \text{(the empty set)}.$$

There are no points in this level set. In fact, all the level sets that correspond to slices at heights less than 4 are empty. ■

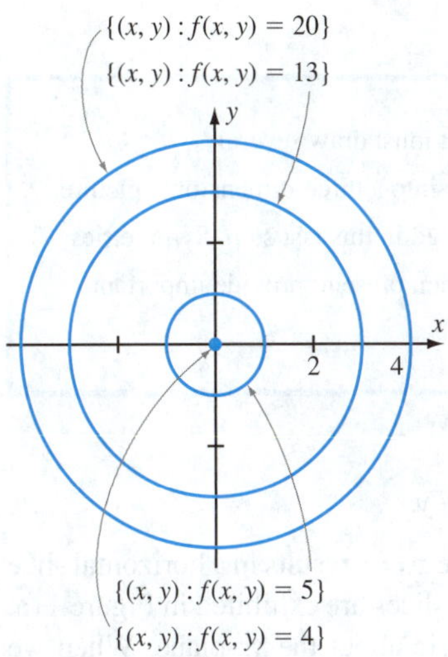

Figure 9

The level sets of a function tell us what the horizontal slices of its graph are. These, in turn, tell us a great deal about the graph of the function.

Example 6 Let $f(x, y) = x^2 + y^2 + 4$, as in Example 5. Use the level sets calculated in Example 5 to plot the horizontal slices at heights 20, 13, 5, 4, and 2. Then plot the graph of f.

Solution As we calculated in Example 5, the level set that corresponds to the slice $z = 20$ is $\{(x, y) : x^2+y^2 = 16\}$. This means that the points $\{(x, y, 20) : x^2+y^2 = 16\}$ lie on the graph of f. This set is a circle with center $(0, 0, 20)$ and radius 4. The set is located at height 20. Likewise, the slice of the graph of f at height 13 is the circle with $(0, 0, 13)$ and radius 3. The slice of the graph at height 5 is the circle with center $(0, 0, 5)$ and radius 1. Finally, the slice of the graph at height 4 is just the point $(0, 0, 4)$. It is important to note that at heights below 4, the slice of the graph is the empty set. This means that the graph does not extend below the plane $z = 4$.

All this information is amalgamated into a picture in Figure 10a. How do we go from a few horizontal slices to the plot of the entire surface? This is a critical step. We piece together the horizontal slices by taking a slice in another direction: Imagine slicing the figure with the yz-plane. The shape of this last slice will tell us a lot about the shape of the figure. The yz-plane has equation $x = 0$. Now, substituting $x = 0$ into $z = f(x, y)$ results in $z = y^2 + 4$, which is the equation of a parabola in the yz-plane (Figure 10b). Because all the nonempty level sets of f are circles with centers at $(0, 0)$, it follows that the graph of f has rotational symmetry about the z-axis. Therefore, the slice we have taken in the yz-plane suffices to confirm that the graph should be as shown in Figure 10c. ■

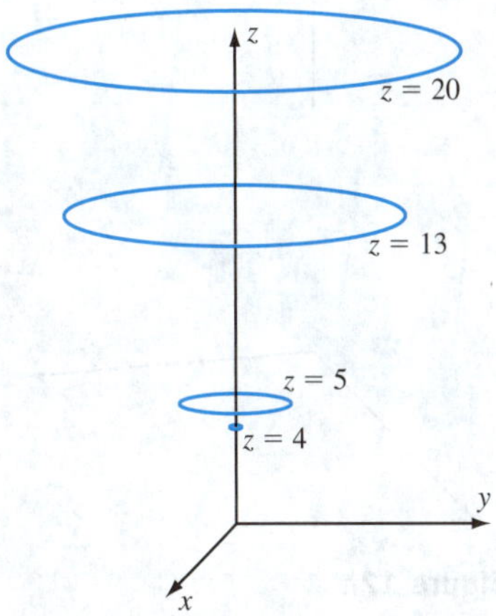

Figure 10a
Four horizontal slices of $z = x^2 + y^2 + 4$

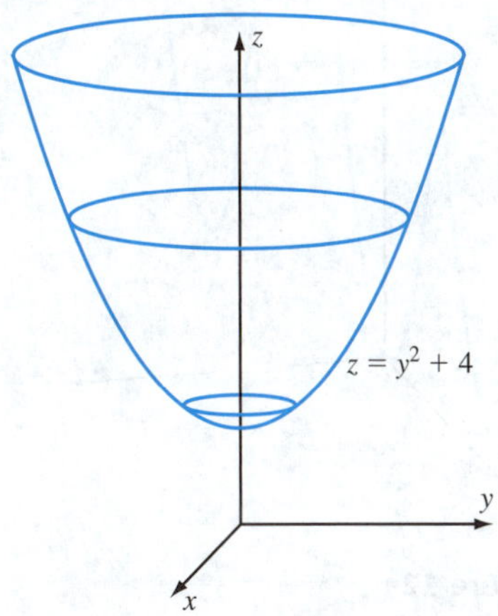

Figure 10b
The slice with the yz-plane is added to the plot.

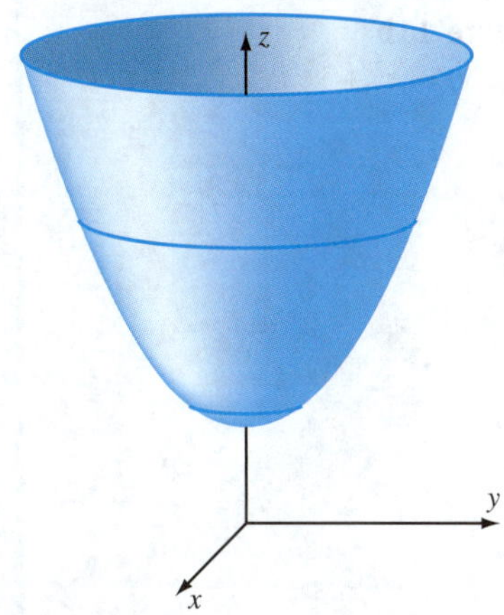

Figure 10c
The plot of $z = x^2 + y^2 + 4$

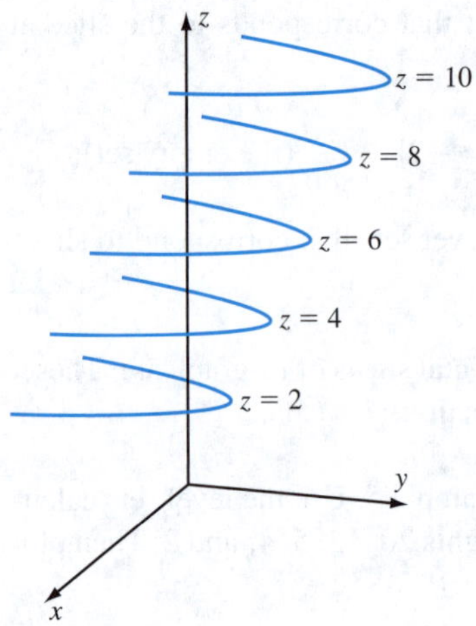

Figure 11a
Horizontal slices of $f(x, y) = x^2 + y$

inSIGHT

Our new graphing procedure has several steps. We first must draw several two-dimensional level sets. Then we must merge these into a three-dimensional picture. One or more slices perpendicular to the *xy*-plane are used in this last step. Symmetries about a plane or rotational symmetry about an axis, when present, provide important information.

Example 7 Sketch the graph of $f(x, y) = x^2 + y$.

Solution Intersecting the graph of f with the plane $z = c$ results in a horizontal slice of the form $\{(x, y, c) : x^2 + y = c\}$. Several of these slices are exhibited in Figure 11a. Notice that they are all parabolas that are symmetric about the *yz*-plane. When we slice the graph of f with the *yz*-plane by setting $x = 0$, we see that the vertices of the parabolas lie on the line $x = 0$, $z = f(0, y) = y$. The plot appears in Figure 11b. ■

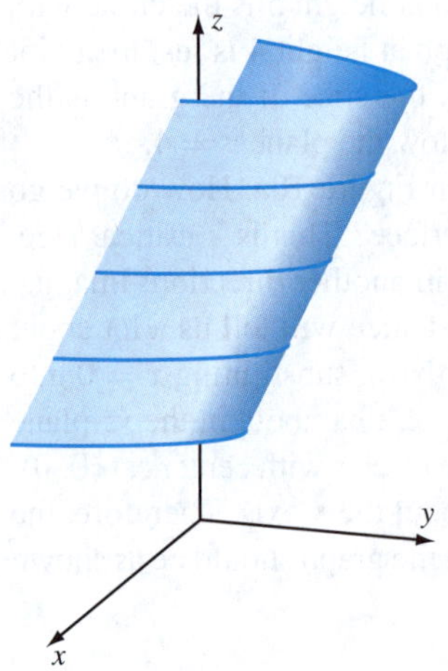

Figure 11b
$f(x, y) = x^2 + y$

inSIGHT

When you graph a function $(x, y) \mapsto f(x, y)$, you may use any slices that are convenient. In Example 7, if we slice the graph of f with planes $y = c$ that are parallel to the *xz*-plane, then we obtain a family of parabolas of the form $y = c$, $z = x^2 + c$ (Figure 12a). These vertical slices can be pieced together to obtain the plot of f (Figure 12b). By comparing Figures 11b and 12b, we see that different slices can provide different perspectives of a surface.

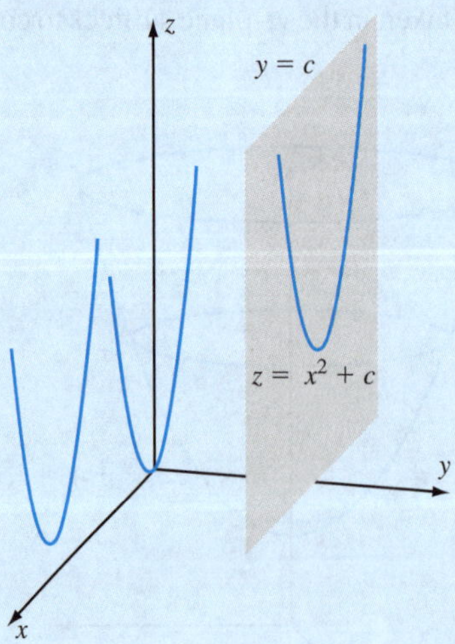

Figure 12a
Vertical slices of $f(x, y) = x^2 + y$

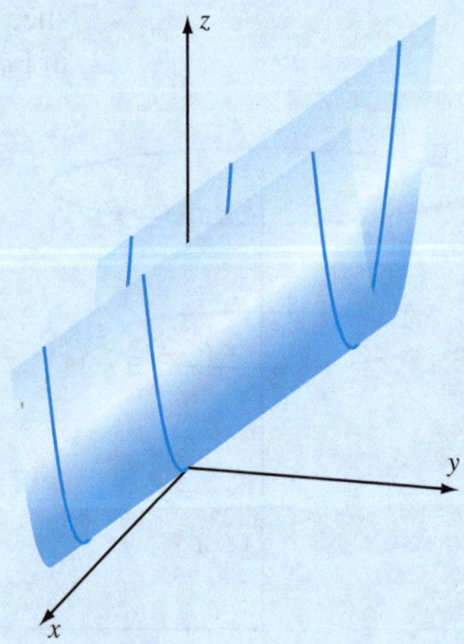

Figure 12b
$f(x, y) = x^2 + y$

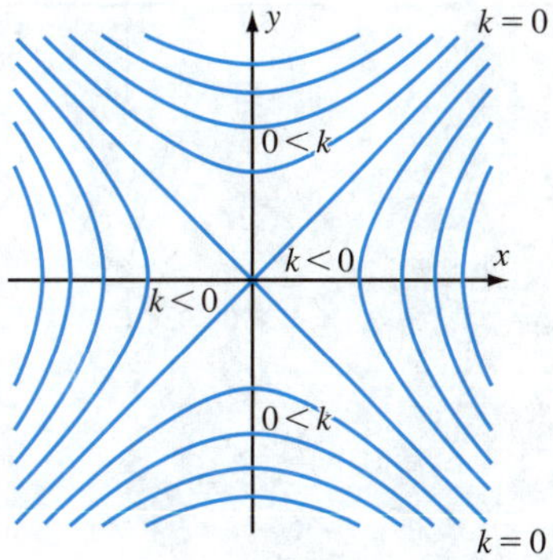

Figure 13
The family of curves $y^2 - x^2 = k$, where k is a constant

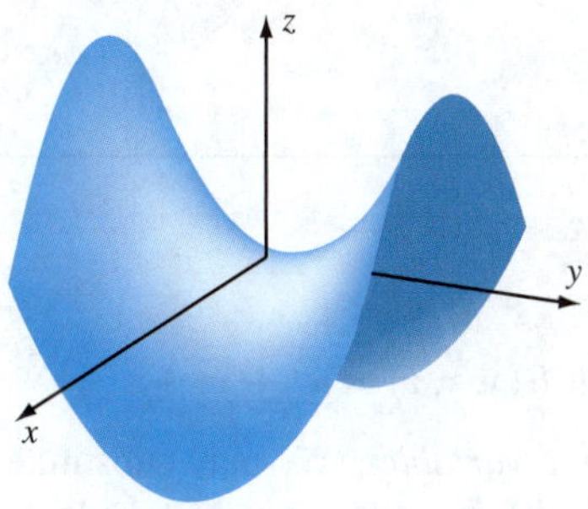

Figure 14
The graph of $z = y^2 - x^2$

Example 8 Sketch the graph of $f(x, y) = y^2 - x^2$.

Solution Figure 13 exhibits several level sets. Notice that they are all hyperbolas. These hyperbolas change orientation at height $c = 0$. In fact, $c = 0$ corresponds to the level set $y^2 - x^2 = 0$ or $(y - x)(y + x) = 0$, which is a pair of crossed lines. The level sets are hyperbolas that open sideways (in the xy-plane) when $c < 0$, and they are hyperbolas that open up and down (in the xy-plane) when $c > 0$. We use the slice corresponding to $x = 0$, which is the parabola $z = y^2$ in the yz-plane, and the slice corresponding to $y = 0$, which is the parabola $z = -x^2$ in the xz-plane, to complete our idea of the graph. The plot of f is shown in Figure 14. ■

More on Level Sets

Level sets occur in many applications of mathematics, including geography, geology, and meteorology. For example, the contours of a topographic map are level sets of the *height function*. Thus, if h is the function that assigns to each point on the map the height above sea level at that point, then the curves $h(x, y) = c$ that are drawn on the map give us an idea of where the mountains, valleys, and ridges are located. An example is shown in Figure 15.

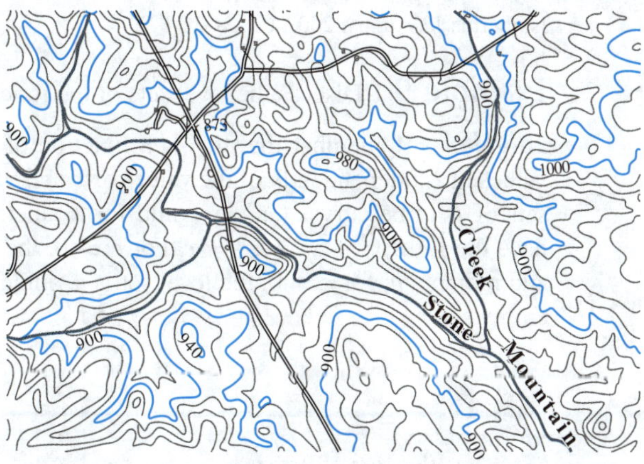

Figure 15
Topographic map of Stone Mountain, Georgia

In newspapers, we often see maps featuring level sets of the temperature function. That is, let $T(x, y)$ be the function that assigns to each point on the map the temperature in degrees Fahrenheit at that point. The curves $T(x, y) = c$, called *isothermal curves,* or *isotherms* for short, give us an idea of the changes in weather, the location of cold pockets, and so on (see Figure 16, next page).

Likewise, *isobars* are the level sets of the barometric pressure function, *isohyets* are level sets of the rainfall function, and so on.

We have seen that, in general, the graph $z = f(x, y)$ of a function of two variables will be a surface. Although we cannot graph a function F of three variables (because the graph would live in four-dimensional space), we can graph its level sets $\{(x, y, z) : F(x, y, z) = c\}$. These too are usually surfaces. A simple example illustrates the idea.

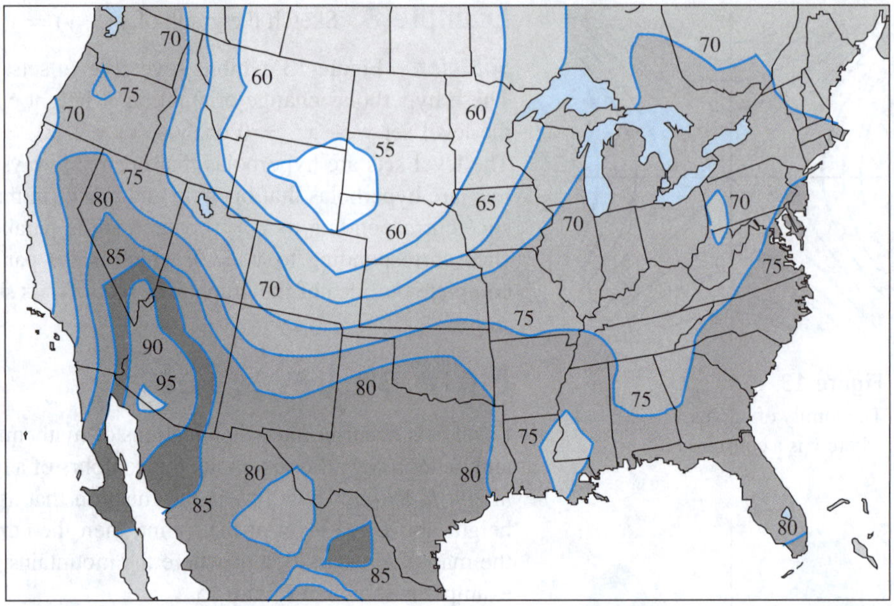

Figure 16
Isotherms 13 August 2002 12 AM EDT

Example 9 Discuss the level sets of the function $F(x, y, z) = x^2 + y^2 + z^2$.

Solution Remember that this is a function of *three variables:* We may substitute values of z independently of the values of x and y. We therefore do not expect the level sets of F to be curves (as we would for a function of two variables). The level sets have the form $x^2 + y^2 + z^2 = c$. For $c > 0$, the level set is the sphere with center 0 and radius $\sqrt{c}$; for $c = 0$, the level set is a single point; for $c < 0$, the level set is the empty set. ■

A LOOK BACK

To understand the different types of functions that we have studied, keep in mind the geometry that each type of function describes. The graphs of scalar-valued functions of one variable are planar curves that meet the Vertical Line Test. Vector-valued functions of one variable describe curves in the plane or in space. The graphs of scalar-valued functions of two variables are surfaces in space. The level sets of scalar-valued functions of three variables describe surfaces in three-dimensional space.

quickquiz

1. What algebraic operations are respected by functions of two or more variables?
2. What geometric device do we use for graphing a function of two variables?
3. What geometric object is the graph of a function of two variables?
4. Why is it impractical to graph a function of three or more variables?

EXERCISES

Problems for Practice

In Exercises 1–8, express the quantity that is described as a function of two or more variables. Describe the domain.

1. The volume $V(h, r)$ of a cylinder in terms of its height h and its radius r
2. The surface area $S(h, r)$ of a cylinder in terms of its height h and its radius r
3. The volume $V(h, r)$ of a cone in terms of its height h and its radius r
4. The area $A(r, \theta)$ of a circular sector of radius r and interior angle θ
5. The area $A(a, b)$ under the graph of $y = x^2$ and over the interval $[a, b]$
6. The surface area $S(x, y, V)$ of a rectangular box in terms of two side lengths, x and y, and its volume V
7. The mass $m(M, \tau, T)$ of a radioactive substance as a function of the initial mass M, the half-life τ, and the amount of time T that has elapsed since the initial measurement
8. The magnitude $a_N(\kappa, v)$ of centripetal acceleration of a moving body at a moment when its speed is v and the curvature of its trajectory is κ

Let $f(x, y) = \exp(x - y)$, $g(x, y) = x^2 - y^3$, $h(x, y, z) = xy^{z+2}$, and $k(x, y, z) = y/(x + z)$. Calculate the given quantity in Exercises 9–12.

9. $(2f + g)(5, 2)$ **10.** $(f \cdot g)(3, 2)$
11. $(h - k/3)(1/2, 3, 0)$ **12.** $(h/k)(3, 4, -1)$

Let $f(x, y) = 3xy/(x^2 + y^2)$, $g(x, y) = x^2 + 2y$, and $\phi(t) = \sqrt{t}$. In Exercises 13–18, calculate the indicated quantity.

13. $(\phi \circ f)(3, 4)$
14. $(\phi \circ g)(4, 10)$
15. $(\phi \circ (2f + 11g))(1, 1)$
16. $((\phi \circ f) \cdot g)(1, 2)$
17. $((2\phi) \circ (f \cdot g))(3, 4)$
18. $(((2\phi) \circ f) \cdot ((2\phi) \circ g))(3, 4)$

In Exercises 19–28, sketch the level sets of f that correspond to horizontal slices at heights $-6, -4, -2, 0, 2, 4,$ and 6.

19. $f(x, y) = 3x$
20. $f(x, y) = x + 2y$
21. $f(x, y) = 3x - 2y$
22. $f(x, y) = 4x^2 - y$
23. $f(x, y) = 1 + x - y^2$
24. $f(x, y) = x^2 + y^2/4$
25. $f(x, y) = 1 - x^2/9 - y^2$
26. $f(x, y) = x^2 - 2x + y^2$
27. $f(x, y) = x^2 - y^2$
28. $f(x, y) = 2 + y^2 - x^2$

In Exercises 29–38, a function f of two variables is given. Sketch several level sets of f. Then sketch the graph of $z = f(x, y)$.

29. $f(x, y) = 1 + x$
30. $f(x, y) = 6 - 2x - 3y$
31. $f(x, y) = x^2 + y^2 - 1$
32. $f(x, y) = 2 - (x^2 + y^2)$
33. $f(x, y) = \sqrt{x^2 + y^2}$
34. $f(x, y) = 3 - \sqrt{x^2 + y^2}$
35. $f(x, y) = \sqrt{1 - x^2 - y^2}$
36. $f(x, y) = 1 + \sqrt{4 - x^2 - y^2}$
37. $f(x, y) = y^2 + 2x + x^2$
38. $f(x, y) = x^2 + y^2/4 + 6$

In Exercises 39–44, sketch level sets corresponding to heights $-8, -4, 0, 4,$ and 8 for the given function F.

39. $F(x, y, z) = x + y + z$
40. $F(x, y, z) = 10 - x - y$
41. $F(x, y, z) = 12 - z^2$
42. $F(x, y, z) = x^2 + y^2 + z^2 - 8$
43. $F(x, y, z) = x^2 + 2x + y^2 + 4y + z^2 + 6z - 25$
44. $F(x, y, z) = x^2 + y^2 - 12$

Further Theory and Practice

45. Let $f(x, y) = xy/\sqrt{x^2 + y^2}$. Calculate $f(f(x, y), f(x, y))$.
46. Match the functions to the correct set of level curves in Figure 17, next page.
    - **a.** $f(x, y) = x^2 - y$
    - **b.** $f(x, y) = xy$
    - **c.** $f(x, y) = |x + y|$
    - **d.** $f(x, y) = 2x^2 + y^2$
    - **e.** $f(x, y) = x^2 - y^2$
    - **f.** $f(x, y) = (2x - y)^2$

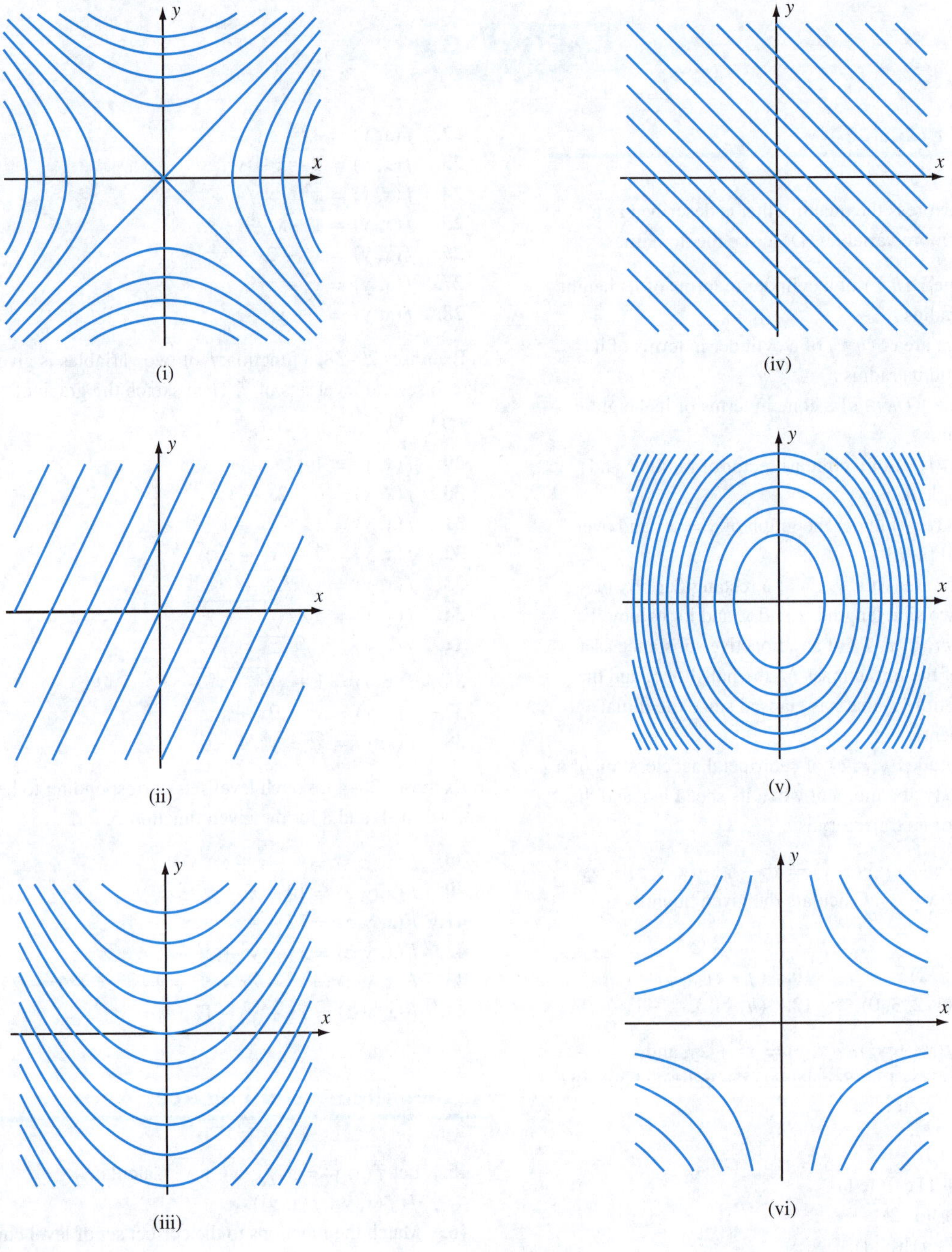

Figure 17

47. How do the level sets of $f_1(x, y) = x - y$ and $f_2(x, y) = (x - y)^2$ differ?

48. Calculate the time $T(h, v_0)$ until ground impact of an object propelled downward from a height h with initial speed v_0. (Assume gravity is the only force present.)

49. Calculate the x-intercept $a(h, c)$ of the tangent line to the curve $x \mapsto h + e^x$ at $x = c$.

50. Calculate the y-intercept $b(k, c)$ of the tangent line to the curve $y = k/x$ at $x = c$.

51. Let $f(x, y) = x^2 + y^2$ and $g(x, y) = 4 - 2x^2 - 2y^2$. Sketch the set $S = \{(x, y) : f(x, y) > g(x, y)\}$.

52. Let $f(x, y) = x^2 + y^2$. Sketch the set $S = \{(x, y) : f(x, y) \geq f(x, y)^2\}$.

53. Let $f(x, y) = x^2 - y + 4$. Sketch the set $S = \{(x, y) : 2 \leq f(x, y) < 4\}$.

In Exercises 54–59, describe the level sets of the function f.

54. $f(x, y) = \ln(xy)$

55. $f(x, y) = \sin(y - x^2)$

56. $f(x, y) = y \sin(x)$

57. $f(x, y) = \sin(x^2 + y^2)$

58. $f(x, y) = y + \sin(x)$

59. $f(x, y) = y^2/(1 + x^2)$

60. Figure 18 shows typical level curves of a function f. Answer the following questions about f, stating the reasons for your answers.

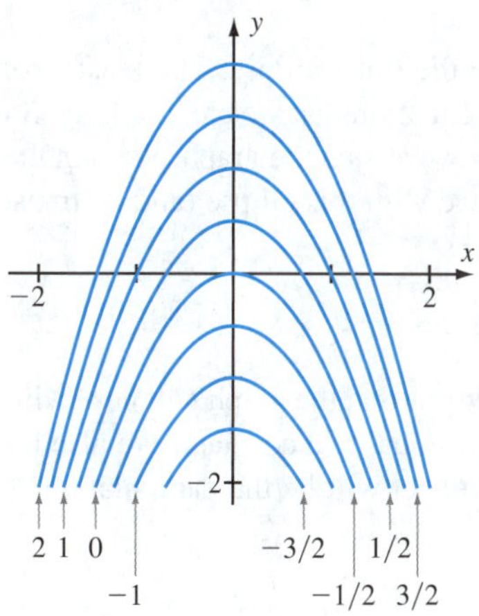

Figure 18
Level curves of f

a. Does $f(0, 1) = f(1, 0)$?

b. Does $f(1/2, 0) = f(0, 1/2)$?

c. Which is greater: $f(-1, 1)$ or $f(1/2, 1)$?

d. As t increases, does $f(-1/2 - t, -3/2 + t)$ increase or decrease?

e. On what interval is $f(x, -1)$ an increasing function of x?

f. On what interval is $f(-1/2, y)$ an increasing function of y?

61. Express the function $f(x, y) = \sum_{j=0}^{\infty} (x/y)^j$ without using sigma notation. What is its domain?

62. What is the domain of the function $f(x, y) = \sum_{n=1}^{\infty} n^{x+y}$? Sketch its level sets.

63. Give an example of a function $f(x, y)$ such that for each fixed y_0, the function $x \mapsto f(x, y_0)$ is a parabola, while for each fixed x_0, the function $y \mapsto f(x_0, y)$ is a cubic.

64. Parameterize the level curves of $f(x, y) = x^2 + 2y^2$.

65. Parameterize the level curves of $f(x, y) = x^2 - 2xy + 2y^2$.

Calculator/Computer Exercises

In Exercises 66–69, a function f and a viewing window R are given. Plot the level curves of the function f in the given viewing window. Use the horizontal slices $z = k/4$ for $k = -3, -2, \ldots, 3$.

66. $f(x, y) = x^3 + x^2y^4$, $R = [-1, 1] \times [-3, 3]$

67. $f(x, y) = x^2 + xy + y^3$, $R = [-1.2, 1] \times [-1, 1]$

68. $f(x, y) = (1 - x + y)/(1 + x^2 + y^2)$, $R = [-6, 4] \times [-4, 6]$

69. $f(x, y) = 2xy^2 + x^4y$, $R = [-2, 2] \times [-2, 2]$

13.2 Cylinders and Quadric Surfaces

You already know from your experience with functions of one variable that it requires some practice to feel comfortable with graphing. We have to do a lot of graphing and encounter a great many different functions and pictures before we can readily visualize ideas. The purpose of this section is to provide that vital experience for functions of two variables.

Some of the graphs that we will encounter in this section will *not* be the graphs of functions. Part of what we will learn is to distinguish which graphs come from functions. We begin now with the simplest graphs that we encounter in the theory of several variables.

Cylinders

An equation in three-dimensional space has a graph that is said to be a *cylinder* if one (or more) of the variables x, y, z does not appear in the equation. The justification for the term "cylinder" is apparent from the graph of the surface $\{(x, y, z) : x^2 + y^2 = 1\}$. In Example 2 from Section 12.1, we deduced that the plot of this equation in xyz-space is a cylinder (in the ordinary sense of the word). The next examples show that the mathematical meaning of "cylinder" is quite a bit more general.

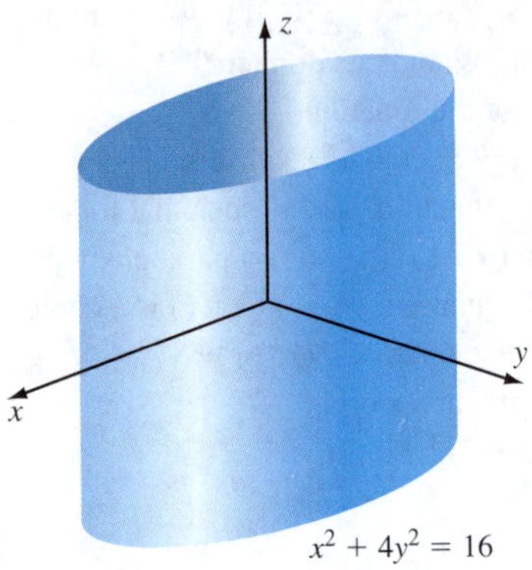

Figure 1
Elliptic cylinder

Example 1 Sketch the set of points in three-dimensional space satisfying the equation $x^2 + 4y^2 = 16$.

Solution We exploit the missing variable by taking slices that are perpendicular to its axis. In this case, z is missing, so we calculate the equations of some level sets $z = c$:

$$\begin{aligned} z &= -1 \qquad & x^2 + 4y^2 &= 16, \\ z &= 0 \qquad & x^2 + 4y^2 &= 16, \\ z &= 2 \qquad & x^2 + 4y^2 &= 16. \end{aligned}$$

Notice that all the level sets of the graph are the same ellipse. The reason for this is simple: The variable z does not appear in the equation that we are graphing. Substituting in different values of z has no effect. Now we sketch the graph. We plot the ellipse in the xy-plane and form the cylindrical surface with this ellipse (and its translates up and down) as cross sections. Refer to Figure 1. ■

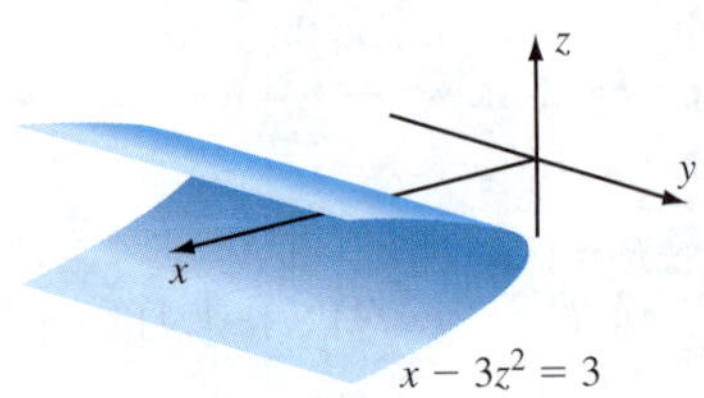

Figure 2

Example 2 Sketch the graph of $x - 3z^2 = 3$.

Solution The variable y is missing, so we know that the graph will be a cylinder and that we should use slices obtained by setting y equal to a constant. We also know that no matter what value c we take for y, the level set will be the parabola $x = 3 + 3z^2$. The resulting graph is sketched in Figure 2. ■

Quadric Surfaces

Quadratic equations provide a rich variety of examples on which to hone our technique at sketching graphs. We will only discuss equations of the form

$$Ax^2 + By^2 + Cz^2 + Dx + Ey + Fz = G. \tag{13.1}$$

The graphs of equations with terms like xy, yz, and xz in them are obtained by rotating the graphs of equations of the form in equation (13.1).

Ellipsoids

An *ellipsoid* is the set of points in space satisfying an equation of the form

$$\frac{x^2}{\alpha^2} + \frac{y^2}{\beta^2} + \frac{z^2}{\gamma^2} = 1$$

with α, β, and γ all positive constants. The distinguishing feature of such an equation is that all of its intersections with planes parallel to the coordinate planes are ellipses. More precisely, if we set $x = c$, then we obtain the set

$$\frac{y^2}{\beta^2} + \frac{z^2}{\gamma^2} = 1 - \frac{c^2}{\alpha^2}.$$

This is an ellipse if $|c| < \alpha$, a point (which may be regarded as a degenerate ellipse) if $|c| = \alpha$, and empty otherwise. Similarly, if we set $y = c$, then we obtain the set

$$\frac{x^2}{\alpha^2} + \frac{z^2}{\gamma^2} = 1 - \frac{c^2}{\beta^2}.$$

This is an ellipse if $|c| < \beta$, a point if $|c| = \beta$, and empty otherwise. The intersections with the planes $z = c$ are similar. It is worth noting that when x is replaced by $-x$ or y by $-y$ or z by $-z$, the equation is unchanged. Thus, the graph will be symmetric in each coordinate plane. We put this information together into Figure 3. Notice that, when α, β, and γ are distinct, the three axes of the ellipsoid have different lengths. When they are all equal, the ellipsoid is in fact a sphere.

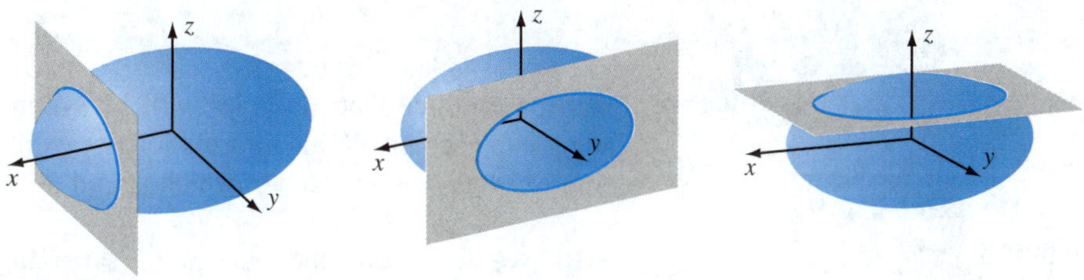

Figure 3
Intersections of the ellipsoid $\frac{x^2}{\alpha^2} + \frac{y^2}{\beta^2} + \frac{z^2}{\gamma^2} = 1$ with planes parallel to the coordinate planes are ellipses.

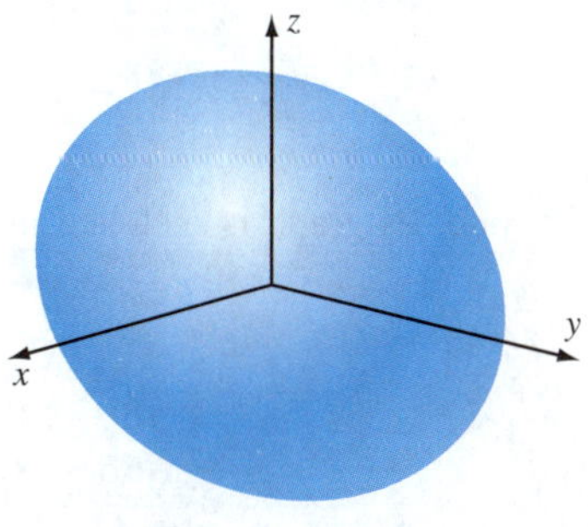

$4x^2 + y^2 + 2z^2 = 4$

Figure 4

Example 3 Sketch the set of points satisfying the equation $4x^2 + y^2 + 2z^2 = 4$.

Solution We divide through by 4 to write the equation in standard form:

$$\frac{x^2}{1^2} + \frac{y^2}{2^2} + \frac{z^2}{(\sqrt{2})^2} = 1.$$

According to our discussion, this is the equation of an ellipsoid. Its graph is sketched in Figure 4. Observe that it is easy to read off from the standard form of the equation that the intercepts of the surface are $(\pm 1, 0, 0)$, $(0, \pm 2, 0)$, and $(0, 0, \pm\sqrt{2})$. ■

Elliptic Cones

Elliptic cones arise from equations of the form

$$\frac{x^2}{\alpha^2} + \frac{y^2}{\beta^2} = \frac{z^2}{\gamma^2} \quad \text{or} \quad \frac{y^2}{\beta^2} + \frac{z^2}{\gamma^2} = \frac{x^2}{\alpha^2} \quad \text{or} \quad \frac{z^2}{\gamma^2} + \frac{x^2}{\alpha^2} = \frac{y^2}{\beta^2}$$

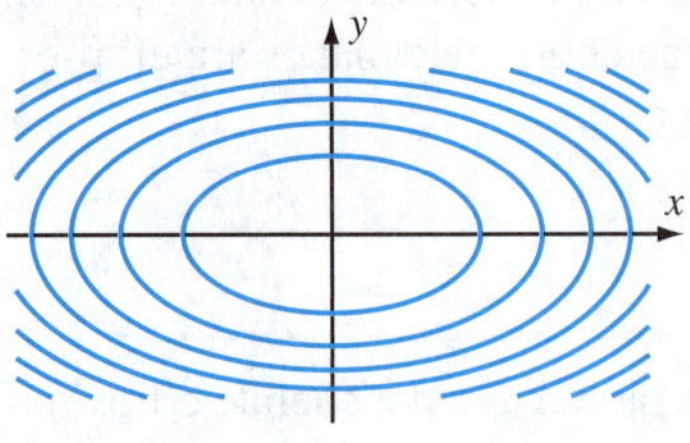

Figure 5
Some level sets of $\frac{x^2}{\alpha^2} + \frac{y^2}{\beta^2} = \frac{c^2}{\gamma^2}$

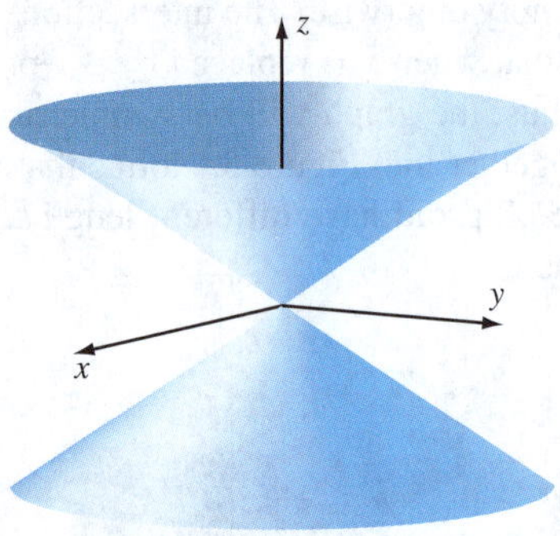

$$\frac{x^2}{\alpha^2} + \frac{y^2}{\beta^2} = \frac{z^2}{\gamma^2}$$

Figure 6
An elliptic cone

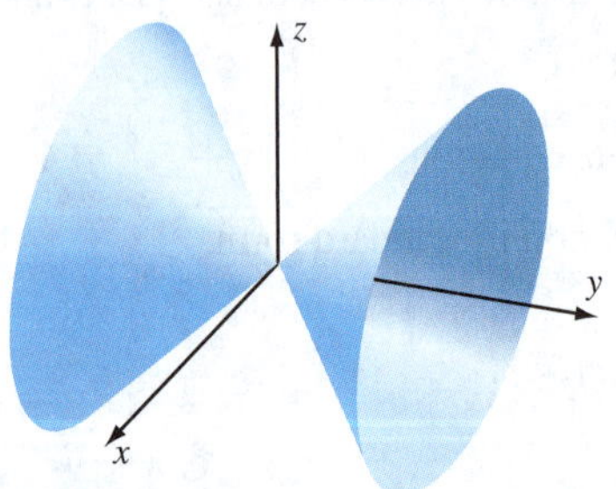

Figure 7
The elliptic cone $x^2 + 2z^2 = 2y^2$

with α, β, and γ positive. We will analyze the first of these equations. The remaining two can be handled in a similar way.

When $z = c$, the level set of $x^2/\alpha^2 + y^2/\beta^2 = z^2/\gamma^2$ has the form

$$\frac{x^2}{\alpha^2} + \frac{y^2}{\beta^2} = \frac{c^2}{\gamma^2}.$$

This is an ellipse whose axes have ratio $\alpha : \beta$. As $|c|$ becomes larger, the ellipse becomes larger, but the ratio of the axes remains the same. Some level curves are sketched in Figure 5.

How do the z–level sets stack up? To find out, we take some slices parallel to the xz- and yz-planes. Setting $x = 0$, that is, slicing with the yz-coordinate plane, we obtain $y^2/\beta^2 = z^2/\gamma^2$, or

$$y = \pm\frac{\beta}{\gamma}z,$$

which describes a pair of lines. Likewise setting $y = 0$, that is, slicing with the xz-coordinate plane, we obtain the two lines described by

$$x = \pm\frac{\alpha}{\gamma}z.$$

Finally, we notice that the graph will be symmetric in all three coordinate planes. Taking into account all this information, we exhibit the graph in Figure 6. Note that it comprises two *nappes*, each of which is said to be a cone in everyday language.

Example 4 Sketch the set of points satisfying the equation $x^2 + 2z^2 = 2y^2$.

Solution We divide through by 2 to write the equation in standard form:

$$\frac{x^2}{(\sqrt{2})^2} + \frac{z^2}{1^2} = \frac{y^2}{1^2}.$$

According to our discussion, this surface is an elliptic cone. Its graph is sketched in Figure 7. ■

Elliptic Paraboloids

Elliptic paraboloids arise from equations of the form

$$\frac{x^2}{\alpha^2} + \frac{y^2}{\beta^2} = \frac{z}{\gamma} \quad \text{or} \quad \frac{y^2}{\beta^2} + \frac{z^2}{\gamma^2} = \frac{x}{\alpha} \quad \text{or} \quad \frac{z^2}{\gamma^2} + \frac{x^2}{\alpha^2} = \frac{y}{\beta}.$$

We assume as usual that $\alpha, \beta, \gamma > 0$. Again, we will discuss only the first of these equations; the others yield to a similar treatment. As in the case of the elliptic cone, the z–level sets of $x^2/\alpha^2 + y^2/\beta^2 = z/\gamma$ are ellipses of the same eccentricity:

$$\frac{x^2}{\alpha^2} + \frac{y^2}{\beta^2} = \frac{c}{\gamma}.$$

As $c > 0$ becomes larger, the ellipses get larger. When $c < 0$, the level set is empty.

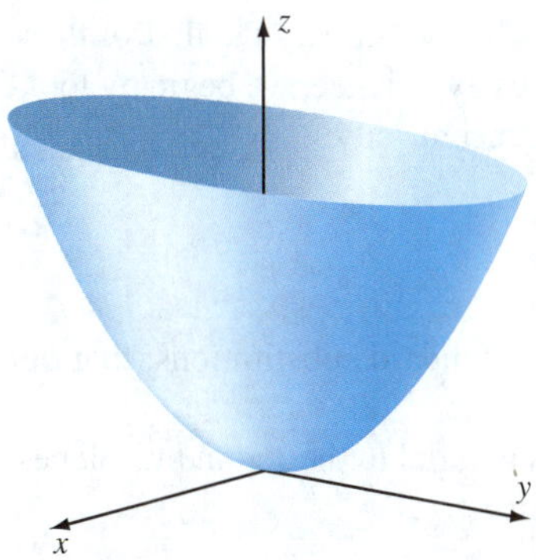

Figure 8
$\frac{x^2}{\alpha^2} + \frac{y^2}{\beta^2} = \frac{z}{\gamma}$;
Elliptic paraboloid

Next we look at slices parallel to the xz- and yz-planes to see how to stack the ellipses. Setting $x = 0$, that is, slicing with the yz-coordinate plane, we obtain $y^2/\beta^2 = z/\gamma$, or

$$z = \gamma \frac{y^2}{\beta^2},$$

which is an upward-opening parabola. Likewise setting $y = 0$, that is, slicing with the xz-coordinate plane, we obtain $x^2/\alpha^2 = z/\gamma$, or

$$z = \gamma \frac{x^2}{\alpha^2}.$$

This, too, is an upward-opening parabola. Finally, we notice that the graph will be symmetric in the xz-coordinate plane (replacing y with $-y$ results in no change) and in the yz-coordinate plane (replacing x with $-x$ results in no change). However, it will *not* be symmetric in the xy-coordinate plane. The resulting sketch appears in Figure 8.

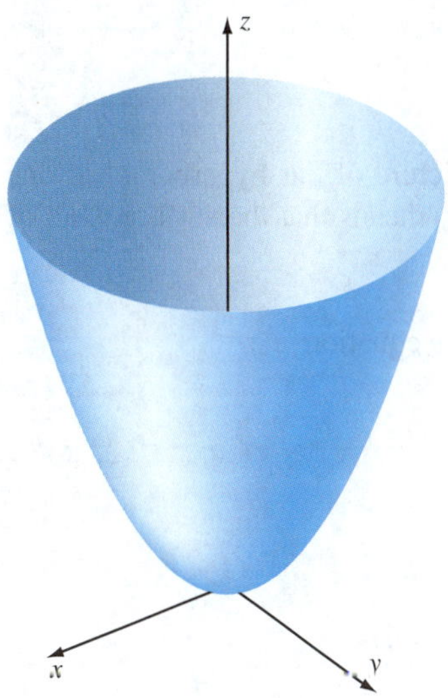

Figure 9
$\frac{x^2}{\alpha^2} + \frac{y^2}{\alpha^2} = \frac{z}{\gamma}$;
Circular paraboloid

inSIGHT

The special case of the elliptic paraboloid when $\alpha = \beta$ is called the *circular* paraboloid. For this surface, the z–level sets are circles. The sketch is in Figure 9.

Circular paraboloids are of interest because they are the model for the shape of radio, television, and radar antennas. The reflecting properties of paraboloids are also crucial to the construction of reflecting telescopes.

Example 5 Sketch the set of points satisfying the equation $x^2 + 2z^2 = y$.

Solution We divide through by 2 to put the equation in standard form:

$$\frac{x^2}{(\sqrt{2})^2} + \frac{z^2}{1^2} = \frac{y}{2}.$$

According to our discussion, the locus of this equation is an elliptic paraboloid. Its graph is shown in Figure 10. ■

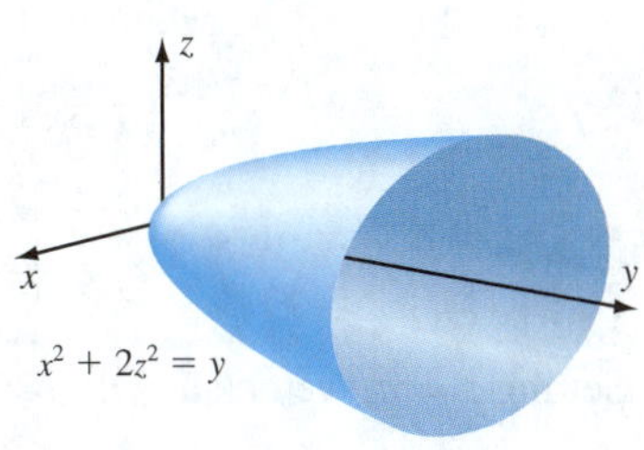

Figure 10
Elliptic paraboloid

Hyperboloids of One Sheet

The hyperboloid of one sheet arises from the equations

$$\frac{x^2}{\alpha^2} + \frac{y^2}{\beta^2} - \frac{z^2}{\gamma^2} = 1 \quad \text{or} \quad \frac{y^2}{\beta^2} + \frac{z^2}{\gamma^2} - \frac{x^2}{\alpha^2} = 1 \quad \text{or} \quad \frac{z^2}{\gamma^2} + \frac{x^2}{\alpha^2} - \frac{y^2}{\beta^2} = 1.$$

We assume that α, β, and γ are positive. Again, we will discuss only the first of these equations; the other two are similar. Following previous experience, we begin by looking at z–level sets. These will be ellipses of the same eccentricity:

$$\frac{x^2}{\alpha^2} + \frac{y^2}{\beta^2} = \frac{c^2}{\gamma^2} + 1.$$

They increase in size as $|c|$ increases. We see, by the standard substitutions, that our figure will be symmetric in all three coordinate axes.

To see how the ellipses stack up, we look at slices parallel to the xz- and yz-planes. We slice with the yz-plane by setting $x = 0$ and obtain

$$\frac{y^2}{\beta^2} - \frac{z^2}{\gamma^2} = 1.$$

This is a hyperbola with axis the z-axis. Also, we slice by the xz-plane, setting $y = 0$, to obtain

$$\frac{x^2}{\alpha^2} - \frac{z^2}{\gamma^2} = 1.$$

This too is a hyperbola with axis the z-axis. The picture of our hyperboloid of one sheet is in Figure 11. The phrase "one sheet" simply means that the surface is all of one piece.

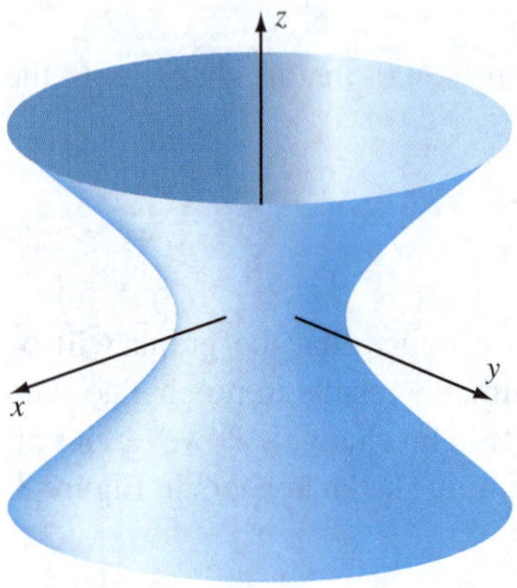

$$\frac{x^2}{\alpha^2} + \frac{y^2}{\beta^2} - \frac{z^2}{\gamma^2} = 1$$

Figure 11
Hyperboloid of one sheet

Example 6 Sketch the set of points satisfying the equation

$$3x^2 - \frac{y^2}{2} + z^2 = 1.$$

Solution We put the equation in standard form:

$$\frac{x^2}{(1/\sqrt{3})^2} + \frac{z^2}{1^2} - \frac{y^2}{(\sqrt{2})^2} = 1.$$

According to our discussion, this surface is a hyperboloid of one sheet. Its graph is shown in Figure 12. ■

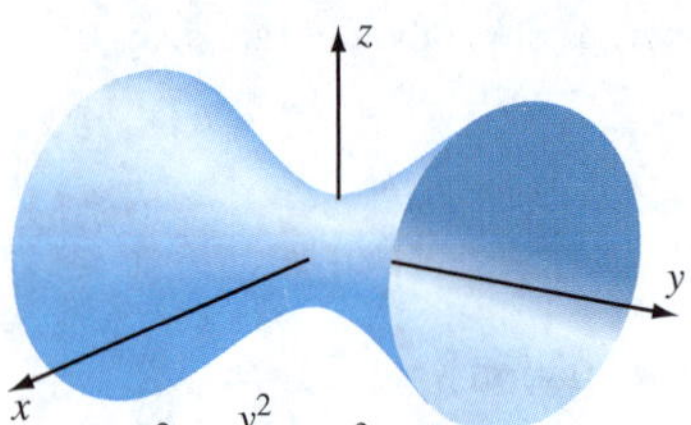

$$3x^2 - \frac{y^2}{2} + z^2 = 1$$

Figure 12
Hyperboloid of one sheet

Hyperboloids of Two Sheets

Consider the equation

$$\frac{z^2}{\gamma^2} - \frac{x^2}{\alpha^2} - \frac{y^2}{\beta^2} = 1.$$

It is convenient to begin by considering z–level sets. Setting $z = c$, we obtain

$$\frac{x^2}{\alpha^2} + \frac{y^2}{\beta^2} = \frac{c^2}{\gamma^2} - 1.$$

If $|c| \geq \gamma$, then this level set is an ellipse (the ellipses have eccentricity independent of c and increase in size with $|c|$); but if $|c| < \gamma$, then the level set is empty.

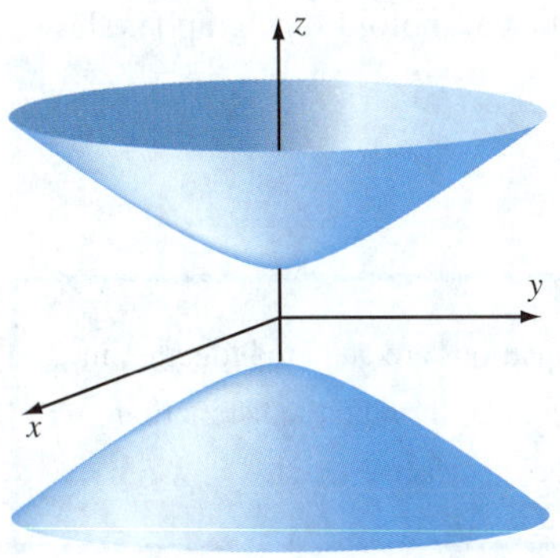

$$\frac{z^2}{\gamma^2} - \frac{x^2}{\alpha^2} - \frac{y^2}{\beta^2} = 1$$

Figure 13
Hyperboloid of two sheets

As usual, we now slice by the coordinate planes and see that the slice with the yz-plane is a hyperbola with axis the z-axis:

$$\frac{z^2}{\gamma^2} - \frac{y^2}{\beta^2} = 1.$$

Similarly, the slice with the xz-plane is a hyperbola with axis the z-axis. The picture of our hyperboloid of two sheets is in Figure 13. By analogous reasoning, we see that the equations

$$\frac{x^2}{\alpha^2} - \frac{y^2}{\beta^2} - \frac{z^2}{\gamma^2} = 1 \quad \text{and} \quad \frac{y^2}{\beta^2} - \frac{z^2}{\gamma^2} - \frac{x^2}{\alpha^2} = 1$$

are also hyperboloids of two sheets.

Example 7 Sketch the set of points satisfying $x^2 - 2z^2 - 4y^2 = 4$.

Solution We divide by 4 to put the equation in standard form:

$$\frac{x^2}{2^2} - \frac{y^2}{1^2} - \frac{z^2}{\left(\sqrt{2}\right)^2} = 1.$$

According to our discussion, this surface is a hyperboloid of two sheets. Its graph is shown in Figure 14. ■

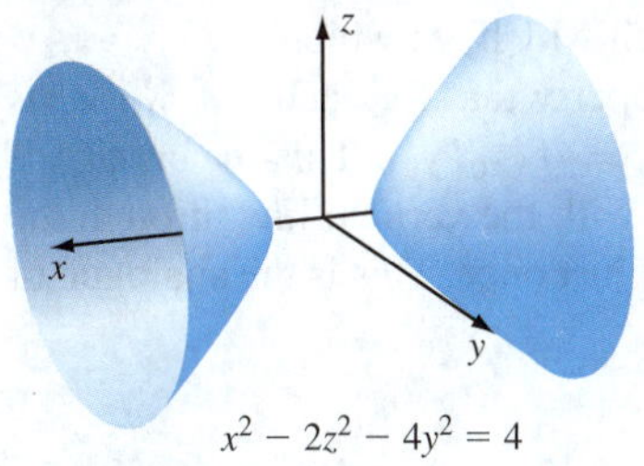

$$x^2 - 2z^2 - 4y^2 = 4$$

Figure 14
Hyperboloid of two sheets

Hyperbolic Paraboloids

Finally, we consider the equation

$$\frac{x^2}{\alpha^2} - \frac{y^2}{\beta^2} = \frac{z}{\gamma}.$$

As usual, α, β, and γ are positive. As long as $c \neq 0$, the $z = c$ level sets are hyperbolas:

$$\frac{x^2}{\alpha^2} - \frac{y^2}{\beta^2} = \frac{c}{\gamma}.$$

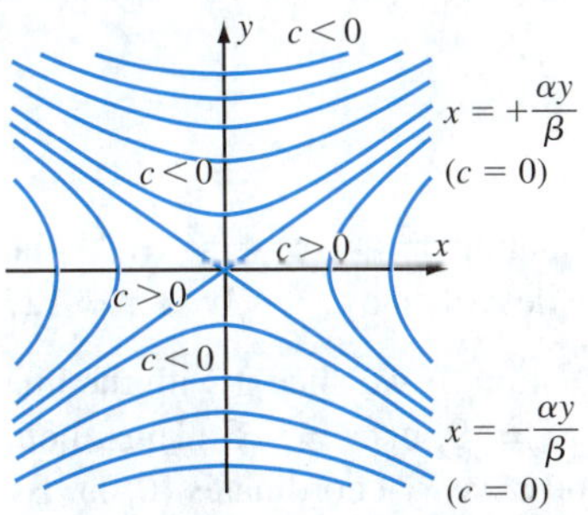

Figure 15
Level curves of $\frac{x^2}{\alpha^2} - \frac{y^2}{\beta^2} = \frac{z}{\gamma}$

When $c = 0$, the level set is $x^2/\alpha^2 - y^2/\beta^2 = 0$, or $x = \pm\alpha y/\beta$, which is a pair of lines. Figure 15 shows the family of level sets for $c < 0$, $c = 0$, and $c > 0$. The slices corresponding to $x = 0$ and $y = 0$ are parabolas. The final picture is in Figure 16. The resulting surface is commonly known as a *saddle surface*. It is of particular interest when we solve maximum-minimum problems, for the origin looks like a local minimum in the x-direction and like a local maximum in the y-direction.

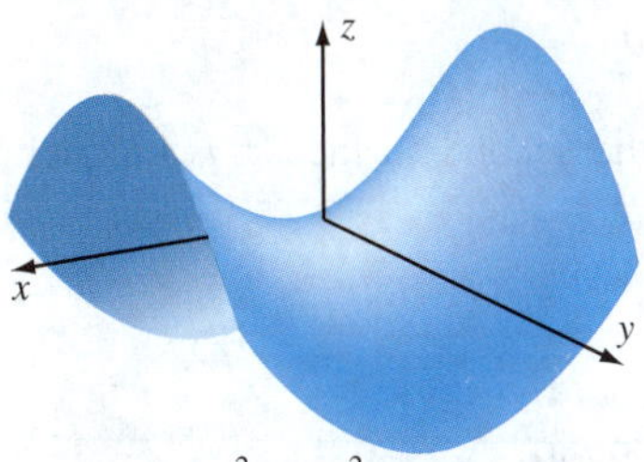

$$\frac{x^2}{\alpha^2} - \frac{y^2}{\beta^2} = \frac{z}{\gamma}$$

Figure 16
Hyperbolic paraboloid (saddle surface)

Example 8 Sketch the set of points satisfying the equation

$$z = 2y^2 - 4x^2.$$

Solution We divide through by 4 to put the equation in standard form:

$$\frac{y^2}{\left(\sqrt{2}\right)^2} - \frac{x^2}{1^2} = \frac{z}{4}.$$

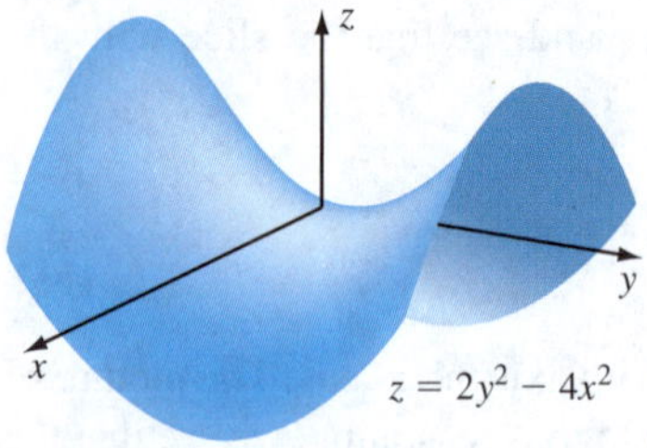

Figure 17
Hyperbolic paraboloid

According to our discussion, the surface is a hyperbolic paraboloid. The graph is shown in Figure 17. ■

inSIGHT

We have studied six kinds of quadric surfaces. Their equations are gathered together in this chapter's Summary of Key Topics. It is a good idea to go through the list, noting similarities and differences between the equations of these surfaces.

Recognizing the Graph of a Function

Many of the quadric surfaces we have discussed in this section are *not* the graphs of functions, although a few of them are. How do we tell which are which?

A function $(x, y) \mapsto f(x, y)$ has the crucial property that for each (x_0, y_0) in its domain, there is precisely one value z_0 such that $z_0 = f(x_0, y_0)$. This means that if (x_0, y_0) is in the domain of f and we cut the graph with the vertical line through the point $(x_0, y_0, 0)$, then the line will intersect the graph just once. This is the analogue of the Vertical Line Test for functions of one variable:

A surface in space is the graph of a function $z = f(x, y)$ if and only if every vertical line intersects the surface *at most* once.

Example 9 Is the set of points satisfying the equation $x^2 + 3y^2 + z^2 = 12$ the graph of a function of x and y? What about the surface described by $x^2+3y^2+z = 12$?

Solution The first surface is *not* the graph of a function. If we slice it with the line $x = 0$, $y = 1$, we see that the equation becomes $z^2 = 9$, or $z = \pm 3$. Thus, both the points $(0, 1, 3)$ and $(0, 1, -3)$ lie on the graph and have xy-coordinates $(0, 1)$. By contrast, the surface $x^2 + 3y^2 + z = 12$ *is* the graph of a function, because once x_0, y_0 are specified, z must be

$$z = 12 - (x_0)^2 - (y_0)^2$$

for (x_0, y_0, z) to lie on the surface. In the case of the second surface, z *is* uniquely determined by x and y. For the first, *it is not.* ■

quickquiz

1. What is a quadric surface?
2. How do we recognize the equation of an ellipsoid?
3. How do we recognize the equation of a paraboloid?
4. Turn to the list of quadric surfaces given in this chapter's Summary of Key Topics. Without looking at the formulas that appear to the left of the named surfaces, write the equations of each surface.

EXERCISES

Problems for Practice

In Exercises 1–14, the equation of a cylinder in xyz-space is given. Sketch its plot.

1. $x + y = 5$
2. $4x^2 + z^2 = 4$
3. $4x^2 + z = 4$
4. $y^2 + 2z^2 = 4$
5. $x^2 + z = -4$
6. $y - x^2 = 9$
7. $x^2 + y = 4$
8. $z^2 + y^2 = 2y$
9. $x^2 - 2y^2 = 1$
10. $4x^2 - 2y^2 = 0$
11. $y^2 - 4z^2 = 1$
12. $z = \sin(y)$
13. $x - 2xy = 1$
14. $z^2 - y^2 = 1$

Sketch some slices, and then give a complete sketch of each quadric surface in Exercises 15–30. Identify each quadric surface as to type (cylinder, paraboloid, etc.). Is the surface the graph of a function of x and y?

15. $x^2 - 4y^2 + 9z^2 = 36$
16. $x^2 + y^2 = z^2$
17. $y - x^2 = z^2$
18. $x^2 - 9z^2 - 4y^2 = 16$
19. $4x^2 + y^2 + 9z^2 = 25$
20. $2x^2 + 4z^2 = y$
21. $x^2 + y^2 + z = 4$
22. $y^2 + x^2 - z = 4$
23. $4x^2 + y^2 = z$
24. $4x^2 + y^2 + 16z^2 = 16$
25. $y^2 - z^2 - x^2 = 1$
26. $z^2 - 4x^2 = y$
27. $y^2 + 8z^2 = x^2$
28. $z^2 - 4x^2 - 9y^2 = 36$
29. $y^2 - x - z^2 = 0$
30. $x^2 - y^2 - 2z^2 = 6$

Further Theory and Practice

31. True or false: If every level curve of a surface is the graph of a quadratic equation, then the surface is a quadric surface. Give reasons for your answer.
32. If a particle bounces off a smooth surface, then the angle of the bounce is determined just as it would be if the particle bounces off the tangent plane at that point: The incoming angle with the tangent plane is the same as the outgoing angle. Now consider a circular paraboloid $z = x^2 + y^2$. Show that there is a point $P = (0, 0, p)$ on the positive z-axis with the property that any particle falling vertically through the paraboloid and striking the wall of the paraboloid will bounce to the point P. Thus, P plays the role of a focal point. This is why parabolic mirrors and antennas are important. (*Hint:* Reduce the problem to a two-dimensional problem with a slice, and use the tangent line to see how the particle bounces.)
33. Prove that the circular paraboloid is the only surface that has circular symmetry in the x- and y-variables and that has the property described in Exercise 32.

The projection in the xy-plane of a space curve $\mathcal{C}$ is the set $\{(x, y) : (x, y, z) \in \mathcal{C} \text{ for some } z\}$. In Exercises 34 and 35, find the projection in the xy-plane of the given space curve $\mathcal{C}$.

34. $\mathcal{C}$ is the intersection of the elliptic paraboloids $z = x^2 + 2y^2 - 8$ and $z = 12 - 4x^2 - 3y^2$.
35. $\mathcal{C}$ is the intersection of the hyperbolic paraboloid $z = x^2/2 - y^2$ and the elliptic paraboloid $z = 1 - x^2/2 - 5y^2/4$.

Calculator/Computer Exercises

In Exercises 36–39, plot the given quadric surface.

36. $x^2 + 2xy + 2y^2 + z^2 = 1$
37. $x^2 + 2xy + 2y^2 - z^2 = 0$
38. $x^2 + 2xy + 2y^2 - z^2 = 1$
39. $x^2 + 2xy + 2y^2 - z^2 = -1$
40. Plot $z = x^2 + y^2$ and $2x + 2y + z = 2$ for $-4 \le x \le 3$, $-4 \le y \le 3$. Parameterize the curve of intersection. Add the graph of this curve to your plot.
41. Plot $z^2 = 4(x^2 + y^2)$ for $0 \le z \le 6$. Plot the plane $z = \sqrt{2}(1 - x)$ for $-6/\sqrt{2} + 1 \le x \le 1$, $-3 \le y \le 3$. Display the two plots in the same viewing box. Parameterize the curve of intersection, graph it, and add it to your previous plot.

13.3 Limits and Continuity

Now that we have treated functions and graphs, we can discuss limits. Recall from Chapter 2 that the limit at c of a function of *one variable* is the value that we *expect* the function to take at c. The same philosophy will prevail for functions of several variables, but there is an important difference. In one variable, the domain is linear, and we decide what we "expect" the function will do by approaching c from the left and from the right. In two variables, however, the domain is planar, and we decide what we "anticipate" the function will do at c by approaching c *along all possible paths that tend to c*. See Figure 1. In a sense, then, it is more difficult for a function of two variables to have a limit.

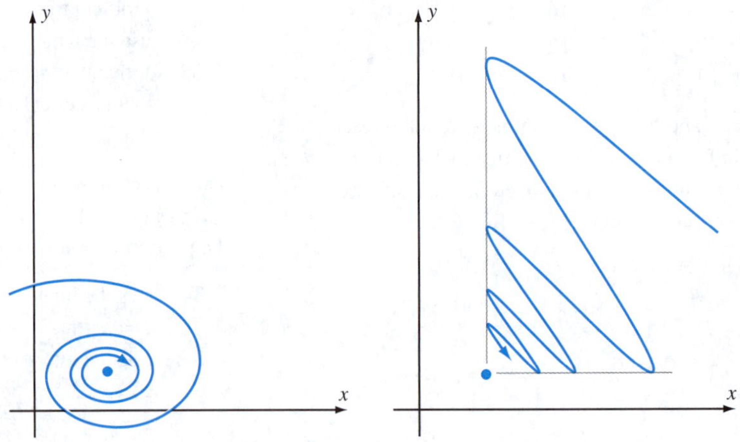

Figure 1
Two possible paths that tend to a planar point

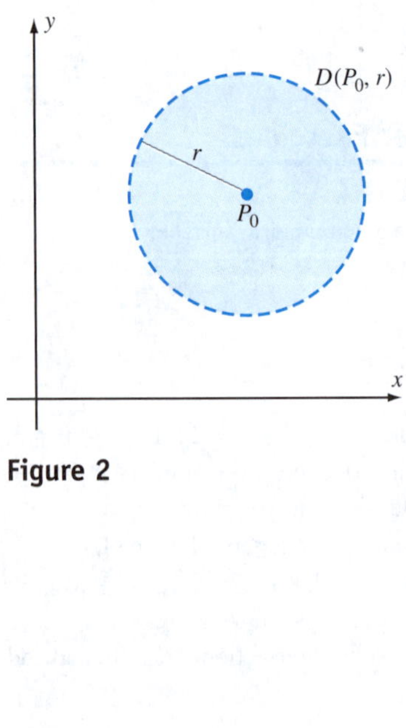

Figure 2

If $P_1 = (x_1, y_1)$ and $P_2 = (x_2, y_2)$ are points in the plane, then it is convenient to denote the distance between them with the notation $d(P_1, P_2)$. Notice that this distance is the magnitude of the vector $\overrightarrow{P_1P_2}$:

$$d(P_1, P_2) = \left\| \overrightarrow{P_1P_2} \right\| = \sqrt{(x_2 - x_1)^2 + (y_2 - y_1)^2}.$$

If $P_0 = (x_0, y_0)$ is a point in the plane and r is a positive number, then we will let $D(P_0, r)$ denote the open disk with positive radius r that is centered at P_0—that is, $D(P_0, r) = \{P \in \mathbb{R}^2 : d(P, P_0) < r\}$. Because of the strict inequality, the boundary of the disk is not included in $D(P_0, r)$. See Figure 2.

Limits

In many of the constructions of calculus, we must investigate the limit of a function at a point that is not in its domain. For example, we learned in Chapter 3 that the formula $\frac{d}{dx}\sin(x) = \cos(x)$ depends entirely on establishing the limit $\lim_{x\to 0} \sin(x)/x = 1$. Yet the point 0 is not in the domain of the expression $\sin(x)/x$ that is under consideration. To handle similar situations in the plane, it is convenient to refer to the "punctured" disk $D_*(P_0, r)$ that is formed from the open disk $D(P_0, r)$ by removing the center: $D_*(P_0, r) = \{P \in \mathbb{R}^2 : 0 < d(P, P_0) < r\}$. See Figure 3.

Definition

Suppose that f is a function of two variables. Let $P_0 = (x_0, y_0)$ be a point in the plane such that every punctured disk $D_*(P_0, r)$ intersects the domain of f. We say that the real number ℓ is the *limit* of $f(P)$ as $P = (x, y)$ approaches (or tends to) P_0, and we write

$$\lim_{P \to P_0} f(P) = \ell, \qquad \text{or equivalently,} \qquad \lim_{(x,y)\to(x_0,y_0)} f(x, y) = \ell,$$

if for any $\epsilon > 0$, there is a $\delta > 0$ such that $|f(P) - \ell| < \epsilon$ for all points P in the domain of f with $0 < d(P, P_0) < \delta$. Stated another way, if an element (x, y) of the domain of f belongs to $D_*(P_0, \delta)$, then $f(x, y)$ belongs to the interval $(\ell - \epsilon, \ell + \epsilon)$. Notice that the definition does not require P_0 to be in the domain of f for us to consider $\lim_{P\to P_0} f(P)$.

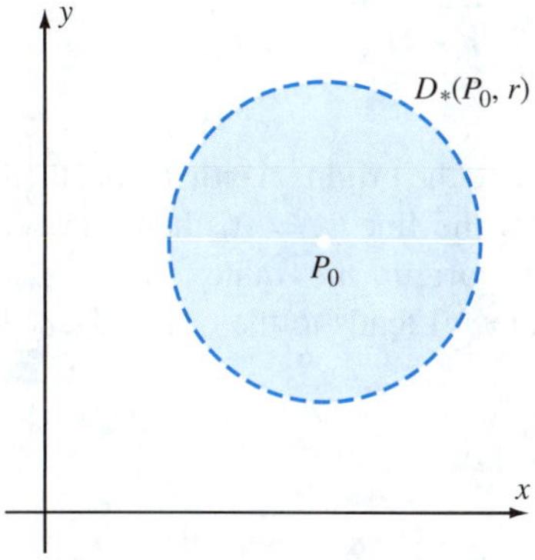

Figure 3

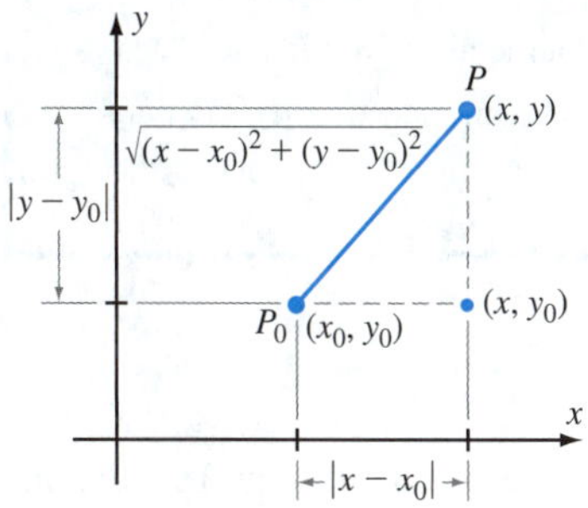

Figure 4

inSIGHT

Compare this definition of limit with the rigorous definition of "limit" in Section 2.2. The principal difference is that the distance between $P = (x, y)$ and $P_0 = (x_0, y_0)$ is measured using $d(P, P_0) = \sqrt{(x - x_0)^2 + (y - y_0)^2}$ rather than | |. What does this mean? Since the hypotenuse of the right triangle in Figure 4 is at least as long as each leg, we have $d(P, P_0) \geq |x - x_0|$ and $d(P, P_0) \geq |y - y_0|$. These inequalities tell us that when P is close to P_0, x is close to x_0 *and*, simultaneously, y is close to y_0. Also, the Triangle Inequality tells us that $d(P, P_0) = \|\overrightarrow{P_0P}\| \leq |x - x_0| + |y - y_0|$. This inequality tells us that when both $|x - x_0|$ and $|y - y_0|$ are small, the distance $d(P, P_0)$ must also be small. In summary, $d(P, P_0) = \sqrt{(x - x_0)^2 + (y - y_0)^2}$ is small precisely when x is close to x_0 *and*, simultaneously, y is close to y_0.

Example 1 Define $f(x, y) = x^2 + y^2$. Verify that $\lim_{(x,y)\to(0,0)} f(x, y) = 0$.

Solution Let $\ell = 0$, $P = (x, y)$, and $P_0 = (0, 0)$. Suppose that $\epsilon > 0$. Then

$$|f(P) - \ell| = |f(x, y) - 0| = |x^2 + y^2|.$$

Thus, $|f(P) - \ell| < \epsilon$ if $|x^2 + y^2| < \epsilon$. But this last inequality holds when $d(P, P_0) = \sqrt{(x - 0)^2 + (y - 0)^2} < \sqrt{\epsilon}$. Unwinding these observations, we see that we should set $\delta = \sqrt{\epsilon}$. It follows that if $0 < d(P, P_0) < \delta$, then $|f(P) - \ell| < \epsilon$. ■

Example 2 Define

$$f(x, y) = \begin{cases} \dfrac{xy}{x^2 + y^2}, & \text{if } (x, y) \neq (0, 0) \\ 0, & \text{if } (x, y) = (0, 0) \end{cases}.$$

Discuss the limiting behavior of $f(x, y)$ as $(x, y) \to (0, 0)$.

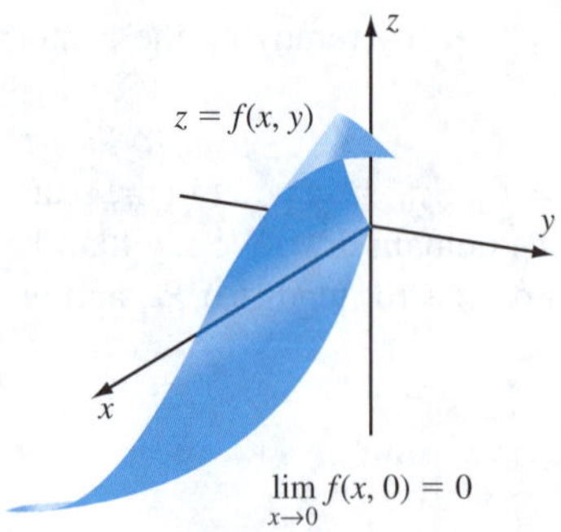

Figure 5

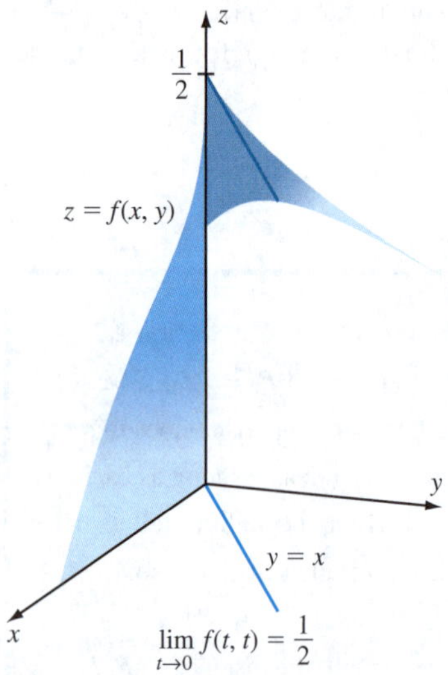

Figure 6

Solution Notice that $\lim_{x\to 0} f(x, 0) = \lim_{x\to 0} 0 = 0$. So, $f(x, y)$ approaches 0 as (x, y) approaches the origin along the x-axis. See Figure 5. Also, $\lim_{y\to 0} f(0, y) = \lim_{y\to 0} 0 = 0$. Thus, $f(x, y)$ approaches 0 as (x, y) approaches the origin along the y-axis. However, we find a different limiting behavior when we consider points $(x, y) = (t, t)$ that approach the origin diagonally as t approaches 0. In this case, we have

$$\lim_{t\to 0} f(t, t) = \lim_{t\to 0} \frac{t^2}{t^2 + t^2} = \lim_{t\to 0} \frac{1}{2} = \frac{1}{2}.$$

See Figure 6. In summary, if (x, y) is sufficiently close to the origin on either axis, then $f(x, y)$ is near 0, but if (x, y) is near the origin on the line $y = x$, then $f(x, y)$ is near $1/2$. For the limit of $f(x, y)$ to exist at the origin, the values of f must approach the same number no matter how the point (x, y) tends to the origin. Hence $\lim_{(x,y)\to(0,0)} f(x, y)$ does not exist. ■

inSIGHT

Although the function f in Example 2 happens to have been defined at $P_0 = (0, 0)$, we do not actually use the defined value $f(P_0)$ when we investigate the existence of $\lim_{P\to P_0} f(P)$. That is because the definition of "limit" refers to the punctured disk $D_*(P_0, \delta)$ from which the point P_0 has been removed. Our analysis would have been exactly the same had $f(P_0)$ been assigned a different value or, indeed, had $f(P_0)$ not been defined at all.

Continuity

The definition of continuity for a function of two variables is no different from the one in Section 2.3 for functions of one variable, except that we now use our new notion of limit. The philosophy is the same: We require that the value that we *anticipate* f will take at P_0 be the value that f *actually* takes:

Definition Suppose f is a function of two variables that is defined at a point $P_0 = (x_0, y_0)$. If $f(x, y)$ has a limit as (x, y) approaches (x_0, y_0) and if

$$\lim_{(x,y)\to(x_0,y_0)} f(x, y) = f(x_0, y_0),$$

then we say that f is continuous at P_0. If f is not continuous at a point in its domain, then we say that f is *discontinuous* there.

Notice that we only discuss the continuity of a function at points of its domain. The function $f(x, y) = x^2 + y^2$ from Example 1 *is* continuous at $(0, 0)$ because $\lim_{(x,y)\to(0,0)} f(x, y) = 0 = f(0, 0)$. However, the function

$$f(x, y) = \begin{cases} \dfrac{xy}{x^2 + y^2}, & \text{if } (x, y) \neq (0, 0) \\ 0, & \text{if } (x, y) = (0, 0) \end{cases}$$

from Example 2 is *not* continuous at $(0, 0)$ because $\lim_{(x,y)\to(0,0)} f(x, y)$ does not exist, so it certainly cannot equal $f(0, 0)$.

Example 3 Suppose that

$$f(x, y) = \begin{cases} \dfrac{(x^2 - y^2)^2}{x^2 + y^2}, & \text{if } (x, y) \neq (0, 0) \\ 1, & \text{if } (x, y) = (0, 0) \end{cases}.$$

Is f continuous at $(0, 0)$?

Solution Observe that, for $(x, y) \neq (0, 0)$,

$$0 \leq |f(x, y)| = \frac{(x^2 - y^2)^2}{x^2 + y^2} \leq \frac{(x^2 + y^2)^2}{x^2 + y^2} = x^2 + y^2.$$

In Example 1, we proved that $\lim_{(x,y)\to(0,0)}(x^2 + y^2) = 0$. It follows that $f(x, y)$ also tends to 0 as (x, y) approaches $(0, 0)$. Therefore,

$$\lim_{(x,y)\to(0,0)} f(x, y) = 0 \neq 1 = f(0, 0).$$

We conclude that f is not continuous at $(0, 0)$. ■

Rules for Limits

Fortunately, there are a number of rules about limits of functions, and about continuous functions, that often make it easy to check limits and continuity in practice. These rules are similar to those that we learned in Chapter 2 for functions of one variable. We record the new rules here for reference.

Theorem 1 Suppose that f and g are functions of two variables such that

$$\lim_{(x,y)\to(x_0,y_0)} f(x, y) \quad \text{and} \quad \lim_{(x,y)\to(x_0,y_0)} g(x, y)$$

exist. Then

a. $\lim_{(x,y)\to(x_0,y_0)}(f + g)(x, y) = \lim_{(x,y)\to(x_0,y_0)} f(x, y) + \lim_{(x,y)\to(x_0,y_0)} g(x, y)$,
b. $\lim_{(x,y)\to(x_0,y_0)}(f - g)(x, y) = \lim_{(x,y)\to(x_0,y_0)} f(x, y) - \lim_{(x,y)\to(x_0,y_0)} g(x, y)$,
c. $\lim_{(x,y)\to(x_0,y_0)}(f \cdot g)(x, y) = (\lim_{(x,y)\to(x_0,y_0)} f(x, y)) \cdot (\lim_{(x,y)\to(x_0,y_0)} g(x, y))$,
d. $\lim_{(x,y)\to(x_0,y_0)}(f/g)(x, y) = \lim_{(x,y)\to(x_0,y_0)} f(x, y)/\lim_{(x,y)\to(x_0,y_0)} g(x, y)$ provided that $\lim_{(x,y)\to(x_0,y_0)} g(x, y) \neq 0$, and
e. $\lim_{(x,y)\to(x_0,y_0)}(\lambda \cdot f(x, y)) = \lambda \cdot (\lim_{(x,y)\to(x_0,y_0)} f(x, y))$ for any constant λ.

Example 4 Define $f(x, y) = (x+y+1)/(x^2-y^2)$. What is the limiting behavior of f as (x, y) tends to $(1, 2)$?

Solution Since

$$\lim_{(x,y)\to(1,2)} (x^2 - y^2) = \lim_{(x,y)\to(1,2)} x^2 - \lim_{(x,y)\to(1,2)} y^2 = 1 - 4 = -3,$$

which is nonzero, we have

$$\lim_{(x,y)\to(1,2)} \frac{x+y+1}{x^2-y^2} = \frac{\lim_{(x,y)\to(1,2)}(x+y+1)}{\lim_{(x,y)\to(1,2)}(x^2-y^2)}$$
$$= \frac{\lim_{(x,y)\to(1,2)} x + \lim_{(x,y)\to(1,2)}(y+1)}{-3} = \frac{1+2+1}{-3} = -\frac{4}{3}.$$

Example 5 Evaluate the limit

$$\lim_{(x,y)\to(3,-2)} \frac{2x^2+5xy+3y^2}{2x+3y}.$$

Solution Notice that as $(x, y) \to (3, -2)$, both the numerator and the denominator tend to 0. Using our experience from Chapter 2, we factor the numerator to see if anything cancels. Our limit problem then becomes

$$\lim_{(x,y)\to(3,-2)} \frac{2x^2+5xy+3y^2}{2x+3y} = \lim_{(x,y)\to(3,-2)} \frac{(2x+3y)(x+y)}{2x+3y}$$
$$= \lim_{(x,y)\to(3,-2)} (x+y) = 1.$$

The usual rules for continuity follow from the limit rules of Theorem 1.

Theorem 2 Suppose that f and g are functions that are continuous at $P_0 = (x_0, y_0)$. Then $f + g$, $f - g$, and $f \cdot g$ are also continuous at P_0. If $g(P_0) \neq 0$, then f/g is also continuous at P_0. If ϕ is a function of one variable that is continuous at $f(P_0)$, then $\phi \circ f$ is continuous at P_0.

Example 6 Define

$$g(x, y) = y^3 \sin(x) - \frac{\cos(y)}{1+xy^2}.$$

Discuss the continuity of g at $(-\pi/2, \pi)$.

Solution Observe that

$$\lim_{(x,y)\to(-\pi/2,\pi)} (1+xy^2) = 1 - \frac{\pi^3}{2} \quad \text{and} \quad \lim_{(x,y)\to(-\pi/2,\pi)} \cos(y) = -1.$$

Here we are using the continuity properties of the one-variable functions x, y^2, $\cos(y)$, and 1, together with the rules for limits. Thus

$$\lim_{(x,y)\to(-\pi/2,\pi)} \frac{\cos(y)}{1+xy^2} = \frac{-1}{1-\pi^3/2} = \frac{-2}{2-\pi^3}.$$

A similar calculation shows that

$$\lim_{(x,y)\to(-\pi/2,\pi)} y^3 \sin(x) = -\pi^3.$$

It follows from Theorem 1b that

$$\lim_{(x,y)\to(-\pi/2,\pi)} g(x,y) = -\pi^3 + \frac{2}{2-\pi^3}.$$

But we may also calculate that

$$g\left(-\frac{\pi}{2}, \pi\right) = -\pi^3 + \frac{2}{2-\pi^3}.$$

We conclude that the function g is continuous at $(-\pi/2, \pi)$. ■

Example 7 Discuss the continuity of the function $g(x, y) = \sin(xy^2 - x)$.

Solution The function is continuous at every (x, y), since $g = \phi \circ f$ where f and ϕ are the continuous functions defined by $f(x, y) = xy^2 - x$ and $\phi(u) = \sin(u)$. ■

We conclude by noting that the entire discussion of this section applies to functions of three variables x, y, z with no change (except for notation). In particular, all of the rules and remarks are still true.

quickquiz

1. How do we define the limit of a function of two variables?
2. Suppose that $f(x, y)$ has limit ℓ as (x, y) tends to P_0. What additional conditions must be true for f to be continuous at P_0?
3. What algebraic operations are respected by the limiting process for functions of two variables?
4. What algebraic operations are respected by continuous functions of two variables?

EXERCISES

Problems for Practice

In Exercises 1–14, calculate the limit.

1. $\lim_{(x,y)\to(5,7)} x(2xy-3)^2$
2. $\lim_{(x,y)\to(-1,6)} \sin\left(\frac{\pi xy}{4}\right)$
3. $\lim_{(x,y)\to(4,2)} \sqrt{x^2y - x}$
4. $\lim_{(x,y)\to(1,-1)} \frac{x^2-y^2}{x+y}$
5. $\lim_{(x,y)\to(-1,4)} \ln(y^2+xy)$
6. $\lim_{(x,y)\to(4,-8)} x^{-1/2}y^{1/3}$
7. $\lim_{(x,y)\to(1/4,1/4)} \frac{\arccos(x-y)}{e^{x-y}}$
8. $\lim_{(x,y)\to(\pi/2,2)} \arctan\left(\cos\left(\frac{x}{2}\right)\sqrt{y}\right)$
9. $\lim_{(x,y)\to(0,\pi/2)} \sin(y\cos(x)) + \cos(x+\cos(y))$
10. $\lim_{(x,y)\to(0,0)} \frac{\sin(x+y)}{(x+y)}$
11. $\lim_{(x,y)\to(1,2)} \frac{(x-1)(y^2-4)}{(y-2)(x^2-1)}$
12. $\lim_{(x,y)\to(0,3)} \frac{y\cdot\sin(x)}{x(y^2+1)}$
13. $\lim_{(x,y)\to(6,3)} \frac{x^2-xy-2y^2}{x-2y}$
14. $\lim_{(x,y)\to(0,0)} \frac{1-\exp(x^2+y^2)}{x^2+y^2}$

Each expression in Exercises 15–20 does *not* have a limit at (0, 0). Let (x, y) approach the origin along two different paths, as in Example 2, to demonstrate the nonexistence of the limit.

15. y/x
16. $y(x^2+y^2)^{-1/2}$
17. $xy^2/|xy^2|$
18. $(x^2+y^2)/(x^2-y^2)$
19. $(x+y)/(x-y)$
20. $(x+y)/(x+2y)$

Further Theory and Practice

In Exercises 21–24, determine whether the function is continuous at (0, 0).

21.

$$f(x, y) = \begin{cases} \dfrac{\sin(x^2 + y^2)}{x^2 + y^2}, & (x, y) \neq (0, 0) \\ 1, & (x, y) = (0, 0) \end{cases}$$

22.

$$f(x, y) = \begin{cases} \dfrac{(x^2 + y^2)^2}{x^4 + y^4}, & (x, y) \neq (0, 0) \\ 1, & (x, y) = (0, 0) \end{cases}$$

23.

$$f(x, y) = \begin{cases} \dfrac{y^2 - 2x^2}{x^2 + y^2}, & (x, y) \neq (0, 0) \\ 1, & (x, y) = (0, 0) \end{cases}$$

24.

$$f(x, y) = \begin{cases} \dfrac{x^4 + 3y^4}{x^2 + y^2}, & (x, y) \neq (0, 0) \\ 0, & (x, y) = (0, 0) \end{cases}$$

25. If ϕ and ψ are continuous functions, show that $(x, y) \mapsto \phi(x) + \psi(y)$ is continuous.

26. If f is continuous at (x_0, y_0), show that $\phi(x) = f(x, y_0)$ is continuous at x_0 and that $\psi(y) = f(x_0, y)$ is continuous at y_0.

27. If $\phi(x) = f(x, y_0)$ is continuous at x_0 and $\psi(y) = f(x_0, y)$ is continuous at y_0, is it true that f is continuous at (x_0, y_0)? Give a reason for your answer.

28. Calculate $\lim_{(x,y)\to(1,0)} (1 + xy)^{1/xy}$.

29. Calculate $\lim_{(x,y)\to(0,0)} x^2 \ln(x^2 + y^2)$.

30. Let $f(x, y) = 2xy/(x^2 + y^2)$. Calculate $\lim_{y\to 0} f(0, y)$. Also, for any slope $-\infty < m < \infty$, calculate $\lim_{x\to 0} f(x, mx)$. Show that $f(x, y)$ has a limit as (x, y) approaches (0, 0) along any straight line path approaching the origin and that the limit can be any number in the interval $[-1, 1]$.

31. Let $f(x, y) = xy^2/(x^2 + y^4)$. This function does not have a limit as (x, y) tends to (0, 0). However, you cannot detect this by looking only at straight line paths approaching the origin. Calculate $\lim_{y\to 0} f(0, y)$. Also, for any slope $-\infty < m < \infty$, calculate $\lim_{x\to 0} f(x, mx)$. Deduce that along any straight line path approaching the origin, the limit is 0. However, show that there is a different limit as (x, y) approaches (0, 0) along the parabolic path $x = y^2$.

32. Let $f(x, y) = xy^3/(x^2 + y^6)$. Prove that along any straight line path approaching the origin, the limit of $f(x, y)$ is 0. However, show that there is another path along which $f(x, y)$ has a different limit as (x, y) approaches (0, 0). Therefore, $f(x, y)$ does not have a limit as (x, y) tends to (0, 0).

Calculator/Computer Exercises

33. Plot $f(x, y) = xy/(x^2 + y^2)$ for (x, y) in a rectangle centered at (0, 0). Use your plot to verify visually that f does not have a limit at (0, 0).

34. Plot $f(x, y) = (x^2 - y^2)/(x^2 + y^2)$ for (x, y) in a rectangle centered at (0, 0). Use your plot to verify visually that f does not have a limit at (0, 0).

35. Plot $f(x, y) = (x + y)^2/(x^2 + y^2)$ for (x, y) in a rectangle centered at (0, 0). Use your plot to verify visually that f does not have a limit at (0, 0).

36. Plot $f(x, y) = xy^2/(x^2 + y^4)$ for (x, y) in a rectangle centered at (0, 0). Use your plot to verify visually that f does not have a limit at (0, 0). Note, however, that the limit *does exist* and does equal zero along straight lines through the origin.

13.4 Partial Derivatives

In this section, we will learn how to determine the rate at which a function f of two variables changes. To do so, we will need an appropriate notion of differentiation. Since there are *two* independent variables, we shall develop a procedure for differentiating in each variable separately. The geometric idea behind these new differentiation techniques is simple: We take slices of the graph of f. You are already familiar with the strategy of intersecting the graph of $z = f(x, y)$ with planes. Up until now, we have relied primarily on horizontal planes as an aid in plotting. Such planes, however, are

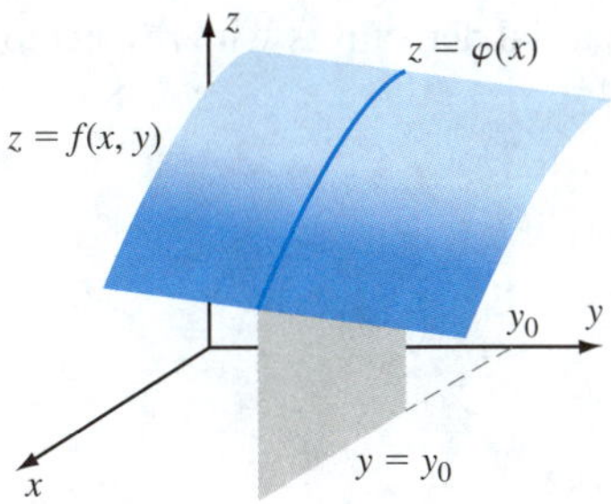

Figure 1

of no use for our current purposes, because the values of $f(x, y)$ do not change on the level curves that they produce. Instead, we will now intersect the graph of $z = f(x, y)$ with planes that are parallel to the xz- and yz-coordinate planes.

Let $P_0 = (x_0, y_0)$ be a point in the plane. Suppose that f is a function whose domain contains a disk centered at P_0. If we fix y_0 and allow only x to vary, then we obtain a function $\varphi(x) = f(x, y_0)$ of one variable. The graph of $z = \varphi(x)$ can be realized by intersecting the graph of $z = f(x, y)$ with the plane $y = y_0$, as shown in Figure 1. Ordinarily, the graph of $z = \varphi(x)$ would be plotted in the xz-plane, which is the plane $y = 0$ in xyz-space. Here it is realized in the plane $y = y_0$, which is parallel to the xz-plane.

Because φ is a function of one variable, we can differentiate it in the ordinary way. In addition, the number

$$\left.\frac{d\varphi}{dx}(x)\right|_{x=x_0} = \lim_{\Delta x \to 0} \frac{\varphi(x_0 + \Delta x) - \varphi(x_0)}{\Delta x} = \lim_{\Delta x \to 0} \frac{f(x_0 + \Delta x, y_0) - f(x_0, y_0)}{\Delta x}$$

has its usual interpretation—the instantaneous rate of change of φ at x_0. Since φ is just the function f in which the y-variable has been fixed at y_0, the value of $\varphi'(x_0)$ is the instantaneous rate of change of f at (x_0, y_0) as x changes and $y = y_0$ remains fixed.

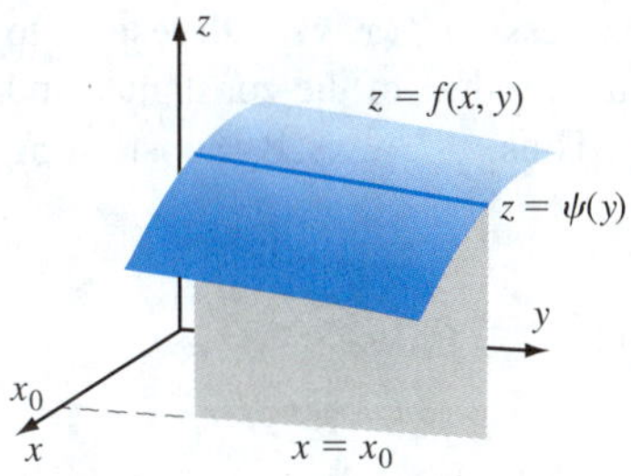

Figure 2

In an analogous manner, we can fix $x = x_0$, define $\psi(y) = f(x_0, y)$, and use

$$\left.\frac{d\psi}{dy}(y)\right|_{y=y_0} = \lim_{\Delta y \to 0} \frac{\psi(y_0 + \Delta y) - \psi(y_0)}{\Delta y} = \lim_{\Delta y \to 0} \frac{f(x_0, y_0 + \Delta y) - f(x_0, y_0)}{\Delta y}$$

to tell us the instantaneous rate of change of f at (x_0, y_0) as y changes and $x = x_0$ remains fixed. See Figure 2 for the geometry. These considerations give us a way to define differentiation for a function of two variables.

Definition

Let $P_0 = (x_0, y_0)$ be a point in the plane. Suppose that f is a function defined on a disk $D(P_0, r)$. We say that f is differentiable in the x-variable at P_0 if

$$\lim_{\Delta x \to 0} \frac{f(x_0 + \Delta x, y_0) - f(x_0, y_0)}{\Delta x}$$

exists. We call this limit the *partial derivative of f with respect to x at the point P_0* and denote it by

$$\frac{\partial f}{\partial x}(P_0).$$

Similarly, we say that f is differentiable in the y-variable at P_0 if

$$\lim_{\Delta y \to 0} \frac{f(x_0, y_0 + \Delta y) - f(x_0, y_0)}{\Delta y}$$

exists. We call this limit the *partial derivative* of f *with respect to y at the point P_0* and denote it by

$$\frac{\partial f}{\partial y}(P_0).$$

Example 1 Define $f(x, y) = xy^2$. Calculate the partial derivatives with respect to x and with respect to y. What are

$$\frac{\partial f}{\partial x}(1, 3) \quad \text{and} \quad \frac{\partial f}{\partial y}(1, 3)?$$

Compare these values with the derivatives

$$\frac{d\varphi}{dx}(1) \quad \text{and} \quad \frac{d\psi}{dy}(3)$$

where φ and ψ are functions of one variable defined by $\varphi(x) = f(x, 3)$ and $\psi(y) = f(1, y)$.

Solution When we calculate $\frac{\partial f}{\partial x}$, we differentiate the expression $f(x, y)$ with respect to x, treating y and any expression depending only on y as if it were a constant. In the case of xy^2, that means we differentiate with respect to x the product of the constant y^2 and the variable x. The result is y^2 times 1. Thus, $\frac{\partial f}{\partial x} = y^2$. Similarly, when we calculate $\frac{\partial f}{\partial y}$, we think of x as a constant and differentiate the expression $f(x, y)$ with respect to y. In other words, we differentiate with respect to y the product of the constant x and the expression y^2. The result is the constant x times $2y$. Thus, $\frac{\partial f}{\partial y} = 2xy$. By substituting $x = 1$ and $y = 3$, we obtain

$$\frac{\partial f}{\partial x}(1, 3) = 9 \quad \text{and} \quad \frac{\partial f}{\partial y}(1, 3) = 6.$$

Next, consider the function $\varphi(x) = f(x, 3) = x(3)^2 = 9x$. We calculate $\varphi'(x) = 9$. Therefore, $\varphi'(1) = 9$, which is the same as $\frac{\partial f}{\partial x}$ at $(1, 3)$. Similarly, $\psi(y) = f(1, y) = 1 \cdot y^2$ and

$$\psi'(3) = \frac{d}{dy}(y^2)\bigg|_{y=3} = 2y\big|_{y=3} = 6,$$

which equals $\frac{\partial f}{\partial y}(1, 3)$. ■

All of our familiar rules for differentiation—the Product Rule, the Quotient Rule, the Sum-Difference Rule, the Chain Rule, and so on—are still valid for partial derivatives. This is so because partial differentiation is simply the differentiation that we already know applied one variable at a time.

Example 2 Calculate the partial derivatives $\frac{\partial f}{\partial x}$ and $\frac{\partial f}{\partial y}$ for $f(x, y) = \sqrt{x^3 + y^2}$.

Solution We have

$$\frac{\partial f}{\partial x}(x, y) = \frac{\partial}{\partial x}(x^3 + y^2)^{1/2} \overset{\text{Chain Rule}}{=} \frac{1}{2}(x^3 + y^2)^{-1/2}\frac{\partial}{\partial x}(x^3 + y^2) = \frac{3}{2}x^2(x^3 + y^2)^{-1/2}.$$

Also,

$$\frac{\partial f}{\partial y}(x, y) = \frac{\partial}{\partial y}(x^3 + y^2)^{1/2} \overset{\text{Chain Rule}}{=} \frac{1}{2}(x^3 + y^2)^{-1/2} \frac{\partial}{\partial y}(x^3 + y^2) = y(x^3 + y^2)^{-1/2}.$$

There are several common notations for the partial derivatives of a function $(x, y) \mapsto f(x, y)$ of two variables. Thus,

$$\frac{\partial f}{\partial x},\ D_x f,\ f_x,\ f_1, \text{ and } D_1 f$$

are all used for the partial derivative of f with respect to x. Similarly, $\frac{\partial f}{\partial y}$, $D_y f$, f_y, f_2, and $D_2 f$ all stand for the partial derivative of f with respect to y.

Example 3 Calculate f_x and f_y for the function f defined by

$$f(x, y) = \ln(x) \cdot e^{x\cos(y)}.$$

Solution We calculate f_x using the Product Rule and the Chain Rule:

$$\begin{aligned} f_x(x, y) &= \left(\frac{\partial}{\partial x}\ln(x)\right) e^{x\cos(y)} + \ln(x)\frac{\partial}{\partial x}e^{x\cos(y)} \\ &= \frac{1}{x} \cdot e^{x\cos(y)} + \ln(x)\big(e^{x\cos(y)} \cdot \cos(y)\big). \end{aligned}$$

When we calculate a partial derivative with respect to y, we treat x as a constant. Therefore, $\ln(x)$ is treated as a constant when calculating f_y. We do not need the Product Rule for this calculation:

$$f_y(x, y) = \ln(x)\frac{\partial}{\partial y}\left(e^{x\cos(y)}\right) = \ln(x)e^{x\cos(y)}\frac{\partial}{\partial y}(x\cos(y)) = -\ln(x)e^{x\cos(y)} \cdot x\sin(y).$$

Example 4 At time $t = 0$, a string is at rest along the x-axis from 0 to 1. It may vibrate up and down in the y-direction. If the displacement of the string at point x and time t is given by the formula $y(x, t) = \sin(\pi x)\sin(2t)$, then at the point $x = 1/4$, what is the instantaneous rate of change of y with respect to time?

Solution We calculate

$$\frac{\partial}{\partial t}y(x, t) = \frac{\partial}{\partial t}(\sin(\pi x)\sin(2t)) = 2\sin(\pi x)\cos(2t).$$

The *instantaneous rate of change with respect to time* of the string's displacement from equilibrium at the point $x = 1/4$ is $2\sin(\pi/4)\cos(2t)$, or $\sqrt{2}\cos(2t)$.

inSIGHT

Example 4 demonstrates that the partial derivative with respect to a variable can be interpreted as the *rate of change* with respect to that variable. Because many physical quantities depend on several parameters, we need to have this new multivariate way to measure rates of change. For instance, if heat is being applied to the ends of a rod, then the temperature T at a point x on the rod depends both on x and on the time t. The partial derivatives $\frac{\partial T}{\partial x}$ and $\frac{\partial T}{\partial t}$ tell us about the rates of change when one of the two independent variables is held constant.

Functions of Three Variables

The notion of partial derivative also applies to functions of three (or more) variables. If $F(x, y, z)$ is an expression involving three independent variables x, y, and z, then we calculate $\frac{\partial F}{\partial x}$ by holding y and z constant and differentiating with respect to x. The partial derivatives in y and z are calculated similarly.

Example 5 Calculate the partial derivatives of $F(x, y, z) = xz\sin(y^2z)$ with respect to x, with respect to y, and with respect to z.

Solution When we hold y and z constant, the expression $z\sin(y^2z)$ is a constant, and $f(x, y, z)$ is this constant times x. Therefore, $\frac{\partial F}{\partial x} = z\sin(y^2z)$. Next, hold x and z constant to calculate the partial derivative with respect to y:

$$\begin{aligned}\frac{\partial F}{\partial y}(x, y, z) &= \frac{\partial}{\partial y}(xz\sin(y^2z))\\ &= xz\frac{\partial}{\partial y}(\sin(y^2z)) && \text{The constant slides outside the derivative}\\ &= xz\cos(y^2z)\frac{\partial}{\partial y}(y^2z) && \text{Using the Chain Rule}\\ &= xz^2\cos(y^2z)\frac{\partial}{\partial y}(y^2) && \text{The constant slides outside the derivative}\\ &= 2xyz^2\cos(y^2z).\end{aligned}$$

We calculate $\frac{\partial F}{\partial z}$ in a similar way, but for this partial derivative, we require the Product Rule:

$$\frac{\partial F}{\partial z}(x, y, z) = \frac{\partial}{\partial z}(xz\sin(y^2z)) = x\frac{\partial}{\partial z}(z\sin(y^2z)) = x\sin(y^2z) + xz\frac{\partial}{\partial z}(\sin(y^2z)).$$

Continuing the calculation with an application of the Chain Rule, we have

$$\frac{\partial F}{\partial z}(x, y, z) = x\sin(y^2z) + xz\cos(y^2z)\frac{\partial}{\partial z}(y^2z) = x\sin(y^2z) + xy^2z\cos(y^2z).$$ ■

For a function $(x, y, z) \mapsto F(x, y, z)$ of three variables, there are several alternative notations for the partial derivative with respect to x:

$$\frac{\partial F}{\partial x}, D_x F, F_x, F_1, D_1 F.$$

Similarly, $\frac{\partial F}{\partial y}$, $D_y F$, F_y, F_2, and $D_2 F$ can be used interchangeably for the partial derivative with respect to y. Likewise, $\frac{\partial F}{\partial z}$, $D_z F$, F_z, F_3, and $D_3 F$ all refer to the partial derivative with respect to z.

Higher Partial Derivatives

Since the partial derivative of a given function $f(x, y)$ is a new function, we may attempt to differentiate that function again. The result, if it exists, is called a *second partial derivative*. Some examples are

$$\frac{\partial^2 f}{\partial x^2}, \quad \text{which means} \quad \frac{\partial}{\partial x}\left(\frac{\partial}{\partial x} f\right);$$
$$\frac{\partial^2 f}{\partial y^2}, \quad \text{which means} \quad \frac{\partial}{\partial y}\left(\frac{\partial}{\partial y} f\right);$$
$$\frac{\partial^2 f}{\partial x \partial y}, \quad \text{which means} \quad \frac{\partial}{\partial x}\left(\frac{\partial}{\partial y} f\right);$$
$$\frac{\partial^2 f}{\partial y \partial x}, \quad \text{which means} \quad \frac{\partial}{\partial y}\left(\frac{\partial}{\partial x} f\right).$$

The last two of these higher-order partial derivatives involve differentiation with respect to both x and y. Therefore, they are sometimes called *mixed partial derivatives*.

Of course, we can also use the other partial derivative notations for second derivatives. For example, $\frac{\partial^2 f}{\partial x^2}$ can be written as

$$D_{xx} f, f_{xx}, f_{1,1}, \text{ or } D_{1,1} f;$$

$\frac{\partial^2 f}{\partial y \partial x}$ may be written as

$$D_{xy} f, f_{xy}, f_{1,2}, \text{ or } D_{1,2} f;$$

and so on.

Example 6 Calculate all the second partial derivatives of $f(x, y) = xy - y^3 + x^2 y^4$.

Solution We have

$$\frac{\partial^2 f}{\partial x^2}(x, y) = \frac{\partial}{\partial x}\left(\frac{\partial f}{\partial x}(x, y)\right) = \frac{\partial}{\partial x}(y + 2xy^4) = 2y^4.$$

Also,

$$\frac{\partial^2 f}{\partial y^2} = \frac{\partial}{\partial y}\left(\frac{\partial f}{\partial y}(x, y)\right) = \frac{\partial}{\partial y}(x - 3y^2 + 4x^2 y^3) = -6y + 12x^2 y^2.$$

And

$$\frac{\partial^2 f}{\partial x \partial y} = \frac{\partial}{\partial x}\left(\frac{\partial f}{\partial y}(x, y)\right) = \frac{\partial}{\partial x}(x - 3y^2 + 4x^2y^3) = 1 + 8xy^3.$$

Finally,

$$\frac{\partial^2 f}{\partial y \partial x} = \frac{\partial}{\partial y}\left(\frac{\partial f}{\partial x}(x, y)\right) = \frac{\partial}{\partial y}(y + 2xy^4) = 1 + 8xy^3.$$ ■

inSIGHT

A remarkable thing occurred in Example 6. The first order partial derivatives turned out to be

$$\frac{\partial f}{\partial x}(x, y) = y + 2xy^4 \quad \text{and} \quad \frac{\partial f}{\partial y}(x, y) = x - 3y^2 + 4x^2y^3.$$

At first glance, these expressions do not appear to have much in common. Yet, when we used $\frac{\partial f}{\partial x}$ and $\frac{\partial f}{\partial y}$ to calculate the mixed partial derivatives $\frac{\partial^2 f}{\partial y \partial x}$ and $\frac{\partial^2 f}{\partial x \partial y}$, respectively, we discovered that

$$\frac{\partial^2 f}{\partial x \partial y} = 1 + 8xy^3 = \frac{\partial^2 f}{\partial y \partial x}.$$

No matter which order we choose to differentiate f with respect to x and y, we obtain the same mixed partial derivative. Is this a coincidence or a particular case of something that is true all the time? The answer is that if we restrict our attention to a class of reasonably nice functions, then the order of differentiation is, indeed, irrelevant. The class of "nice" functions that we will study is described in the next definition.

Definition A function $(x, y) \mapsto f(x, y)$ is called *continuously differentiable* on an open disk $D(P, r)$ if f and its partial derivatives

$$f_x \quad \text{and} \quad f_y$$

are continuous on $D(P, r)$. The function f is called *twice continuously differentiable* on $D(P, r)$ if f and its partial derivatives

$$f_x, f_y, f_{xx}, f_{yy}, f_{xy}, \text{and } f_{yx}$$

are all continuous on $D(P, r)$.

Theorem 1 If $(x, y) \mapsto f(x, y)$ is twice continuously differentiable on an open disk $D(P, r)$, then its mixed partial derivatives are equal in the disk:

$$f_{xy}(x, y) = f_{yx}(x, y), \qquad (x, y) \in D(P, r).$$

inSIGHT

The proof of Theorem 1 is discussed in Exercise 68. Does Theorem 1 mean that we have to check that every function we encounter is twice continuously differentiable? Fortunately, the answer is "no." The functions that we encounter in practice are built up with elementary operations from x, y, $u \mapsto \sin(u)$, $u \mapsto \ln(u)$, and so on. Each of these is twice continuously differentiable wherever it is defined, and the elementary operations preserve this property (as long as we never divide by 0). So, for the functions we will be using in the rest of this book, the real meaning of Theorem 1 is that we *can* differentiate in any order we like.

The class of functions that admit switching of the order of differentiation is in fact somewhat larger than the class of twice continuously differentiable functions—but this is a matter for an advanced analysis course and need not concern us.

Concluding Remarks about Partial Differentiation

The derivatives that we learned about in Chapter 3—namely, derivatives of functions of one variable—are sometimes called *ordinary derivatives* to distinguish them from the partial derivatives that we discuss in this chapter. When we compute an ordinary derivative, we write $\frac{d}{dx}$; but if we compute a partial derivative, then we write $\frac{\partial}{\partial x}$ or $\frac{\partial}{\partial y}$ (or one of the other notations we have discussed). Sometimes a partial derivative can turn out to be an ordinary derivative. For example, if g is a function of one variable, then we have

$$\frac{\partial^2}{\partial x \partial y}\big(g(x)y + 4\sin(x)\big) = \frac{\partial}{\partial x}\left(\frac{\partial}{\partial y}\big(g(x)y + 4\sin(x)\big)\right) = \frac{d}{dx}g(x).$$

In some applications, it is necessary to calculate derivatives of order higher than two. For example,

$$\frac{\partial}{\partial x}\left(\frac{\partial^2 f}{\partial x \partial y}\right)$$

is a third derivative obtained by applying $\frac{\partial}{\partial x}$ to the already familiar quantity $\frac{\partial^2 f}{\partial x \partial y}$. In general, the expression

$$\frac{\partial^m f}{\partial x^j \partial y^k},$$

where $j + k = m$, denotes a derivative of order m: Differentiation in x is applied j times, and differentiation in y is applied k times. To guarantee that the differentiations may be applied in any order with the same result, we usually assume that the function is m times continuously differentiable (all partial derivatives up to and including order m exist and are continuous).

quickquiz

1. How do we define the partial derivative in x of a function $f(x, y)$ of two variables?
2. Calculate $f_x(x, y)$ and $f_y(x, y)$ for $f(x, y) = \sin(x) + \cos(2xy)$.
3. Discuss the order of differentiation when computing higher derivatives.
4. Calculate $f_{xy}(x, y)$ where $f(x, y) = \sin(x \ln(y))$.

EXERCISES

Problems for Practice

For Exercises 1–10, a function f and a point $P_0 = (x_0, y_0)$ are given. Calculate $\varphi(x) = f(x, y_0)$, $\psi(y) = f(x_0, y)$, $\varphi'(x_0)$, and $\psi'(y_0)$. Then calculate $\frac{\partial f}{\partial x}(x_0, y_0)$ and $\frac{\partial f}{\partial y}(x_0, y_0)$. Verify that $\varphi'(x_0) = \frac{\partial f}{\partial x}(x_0, y_0)$ and $\psi'(y_0) = \frac{\partial f}{\partial y}(x_0, y_0)$.

1. $f(x, y) = 2x + y^3$, $P_0 = (5, 1)$
2. $f(x, y) = \pi + \sin(x) - \cos(y)$, $P_0 = (\pi/4, \pi/3)$
3. $f(x, y) = 7 - 3y^4$, $P_0 = (2, -1)$
4. $f(x, y) = 1 + 3xy^2 - y$, $P_0 = (1, 2)$
5. $f(x, y) = \cos(xy^2)$, $P_0 = (\pi, 1/2)$
6. $f(x, y) = x\exp(y)$, $P_0 = (1, 0)$
7. $f(x, y) = x\ln(xy)$, $P_0 = (1, e)$
8. $f(x, y) = x^2/(x + 2y)$, $P_0 = (4, 2)$
9. $f(x, y) = x^y$, $P_0 = (2, 2)$
10. $f(x, y) = (2x - y)/(1 + 3x + 3y)$, $P_0 = (3, -2)$

In Exercises 11–20, a function $(x, y) \mapsto f(x, y)$ is given. Calculate f_{xx}, f_{yy}, f_{xy}, and f_{yx}. Verify that the mixed partial derivatives are equal.

11. $f(x, y) = x^2y - yx + y^5$
12. $f(x, y) = x/(x + y)$
13. $f(x, y) = e^{xy}\cos(x)$
14. $f(x, y) = \tan(x^2y)$
15. $f(x, y) = \sin(x)/\cos(y)$
16. $f(x, y) = \sin(xy)$
17. $f(x, y) = \ln(xy - y^2)$
18. $f(x, y) = y^2\ln(x + y)$
19. $f(x, y) = \cos(x)\sin(y)$
20. $f(x, y) = (x - y)/(x + y)$

For the functions in Exercises 21–30, calculate $D_y f$, $D_{yy} f$, $D_{yx} f$, f_{12}, and f_{11}. Verify that $D_{yx} f = f_{12}$.

21. $f(x, y) = x\sec(y)$
22. $f(x, y) = (x - y)/(1 - xy)$
23. $f(x, y) = \sqrt{\sin(xy)}$
24. $f(x, y) = \tan(x^2 + y^2)$
25. $f(x, y) = (\sin(x) - \cos(y))^5$
26. $f(x, y) = \exp(x - 2y)$
27. $f(x, y) = (1 + x^2)(x + 3y)^6$
28. $f(x, y) = 1/(4 + x^2y^2)^3$
29. $f(x, y) = \sqrt{1 + 3x^2 + y^4}$
30. $f(x, y) = y\sin(x^2y)$

In Exercises 31–40, a function $(x, y, z) \mapsto f(x, y, z)$ is given. Calculate f_x, f_y, f_z, f_{xy}, f_{xz}, f_{yz}, f_{xx}, f_{yy}, and f_{zz}.

31. $f(x, y, z) = 2x^3 + y^5z^8$
32. $f(x, y, z) = x^3y - 3z\ln(y)$
33. $f(x, y, z) = (x + 2y + 3z)^5$
34. $f(x, y, z) = \sin(xyz)$
35. $f(x, y, z) = \exp(xy - yz)$
36. $f(x, y, z) = x^2/(y^2 + z^2)$
37. $f(x, y, z) = y\sin(\ln(xz))$
38. $f(x, y, z) = \ln(x^2y + 2zy^3)$
39. $f(x, y, z) = (x^2 + y^3 + z^4)^3$
40. $f(x, y, z) = \exp(z\cos(x)\sin(y))$

In Exercises 41–44, a function $(x, y, z) \mapsto f(x, y, z)$ is given. Calculate f_{xyz}, f_{yxz}, f_{xzy}, f_{zxy}, f_{yzx}, and f_{zyx}. (All six calculations should result in the same final expression, but each will involve different intermediate results.)

41. $f(x, y, z) = xy^2z^3$
42. $f(x, y, z) = xy + 2xz$
43. $f(x, y, z) = (y + 3z)/x$
44. $f(x, y, z) = (x + 2y + 3z)^{-2}$

Further Theory and Practice

45. Express the volume V of a cone as a function of height h and radius r. What is the rate of change of volume with respect to radius? With respect to height?

46. Express the surface area A of a box in terms of length x, width y, and height z. What is the rate of change of surface area with respect to length?

47. The pressure P, volume V, and temperaure T of a sample of an ideal gas are related by the equation $PV = kT$ for a constant k. Express each variable, P, V, and T, as a function of the other two variables. Then show that

$$\frac{\partial P}{\partial V} \cdot \frac{\partial V}{\partial T} \cdot \frac{\partial T}{\partial P} = -1.$$

(*Note:* The partial differentials cannot be formally cancelled.)

48. Show that $u(x, y) = x \sin(x/y)$ satisfies the equation $xu_x(x, y) + yu_y(x, y) = u(x, y)$.

49. Show that $u(x, y) = \ln(\exp(x) + \exp(y))$ satisfies the equation $u_x(x, y) + u_y(x, y) = 1$.

50. Show that $u(x, y) = \arctan(y/x)$, $u(x, y) = \ln(y) - \ln(x)$, and $u(x, y) = 2xy/(x^2 + y^2)$ all satisfy the equation $xu_x(x, y) + yu_y(x, y) = 0$.

The partial differential equation

$$\frac{\partial^2 u}{\partial x^2} + \frac{\partial^2 u}{\partial y^2} = 0$$

is called *Laplace's equation*. It describes, for example, the steady state heat distribution in a thin metal plate. A function $u(x, y)$ that satisfies Laplace's equation is said to be *harmonic*. Exercises 51–58 concern solutions of Laplace's equation.

51. Which of the following functions is harmonic?

a. $u(x, y) = x^2 - y^2$
b. $u(x, y) = xy$
c. $u(x, y) = e^x \cos(y)$
d. $u(x, y) = x^2 + y^2$
e. $u(x, y) = x^3$
f. $u(x, y) = e^{-x} \sin(y)$

52. Verify that the function

$$P(x, y) = \frac{1}{2\pi} \cdot \frac{1 - (x^2 + y^2)}{(x - 1)^2 + y^2},$$

which is called the *Poisson kernel* for the open unit disk $D(0, 1)$, is harmonic.

53. Verify that the function

$$P(x, y) = \frac{1}{\pi} \cdot \frac{y}{x^2 + y^2},$$

which is called the Poisson kernel for the upper half plane $\{(x, y) : y > 0\}$, is harmonic.

54. Determine all nonzero solutions of Laplace's equation having the form $u(x, y) = Ax^2 + Bxy + Cy^2$. These are the *spherical harmonics* of degree two.

55. Suppose that A and E are given constants. Determine B and C so that $u(x, y) = Ax^3 + Bx^2y + Cxy^2 + Ey^3$ satisfies Laplace's equation. In this case, we call u a spherical harmonic of degree three.

56. Let (x_0, y_0) be any fixed point in the plane. Show that $u(x, y) = \ln((x - x_0)^2 + (y - y_0)^2)$ is harmonic.

57. Let $v_n(x, y) = (x^2 + y^2)^n$. Show that

$$\left(\frac{\partial^2}{\partial x^2} + \frac{\partial^2}{\partial y^2}\right) v_n(x, y) = \frac{4n^2}{x^2 + y^2} v_n(x, y).$$

Deduce that $v_n(x, y)$ cannot be harmonic for any $n \neq 0$.

58. Let $B_*(0, \infty) = \{(x, y, z) \in \mathbb{R}^3 : (x, y, z) \neq (0, 0, 0)\}$. Set $v_n(x, y, z) = (x^2 + y^2 + z^2)^n$. Show that

$$\left(\frac{\partial^2}{\partial x^2} + \frac{\partial^2}{\partial y^2} + \frac{\partial^2}{\partial z^2}\right) v_n(x, y, z) = \frac{2n(2n+1)}{x^2 + y^2 + z^2} v_n(x, y, z).$$

For what values of n does $v_n(x, y, z)$ satisfy Laplace's equation in $B_*(0, \infty)$?

59. If $u(x, t)$ denotes the temperature at position x in a long thin rod at time t, then u satisfies the *heat equation*

$$\frac{\partial u}{\partial t} = \alpha^2 \frac{\partial^2 u}{\partial x^2}.$$

The number α is a constant determined by the rod's heat conductivity properties. Show that the function

$$u_n(x, t) = \exp(-n^2\alpha^2 t) \sin(nx)$$

satisfies the heat equation for each n in $\mathbb{Z}$.

60. Let $u(x, t)$ describe the height above or below the x-axis at time t of the point x on a vibrating string with ends fixed at $x = 0$ and $x = 2\pi$. Using principles of elementary mechanics, it can be shown that u satisfies the *wave equation*

$$\frac{\partial^2 u}{\partial t^2} = c^2 \frac{\partial^2 u}{\partial x^2}$$

where c is a constant that depends on the properties of the string. Show that the function

$$u_m(x, t) = A_m \sin(cmt) \sin(mx)$$

satisfies the wave equation for all m in $\mathbb{Z}$.

Suppose that $f_x(x, y) = 2x$ and $f_y(x, y) = 3y^2$. If we integrate the first equation with respect to x, then we obtain $f(x, y) = x^2 + g(y)$. Notice that the constant of integration is a function of y because y is held constant when f_x is calculated. If we differentiate this last equation with respect to y, holding x fixed, we obtain $f_y(x, y) = g'(y)$. But we are given $f_y(x, y) = 3y^2$. It follows that $g'(y) = 3y^2$ and $g(y) = y^3 + C$ where C is a constant of integration. If we substitute this expression for $g(y)$ into our formula for $f(x, y)$, then we see that $f(x, y) = x^2 + y^3 + C$. In Exercises 61–64, use this method to find a function $f(x, y)$ that satisfies the equation.

61. $f_x(x, y) = 2xy$, $f_y(x, y) = x^2 + 1$, $f(2, 1) = 4$

62. $f_x(x, y) = 2x + 2$, $f_y(x, y) = 2y$, $f(1, 1) = 6$

63. $f_x(x, y) = 2x$, $f_y(x, y) = 1 - 2y$, $f(0, 0) = 3$

64. $f_x(x, y) = 3(x + y)^2 - 1$, $f_y(x, y) = 3(x + y)^2 + 2$, $f(0, 0) = -2$

65. Let

$$P(x, y) = e^{-xy} \frac{\partial^{n+m}}{\partial x^n \partial y^m} e^{xy}.$$

Show that $P(x, y)$ is a polynomial of total degree $m + n$ in x and y. In fact,

$$P(x, y) = \sum_{k=0}^{\infty} k! \binom{n}{k} \binom{m}{k} x^{m-k} y^{n-k}$$

where the binomial coefficient $\binom{p}{q}$ is taken to be 0 if $q > p$.

66. Define

$$f(x, y) = \begin{cases} \dfrac{xy(x^2 - y^2)}{x^2 + y^2}, & \text{if } (x, y) \neq (0, 0) \\ 0, & \text{if } (x, y) = (0, 0) \end{cases}.$$

Calculate f_x and f_y. (*Note:* To calculate these at $(0, 0)$, you will have to resort to the definition of "partial derivative," using a limit, and you will have to calculate the limit by hand.) Now, calculate $f_{xy}(0, 0)$ and $f_{yx}(0, 0)$. They are unequal. Explain why Theorem 1 is not contradicted.

67. In Chapter 3, we observed that if a function of one variable is differentiable at a point, then it is continuous at that point. However, the function

$$f(x, y) = \begin{cases} \dfrac{xy}{x^2 + y^2}, & \text{if } (x, y) \neq (0, 0) \\ 0, & \text{if } (x, y) = 0 \end{cases}$$

has the property that f_x and f_y exist at $(0, 0)$, yet f is discontinuous at $(0, 0)$. Prove these assertions.

68. Complete the following outline to obtain a proof of Theorem 1.

a. Fix a base point (x_0, y_0). Expand f in a Taylor expansion of order 1 *in the x-variable* about the point (x_0, y).

b. Expand the Taylor polynomial P_1 from part a in a first order Taylor expansion *in the y-variable* about the point y_0.

c. Repeat parts a and b with the roles of x and y reversed.

d. Subtract the two formulas for $f(x, y)$ that you obtained in parts b and c, cancelling like terms.

e. Divide through by $(x - x_0)(y - y_0)$, and let $(x, y) \to (0, 0)$ to obtain the result.

Calculator/Computer Exercises

In Exercises 69 and 70, use numerical differentiation to calculate $f_x(P_0)$ and $f_y(P_0)$ for the function f and the point P_0.

69. $f(x, y) = (\sqrt{x} + e^{y-x})/\sqrt{3 + x^2 + y^4}$, $P_0 = (4, 3)$

70. $f(x, y) = \sqrt{x^2 \cos^2(y) + y^4}/\sqrt{1 + y^2 \cos^2(x) + x^4}$, $P_0 = (\pi/4, 3\pi/4)$

71. Use numerical differentiation to calculate $f_{xx}(P_0)$ and $f_{yy}(P_0)$ for the function f and the point P_0 in Exercise 69.

72. Use numerical differentiation to calculate $f_{xx}(P_0)$ and $f_{yy}(P_0)$ for the function f and the point P_0 in Exercise 70.

13.5 Differentiability and the Chain Rule

Partial derivatives tell us how a multivariable function changes one variable at a time. In practice, it often happens that two or more of the independent variables are simultaneously changing. In this section, we will learn how to analyze the rate of change of a function when more than one of its variables is changing. Throughout this discussion, the notation $a \cdot b$ will be used to signify the product of scalars; there are no dot products of vectors in this section.

Let us review what we know for a function ϕ of one variable that is differentiable at c. By definition of $\phi'(c)$, the quantity

$$\epsilon = \frac{\phi(x) - \phi(c)}{x - c} - \phi'(c)$$

tends to 0 as x tends to c. After multiplying by $x - c$ and isolating $\phi(x)$, we obtain

$$\phi(x) = \phi(c) + \phi'(c) \cdot (x - c) + \epsilon \cdot (x - c) \qquad \text{where} \qquad \lim_{x \to c} \epsilon = 0. \qquad \textbf{(13.2)}$$

Equation (13.2) is the essence of the differential approximation: When x is near c, each factor of the product $\epsilon \cdot (x - c)$ is small. As $x \to c$, therefore, the expression $\epsilon \cdot (x - c)$ becomes negligible compared with the other terms in the equation. This idea is key to the concept of differentiability of functions of two or more variables. For simplicity, we will discuss the case of a function of two variables—functions of three or more variables are treated analogously.

Definition Suppose that $P_0 = (x_0, y_0)$ is the center of an open disk $D(P_0, r)$ on which a function f of two variables is defined. Suppose that $f_x(P_0)$ and $f_y(P_0)$ both exist. We say that f is *differentiable* at the point P_0 if we can express $f(x, y)$ by the formula

$$f(x, y) = f(P_0) + f_x(P_0) \cdot (x - x_0) + f_y(P_0) \cdot (y - y_0) + \epsilon_1 \cdot (x - x_0) + \epsilon_2 \cdot (y - y_0) \qquad \textbf{(13.3)}$$

where

$$\lim_{(x,y)\to(x_0,y_0)} \epsilon_1 = 0 \qquad \text{and} \qquad \lim_{(x,y)\to(x_0,y_0)} \epsilon_2 = 0.$$

We say that f is differentiable on a set if it is differentiable at each point of the set.

It is clear that each of the last four summands on the right side of equation (13.3) tends to 0 as (x, y) tends to (x_0, y_0). Therefore, $\lim_{(x,y)\to(x_0,y_0)} f(x, y) = f(P_0)$ if f is differentiable at P_0. We state this observation as a theorem.

Theorem 1 If f is differentiable at P_0, then f is continuous at P_0.

It is reasonable to wonder whether the existence of both partial derivatives at a point is enough to ensure the differentiability of the function at the point. As the following example will show, the answer is "no!"

Example 1 Show that the function f, defined by

$$f(x, y) = \begin{cases} \dfrac{xy}{x^2 + y^2}, & \text{if } (x, y) \neq (0, 0) \\ 0, & \text{if } (x, y) = (0, 0) \end{cases},$$

is not differentiable at the origin even though both partial derivatives $f_x(0, 0)$ and $f_y(0, 0)$ exist.

Solution We considered this function in Example 2 of Section 13.3. We observed that $\lim_{(x,y)\to(0,0)} f(x, y)$ does not exist. Therefore, f cannot be continuous at $(0, 0)$, by definition. In view of Theorem 1, this rules out differentiability at (0, 0). On the other hand,

$$\lim_{\Delta x\to 0}\frac{f(0+\Delta x,0)-f(0,0)}{\Delta x}=\lim_{\Delta x\to 0}\frac{1}{\Delta x}\left(\frac{(\Delta x)\cdot 0}{(\Delta x)^2+0^2}-0\right)=\lim_{\Delta x\to 0}\frac{0}{\Delta x}=0.$$

Therefore, $f_x(0, 0)$ exists and is 0. A symmetric argument shows that $f_y(0, 0)$ exists and is 0 as well. ■

inSIGHT

In Example 1, we must use the definitions of the partial derivatives f_x and f_y to calculate $f_x(0, 0)$ and $f_y(0, 0)$. We cannot apply the Quotient Rule to $xy/(x^2 + y^2)$ at the point (0, 0) because the denominator would be 0.

Example 1 shows that we cannot take differentiability for granted—it must be checked. Unfortunately, equation (13.3) is unwieldy to work with. Our next theorem provides what we need—sufficient conditions for differentiability that are usually easy to verify.

Theorem 2 Suppose that $P_0 = (x_0, y_0)$ is the center of an open disk $D(P_0, r)$ on which a function f of two variables is defined. If both $f_x(x, y)$ and $f_y(x, y)$ exist and are continuous on $D(P_0, r)$, then f is differentiable at P_0. In other words, if f is continuously differentiable at P_0 (as defined in Section 13.4), then f is differentiable at P_0.

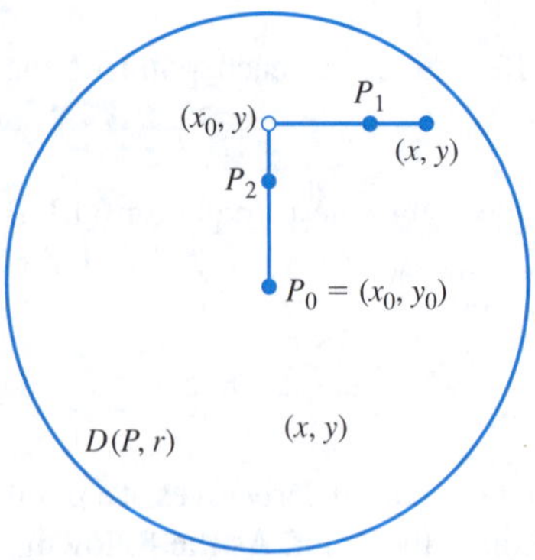

Figure 1

Proof We start with the identity

$$f(x, y) = f(x_0, y_0) + (f(x, y) - f(x_0, y)) + (f(x_0, y) - f(x_0, y_0)). \qquad \textbf{(13.4)}$$

We apply the Mean Value Theorem to the function of one variable $x \mapsto f(x, y)$. (Since we are holding y fixed, the derivative of this function is the partial derivative f_x.) The Mean Value Theorem tells us that there is a point P_1 on the line segment joining (x_0, y) to (x, y) such that $f(x, y) - f(x_0, y) = f_x(P_1) \cdot (x - x_0)$. See Figure 1. A similar application of the Mean Value Theorem tells us that there is a point P_2 on the line segment joining (x_0, y_0) to (x_0, y) such that $f(x_0, y) - f(x_0, y_0) = f_y(P_2) \cdot (y - y_0)$. We can therefore rewrite equation (13.4) as

$$f(x, y) = f(x_0, y_0) + f_x(P_1) \cdot (x - x_0) + f_y(P_2) \cdot (y - y_0). \qquad \textbf{(13.5)}$$

Let $\epsilon_1 = f_x(P_1) - f_x(P_0)$ and $\epsilon_2 = f_y(P_2) - f_y(P_0)$. Then $f_x(P_1) = f_x(P_0) + \epsilon_1$ and $f_y(P_2) = f_y(P_0) + \epsilon_2$. If we substitute these formulas into equation (13.5), we obtain

$$f(x, y) = f(x_0, y_0) + f_x(P_0) \cdot (x - x_0) + f_y(P_0) \cdot (y - y_0) + \epsilon_1 \cdot (x - x_0) + \epsilon_2 \cdot (y - y_0).$$

Furthermore, P_1 and P_2 both tend to P_0 as (x, y) does (refer to Figure 1). By the assumed continuity of f_x and f_y, we have

$$\lim_{(x,y)\to(x_0,y_0)} \epsilon_1 = \lim_{(x,y)\to(x_0,y_0)} (f_x(P_1) - f_x(P_0)) = 0$$

and

$$\lim_{(x,y)\to(x_0,y_0)} \epsilon_2 = \lim_{(x,y)\to(x_0,y_0)} (f_y(P_2) - f_y(P_0)) = 0.$$

In conclusion, we have established equation (13.3), with ϵ_1 and ϵ_2 having the required limit properties. This proves the differentiability of f at P_0. ∎

Example 2 Show that $f(x, y) = y/(1+x^2)$ is differentiable on the entire xy-plane.

Solution We have $f_x(x, y) = -2xy/(1 + x^2)^2$ and $f_y(x, y) = 1/(1 + x^2)$. Since these partial derivatives exist and are continuous at each point of the plane, we may apply Theorem 2 to obtain the desired conclusion. ∎

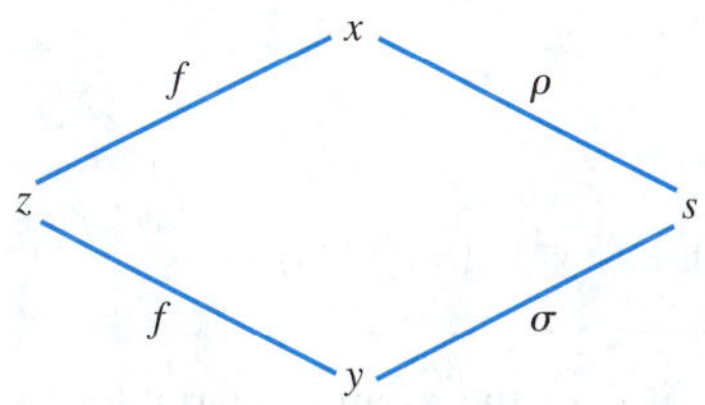

$z = f(x, y),\ x = \rho(s),\ y = \sigma(s)$

Figure 2

The Chain Rule

Suppose that $z = f(x, y)$ is a function of the two variables x and y. Suppose also that $x = \rho(s)$ and $y = \sigma(s)$ are functions of another variable s. The situation is represented schematically in Figure 2. We see that z depends on x and y, each of which, in turn, depends on s. So, z is, indirectly, a function of s, and we may ask for the value of $\frac{dz}{ds}$. Naturally, the answer is given in terms of a Chain Rule.

Theorem 3 Let $z = f(x, y)$ be a differentiable function of x and y. Suppose that $x = \rho(s)$ and $y = \sigma(s)$ are differentiable functions of s. Then $z = f(\rho(s), \sigma(s))$ is a differentiable function of s and

$$\frac{dz}{ds} = \frac{\partial f}{\partial x} \cdot \frac{d\rho}{ds} + \frac{\partial f}{\partial y} \cdot \frac{d\sigma}{ds}.$$

In more schematic form,

$$\frac{dz}{ds} = \frac{\partial z}{\partial x} \cdot \frac{dx}{ds} + \frac{\partial z}{\partial y} \cdot \frac{dy}{ds}. \tag{13.6}$$

Proof Let $x = \rho(s + \Delta s)$, $y = \sigma(s + \Delta s)$, $x_0 = \rho(s)$, and $y_0 = \sigma(s)$. Then equation (13.3) becomes

$$f(\rho(s+\Delta s), \sigma(s+\Delta s)) - f(\rho(s), \sigma(s)) = \left(\frac{\partial f}{\partial x}(x, y) + \epsilon_1\right) \cdot (\rho(s+\Delta s) - \rho(s)) + \left(\frac{\partial f}{\partial y}(x, y) + \epsilon_2\right) \cdot (\sigma(s+\Delta s) - \sigma(s)),$$

from which we obtain

$$\frac{f(\rho(s+\Delta s), \sigma(s+\Delta s)) - f(\rho(s), \sigma(s))}{\Delta s} = \left(\frac{\partial f}{\partial x}(x, y) + \epsilon_1\right) \cdot \left(\frac{\rho(s+\Delta s) - \rho(s)}{\Delta s}\right) + \left(\frac{\partial f}{\partial y}(x, y) + \epsilon_2\right) \cdot \left(\frac{\sigma(s+\Delta s) - \sigma(s)}{\Delta s}\right).$$

On letting Δs tend to 0, the left side of this equation tends to $\frac{dz}{ds}$. The quotients on the right tend to $\frac{d\rho}{ds}$ and $\frac{d\sigma}{ds}$, while the coefficients ϵ_1 and ϵ_2 both tend to 0. The limiting form of the equation is therefore

$$\frac{dz}{ds} = \frac{\partial f}{\partial x} \cdot \frac{d\rho}{ds} + \frac{\partial f}{\partial y} \cdot \frac{d\sigma}{ds},$$

as asserted. ■

The only way to learn this Chain Rule, and the others that will be presented later in this section, is to study the examples that follow and do several yourself.

Example 3 Define $z = f(x, y) = x^2 + y^3$, $x = \sin(s)$, and $y = \cos(s)$. Calculate $\frac{dz}{ds}$.

Solution The Chain Rule tells us that

$$\frac{dz}{ds} = \frac{\partial z}{\partial x} \cdot \frac{dx}{ds} + \frac{\partial z}{\partial y} \cdot \frac{dy}{ds} = (2x) \cdot \cos(s) + (3y^2) \cdot (-\sin(s)).$$

This is not a satisfactory form for our answer, since it uses the x- and y-variables to express a rate of change with resepect to s. Therefore, we substitute for x and y. The result is

$$\frac{dz}{ds} = 2(\sin(s)) \cdot \cos(s) + 3(\cos(s))^2 \cdot (-\sin(s)) = \sin(s)\cos(s)(2 - 3\cos(s)).$$ ■

inSIGHT

You may think that we could have done Example 3 without the new Chain Rule, and you are right. Why not just substitute the formulas for x and y directly into the formula for z? Namely,

$$z = f(x, y) = x^2 + y^3 = (\sin(s))^2 + (\cos(s))^3.$$

In this way, we have expressed z explicitly as a function of s so that we may differentiate it directly. (Verify that this method results in the same answer obtained in Example 3.) In spite of the fact that the Chain Rule was not absolutely necessary in Example 3, you will find that it is often useful; it is also an excellent device for organizing complicated differentiation problems.

The Chain Rule for Two or More Independent Variables

Frequently, we are faced with a function $(x, y) \mapsto f(x, y)$ for which each variable x, y depends on more than one independent variable. To begin, let us suppose that $x = \rho(s, t)$ and $y = \sigma(s, t)$. The Chain Rule for this setup is as follows.

Theorem 4 Let $z = f(x, y)$ be a differentiable function of x and y. Furthermore, assume that $x = \rho(s, t)$ and $y = \sigma(s, t)$ are differentiable functions of s and t (as represented in Figure 3). Then the composition $z = f(\rho(s, t), \sigma(s, t))$ is a differentiable function of s and t and

$$\frac{\partial z}{\partial s} = \frac{\partial f}{\partial x} \cdot \frac{\partial \rho}{\partial s} + \frac{\partial f}{\partial y} \cdot \frac{\partial \sigma}{\partial s},$$
$$\frac{\partial z}{\partial t} = \frac{\partial f}{\partial x} \cdot \frac{\partial \rho}{\partial t} + \frac{\partial f}{\partial y} \cdot \frac{\partial \sigma}{\partial t}.$$

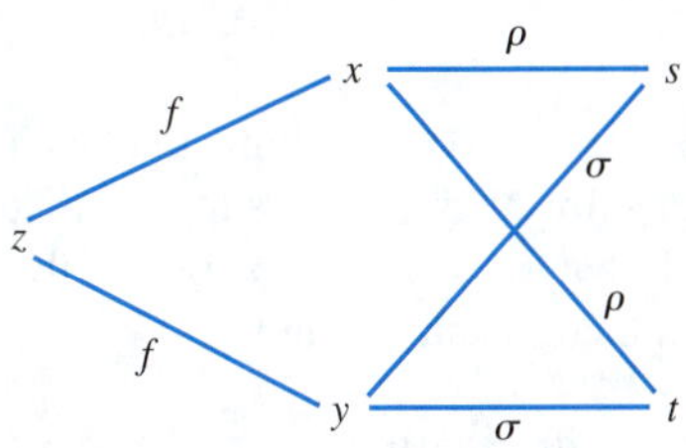

$z = f(x, y),\ x = \rho(s, t),\ y = \sigma(s, t)$

Figure 3

In a more schematic form,

$$\frac{\partial z}{\partial s} = \frac{\partial z}{\partial x} \cdot \frac{\partial x}{\partial s} + \frac{\partial z}{\partial y} \cdot \frac{\partial y}{\partial s}, \tag{13.7}$$
$$\frac{\partial z}{\partial t} = \frac{\partial z}{\partial x} \cdot \frac{\partial x}{\partial t} + \frac{\partial z}{\partial y} \cdot \frac{\partial y}{\partial t}. \tag{13.8}$$

The proof of this second Chain Rule is similar to that of the first. Notice that the form of the Chain Rule when there are two independent variables is similar to the form when there is just one variable: We treat each of the variables s and t one at a time.

Example 4 Consider the functions $z = f(x, y) = \ln(x + y^2)$, $x = s \cdot e^t$, and $y = t \cdot e^{-s}$. Calculate $\frac{\partial z}{\partial s}$ and $\frac{\partial z}{\partial t}$.

Solution In this example, x and y depend on both s and t. Let us differentiate z with respect to s (in effect, treating the independent variable t as a constant). We have

$$\frac{\partial z}{\partial s} = \frac{\partial z}{\partial x} \cdot \frac{\partial x}{\partial s} + \frac{\partial z}{\partial y} \cdot \frac{\partial y}{\partial s}.$$

Now we calculate

$$\frac{\partial z}{\partial s} = \left(\frac{1}{x + y^2}\right) \cdot e^t + \left(\frac{2y}{x + y^2}\right) \cdot (t \cdot (-e^{-s})).$$

Finally, we substitute for x and y to obtain

$$\frac{\partial z}{\partial s} = \left(\frac{1}{s \cdot e^t + t^2 \cdot e^{-2s}}\right) \cdot e^t + \left(\frac{2t \cdot e^{-s}}{s \cdot e^t + t^2 \cdot e^{-2s}}\right) \cdot (-t \cdot e^{-s}).$$

A similar calculation (holding s constant and differentiating in t) yields

$$\frac{\partial z}{\partial t} = \underbrace{\left(\frac{1}{s \cdot e^t + t^2 \cdot e^{-2s}}\right)}_{\frac{\partial z}{\partial x}} \cdot \underbrace{(s \cdot e^t)}_{\frac{\partial x}{\partial t}} + \underbrace{\left(\frac{2t \cdot e^{-s}}{s \cdot e^t + t^2 \cdot e^{-2s}}\right)}_{\frac{\partial z}{\partial y}} \cdot \underbrace{(e^{-s})}_{\frac{\partial y}{\partial t}}.$$

■

Functions of Three or More Variables

There is a version of the Chain Rule for functions of any number of variables. We formulate it as follows.

Theorem 5 Let $z = f(x_1, \ldots, x_N)$ be a function of the N variables $x_1, \ldots, x_N$. Assume that each variable x_j depends in turn on the variables $t_1, \ldots, t_M$. In other words, suppose that for each j, there is a function σj of M variables such that $x_j = \sigma_j(t_1, \ldots, t_M)$. If the functions $f, \sigma_1, \ldots, \sigma_N$ are differentiable, then so is the composition, and

$$\frac{\partial z}{\partial t_k} = \frac{\partial f}{\partial x_1} \cdot \frac{\partial \sigma_1}{\partial t_k} + \frac{\partial f}{\partial x_2} \cdot \frac{\partial \sigma_2}{\partial t_k} + \cdots + \frac{\partial f}{\partial x_N} \cdot \frac{\partial \sigma_N}{\partial t_k}$$

for $k = 1, \ldots, N$. Schematically, we may write this as

$$\frac{\partial z}{\partial t_k} = \frac{\partial z}{\partial x_1} \cdot \frac{\partial x_1}{\partial t_k} + \frac{\partial z}{\partial x_2} \cdot \frac{\partial x_2}{\partial t_k} + \cdots + \frac{\partial z}{\partial x_N} \cdot \frac{\partial x_N}{\partial t_k}.$$

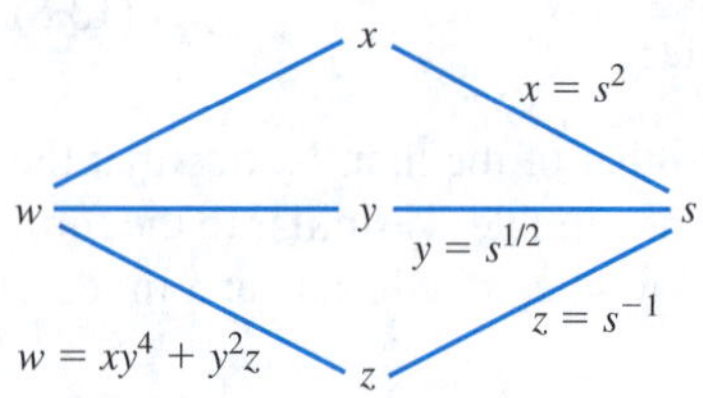

Figure 4

Example 5 Consider the function $w = xy^4 + y^2 z$ and assume that $x = s^2$, $y = s^{1/2}$, and $z = s^{-1}$. Calculate $\frac{dw}{ds}$.

Solution Refer to Figure 4. We have

$$\frac{dw}{ds} = \frac{\partial w}{\partial x} \cdot \frac{dx}{ds} + \frac{\partial w}{\partial y} \cdot \frac{dy}{ds} + \frac{\partial w}{\partial z} \cdot \frac{dz}{ds}.$$

Therefore,

$$\frac{dw}{ds} = y^4 \cdot (2s) + (4xy^3 + 2yz) \cdot \left(\frac{1}{2} s^{-1/2}\right) + y^2 \cdot (-s^{-2}).$$

Substituting for x, y, z and simplifying gives

$$\frac{dw}{ds} = (s^{1/2})^4 \cdot (2s) + (2s^2 s^{3/2} + s^{1/2} s^{-1}) \cdot s^{-1/2} + (s^{1/2})^2 \cdot (-s^{-2}) = 4s^3.$$

We may verify this result by substituting the expressions for x, y, and z into w. It turns out that $w = s^4 + 1$, and the calculation $\frac{dw}{ds} = 4s^3$ follows immediately. ■

Example 6 Define $w = 3xy + z^2 x$, and suppose that $x = t \cdot \cos(s)$, $y = t \cdot \sin(s)$, and $z = s$. Calculate $\frac{\partial w}{\partial t}$.

Solution The Chain Rule tells us that

$$\frac{\partial w}{\partial t} = \frac{\partial w}{\partial x} \cdot \frac{\partial x}{\partial t} + \frac{\partial w}{\partial y} \cdot \frac{\partial y}{\partial t} + \frac{\partial w}{\partial z} \cdot \frac{\partial z}{\partial t}.$$

Therefore,

$$\begin{aligned} \frac{\partial w}{\partial t} &= (3y + z^2) \cdot \cos(s) + (3x) \cdot \sin(s) + (2zx) \cdot 0 \\ &= (3t \sin(s) + s^2) \cdot \cos(s) + 3t \cos(s) \sin(s). \end{aligned}$$ ■

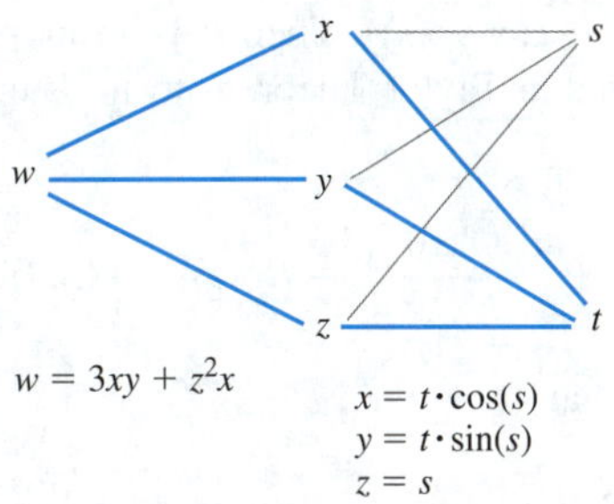

Figure 5

INSIGHT

A schematic diagram can help you keep all terms straight, even in complicated examples. Figure 5 shows how this works in Example 6. There are three "paths" from w to t, which tells us that the formula for $\frac{\partial w}{\partial t}$ has three summands. Each path has two segments, which tells us that each summand is the product of two factors. Each factor is obtained by differentiating the variable that appears on the left of the segment with respect to the variable that appears on the right.

Taylor's Formula in Several Variables

The basic ideas connected with Taylor's formula for functions of one variable were developed in Chapter 10. We now wish to consider an analogous formula for functions of two variables. Whereas the Taylor formula in one variable was specified on an interval centered at the point about which the formula was expanded, we now formulate the theorem on a rectangle $I = \{(x, y) : |x - x_0| < r_1, |y - y_0| < r_2\}$ that is centered about the base point of expansion (x_0, y_0).

Theorem 6 Let N be a nonnegative integer. Suppose f has continuous partial derivatives of all orders, up to and including $N+1$, on a rectangle I centered at $P_0 = (x_0, y_0)$. Suppose that (x, y) is a point in I. Set $h = x - x_0$ and $k = y - y_0$. Then

$$\begin{aligned} f(x, y) = {} & f(P_0) + f_x(P_0)h + f_y(P_0)k \\ & + \frac{1}{2!}(f_{xx}(P_0)h^2 + 2f_{xy}(P_0)hk + f_{yy}(P_0)k^2) \\ & + \frac{1}{3!}(f_{xxx}(P_0)h^3 + 3f_{xxy}(P_0)h^2k + 3f_{xyy}(P_0)hk^2 + f_{yyy}(P_0)k^3) \\ & + \\ & \vdots \\ & + \\ & + \frac{1}{N!}\left(f_{xx\ldots x}(P_0)h^N + Nf_{xx\ldots xy}(P_0)h^{N-1}k + \binom{N}{2} f_{xx\ldots xyy}(P_0)h^{N-2}k^2 \right. \\ & \qquad \left. + \cdots + Nf_{xyy\ldots y}(P_0)hk^{N-1} + f_{yy\ldots y}(P_0)k^N \right) \\ & + R_N(x, y). \end{aligned}$$

The coefficient of $h^p k^q$ is

$$\frac{\binom{p+q}{p}}{(p+q)!} f\underbrace{x\ldots x}_{p}\underbrace{y\ldots y}_{q}(P_0).$$

The remainder term $R_N(x, y)$ is the sum of monomials $C_{p,q}h^p k^q$ where $C_{p,q}$ is a constant and $p + q = N + 1$.

Proof Define a function $F(t) = f(x_0 + t(x - x_0), y_0 + t(y - y_0))$. Then F is a function of one variable, and we may apply the one-variable Taylor Theorem to it. The result is that

$$F(t) = F(0) + \frac{F'(0)}{1!}t + \frac{F''(0)}{2!}t^2 + \frac{F'''(0)}{3!}t^3 + \cdots + \frac{F^{(N)}(0)}{N!}t^N + R_N(t). \quad \textbf{(13.9)}$$

We use the Chain Rule from Theorem 3 to calculate that

$$F'(0) = f_x(P_0)(x - x_0) + f_y(P_0)(y - y_0).$$

A similar calculation shows that

$$F''(0) = f_{xx}(P_0)(x - x_0)^2 + 2f_{xy}(P_0)(x - x_0)(y - y_0) + f_{yy}(P_0)(y - y_0)^2.$$

Calculating all the derivatives of F up to and including order N, substituting them into equation (13.9), and setting $t = 1$ yields the desired formula. By inspecting the error term of Taylor's Theorem (from Section 10.5), we see that the present remainder term involves powers of $x - x_0$ and $y - y_0$ that are of order $N + 1$. ■

Example 7 Use Taylor's formula to find a quadratic polynomial that approximates the function $f(x, y) = \cos(x)\cos(y)$ near the origin.

Solution We take $N = 2$ and base point $(x_0, y_0) = (0, 0)$ in Theorem 6. Given this chosen base point, the increments $h = x - x_0$ and $k = y - y_0$ simplify to x and y, respectively. With these specifications of parameters, Theorem 6 states that

$$\begin{aligned} f(x, y) = {} & f(0,0) + (f_x(0,0)x + f_y(0,0)y) \\ & + \frac{1}{2!}(f_{xx}(0,0)x^2 + 2f_{xy}(0,0)xy + f_{yy}(0,0)y^2) + R_3(x, y). \end{aligned}$$

For $f(x, y) = \cos(x)\cos(y)$ we calculate

$$\begin{aligned} f(0,0) &= \cos(x)\cos(y)\big|_{(0,0)} = 1 \\ f_x(0,0) &= -\sin(x)\cos(y)\big|_{(0,0)} = 0 \\ f_y(0,0) &= -\cos(x)\sin(y)\big|_{(0,0)} = 0 \\ f_{xx}(0,0) &= -\cos(x)\cos(y)\big|_{(0,0)} = -1 \\ f_{xy}(0,0) &= \sin(x)\sin(y)\big|_{(0,0)} = 0 \\ f_{yy}(0,0) &= -\cos(x)\cos(y)\big|_{(0,0)} = -1. \end{aligned}$$

Substituting these values into our second order Taylor formula and dropping the remainder term, we obtain the approximation

$$\cos(x)\cos(y) \approx 1 - \frac{1}{2}(x^2 + y^2).$$ ■

Example 8 Use Taylor's formula to find a quadratic polynomial $T(x, y)$ that approximates the function $f(x, y) = xy + 3/x + 9/y$ near the point $(1, 3)$.

Solution We begin by calculating the first and second order partial derivatives of $f(x, y)$: $f_x(x, y) = y - 3/x^2$, $f_y(x, y) = x - 9/y^2$, $f_{xy}(x, y) = 1$, $f_{xx}(x, y) = 6/x^3$, and $f_{yy}(x, y) = 18/y^3$. Thus, at the base point $P_0 = (1, 3)$, we have $f_x(1, 3) = 3 - 3/1^2 = 0$, $f_y(1, 3) = 1 - 9/3^2 = 0$, $f_{xy}(1, 3) = 1$, $f_{xx}(1, 3) = 6/1^3 = 6$, and $f_{yy}(1, 3) = 18/3^3 = 2/3$. The required order two Taylor polynomial is therefore

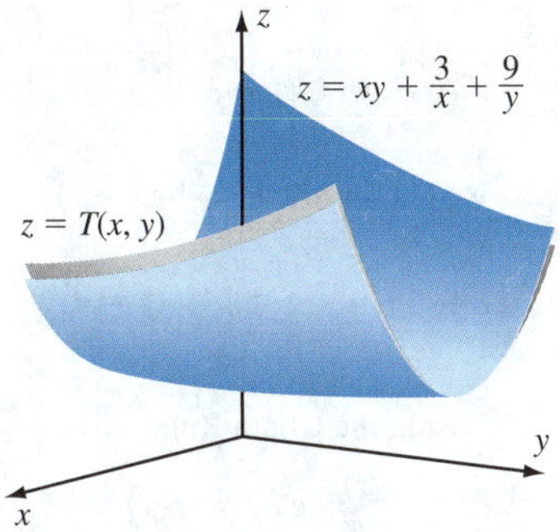

Figure 6

$$f(1, 3) + (f_x(1, 2)(x - 1) + f_y(1, 3)(y - 3)) + \frac{1}{2!}(f_{xx}(1, 3)(x - 1)^2 + 2f_{xy}(1, 3)(x - 1)(y - 3) + f_{yy}(1, 3)(y - 3)^2),$$

or

$$\begin{aligned} T(x, y) &= 9 + \frac{1}{2!}\left(6(x - 1)^2 + 2(x - 1)(y - 3) + \frac{2}{3}(y - 3)^2\right) \\ &= 9 + 3(x - 1)^2 + (x - 1)(y - 3) + \frac{1}{3}(y - 3)^2. \end{aligned}$$

Figure 6 shows the graphs of f and T in the viewing box $[0.9, 1.1] \times [2.9, 3.1] \times [8.985, 9.05]$.

inSIGHT

When we graph a function of one variable in a small window containing a maximum or minimum value, we expect to have visual evidence of the extremum. However, extrema of functions of two variables can be difficult to spot, especially when the graph is observed from only one direction. In Example 8, we complete the square in the expression for $T(x, y)$ to obtain

$$\begin{aligned} T(x, y) &= 9 + 3\left((x - 1)^2 + \frac{1}{3}(x - 1)(y - 3) + \frac{1}{36}(y - 3)^2 + \left(\frac{1}{9} - \frac{1}{36}\right)(y - 3)^2\right) \\ &= 9 + 3\left(x - 1 + \frac{1}{6}(y - 3)\right)^2 + \frac{1}{4}(y - 3)^2. \end{aligned}$$

Because $T(1, 3) = 9$, we see that $T(x, y)$ is the sum of $T(1, 3)$ and two nonnegative quantities. We conclude that T has an absolute minimum value of 9 at the point $(1, 3)$. Since $f(x, y)$ differs from $T(x, y)$ by a remainder term that is small when the point (x, y) is close to the base point $(1, 3)$, we expect f to have a local minimum at the point $(1, 3)$—and it does. We will return to these ideas in Section 13.8.

quickquiz

1. What conditions ensure that $(x, y) \mapsto f(x, y)$ is differentiable at (x_0, y_0)?
2. State the Chain Rule for a function $f(x, y)$ with x and y each depending on a variable s.

3. Calculate $\frac{\partial z}{\partial t}$ when $z = \cos(x) - \sin(y)$ and $x = s^2 t$, $y = s^2 + t$.
4. State a quadratic polynomial that approximates the function $\ln(1 + x + y^2)$ near the origin.

EXERCISES

Problems for Practice

In Exercises 1–10, calculate $\frac{dz}{ds}$ by the following two methods:

a. Use the Chain Rule.

b. Substitute the formulas for x and y into the formula for z, and calculate the derivative directly.

1. $z = f(x, y) = x^2 y - y^3$, $x = \rho(s) = s^3 - 1$, $y = \sigma(s) = s^{-1}$

2. $z = f(x, y) = x\sin(y)$, $x = \rho(s) = e^s$, $y = \sigma(s) = e^{-2s}$

3. $z = f(x, y) = x/(x^2 + y^2)$, $x = \rho(s) = \cos(s)$, $y = \sigma(s) = \sin(s)$

4. $z = f(x, y) = xe^y$, $x = \rho(s) = s^2 - s$, $y = \sigma(s) = s^3 + s$

5. $z = f(x, y) = \ln(x^3 + y)$, $x = \rho(s) = s^{1/2}$, $y = \sigma(s) = s^{3/2}$

6. $z = f(x, y) = \tan(x^3 y)$, $x = \rho(s) = s \cdot e^s$, $y = \sigma(s) = s^2 e^s$

7. $z = f(x, y) = e^{(2x-3y)}$, $x = \rho(s) = \cos(s)$, $y = \sigma(s) = \sin(s)$

8. $z = f(x, y) = x(y - 1)$, $x = \rho(s) = \tan(s)$, $y = \sigma(s) = \cot(s)$

9. $z = f(x, y) = (x + y)/(x - y)$, $x = \rho(s) = \cos(s^2)$, $y = \sigma(s) = \sin(s^2)$

10. $z = f(x, y) = xy^2$, $x = s\ln(s)$, $y = \sigma(s) = s^3$

In Exercises 11–18, use the Chain Rule to calculate $\frac{\partial z}{\partial s}$ and $\frac{\partial z}{\partial t}$.

11. $z = f(x, y) = x^2 - y^7$, $x = \rho(s, t) = \sin(3st)$, $y = \sigma(s, t) = \cos(3st)$

12. $z = f(x, y) = \ln(3x - y^2)$, $x = \rho(s, t) = s^2 - 4t^3$, $y = \sigma(s, t) = s^3 t^{-3}$

13. $z = f(x, y) = e^{3xy}$, $x = \rho(s, t) = s/t$, $y = \sigma(s, t) = t/s$

14. $z = f(x, y) = x/y$, $x = \rho(s, t) = \tan(3st)$, $y = \sigma(s, t) = \cot(3st)$

15. $z = f(x, y) = \sqrt{x^2 - y^2}$, $x = \rho(s, t) = (s^2 - t^2)^{-1}$, $y = \sigma(s, t) = (s^2 + t^2)^{-1}$

16. $z = f(x, y) = 1/(x + y)$, $x = \rho(s, t) = e^{s-t}$, $y = \sigma(s, t) = e^{8t-5s}$

17. $z = f(x, y) = \exp(2x - y)$, $x = \rho(s, t) = \ln(s - t)$, $y = \sigma(s, t) = \ln(2s + 3t)$

18. $z = f(x, y) = (x + 3y)/(y - 5x)$, $x = \rho(s, t) = e^{5st}$, $y = \sigma(s, t) = e^{-4st}$

In Exercises 19–26, calculate $\frac{dw}{ds}$ using the Chain Rule.

19. $w = f(x, y, z) = xyz^2$, $x = \rho(s) = e^s$, $y = \sigma(s) = s^3$, $z = \tau(s) = (s - 2)^3$

20. $w = f(x, y, z) = (x^2 + y)/z$, $x = \rho(s) = \cos(s)$, $y = \sigma(s) = \sin(s)$, $z = \tau(s) = \tan(s)$

21. $w = f(x, y, z) = \sqrt{x^2 - y^3 + z}$, $x = \rho(s) = \ln(s)$, $y = \sigma(s) = s^{-1}$, $z = \tau(s) = \tan(s^2)$

22. $w = f(x, y, z) = yz\sin(x)$, $x = \rho(s) = s^5$, $y = \sigma(s) = s^7$, $z = \tau(s) = s^9 + 3$

23. $w = f(x, y, z) = \sin(xyz)$, $x = \rho(s) = e^{3s}$, $y = \sigma(s) = s^{-4}$, $z = \tau(s) = s^7$

24. $w = f(x, y, z) = \ln(x + y^2 + z^3)$, $x = \rho(s) = s^2$, $y = \sigma(s) = (s^2 + 3)^{1/2}$, $z = \tau(s) = (s^2 + 8)^{1/3}$

25. $w = f(x, y, z) = x^2/(y^2 + z^2)$, $x = \rho(s) = 8^s$, $y = \sigma(s) = 2^s$, $z = \tau(s) = 4^s$

26. $w = f(x, y, z) = x^3\ln(z^4 + y)$, $x = \rho(s) = \sin^3(s)$, $y = \sigma(s) = s^{-4}$, $z = \tau(s) = s^{-1}$

In Exercises 27–32, give the Taylor polynomial for f of the specified order m about the given point P_0.

27. $f(x, y) = y\ln(x) - \sin(xy)$, $m = 2$, $P_0 = (1, \pi)$

28. $f(x, y) = \cos(x^2 y)$, $m = 2$, $P_0 = (0, 0)$

29. $f(x, y) = 1/(1 + xy)$, $m = 3$, $P_0 = (0, 0)$

30. $f(x, y) = e^{x+2y}$, $m = 3$, $P_0 = (0, 0)$

31. $f(x, y) = \tan(2x - y)$, $m = 2$, $P_0 = (\pi/4, \pi/4)$

32. $f(x, y) = \sqrt{2 + xy}$, $m = 2$, $P_0 = (1, 2)$

Further Theory and Practice

33. Suppose that $z = h(x, y)$, $x = \rho(s, t)$, $y = \sigma(s, t)$, $s = \alpha(u, v)$, and $t = \beta(u, v)$. State a Chain Rule for $\frac{\partial z}{\partial u}$ and $\frac{\partial z}{\partial v}$.

34. A rectangular box has sides that, due to temperature fluctuations, undergo changes in dimension. At a given moment, the length increases at a rate of 0.01 cm/s, the width decreases at a rate of 0.02 cm/s, and the height increases at a rate of 0.005 cm/s. At this particular moment, the length is 15 cm, the width is 8 cm, and the height is 5 cm. Is the volume increasing or decreasing at this moment?

35. In Exercise 34, is the surface area increasing or decreasing?

36. Given that $z = f(x, y)$, $x = \rho(s, t)$, and $y = \sigma(s, t)$, find formulas for $\frac{\partial^2 z}{\partial s^2}$, $\frac{\partial^2 z}{\partial t^2}$, and $\frac{\partial^2 z}{\partial s \partial t}$.

37. Let ϕ be a continuously differentiable function of one variable, and let f be a differentiable function of two variables. Prove that if $g(x, y) = \phi(f(x, y))$, then

$$g_x(x, y) = \phi'(f(x, y)) f_x(x, y)$$

and

$$g_y(x, y) = \phi'(f(x, y)) f_y(x, y).$$

Draw a schematic diagram for this version of the Chain Rule. Use the function $g(x, y) = \sin(x^2 \ln(y))$ to illustrate this Chain Rule.

38. Let ϕ and ψ be continuously differentiable functions of one variable. Suppose that for each x and t there is a unique value $f(x, t)$ for which $f(x, t) = t + x\phi(f(x, t))$. Let $v(x, t) = \psi(f(x, t))$. Show that

$$\frac{\partial v}{\partial x} = \phi(f(x, t)) \frac{\partial v}{\partial t}.$$

39. Let a and b be constants. Suppose that ϕ is a continuously differentiable function of one variable. Show that $u(x, y) = \phi(ax + by)$ satisfies the equation

$$b \frac{\partial u}{\partial x}(x, y) = a \frac{\partial u}{\partial y}(x, y).$$

40. Let ϕ be a continuously differentiable function of one variable. Show that $u(x, y) = \phi(xy)$ satisfies the equation

$$x \frac{\partial u}{\partial x}(x, y) = y \frac{\partial u}{\partial y}(x, y).$$

41. Let ϕ be a continuously differentiable function of one variable. Show that $v(x, y) = \phi(x^2 + y^2)$ satisfies

$$y \frac{\partial v}{\partial x}(x, y) = x \frac{\partial v}{\partial y}(x, y).$$

42. If $y(x, t)$ represents the displacement of a vibrating string at time t and position x, then $y(x, t)$ satisfies the wave equation

$$\frac{\partial^2 y}{\partial x^2}(x, t) = \frac{1}{c^2} \cdot \frac{\partial^2 y}{\partial t^2}(x, t)$$

where c is a positive constant that depends on the tension in the string. If ϕ and ψ are any twice differentiable functions of a single variable, then show that

$$y(x, t) = \phi(x + ct) + \psi(x - ct)$$

is a solution of the wave equation. Notice that

$$\psi((x + c\Delta t) - c(t + \Delta t)) = \psi(x - ct).$$

What physical interpretation of the term $\psi(x - ct)$ results from this observation? What is the interpretation of $\phi(x + ct)$?

43. Let $\xi = x + ct$ and $\eta = x - ct$. Suppose that y is a twice continuously differentiable function of two variables such that $y_{xx}(x, t) = (1/c^2) y_{tt}(x, t)$. Show that

$$v(\xi, \eta) = y\left(\frac{\xi + \eta}{2}, \frac{\xi - \eta}{2c}\right)$$

satisfies $v_{\xi\eta}(\xi, \eta) = 0$. Solve this differential equation and deduce that

$$y(x, t) = \phi(x + ct) + \psi(x - ct)$$

for some twice continuously differentiable functions ϕ and ψ.

44. Let $x = r\cos(\theta)$ and $y = r\sin(\theta)$. (Here r is the distance of $P = (x, y)$ to the origin and θ is the angle that the position vector OP makes with the positive x-axis.) Given a twice continuously differentiable function u of two variables, let $v(r, \theta) = u(r\cos(\theta), r\sin(\theta))$. If u satisfies Laplace's equation $u_{xx}(x, y) + u_{yy}(x, y) = 0$, then what partial differential equation does v satisfy?

45. Let c be a constant, and let f be a differentiable function of two variables. Suppose that the equation $f(x, y) = c$ defines y implicitly as a differentiable function of x. Show that $y'(x) = -f_x(x, y)/f_y(x, y)$ at any point at which the denominator is nonzero.

46. Let c be a constant, and let F be a differentiable function of three variables. Suppose that the equation $F(x, y, z) = c$ defines z implicitly as a differentiable function of x and y. Show that

$$\frac{\partial z}{\partial x}(x, y) = -\frac{F_x(x, y, z)}{F_z(x, y, z)} \quad \text{and} \quad \frac{\partial z}{\partial y}(x, y) = -\frac{F_y(x, y, z)}{F_z(x, y, z)}$$

at any point at which the denominator is nonzero.

47. Let $f(x, y)$ be a continuously differentiable function of two variables. Calculate the derivative with respect to x of $f(f(f(x, y)), f(x, f(x, y)))$.

48. Let $\Delta u(x, y, z) = u_{xx}(x, y, z) + u_{yy}(x, y, z) + u_{zz}(x, y, z)$. (This expression is called the *Laplacian of* u.) For $r = \sqrt{x^2 + y^2 + z^2} > 0$, let

$$v(x, y, z) = \frac{1}{r} u\left(\frac{x}{r^2}, \frac{y}{r^2}, \frac{z}{r^2}\right).$$

Calculate $\Delta v(x, y, z)$ in terms of $\Delta u(x, y, z)$.

Computer/Calculator Exercises

In Exercises 49–52, functions $F(x, y)$, $\rho(s, t)$, and $\sigma(s, t)$ are given. Let $f(s, t) = F(\rho(s, t), \sigma(s, t))$. For the given point (s_0, t_0), set $x_0 = \rho(s_0, t_0)$ and $y_0 = \sigma(s_0, t_0)$. Use central difference quotients to evaluate $f_s(s_0, t_0)$ and $f_t(s_0, t_0)$. Also, use central difference quotients to evaluate $\rho_s(s_0, t_0)$, $\rho_t(s_0, t_0)$, $\sigma_s(s_0, t_0)$, $\sigma_t(s_0, t_0)$, $F_x(x_0, y_0)$, and $F_y(x_0, y_0)$. Then use these values, together with the Chain Rule, to obtain alternative evaluations of $f_s(s_0, t_0)$ and $f_t(s_0, t_0)$.

49. $F(x, y) = (x + y)/(x + 1)$, $\rho(s, t) = \sqrt{st + t}$, $\sigma(s, t) = 2^{s-t}$, $(s_0, t_0) = (3, 1)$

50. $F(x, y) = xe^{2y-x}$, $\rho(s, t) = s/(1 + t)$, $\sigma(s, t) = s^t$, $(s_0, t_0) = (2, 0)$

51. $F(x, y) = \ln(1 + x^4 + y^2)$, $\rho(s, t) = (2s + t)^{1/3}$, $\sigma(s, t) = \tan(s + t)$, $(s_0, t_0) = (1, -1)$

52. $F(x, y) = \sin(2x + 6y)$, $\rho(s, t) = \arcsin(s/t)$, $\sigma(s, t) = \arcsin((2s - 1)/t^2)$, $(s_0, t_0) = (1, -\sqrt{2})$

13.6 Gradients and Directional Derivatives

Recall that a *direction* is another word for a *unit vector*. In this section, we want to discuss the notion of differentiating a function of several variables in a given direction. For motivation, look at the topographic map in Figure 1. The level sets are for the height function f (measured in feet above sea level). They are plotted at 20 ft increments. Imagine you are standing at the point P, which is the origin of the superimposed coordinate axes. The level curve that passes through P tells us that P is situated 1040 ft above sea level. If you look east, the height is increasing; so, the derivative of f in that direction should be positive. If you look south or north (along the level curve), the height is not changing; so, the derivatives of f in those directions should be 0. If you look west, the height is decreasing; so, the derivative in that direction should be negative.

At what rate does your elevation change if you set out from P in a given direction? In which direction does elevation increase most rapidly? Decrease most rapidly? This section introduces methods for answering these questions.

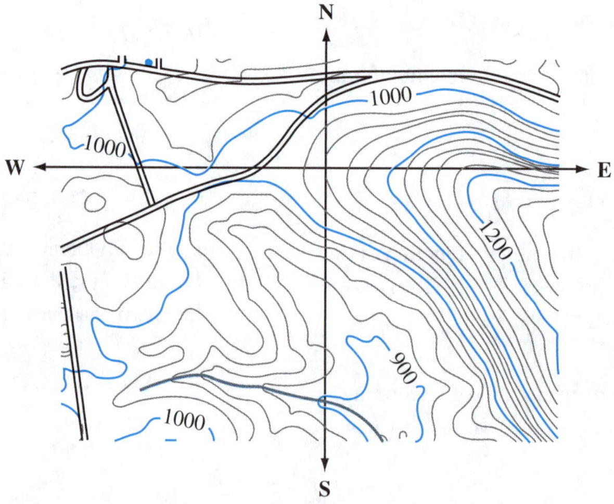

Figure 1
Stone Mountain, Georgia

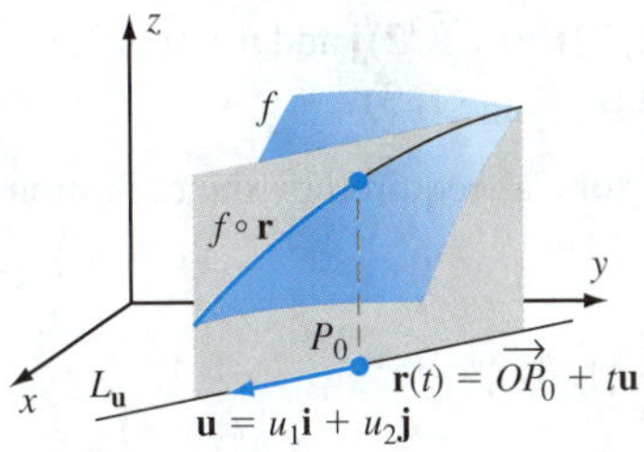

Figure 2

The Directional Derivative

Let $\mathbf{u} = u_1\mathbf{i} + u_2\mathbf{j}$ be a direction vector and $P_0 = (x_0, y_0)$ a point in the xy-plane. Let $L_\mathbf{u}$ denote the line in the xy-plane that passes through P_0 and has direction $\mathbf{u}$. We parameterize $L_\mathbf{u}$ by the vector-valued function $\mathbf{r}(t) = \overrightarrow{OP_0} + t\mathbf{u}$. Since $\|\mathbf{r}'(t)\| = \|\mathbf{u}\| = 1$, we note that $t \mapsto \mathbf{r}(t)$ is the arc length parameterization of $L_\mathbf{u}$ that passes through P_0 at $t = 0$. See Figure 2.

Let $(x, y) \mapsto f(x, y)$ be a differentiable function defined on a disk centered at P_0. The first step toward calculating the rate of change of f at the point P_0 in the direction $\mathbf{u} = u_1\mathbf{i} + u_2\mathbf{j}$ is to restrict f to $L_\mathbf{u}$ by forming the composition $f \circ \mathbf{r}$:

$$(f \circ \mathbf{r})(t) = f(\overrightarrow{OP_0} + t\mathbf{u}) = f(x_0 + tu_1, y_0 + tu_2).$$

This is a scalar-valued function of one variable. Its plot can be obtained by intersecting the surface $z = f(x, y)$ with the plane that contains the line $L_\mathbf{u}$ and that is perpendicular to the xy-plane. Refer to Figure 2. The rate of change that we seek is obtained from the derivative of $f \circ \mathbf{r}$.

Definition The *directional derivative* of the function f in the direction $\mathbf{u} = u_1\mathbf{i} + u_2\mathbf{j}$ at the point P_0 is defined to be

$$D_\mathbf{u} f(P_0) = \frac{d}{dt} f(\overrightarrow{OP_0} + t\mathbf{u})\bigg|_{t=0} = \frac{d}{dt} f(x_0 + tu_1, y_0 + tu_2)\bigg|_{t=0}.$$

Now that we have defined the directional derivative, how can we calculate it? We use the Chain Rule to differentiate the function $z = f \circ \mathbf{r}$ with respect to t:

$$\frac{dz}{dt}\bigg|_{t=0} = \frac{\partial f}{\partial x}(P_0)\frac{dx}{dt}\bigg|_{t=0} + \frac{\partial f}{\partial y}(P_0)\frac{dy}{dt}\bigg|_{t=0} = \frac{\partial f}{\partial x}(P_0)u_1 + \frac{\partial f}{\partial y}(P_0)u_2.$$

Now we can summarize what we have found.

Theorem 1 Let P_0 be a point in the plane, $\mathbf{u} = u_1\mathbf{i} + u_2\mathbf{j}$ a unit vector, and f a differentiable function on a disk centered at P_0. Then the directional derivative of f at P_0 in the direction $\mathbf{u}$ is given by the formula

$$D_\mathbf{u} f(P_0) = \frac{\partial f}{\partial x}(P_0)u_1 + \frac{\partial f}{\partial y}(P_0)u_2.$$

inSIGHT

Notice that if $\mathbf{u} = \mathbf{i}$, then $u_1 = 1$ and $u_2 = 0$. Therefore, $D_\mathbf{i} f(P_0) = f_x(P_0)$. Similarly, $D_\mathbf{j} f(P_0) = f_y(P_0)$. This tells us that the partial derivatives $f_x(P_0)$ and $f_y(P_0)$ are themselves directional derivatives: They correspond to the directions of the positive x-axis and positive y-axis, respectively. From this point of view, the directional derivative may be regarded as a generalization of the concept of partial derivative.

Example 1 Let $f(x, y) = 1 - x^2 y$. Let $\mathbf{v} = (\sqrt{2}/2)\mathbf{i} + (\sqrt{2}/2)\mathbf{j}$ and $\mathbf{u} = (1/2)\,\mathbf{i} - (\sqrt{3}/2)\mathbf{j}$. Calculate $D_{\mathbf{v}} f(1, 3)$ and $D_{\mathbf{u}} f(1, 3)$. What is $D_{-\mathbf{u}} f(1, 3)$?

Solution To begin, notice that **u** and **v** are unit vectors, as required by the definition of the directional derivative. We calculate that

$$\frac{\partial f}{\partial x}(1, 3) = -2xy|_{(1,3)} = -6 \quad \text{and} \quad \frac{\partial f}{\partial y}(1, 3) = -x^2|_{(1,3)} = -1.$$

Therefore

$$D_{\mathbf{v}} f(1, 3) = \frac{\partial f}{\partial x}(1, 3)v_1 + \frac{\partial f}{\partial y}(1, 3)v_2 = (-6)\,\frac{\sqrt{2}}{2} + (-1)\,\frac{\sqrt{2}}{2} = -7\frac{\sqrt{2}}{2}.$$

Likewise,

$$D_{\mathbf{u}} f(1, 3) = \frac{\partial f}{\partial x}(1, 3)u_1 + \frac{\partial f}{\partial y}(1, 3)u_2 = (-6)\,\frac{1}{2} + (-1)\left(-\frac{\sqrt{3}}{2}\right) = -\left(3 - \frac{\sqrt{3}}{2}\right).$$

Similarly,

$$\begin{aligned} D_{-\mathbf{u}} f(1, 3) &= \frac{\partial f}{\partial x}(1, 3)\,(-u_1) + \frac{\partial f}{\partial y}(1, 3)\,(-u_2) \\ &= -\left(\frac{\partial f}{\partial x}(1, 3)u_1 + \frac{\partial f}{\partial y}(1, 3)u_2\right) = -D_{\mathbf{u}} f(1, 3) = 3 - \frac{\sqrt{3}}{2}. \end{aligned}$$

inSIGHT

The last computation in Example 1 shows that $D_{-\mathbf{u}} f(P_0) = -D_{\mathbf{u}} f(P_0)$. There is a simple geometric explanation for this. Remember that the sign of a derivative indicates whether the function increases (positive derivative) or decreases (negative derivative). Now think of a mountain path while looking at Figure 2. Follow the path downward. Relative to the ground, your path has a heading **u**. Since you are descending, your height function has a negative rate of change. Now follow the same path upward by reversing your ground direction from **u** to $-\mathbf{u}$. The slope of the path is the same, but since you are now ascending, the rate of change of height reverses in sign.

Example 2 Let $f(x, y) = 1 + 2x + y^3$. What is the directional derivative of f at $P = (2, 1)$ in the direction from P to $Q = (14, 6)$?

Solution We calculate $f_x(x, y) = 2$ and $f_y(x, y) = 3y^2$. Therefore $f_x(P) = 2$ and $f_y(P) = 3$. The next step is to compute the required direction. The initial candidate, $\overrightarrow{PQ} = 12\mathbf{i} + 5\mathbf{j}$, has length $\sqrt{12^2 + 5^2}$, which simplifies to 13. Since $\overrightarrow{PQ}$ is not a unit vector, we do not use it in the formula for the directional derivative. Instead, we use the unit vector in the direction of P to Q, namely, $\mathbf{u} = (1/13)\overrightarrow{PQ} = (12/13)\mathbf{i} + (5/13)\mathbf{j}$. The required directional derivative is $D_{\mathbf{u}} f(P) = (2)(12/13) + (3)(5/13) = 3$.

The Gradient

Before proceeding, we introduce a notation that we will use to organize our calculations.

Definition Let f be a differentiable function of two variables. The *gradient function* of f is the vector-valued function ∇f defined by

$$\nabla f(P) = \left(\frac{\partial f}{\partial x}(P)\right)\mathbf{i} + \left(\frac{\partial f}{\partial y}(P)\right)\mathbf{j}.$$

In this notation, ∇f is a vector-valued function created from the scalar-valued function f, and $\nabla f(P)$ is the vector obtained by evaluating the function ∇f at point P. We say that $\nabla f(P)$ is the *gradient* of f at the point P. Sometimes the alternative notation grad(f) is used to denote the gradient function of f.

We can use the gradient to rewrite the directional derivative as a dot product:

$$D_{\mathbf{u}} f(P) = \nabla f(P) \cdot \mathbf{u}.$$

Notice that if $\nabla f(P) = \vec{\mathbf{0}}$—that is, if $f_x(P) = 0$ and $f_y(P) = 0$—then $D_{\mathbf{u}} f(P) = 0$ for all unit vectors $\mathbf{u}$. In other words, if the gradient at P is 0, then the rate of change of f at P is 0 in all directions.

Example 3 Let $f(x, y) = x\sin(y)$. Calculate $\nabla f(x, y)$. If $\mathbf{u} = (-3/5)\mathbf{i} + (4/5)\mathbf{j}$, what is $D_{\mathbf{u}} f(2, \pi/6)$?

Solution We have

$$\nabla f(x, y) = \left(\frac{\partial f}{\partial x}(x, y)\right)\mathbf{i} + \left(\frac{\partial f}{\partial y}(x, y)\right)\mathbf{j} = \sin(y)\mathbf{i} + x\cos(y)\mathbf{j}.$$

Therefore, $\nabla f(2, \pi/6) = \sin(\pi/6)\mathbf{i} + 2\cos(\pi/6)\mathbf{j} = (1/2)\mathbf{i} + \sqrt{3}\mathbf{j}$ and

$$D_{\mathbf{u}} f\left(2, \frac{\pi}{6}\right) = \nabla f\left(2, \frac{\pi}{6}\right) \cdot \mathbf{u} = \left(\frac{1}{2}\mathbf{i} + \sqrt{3}\mathbf{j}\right) \cdot \left(-\frac{3}{5}\mathbf{i} + \frac{4}{5}\mathbf{j}\right) = -\frac{3}{10} + \frac{4\sqrt{3}}{5}.$$

The Directions of Greatest Increase and Decrease

Suppose now that the point P_0 and the function f are fixed. Which direction $\mathbf{u}$ will result in the greatest possible value for $D_{\mathbf{u}} f(P_0)$? Which direction $\mathbf{u}$ will result in the least possible value for $D_{\mathbf{u}} f(P_0)$? Since $D_{\mathbf{u}} f(P_0)$ measures the rate of change of f at P_0 in the direction $\mathbf{u}$, we can pose our questions in this way: What is the direction of greatest increase for the function f at P_0? What is the direction of greatest decrease for the function f at P_0?

First of all, let us exclude from consideration points at which $\nabla f(P_0) = \vec{\mathbf{0}}$: We have already noticed that all directional derivatives at P_0 are 0 in this case. We therefore assume that $\nabla f(P_0) \neq \vec{\mathbf{0}}$. With this assumption, we are guaranteed that there is a well-defined angle θ between the nonzero vectors $\nabla f(P)$ and $\mathbf{u}$. Expressing the dot product of these vectors in terms of θ, we have

$$D_{\mathbf{u}} f(P_0) = \nabla f(P_0) \cdot \mathbf{u} = \|\nabla f(P_0)\| \|\mathbf{u}\| \cos(\theta) = \|\nabla f(P_0)\| \cos(\theta).$$

Remember that f and P_0 are fixed. We are trying to select a direction $\mathbf{u}$ to make $\|\nabla f(P_0)\| \cos(\theta)$ as big or as small as possible. In other words, we wish to determine θ so that $\cos(\theta)$ is as big or as small as possible. Clearly $\theta = 0$ gives the greatest value for $\cos(\theta)$, namely, 1. This value of θ occurs when $\mathbf{u}$ is the direction of the gradient $\nabla f(P_0)$. The value $\theta = \pi$ gives the least value of $\cos(\theta)$, namely, -1. This value of θ occurs when $\mathbf{u}$ is the opposite direction of the gradient $\nabla f(P_0)$. We state these observations as a theorem.

Theorem 2 Let P_0 be a fixed point in the plane. Suppose that f is a differentiable function at P_0. Then $D_{\mathbf{u}} f(P_0)$ is maximal when the unit vector $\mathbf{u}$ is the direction of the gradient $\nabla f(P_0)$; for this choice of $\mathbf{u}$, the directional derivative is $D_{\mathbf{u}} f(P_0) = \|\nabla f(P_0)\|$. Also, $D_{\mathbf{u}} f(P_0)$ is minimal when $\mathbf{u}$ is opposite in direction to $\nabla f(P_0)$; for this choice of $\mathbf{u}$, the directional derivative is $D_{\mathbf{u}} f(P_0) = -\|\nabla f(P_0)\|$.

inSIGHT

The maximal value of $D_{\mathbf{u}} f(P_0)$ is the greatest rate of increase of f. The minimal value of $D_{\mathbf{u}} f(P_0)$ is the most negative $D_{\mathbf{u}} f(P_0)$ can be—it represents the greatest rate of decrease of f. See the discussion in the *Insight* for Example 1 for an explanation, geometrically, of why the two directions that give these extreme values are opposite to each other.

Example 4 At the point $P_0 = (-2, 1)$, what is the direction that results in the greatest increase for $f(x, y) = x^2 + y^2$, and what is the direction of greatest decrease? What are the greatest and least values of the directional derivative at P_0?

Solution The gradient of f at (x, y) is $\nabla f(x, y) = 2x\mathbf{i} + 2y\mathbf{j}$. At P_0, this gradient is equal to $-4\mathbf{i} + 2\mathbf{j}$. The direction of greatest increase is therefore

$$\mathbf{u} = \frac{1}{\|\nabla f(P_0)\|} \nabla f(P_0) = \frac{1}{\sqrt{20}}(-4\mathbf{i} + 2\mathbf{j}) = -\frac{2}{\sqrt{5}}\mathbf{i} + \frac{1}{\sqrt{5}}\mathbf{j}.$$

The corresponding greatest value for the directional derivative at P_0 is

$$D_{\mathbf{u}} f(P_0) = \|\nabla f(P_0)\| = \sqrt{20} = 2\sqrt{5}.$$

The direction of greatest decrease is the negative of the vector $\mathbf{u}$, namely, $(2/\sqrt{5})\mathbf{i} - (1/\sqrt{5})\mathbf{j}$. The corresponding least value for the directional derivative at P_0 is $-\|\nabla f(P_0)\| = -2\sqrt{5}$. ■

The Gradient and Level Curves

We turn to another important application of the gradient. Suppose f is a function that is differentiable at a point P_0 in the plane. Let c denote the value $f(P_0)$, and let $\mathbf{r} = \langle r_1, r_2 \rangle$ be the arc length parameterization of the level curve $\mathcal{C} = \{(x, y) : f(x, y) = c\}$ for which $\mathbf{r}(0) = P_0$. Then $\mathbf{T} = \mathbf{r}'(0)$ is the unit tangent vector to $\mathcal{C}$ at P_0. Because $\mathbf{r}(s)$ is a point on $\mathcal{C}$ for all s, we see that $z = (f \circ \mathbf{r})(s) = f(\mathbf{r}(s)) = c$ is constant. It follows that $\frac{dz}{ds} = 0$ holds for all s. In particular, at $s = 0$, we have

$$\begin{aligned} D_{\mathbf{T}} f(P_0) &= \nabla f(P_0) \cdot \mathbf{T} = \nabla f(P_0) \cdot \mathbf{r}'(0) \\ &= \frac{\partial f}{\partial x}(P_0) r_1'(0) + \frac{\partial f}{\partial y}(P_0) r_2'(0) \overset{\text{Chain Rule}}{=} \left. \frac{dz}{ds} \right|_{s=0} = 0. \end{aligned}$$

We now gather together some conclusions that we can draw from this equation.

Theorem 3 Suppose f is differentiable at P_0. Let $\mathbf{T}$ be a unit tangent vector to the level curve of f at P_0. Then

a. $D_{\mathbf{T}}f(P_0) = 0$, and
b. $\nabla f(P_0)$ is perpendicular to $\mathbf{T}$.

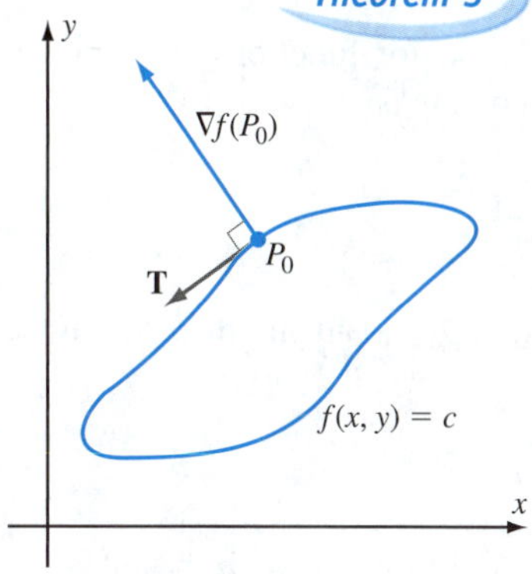

Figure 3

In words, part a tells us that the instantaneous rate of change of f at P_0 is 0 in the direction of the tangent to the level curve. Part b tells us that the gradient $\nabla f(P_0)$ is normal to the level curve of f at P_0. See Figure 3.

inSIGHT

In combination, Theorems 2 and 3 tell us that a direction perpendicular to a level curve of f at P is either the direction in which f has the greatest increase at P or the direction in which f has the greatest decrease at P. This principle is actually quite familar to us. Imagine ascending the ramp shown in Figure 4. If we begin at point P, then our instinct is to walk toward Q in the direction that is perpendicular to the level curves, rather than toward another point, such as R.

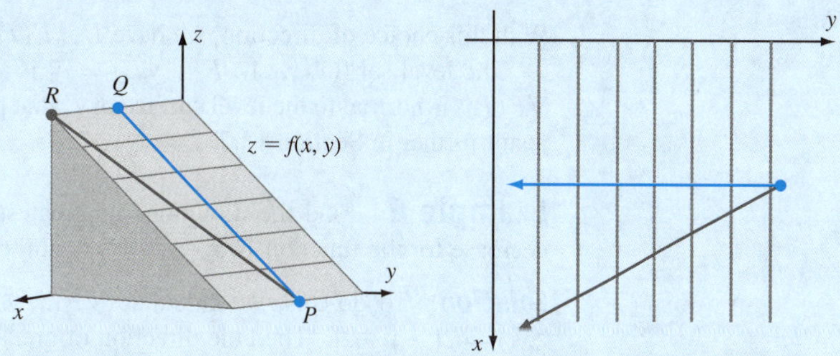

Figure 4
The direction perpendicular to the level curves of f is the direction of greatest increase or the direction of greatest decrease.

Example 5 Consider the curve $\mathcal{C}$ in the xy-plane that is the graph of the equation $x^2 + 6y^4 = 10$. Find the line that is normal to the curve at the point (2, 1).

Solution Let $P_0 = (2, 1)$. We may regard $\mathcal{C}$ as the level curve $f(x, y) = 10$ where $f(x, y) = x^2 + 6y^4$. We calculate

$$\nabla f(P_0) = (2x\mathbf{i} + 24y^3\mathbf{j})|_{(x,y)=(2,1)} = 4\mathbf{i} + 24\mathbf{j} = 4(\mathbf{i} + 6\mathbf{j}).$$

According to Theorem 3, this gradient vector is normal to $\mathcal{C}$ at P_0. We may therefore write the normal line as $(x - 2)/1 = (y - 1)/6$, or equivalently, $y = 6(x - 2) + 1$. See Figure 5. ■

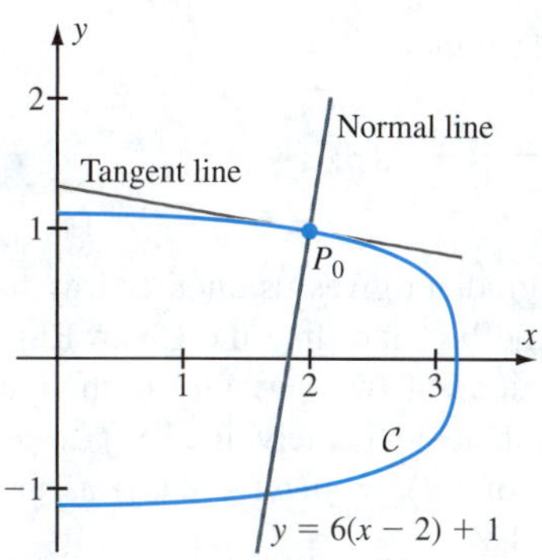

Figure 5
A normal line to the graph of $x^2 + 6y^4 = 10$

Functions of Three or More Variables

All of the ideas introduced in this section are also valid for functions $(x, y, z) \mapsto F(x, y, z)$ of three variables. The gradient of such a function is

$$\nabla F = \frac{\partial F}{\partial x}\mathbf{i} + \frac{\partial F}{\partial y}\mathbf{j} + \frac{\partial F}{\partial z}\mathbf{k}.$$

The directional derivative of F at the point $P_0 = (x_0, y_0, z_0)$ in the direction $\mathbf{u} = \langle u_1, u_2, u_3 \rangle$ is

$$D_{\mathbf{u}}F(P_0) = \nabla F(P_0) \cdot \mathbf{u}.$$

The direction of greatest rate of increase of the function f at the point P_0 is

$$\mathbf{u} = \frac{1}{\|\nabla F(P_0)\|}\nabla F(P_0).$$

With this choice of direction $\mathbf{u}$, we have $D_{\mathbf{u}}F(P_0) = \|\nabla F(P_0)\|$. The direction of greatest rate of decrease of the function F at the point P_0 is

$$-\mathbf{u} = -\frac{1}{\|\nabla F(P_0)\|}\nabla F(P_0).$$

With this choice of direction, we have $D_{-\mathbf{u}}F(P_0) = -\|\nabla F(P_0)\|$.

The level set $\{(x, y, z) : F(x, y, z) = c\}$ is a surface. At the point P_0, the gradient $\nabla F(P_0)$ is normal to the level surface of F that passes through P_0. We will discuss this result further in Section 13.7.

Example 6 Find the directions of greatest rate of increase and greatest rate of decrease for the function $F(x, y, z) = xyz$ at the point $(-1, 2, 1)$.

Solution To do this, we calculate $\nabla F(x, y, z) = yz\mathbf{i} + xz\mathbf{j} + xy\mathbf{k}$ and $\nabla F(-1, 2, 1) = 2\mathbf{i} - \mathbf{j} - 2\mathbf{k}$. Then the direction of greatest rate of increase is

$$\frac{1}{\|\nabla F(-1, 2, 1)\|}\nabla F(-1, 2, 1) = \frac{1}{\sqrt{2^2 + (-1)^2 + (-2)^2}}(2\mathbf{i} - \mathbf{j} - 2\mathbf{k}) = \frac{2}{3}\mathbf{i} - \frac{1}{3}\mathbf{j} - \frac{2}{3}\mathbf{k}.$$

The direction of greatest rate of decrease at $(-1, 2, 1)$ is then

$$-\frac{1}{\|\nabla F(-1, 2, 1)\|}\nabla F(-1, 2, 1) = -\frac{2}{3}\mathbf{i} + \frac{1}{3}\mathbf{j} + \frac{2}{3}\mathbf{k}.$$

■

We conclude this section by observing that the gradient gives us another way to think about the Chain Rule that makes it look a little bit more like the Chain Rule of one variable. Let us illustrate this point for functions of two variables (although the point is valid for functions of any number of variables)—namely, if $z = f(x, y)$ where $x = \rho(t)$ and $y = \sigma(t)$, then we can think of $\mathbf{r}(t) = \rho(t)\mathbf{i} + \sigma(t)\mathbf{j}$ as the parameterization of a curve. The Chain Rule tells us that

$$\frac{dz}{dt} = \frac{\partial f}{\partial x}\frac{d\rho}{dt} + \frac{\partial f}{\partial y}\frac{d\sigma}{dt},$$

which we can write as

$$\frac{dz}{dt} = (\nabla f) \cdot \left(\frac{d\mathbf{r}}{dt}\right). \tag{13.10}$$

We see that the gradient lets us write the Chain Rule in the more familiar form of a product, provided we interpret the dot in equation (13.10) as the dot product. Note that this last notation for the Chain Rule looks the same when more independent variables are present. For instance, if $w = F(x, y, z)$ where $x = \sigma_1(t)$, $y = \sigma_2(t)$, $z = \sigma_3(t)$, then we can think of $\mathbf{r}(t) = \sigma_1(t)\mathbf{i} + \sigma_2(t)\mathbf{j} + \sigma_3(t)\mathbf{k}$ as the parameterization of a curve in xyz-space. In this case, the Chain Rule is

$$\frac{dw}{dt} = \frac{\partial F}{\partial x}\frac{d\sigma_1}{dt} + \frac{\partial F}{\partial y}\frac{d\sigma_2}{dt} + \frac{\partial F}{\partial z}\frac{d\sigma_3}{dt},$$

which can be written as

$$\frac{dw}{dt} = (\nabla F) \cdot \left(\frac{d\mathbf{r}}{dt}\right). \tag{13.11}$$

quickquiz

1. Define the directional derivative of a function $f(x, y)$ at a point P in the direction $\mathbf{v}$.
2. When $\mathbf{v} = \mathbf{i}$ or $\mathbf{v} = \mathbf{j}$, what does the directional derivative signify?
3. What is the gradient of a function $f(x, y)$?
4. Given a function f and a point P of its domain, which direction indicates the direction of greatest decrease? Greatest increase?

EXERCISES

Problems for Practice

Calculate the gradient of each function in Exercises 1–10.

1. $f(x, y) = x\sin(y)$
2. $f(x, y) = x\ln(x + 3y)$
3. $f(x, y) = \tan(xy^2)$
4. $f(x, y, z) = xy^2/z^3$
5. $f(x, y, z) = \cos(xy)\sin(yz)$
6. $f(x, y) = x^{3/2}/y^{5/2} - xy$
7. $f(x, y) = (xy^2)/\cos(y)$
8. $f(x, y) = \sqrt{y}\cot(x+y)$
9. $f(x, y, z) = z \cdot \sin(xy)$
10. $f(x, y, z) = xy^2\sin(yz^2)$

In Exercises 11–22, calculate the directional derivative of f at P_0 in the direction $\mathbf{u}$.

11. $f(x, y) = x^2y - y^3x$, $P_0 = (2, -3)$, $\mathbf{u} = \langle -6/10, 8/10\rangle$
12. $f(x, y) = x\ln(x^2 + y^2)$, $P_0 = (2, 4)$, $\mathbf{u} = \langle 4/5, 3/5\rangle$
13. $f(x, y) = \cos(x^2 - y^2)$, $P_0 = (\pi^{1/2}, \pi^{1/2}/2)$, $\mathbf{u} = \langle 1/\sqrt{2}, -1/\sqrt{2}\rangle$
14. $f(x, y) = (x + y)/(x - y)$, $P_0 = (4, 7)$, $\mathbf{u} = \langle \sqrt{3}/2, -1/2\rangle$
15. $f(x, y) = e^{4x-7y}$, $P_0 = (0, \ln(2))$, $\mathbf{u} = \langle 1, 0\rangle$
16. $f(x, y) = \tan(x/y)$, $P_0 = (\pi/4, 1)$, $\mathbf{u} = \langle -12/13, 5/13\rangle$
17. $f(x, y) = 1/(x^2 + y^2)$, $P_0 = (2, 2)$, $\mathbf{u} = \langle 0, 1\rangle$
18. $f(x, y) = (x - y)\ln(x + y)$, $P_0 = (1, 1)$, $\mathbf{u} = \langle \sqrt{3/2}, -1/2\rangle$
19. $f(x, y) = \sin(xy)/(x + y)$, $P_0 = (0, 1)$, $\mathbf{u} = \langle 1/\sqrt{2}, -1/\sqrt{2}\rangle$
20. $f(x, y) = e^{xy}e^{-y}$, $P_0 = (1, \ln(2))$, $\mathbf{u} = \langle 0, -1\rangle$
21. $f(x, y) = \sin^3(x + y)$, $P_0 = (\pi/3, -\pi/6)$, $\mathbf{u} = \langle -1, 0\rangle$
22. $f(x, y) = (x^2 + y^3)^4$, $P_0 = (1, 1)$, $\mathbf{u} = \langle -1/2, \sqrt{3}/2\rangle$

In Exercises 23–32, find the following:

a. the direction $\mathbf{u}$ of greatest rate of increase for the function f at the point P_0
b. the directional derivative of f at P_0 in the direction of $\mathbf{u}$

c. the direction $\mathbf{v}$ of greatest rate of decrease for f at the point P_0

d. the directional derivative of f at P_0 in the direction of $\mathbf{v}$

23. $f(x, y) = x^3y^2 - yx$, $P_0 = (2, 1/2)$

24. $f(x, y) = \cos(x - y^2)$, $P_0 = (\pi, \pi^{1/2}/2)$

25. $f(x, y) = \ln(x + y)/(x + y)$, $P_0 = (1, 1)$

26. $f(x, y) = x^2 \tan(y)$, $P_0 = (1, \pi/4)$

27. $f(x, y) = \sin(x + y)/\cos(x - y)$, $P_0 = (\pi/2, \pi/3)$

28. $f(x, y) = \tan(x^2y)$, $P_0 = (1, \pi/3)$

29. $f(x, y) = e^{-x} \cos(y)$, $P_0 = (1, \pi)$

30. $f(x, y) = x/(x - y)$, $P_0 = (4, 1)$

31. $f(x, y) = (x^2 + y)^5$, $P_0 = (1, 1)$

32. $f(x, y) = \sec(x - y)$, $P_0 = (\pi/2, \pi/3)$

In Exercises 33–38, find the directional derivative of the function F at the point P_0 in the direction $\mathbf{u}$.

33. $F(x, y, z) = xyz^3$, $P_0 = (2, 1, -3)$, $\mathbf{u} = \langle 2/3, -2/3, 1/3\rangle$

34. $F(x, y, z) = x/(y + z)$, $P_0 = (4, 2, 2)$, $\mathbf{u} = \langle 3/5, \sqrt{7}/5, -3/5\rangle$

35. $F(x, y, z) = y \cdot \sin(xz)$, $P_0 = (1/2, 3, \pi)$, $\mathbf{u} = \langle\sqrt{3}/8, -6/8, 5/8\rangle$

36. $F(x, y, z) = x/yz$, $P_0 = (6, 3, 2)$, $\mathbf{u} = \langle 5/7, -2\sqrt{2}/7, -4/7\rangle$

37. $F(x, y, z) = \ln(xy + z^3)$, $P_0 = (1, 1, 1)$, $\mathbf{u} = \langle -1/2, 3/4, \sqrt{3}/4\rangle$

38. $F(x, y, z) = \sin(x - z)/\cos(y + z)$, $P_0 = (\pi/6, \pi/3, \pi/2)$, $\mathbf{u} = \langle 1/2, -1/2, 1/\sqrt{2}\rangle$

In Exercises 39–42, find the following:

a. the direction $\mathbf{u}$ of greatest rate of increase for the function F at the point P_0

b. the directional derivative of F at P_0 in the direction of $\mathbf{u}$

c. the direction $\mathbf{v}$ of greatest rate of decrease for F at the point P_0

d. the directional derivative of F at P_0 in the direction of $\mathbf{v}$

39. $F(x, y, z) = x^2(y^3 - zy)$, $P_0 = (2, -1, 4)$

40. $F(x, y, z) = \ln(x - y)/z$, $P_0 = (2, 1, 3)$

41. $F(x, y, z) = \sin(x^2 - y + z)$, $P_0 = (0, \pi, 2\pi)$

42. $F(x, y, z) = \sin(2x) + 12\cos(yz)$, $P_0 = (\pi/6, 1, \pi/6)$

Further Theory and Practice

43. Let $f(x, y)$ be a continuously differentiable function, and suppose that $\nabla f(P) = 0$. What does this say about the direction of greatest increase or greatest decrease of f at P? Give examples.

In Exercises 44–47, find the directional derivative of the given function f at P in the direction from P to Q.

44. $f(x, y) = x^2 + xy$, $P = (1, 2)$, $Q = (3, 4)$

45. $f(x, y) = x\sqrt{1 + xy}$, $P = (2, 4)$, $Q = (5, 8)$

46. $f(x, y) = \exp(8x/y^2 + y/x)$, $P = (1, -2)$, $Q = (-5, 6)$

47. $f(x, y) = \tan(xy^2)$, $P = (\pi, 1/2)$, $Q = (-3\pi, 1/2 + 3\pi)$

In Exercises 48–51, find $\frac{\partial f}{\partial x}(P)$ and $\frac{\partial f}{\partial y}(P)$ using the given information.

48. $\mathbf{u} = \langle 0, 1\rangle$, $\mathbf{v} = \langle 1/\sqrt{2}, 1/\sqrt{2}\rangle$, $D_{\mathbf{u}}f(P) = 2\sqrt{2}$, $D_{\mathbf{v}}f(P) = 3\sqrt{2}$

49. $\mathbf{u} = \langle 3/5, 4/5\rangle$, $\mathbf{v} = \langle 1/\sqrt{2}, 1/\sqrt{2}\rangle$, $D_{\mathbf{u}}f(P) = 1$, $D_{\mathbf{v}}f(P) = \sqrt{2}$

50. $\mathbf{u} = \langle 5/13, 12/13\rangle$, $\mathbf{v} = \langle -12/13, 5/13\rangle$, $D_{\mathbf{u}}f(P) = 19$, $D_{\mathbf{v}}f(P) = 22$

51. $\mathbf{u} = \langle 1/2, -\sqrt{3}/2\rangle$, $\mathbf{v} = \langle 1/\sqrt{2}, 1/\sqrt{2}\rangle$, $D_{\mathbf{u}}f(P) = 3$, $D_{\mathbf{v}}f(P) = 0$

52. Let $f(x, y) = x^2 - y^3$. At the point $P = (3, 7)$, which direction(s) $\mathbf{u}$ will have the property that $D_{\mathbf{u}}f(P) = 8$?

53. Let $f(x, y) = xy - y^4$ and $P = (8, -2)$. Explain why there is no direction $\mathbf{u}$ such that $D_{\mathbf{u}}f(P) = 50$.

54. You are standing at the bottom, $(0, 0, 0)$, of a hole that is shaped like the graph of $f(x, y) = x^4 + y^2 + x^2y^6$. You need to climb out of the hole. What is the direction of steepest ascent? What is the direction of slowest ascent?

55. A heat-seeking missile can only sense heat in its immediate vicinity. In an effort to find the source of the heat, the missile always moves in the direction of greatest increase in temperature. If the temperature at any point in space is given by $T(x, y, z) = 1000 - 10(x - 90)^2 - 2(y - 12)^4 - z^2$ and the missile is currently at $(100, 15, 8)$, in what direction will it move next?

56. Laplace's equation for a function of two variables is given by

$$\Delta f(x, y) = \left(\frac{\partial^2}{\partial x^2} + \frac{\partial^2}{\partial y^2}\right) f(x, y) = 0.$$

Let $\mathbf{u}$ and $\mathbf{v}$ be any pair of orthogonal unit vectors in the plane. Prove that Laplace's equation is also given by

$$D_{\mathbf{u}}(D_{\mathbf{u}}f) + D_{\mathbf{v}}(D_{\mathbf{v}}f)(x, y) = 0.$$

57. If $\mathbf{f}(x, y) = f_1(x, y)\mathbf{i} + f_2(x, y)\mathbf{j}$ is a vector-valued function of two variables, then we define the divergence of $\mathbf{f}$ to be

$$\operatorname{div} \mathbf{f}(x, y) = \frac{\partial f_1}{\partial x}(x, y) + \frac{\partial f_2}{\partial y}(x, y).$$

Show that if φ is a twice continuously differentiable function, then

$$\operatorname{div}(\operatorname{grad}(\varphi)) = \Delta\varphi.$$

58. Let f be a continuously differentiable function defined on a planar region. Suppose that P is a point in the domain of f. Show that the set of all possible values for the directional derivative of f at P forms an interval. What are the endpoints of this interval?

59. Let f be a twice continuously differentiable function with P_0 in its domain. Let $\mathbf{u} = u_1\mathbf{i} + u_2\mathbf{j}$ and $\mathbf{v} = v_1\mathbf{i} + v_2\mathbf{j}$ be directions. Give formulas for $D_{\mathbf{u}}(D_{\mathbf{u}} f)(P_0)$, $D_{\mathbf{v}}(D_{\mathbf{v}} f)(P_0)$, and $D_{\mathbf{u}}(D_{\mathbf{v}} f)(P_0)$. What does $D_{\mathbf{u}}(D_{\mathbf{u}} f)(P_0)$ tell us about the concavity of the graph of f?

Computer/Calculator Exercises

60. In the viewing window $[-1, 3.2] \times [-2.2, 2.2]$, plot the level curve of $f(x, y) = 4x + y - x^2/8 - x^4/8 + 4y^2 - y^4$ at height 1. Calculate and sketch the directions of the gradients at the six points of intersection with the coordinate axes.

61. In the viewing window $[-1.25, 4.5] \times [-5, 3/4]$, plot the level curves L_c of $f(x, y) = 1 - x^2y/2 - x^2 - y^2 - 2y$ that correspond to values $f(x, y) = c$ where $c = -1/2, 1/2, 3/2, 5/2$, and $7/2$. Calculate and sketch the directions of the gradients at the following points.

a. $(3, -3/2)$ on $L_{-1/2}$

b. $(2, -2 - \sqrt{2}/2)$ and $(2, -2 + \sqrt{2}/2)$ on $L_{1/2}$

c. $(1, -1)$ and $(1, -3/2)$ on $L_{3/2}$

d. $(3, -3)$ and $(3, -7/2)$ on $L_{5/2}$

e. $(4, -5 + \sqrt{26}/2)$ on $L_{7/2}$

13.7 Tangent Planes

If f is a differentiable function of one variable, then its derivative at a point x tells us the slope of the tangent line at that point. Now we will learn that the gradient of a function of two variables tells us about the tangent lines to the graph of f. We shall see that the gradient plays the role of a "total derivative."

But we have begged an important question: *What* geometric object should the tangent to the graph of a function of two variables be? Look at Figure 1, which exhibits the graph of $f(x, y) = 1 + x^2 + y^2$. There are a great many tangent lines at the point $(0, 0, 1)$, as the figure illustrates. The union of these tangent lines suggests that a *tangent plane* is the geometric object we seek. See Figure 2.

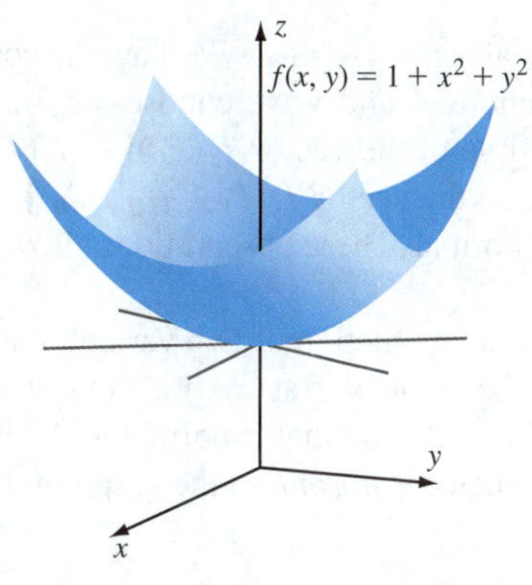

Figure 1
Some tangent lines at the point $(0, 0, 1)$

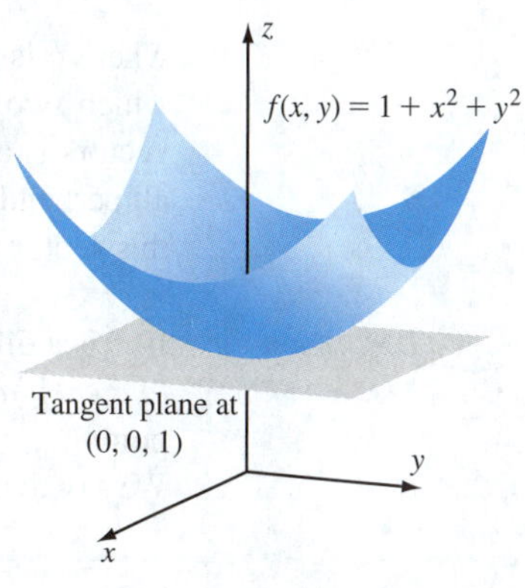

Figure 2
The union of all tangent lines at a point is a tangent plane.

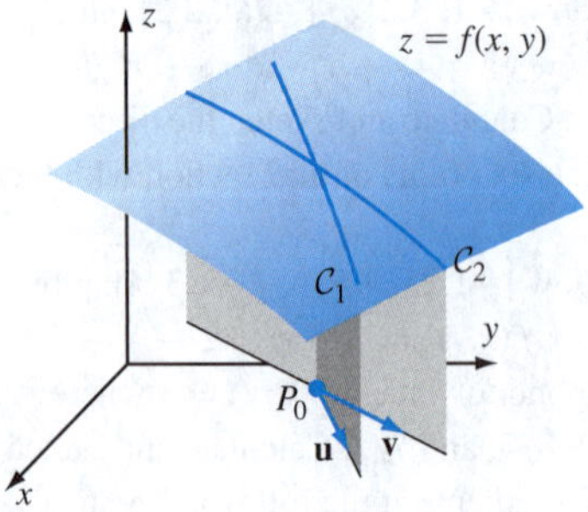

Figure 3

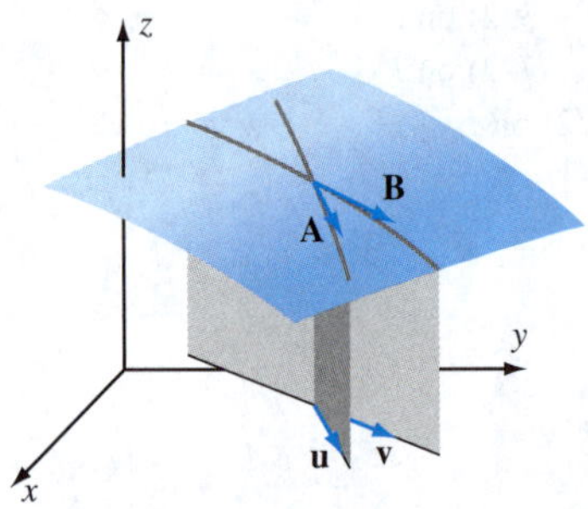

Figure 4
Tangent vectors
$\mathbf{A} = u_1\mathbf{i} + u_2\mathbf{j} + D_{\mathbf{u}}f(P_0)\mathbf{k}$ and
$\mathbf{B} = v_1\mathbf{i} + v_2\mathbf{j} + D_{\mathbf{v}}f(P_0)\mathbf{k}$

Now consider an arbitrary differentiable function f of two variables. Fix a point $P_0 = (x_0, y_0)$ in the domain of f. Also fix two nonparallel directions $\mathbf{u} = u_1\mathbf{i} + u_2\mathbf{j}$ and $\mathbf{v} = v_1\mathbf{i} + v_2\mathbf{j}$ in the xy-plane. Then the curves $\mathcal{C}_1$ and $\mathcal{C}_2$ that are parameterized by

$$r_{\mathbf{u}}(t) = (x_0 + tu_1)\mathbf{i} + (y_0 + tu_2)\mathbf{j} + f(x_0 + tu_1, y_0 + tu_2)\mathbf{k}$$

and

$$r_{\mathbf{v}}(t) = (x_0 + tv_1)\mathbf{i} + (y_0 + tv_2)\mathbf{j} + f(x_0 + tv_1, y_0 + tv_2)\mathbf{k}$$

are subsets of the graph of f. Each curve passes through the point P_0 when $t = 0$. The geometry is shown in Figure 3. We know that the vectors $r'_{\mathbf{u}}(0)$ and $r'_{\mathbf{v}}(0)$ are tangent to $\mathcal{C}_1$ and $\mathcal{C}_2$, respectively. We use the Chain Rule to calculate these two tangent vectors:

$$r'_{\mathbf{u}}(0) = u_1\mathbf{i} + u_2\mathbf{j} + \left(\frac{d}{dt} f(x_0 + tu_1, y_0 + tu_2)\bigg|_{t=0} \right)\mathbf{k} = u_1\mathbf{i} + u_2\mathbf{j} + D_{\mathbf{u}}f(P_0)\mathbf{k}$$

and

$$r'_{\mathbf{v}}(0) = v_1\mathbf{i} + v_2\mathbf{j} + \left(\frac{d}{dt} f(x_0 + tv_1, y_0 + tv_2)\bigg|_{t=0} \right)\mathbf{k} = v_1\mathbf{i} + v_2\mathbf{j} + D_{\mathbf{v}}f(P_0)\mathbf{k}.$$

See Figure 4. Let us use the cross product to calculate a vector $\mathbf{N}$ that is normal to both of these tangent vectors. We obtain

$$\mathbf{N} = (u_1\mathbf{i}+u_2\mathbf{j}+D_{\mathbf{u}}f(P_0)\mathbf{k})\times(v_1\mathbf{i}+v_2\mathbf{j}+D_{\mathbf{v}}f(P_0)\mathbf{k}) = \det\left(\begin{bmatrix} \mathbf{i} & \mathbf{j} & \mathbf{k} \\ u_1 & u_2 & D_{\mathbf{u}}f(P_0) \\ v_1 & v_2 & D_{\mathbf{v}}f(P_0) \end{bmatrix}\right),$$

or equivalently,

$$\mathbf{N} = (u_2D_{\mathbf{v}}f(P_0) - v_2D_{\mathbf{u}}f(P_0))\mathbf{i} - (u_1D_{\mathbf{v}}f(P_0) - v_1D_{\mathbf{u}}f(P_0))\mathbf{j} + (u_1v_2 - u_2v_1)\mathbf{k}.$$

After substituting $D_{\mathbf{u}}f(P_0) = u_1f_x(P_0) + u_2f_y(P_0)$ and $D_{\mathbf{v}}f(P_0) = v_1f_x(P_0) + v_2f_y(P_0)$ into the formula for $\mathbf{N}$, we find that the resulting expression simplifies to

$$\mathbf{N} = c(f_x(P_0)\mathbf{i} + f_y(P_0)\mathbf{j} - \mathbf{k})$$

where c is the scalar $u_2v_1 - v_2u_1$. We have discovered a remarkable fact: No matter which two directions $\mathbf{u}$ and $\mathbf{v}$ we choose, any normal to the corresponding tangent vectors is a multiple of the vector $f_x(P_0)\mathbf{i} + f_y(P_0)\mathbf{j} - \mathbf{k}$. Since the tangent lines are all perpendicular to the vector $f_x(P_0)\mathbf{i} + f_y(P_0)\mathbf{j} - \mathbf{k}$, they must form a plane that has this vector as its normal. These observations allow us to make the following definition.

Definition If f is a differentiable function of two variables and (x_0, y_0) is in its domain, then the *tangent plane* to the graph of f at $(x_0, y_0, f(x_0, y_0))$ is the plane that passes through the point $(x_0, y_0, f(x_0, y_0))$ and that is normal to the vector $f_x(x_0, y_0)\mathbf{i} + f_y(x_0, y_0)\mathbf{j} - \mathbf{k}$. We say that this vector is *normal* to the graph of f at the point $(x_0, y_0, f(x_0, y_0))$.

Using equation (11.26), we see that the Cartesian equation of the tangent plane to the graph of f at the point $(x_0, y_0, f(x_0, y_0))$ is

$$f_x(x_0, y_0)(x - x_0) + f_y(x_0, y_0)(y - y_0) - (z - f(x_0, y_0)) = 0. \qquad \textbf{(13.12)}$$

The equivalent equation,

$$z = f(x_0, y_0) + f_x(x_0, y_0)(x - x_0) + f_y(x_0, y_0)(y - y_0), \tag{13.13}$$

has the advantage of giving the formula for z explicitly.

inSIGHT

When we use equations (13.12) and (13.13), we calculate the derivatives $f_x(x, y)$ and $f_y(x, y)$, and then we evaluate them at the point of tangency. Consequently, the coefficients $f_x(x_0, y_0)$ and $f_y(x_0, y_0)$ are constants. When our calculation of equation (13.12) or (13.13) is complete, each of the variables x and y appears at most once, has power 1, and is multiplied by a (possibly zero) constant. Equation (13.13), for example, has the form

$$\underset{\substack{\uparrow \\ \text{Variable}}}{z} = \underbrace{f(x_0, y_0)}_{\text{Constant}} + \underbrace{f_x(x_0, y_0)}_{\text{Constant}} (\underset{\substack{\uparrow \\ \text{Variable}}}{x} - \underbrace{x_0}_{\text{Constant}}) + \underbrace{f_y(x_0, y_0)}_{\text{Constant}} (\underset{\substack{\uparrow \\ \text{Variable}}}{y} - \underbrace{y_0}_{\text{Constant}}).$$

Example 1 Find the equation of the tangent plane to the graph of $f(x, y) = 2x - 3xy^3$ at the point $(2, -1, 10)$.

Solution We calculate $f_x(2, -1) = (2 - 3y^3)|_{(2,-1)} = 5$ and $f_y(2, -1) = (-9xy^2)|_{(2,-1)} = -18$. Also, $z_0 = f(2, -1) = 10$. According to equation (13.12), the equation of the tangent plane is $5(x - 2) + (-18)(y - (-1)) - (z - 10) = 0$, or $5x - 18y - z = 18$. ■

At this point in the theory, the variables x, y, and z do not have symmetric roles in the formula for the tangent plane to the graph of $z = f(x, y)$. A partial explanation is given by the fact that z is a dependent variable, whereas x and y are independent variables. However, a more geometric explanation is provided by the following discussion of level surfaces.

Level Surfaces

In Section 13.6, we learned that the gradient $\nabla f(x_0, y_0)$ of a function of two variables is perpendicular to the level curve of f that passes through the point (x_0, y_0). An analogous principle holds true for a function F of three variables. To understand why, fix a point $Q_0 = (x_0, y_0, z_0)$ in the domain of F. Let $\mathcal{S} = \{(x, y, z) : F(x, y, z) = c\}$ denote the level surface of F that passes through Q_0. We will show that $\nabla F(Q_0)$ is perpendicular to $\mathcal{S}$ at Q_0 by showing that it is perpendicular to all curves on $\mathcal{S}$ that pass through Q_0. Indeed, if $\mathbf{r} = r_1\mathbf{i} + r_2\mathbf{j} + r_3\mathbf{k}$ is a parameterization of a curve $\mathcal{C}$ that passes through Q_0 at $t = t_0$, then

$$\left.\frac{d}{dt}F(\mathbf{r}(t))\right|_{t=t_0} = \nabla F(Q_0) \cdot \mathbf{r}'(t_0), \tag{13.14}$$

by equation (13.11). But if $\mathcal{C}$ is a subset of the level set $\mathcal{S}$, then $F(\mathbf{r}(t))$ is the constant c for all values of t. The left side of equation (13.14) is therefore 0. It follows that $\nabla F(Q_0)$ is perpendicular to the velocity vector $\mathbf{r}'(t_0)$. Since this velocity vector is tangent to $\mathcal{C}$ at Q_0, we conclude that $\nabla F(Q_0)$ is perpendicular to $\mathcal{C}$. We have proved the following theorem.

Theorem 1 If F is a differentiable function of three variables, then $\nabla F(x_0, y_0, z_0)$ is perpendicular to the level surface of F at (x_0, y_0, z_0).

We can now tie together the idea of tangent plane with our new insight into level surfaces. Suppose that $(x, y) \mapsto f(x, y)$ is a function of two variables. We want to know a normal to the tangent plane to the graph at the point $Q_0 = (x_0, y_0, f(x_0, y_0))$. Define

$$F(x, y, z) = f(x, y) - z.$$

Then the graph of f is just the set $\{(x, y, z) : F(x, y, z) = 0\}$; that is, the graph of f is a level surface of F. Therefore, $\nabla F(Q_0)$ is perpendicular to the graph of f. But

$$\nabla F(Q_0) = \langle f_x(x_0, y_0), f_y(x_0, y_0), -1\rangle.$$

Thus, we have rediscovered the fact that $\langle f_x(x_0, y_0), f_y(x_0, y_0), -1\rangle$ is a normal to the graph of f at the point $(x_0, y_0, f(x_0, y_0))$.

Example 2 Find the tangent plane to the surface $x^2 + 4y^2 + 8z^2 = 13$ at the point $(1, -1, 1)$.

Solution We do *not* need to solve for z as a function of x and y. Instead, we define the function of three variables $F(x, y, z) = x^2 + 4y^2 + 8z^2$. Then our surface is just the level set $F(x, y, z) = 13$. By Theorem 1, a normal to the surface $F(x, y, z) = 13$ at the point $(1, -1, 1)$ is given by

$$\nabla F(1, -1, 1) = (2x\mathbf{i} + 8y\mathbf{j} + 16z\mathbf{k})|_{(x,y,z)=(1,-1,1)} = 2\mathbf{i} - 8\mathbf{j} + 16\mathbf{k}.$$

Since the tangent plane also passes through the point $(x_0, y_0, z_0) = (1, -1, 1)$, we may write the equation of the plane as $2(x-1) - 8(y-(-1)) + 16(z-1) = 0$. This equation simplifies to $2x - 8y + 16z = 26$. ■

Example 3 Let $\mathcal{S}$ be the surface defined by the equation $z = \sqrt{(5 + x^2 + 2y^4)/(y^2 + x^4)}$. Find the tangent plane to $\mathcal{S}$ at the point $(1, -1, 2)$.

Solution Although z is given explicitly as a function of x and y, the calculation of the derivatives $\frac{\partial z}{\partial x}$ and $\frac{\partial z}{\partial y}$ involves a lengthy application of the Quotient Rule. We may avoid this difficulty by observing that $\mathcal{S}$ is a subset of a level surface of a function that is easy to differentiate. Indeed, every point (x, y, z) of $\mathcal{S}$ satisfies the equation $z^2 = (5 + x^2 + 2y^4)/(y^2 + x^4)$, or equivalently, $5 + x^2 + 2y^4 - (y^2 + x^4)z^2 = 0$. Therefore $\mathcal{S}$ is a subset of a level set of $F(x, y, z) = 5 + x^2 + 2y^4 - (y^2 + x^4)z^2$. We calculate $\nabla F(x, y, z) = (2x - 4x^3z^2)\mathbf{i} + (8y^3 - 2yz^2)\mathbf{j} - 2(y^2 + x^4)z\mathbf{k}$. A normal to $\mathcal{S}$ at the point $(1, -1, 2)$ is therefore given by

$$\begin{aligned}\nabla F(1, -1, 2) &= ((2x - 4x^3z^2)\mathbf{i} + (8y^3 - 2yz^2)\mathbf{j} - 2(y^2 + x^4)z\mathbf{k})|_{(x,y,z)=(1,-1,2)} \\ &= -14\mathbf{i} - 8\mathbf{k}.\end{aligned}$$

Since the tangent plane we seek passes through the point $(x_0, y_0, z_0) = (1, -1, 2)$, we may write its equation as $(-14)(x-1) + (0)(y-(-1)) + (-8)(z-2) = 0$. This equation simplifies to $7x + 4z = 15$. ■

Normal Lines

The normal line to a surface at a point $Q_0 = (x_0, y_0, f(x_0, y_0))$ on the surface is the line with direction vector given by a normal to the surface at Q_0 and passing through Q_0.

Example 4 Find the normal line to the graph of $f(x, y) = -y^2 - x^3 + xy^2$ at the point $(1, 4, -1)$.

Solution We will solve this problem in two ways. A normal to the graph at Q_0 is

$$\langle f_x(x, y), f_y(x, y), -1\rangle|_{x=1, y=4} = \langle -3x^2 + y^2, -2y + 2xy, -1\rangle|_{x=1, y=4} = \langle 13, 0, -1\rangle.$$

Since the desired line also passes through $Q_0 = (1, 4, f(1, 4)) = (1, 4, -1)$, we may write the line parametrically as

$$x = 1 + 13t, \quad y = 4, \quad z = -1 + (-1)t.$$

The symmetric equations are

$$\frac{x-1}{13} = \frac{z+1}{-1}, \quad y = 4.$$

As an alternative, we can obtain the same result by using the equation

$$F(x, y, z) = f(x, y) - z = -y^2 - x^3 + xy^2 - z$$

to define a function F of three variables. Then the level set $\{(x, y, z) : F(x, y, z) = 0\}$ is precisely the graph of f. Therefore, a normal to the graph of f is

$$\nabla F(1, 4, -3) = ((-3x^2 + y^2)\mathbf{i} + (-2y + 2xy)\mathbf{j} - 1\mathbf{k})|_{x=1, y=4, z=f(1,4)} = 13\mathbf{i} - \mathbf{k}.$$

The rest of the computation continues exactly as in the first solution. ■

Numerical Approximations Using the Tangent Plane

As we learned in Section 3.9, we can use the tangent line as a means for approximating a differentiable function of one variable. Now we apply the same idea to functions of two variables. Let $\mathcal{T}$ be the tangent plane to the graph of a differentiable function $(x, y) \mapsto f(x, y)$ at the point $(x_0, y_0, f(x_0, y_0))$. Equation (13.13) tells us that $\mathcal{T}$ is the graph of the function

$$L(x, y) = f(P_0) + f_x(P_0)(x - x_0) + f_y(P_0)(y - y_0) \qquad \textbf{(13.15)}$$

where $P_0 = (x_0, y_0)$. Recall that the terms $f(P_0)$, $f_x(P_0)$, and $f_y(P_0)$ are just constants. Once we have calculated these three fixed numbers, we need only perform two multiplications and a few additions to compute $L(x, y)$ for any x and y. What is significant is that this particularly simple function gives a good approximation to $f(x, y)$ for x and y close to x_0 and y_0, respectively. This can be seen from equation (13.3), which states that

$$f(x, y) = f(P_0) + f_x(P_0)(x - x_0) + f_y(P_0)(y - y_0) + \epsilon_1 \cdot (x - x_0) + \epsilon_2 \cdot (y - y_0).$$

Notice that the first three summands on the right side exactly constitute the formula for $L(x, y)$. We conclude that

$$f(x, y) = L(x, y) + \epsilon_1 \cdot (x - x_0) + \epsilon_2 \cdot (y - y_0).$$

Finally, recall that ϵ_1 and ϵ_2 both tend to 0 as x and y tend to x_0 and y_0, respectively. This means that $f(x, y) - L(x, y)$ is the sum of two terms, $\epsilon_1 \cdot (x - x_0)$ and $\epsilon_2 \cdot (y - y_0)$, each of which becomes small compared with the increments $\Delta x = (x - x_0)$ and $\Delta y = (y - y_0)$ as x and y tend to x_0 and y_0. We summarize our analysis in the following theorem.

Theorem 2 Let f be a differentiable function of two variables on a rectangular region centered at $P_0 = (x_0, y_0)$. Let L be the linear function of two variables defined by

$$L(x, y) = f(P_0) + f_x(P_0)(x - x_0) + f_y(P_0)(y - y_0).$$

Then for every point (x, y) near the fixed point P_0, we have the approximation $f(x, y) \approx L(x, y)$. The difference $f(x, y) - L(x, y)$ between the exact value $f(x, y)$ and the approximating value $L(x, y)$ is equal to $\epsilon_1 \cdot (x - x_0) + \epsilon_2 \cdot (y - y_0)$ where ϵ_1 and ϵ_2 both tend to 0 as x and y tend to x_0 and y_0, respectively.

Definition The expression $L(x, y)$ defined by equation (13.15) is called the *linear approximation* (or the *tangent plane approximation*) to $f(x, y)$ at the point P_0.

Nowadays, with calculators and computers so common, we can easily compute the expression $f(x, y)$ when it is given by an explicit formula that involves x and y. What is the point of an approximation? The answer is that $f(x, y)$ may involve complicated combinations and compositions of expressions, such as e^x, $\ln(x)$, $\tan(y)$, that are difficult to analyze. Because the expression $L(x, y)$ is either a constant or a first-degree polynomial in x and y, it is always easy to work with.

It is natural to ask how good the approximation given by the theorem is. After all, we do not have exact expressions for the terms ϵ_1 and ϵ_2 that appear in the error $f(x, y) - L(x, y)$. In fact, by using the Taylor formula derived in Section 13.5, we can see that on a rectangular region

$$R = \{(x, y) : |x - x_0| \le r_1, |y - y_0| \le r_2\},$$

the difference between the given function f and its linear approximation L at a point $(x, y) \in R$ satisfies the bound

$$|f(x, y) - L(x, y)| \le \frac{M}{2} \cdot \left(|x - x_0| + |y - y_0|\right)^2, \qquad \textbf{(13.16)}$$

where M is any number that is larger than the absolute values of all the second derivatives $|f_{xx}|$, $|f_{yy}|$, $|f_{xy}|$ on the region R.

Example 5 Let f be defined by $f(x, y) = \sqrt{x^2 + y^4}$. Using $P_0 = (3, 2)$, what is the tangent plane approximation $L(2.9, 2.1)$ of $f(2.9, 2.1)$?

Solution We use formula (13.15) with $(x, y) = (2.9, 2.1)$ and $(x_0, y_0) = (3, 2)$. The value $f(3, 2)$ is easily computed to be 5. Also

$$f_x(3, 2) = x(x^2 + y^4)^{-1/2}\big|_{(3,2)} = \frac{3}{5} \quad \text{and} \quad f_y(3, 2) = 2y^3(x^2 + y^4)^{-1/2}\big|_{(3,2)} = \frac{16}{5}.$$

Therefore, according to Theorem 2,

$$\begin{aligned} f(2.9, 2.1) &\approx f(3, 2) + f_x(3, 2) \cdot (2.9 - 3) + f_y(3, 2) \cdot (2.1 - 2) \\ &= 5 + \left(\frac{3}{5}\right) \cdot (-0.1) + \left(\frac{16}{5}\right) \cdot (0.1) = 5.26. \end{aligned}$$

The true value of $f(2.9, 2.1)$ is $5.278\ldots$, so our approximation is accurate to one decimal place. ■

inSIGHT

If we assume that the rectangular region R on which Example 5 takes place is $R = \{(x, y) : |x - 3| \le 0.1, |y - 2| \le 0.1\}$, then we may calculate that the second derivatives of f are not larger than $M = 10$ in absolute value. Thus, estimate (13.16) for the maximal size of the error tells us that it cannot be larger than

$$\frac{M}{2} \cdot (|-0.1| + |0.1|)^2 = 0.2.$$

In fact, the accuracy that we actually achieved is even better than this prediction. The error bound given by estimate (13.16) is a "worst case" value.

A Restatement of Theorem 2 Using Increments

Before we proceed with our discussion of error analysis, we formulate Theorem 2 in a more intuitive fashion. Since x is a point that is near x_0, let us write $x = x_0 + \Delta x$. Similarly, we write $y = y_0 + \Delta y$. This simple change of notation makes the change in x and the change in y more explicit. Continuing with this theme, we let $f(x, y) = f(x_0, y_0) + \Delta f$. We can now express Theorem 2 as follows.

Theorem 3 Let f be a differentiable function of two variables, and let $P_0 = (x_0, y_0)$ be a point of its domain. Then the difference in the values of f at a point (x_0, y_0) and a nearby point $(x, y) = (x_0 + \Delta x, y_0 + \Delta y)$ is given approximately by

$$\Delta f \approx f_x(P_0)\Delta x + f_y(P_0)\Delta y = \nabla f(P_0) \cdot \langle \Delta x, \Delta y \rangle.$$

Using differentials, we write the approximation $\Delta f \approx f_x(P_0)\Delta x + f_y(P_0)\Delta y$ in the form

$$df = f_x(P_0)\,dx + f_y(P_0)\,dy. \tag{13.17}$$

The quantity on the right side of equation (13.17) is called the *total differential* of f at the point P_0. Let us see how to use this quantity in error analysis.

Example 6 A machinist is making a bearing that is in the shape of a small cone. It is being manufactured from a costly titanium-vanadium alloy. The bearing is to have base of radius 3 mm and height 6 mm. The manufacturer can only allow errors of 1% in the amount of material used in each bearing. How large an error is allowable in the dimensions of the bearing?

Solution The volume of the cone is $V(r, h) = \pi r^2 h/3$. The maximum allowable error is 1% of volume, or

$$\text{Allowable error} = (0.01) \cdot \frac{1}{3}\pi r^2 h\Big|_{r=3,h=6} = 0.18\pi.$$

According to Theorem 2,

$$\begin{aligned}\Delta V \approx V_r(3,6)\Delta r + V_h(3,6)\Delta h &= \frac{2}{3}\pi r h\Big|_{r=3,h=6}\Delta r + \frac{1}{3}\pi r^2\Big|_{r=3,h=6}\Delta h\\ &= 12\pi\Delta r + 3\pi\Delta h.\end{aligned}$$

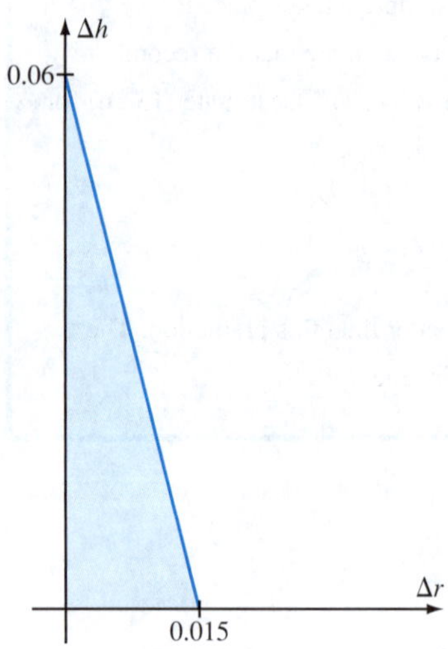

Figure 5
The region $4\Delta r + \Delta h \le 0.06$

It is therefore required that Δr and Δh satisfy $12\pi\Delta r + 3\pi\Delta h \le 0.18\pi$, or

$$4\Delta r + \Delta h \le 0.06.$$

How do we interpret this inequality? Refer to Figure 5 for a schematic of Δr and Δh. If a given bearing has *no error* in the height, then the formula reduces to $\Delta r \le 0.015$. Thus, a *maximum error* of 0.015 mm can be allowed in the radius of the bearing if there is no error in the height. Similarly, if there is no error in the radius, then the formula becomes $\Delta h \le 0.06$. We see that a maximum error of 0.06 mm in the height can be allowed if there is no error in the radius. It is more realistic to suppose that the machinist can keep the height error down to 0.01 mm and ask how much tolerance can be allowed in the radius. To apply our error estimate, we set $\Delta h = 0.01$ to obtain $4\Delta r + 0.01 \le 0.06$, or $\Delta r \le 0.0125$. Thus, the greatest variation in radius that the manufacturer can allow is 0.0125 mm. ■

Functions of Three or More Variables

The idea of linear approximation applies to functions of three or more variables. If $(x, y, z) = (x_0 + \Delta x, y_0 + \Delta y, z_0 + \Delta z)$ is near the point $Q_0 = (x_0, y_0, z_0)$, then the linear approximation of $F(x, y, z)$ is

$$F(x, y, z) \approx F(Q_0) + F_x(Q_0)(x - x_0) + F_y(Q_0)(y - y_0) + F_z(Q_0)(z - z_0).$$

Stated another way, the difference ΔF between the values of F at (x, y, z) and $Q_0 = (x_0, y_0, z_0)$ is approximately

$$\Delta F \approx F_x(Q_0)\Delta x + F_y(Q_0)\Delta y + F_z(Q_0)\Delta z.$$

Using the notation of differentials, this approximation is written as

$$dF = F_x(Q_0)\,dx + F_y(Q_0)\,dy + F_z(Q_0)\,dz, \qquad \textbf{(13.18)}$$

and the right side is called the *total differential* of F at Q_0.

Example 7 The initial mass of a bacterial sample is measured to be $m = 3.00$ mg. The mass grows according to the exponential law $m\exp(kt)$. One hour after the initial measurement, a second measurement $M = 3.30$ mg is taken. The two measurements allow the researchers to determine the growth constant: $k = \ln(3.3/3)$. It is therefore predicted that 8 hr after the initial measurement, the mass will be $3.00\exp(8\ln(3.3/3)) = 6.43$ mg. By about how much can the actual mass exceed this figure if (a) the two mass measurements were as much as 0.05 mg off and (b) the time measurement $t = 1$ hr was as much as 0.01 hr in error?

Solution If an exact initial mass measurement m was made and if an exact mass measurement M was made exactly T hours after the first, then we would have $M = m\exp(kt)$ or $k = \ln(M/m)/T$. The mass W at a fixed instant of time t hours after the initial measurement would be

$$W(m, M, T) = m\exp(kt) = m\exp\left(\frac{t}{T}\ln\left(\frac{M}{m}\right)\right) = m\exp\left(\ln\left(\left(\frac{M}{m}\right)^{t/T}\right)\right)$$
$$= m\left(\frac{M}{m}\right)^{t/T}.$$

In particular, 8 hr after the initial measurement, the mass would be

$$W(m, M, T) = m\left(\frac{M}{m}\right)^{8/T}.$$

Using the values $m_0 = 3.00$, $M_0 = 3.30$, and $T_0 = 1.00$, we have $W(3.00, 3.30, 1) = 6.43$, as announced in the statement of the problem. We calculate

$$W_m(3.0, 3.3, 1) = \frac{\partial}{\partial m}\left(m\left(\frac{M}{m}\right)^{8/T}\right)\bigg|_{m=3.0, M=3.3, T=1}$$
$$= \left(\frac{M}{m}\right)^{8/T} - 8\frac{(M/m)^{8/T}}{T}\bigg|_{m=3.0, M=3.3, T=1} = -15.005,$$

$$W_M(3.0, 3.3, 1) = \frac{\partial}{\partial M}\left(m\left(\frac{M}{m}\right)^{8/T}\right)\bigg|_{m=3.0, M=3.3, T=1}$$
$$= 8m\frac{(M/m)^{8/T}}{TM}\bigg|_{m=3.0, M=3.3, T=1} = 15.59,$$

and

$$W_T(3.0, 3.3, 1) = \frac{\partial}{\partial T}\left(m\left(\frac{M}{m}\right)^{8/T}\right)\Bigg|_{m=3.0, M=3.3, T=1}$$

$$= -8m\frac{(M/m)^{8/T}}{T^2}\ln\left(\frac{M}{m}\right)\Bigg|_{m=3.0, M=3.3, T=1} = -4.903.$$

The error ΔW is approximately

$$\Delta W \approx \frac{\partial W}{\partial m}(3.0, 3.3, 1)\Delta m + \frac{\partial W}{\partial M}(3.0, 3.3, 1)\Delta M + \frac{\partial W}{\partial T}(3.0, 3.3, 1)\Delta T$$
$$= (-15.005)\Delta m + (15.590)\Delta M + (-4.903)\Delta T.$$

Now the given bounds for the measurement errors are $-0.05 \le \Delta m \le 0.05$, $-0.05 \le \Delta M \le 0.05$, and $-0.01 \le \Delta T \le 0.01$. In the worst case, the signs of Δm, ΔM, and ΔT will cause the three summands of ΔW to have the same sign. When this happens, the individual measurement errors will accumulate and not cancel. Thus, if $\Delta m = -0.05$, $\Delta M = +0.05$, and $\Delta T = -0.05$, then

$$\Delta W \approx (-15.005)(-0.05) + (15.590)(0.05) + (-4.903)(-0.01) = 1.57878,$$

which is about one-fourth of the predicted value. In conclusion, our error analysis tells us that in the worst case, the error can be about 25% of the predicted value. ■

A LOOK BACK

In Section 3.9, we learned the linear approximation, $\Delta f \approx f'(c)\Delta x$, of a function f of one variable. Equation (3.28), $df = f'(c)\,dx$, expresses this linear approximation in the language of differentials. Equations (13.17) and (13.18) are, respectively, the two-variable and three-variable analogues of equation (3.28).

quickquiz

1. What is a normal vector to the graph of $z = x^2 + 2y$ at $(3, -2, 5)$?
2. What is the Cartesian equation of the plane that is tangent to the graph of $z = x^2 + 2y$ at the point $(3, -2, 5)$?
3. What is the Cartesian equation of the plane that is tangent to the graph of $xz + 2y + z^3 = 1$ at the point $(2, -1, 1)$?
4. If $f(x, y) = \sqrt{x^2 + 3y}$, what is the tangent plane approximation to $f(4.1, 2.8)$ if the base point $(4, 3)$ is used?

EXERCISES

Problems for Practice

In Exercises 1–12, a differentiable function $(x, y) \mapsto f(x, y)$ and a point P_0 in the domain of f are given. Calculate a normal vector to the graph of f at the point $(P_0, f(P_0))$. Then write the equation of the tangent plane to the graph of f at the point $(P_0, f(P_0))$.

1. $f(x, y) = xy - y^3 + x^2$, $P_0 = (1, 4)$
2. $f(x, y) = \sin(x - y)$, $P_0 = (\pi/2, \pi/3)$
3. $f(x, y) = \ln(2 + x^2 + y^2)$, $P_0 = (1, 2)$
4. $f(x, y) = e^{2x-3y}$, $P_0 = (\ln(2), \ln(3))$
5. $f(x, y) = x/(x + y)$, $P_0 = (2, 1)$
6. $f(x, y) = \sin(x)\cos(y)$, $P_0 = (\pi/3, \pi/4)$
7. $f(x, y) = xye^{x+y}$, $P_0 = (1, 2)$
8. $f(x, y) = \sin(x)/\cos(y)$, $P_0 = (\pi/3, \pi/3)$
9. $f(x, y) = (x^2 + y^2)/(x + y)$, $P_0 = (1, 1)$
10. $f(x, y) = (x + y + 1)^{1/3}$, $P_0 = (6, 1)$
11. $f(x, y) = (x - y)^{-1/3}$, $P_0 = (9, 1)$
12. $f(x, y) = \cot(x/y)$, $P_0 = (\pi/2, 2)$

In Exercises 13–20, find a normal to the surface at the point Q_0. Then find the equation of the tangent plane to the surface at the point Q_0.

13. $x^2 - 4y^2 + z^2 = -14$, $Q_0 = (1, 2, 1)$
14. $xyz = 8$, $Q_0 = (4, 1, 2)$
15. $y/(x + z) = 2$, $Q_0 = (1, 8, 3)$
16. $xy\sin(z) = 5$, $Q_0 = (5, 2, \pi/6)$
17. $\ln(1 + x + y + z) = 3$, $Q_0 = (e^2 + 2, e^3 - e^2, -3)$
18. $(x + y)(z - y^2) = -14$, $Q_0 = (-2, 4, 9)$
19. $(x + y + z^2)^6 = 1$, $Q_0 = (3, -3, 1)$
20. $\cos(xyz) = 0$, $Q_0 = (4, \pi, 1/8)$

In Exercises 21–26, write the equation of the normal line to the given surface at the given point.

21. Graph of $f(x, y) = x^3 - y^5$, $(2, 1, 7)$
22. Graph of $(x + y - z^2)^3 = 8$, $(2, 1, 1)$
23. Graph of $\cos(x + y + z) = 0$, $(\pi/4, 0, \pi/4)$
24. Graph of $f(x, y) = y^2/(x - y)$, $(1, 2, -4)$
25. Graph of $f(x, y) = \cos(x) - \sin(y)$, $(0, \pi, 1)$
26. Graph of $f(x, y) = x\sin(y)$, $(\pi/2, \pi/2, \pi/2)$

In Exercises 27–30, calculate the linear approximation to $f(P)$ using the tangent plane at $(P_0, f(P_0))$.

27. $f(x, y) = \sqrt{x^4 + y^2}$, $P = (2.1, 2.9)$, $P_0 = (2, 3)$
28. $f(x, y) = x/\sqrt{1 + x^2 + y^2}$, $P = (1.9, 2.1)$, $P_0 = (2, 2)$
29. $f(x, y) = x\cos(2x - y)$, $P = (0.9, 2.2)$, $P_0 = (1, 2)$
30. $f(x, y) = \ln(1 + \sqrt{x} - y)$, $P = (9.3, 3.2)$, $P_0 = (9, 3)$
31. A triangular plot of land is to have base 1000 ft and height 500 ft. Suppose that surveying errors are kept within 1%. Use the total differential to estimate the greatest error in area.
32. A rectangular beam with a cross section of width w and thickness t is known to have stiffness $\alpha \cdot w^3 \cdot t$ where α is a physical constant. The beam is cut to have dimensions 10 cm (width) and 5 cm (thickness). Assuming neither cut is off by more than 1%, use the total differential to estimate how much the stiffness of the beam can vary.
33. A rectangular box is to be built with length 6 cm, width 4 cm, and height 3 cm. Suppose the measurement error can be kept accurate to 0.5%. Use the total differential to estimate the greatest error in volume.
34. A machinist's jig is to be made in the shape of an isosceles triangle. The hypotenuse is to measure 12 cm and the legs 8 cm. However, the man building the jig is a novice, and his measurements are off by 0.1 cm. Estimate the resulting error in the area of the jig.

Further Theory and Practice

In Exercises 35–38, a function f and a point P_0 are given. Also given is an incorrect equation for the tangent plane to the graph of f at P_0. Explain what error has been made, and state the correct equation for the tangent plane.

35. $f(x, y) = 1 + 3x^2 + 2y^3$, $P_0 = (1, -1)$, $z = 6(x - 1) + 6(y + 1)$
36. $f(x, y) = 5 + x^2 + y^3$, $P_0 = (2, 1)$, $z = 10 + 2x(x - 2) + 3y^2(y - 1)$
37. $f(x, y) = 3 + y + xy$, $P_0 = (0, -3)$, $z = yx + (x + 1)(y + 3)$
38. $f(x, y) = 7x + 4/y$, $P_0 = (1, 2)$, $z = 9 + 7x - y$
39. Let θ be any number. Show that $\mathbf{r}_\theta(t) = t\cos(\theta)\mathbf{i} + t\sin(\theta)\mathbf{j} + t\mathbf{k}$ is a curve that passes through the origin and that lies on the surface $z^2 = x^2 + y^2$. Show that the zero vector is the only vector perpendicular to $\mathbf{r}'_\theta(0)$ for every θ. Use this observation

to discuss the concept of normal vector and tangent plane to the surface $z^2 = x^2 + y^2$. Sketch the surface. What, if anything, does Theorem 1 tell us when we apply it to the differentiable function $F(x, y, z) = x^2 + y^2 - z^2$ at the origin?

40. Define the angle between two surfaces at a point of intersection to be the angle between their tangent planes, which is the same thing as the angle between their normals. What is the angle between the surfaces $x^2 + y^2 + z^2 = 16$ and $x - y + 2z = 0$ at their points of intersection? (*Hint:* There are many points of intersection.)

41. Refer to Exercise 40 for terminology. Calculate the angle between the surfaces $x^2 - 3xz + 4y = 0$ and $y^2 + 3xyz - 8x = 0$ at the point $(0, 0, 0)$.

42. Calculate the angle between the surface $z = x^2 - 4y$ and the line $x = 2t$, $y = -t + 1$, $z = 4t + 12$ at their two points of intersection. (*Hint:* This is the same as the angle between the direction vector for the line and the tangent plane to the surface.)

43. Calculate the angle between the line $x = 3t + 6$, $y = 4t + 8$, $z = 5t + 5$ and the sphere $x^2 + y^2 + z^2 = 25$ at their two points of intersection.

44. Fix a twice continuously differentiable function $(x, y) \mapsto f(x, y)$, a point $P_0 = (x_0, y_0)$ in the domain of f, and a direction $\mathbf{u} = u_1\mathbf{i} + u_2\mathbf{j}$. Consider the function $\phi(t) = f(P_0 + t\mathbf{u})$. Write out the first order Taylor expansion of ϕ about $t = 0$. What do the zero and first order terms that you see in this expansion have to do with the equation for the tangent plane to the graph of f at P_0?

45. A solid bead (with no hole drilled through it) of radius 4 mm is to be made of an alloy that is 60% gold. If the volume of the gold used is not to be greater than 1% more than intended, then approximately what errors in size and alloy percentage can be permitted?

46. Let S be any sphere and ℓ any line that intersects that sphere in two points. Show that the angle between the line and the sphere is the same at the two points of intersection.

Calculator/Computer Exercises

In Exercises 47–50, plot the function f in a suitable viewing cube containing the given point P_0. On the same set of axes, plot the tangent plane at the point P_0. Zoom in on the given point until it becomes apparent that the tangent plane closely approximates the graph of f.

47. $f(x, y) = 3xy^2/(1 + x^2 + y^4)$, $P_0 = (2, 1, 1)$

48. $f(x, y) = (3x - y)^2/\sqrt{6 + x^2 + y^4}$, $P_0 = (3, 1, 16)$

49. $f(x, y) = (x^2 - y)e^{2x-y}$, $P_0 = (2, 3, e)$

50. $f(x, y) = 17\ln(1 + x^2)/(1 + x^2y^4)$, $P_0 = (1, 2, \ln(2))$

13.8 Maximum-Minimum Problems

Throughout this section, we assume that f is defined on a planar set $\mathcal{G}_f$. We further assume that each point in $\mathcal{G}_f$ is the center of a disk of positive radius that is entirely contained in $\mathcal{G}_f$. A point P_0 in $\mathcal{G}_f$ is called a *local minimum* for f if there is a disk $D(P_0, r)$ with center P_0 and radius $r > 0$ such that

$$f(P_0) \leq f(x, y) \quad \text{for all } (x, y) \in D(P_0, r).$$

If $f(P_0) \leq f(x, y)$ for all (x, y) in $\mathcal{G}_f$, then P_0 is called an *absolute minimum* for f. Similarly, a point P_0 in $\mathcal{G}_f$ is called a *local maximum* for f if there is an $r' > 0$ such that

$$f(P_0) \geq f(x, y) \quad \text{for all } (x, y) \in D(P_0, r').$$

If $f(P_0) \geq f(x, y)$ for all (x, y) in $\mathcal{G}_f$, then P_0 is called an *absolute maximum* for f. A point that is either a local minimum or a local maximum is said to be a *local extremum*.

The Analogue of Fermat's Theorem

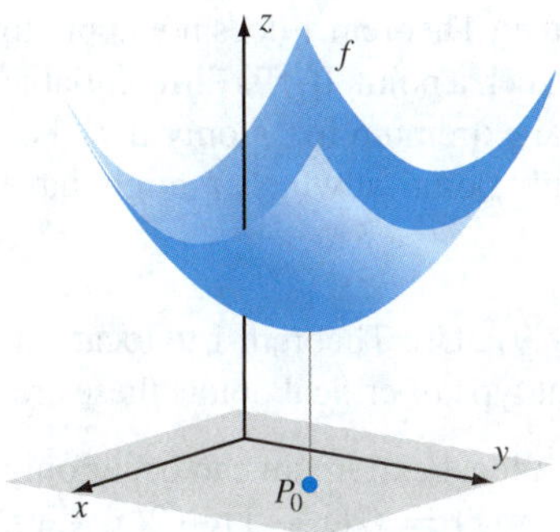

Figure 1a
Local minimum at $P_0 = (x_0, y_0)$

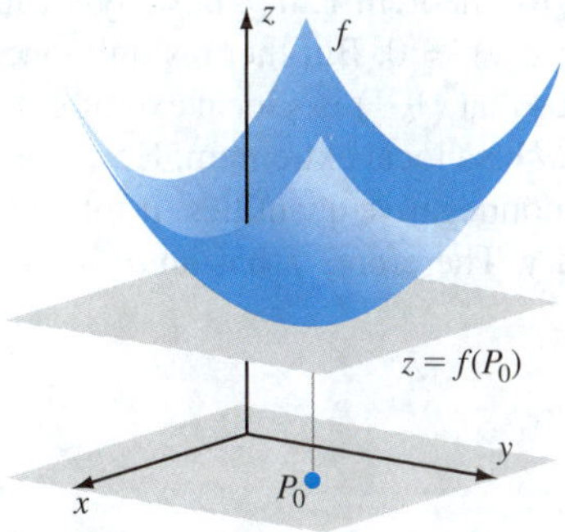

Figure 1b
Horizontal tangent plane at $(x_0, y_0, f(P_0))$

Pictorially, we expect a local minimum $P_0 = (x_0, y_0)$ to correspond to a point $(x_0, y_0, f(P_0))$ where the graph of f looks like Figure 1a. We expect a local maximum to correspond to a point $(x_0, y_0, f(P_0))$ where the graph of f looks like Figure 2a. Figures 1b and 2b suggest that if P_0 is a local extremum of f, then the *tangent plane* to the graph of f at $(x_0, y_0, f(P_0))$ will be parallel to the xy-plane. But this would happen only if the normal vector to the tangent plane, $\langle f_x(P_0), f_y(P_0), -1\rangle$, were a multiple of $\langle 0, 0, 1\rangle$. In other words, the geometry indicates that $f_x(P_0) = 0$ and $f_y(P_0) = 0$ must both hold if P_0 is to be a local extremum.

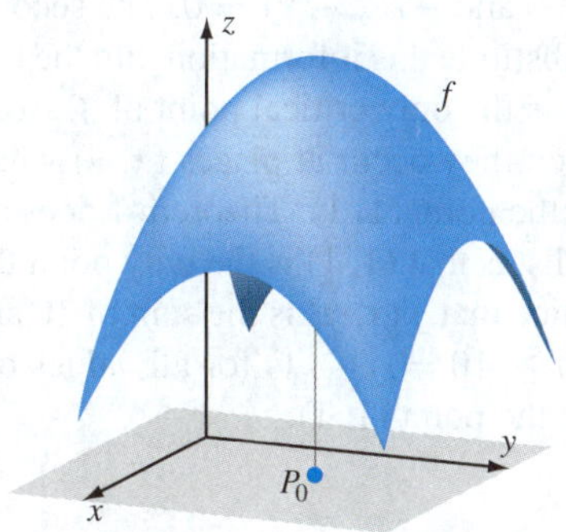

Figure 2a
Local maximum at $P_0 = (x_0, y_0)$

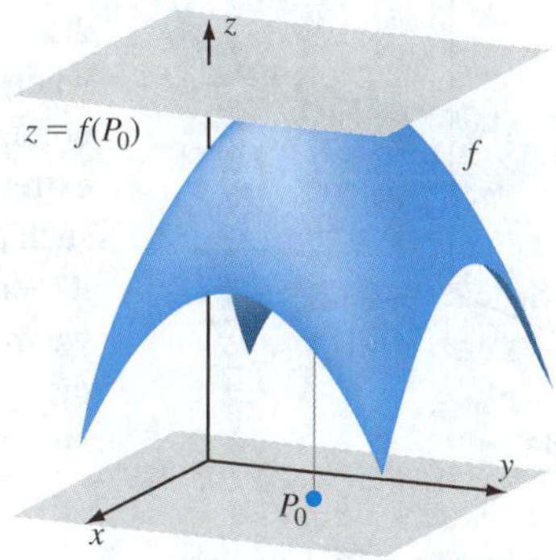

Figure 2b
Horizontal tangent plane at $(x_0, y_0, f(P_0))$

Let us confirm these conjectures, supposing that both $f_x(P_0)$ and $f_y(P_0)$ exist. Notice that if $P_0 = (x_0, y_0)$ is a local minimum for f, then the one-variable function $\varphi(x) = f(x, y_0)$ has a local minimum at $x = x_0$. It follows from the one-variable theory that $\varphi'(x_0) = 0$. But $\varphi'(x_0) = f_x(x_0, y_0)$, as we saw in Section 13.4. It follows that $f_x(x_0, y_0) = 0$ if (x_0, y_0) is a local minimum for f. By similar reasoning with $\psi(y) = f(x_0, y)$, we have $f_y(x_0, y_0) = \psi'(y_0) = 0$. An analogous argument shows that at a local maximum (x_0, y_0), it is also the case that $f_x(x_0, y_0) = 0$ and $f_y(x_0, y_0) = 0$. We summarize our findings as follows.

Theorem 1 Suppose that $(x, y) \mapsto f(x, y)$ is a function of two variables for which both partial derivatives $f_x(P_0)$ and $f_y(P_0)$ exist.

a. If P_0 is a local minimum for f, then $f_x(P_0) = 0$ and $f_y(P_0) = 0$.
b. If P_0 is a local maximum for f, then $f_x(P_0) = 0$ and $f_y(P_0) = 0$.

In short, at a local extremum P_0 we have $\nabla f(P_0) = \vec{\mathbf{0}}$.

Definition We say that P_0 is a *critical point* for f if

1. both partial derivatives exist at P_0 and $\nabla f(P_0) = \vec{\mathbf{0}}$, or
2. f fails to have either a partial derivative with respect to x at P_0 or a partial derivative with respect to y at P_0.

If f does not have both partial derivatives at P_0, then Theorem 1 does not apply to f at P_0. It is possible for a local extremum to occur at such a point. If f is differentiable at P_0, then according to Theorem 1, P_0 can be a local extremum for f only if P_0 is a critical point. Thus, the critical points of f are the only points at which f might have a local extremum.

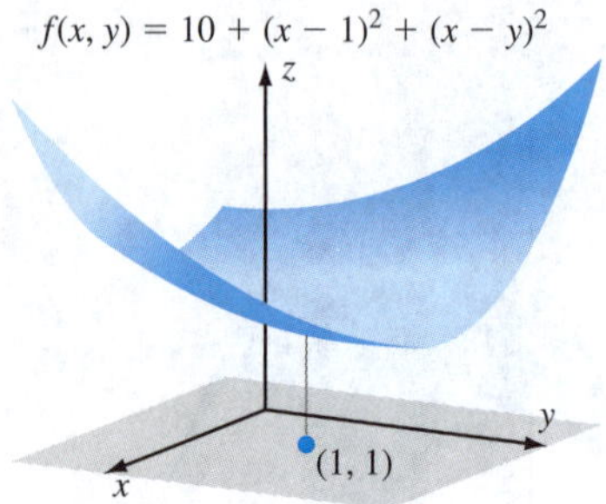

Figure 3a

Example 1 Let $f(x, y) = 10 + (x-1)^2 + (x-y)^2$. Use Theorem 1 to locate all points that might be local extrema for f. Identify what type of critical points these are.

Solution Because f is everywhere differentiable, a local extremum can occur only at a point at which $\nabla f(x, y) = \vec{\mathbf{0}}$. We calculate $\nabla f(x, y) = \langle 2(x-1) + 2(x-y), -2(x-y)\rangle$. We must therefore determine all points (x, y) where *both* $2(x-1) + 2(x-y) = 0$ and $-2(x-y) = 0$. The second of these equations tells us that $y = x$; when we substitute this information into the first equation, we obtain $x = 1$. Therefore, point $(1, 1)$ is the only critical point of f. According to Theorem 1, if f has any local extrema, then they occur at places (x, y) where $\nabla f(x, y) = 0$. But there is only one such place: the point $(1, 1)$. Theorem 1 does not tell us that $(1, 1)$ is a local extremum; it merely tells us that $(1, 1)$ is the only point that *could* be a local extremum. However, we may notice that $f(x, y)$ is the sum of 10 and two nonnegative quantities. It follows that $f(x, y) \geq 10 = f(1, 1)$ for all values of x and y. Therefore, f has an *absolute minimum* at the point $(1, 1)$. ∎

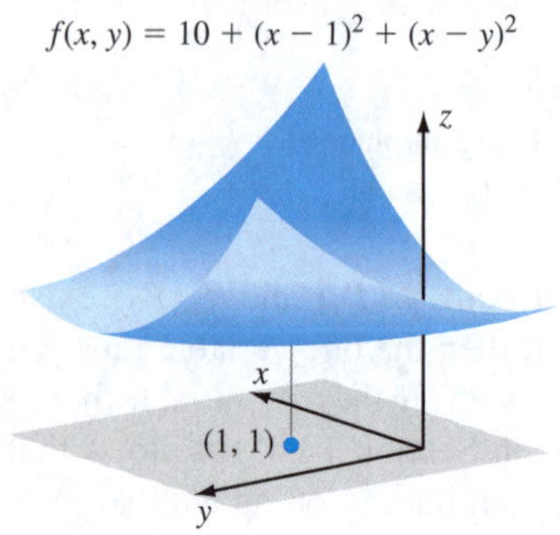

Figure 3b

inSIGHT

Theorem 1 is analogous to Fermat's Theorem from Chapter 4. It gives us a way to find points that are *candidates* for local maxima and minima. Once we isolate the candidates, our next job is to identify them as local maxima, local minima, *or neither*. Figure 3a illustrates the absolute minimum of Example 1, viewed from a standard perspective. As Figure 3b demonstrates, it may be possible to rotate the plot to obtain a more convincing view. Nevertheless, it is desirable to have an analytic procedure for establishing a local extremum. Later in this section, we will learn the analogue of the Second Derivative Test for Local Extrema.

Saddle Points

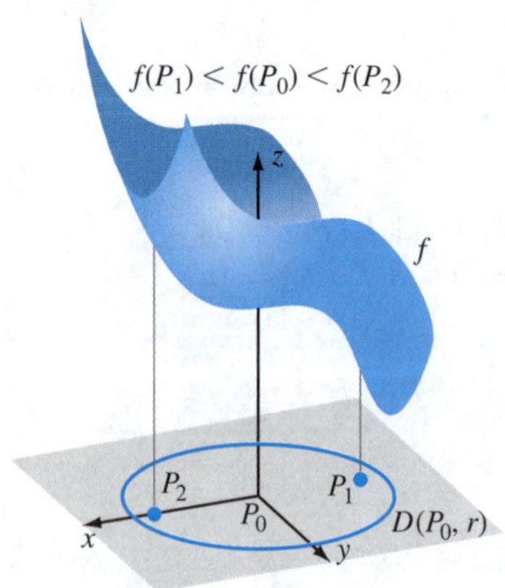

Figure 4
Critical point P_0 that is not a local extremum

As in the one-variable situation, a critical point need not be a local extremum. For example, the point $P_0 = (0, 0)$ is a critical point for $f(x, y) = 1 + x^3 + y^2$. However, if r is any positive number, then the disk $D(P_0, r)$ contains the points $P_1 = (-r/2, 0)$ and $P_2 = (r/2, 0)$. The inequalities

$$f(P_1) = 1 - \frac{r^3}{8} < 1 = f(P_0) < 1 + \frac{r^3}{8} = f(P_2)$$

rule out P_0 as either a local minimum or a local maximum (see Figure 4). The most important instance of what a critical point can be, other than a local maximum or a local minimum, is illustrated in the next example.

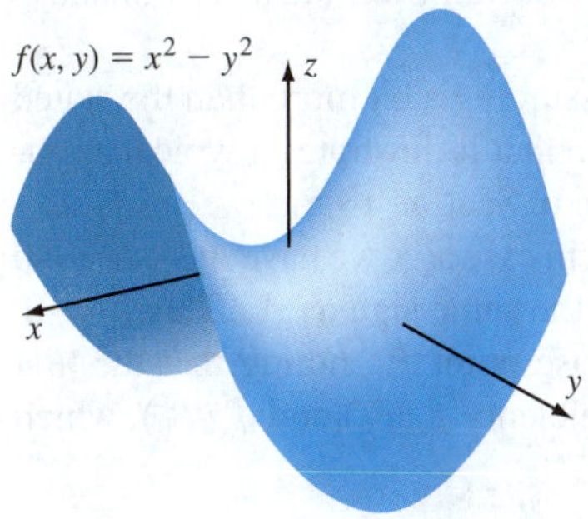

Figure 5a
Saddle surface

Example 2 Locate and analyze the critical points of $f(x, y) = x^2 - y^2$.

Solution The graph of f appears in Figure 5a. (Refer to Section 13.2 for details on plotting this surface.) Since $\nabla f(x, y) = \langle 2x, -2y \rangle$ for every point (x, y), the only critical point is $(0, 0)$. But this point is neither a local maximum nor a local minimum. If we restrict attention to the x-direction by setting $y = 0$, then we have $f(x, 0) = x^2$. Thus, the curve formed by intersecting the graph of f and the plane $y = 0$ is the parabola that consists of the points $(x, 0, x^2)$. A minimum occurs when $x = 0$ (see Figure 5b). If we restrict attention to the y-direction by setting $x = 0$, then we have $f(0, y) = -y^2$. The curve that is formed by intersecting the graph of f and the plane $x = 0$ is a parabola that has maximum value when $y = 0$. It follows that $(0, 0)$ is neither a local maximum nor a local minimum for the function $f(x, y)$ of two variables, in the sense introduced at the beginning of this section. Our surface, concave up in one direction and concave down in the perpendicular direction, looks like a saddle. ■

Definition

We call P_0 a *saddle point* for the function f if f has a maximum value at P_0 in one direction and a minimum value at P_0 in another direction. More precisely, P_0 is a saddle point for f if there are unit vectors $\mathbf{u}$ and $\mathbf{v}$ in the plane and a number $\delta > 0$ such that

$$\underbrace{f(P_0) < f(\overrightarrow{OP_0} + t\mathbf{u}) \text{ for all } 0 < |t| < \delta}_{f \text{ has a local minimum at } P_0 \text{ in the direction } \mathbf{u}.}$$

and

$$\underbrace{f(P_0) > f(\overrightarrow{OP_0} + t\mathbf{v}) \text{ for all } 0 < |t| < \delta.}_{f \text{ has a local maximum at } P_0 \text{ in the direction } \mathbf{v}.}$$

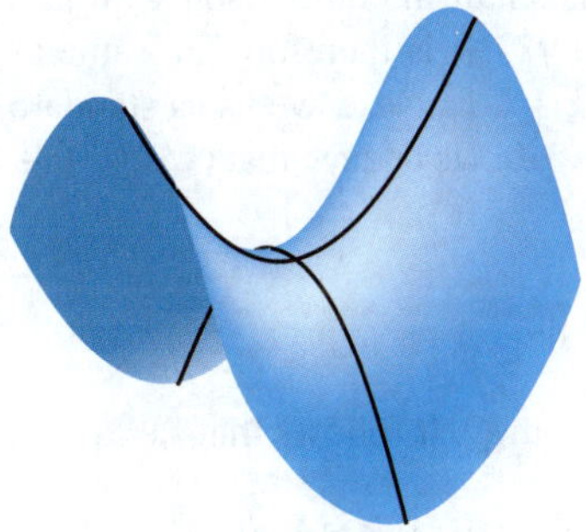

Figure 5b

The Second Derivative Test for Local Extrema

The phenomenon of saddle points makes the analysis of critical points in the plane somewhat more complicated than in the one-variable case. The following construction is frequently useful for streamlining the analysis.

Definition

Let $(x, y) \mapsto f(x, y)$ be a twice continuously differentiable function. The scalar-valued function defined by

$$\text{Discr}(f, P_0) = \det\left(\begin{bmatrix} f_{xx}(P_0) & f_{xy}(P_0) \\ f_{yx}(P_0) & f_{yy}(P_0) \end{bmatrix}\right) = \frac{\partial^2 f}{\partial x^2}(P_0)\frac{\partial^2 f}{\partial y^2}(P_0) - \left(\frac{\partial^2 f}{\partial x \partial y}(P_0)\right)^2$$

is called the *discriminant* of f.

Theorem 2

Second Derivative Test Let $(x, y) \mapsto f(x, y)$ be a twice continuously differentiable function, and suppose that P_0 is a critical point for f. Let $\text{Discr}(f, P_0)$ be the discriminant of f at P_0.

a. If $\text{Discr}(f, P_0) > 0$, then $f_{xx}(P_0)$ and $f_{yy}(P_0)$ have the same sign.
   i. If $f_{xx}(P_0)$ and $f_{yy}(P_0)$ are both positive, then P_0 is a local minimum for f.
   ii. If $f_{xx}(P_0)$ and $f_{yy}(P_0)$ are both negative, then P_0 is a local maximum for f.

b. If $\text{Discr}(f, P_0) < 0$, then P_0 is a saddle point for f.
c. If $\text{Discr}(f, P_0) = 0$, then we can draw no conclusion from the discriminant.

Proof We will prove just one part of this theorem, assuming a bit more than the stated hypotheses in doing so. The analysis is noteworthy in that it illustrates a typical application of the two-variable Taylor Theorem (Theorem 6, Section 13.5).

Let $A = f_{xx}(P_0)$, $B = f_{xy}(P_0)$, and $C = f_{yy}(P_0)$. In case a(i), we have $AC - B^2 > 0$ and $A > 0$. Since $AC > B^2 \geq 0$, we see that C has the same sign as A. Thus $C > 0$. We use Taylor's Theorem to expand f about the base point P_0, noting that the first powers of $h = (x - x_0)$ and $k = (y - y_0)$ have coefficients $f_x(P_0)$ and $f_y(P_0)$, which are zero. We obtain

$$f(x, y) = f(P_0) + \frac{1}{2}(Ah^2 + 2Bhk + Ck^2) \\ + (\text{terms involving } h^\alpha k^\beta)$$

where $\alpha + \beta > 2$. We can suppose that (x, y) is so close to P_0 that the higher powers of h and k are negligible compared with the quadratic terms. (For example, if $|h|$ is one-hundredth, then $|h|^3$ is one-millionth, which is small in comparison with $|h|^2$, which is one-ten-thousandth.) The sign of $f(x, y) - f(P_0)$ is therefore the same as the sign of the quadratic expression $Q = Ah^2 + 2Bhk + Ck^2$. Next, we use a standard inequality for quadratics. Because $(\sqrt{A}|h| - \sqrt{C}|k|)^2 \geq 0$, we deduce that $(\sqrt{A}|h|)^2 - 2\sqrt{A}\sqrt{C}|h||k| + (\sqrt{C}|k|)^2 \geq 0$, or

$$Ah^2 + Ck^2 \geq 2\sqrt{A}\sqrt{C}|h||k|.$$

The sum on the left side of this inequality also appears in Q. It follows that

$$Q \geq 2\sqrt{A}\sqrt{C}|h||k| + 2Bhk \geq 2\left(\sqrt{A}\sqrt{C} - |B|\right)|h|\,|k| \geq 0.$$

Since $f(x, y) - f(P_0)$ and Q have the same sign, we conclude that $f(x, y) \geq f(P_0)$, proving that P_0 is a local minimum. ■

Example 3 Locate all local maxima, local minima, and saddle points for the function

$$f(x, y) = 2x^2 + 3xy + 4y^2 - 5x + 2y + 3.$$

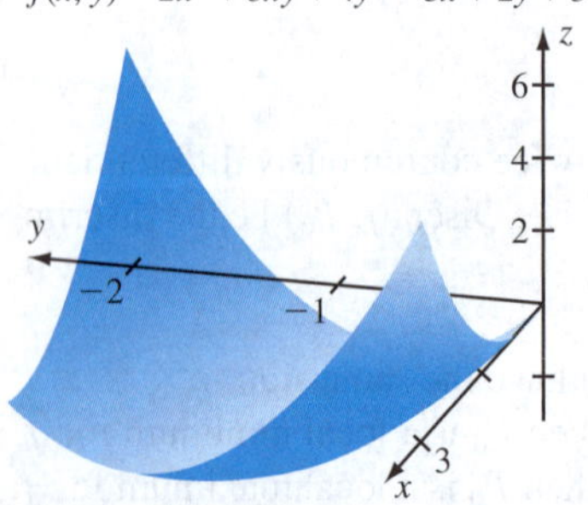

Figure 6a

Solution We calculate that $f_x(x, y) = 4x + 3y - 5$, $f_y(x, y) = 3x + 8y + 2$, $f_{xx}(x, y) = 4$, $f_{yy}(x, y) = 8$, and $f_{xy}(x, y) = f_{yx}(x, y) = 3$. The critical points will be determined by the simultaneous solution of the equations

$$f_x(x, y) = 4x + 3y - 5 = 0 \quad \text{and} \quad f_y(x, y) = 3x + 8y + 2 = 0.$$

We find that $x = 2$ and $y = -1$; hence, $P_0 = (2, -1)$ is the only critical point. Since

$$\text{Discr}(f, P_0) = \det\left(\begin{bmatrix} f_{xx}(P_0) & f_{xy}(P_0) \\ f_{yx}(P_0) & f_{yy}(P_0) \end{bmatrix}\right) = \det\left(\begin{bmatrix} 4 & 3 \\ 3 & 8 \end{bmatrix}\right) = 23 > 0,$$

we see that we are in case a of Theorem 2. Since $f_{xx}(2, -1) > 0$ and $f_{yy}(2, -1) > 0$, we know that (i) applies, and $(2, -1)$ is therefore a local minimum. Figure 6a shows a

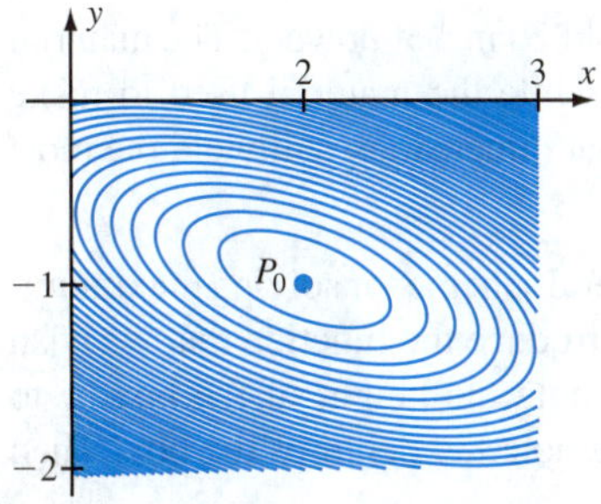

Figure 6b
Level curves of Figure 6a. Local minimum at $P_0 = (2, -1)$

plot of f, and Figure 6b shows several level curves near the minimum. The pattern of curves enclosing the local extremum is typical. ■

inSIGHT

The key to success with maximum-minimum problems is to be systematic. The Second Derivative Test requires several pieces of information, and the only way to keep track of them all is to follow the recipe given in Example 3.

In case a of Theorem 2, the signs of $f_{xx}(P_0)$ and $f_{yy}(P_0)$ are the same. Therefore, we need examine only one of them to determine the nature of the local extremum. However, because both derivatives must be calculated in the course of obtaining the discriminant, this observation does not constitute a real shortcut. In Example 3 and the examples that follow, we state both $f_{xx}(P_0)$ and $f_{yy}(P_0)$ for emphasis.

Example 4 Locate and identify the critical points of the function $f(x, y) = 2x^3 - 2y^3 - 4xy + 5$.

Solution We calculate that $f_x = 6x^2 - 4y$, $f_y = -6y^2 - 4x$, $f_{xx} = 12x$, $f_{yy} = -12y$, and $f_{xy} = f_{yx} = -4$. The critical points will be the simultaneous solutions of the equations

$$6x^2 - 4y = 0 \qquad \text{and} \qquad -6y^2 - 4x = 0.$$

The first equation yields $y = (3/2)x^2$, which we may substitute into the second equation to obtain $-6((3/2)x^2)^2 - 4x = 0$, or equivalently, $27x^4 + 8x = 0$. We factor this as $x(27x^3 + 8) = 0$. It follows that $x = 0$ and $x = -2/3$ are the only possibilities. We return to the rewritten first equation, $y = (3/2)x^2$, to find that the value of y corresponding to $x = 0$ is 0, and the value of y corresponding to $x = -2/3$ is $2/3$. Thus, our critical points are $P_0 = (0, 0)$ and $P_1 = (-2/3, 2/3)$. To identify them, we need to examine the discriminant. For $P = (x, y)$, we have

$$\text{Discr}(f, P) = \det\left(\begin{bmatrix} f_{xx}(P) & f_{xy}(P) \\ f_{yx}(P) & f_{yy}(P) \end{bmatrix}\right) = \det\left(\begin{bmatrix} 12x & -4 \\ -4 & -12y \end{bmatrix}\right) = -144xy - 16.$$

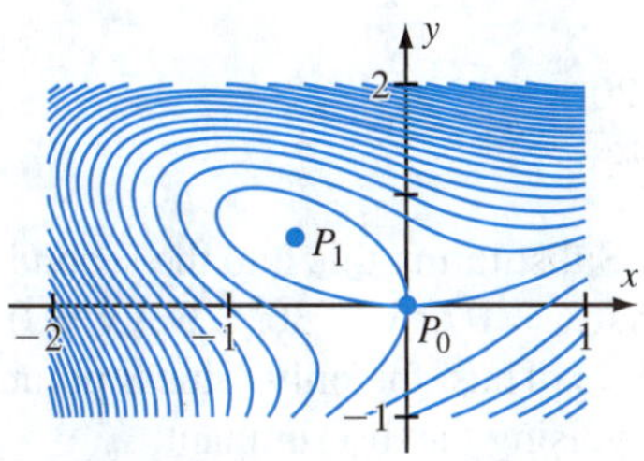

Figure 7
Saddle at $P_0 = (0, 0)$, Local maximum at $P_1 = (-2/3, 2/3)$

We see that $\text{Discr}(f, P_0) = -16 < 0$ and conclude that $P_0 = (0, 0)$ is a saddle point. Next, we consider $P_1 = (-2/3, 2/3)$. Since $\text{Discr}(f, P_1) = 48 > 0$, Theorem 2a applies. We calculate $f_{xx}(-2/3, 2/3) = -8$ and $f_{yy}(-2/3, 2/3) = -8$. These are both negative, so we find ourselves in case (ii); that is, $(-2/3, 2/3)$ is a local maximum. Level curves near the critical points are shown in Figure 7. The pattern of curves that enclose the local maximum is similar to what we saw near the local minimum of Example 3. The "crossing" that appears at the saddle point is typical for that type of critical point. ■

Applied Maximum-Minimum Problems

Certainly part of the interest of learning to find local maxima and minima is using them to solve practical problems.

Example 5 A rectangular box with a top is to hold 20 in.^{3} of powder. The material used to make the top and bottom costs \$0.02/in.^{2}, while the material used to make the front and back and the sides costs \$0.03/in.^{2} What dimensions will yield the most economical box?

Solution To solve this problem, we follow the general scheme for solving maximum-minimum problems introduced in Section 4.3. We introduce the function that we wish to minimize, namely, *cost*. We use variable z to denote the height of the box, x to represent the width of its front and back, and y to denote the depth of the box. Then the following information becomes clear:

1. Front and back each have area xz in.^{2} and cost $3xz$ cents.
2. Sides each have area yz in.^{2} and cost $3yz$ cents.
3. Top and bottom each have area xy in.^{2} and cost $2xy$ cents.

It follows that the total cost is given by

$$\text{Cost} = 2(3xz) + 2(3yz) + 2(2xy) = 6xz + 6yz + 4xy.$$

Notice that the factors of 2 appear because there are two of each item.

If either $x = 0$ or $y = 0$, then the box degenerates and can hold no powder. So we will assume in what follows that $x > 0$ and $y > 0$. Just as in the examples in Section 4.3, we now use the information provided to eliminate one of the variables—namely, volume $= xyz = 20$, or $z = 20/xy$. Substituting this into the formula for C gives

$$C(x, y) = \frac{120}{y} + \frac{120}{x} + 4xy \quad (0 < x, y < \infty)$$

as the function we are required to minimize. We can now apply the methods developed in this section to analyze $C(x, y)$. We have

$$\frac{\partial C}{\partial x}(x, y) = -\frac{120}{x^2} + 4y, \qquad \frac{\partial C}{\partial y}(x, y) = -\frac{120}{y^2} + 4x,$$

and

$$\frac{\partial^2 C}{\partial x^2}(x, y) = \frac{240}{x^3}, \qquad \frac{\partial^2 C}{\partial y^2} = \frac{240}{y^3}, \qquad \frac{\partial^2 C}{\partial x \partial y}(x, y) = \frac{\partial^2 C}{\partial y \partial x}(x, y) = 4.$$

The critical points are the simultaneous solutions of

$$-\frac{120}{x^2} + 4y = 0 \quad \text{and} \quad -\frac{120}{y^2} + 4x = 0.$$

We solve the first equation for y to obtain $y = 30/x^2$. Substituting this into the second equation gives $-x^4 + 30x = 0$. As a result, we find that $x = 0$ or $x = 30^{1/3}$. But $x = 0$ does not correspond to a point in the domain of $C(x, y)$. Thus, the only critical point is $P_0 = (30^{1/3}, 30^{1/3})$. We analyze this critical point by using the discriminant,

$$\text{Discr}(f, P_0) = \det\left(\begin{bmatrix} 240/x^3 & 4 \\ 4 & 240/y^3 \end{bmatrix}\right) = \frac{57600}{x^3 y^3} - 16.$$

At our critical point, $\text{Discr}(f, P_0) = 48$. Therefore, P_0 is either a local maximum or a local minimum. But since $C_{xx}(30^{1/3}, 30^{1/3})$ and $C_{yy}(30^{1/3}, 30^{1/3})$ are both positive, the point P_0 is a local minimum. Next, observe that if either x or y tends to 0, then $C(x, y) = 120/y + 120/x + 4xy$ tends to infinity. This is not surprising. If, for example, y tends to 0, then $xz = 20/y$ tends to infinity. So, the areas of the front and back of the box become infinite, as does the cost of material. Since $C(x, y)$ tends to infinity as (x, y) tends to a point on the boundary of its domain, we deduce that the single local extremum in the domain of C must be an absolute minimum. Thus, the optimal dimensions are

$$x = 30^{1/3}, \qquad y = 30^{1/3}, \qquad z = 20 \cdot 30^{-2/3}.$$

We now consider a strategy that often eases the computational burden of locating extrema. Suppose that ϕ is an increasing function of one variable and that f is a differentiable function of two variables. Set $g = \phi \circ f$. Because increasing functions preserve inequalities, we have $g(P_0) = \phi(f(P_0)) \leq \phi(f(P)) = g(P)$ if and only if $f(P_0) \leq f(P)$. Thus, P_0 is a local minimum for g if and only if it is a local minimum for f. The same reasoning is true for local maxima. We can benefit from this observation as follows: By composing f with a suitable increasing function ϕ, we may create a function that is easier to analyze. The next example illustrates this idea.

Example 6 Find the point on the plane $3x + 2y + z = 6$ that is nearest to the origin.

Solution The function to be minimized is the distance to the origin, $f(x, y, z) = \sqrt{x^2 + y^2 + z^2}$, where $3x + 2y + z = 6$. Because the partial derivatives of f are ratios with the term $\sqrt{x^2 + y^2 + z^2}$ in the denominator, it appears that it will be algebraically simpler to compose f with the increasing function $\phi(u) = u^2$ and minimize the *square* of the distance to the origin: $f(x, y, z)^2 = x^2 + y^2 + z^2$. Let us do that. We make a substitution to eliminate one of the variables. Since we are only considering points in the plane $3x + 2y + z = 6$, we see that $z = 6 - 2y - 3x$. Substituting this expression for z into $x^2 + y^2 + z^2$ yields the function we want to minimize:

$$g(x, y) = x^2 + y^2 + (6 - 2y - 3x)^2 = 10x^2 + 12xy + 5y^2 - 36x - 24y + 36.$$

Now

$$g_x(x, y) = 20x + 12y - 36, \qquad g_y(x, y) = 12x + 10y - 24$$

and

$$g_{xx}(x, y) = 20, \qquad g_{yy}(x, y) = 10, \qquad g_{xy}(x, y) = g_{yx}(x, y) = 12.$$

The only critical point is the simultaneous solution of the equations

$$20x + 12y - 36 = 0 \qquad \text{and} \qquad 12x + 10y - 24 = 0,$$

which is $P_0 = (9/7, 6/7)$. We calculate that

$$\text{Discr}(g, P_0) = \det\left(\begin{bmatrix} 20 & 12 \\ 12 & 10 \end{bmatrix}\right) = 20 \cdot 10 - 12 \cdot 12 = 56 > 0.$$

Thus, our critical point is either a relative maximum or minimum. Since $g_{xx}(9/7, 6/7)$ and $g_{yy}(9/7, 6/7)$ are both positive, the point P_0 is a relative minimum. The corresponding point on the plane is

$$x = \frac{9}{7}, \qquad y = \frac{6}{7}, \qquad z = 6 - 2y - 3x = \frac{3}{7}.$$

Thus, $(9/7, 6/7, 3/7)$ is the point in the plane that is nearest to the origin. ■

Least Squares Lines

Table 1 lists the weights and pulse rates of ten land mammals. The x column represents the natural logarithms of the weights, and the y column represents the natural logarithms of the pulse rates.

Mammal	Weight (g)	Pulse Rate (per minute)	x	y
Mouse	25	670	3.2189	6.5073
Rat	200	420	5.2983	6.0403
Guinea pig	300	300	5.7038	5.7038
Rabbit	2000	205	7.6009	5.3230
Dog (small)	5000	120	8.5172	4.7875
Dog (large)	30000	85	10.309	4.4427
Sheep	50000	70	10.820	4.2485
Human	70000	72	11.156	4.2767
Horse	450000	38	13.017	3.6376
Elephant	3000000	48	14.914	3.8712

Table 1
Source: A. J. Clark, *Comparative Physiology of the Heart* (Macmillan, 1972).

Figure 8 shows a scatter plot of the ten data points (x, y). The plotted points certainly suggest a straight line relationship, but what line should we use? The ten data points give rise to 45 pairs of points, each of which determines a straight line. Whichever line we select, it will not pass through the other eight points. To choose between these possible lines, and other candidate lines that may not pass through any of the points, we need a measure of how well a line ℓ "fits" the data. To each data point not on ℓ, there is an error. If we quantify this error and sum over all data points (having made sure that all errors are taken to be positive so that there is no cancellation when we sum), then we obtain a measure of how well our line ℓ fits the data. The line for which the total error is minimized is called a *best-fitting line*.

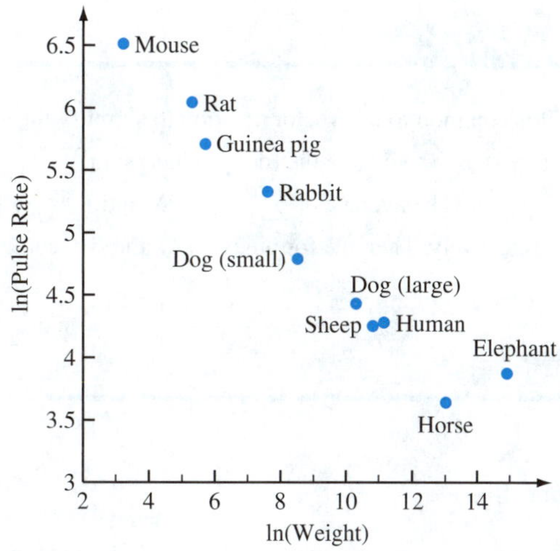

Figure 8

The first question, then, is how do we measure the amount by which a line $y = mx + b$ misses a data point (x_j, y_j)? One possibility is to use the absolute value $|y_j - (mx_j + b)|$ of the vertical distance that separates the point from the line. The absolute value is necessary so that errors on different sides of the line do not cancel. However, summing the absolute errors results in a function that is not differentiable, preventing us from using calculus to identify the minimum. Instead, we use the *sum of the squares of the errors (SSE)*:

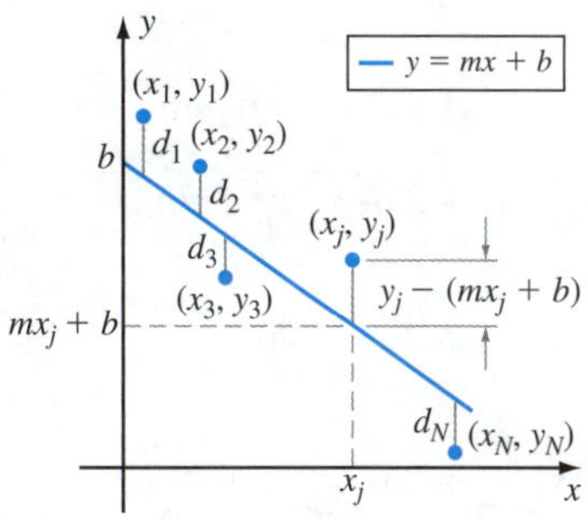

Figure 9
The least squares line minimizes $d_1^2 + d_2^2 + \cdots + d_N^2$.

$$SSE(m, b) = \sum_{j=1}^{N} (y_j - (mx_j + b))^2. \tag{13.19}$$

Although this function seems complicated, it is just a quadratic in the variables m and b. The best-fitting line we obtain when we minimize SSE is called a *least squares line*. The term *regression line* is also used. See Figure 9.

Theorem 3

Suppose that N is an integer greater than or equal to 2. Given N observations (x_1, y_1), $(x_2, y_2), \ldots, (x_N, y_N)$, the least squares line is $y = mx + b$ where

$$m = \frac{N \sum_{j=1}^{N} x_j y_j - \left(\sum_{j=1}^{N} x_j\right)\left(\sum_{j=1}^{N} y_j\right)}{N \sum_{j=1}^{N} x_j^2 - \left(\sum_{j=1}^{N} x_j\right)^2} \tag{13.20}$$

and

$$b = \frac{1}{N}\left(\sum_{j=1}^{N} y_j - m \sum_{j=1}^{N} x_j\right). \tag{13.21}$$

inSIGHT

It is common to use vector notation to simplify the appearance of these formulas. Thus, $\mathbf{x}\cdot\mathbf{y}$ is used to denote the (dot product) sum $\sum_{j=1}^{N} x_j y_j$ and $\mathbf{x}\cdot\mathbf{x}$ is used for the sum $\sum_{j=1}^{N} x_j^2$. The averages $\left(\sum_{j=1}^{N} x_j\right)/N$ and $\left(\sum_{j=1}^{N} y_j\right)/N$ are denoted by $\overline{x}$ and $\overline{y}$ respectively. Then the formulas for m and b become

$$m = \frac{\mathbf{x}\cdot\mathbf{y} - N\overline{x}\,\overline{y}}{\mathbf{x}\cdot\mathbf{x} - N\overline{x}\,\overline{x}} \qquad \text{and} \qquad b = \overline{y} - m\overline{x}. \tag{13.22}$$

Proof Since

$$\frac{\partial}{\partial m}(y_j - (mx_j + b))^2 = -2x_j(y_j - mx_j - b) \qquad \text{and}$$
$$\frac{\partial}{\partial b}(y_j - (mx_j + b))^2 = -2(y_j - mx_j - b),$$

we have

$$\frac{\partial}{\partial m}SSE(m, b) = -2\sum_{j=1}^{N} x_j(y_j - mx_j - b) = -2\sum_{j=1}^{N} x_j y_j + 2m\sum_{j=1}^{N} x_j^2 + 2b\sum_{j=1}^{N} x_j$$

and

$$\frac{\partial}{\partial b}SSE(m, b) = -2\sum_{j=1}^{N}(y_j - mx_j - b) = -2\sum_{j=1}^{N} y_j + 2m\sum_{j=1}^{N} x_j + 2Nb.$$

To find the critical point, we set $SSE_m(m, b) = 0$ and $SSE_b(m, b) = 0$. Using the simplified vector notation, we obtain

$$-2\mathbf{x}\cdot\mathbf{y} + 2m\mathbf{x}\cdot\mathbf{x} + 2Nb\overline{x} = 0 \tag{13.23}$$

and

$$-2N\overline{y} + 2mN\overline{x} + 2Nb = 0. \tag{13.24}$$

Equation (13.24) simplifies to $b = \overline{y} - m\overline{x}$, which is one of the required equations. If we replace b with $\overline{y} - m\overline{x}$ in equation (13.23) and solve for m, then we obtain the other required equation after a little algebraic simplification. An application of the Second Derivative Test, outlined in Exercise 37, shows that the critical point we have found is a local minimum. From the nature of the function SSE, we can reason that the critical point corresponds to an absolute minimum. ■

Example 7 Use the ten observations in Table 1 to formulate an explicit relationship between the body weight and the pulse rate of land mammals.

Solution We calculate

$$\begin{aligned}\mathbf{x}\cdot\mathbf{y} &= \ln(25)\ln(670) + \ln(200)\ln(420) + \ln(300)\ln(300)\\ &\quad + \cdots + \ln(3000000)\ln(48) = 411.283,\\ \mathbf{x}\cdot\mathbf{x} &= \ln(25)^2 + \ln(200)^2 + \ln(300)^2 + \cdots + \ln(3000000)^2 = 940.96,\\ \overline{x} &= \frac{1}{10}(\ln(25) + \ln(200) + \ln(300) + \cdots + \ln(3000000)) = 9.056,\end{aligned}$$

and

$$\overline{y} = \frac{1}{10}(\ln(670) + \ln(420) + \ln(300) + \cdots + \ln(48)) = 4.88.$$

The equation of the least squares line is $y = mx + b$ where

$$m = \frac{\mathbf{x}\cdot\mathbf{y} - 10\overline{x}\overline{y}}{\mathbf{x}\cdot\mathbf{x} - \overline{x}\overline{x}} = \frac{411.283 - 10(9.056)(4.88)}{940.96 - 10(9.056)(9.056)} = -0.254$$

and

$$b = \overline{y} - m\overline{x} = 4.88 - (-0.254)(9.056) = 7.18.$$

In Figure 10, this line is superimposed on the scatter plot of the data. Plots such as this are often called log-log plots because we plotted the log of the dependent variable p, the pulse rate (measured in beats per minute), as a function of the log of the independent variable w, the body weight (measured in grams). In terms of these variables, the equation we have found is $\ln(p) = -0.254\ln(w) + 7.18$, or

$$p = \frac{1312.91}{w^{0.254}}.$$

Roughly speaking, pulse rate is inversely proportional to the fourth root of weight.

The first "mouse-to-elephant curve," such as the line in Figure 10, was discovered by the biologist Max Kleiber in the 1930s. In 1947, Kleiber found that metabolic rate is proportional to the $(3/4)$th power of weight. An unsolved problem in physiology has

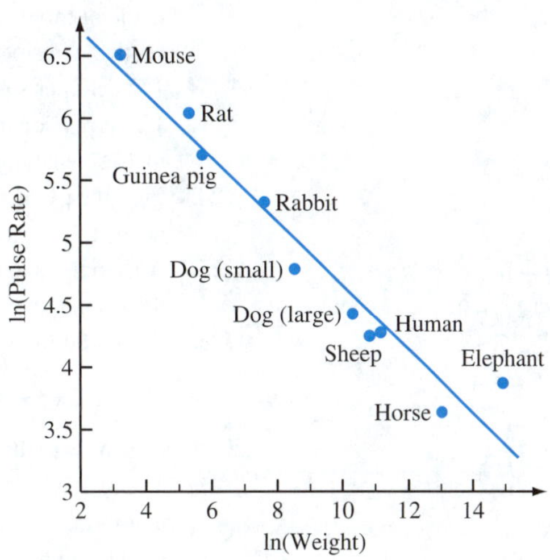

Figure 10

been to convincingly explain the powers that appear in these laws. The mathematical theory of fractals is the basis of an explanation that has recently been advanced. ■

A LOOK BACK

In Section 1.3, we used elementary algebra to find the least squares line that passes through a specified point. That task is a one-variable problem in which only the slope m must be determined. In practice, however, it rarely is the case that there is a preferred observation through which to pass the best-fitting line. Now that we are able to minimize a function of two variables, we can determine the least squares line without making any assumption about the line.

quickquiz

1. What is a saddle point for a function of two variables?
2. If $f(x, y)$ is defined on an open subset of the plane, then at what points can f have a local extremum?
3. True or false: (a) If $\nabla f(P_0) = \vec{\mathbf{0}}$, then f has a local extremum at P_0. (b) If f has a local extremum at P_0, then $\nabla f(P_0) = \vec{\mathbf{0}}$.
4. What is the Second Derivative Test for extrema of a function of two variables?

EXERCISES

Problems for Practice

In Exercises 1–20, find all the critical points of function f. For each critical point P, determine whether P is a local minimum, a local maximum, or a saddle point of f.

1. $f(x, y) = -2x^2 - 2y^2 + 4x + 8y + 7$
2. $f(x, y) = x^2 - 4xy + y^2 + 6x + 8y + 6$
3. $f(x, y) = x^3 - y^3 - 6xy + 8$
4. $f(x, y) = x\cos(y)$
5. $f(x, y) = 3x^2 + 2y^3 - 6xy$
6. $f(x, y) = 3xy + 3/x + 2/y$
7. $f(x, y) = (3 + x + y)xy$
8. $f(x, y) = x^2 + y^2 + 4x - 8y + 7$
9. $f(x, y) = 2x^2 + 4y^2 - 12x - 24y + 6$
10. $f(x, y) = x^2 + 2y^2 - 4xy + 6x + 12y + 9$
11. $f(x, y) = (x - 2)(y + 1)x$
12. $f(x, y) = x^2 - 3y^2 + 6xy + 3x + 5y + 4$
13. $f(x, y) = xy^2 + yx^2 + 8xy + 6$
14. $f(x, y) = 3y^2 + 6x^2 - 2xy + 4x + 3y + 7$
15. $f(x, y) = y^3 + 3x^2y + 3x^2 - 15y + 6$
16. $f(x, y) = -2y^3 + 3x^2 + 6y^2 + 6xy$
17. $f(x, y) = x^4 + y^4 - 8xy$
18. $f(x, y) = \sin(x) + \cos(y)$
19. $f(x, y) = \ln(1 + x^2 + y^2)$
20. $f(x, y) = y^2 - 4xy + x^2 + 5y - 2x + 7$
21. Find three positive numbers whose sum is 100 and whose product is as large as possible.
22. Which point on the plane $x + 2y + 2z = 6$ is the nearest to the origin?
23. A box in the shape of a rectangular parallelepiped is made of three types of material. The material for the top costs 2 cents per square inch, that for the sides costs 1/2 cent per square inch, and that for the bottom costs 1 cent per square inch. If the box is to hold 64 cubic inches, what dimensions will result in the cheapest box?
24. Find the point(s) on the surface $z = x^2 + 8y^2$ that are nearest to the point $(0, 0, 1)$.
25. A triangle is to enclose area 100. What dimensions for the triangle result in the least perimeter?
26. Which point on the sphere

$$(x - 4)^2 + (y - 2)^2 + (z - 3)^2 = 1$$

is nearest to the origin?

27. The following table records several paired values of automobile mileage (x), measured in thousands of miles, and hydrocarbon emissions per mile (y), measured in grams.

x	5.013	10.124	15.060	24.899	44.862
y	0.270	0.277	0.282	0.310	0.345

Plot these points. Determine the regression line for this data. If this pattern continues for higher-mileage cars, about how many grams of hydrocarbons per mile would a car with 100000 miles emit?

28. The table below displays total annual American income (x) and consumption (y) in billions of dollars for the years 1985–1991.

	Income (x)	Consumption (y)
1985	3325.3	2629.0
1986	3526.2	2797.4
1987	3776.6	3009.4
1988	4070.8	3296.1
1989	4384.3	3523.1
1990	4679.8	3748.4
1991	4834.4	3887.7

Plot these points. Determine the regression line for this data.

29. In the following table, x represents cigarette consumption per adult (in 100s) for the year 1962 for eight countries. The variable y represents mortality per 100000 due to heart disease in 1962.

	x	y
Australia	32.2	238.1
Belgium	17.0	118.1
Canada	33.5	211.6
Great Britain	27.9	194.1
Ireland	27.7	187.3
Netherlands	18.14	124.7
United States	39	256.9
West Germany	18.9	150.3

Plot the eight points (x, y). Find the least squares line $y = mx + b$ that could be used to model the relationship between cigarette consumption and mortality due to heart disease.

Further Theory and Practice

30. A 20 in. piece of wire is to be cut into three pieces. From one piece is made a square. From the second is made a rectangle with length equal to twice its width. From the third is made an equilateral triangle. How should the wire be cut so that the sum of the three areas is a maximum?

31. A radioactive isotope of gold is used to diagnose arthritis. Let $y(t)$ denote the blood serum gold at time t (measured in days) as a fraction of the initial dose at time $t = 0$. Measured values of $y(t)$ for $t = 1, 2, \ldots, 8$ are as shown in the table.

t	Serum gold
0	1.0
1	0.91
2	0.77
3	0.66
4	0.56
5	0.49
6	0.43
7	0.38
8	0.34

Plot the points $(t, \ln(y))$. Find a least squares line of the form $\ln(y) = b - mt$. Use the model to predict the fraction of the initial dose that remains in the blood serum after 10 days.

32. If $P_1 = (x_1, y_1, z_1)$, $P_2 = (x_2, y_2, z_2)$, $\ldots$, $P_N = (x_N, y_N, z_N)$ are points in space, then their "fit" to the plane $z = Ax + By + D$ is measured by

$$S = \sum_{j=1}^{N} (z_j - (Ax_j + By_j + D))^2.$$

The plane of best fit is that plane that minimizes the expression S. Find, in terms of $P_1, \ldots, P_N$, the plane of best fit.

33. Consider a quadratic function of the form

$$f(x, y) = Ax^2 + Cy^2 + Dx + Ey + F.$$

Explain how, by completing squares, one can immediately identify any local maxima or minima.

34. Apply the method of Exercise 33 to analyze the following quadratic functions.

a. $f(x, y) = x^2 + 3y^2 - 6x + 8y - 7$

b. $f(x, y) = -3x^2 + 6x - 8y - 5y^2 + 6$

c. $f(x, y) = 4x^2 - 7y^2 + 4x - 7y + 3$

d. $f(x, y) = 5x^2 + 8y^2 - 10x + 24y + 2$

35. If $F(x, y, z) = 0$ is a surface in space, P is a point that does not lie on the surface, and Q is the point *on* the surface that is nearest to P, then prove that the line through P and Q is orthogonal to the surface at the point Q.

36. Let $f(x, y) = (x^2 - y)(3x^2 - y) = 3x^4 - 4x^2y + y^2$.

a. Show that $(0, 0)$ is a critical point. Show that the Second Derivative Test is inconclusive, however.

b. Plot $y = x^2$ and $y = 3x^2$ in the xy-plane. These two curves divide the plane into three regions. The function f takes positive values on two of the regions and negative values on the other. Label these regions $+$ and $-$ accordingly.

c. Referring to your plot in part b, show that the restriction of f to any straight line through $(0, 0)$ has a minimum there.

d. Surprisingly, in view of part c, (0,0) is a saddle point for f. Show this by using your plot to find a parabolic curve $\mathcal{C}$ such that f restricted to $\mathcal{C}$ has a maximum value at $(0, 0)$.

37. Let $SSE(m, b)$ be defined by equation (13.19).

a. Show that for any point P,

$$\text{Discr}\,(SSE, P) = 4\left(N\sum_{j=1}^{N} x_j^2 - 2\sum_{1\le i<j\le N} x_i x_j\right).$$

b. Show that

$$N\sum_{j=1}^{N} x_j^2 = \sum_{j=1}^{N} x_j^2 + \sum_{i=1}^{N}\sum_{j=i+1}^{N}\left(x_i^2 + x_j^2\right).$$

c. Deduce that $\text{Discr}\,(SSE, P) \ge 4\sum_{j=1}^{N} x_j^2 > 0$. Then apply the Second Derivative Test to show that formulas (13.20) and (13.21) define a point at which SSE has a local minimum.

Calculator/Computer Exercises

38. Plot $x^4 + y^2 - 10xy + y$ for $3 \le x \le 4$ and $16 \le y \le 18$. Identify the coordinates of the local minimum that you see in the plot. Use the Second Derivative Test to verify this local minimum.

39. Plot $x^4 - y^5 + x^2y + x$ for $-1 \le x \le 0$ and $0 \le y \le 1$. Identify the coordinates of the saddle point that you see in the plot. Use the Second Derivative Test to verify that this critical point is not an extremum.

40. Plot $y^3 - xy - x^5$ for $0 \le x \le 1, -1 \le y \le 0$. Identify the coordinates of the local maximum that you see in the plot. Use the Second Derivative Test to verify this local maximum.

41. Plot $\ln(1 + x^4) - \sin(y)$ for $-1 \le x \le 1$ and $1 \le y \le 2$. Identify the coordinates of a critical point that you see in the plot. Verify that the Second Derivative Test is inconclusive at this critical point. What does your plot tell you about the behavior?

13.9 Lagrange Multipliers*

In many applications, when we are solving a maximum-minimum problem for a given function, we are not interested in the entire domain of the function. For example, every point on a mountain has a height $f(x, y)$ above its ground coordinates (x, y). But when we walk across the mountain and speak of the highest point of our walk, we are only interested in the points along the path that we follow. This situation is illustrated in Figure 1a with a path whose ground coordinates satisfy the equation $y = c$ for some constant c. If we set $g(x, y) = y$, then we can express the constraint on x and y as $g(x, y) = c$. To find the high point of our mountain path, we must maximize the expression $f(x, y)$ subject to the constraint that $g(x, y) = c$ for a given constant c.

*Knowledge of this section is not required for subsequent sections.

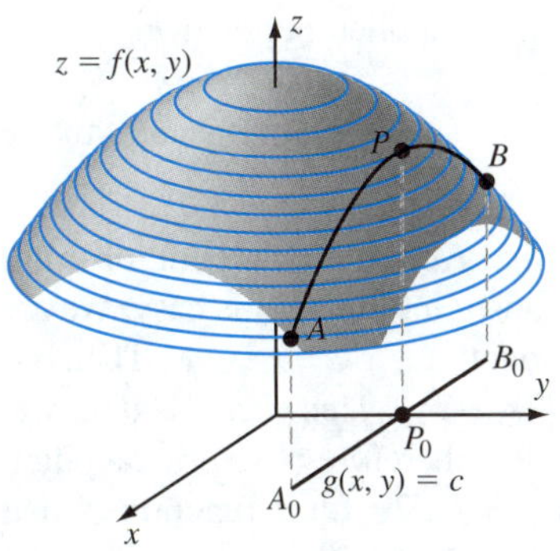

Figure 1a

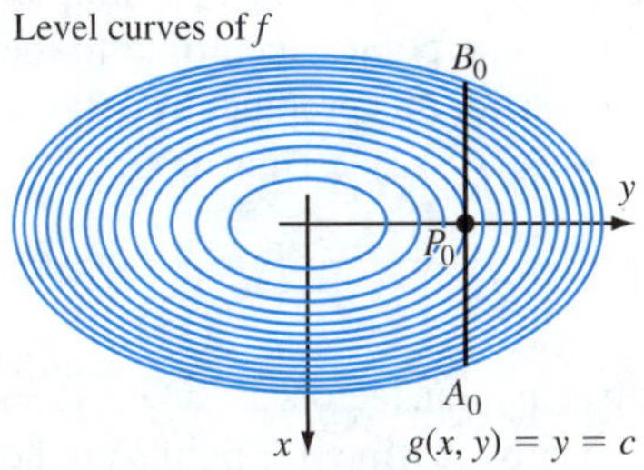

Figure 1b

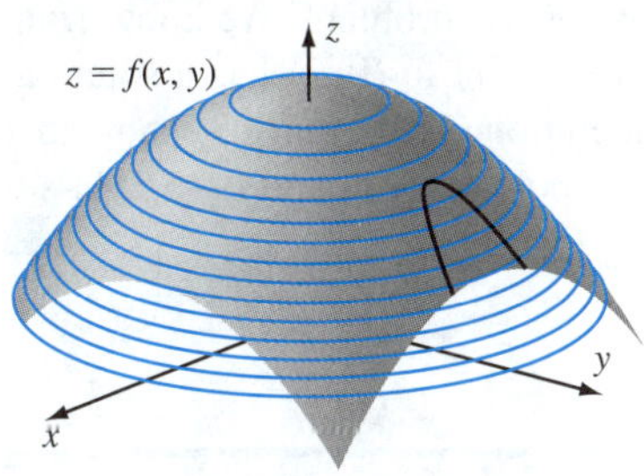

Figure 2a

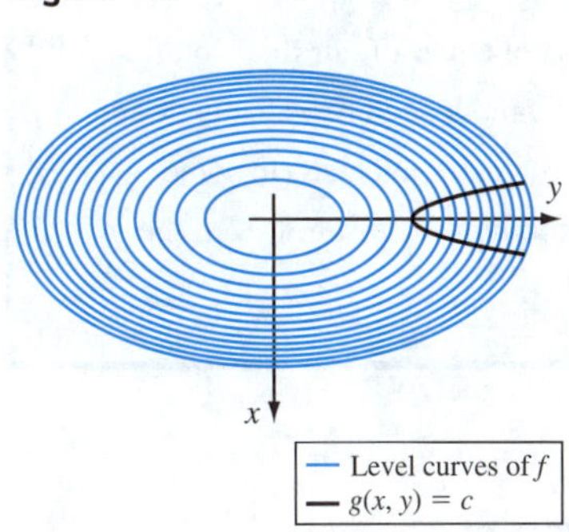

Figure 2b

In this section, we study an elegant mathematical procedure for determining the extrema of a function subject to a constraint. We begin by finding what might be called *critical points for $f(x, y)$ subject to the constraint $g(x, y) = c$.* These points are analogous to the critical points that we studied in earlier parts of this book; that is, we locate the critical points by finding the points where derivatives involving f and g are 0. These are the potential extrema. We then examine the roots to determine the extremum that we seek.

The example of a mountain path provides us with a good geometric understanding of the technique. Look again at Figure 1a and imagine yourself traversing the path from point A to point B. By following this path, you are constraining your x- and y-coordinates to the curve $g(x, y) = c$ in the xy-plane. As you start out from point A toward point B, you begin to ascend the mountain. Your projection into the xy-plane moves along the curve $g(x, y) = c$ from point A_0 toward point B_0, *crossing the level curves of f* (as seen in Figure 1b). That crossing behavior corresponds to passing from one level to another on the mountain. At the moment you pass through the high point P along your path, you break the pattern of passing to a higher level. That suggests that the curve $g(x, y) = c$ will *touch but not cross* the level curve of f at the corresponding point P_0 in the xy-plane. Indeed, Figure 1b shows that the curve $g(x, y) = c$ is tangent at P_0 to the level curve of f through P_0. The same holds true even when the constraint curve is not a straight line (Figure 2).

What we need, then, is an algebraic condition that tells us where $g(x, y) = c$ is tangent to the level curve of f through (x, y). Two curves are tangent at a point of intersection P_0 if and only if their normal vectors at the point are parallel. But we already know that $\nabla g(P_0)$ is normal to the curve $g(x, y) = c$ at P_0 and that $\nabla f(P_0)$ is normal to the level curve of f at P_0. So we want to find all points P_0 at which the vectors $\nabla g(P_0)$ and $\nabla f(P_0)$ are parallel. Finally, we recall that vectors $\nabla g(P_0)$ and $\nabla f(P_0)$ are parallel if and only if either

$$\nabla f(P_0) = \lambda \nabla g(P_0)$$

for some scalar λ or

$$\nabla g(P_0) = \vec{0}.$$

The critical points of $f(x, y)$ subject to the constraint $g(x, y) = c$ are the places on the constraint curve at which one of these two equations holds. The factor λ is known as a *Lagrange multiplier*. The procedure for finding critical points by solving the equation $\nabla f(x, y) = \lambda \nabla g(x, y)$ is called the *Method of Lagrange Multipliers.*

inSIGHT

It is important to understand that a critical point of f in the sense of Section 13.8 is not, in general, a critical point of $f(x, y)$ subject to the constraint $g(x, y) = c$. The surface $z = f(x, y)$ that is plotted in Figures 1 and 2 has a horizontal tangent plane only at the point $(0, 0)$. That point is the only critical point of $f(x, y)$, just as P_0 is the only critical point of $f(x, y)$ subject to the constraint $g(x, y) = c$. There is no connection between the set of critical points of f and the set of critical points of f subject to the constraint $g(x, y) = c$.

Let us see how the Method of Lagrange Multipliers is applied in practice.

Example 1 Find the point on the hyperbola $x^2 - y^2 = 4$ that is nearest to the point $(0, 2)$.

Solution Our first job is to identify the function that we are maximizing or minimizing and to identify the constraint function. The constraint function is easy. We are only interested in points on the curve $x^2 - y^2 = 4$. So let $g(x, y) = x^2 - y^2$. Then our constraint is $g(x, y) = 4$. We calculate $\nabla g(x, y) = \langle 2x, -2y \rangle$. This vector is $\vec{\mathbf{0}}$ only at the point $(0, 0)$, which is not on the constraint curve. It is therefore not a critical point.

We turn to the Lagrange multiplier equation $\nabla f = \lambda \nabla g$ for a function f that we will now specify. Rather than minimize the distance to $(0, 2)$, let us employ the strategy discussed in Section 13.8 to minimize the square of the distance to $(0, 2)$, namely, $f(x, y) = x^2 + (y - 2)^2$. We do this because the expression $f(x, y)$ is simpler than $\sqrt{x^2 + (y - 2)^2}$, but the two are minimized at the same point. According to the Method of Lagrange Multipliers, we find the critical points for this problem by solving the equation $\nabla f(x, y) = \lambda \nabla g(x, y)$, or $\langle 2x, 2y - 4 \rangle = \lambda \langle 2x, -2y \rangle$. Thus,

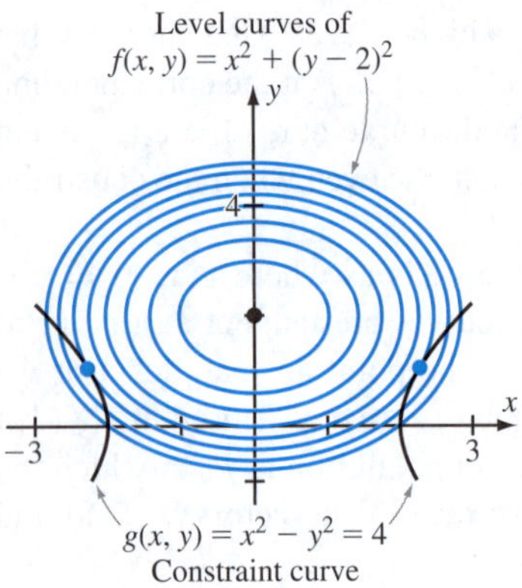

Figure 3
Critical points at $(-\sqrt{5}, 1)$ and $(\sqrt{5}, 1)$

$$2x = \lambda 2x \qquad \text{and} \qquad 2y - 4 = \lambda(-2y).$$

We need to solve these equations for λ, x, and y. The first equation factors as $2x(\lambda - 1) = 0$; hence, its solutions are $x = 0$ or $\lambda = 1$. But $x = 0$ is not possible for points on the curve $g(x, y) = 4$. Therefore, $\lambda = 1$ is the only possible solution. Substituting this into the second equation gives $y = 1$. The points on the curve $g(x, y) = 4$ with $y = 1$ are $(\sqrt{5}, 1)$ and $(-\sqrt{5}, 1)$. These are the critical points for our problem. We know from common sense that the problem of finding the nearest point to $(0, 2)$ has a solution. Which of the critical points is it? Symmetry considerations (see Figure 3) show that our two critical points are equidistant from $(0, 2)$. So both $(\sqrt{5}, 1)$ and $(-\sqrt{5}, 1)$ are solutions to our constrained extremal problem. ■

inSIGHT

Notice that, in the process of applying the Method of Lagrange Multipliers in Example 1, we solve for λ, but the actual value of λ is not a part of our final solution to the problem. Although the parameter λ is part of the technique, and its value is often crucial to the solution process, its value is usually of no greater interest. (In some applications in economics and other constrained extremum problems, however, the actual value of λ does have significance.)

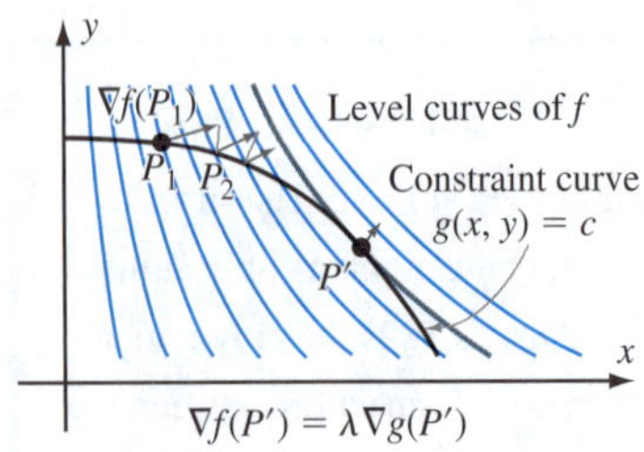

Figure 4

Why the Method of Lagrange Multipliers Works

Let us say a few more words about why the Lagrange multiplier technique works. Look at Figure 4, which illustrates the level curve $g(x, y) = c$. Our job is to find the maxima and minima of f *among those points that lie on this curve*. Imagine that we are standing at a point P_1 on the curve and looking for a maximum. If we are not already at a maximum, then we should move to a different point in our search. In which direction should we go? Ideally, we should go in the direction of $\nabla f(P_1)$, since that is the direction of most rapid increase of f. The trouble is that $\nabla f(P_1)$ probably points out of the curve (as happens in Figure 4). The next best thing is to move in the direction of the *projection of* $\nabla f(P_1)$ *onto the tangent to the curve* (as shown in Figure 4). Thus, we move a little bit in that direction to a point P_2 and repeat our program: If we are not already at a maximum, then we move in the direction of the projection of $\nabla f(P_2)$ onto the tangent to the curve at P_2. When we find the maximum P' we seek, this procedure breaks down. There is no direction to move that will increase the value of f. Thus, $\nabla f(P')$ must have zero projection into the tangent to the curve. In other words, at a maximum P', we have that $\nabla f(P')$ is normal to the curve. But $\nabla g(P')$ is normal to the curve $g(x, y) = c$ at P', so $\nabla f(P')$ and $\nabla g(P')$ are parallel. Thus, either $\nabla g(P') = \vec{0}$ or $\nabla f(P') = \lambda \nabla g(P')$ for some constant λ. The same reasoning works at a minimum. This explains why we declare places at which $\nabla f(P') = \lambda \nabla g(P')$ to be critical points for this constrained extremal problem.

We state the results of our investigation as a theorem. A rigorous proof is outlined in Exercise 45.

Theorem 1

Method of Lagrange Multipliers Suppose that $(x, y) \mapsto f(x, y)$ and $(x, y) \mapsto g(x, y)$ are differentiable functions. Let c be a constant. If f has an extreme value at a point P' on the constraint curve $g(x, y) = c$, then either $\nabla g(P') = \vec{0}$ or there is a constant λ such that $\nabla f(P') = \lambda \nabla g(P')$.

inSIGHT

As noted, the points P' for which $\nabla g(P') = \vec{0}$ must be considered, along with the solutions of the Lagrange multiplier equation $\nabla f(P') = \lambda \nabla g(P')$, as potential extrema. Also, the Method of Lagrange Multipliers fails at points P' where either $\nabla f(P')$ or $\nabla g(P')$ is undefined. As with the theory of extrema of one variable, these situations must be handled with ad hoc techniques. We shall not treat those techniques here.

Example 2 Find the maximum and minimum values of the function $f(x, y) = 2x^2 - y^2$ on the ellipse $x^2 + 2(y - 1)^2 = 2$.

Solution Let $g(x, y) = x^2 + 2(y-1)^2$. Our job is to maximize the function f subject to the constraint $g(x, y) = 2$. Notice that $\nabla g(x, y) = \langle 2x, 4(y - 1)\rangle$, which is equal to $\vec{0}$ only at the point $(0, 1)$. Because this point is not on the constraint curve $g(x, y) = 2$, it is not a critical point for our problem. The only critical points will be the solutions of the equation $\nabla f(x, y) = \lambda \nabla g(x, y)$ that lie on the constraint curve. Therefore, we want to solve $\langle 4x, -2y\rangle = \lambda\langle 2x, 4(y - 1)\rangle$, or

$$4x = \lambda 2x \tag{13.25}$$

and

$$-2y = \lambda 4(y - 1). \tag{13.26}$$

Equation (13.25) factors as $2x(2 - \lambda) = 0$; it is valid when either $\lambda = 2$ or $x = 0$.

1. If $x = 0$, then the corresponding points on the curve $g(x, y) = 2$ are $(0, 0)$ and $(0, 2)$. We still must determine whether equation (13.26) is satisfied at either of these two points. In fact, by taking $\lambda = 0$, equation (13.26) holds for the point $(0, 0)$. Similarly, we find that the point $(0, 2)$ satisfies equation (13.26) for $\lambda = -1$. Therefore, $(0, 0)$ and $(0, 2)$ are both critical points.
2. If $\lambda = 2$, then equation (13.26) yields $y = 4/5$. The corresponding points on the curve $g(x, y) = 2$ are $(4\sqrt{3}/5, 4/5)$ and $(-4\sqrt{3}/5, 4/5)$.

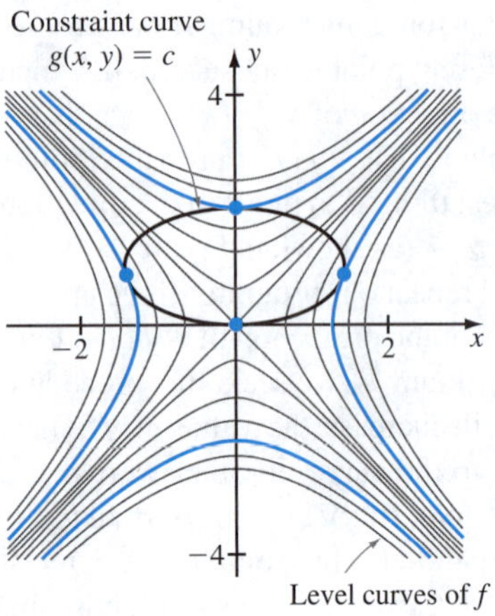

Figure 5

We have found four critical points:

$$(0, 0), \qquad (0, 2), \qquad (4\sqrt{3}/5, 4/5), \qquad \text{and} \qquad (-4\sqrt{3}/5, 4/5).$$

The maximum we seek will occur at one (or more) of these four points. The values of f at these four points are 0, -4, 80/25, and 80/25, respectively. We conclude that the maximum *value* of f, when restricted to the curve $g(x, y) = 2$, is 80/25. This maximum value is *attained* at the points $(4\sqrt{3}/5, 4/5)$ and $(-4\sqrt{3}/5, 4/5)$. The minimum value of f, when restricted to the curve $g(x, y) = 2$, is -4. This minimum value is *attained* at the point $(0, 2)$. Figure 5 shows the location of the three extrema on the constraint curve. The critical point $(0, 0)$ yields neither a maximum nor a minimum. There is no contradiction in this: As in the extremization topics that were treated earlier, a critical point is a *candidate* for an extremum, but it may turn out not to be an extremum. ■

inSIGHT

The algebra in Lagrange multiplier problems is often tricky. If we are not careful, it is easy to overlook or lose solutions. When we solve a constrained extremal problem with the Method of Lagrange Multipliers in two variables x and y, we set $\nabla f(x, y) = \lambda \nabla g(x, y)$. This leads to *two equations* in the *three unknowns* x, y, and λ. However, the extremum we seek must lie on the constraint curve, which means there is an important third equation $g(x, y) = c$. It is a heuristic principle of algebra that when the number of unknowns equals the number of equations, the system will have a finite number of solutions. The unknown λ will always appear in the first two equations in a simple fashion because of the form of the Lagrange multiplier condition. Typically, we may solve for λ in each of these equations and equate the two results. This leads to a single equation relating x and y. The constraint gives a second equation relating x and y. We attempt to solve these two equations simultaneously by eliminating one of the variables.

Our next example emphasizes that it is important for us to know that an extremum exists before we leap to any conclusion about a candidate located by the Method of Lagrange Multipliers.

Example 3 What are the extreme values of $f(x, y) = x^2 + 2y + 16$, subject to the constraint $(x + y)^2 = 1$?

Solution The vector equation $\nabla f = \lambda \nabla g$ gives us the two scalar equations $2x = 2\lambda(x + y)$ and $2 = 2\lambda(x + y)$. Because the right sides of these two equations are the same, we see that $2x = 2$, or $x = 1$. Substituting this value of x into the constraint equation $(x + y)^2 = 1$, we find that $y = 0$ or $y = -2$. Thus, $(1, 0)$ and $(1, -2)$ are the only critical points for $f(x, y)$ subject to the constraint $(x + y)^2 = 1$. Because $17 = f(1, 0) > f(1, -2) = 13$, it is tempting to think that we have located a maximum at $(1, 0)$. That conclusion would be wrong, however. Notice that the constraint condition tells us that either $x + y = 1$ or $x + y = -1$. Therefore, on the constraint curve, either $y = 1 - x$ or $y = -1 - x$. Because $f(x, 1-x) = x^2 + 2(1-x) + 16 = (x-1)^2 + 17$ and $f(x, -1-x) = x^2 + 2(-1-x) + 16 = (x-1)^2 + 13$, we see that, subject to the given constraint, f can assume arbitrarily large values. Therefore, f does not have an extreme value at the point $(1, 0)$. On the other hand, our formulas for $f(x, 1-x)$ and $f(x, -1-x)$ show that $f(x, y) \geq 13 = f(1, -2)$ at points on the constraint curve. Therefore, subject to the constraint, f has a minimum value of 13, which occurs at $(1, -2)$. See Figure 6. ■

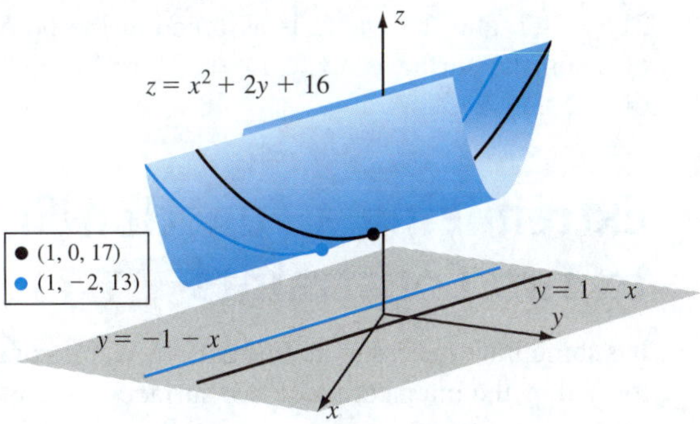

Figure 6

Lagrange Multipliers and Functions of Three Variables

The Method of Lagrange Multipliers works in much the same way for functions of three variables. If we wish to extremize $F(x, y, z)$ subject to the constraint $G(x, y, z) = c$, then we must find all points $P = (x, y, z)$ on the constraint curve for which the vector equation $\nabla F(P) = \lambda \nabla G(P)$ holds. Any point P where $\nabla G(P) = \vec{\mathbf{0}}$ or where $\nabla F(P)$ does not exist or where $\nabla G(P)$ does not exist is also included among the critical points.

Example 4 The temperature at a point (x, y, z) in space is given by $F(x, y, z) = 8x - 4y + 2z$. Find the maximum and minimum temperatures on the sphere $x^2 + y^2 + z^2 = 21$.

Solution Let $G(x, y, z) = x^2 + y^2 + z^2$. Thus, we are finding extrema of F subject to the constraint $G(x, y, z) = 21$. We first observe that F and G are differentiable and that ∇G is 0 only at the origin, which is not on the constraint surface. We therefore find all the critical points by setting $\nabla F(x, y, z) = \lambda \nabla G(x, y, z)$. Writing out the gradients more explicitly gives $\langle 8, -4, 2 \rangle = \lambda \langle 2x, 2y, 2z \rangle$, or

$$8 = \lambda 2x, \qquad -4 = \lambda 2y, \qquad \text{and} \qquad 2 = \lambda 2z.$$

The first and third equations yield

$$\frac{4}{x} = \lambda = \frac{1}{z}, \qquad \text{or} \qquad x = 4z.$$

The second and third equations yield

$$\frac{-2}{y} = \lambda = \frac{1}{z}, \qquad \text{or} \qquad y = -2z.$$

We may substitute these into the constraint equation $G(x, y, z) = 21$. The result of these substitutions is $(4z)^2 + (-2z)^2 + z^2 = 21$. Therefore, $z = \pm 1$. If $z = 1$, then $x = 4$ and $y = -2$; so we have found the critical point $(4, -2, 1)$. Likewise, $z = -1$ yields the critical point $(-4, 2, -1)$. Since $F(4, -2, 1) = 42$ and $F(-4, 2, -1) = -42$, we conclude the following: The minimum value of F on the surface $G(x, y, z) = 21$ is -42, and this value is assumed at the point $(-4, 2, -1)$; the maximum value of F on the surface $G(x, y, z) = 21$ is 42, and this value is assumed at the point $(4, -2, 1)$. ■

Extremizing a Function Subject to Two Constraints

It is sometimes necessary to find extreme values of an expression $F(x, y, z)$ on the curve formed by the intersection of two surfaces $G(x, y, z) = c_1$ and $H(x, y, z) = c_2$. In this case, we use two Lagrange multipliers λ and μ. The Lagrange multiplier equation is then

$$\nabla F(P) = \lambda \nabla G(P) + \mu \nabla H(P). \tag{13.27}$$

By extracting components, we obtain from this vector equation *three* scalar equations in the *five* unknowns x, y, z, λ, and μ. The other two equations that we need are the constraint equations $G(x, y, z) = c_1$ and $H(x, y, z) = c_2$. The reason this procedure works is discussed in Exercise 46. However, the algebra required to solve the resulting system of five equations in five unknowns is usually tedious and often complicated.

Example 5 Let C be the curve that results when the plane $2x + y + z = 11/5$ intersects the surface $x^2 + y^2 + z = 1$. What point on C has the largest z-value?

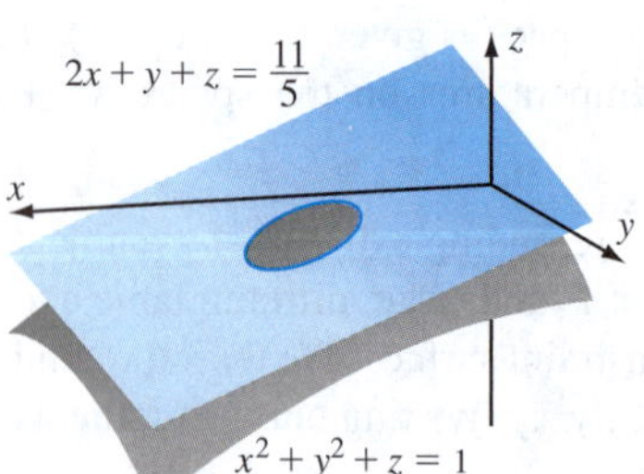

Figure 7

Solution Since $z = 1 - (x^2 + y^2) \le 1$ for each point (x, y, z) on C, we deduce that there is a maximum value of z on C. Figure 7, which shows the two surfaces, also makes it clear that our problem has a solution. It appears from the figure that there is also a minimum value of z. We should therefore expect two solutions of the Lagrange multiplier equations. The function to maximize is $F(x, y, z) = z$. The constraints are $G(x, y, z) = 2x + y + z = 11/5$ and $H(x, y, z) = x^2 + y^2 + z = 1$. Equation (13.27) becomes $\langle 0, 0, 1 \rangle = \lambda \langle 2, 1, 1 \rangle + \mu \langle 2x, 2y, 1 \rangle$, or

$$0 = 2\lambda + 2\mu x, \qquad 0 = \lambda + 2\mu y, \qquad \text{and} \qquad 1 = \lambda + \mu.$$

The third equation gives us $\lambda = 1 - \mu$. Substituting this into $0 = 2\lambda + 2\mu x$ and simplifying, we obtain $\mu(1 - x) = 1$. This equation tells us that $\mu \ne 0$ and $1 - x = 1/\mu$. Similarly, by replacing λ with $1 - \mu$ in $0 = \lambda + 2\mu y$, we obtain $\mu(1 - 2y) = 1$. Therefore, $1 - x = 1/\mu = 1 - 2y$, or $x = 2y$. We can now eliminate x from the two constraints: $G(2y, y, z) = 2(2y) + y + z = 11/5$, or $z = 11/5 - 5y$, and $H(2y, y, z) = (2y)^2 + y^2 + z = 1$. Replacing z with $11/5 - 5y$ in this last equation gives us $5y^2 + (11/5 - 5y) = 1$, or $5y^2 - 5y + 6/5 = 0$. Using the quadratic formula, we have

$$y = \frac{-(-5) \pm \sqrt{(-5)^2 - 4(5)(6/5)}}{2(5)} = \frac{5 \pm 1}{10} = \frac{2}{5}, \frac{3}{5}.$$

But we have found that $z = 11/5 - 5y$. Therefore, the largest value of z on the curve of intersection is $11/5 - 5(2/5) = 1/5$. The least value is $11/5 - 5(3/5) = -4/5$. ■

quickquiz

1. What is a constrained extremal problem?
2. What is the Method of Lagrange Multipliers?
3. What is the role of the multiplier λ in the Lagrange multiplier technique?
4. What is the difference between the role of the function f to be extremized and the function g that defines the constraint in the Lagrange technique?

EXERCISES

Problems for Practice

In Exercises 1–12, find the extrema of $f(x, y)$ subject to the constraint $g(x, y) = c$.

1. $f(x, y) = 2x - 3y + 6$, $g(x, y) = x^2 + 2y^2 = 4$
2. $f(x, y) = 3x - 4y$, $g(x, y) = 4x^2 + y^2 = 7$
3. $f(x, y) = (x + 1)^2 + y^2$, $g(x, y) = x^2 + 4y^2 = 16$
4. $f(x, y) = x^2 - y^2$, $g(x, y) = x^2 + (y - 2)^2 = 9$
5. $f(x, y) = 3xy^2 - 24$, $g(x, y) = x^2 + y^2 = 16$
6. $f(x, y) = xy$, $g(x, y) = x^2 + 9y^2 = 18$
7. $f(x, y) = x + y^2$, $g(x, y) = x^2 + y^2 = 9$
8. $f(x, y) = y^2 - x^2$, $g(x, y) = y^2 + 2x^2 = 4$
9. $f(x, y) = 4x^2 + 4y^2$, $g(x, y) = x^4 + y^4 = 16$
10. $f(x, y) = \sin^2(x) + \sin^2(y)$, $g(x, y) = x + y = \pi$
11. $f(x, y, z) = xyz$, $g(x, y, z) = x^2 + y^2 + z^2 = 4$
12. $f(x, y, z) = 4x - 7y + 6z$,
 $g(x, y, z) = x^2 + 7y^2 + 12z^2 = 84$

In Exercises 13–25, use the Method of Lagrange Multipliers to solve the stated problem.

13. Minimize the function $f(x, y) = x^2 + 2y^2 + 9$ subject to the constraint $2x - 6y = 5$.
14. Maximize the function $f(x, y) = x - y^2$ subject to the constraint $2x + y^2 = 4$.
15. Minimize $x^2 + y^2 + 4y$ subject to the constraint $x^2 + 2y^2 = 8$.
16. Maximize the product xy subject to the condition that $6x^2 + y^2 = 8$.
17. Maximize $x^3 + 2y$ subject to the condition that $x^2 + y^2 = 4/3$.
18. A window is to be constructed in the shape of a rectangle surmounted by an isosceles triangle. Building codes require the total area of the window to be 6 ft^2, but the material used in the frame is very costly.
 What dimensions will minimize the perimeter of the window?
19. Minimize the surface area $2\pi r^2 + 2\pi rh$ of a cylindrical can of height h and radius r, subject to the constraint that the volume $(\pi r^2 h)$ of the cylinder is equal to a fixed positive constant V_0.
20. Minimize $x^2 - y^2$subject to the constraint $x^2 + 2y^2 = 8$.
21. Maximize $x^4 + y^4 + z^4$ on the sphere $x^2 + y^2 + z^2 = 12$.
22. Find the point on the plane $x - 3y + 5z = 6$ that is nearest to the origin.
23. Find the point on the ellipsoid $x^2 + 2y^2 + 4z^2 = 4$ that is nearest to $(1, 0, 0)$.
24. The temperature of any point (x, y, z) in space is given by

$$T(x, y, z) = 2x - 6y + 5z.$$

 Find the greatest and least temperatures on the surface $x^2 + 6y^2 + 4z^2 = 24$.

25. Maximize the product xy^2z subject to the constraint $x^2 + 3y^2 + 2z^2 = 64$.

Further Theory and Practice

26. By using the Method of Lagrange Multipliers, find two level curves of $f(x, y) = 5x + 4y$ that are tangent to the hyperbola $x^2 - y^2 = 1$. Does an extreme value occur at either point of tangency?

27. What is the largest possible y-coordinate of a point on the curve $3x^2 + 2xy + 3y^2 = 24$?

28. The temperature on an ellipsoid $2x^2 + y^2 + 2z^2 = 8$ is given by the formula

$$T(x, y, z) = 8z^2 + 4xy - 12y + 200.$$

What is the location of the hottest and coldest points on the ellipsoid?

29. A certain county consists of the region $\{(x, y) : 4x^2 + 2y^2 \le 16\}$. The altitude at any point of this county is given by $a(x, y) = 80x - 70y + 150$. What are the highest and lowest points in this county?

30. If α, β, and γ are the angles of a triangle, how large can $\sin(\alpha)\sin(\beta)\sin(\gamma)$ be?

31. Some cylindrical cans, such as containers of juice concentrate, are manufactured with metal tops and bottoms but cardboard sides. Suppose that the cost per unit area of the metal is k times that of the cardboard. Minimize the cost of materials of such a can if its volume is to equal a fixed positive constant V_0.

32. A capsule has the shape of a cylinder that is "capped" at each end by a hemisphere. The cylinder has height $h > 0$ and radius $r > 0$. Is it possible to maximize or minimize the surface area $4\pi r^2 + 2\pi rh$ subject to the constraint that the volume of the capsule is equal to a fixed positive constant V_0?

33. Minimize $x^2 + y^2 + z^2$ subject to the constraints $x + y + z = 6$ and $x + 2y - 3z = 12$.

34. Minimize $f(x, y, z) = x^2 + y^2$ subject to the constraints $x + y + z = 6$ and $x + 3y - z = 12$.

35. Maximize $f(x, y, z) = z$ subject to the constraints $x + 2y + 3z = 6$ and $5(x^2 + y^2 + z^2) = 14$.

36. What is the minimum x-coordinate a point can have if it is on the intersection of the plane $2x + y + z = 2$ and the ellipsoid $x^2 + y^2/4 + z^2 = 1$?

In economics, a *utility function* $f(x, y)$ quantifies the satisfaction a consumer derives from x units of one item and y units of a second item. The level curves of f are called *indifference curves*. The consumer's *budget line* is the line segment in the first quadrant of the xy-plane that represents how many of the two items can be purchased for a total amount T. Exercises 37 and 38 concern these concepts.

37. A Cobb-Douglas utility function has the form $f(x, y) = Cx^p y^q$ where C, p, and q are positive constants. If the first item costs A per unit and the second costs B per unit and if the consumer has a total amount T that he can spend on the two items, then for what values of x and y is the consumer's utility maximized?

38. Using $A = 15$, $B = 10$, and $T = 120$, plot the consumer's budget line in the viewing window $[0, 12] \times [0, 12]$. In this viewing window, add the plots (or sketches) of several indifference curves of the Cobb-Douglas utility function described in Exercise 37. Use the values $C = 1$, $p = 3/4$, and $q = 1/4$. Include the indifference curve on which the consumer's maximum utility lies. What is the relationship of this indifference curve to the budget line?

39. A company allocates an amount T to produce an item. The available capital can be divided into labor costs and plant costs. Let $f(x, y)$ be the number of units of an item that can be produced with x units of labor and y units of plant expenditures. Suppose that the unit costs of these outlays are a and b, respectively. Show that when production is maximized, $f_x/f_y = a/b$. This equation can be stated as the following economic principle: When labor and plant costs are allocated optimally, the ratio of their marginal productivities is equal to the ratio of their unit costs.

40. Suppose that f and g are differentiable functions of two variables. Suppose also that for every c, there is a unique point $(x(c), y(c))$ that maximizes f subject to the constraint $g(x, y) = c$. Assume that $\nabla g(x(c), y(c)) \neq \vec{\mathbf{0}}$, and let $\lambda(c)$ be the Lagrange multiplier defined by $\nabla f(x(c), y(c)) = \lambda(c)\nabla g(x(c), y(c))$. Let $\mathbf{r}(c) = \langle x(c), y(c)\rangle$. Set $M(c) = f(x(c), y(c))$, the maximum value of f subject to the constraint $g(x, y) = c$.

a. Show that $1 = \nabla g(x(c), y(c)) \cdot \mathbf{r}'(c)$.

b. Show that $\lambda(c) = \nabla f(x(c), y(c)) \cdot \mathbf{r}'(c)$.

c. Deduce that $\lambda(c) = M'(c)$.

41. Let $f(x, y) = xy^{3/4}$ be a production function (in the sense of Exercise 39). Suppose that the equation $15x + 12y = c$ describes the allocation of labor and plant costs. For an arbitrary positive constant c, find the point $(x(c), y(c))$ on the constraint curve at which f is maximized. Calculate the maximum value $M(c)$ and the Lagrange multiplier $\lambda(c)$. According to Exercise 40, $\lambda(c)$ is approximately equal to the marginal productivity $M(c+1) - M(c)$. Verify the approximation $\lambda(c) \approx M(c+1) - M(c)$ for $c = 10000$.

42. Suppose that A and B are positive constants, that $f(x, y) = Ax + By$, and that $g(x, y) = x^2 + y^2$. Let $(x(c), y(c))$ be the point on the constraint curve $g(x, y) = c$ at which f is maximized. Let $M(c) = f(x(c), y(c))$. Calculate the Lagrange multiplier $\lambda(c)$, and verify that $\lambda(c) = M'(c)$.

43. *The Milkmaid Problem.* A river, in the shape of a smooth curve, flows near a house H and a barn B. Each morning a milkmaid leaves the house, fills a bucket of water at a point R on the river, then goes to the barn. The distance of this walk is $|\overline{HR}| + |\overline{RB}|$. If the point R_0 minimizes this distance, then show that there is an ellipse with foci B and H that is tangent to the river at R_0.

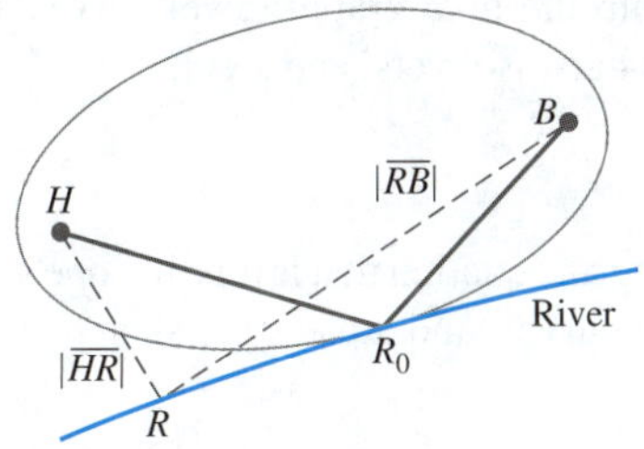

44. Let $f(x, y, z) = ax + by + cz$ be a linear function and let

$$g(x, y, z) = \alpha^2(x - e)^2 + \beta^2(y - f)^2 + \gamma^2(x - g)^2.$$

Prove that the extrema of f on the region

$$\mathcal{R} = \{(x, y, z) : g(x, y, z) \le 1\}$$

will always occur on the boundary of $\mathcal{R}$. Prove that there will always be exactly one maximum and one minimum and that they will occur at diametrically opposite points of the region $\mathcal{R}$.

45. Complete the following outline to obtain a more rigorous treatment of the Method of Lagrange Multipliers. We will assume that ∇f and ∇g exist at the critical points and that ∇g does not vanish at those points. We seek extrema of $f(x, y)$ subject to the constraint $g(x, y) = c$.

a. Let $P_0 = (x_0, y_0)$ satisfy $g(P_0) = c$, and let $\mathbf{r}(t) = x(t)\mathbf{i} + y(t)\mathbf{j}$ be a parametrization of a curve lying in $g(x, y) = c$ that passes through P_0. Say that $\mathbf{r}(t_0) = P_0$.

b. If P_0 is an extremum for f subject to the constraint $g = c$, then the function of one variable $f(\mathbf{r}(t))$ has an extremum at $t = t_0$.

c. We have

$$\frac{d}{dt} f(\mathbf{r}(t))\bigg|_{t=t_0} = 0.$$

d. We conclude that

$$\nabla f(P_0) \perp \mathbf{r}'(t_0).$$

e. Since $\mathbf{r}$ was an arbitrary curve in the level set $g = c$, we conclude that $\nabla f(P_0)$ is normal to the level set $g = c$ at P_0.

Deduce that $\nabla f(P_0) = \lambda \nabla g(P_0)$ for some $\lambda \in \mathbb{R}$.

46. Suppose that $G(x, y, x) = c_1$ and $H(x, y, z) = c_2$ intersect in a smooth curve $\mathcal{C}$. Let $P_0 = (x_0, y_0, z_0)$ belong to $\mathcal{C}$, and let $\mathbf{r}(t) = x(t)\mathbf{i} + y(t)\mathbf{j} + z(t)\mathbf{k}$ be a parametrization of $\mathcal{C}$ that passes through P_0. Say that $\mathbf{r}(t_0) = P_0$. If P_0 is an extremum for $F(x, y, z)$ subject to the constraint $G(x, y, x) = c_1$ and $H(x, y, z) = c_2$, then the function of one variable $F(\mathbf{r}(t))$ has an extremum at $t = t_0$. Deduce that

$$\frac{d}{dt} F(\mathbf{r}(t))\bigg|_{t=t_0} = 0$$

and therefore $\nabla F(P_0) \perp \mathbf{r}'(t_0)$. Conclude that $\nabla F(P_0)$ is in the plane that is normal to $\mathcal{C}$ at P_0. Show that every vector in this normal plane has the form $\lambda \nabla G(P_0) + \mu \nabla H(P_0)$ for some scalars λ and μ.

47. Maximize $x_1 + x_2 + \cdots + x_N$ subject to the constraint $x_1^2 + x_2^2 + \cdots + x_N^2 = 1$. Deduce the inequality

$$\frac{x_1 + x_2 + \cdots + x_N}{N} \le \sqrt{\frac{x_1^2 + x_2^2 + \cdots + x_N^2}{N}}.$$

48. Suppose $x_1, x_2, \ldots, x_N$ are positive numbers. Minimize $x_1 + x_2 + \cdots + x_N$ subject to the constraint $x_1 x_2 \cdots x_N = 1$. Deduce the *Arithmetic-Geometric Mean Inequality:*

$$(x_1 x_2 \cdots x_N)^{1/N} \le \frac{x_1 + x_2 + \cdots + x_N}{N}.$$

Computer/Calculator Exercises

49. Find the largest value of $x \exp(x^2 - xy)$ subject to the constraint $x^2 + y^2 = 1$.

50. Find the point on the curve $x \exp(x^2 - xy) = 1$ that is closest to the origin.

51. Find the minimum value of $(1 + x^2 + xy)/(1 + x^2 + y^4)$ subject to the constraint $x^2 + y^2/4 = 1$.

52. Minimize $\sqrt{x^2 + y^2} + \sqrt{(x-2)^2 + y^2}$ subject to the constraint $xy = 2$.

Summary of Key Topics

Functions of Several Variables (Section 13.1)

A function of two variables has as its domain a set of ordered pairs of numbers; a function of three variables has as its domain a set of ordered triples. In this chapter, these functions are scalar-valued. Functions of several variables may be added, subtracted, and multiplied in the usual way. They may be divided as long as we do not divide by 0. If f is a scalar-valued function of several variables and ϕ is a scalar-valued function of one variable and if the values of f lie in the domain of ϕ, then we may form the composition $\phi \circ f$ to obtain a new scalar-valued function of several variables.

Functions of two variables are graphed in three-dimensional space. The graph of f is a surface consisting of the points $\{(x, y, f(x, y)) : (x, y) \in \text{ domain of } f\}$. In general, a surface in space is the graph of a function if and only if no vertical line intersects the surface more than once. We form the graph by sketching level curves

$$\{(x, y) : f(x, y) = c\}$$

in the plane and then amalgamating them into the final graph. Level curves arise in applications of mathematics as isotherms, isobars, isohyets, and so on.

Cylinders (Section 13.2)

A cylinder is the set of points in space satisfying an equation that is missing one variable. Such surfaces have uniform sections, or level curves, and are easy to sketch.

Quadric Surfaces (Section 13.2)

The quadric surfaces are classified by signature:

$\dfrac{x^2}{\alpha^2} + \dfrac{y^2}{\beta^2} + \dfrac{z^2}{\gamma^2} = 1$	Ellipsoid
$\dfrac{x^2}{\alpha^2} + \dfrac{y^2}{\beta^2} = \dfrac{z^2}{\gamma^2}$	Elliptic cone
$\dfrac{x^2}{\alpha^2} + \dfrac{y^2}{\beta^2} = \dfrac{z}{\gamma}$	Elliptic paraboloid
$\dfrac{x^2}{\alpha^2} + \dfrac{y^2}{\beta^2} - \dfrac{z^2}{\gamma^2} = 1$	Hyperboloid of one sheet
$\dfrac{z^2}{\gamma^2} - \dfrac{x^2}{\alpha^2} - \dfrac{y^2}{\beta^2} = 1$	Hyperboloid of two sheets
$\dfrac{x^2}{\alpha^2} - \dfrac{y^2}{\beta^2} = \dfrac{z}{\gamma}$	Hyperbolic paraboloid

Limits and Continuity (Section 13.3)

If $P = (p_1, p_2)$ is a point in the plane, let $D(P_0, r)$ be the disk of radius $r > 0$ centered at P_0. Let $D_*(P_0, r)$ be the same disk but with the center removed. Let $d(P_0, P)$ denote the distance between the points P and P_0. If the domain of $f(x, y)$ contains $D_*(P_0, r)$, then we say that

$$\lim_{(x,y)\to P_0} f(x, y) = \ell$$

if, for any $\epsilon > 0$, there is a $\delta > 0$ such that

$$0 < d(P_0, P) < \delta \quad \text{implies} \quad |f(x, y) - \ell| < \epsilon.$$

We say that f is continuous at $P_0 = (x_0, y_0)$ if the domain of f contains P_0 and

$$\lim_{(x,y)\to P_0} f(x, y) = f(x_0, y_0).$$

The standard limit rules apply to these new notions of limit; the standard results about continuity also hold.

Partial Derivatives (Section 13.4)

If the domain of f contains a disk $D(P_0, r)$, then we define the partial derivatives of f at $P = (x_0, y_0)$ to be

$$\frac{\partial f}{\partial x}(P_0) = \lim_{\Delta x\to 0} \frac{f(x_0 + \Delta x, y_0) - f(x_0, y_0)}{\Delta x}$$

and

$$\frac{\partial f}{\partial y}(P_0) = \lim_{\Delta y\to 0} \frac{f(x_0, y_0 + \Delta y) - f(x_0, y_0)}{\Delta y},$$

provided these limits exist. The partial differentiation process can be iterated to obtain higher partial derivatives. We will usually work with functions with continuous derivatives, called continuously differentiable functions. The familiar differentiation rules also hold for partial differentiation.

Other notations for the partial derivatives of f are f_x, f_1, f_y, f_2, f_{xy}, f_{12}, $D_x f$, $D_2 f$, and so on.

All of these ideas can be adapted to functions of three variables.

The Chain Rule (Section 13.5)

If $z = f(x, y)$, and in turn $x = \rho(s)$ and $y = \sigma(s)$, with all functions continuously differentiable, then

$$\frac{dz}{ds} = \frac{\partial z}{\partial x} \cdot \frac{dx}{ds} + \frac{\partial z}{\partial y} \cdot \frac{dy}{ds}.$$

A similar Chain Rule applies when $x = \rho(z, t)$ and $y = \sigma(s, t)$:

$$\frac{\partial z}{\partial s} = \frac{\partial z}{\partial x} \cdot \frac{\partial x}{\partial s} + \frac{\partial z}{\partial y} \cdot \frac{\partial y}{\partial s}$$

and

$$\frac{\partial z}{\partial t} = \frac{\partial z}{\partial x} \cdot \frac{\partial x}{\partial t} + \frac{\partial z}{\partial y} \cdot \frac{\partial y}{\partial t}.$$

There is an analogous Chain Rule when f is a function of three or more variables.

Gradients and Directional Derivatives (Section 13.6)

If $f(x, y)$ is a continuously differentiable function, then the gradient of f is

$$\operatorname{grad} f(x, y) = \nabla f(x, y) = f_x(x, y)\mathbf{i} + f_y(x, y)\mathbf{j}.$$

If $\mathbf{u}$ is a unit vector, then the directional derivative of f in the direction $\mathbf{u}$ at the point (x, y) is

$$D_{\mathbf{u}} f(x, y) = \mathbf{u} \cdot \nabla f(x, y).$$

If P_0 is fixed, then the greatest directional derivative of f at P_0 occurs when $\mathbf{u}$ is the direction of $\nabla f(P_0)$. In this case, $D_{\mathbf{u}} f(P_0) = \|\nabla f(P_0)\|$. Also, the least directional derivative of f at P_0 occurs when $\mathbf{u}$ is the direction of $-\nabla f(P_0)$. In this case, $D_{\mathbf{u}} f(P_0) = -\|\nabla f(P_0)\|$.

All of these ideas can be adapted to functions of three variables.

Normal Vectors and Tangent Planes (Section 13.7)

If $f(x, y)$ is a continuously differentiable function and $P_0 = (x_0, y_0)$ is a point in its domain, then a normal vector to the graph at $(P_0, f(P_0))$ is

$$f_x(P_0)\mathbf{i} + f_y(P_0)\mathbf{j} - \mathbf{k}.$$

The tangent plane to the graph at $(P_0, f(P_0))$ will be the plane passing through $(P_0, f(P_0))$ and normal to this vector. It will have equation

$$f_x(P_0) \cdot (x - x_0) + f_y(P_0) \cdot (y - y_0) - (z - f(x_0, y_0)) = 0.$$

For any continuously differentiable function f of several variables, the gradient of f is perpendicular to the level surfaces of f.

Numerical Approximation (Section 13.7)

Because the tangent plane geometrically approximates a graph near the point of contact, we may approximate $f(x, y)$ for (x, y) near $P_0 = (x_0, y_0)$ by the function

$$L(x, y) = f(P_0) + f_x(P_0)(x - x_0) + f_y(P_0)(y - y_0).$$

The error, or rate of approximation, can be estimated using the second derivatives of f. This approximation is useful in numerical calculations and in error analysis.

Critical Points (Section 13.8)

Local maxima and minima are defined as they are for functions of one variable. A point at which a function has a local minimum when approached from one direction and a local maximum when approached from another is called a saddle point. If $f(x, y)$ is continuously differentiable, then a critical point is a point where the gradient vanishes. Let $P_0 = (x_0, y_0)$ be a critical point. If f is twice continuously differentiable, then define

$$\text{Discr}(f, P_0) = \det\left(\begin{bmatrix} f_{xx}(P_0) & f_{xy}(P_0) \\ f_{yx}(P_0) & f_{yy}(P_0) \end{bmatrix}\right) = f_{xx}(P_0) \cdot f_{yy}(P_0) - (f_{xy}(P_0))^2.$$

1. If $\text{Discr}(f, P_0) > 0$, $f_{xx}(P_0) > 0$, and $f_{yy}(P_0) > 0$, then P_0 is a local minimum for f.
2. If $\text{Discr}(f, P_0) > 0$, $f_{yy}(P_0) < 0$, and $f_{yy}(P_0) < 0$, then P_0 is a local maximum for f.
3. If $\text{Discr}(f, P_0) < 0$, then P_0 is a saddle point for f.
4. If $\text{Discr}(f, P_0) = 0$, then we can draw no conclusion.

The ability to find and identify critical points is an aid in graphing and it enables us to solve applied problems about extrema.

Lagrange Multipliers (Section 13.9)

To find the critical points for the problem of extremizing $f(x, y)$ subject to the constraint $g(x, y) = c$, we solve the equation

$$\nabla f(x, y) = \lambda \nabla g(x, y).$$

This procedure will locate those critical points for the problem at which both ∇f and ∇g exist and $\nabla g \neq 0$.

Genesis & Development

The two founders of calculus, Newton and Leibniz, both had occasion to employ partial derivatives. For example, in a letter written to l'Hôpital in 1694, Leibniz introduced the total differential of a function of two variables. Although Leibniz wrote δf and ϑf for the partial derivatives $\frac{\partial f}{\partial x}$ and $\frac{\partial f}{\partial y}$, respectively, the practice of using notation to distinguish between ordinary and partial derivatives did not take hold among the mathematicians who followed. In fact, the notation $\frac{\partial f}{\partial x}$ and $\frac{\partial f}{\partial y}$, introduced by Carl Gustav Jacobi in 1841, did not receive widespread acceptance until the end of the 19th century.

Clairaut's Theorem

Nowadays, we often refer to the identity $f_{xy}(x, y) = f_{yx}(x, y)$ as *Clairaut's Theorem*. Alexis Claude Clairaut (1713–1765), the son of a mathematics teacher and the only one of 20 siblings to reach adulthood, was a mathematical prodigy who published his first paper at the age of 13. When only 16, he initiated the study of space curves, and, as a result, he became the youngest member ever appointed to the prestigious Paris Academy of Sciences. In addition to geometry, Clairaut undertook important research in differential equations, the calculus of variations, mechanics, and astronomy. He participated in an expedition to Lapland to measure 1 degree of longitude, with the aim of verifying Newton's assertion that Earth is an oblate sphere. Clairaut's scientific activity continued unabated until his early death, but not to the exclusion of all other interests. As the contemporary mathematician Charles Bossut (1730–1814) wrote, "[O]ccupied by dinner parties and late-night discussions, burdened by unrestrained womanizing, happy to mix pleasure with business, Clairaut lost his rest, his health, and, finally, at the age of 52, his life."

Before Clairaut, the equality of mixed partial derivatives had been observed in particular cases and was tacitly assumed to be true in general. In 1740, Clairaut published a derivation of the equation $f_{xy}(x, y) = f_{yx}(x, y)$, as did Euler. However, when the standards of mathematical rigor tightened up half a century later, both proofs were judged inadequate. A succession of eminent mathematicians, including Lagrange in 1797 and Cauchy in 1823, advanced alternative demonstrations that were equally flawed. In 1867, the Finnish mathematician Leonard Lorenz Lindelöf (1827–1908) published a critique of the existing faulty proofs. Reinforcing the need for a valid proof, he also exhibited a non-differentiable function f for which $f_{xy}(x, y)$ and $f_{yx}(x, y)$ exist but are unequal. At last, in 1873, Hermann Amandus Schwarz (1843–1921) conclusively established the validity of Clairaut's Theorem under appropriate assumptions on f. Thereafter, work on Clairaut's Theorem was directed toward relaxing the hypotheses. Well-known mathematicians such as Camille Jordan (1838–1922), Ulisse Dini (1845–1918), Axel Harnack (1851–1888), and Giuseppe Peano (1858–1932) all contributed to extending Clairaut's Theorem to a wider class of functions.

The Vibrating String

A large part of the mathematical enterprise of the 18th century was devoted to applying calculus to natural phenomena. One of the problems that first attracted attention was the motion of a vibrating string (Figure 1). We let $y(x, t)$ be the vertical displacement

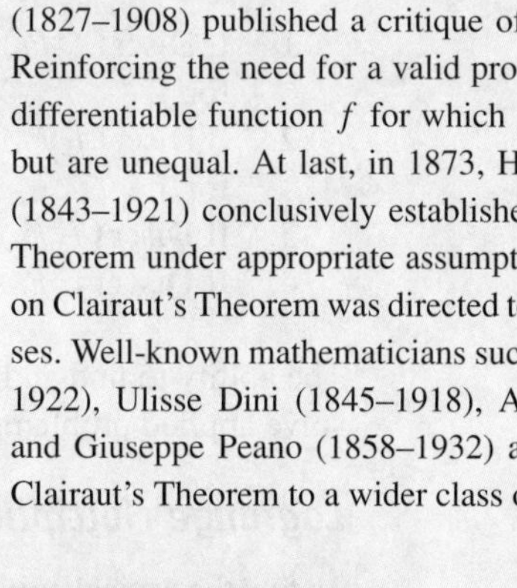

Figure 1
A snapshot of a vibrating string at time t

of the string at point x and time t. An analysis that is quite similar to the investigation of the catenary in Chapter 6 shows that

$$\frac{\partial^2 y}{\partial t^2} = c^2 \frac{\partial^2 y}{\partial x^2}$$

where c^2 is the tension in the string divided by the mass density of the string. This partial differential equation is called the *one-dimensional wave equation*. It appeared for the first time in a paper that Jean le Rond d'Alembert (1717–1783) published in 1747. In the same article, d'Alembert also argued that any solution of the one-dimensional wave equation can be expressed in the form

$$y(x, t) = f_1(x - ct) + f_2(x + ct) \qquad \textbf{(13.28)}$$

where f_1 and f_2 are twice-differentiable functions of one variable. Letting $\Delta x = c\Delta t$, we observe that $f_1((x + \Delta x) - c(t + \Delta t)) = f_1(x - ct)$. This means that the displacement of the string at a point $c\Delta t$ to the right of x is the same displacement that existed at x at a time Δt units earlier. We conclude that f_1 represents a wave that travels to the right with speed c. See Figure 2. Similarly, $f_2((x - \Delta x) + c(t + \Delta t)) = f_2(x + ct)$, so f_2 represents a wave that travels to the left with speed c.

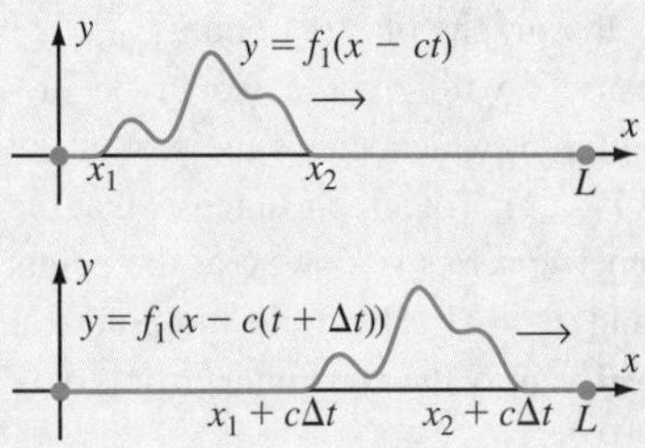

Figure 2
Snapshots at times t and $t + \Delta t$ of a traveling wave on a vibrating string

Six years later, Daniel Bernoulli (1700–1782), son of Johann Bernoulli, discovered an alternative method of solving the wave equation. His idea was to represent the solution as a superposition of fundamental waves (Figure 3). Following that approach, Bernoulli found that the general solution of the wave equation has the form

$$y(x, t) = \sum_{n=0}^{\infty} \left(a_n \cos\left(\frac{n\pi ct}{L}\right) + b_n \sin\left(\frac{n\pi ct}{L}\right) \right) \sin\left(\frac{n\pi x}{L}\right). \qquad \textbf{(13.29)}$$

Now, it often happens that different methods lead to different forms of an answer. Usually these differences can be easily reconciled. In the case of the vibrating string, however, there was no obvious way to relate the solutions of d'Alembert and Bernoulli. The matter was further complicated by Euler, who agreed with the form of d'Alembert's solution (13.28) but disagreed on the type of functions f_1 and f_2 that can appear in the solution. Calculus was left in something of a crisis—three of its leading practitioners had tackled the same problem only to arrive at apparently different answers.

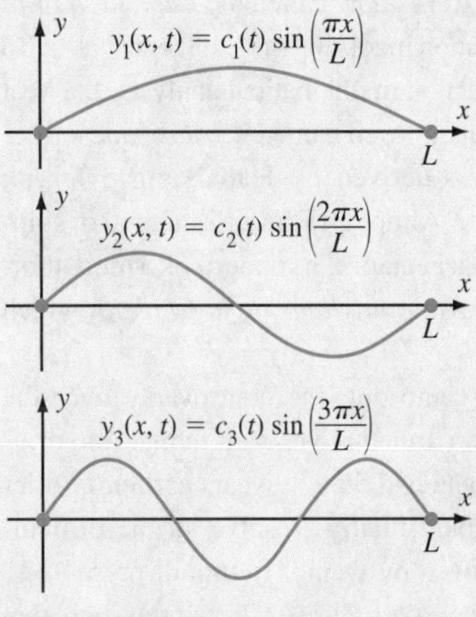

Figure 3
In Daniel Bernoulli's solution of the wave equation, every vibration is the superposition of waves that have fundamental frequencies.

Other Equations of Mathematical Physics

As the 18th century progressed, mathematicians used partial differential equations to study acoustics, hydrodynamics, elasticity, and celestial mechanics. One particular type of expression appeared repeatedly. If y depends on only one space variable x, if z depends on only two space variables x and y, and if u depends on all three space variables x, y, and z, then we write

$$\Delta y = \frac{\partial^2 y}{\partial x^2}, \quad \Delta z = \frac{\partial^2 z}{\partial x^2} + \frac{\partial^2 z}{\partial y^2}, \quad \text{and} \quad \Delta u = \frac{\partial^2 u}{\partial x^2} + \frac{\partial^2 u}{\partial y^2} + \frac{\partial^2 u}{\partial z^2}.$$

Using this notation, d'Alembert's one-dimensional wave equation takes the form $y_{tt} = c^2 \Delta y$. Similarly, the two-dimensional wave equation, which Euler introduced in a 1764 study of the vibrating circular membrane (Figure 4), has the form $z_{tt} = c^2 \Delta z$.

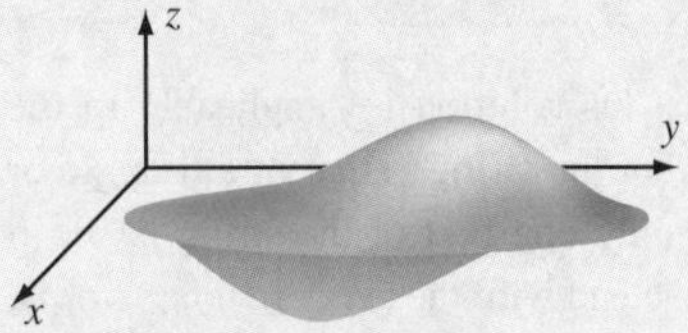

Figure 4
A vibrating circular membrane

Although the expression Δu first appeared in a 1752 paper of Euler that concerned fluid flow, Pierre-Simon Laplace (1749–1827) used the equation $\Delta u = 0$ so extensively in his work on celestial mechanics that $\triangle u$ is now called the *Laplacian* of u. One partial differential equation involving the Laplacian has had a particularly important impact on mathematical analysis: the heat equation $\triangle u = ku_t$. This equation, so named because it describes the conduction of heat, was derived by Jean Baptiste Joseph Fourier (1768–1830) in 1807. Although Fourier's method of solution did not gain immediate acceptance, he undertook a more thorough exposition, titled *Théorie Analytique de la Chaleur,* which he published in 1822.

When Fourier's book came out, the controversy over the vibrating string was still very much alive, even though the original participants in the dispute had died 39 years earlier. Fourier realized that his method could finally "resolve all the difficulties that the analysis employed by Daniel Bernoulli presented." To that end, consider a wave $y(x, t) = f(x - ct)$ such that $y(0, t) = y(L, t) = 0$ for all t. We have, on the one hand, $y(x, 0) = f(x)$. On the other hand, Bernoulli's solution tells us that we can express $y(x, 0)$ by substituting $t = 0$ into the right side of equation (13.29). This leads to the equation

$$f(x) = \sum_{n=1}^{\infty} a_n \sin\left(\frac{n\pi x}{L}\right). \qquad \textbf{(13.30)}$$

Before Fourier's work on heat conduction, most mathematicians did not believe that a general function could admit such an expansion as a trigonometric series. Fourier was convinced that such skepticism would vanish once a concrete relationship between f and the coefficients a_n was made. As he phrased the matter, "Geometers only admit that which they cannot dispute. Of all the derivations of Bernoulli's solution, the most complete is that which consists of actually resolving an arbitrary function into such a series [as in equation (13.30)] and assigning the values of the coefficients." The methods Fourier brought to bear on the heat equation were exactly the right tools for determining the coefficients a_n in equation (13.30):

$$a_n = \frac{2}{L}\int_0^L f(x)\sin\left(\frac{n\pi x}{L}\right)\,dx. \qquad \textbf{(13.31)}$$

Fourier emphasized that his solution is "applicable to the case where the initial figure of the string is that of a triangle or a trapezoid or is such that only one part of the string is set in motion while the other parts blend with the axis." In other words, he asserted that his method remains valid for functions that are defined by different analytic expressions on different parts of their domains. Figure 5 shows the graph of a function f on the interval $[0, 10]$. This function is precisely of the type that Fourier mentioned: partly trapezoidal and partly "blended" with the axis. The trigonometric polynomial $T_5(x) = \sum_{n=1}^{5} a_n \sin(n\pi x/10)$,

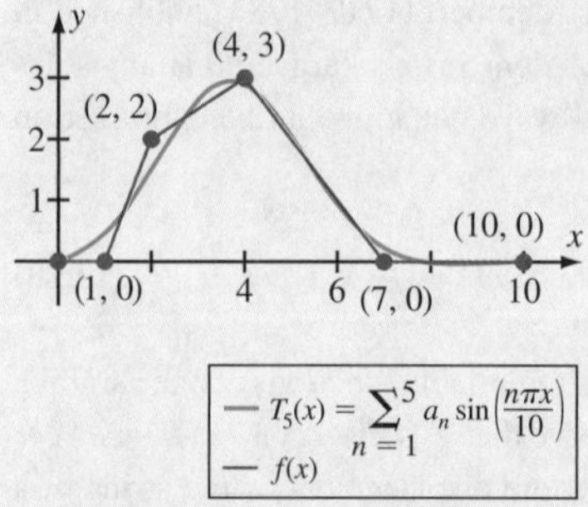

Figure 5

with coefficients specified by equation (13.31), is also shown in Figure 5. Notice that $T_5(x)$ already captures the general shape of the graph of f even though it is composed of a small number of terms. There would be little point displaying an analogous superposition of the graphs of $f(x)$ and $T_{100}(x) = \sum_{n=1}^{100} a_n \sin(n\pi x/10)$, because no differences would be discernible. Indeed, look at Figure 6, which shows the plot of the error function $y = f(x) - T_{100}(x)$. The graph suggests that the convergence of the trigonometric series is slowest at the points $x = 1$, $x = 2$, $x = 4$, and $x = 7$. These are the points at which f is not differentiable. Even so, the maximum error is only about 0.02.

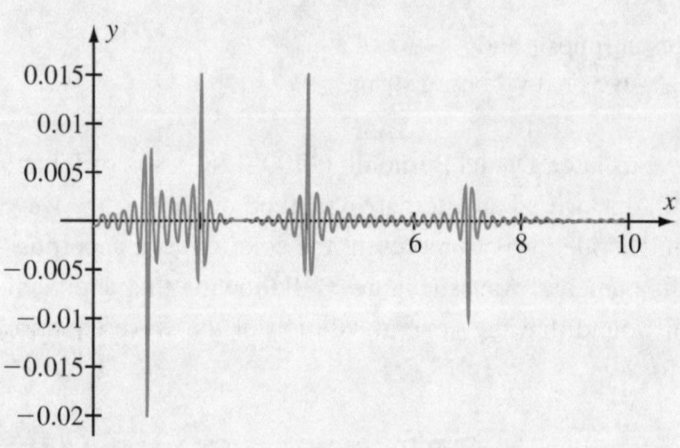

Figure 6

The repercussions of Fourier's work touched nearly every aspect of mathematical analysis. The modern definition of a function as a correspondence of real numbers (as opposed to an analytic expression) ensued in large measure from Fourier's influence. The introduction of Fourier series prompted Niels Henrik Abel (1802–1829) and Peter Gustav Lejeune Dirichlet (1805–1859) to clarify the meaning of convergence of infinite series. By 1850, Fourier series were being used in number theory, far afield from mathematical physics. Since many natural functions of number theory are considerably more irregular than the garden-variety functions of physics, Georg Friedrich Bernhard Riemann (1826–1866) thought it prudent to examine the meaning of equation (13.31) in cases where f has numerous discontinuities. The Riemann integral was the result of his investigations. In 1871, Georg Cantor (1845–1918) studied the uniqueness of representation by trigonometric series. The question he sought to answer was, For what sets X is it true that

$$\sum_{n=1}^{\infty} a_n \sin\left(\frac{n\pi x}{L}\right) = \sum_{n=1}^{\infty} b_n \sin\left(\frac{n\pi x}{L}\right), \quad x \in X,$$

implies $a_n = b_n$ for all n? It was through these studies that Cantor came to create the subject of set theory.

14

Multiple Integrals

PREVIEW

In Chapter 13, we learned how to differentiate functions of two or more variables. Now we will learn how to integrate such functions. To be specific, given a region $\mathcal{R}$ of the xy-plane and a continuous function f defined on $\mathcal{R}$, we will define the integral of f over $\mathcal{R}$. This integral is denoted by $\iint_{\mathcal{R}} f(x, y)\, dA$ and is called a *double integral*. Analogously, if F is a continuous function of three variables that is defined on a solid region $\mathcal{U}$ in xyz-space, then we will define the *triple integral* $\iiint_{\mathcal{U}} F(x, y, z)\, dV$. The main theorem of this chapter tells us that multiple integrals can be evaluated by iterating the familiar process of integration, one variable at a time.

In Chapter 8, we studied certain physical applications of the integral: calculating centers of mass of planar regions, volumes of solids of revolution, and so on. Because we had only single integrals available to us at the time, the scope of our discussion was necessarily limited. For example, when we calculated volumes, we treated only solids with symmetries. We exploited the symmetry to express volume by means of a one-variable integral. In this chapter, we will take a second look at some of the physical interpretations of the integral. As we will see, double and triple integrals allow us to handle these concepts in greater generality.

14.1 Double Integrals over Rectangular Regions

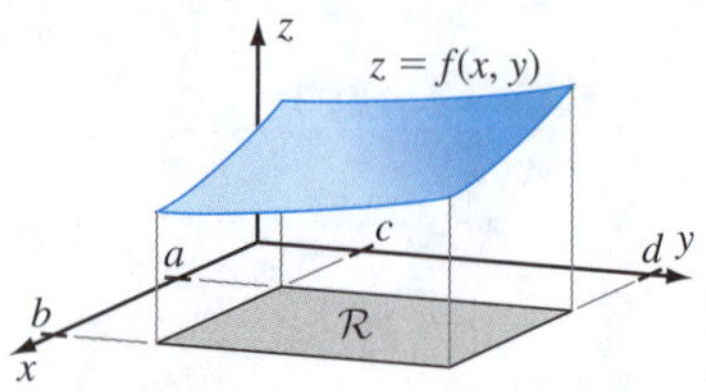

Figure 1

Suppose that $(x, y) \mapsto f(x, y)$ is a positive, continuous function whose domain contains the rectangle $\mathcal{R} = \{(x, y) : a \le x \le b, c \le y \le d\}$. How should we calculate the volume of the solid that lies under the graph of f and above rectangle $\mathcal{R}$? See Figure 1.

We begin by partitioning the domain of the function. A convenient way to do this is to partition the domain of the x-variable:

$$a = x_0 < x_1 < x_2 < \cdots < x_{N-1} < x_N = b;$$

and also to partition the domain of the y-variable:

$$c = y_0 < y_1 < y_2 < \cdots < y_{N-1} < y_N = d.$$

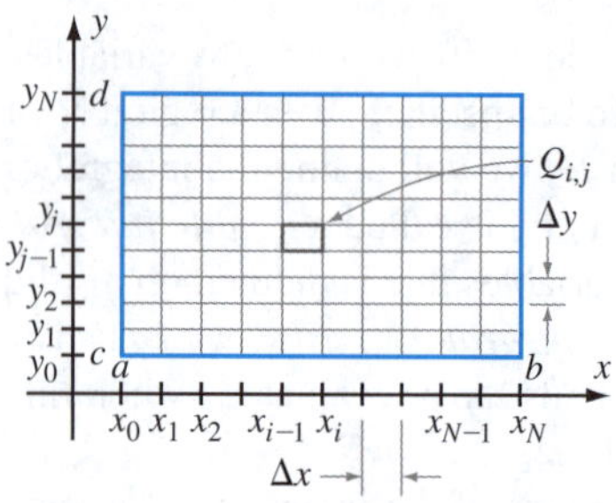

Figure 2

For simplicity, we use only uniform partitions with the same number N of intervals in both the x-direction and the y-direction. This partition breaks up the rectangle $\mathcal{R}$ into smaller rectangles, as shown in Figure 2—that is, to each pair x_{i-1}, x_i in the x-partition and each pair y_{j-1}, y_j in the y-partition, there corresponds a subrectangle $Q_{i,j}$. The side lengths of $Q_{i,j}$ are $\Delta x = x_i - x_{i-1} = (b-a)/N$ and $\Delta y = y_j - y_{j-1} = (d-c)/N$.

Over each subrectangle $Q_{i,j}$, we erect a box with height $f(\xi_{i,j})$ where $\xi_{i,j}$ is some point in $Q_{i,j}$ (see Figure 3). This box has volume

$$V_{i,j} = f(\xi_{i,j})\Delta x \Delta y = f(\xi_{i,j})\Delta A.$$

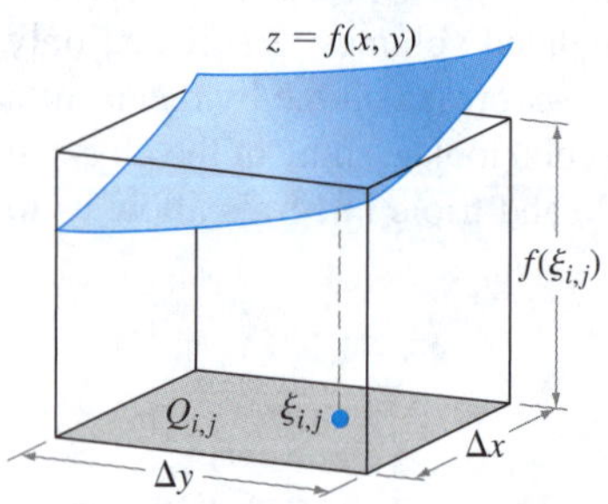

Figure 3
The volume $V_{i,j} = f(\xi_{i,j})\Delta x \Delta y$ of the box of height $f(\xi_{i,j})$ and base area $\Delta A = \Delta x \Delta y$ approximates the volume of the solid under the graph of $z = f(x, y)$ and over the subrectangle $Q_{i,j}$.

Here the symbol ΔA denotes the increment $\Delta x \Delta y$ of (planar) area. The volume V of the solid under the graph of $z = f(x, y)$ and over rectangle $\mathcal{R}$ is then approximated by the sum of the volumes of these boxes, that is,

$$V \approx \sum_{i=1}^{N}\sum_{j=1}^{N} V_{i,j} = \sum_{i=1}^{N}\sum_{j=1}^{N} f(\xi_{i,j})\Delta A. \qquad \textbf{(14.1)}$$

Approximation (14.1) is illustrated for $N = 2$ in Figure 4a and for $N = 4$ in Figure 4b. In Figure 4c, with $N = 16$, we have omitted the graph of $z = f(x, y)$. To see how well the tops of the 16^2 boxes approximate the graph of f, compare Figure 4c with Figure 1.

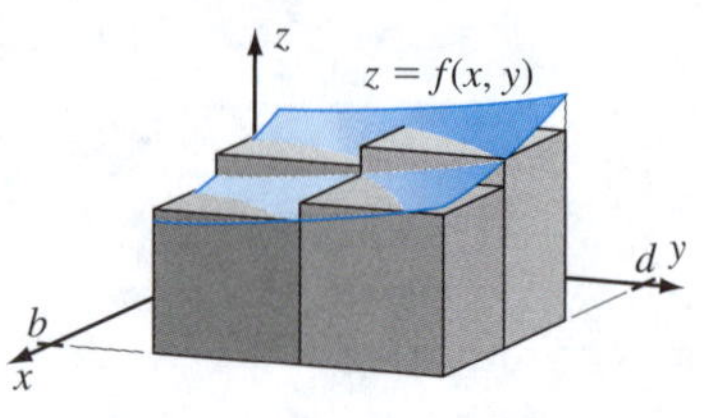

Figure 4a

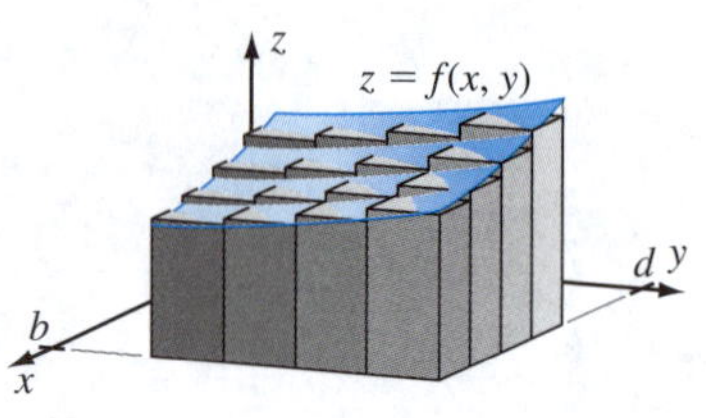

Figure 4b

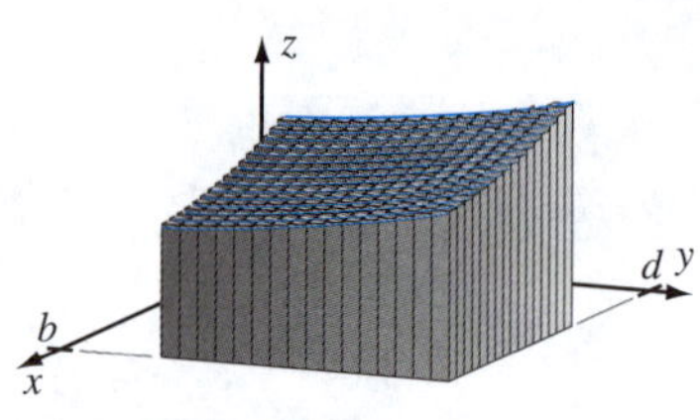

Figure 4c

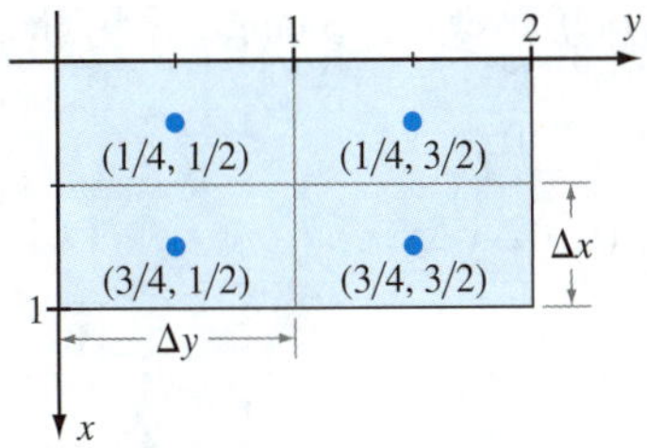

Figure 5
The rectangle [0, 1] × [0, 2] with two equal subdivisions of each side: The midpoints of the subrectangles constitute the choice of points $\{\xi_{i,j}\}$.

Example 1 Use approximation (14.1) with $N = 2$ to estimate the volume of the solid that lies under the surface $z = 6 - x^2 - y^2$ and above the rectangle $[0, 1] \times [0, 2]$ in the xy-plane. Use the midpoint of each subrectangle for the choice of points $\{\xi_{i,j}\}$.

Solution Figure 5 illustrates the four subrectangles that arise from the uniform partitions of the x-interval [0, 1] and y-interval [0, 2] with $N = 2$. It also shows the midpoints of these rectangles. The increment of area is $\Delta A = \Delta x \Delta y = (1)(1/2) = 1/2$. We use $f(x, y) = 6 - x^2 - y^2$ in approximation (14.1) to obtain

$$V \approx \left(6 - \left(\frac{1}{4}\right)^2 - \left(\frac{1}{2}\right)^2\right)\left(\frac{1}{2}\right) + \left(6 - \left(\frac{1}{4}\right)^2 - \left(\frac{3}{2}\right)^2\right)\left(\frac{1}{2}\right)$$
$$+ \left(6 - \left(\frac{3}{4}\right)^2 - \left(\frac{1}{2}\right)^2\right)\left(\frac{1}{2}\right) + \left(6 - \left(\frac{3}{4}\right)^2 - \left(\frac{3}{2}\right)^2\right)\left(\frac{1}{2}\right) = \frac{71}{8}.$$

Figure 6a illustrates the solid, while Figure 6b shows the four approximating boxes. ■

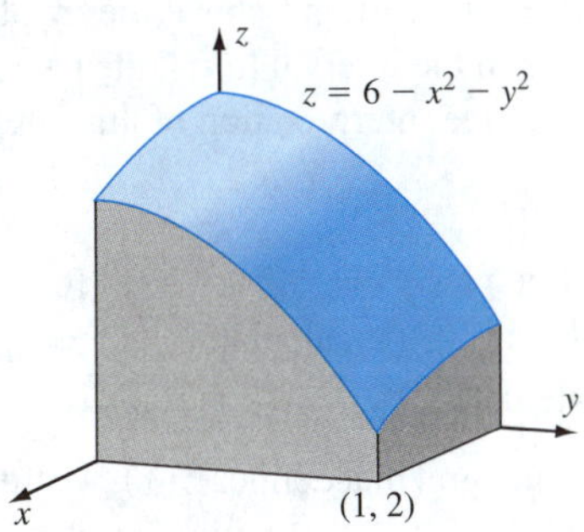

Figure 6a

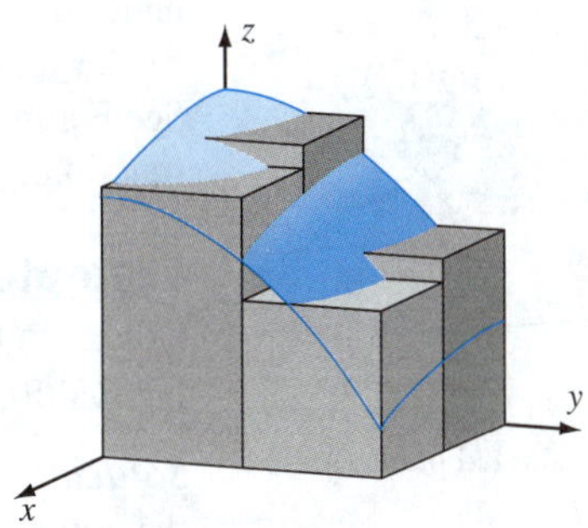

Figure 6b

We call the double sum in approximation (14.1) a *Riemann sum*. As we have seen, these sums can be used to estimate the volume under the graph of f when f is a positive function. However, approximation (14.1) makes perfect sense for any continuous function. Furthermore, it can be shown that the Riemann sums of any continuous function f tend to a limit as N tends to infinity. We may therefore make the following definition.

Definition Let f be a continuous function that is defined on the rectangle $\mathcal{R} = \{(x, y) : a \le x \le b, c \le y \le d\}$. For each positive integer N, let $\{x_0, x_1, \ldots, x_N\}$ and $\{y_0, y_1, \ldots, y_N\}$ be uniform partitions of the intervals $[a, b]$ and $[c, d]$, respectively. Set $\Delta x = (b - a)/N$ and $\Delta y = (d - c)/N$. Let $\xi_{i,j}$ be any point in subrectangle $Q_{i,j} = [x_{i-1}, x_i] \times [y_{j-1}, y_j]$. Then the *double integral* of f over $\mathcal{R}$ is defined to be the limit of the Riemann sums

$$\sum_{i=1}^{N}\sum_{j=1}^{N} f(\xi_{i,j})\Delta x \Delta y = \sum_{i=1}^{N}\sum_{j=1}^{N} f(\xi_{i,j})\Delta A$$

as N tends to infinity. We denote the double integral of f over the rectangle $\mathcal{R}$ by the symbol $\iint_{\mathcal{R}} f(x, y)\, dA$. Thus,

$$\iint_{\mathcal{R}} f(x, y)\, dA = \lim_{N\to\infty} \sum_{i=1}^{N}\sum_{j=1}^{N} f(\xi_{i,j})\Delta x \Delta y. \qquad \textbf{(14.2)}$$

Iterated Integrals

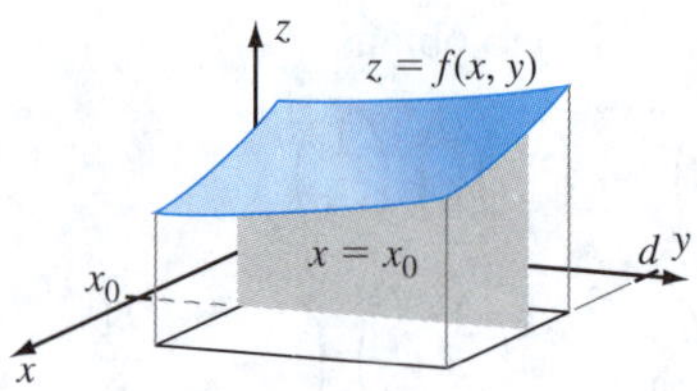

Figure 7
The area of the shaded planar region is $\int_c^d f(x_0, y)\, dy$.

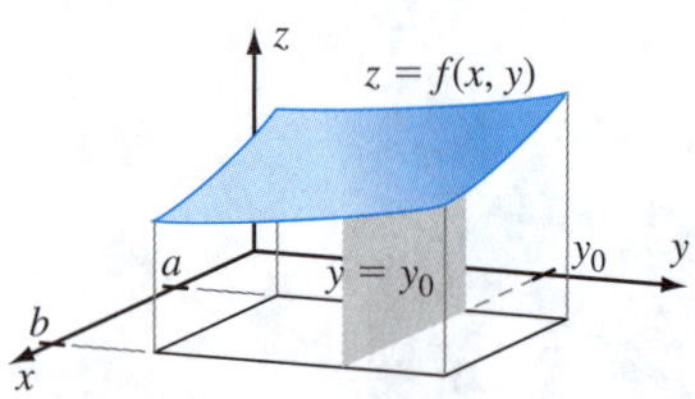

Figure 8
The area of the shaded planar region is $\int_a^b f(x, y_0)\, dx$.

Now that we understand what quantity the double integral $\iint_{\mathcal{R}} f(x, y)\, dA$ signifies, we need to know how to calculate it. The technique is to integrate the integrand $f(x, y)$ one variable at a time. The idea is similar to partial differentiation: One variable of $f(x, y)$ is held constant while we apply an operation of calculus to the other variable. We continue to suppose that f is continuous on the rectangle $\mathcal{R} = \{(x, y) : a \le x \le b, c \le y \le d\} = [a, b] \times [c, d]$. If we fix the value $x = x_0$ in the interval $[a, b]$, then we can form the integral $\int_c^d f(x_0, y)\, dy$. If $y \mapsto f(x_0, y)$ happens to be positive, then the integral $\int_c^d f(x_0, y)\, dy$ represents the area of the region in the $x = x_0$ plane that lies under the slice $z = f(x_0, y)$ and above the xy-plane (see Figure 7). Similarly, if we fix the value $y = y_0$ in the interval $[c, d]$, then we can form the integral $\int_a^b f(x, y_0)\, dx$. See Figure 8 for the area interpretation of this integral when $x \mapsto f(x, y_0)$ is a positive function.

Example 2 Let $f(x, y) = 6 - x^2 - y^2$ for $0 \le x \le 1$ and $0 \le y \le 2$. For each fixed x in the interval $[0, 1]$, calculate $\mathcal{I}(x) = \int_0^2 f(x, y)\, dy$. For each fixed y in the interval $[0, 2]$, calculate $\mathcal{J}(y) = \int_0^1 f(x, y)\, dx$.

Solution In the integral that defines $\mathcal{I}(x)$, we treat the x-variable as a constant. Thus, the antiderivative of $6 - x^2 - y^2$ with respect to the y-variable is $(6 - x^2)y - y^3/3$. Therefore, we have

$$\mathcal{I}(x) = \int_0^2 (6 - x^2 - y^2)\, dy = \left((6 - x^2)y - \frac{y^3}{3}\right)\Bigg|_{y=0}^{y=2} = 2(6 - x^2) - \frac{8}{3} = \frac{28}{3} - 2x^2.$$

Similarly, in the integral that defines $\mathcal{J}(y)$, we treat the y-variable as a constant. Thus, the antiderivative of $6 - x^2 - y^2$ with respect to the x-variable is $(6 - y^2)x - x^3/3$. Therefore, we have

$$\mathcal{J}(y) = \int_0^1 (6 - x^2 - y^2)\, dx = \left((6 - y^2)x - \frac{x^3}{3}\right)\Bigg|_{x=0}^{x=1} = (6 - y^2) - \frac{1}{3} = \frac{17}{3} - y^2.$$

Using the continuity of f, it can be shown that the functions $\mathcal{I}(x) = \int_c^d f(x, y)\, dy$ and $\mathcal{J}(y) = \int_a^b f(x, y)\, dx$ are continuous for $a \le x \le b$ and $c \le y \le d$, respectively. In particular, each of these two functions can be integrated over its domain. Thus, the quantities

$$\int_a^b \left(\int_c^d f(x, y)\, dy\right) dx = \int_a^b \mathcal{I}(x)\, dx$$

and

$$\int_c^d \left(\int_a^b f(x, y)\, dx \right) dy = \int_c^d \mathcal{J}(y)\, dy$$

are well-defined.

Definition The expressions $\int_a^b (\int_c^d f(x, y)\, dy)\, dx$ and $\int_c^d (\int_a^b f(x, y)\, dx)\, dy$ are called *iterated integrals*.

inSIGHT In Example 3, the integrands of the integrals $\int_0^1 \mathcal{I}(x)\, dx$ and $\int_0^2 \mathcal{J}(y)\, dy$ differ, as do the intervals of integration. Nevertheless, Example 3 demonstrates that the two integrals have the same value. In other words, the two iterated integrals $\int_0^1 (\int_0^2 f(x, y)\, dy)\, dx$ and $\int_0^2 (\int_0^1 f(x, y)\, dx)\, dy$ are equal. In the next subsection, we will see that these expressions are equal because they both represent the volume of the solid that lies under the graph of f and over the rectangle $[0, 1] \times [0, 2]$.

In the iterated integral $\int_a^b (\int_c^d f(x, y)\, dy)\, dx$, we first calculate the inner (single) integral $\int_c^d f(x, y)\, dy$. For this computation, we treat the variable x as if it were a constant. Since $\int_c^d f(x, y)\, dy$ is a definite integral in the y-variable, its value does not depend on y, but it *does* depend on x. We integrate this function of x to complete the calculation of the interated integral. We handle the integral $\int_c^d (\int_a^b f(x, y)\, dx)\, dy$ in an analogous way.

Example 3 Let $f(x, y) = 6 - x^2 - y^2$ for $0 \le x \le 1$ and $0 \le y \le 2$, as in Example 2. Calculate the iterated integrals $\int_0^1 (\int_0^2 f(x, y)\, dy)\, dx$ and $\int_0^2 (\int_0^1 f(x, y)\, dx)\, dy$.

Solution Let $\mathcal{I}(x) = \int_0^2 f(x, y)\, dy$ and $\mathcal{J}(y) = \int_0^1 f(x, y)\, dx$. In Example 2, we showed that $\mathcal{I}(x) = 28/3 - 2x^2$ and $\mathcal{J}(y) = 17/3 - y^2$. Therefore,

$$\int_0^1 \left(\int_0^2 f(x, y)\, dy \right) dx = \int_0^1 \mathcal{I}(x)\, dx = \int_0^1 \left(\frac{28}{3} - 2x^2 \right) dx$$

$$= \left(\frac{28}{3}x - \frac{2}{3}x^3 \right)\Big|_0^1 = \frac{26}{3}$$

and

$$\int_0^2 \left(\int_0^1 f(x, y)\, dx \right) dy = \int_0^2 \mathcal{J}(y)\, dy = \int_0^2 \left(\frac{17}{3} - y^2 \right) dy$$

$$= \left(\frac{17}{3}y - \frac{y^3}{3} \right)\Big|_0^2 = \frac{26}{3}. \quad \blacksquare$$

Using Iterated Integrals to Calculate Double Integrals

We will now show that when f is continuous and positive, the iterated integrals $\int_a^b (\int_c^d f(x, y)\, dy)\, dx$ and $\int_c^d (\int_a^b f(x, y)\, dx)\, dy$ both equal the volume of the solid $\mathcal{U}$ that lies under the graph of f and above the rectangle $\mathcal{R} = [a, b] \times [c, d]$. Let $\mathcal{I}(x) = \int_c^d f(x, y)\, dy$ and $\mathcal{J}(y) = \int_a^b f(x, y)\, dx$. Partition the interval $[a, b]$ into N equal subintervals by means of the points $a = x_0, x_1, \ldots, x_{N-1}, x_N = b$. As before, we set $\Delta x = (b - a)/N$. Figure 9 shows that when Δx is small, the quantity $(\int_c^d f(x_i, y)\, dy)\Delta x$ is a good approximation to the volume of the section of $\mathcal{U}$ that lies between the planes $x = x_{i-1}$ and $x = x_i$. It follows that the sum $\sum_{i=1}^N (\int_c^d f(x_i, y)\, dy)\Delta x$

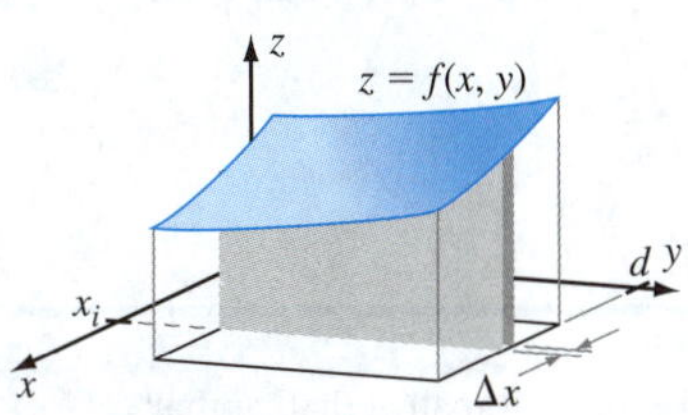

Figure 9
The area of the shaded solid is approximately $(\int_c^d f(x_i, y)\, dy)\Delta x$.

is a good approximation to the volume of $\mathcal{U}$ when Δx is small. Therefore,

$$\lim_{N\to\infty}\sum_{i=1}^{N}\left(\int_c^d f(x_i, y)\,dy\right)\Delta x = \iint_{\mathcal{R}} f(x, y)\,dA. \tag{14.3}$$

On the other hand, the sum $\sum_{i=1}^{N}(\int_c^d f(x_i, y)\,dy)\Delta x$ is a Riemann sum for the Riemann integral $\int_a^b \mathcal{I}(x)\,dx$. Therefore,

$$\lim_{N\to\infty}\sum_{i=1}^{N}\left(\int_c^d f(x_i, y)\,dy\right)\Delta x = \int_a^b \mathcal{I}(x)\,dx = \int_a^b\left(\int_c^d f(x, y)\,dy\right)dx. \tag{14.4}$$

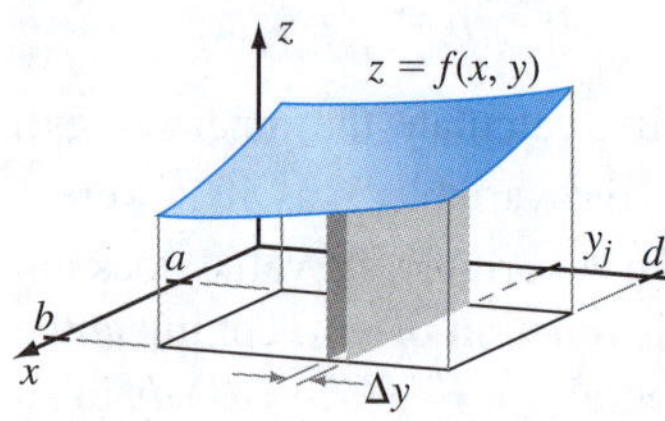

Figure 10
The area of the shaded solid is approximately $(\int_a^b f(x, y_j)\,dx)\Delta y$.

From equations (14.3) and (14.4), we conclude that $\iint_{\mathcal{R}} f(x, y)\,dA = \int_a^b(\int_c^d f(x, y)\,dy)\,dx$.

Figure 10 reveals an analogous situation for the other iterated integral. If we partition the interval $[c, d]$ into N equal subintervals by means of the points $c = y_0, y_1, \ldots, y_{N-1}, y_N = d$, and set $\Delta y = (d - c)/N$, then $(\int_a^b f(x, y_j)\,dx)\Delta y$ is a good approximation to the volume of the section of $\mathcal{U}$ that lies between the planes $y = y_{j-1}$ and $y = y_j$. It follows that the sum $\sum_{j=1}^{N}(\int_a^b f(x, y_j)\,dx)\Delta y$ is a good approximation to the volume of $\mathcal{U}$, and it is also a Riemann sum for $\int_c^d \mathcal{J}(y)\,dy$. Letting N tend to infinity, we obtain

$$\lim_{N\to\infty}\sum_{j=1}^{N}\left(\int_a^b f(x, y_j)\,dx\right)\Delta y = \iint_{\mathcal{R}} f(x, y)\,dA$$

and

$$\lim_{N\to\infty}\sum_{j=1}^{N}\left(\int_a^b f(x, y_j)\,dx\right)\Delta y = \int_c^d \mathcal{J}(y)\,dy = \int_c^d\left(\int_a^b f(x, y)\,dx\right)dy.$$

It follows that $\iint_{\mathcal{R}} f(x, y)\,dA = \int_c^d(\int_a^b f(x, y)\,dx)\,dy$.

Although our discussion has assumed that f is positive, the equalities we have obtained are valid for any continuous function, as our next theorem asserts.

Theorem 1 Let f be continuous on the rectangle $\mathcal{R} = \{(x, y) : a \le x \le b, c \le y \le d\} = [a, b] \times [c, d]$. Then the double integral $\iint_{\mathcal{R}} f(x, y)\,dA$ exists and

$$\iint_{\mathcal{R}} f(x, y)\,dA = \int_a^b\left(\int_c^d f(x, y)\,dy\right)dx = \int_c^d\left(\int_a^b f(x, y)\,dx\right)dy.$$

inSIGHT

If you think back to the Method of Disks in Section 8.1, you will realize that, at that time, we were using a similar idea. We calculated the areas of slices in the y-variable and then integrated out in the x-variable, or vice versa.

Example 4 Use Theorem 1 to calculate the integral of the function $f(x, y) = \cos(x)\sin(y)$ over the rectangle $\mathcal{R} = \{(x, y) : \pi/6 \leq x \leq \pi/2, \pi/4 \leq y \leq \pi/3\}$.

Solution By Theorem 1,

$$\iint_{\mathcal{R}} f(x, y)\, dA = \int_{\pi/6}^{\pi/2} \left(\int_{\pi/4}^{\pi/3} \cos(x)\sin(y)\, dy \right) dx.$$

We evaluate an iterated integral by working from the *inside out*. Notice that the inner integral is an integral in the y-variable. Just as in the theory of partial differentiation, we treat the x-variable as a constant. Thus, the antiderivative of $\cos(x) \cdot \sin(y)$ *with respect to the y-variable* is $\cos(x) \cdot (-\cos(y))$. We have

$$\begin{aligned}\int_{\pi/4}^{\pi/3} \cos(x)\sin(y)\, dy &= \cos(x)(-\cos(y))\Big|_{y=\pi/4}^{y=\pi/3} \\ &= \cos(x)\left(-\cos\left(\frac{\pi}{3}\right)\right) - \cos(x)\left(-\cos\left(\frac{\pi}{4}\right)\right) \\ &= \left(\frac{\sqrt{2}-1}{2}\right)\cos(x).\end{aligned}$$

To finish the calculation of the given double integral, we now do a straightforward integral in one variable. The result is

$$\begin{aligned}\iint_{\mathcal{R}} f(x, y)\, dA &= \int_{\pi/6}^{\pi/2} \left(\left(\frac{\sqrt{2}-1}{2}\right)\cos(x) \right) dx \\ &= \left(\frac{\sqrt{2}-1}{2}\right)\sin(x)\Bigg|_{x=\pi/6}^{x=\pi/2} \\ &= \left(\frac{\sqrt{2}-1}{2}\right)\cdot 1 - \left(\frac{\sqrt{2}-1}{2}\right)\cdot\frac{1}{2} \\ &= \frac{\sqrt{2}-1}{4}.\end{aligned}$$

■

Example 5 Integrate the function $f(x, y) = x^2y - y^3 + 2x$ over the rectangle $\mathcal{R} = \{(x, y) : 1 \leq x \leq 3, -1 \leq y \leq 0\}$.

Solution According to Theorem 1, we have

$$\iint_{\mathcal{R}} f(x, y)\, dA = \int_{-1}^{0} \left(\int_{1}^{3} (x^2y - y^3 + 2x)\, dx \right) dy.$$

We evaluate the inside integral, in the x-variable, first. When calculating the antiderivative in the x-variable, we treat y as a constant. Thus,

$$\begin{aligned}\int_{1}^{3} (x^2y - y^3 + 2x)\, dx &= \left(y\frac{x^3}{3} - y^3x + x^2 \right)\Bigg|_{x=1}^{x=3} \\ &= (9y - 3y^3 + 9) - \left(\frac{1}{3}y - y^3 + 1\right) = \frac{26}{3}y - 2y^3 + 8.\end{aligned}$$

inSIGHT

Theorem 1 gives us a choice. We can calculate $\iint_{\mathcal{R}} f(x, y)\, dA$ by integrating first in the y-variable, then in the x-variable, as in Example 4. Alternatively, we can integrate first with respect to x and then with respect to y, as in Example 5. Sometimes one order is easier, or more effective, than the other.

Therefore,

$$\iint_{\mathcal{R}} f(x, y)\, dA = \int_{-1}^{0} \left(\frac{26}{3}y - 2y^3 + 8\right) dy = \left(\frac{13}{3}y^2 - \frac{1}{2}y^4 + 8y\right)\Bigg|_{y=-1}^{y=0}$$
$$= 0 - \left(\frac{13}{3} - \frac{1}{2} - 8\right) = \frac{25}{6}.$$

Example 6 Consider the integral $\iint_{\mathcal{R}} x\cos(xy)\, dA$, where $\mathcal{R} = \{(x, y) : 0 \le x \le 1, 0 \le y \le \pi\}$. Use Theorem 1 to convert this to an iterated integral. Evaluate the iterated integral by performing the y-integration first.

Solution The antiderivative of $x\cos(xy)$ with respect to y is $\sin(xy)$. Our integral is therefore

$$\iint_{\mathcal{R}} x\cos(xy)\, dA = \int_0^1 \left(\int_0^{\pi} x\cos(xy)\, dy\right) dx = \int_0^1 \left(\sin(xy)\Big|_{y=0}^{y=\pi}\right) dx$$
$$= \int_0^1 (\sin(\pi x) - \sin(0))\, dx = -\frac{1}{\pi}\cos(\pi x)\Bigg|_{x=0}^{x=1}$$
$$= -\frac{1}{\pi}(-1 - 1) = \frac{2}{\pi}.$$

inSIGHT

In Example 6, we could have integrated with respect to x first. However, then we would have had to use the Method of Integration by Parts. Since we chose to do the y-integration first, our calculation was much simpler.

quickquiz

1. What is a double integral over a rectangle?
2. The double sum

$$\sum_{i=1}^{100}\sum_{j=1}^{100}\left(3 + \frac{i}{50}\right)\left(4 + \frac{j}{100}\right)^2\left(\frac{1}{50}\right)\left(\frac{1}{100}\right)$$

is a Riemann sum for the double integral $\iint_{\mathcal{R}}(xy^2)\, dA$ over what rectangle?
3. What is an iterated integral, and how is the concept related to that of a double integral?
4. Evaluate $\iint_{\mathcal{R}}(1 + x^2y)\, dA$, where $\mathcal{R} = \{(x, y) : 1 \le x \le 2, 0 \le y \le 1\}$.

EXERCISES

Problems for Practice

In Exercises 1–4, approximate the volume of the solid that lies under the graph of the given function f and above the given rectangle R in the xy-plane. Use a Riemann sum with $N = 2$ and, for the choice of points $\{\xi_{i,j}\}$, the midpoints of the four subrectangles of R.

1. $f(x, y) = x + 2y$, $R = [0, 2] \times [1, 5]$
2. $f(x, y) = 1 + 6xy^2$, $R = [0, 1] \times [1, 2]$
3. $f(x, y) = 3x^2 + 2y$, $R = [-1, 3] \times [-2, 2]$
4. $f(x, y) = y/x$, $R = [3, 7] \times [2, 4]$

In Exercises 5–10, calculate $\mathcal{I}(x) = \int_c^d f(x, y)\,dy$ and $\mathcal{J}(y) = \int_a^b f(x, y)\,dx$ for the function f and the rectangle $R = [a, b] \times [c, d]$ in the xy-plane.

5. $f(x, y) = x + 2y$, $R = [0, 2] \times [1, 5]$
6. $f(x, y) = 1 + 6xy^2$, $R = [0, 1] \times [1, 2]$
7. $f(x, y) = 3x^2 + 2y$, $R = [-1, 3] \times [-2, 2]$
8. $f(x, y) = y/x$, $R = [3, 7] \times [2, 4]$
9. $f(x, y) = y\exp(xy^2)$, $R = [0, 1] \times [0, 1]$
10. $f(x, y) = \sin(x + 2y)$, $R = [0, 2\pi] \times [0, \pi]$

Evaluate each double integral in Exercises 11–26 by converting it into an iterated integral.

11. $\iint_{\mathcal{R}}(x^2 - y)\,dA$,
$\mathcal{R} = \{(x, y) : -2 \le x \le 5, 1 \le y \le 4\}$
12. $\iint_{\mathcal{R}}(\cos(x) - \sin(y))\,dA$,
$\mathcal{R} = \{(x, y) : 0 \le x \le \pi/2, -\pi/3 \le y \le \pi\}$
13. $\iint_{\mathcal{R}} x\cos(y)\,dA$,
$\mathcal{R} = \{(x, y) : 0 \le x \le \pi, 0 \le y \le \pi\}$
14. $\iint_{\mathcal{R}}(x + 2y)\,dA$,
$\mathcal{R} = \{(x, y) : 0 \le x \le 4, -2 \le y \le 0\}$
15. $\iint_{\mathcal{R}} \exp(x - y)\,dA$,
$\mathcal{R} = \{(x, y) : 0 \le x \le 1, -2 \le y \le 0\}$
16. $\iint_{\mathcal{R}} \sin(x)\tan(y)\,dA$,
$\mathcal{R} = \{(x, y) : 0 \le x \le \pi/4, -\pi/3 \le y \le 0\}$
17. $\iint_{\mathcal{R}}(\cos(x)/y - \cos(y)/x)\,dA$,
$\mathcal{R} = \{(x, y) : \pi \le x \le 2\pi, \pi \le y \le 2\pi\}$
18. $\iint_{\mathcal{R}}(xe^y - ye^x)\,dA$,
$\mathcal{R} = \{(x, y) : 0 \le x \le 1, 1 \le y \le 2\}$
19. $\iint_{\mathcal{R}}(y\sqrt{x} - x\sqrt{y})\,dA$,
$\mathcal{R} = \{(x, y) : 1 \le x \le 4, 1 \le y \le 4\}$
20. $\iint_{\mathcal{R}} x^2/(1 + y^2)\,dA$,
$\mathcal{R} = \{(x, y) : 0 \le x \le 2, 1 \le y \le \sqrt{3}\}$
21. $\iint_{\mathcal{R}}(x/y^2 + y/x^3)\,dA$,
$\mathcal{R} = \{(x, y) : -3 \le x \le -1, -4 \le y \le -2\}$
22. $\iint_{\mathcal{R}} e^x\cos(y)\,dA$,
$\mathcal{R} = \{(x, y) : 0 \le x \le 1, 0 \le y \le \pi/2\}$
23. $\iint_{\mathcal{R}} \cos^2(x)\sin^2(y)\,dA$,
$\mathcal{R} = \{(x, y) : \pi/2 \le x \le \pi, 0 \le y \le \pi\}$
24. $\iint_{\mathcal{R}} y/\sqrt{1 - x^2}\,dA$,
$\mathcal{R} = \{(x, y) : 0 \le x \le 1/2, -1 \le y \le 2\}$
25. $\iint_{\mathcal{R}} \cos^3(x)\sin(y)\,dA$,
$\mathcal{R} = \{(x, y) : 0 \le x \le \pi/2, 0 \le y \le \pi\}$
26. $\iint_{\mathcal{R}} (x + y)^3\,dA$, $\mathcal{R} = \{(x, y) : 1 \le x \le 2, 2 \le y \le 3\}$

Evaluate each double integral in Exercises 27–30 twice by performing iterated integrations in both orders. Remember that your answer should come out the same either way.

27. $\iint_{\mathcal{R}} x^3y^2\,dA$, $\mathcal{R} = \{(x, y) : 0 \le x \le 1, 0 \le y \le 1\}$
28. $\iint_{\mathcal{R}}(x + y)^4 dA$, $\mathcal{R} = \{(x, y) : 0 \le x \le 1, 1 \le y \le 2\}$
29. $\iint_{\mathcal{R}} 1/(x + y)\,dA$,
$\mathcal{R} = \{(x, y) : 0 \le x \le 2, 1 \le y \le e\}$
30. $\iint_{\mathcal{R}} \sin(2x + y)\,dA$,
$\mathcal{R} = \{(x, y) : -\pi/4 \le x \le \pi, \pi/6 \le y \le \pi\}$

Further Theory and Practice

Evaluate each double integral in Exercises 31–34.

31. $\iint_{\mathcal{R}} \ln(xy)\,dA$, $\mathcal{R} = \{(x, y) : 1 \le x \le e, 1 \le y \le e\}$
32. $\iint_{\mathcal{R}} \ln(x^y)\,dA$,
$\mathcal{R} = \{(x, y) : 1 \le x \le e, 1 \le y \le 2e\}$
33. $\iint_{\mathcal{R}} xy\ln(y)\,dA$,
$\mathcal{R} = \{(x, y) : 1 \le x \le e, 1 \le y \le e\}$
34. $\iint_{\mathcal{R}} y^2e^xe^y\,dA$,
$\mathcal{R} = \{(x, y) : -1 \le x \le 1, -2 \le y \le 0\}$

In Exercises 35–38, evaluate the iterated integral.

35. $\int_0^1\left(\int_0^1 xe^{xy}\,dy\right)dx$
36. $\int_0^{1/2}\left(\int_0^{\pi} x\sin(xy)\,dy\right)dx$
37. $\int_1^4\left(\int_1^2 2x^2y\ln(x^2y^2)\,dy\right)dx$
38. $\int_0^{\pi}\left(\int_0^1 xy\cos(xy^2)\,dy\right)dx$

In Exercises 39–42, calculate $\iint_{\mathcal{R}} f(x, y)\, dA$ for function f and rectangle $\mathcal{R}$.

39. $f(x, y) = x^2 \cos(xy)$,
$\mathcal{R} = \{(x, y) : -1 \le x \le 1, -1 \le y \le 1\}$

40. $f(x, y) = y^2 \sin(xy) \cos(xy)$,
$\mathcal{R} = \{(x, y) : 0 \le x \le 1, 0 \le y \le \pi/2\}$

41. $f(x, y) = xy \ln(xy)\, dA$,
$\mathcal{R} = \{(x, y) : 1 \le x \le e, 1 \le y \le e\}$

42. $(x, y) = y \exp(x^2 + y^2)$,
$\mathcal{R} = \{(x, y) : 0 \le x \le 1, -1 \le y \le 1\}$

43. Suppose that ϕ and ψ are continuous functions of one variable on the intervals $[a, b]$ and $[c, d]$, respectively. Show that if $f(x, y) = \phi(x)\psi(y)$, then

$$\int_a^b \left(\int_c^d f(x, y)\, dy \right) dx = \left(\int_a^b \phi(x)\, dx \right) \left(\int_c^d \psi(y)\, dy \right).$$

Let $\mathcal{R}$ denote the rectangle $[0, 1] \times [0, 2]$ in the xy-plane. Given that

$$\int_0^1 x^2(1 - x^2)^{3/2}\, dx = \pi/32$$

and

$$\int_0^2 y^2(4 - y^2)^{1/2}\, dy = \pi,$$

what is $\iint_{\mathcal{R}} (xy)^2 \sqrt{(1 - x^2)^3(4 - y^2)}\, dA$?

44. Suppose that f has continuous partial derivatives on an open set that contains the rectangle $\mathcal{R} = [a, b] \times [c, d]$ in the xy-plane. Evaluate $\iint_{\mathcal{R}} f_{xy}(x, y)\, dA$.

In Exercises 45–48, calculate $\iint_{\mathcal{R}} f(x, y)\, dA$ for the multicase function f and rectangle $\mathcal{R} = [0, 1] \times [0, 1]$ in the xy-plane.

45. $f(x, y) = 0$ if $x \ge y$, and $f(x, y) = 1$ if $x < y$

46. $f(x, y) = 0$ if $x \ge y$, and $f(x, y) = y$ if $x < y$

47. $f(x, y) = x$ if $x \ge y$, and $f(x, y) = y$ if $x < y$

48. $f(x, y) = x - 2y$ if $x \ge 2y$, and $f(x, y) = 0$ if $x < y$

Calculator/Computer Exercises

By using a calculator or computer to implement loops, it is not tedious to estimate a double integral $\iint_{\mathcal{R}} f(x, y)\, dA$ by using approximation (14.1) with a large value of N. Figure 11 illustrates how this can be done using the computer algebra system Maple. In Figure 11, the integrand is given by $f(x, y) = \sqrt{25 - 3x^2 - y^2}$, and the rectangle of integration is $\mathcal{R} = [0, 1] \times [0, 4]$. Each side of the rectangle has been divided into $N = 100$ equal subintervals, and the midpoints of the 10,000 subrectangles have been used for the choice of points $\{\xi_{i,j}\}$. In Exercises 49–52, approximate the double integral $\iint_{\mathcal{R}} f(x, y)\, dA$ for the function f and rectangle $\mathcal{R}$ in the xy-plane. Use $N = 50$, and take the midpoints of the 2500 subrectangles for the choice of points $\{\xi_{i,j}\}$. (In each exercise, the suggested approximation is accurate to at least three decimal places.)

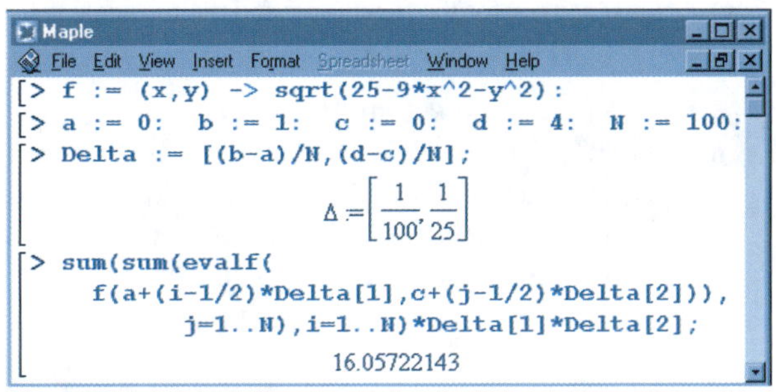

Figure 11

49. $f(x, y) = \cos\left(\sqrt{1 + x} + y\right)$, $\mathcal{R} = [-1, 1] \times [0, 1]$

50. $f(x, y) = \sqrt{1 + x + y^3}$, $\mathcal{R} = [1, 3] \times [0, 1]$

51. $f(x, y) = \exp(-x^2 - y^2)$, $\mathcal{R} = [-1, 1] \times [-1, 1]$

52. $f(x, y) = (1 + x + y)/(1 + x^2 + y^4)$,
$\mathcal{R} = [0, 2] \times [0, 1]$

14.2 Integration over More General Regions

Many integration problems are formulated over planar regions other than rectangles. We need a technique for defining an integral over the interior of a triangle, a circle, an ellipse, or over even more complicated sets in the plane. We will restrict attention to regions bounded by smooth curves. To begin, we need to know how to partition such a set. These are the considerations that we treat in the present section.

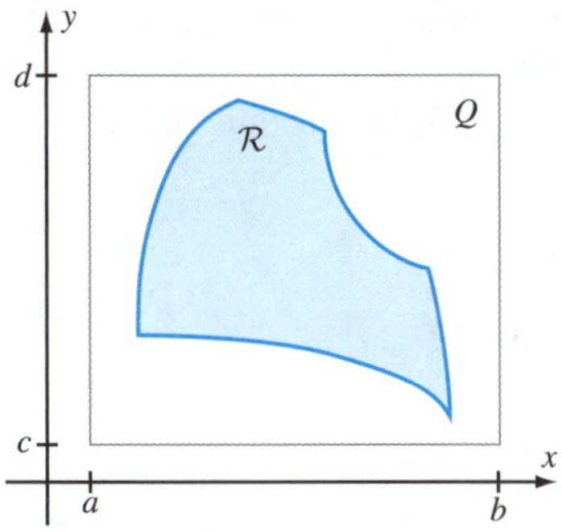

Figure 1

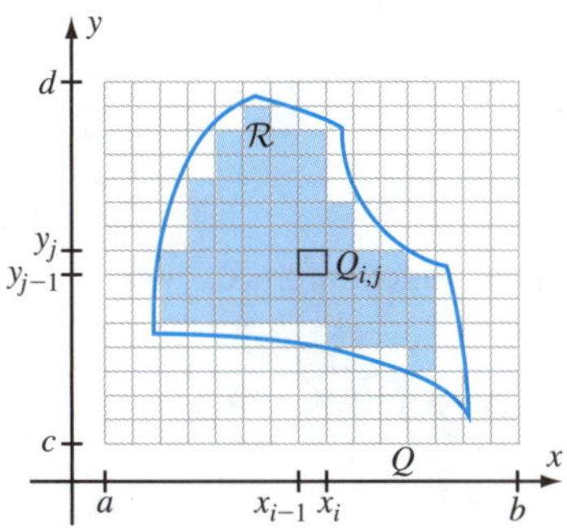

Figure 2

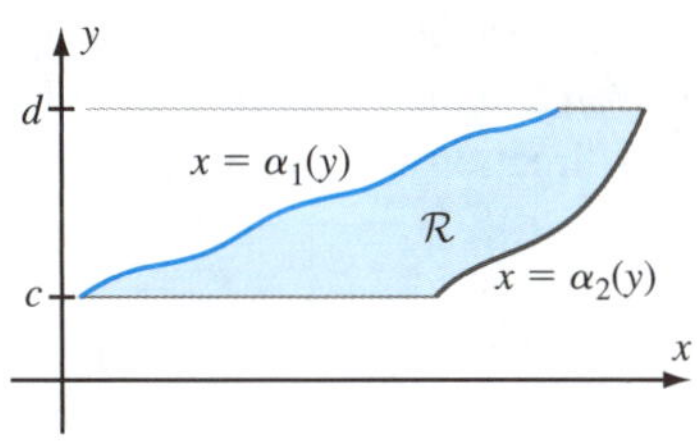

Figure 3
The region $\mathcal{R} = \{(x, y) : c \le y \le d, \alpha_1(y) \le x \le \alpha_2(y)\}$ is x-simple.

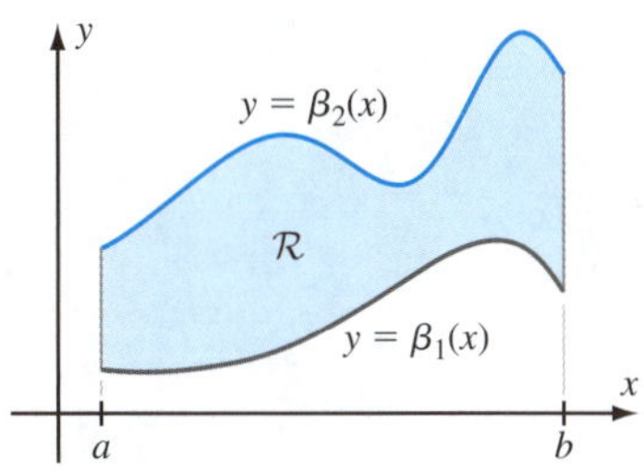

Figure 4
The region $\mathcal{R} = \{(x, y) : a \le x \le b, \beta_1(x) \le y \le \beta_2(x)\}$ is y-simple.

Planar Regions Bounded by Finitely Many Curves

Let $\mathcal{R}$ be a closed region in the plane that is (1) contained in some bounded rectangle Q and (2) bounded by finitely many continuously differentiable curves. Refer to Figure 1. Suppose that $Q = \{(x, y) : a \le x \le b, c \le y \le d\}$. Then we can partition the rectangle Q in the usual way by partitioning the intervals $[a, b]$ and $[c, d]$:

$$a = x_0 < x_1 < x_2 < \cdots < x_{N-1} < x_N = b$$

and

$$c = y_0 < y_1 < y_2 < \cdots < y_{N-1} < y_N = d.$$

For each pair x_{i-1}, x_i in the partition of $[a, b]$ and y_{j-1}, y_j in the partition of $[c, d]$, there is a small subrectangle $Q_{i,j}$, as shown in Figure 2. The only rectangles $Q_{i,j}$ relevant for integration over the region $\mathcal{R}$ are the ones that are completely contained in $\mathcal{R}$. Clearly, the rectangles that lie completely outside $\mathcal{R}$ are irrelevant; the ones that overlap the boundary of $\mathcal{R}$ make a negligible error contribution when the subrectangles of the partition are small.

If f is a continuous function defined on $\mathcal{R}$, then we consider Riemann sums

$$\sum_{Q_{i,j} \subset \mathcal{R}} f(\xi_{i,j}) \Delta x \Delta y = \sum_{Q_{i,j} \subset \mathcal{R}} f(\xi_{i,j}) \Delta A$$

where $\xi_{i,j}$ is any point selected from $Q_{i,j}$ for each i, j.

A theorem from mathematical analysis guarantees that if we let N tend to infinity, then these Riemann sums tend to a limit. We call that limit *the integral of f over the region $\mathcal{R}$*, and we denote it by

$$\iint_{\mathcal{R}} f(x, y)\, dA.$$

The practical issue that we now need to address is how to evaluate integrals over general regions $\mathcal{R}$. To keep things from getting too complicated, we restrict attention to two types of regions: (1) the *x-simple regions,* which are regions of the form

$$\mathcal{R} = \{(x, y) : c \le y \le d, \alpha_1(y) \le x \le \alpha_2(y)\}$$

for some continuously differentiable functions $\alpha_1(y) \le \alpha_2(y)$ on $[c, d]$ (see Figure 3), and (2) the *y-simple regions,* which are regions of the form

$$\mathcal{R} = \{(x, y) : a \le x \le b, \beta_1(x) \le y \le \beta_2(x)\}$$

for some continuously differentiable functions $\beta_1(x) \le \beta_2(x)$ on $[a, b]$ (see Figure 4). Notice that the first type of region is called x-simple because it is spanned from left to right by horizontal line segments. A similar reason motivates the name of the second type of region; it is spanned from bottom to top by vertical line segments.

The following two theorems give a method for evaluating integrals over x-simple and y-simple regions. They tell us that we can evaluate double integrals over simple regions by iterated integrals (just as we did with the rectangles in Section 14.1).

Theorem 1 Suppose that $\alpha_1(y) \le \alpha_2(y)$ for $y \in [c, d]$. Let

$$\mathcal{R} = \{(x, y) : c \le y \le d, \alpha_1(y) \le x \le \alpha_2(y)\}$$

be the corresponding x-simple region. If f is a continuous function on $\mathcal{R}$, then

$$\iint_{\mathcal{R}} f(x, y)\, dA = \int_c^d \left(\int_{\alpha_1(y)}^{\alpha_2(y)} f(x, y)\, dx \right) dy.$$

Theorem 2 Suppose that $\beta_1(x) \le \beta_2(x)$ for $y \in [c, d]$. Let

$$\mathcal{R} = \{(x, y) : a \le x \le b, \beta_1(x) \le y \le \beta_2(x)\}$$

be the corresponding y-simple region. If f is a continuous function on $\mathcal{R}$, then

$$\iint_{\mathcal{R}} f(x, y)\, dA = \int_a^b \left(\int_{\beta_1(x)}^{\beta_2(x)} f(x, y)\, dy \right) dx.$$

You should think back to the Method of Disks that we used to calculate the volumes of solids of revolution. Recall that in those integrations, we would calculate the areas of slices in the y-variable and then integrate out in the x-variable, or vice versa. In Theorem 1, we are, in effect, taking slices in the x-direction and then integrating out in the y-variable. In Theorem 2, we are taking slices in the y-direction and then integrating out in the x-variable.

Example 1 Integrate the function $f(x, y) = 4xy$ over the region $\mathcal{R}$ between the parabola $y = x^2$ and the line $y = 2x + 3$.

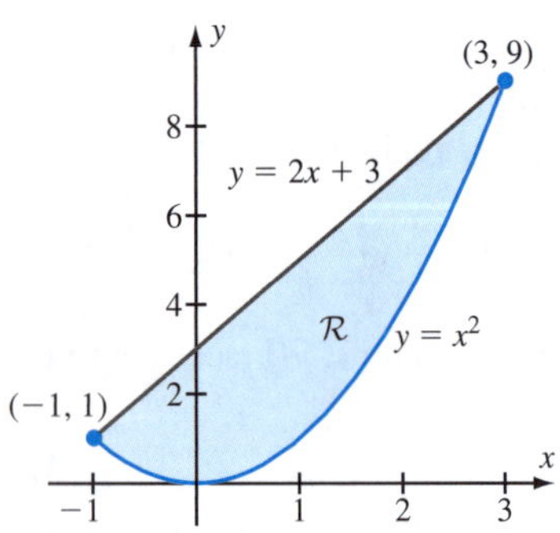

Figure 5
The region $\mathcal{R}$ is y-simple.

Solution Look at Figure 5. The curves intersect at the points $(-1, 1)$ and $(3, 9)$. This is a y-simple region: $\mathcal{R} = \{(x, y) : -1 \le x \le 3, x^2 \le y \le 2x + 3\}$. According to Theorem 2, we have

$$\iint_{\mathcal{R}} 4xy\, dA = \int_{-1}^{3} \left(\int_{x^2}^{2x+3} 4xy\, dy \right) dx = \int_{-1}^{3} \left(2xy^2 \Big|_{y=x^2}^{y=2x+3} \right) dx$$

$$= \int_{-1}^{3} \Big(2x\big((2x + 3)^2 - (x^2)^2\big) \Big) dx.$$

Therefore,

$$\iint_{\mathcal{R}} 4xy\, dA = \int_{-1}^{3} (-2x^5 + 8x^3 + 24x^2 + 18x)\, dx$$

$$= \left(-\frac{2}{6}x^6 + \frac{8}{4}x^4 + \frac{24}{3}x^3 + \frac{18}{2}x^2 \right) \Bigg|_{x=-1}^{x=3} = \frac{640}{3}.$$ ■

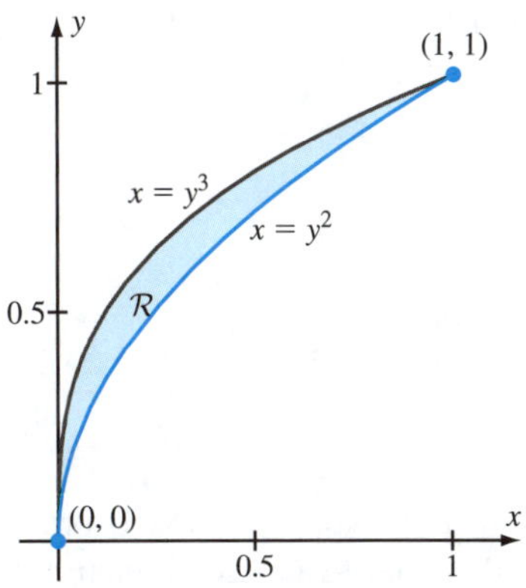

Figure 6
The region $\mathcal{R}$ is x-simple and y-simple.

Example 2 Integrate the function $f(x, y) = 2x - 4y$ over the region $\mathcal{R}$ between the curves $x = y^2$ and $x = y^3$. Refer to Figure 6.

Solution We first notice that the curves intersect at the points $(0, 0)$ and $(1, 1)$. The region $\mathcal{R} = \{(x, y) : 0 \le y \le 1, y^3 \le x \le y^2\}$ is x-simple. According to Theorem 1, the value of the integral is

$$\iint_{\mathcal{R}} (2x - 4y)\, dA = \int_0^1 \left(\int_{y^3}^{y^2} (2x - 4y)\, dx \right) dy = \int_0^1 \left((x^2 - 4xy)\Big|_{x=y^3}^{x=y^2} \right) dy$$
$$= \int_0^1 ((y^4 - 4y^3) - (y^6 - 4y^4))\, dy.$$

The last integrand simplifies to $-y^6 + 5y^4 - 4y^3$ and therefore

$$\iint_{\mathcal{R}} (2x - 4y)\, dA = \int_0^1 (-y^6 + 5y^4 - 4y^3)\, dy = \left(-\frac{1}{7}y^7 + y^5 - y^4 \right)\Big|_{y=0}^{y=1} = -\frac{1}{7}.$$

INSIGHT

Notice that the region $\mathcal{R}$ in Example 2 is also y-simple; indeed,

$$\mathcal{R} = \{(x, y) : 0 \le x \le 1, x^{1/2} \le y \le x^{1/3}\}.$$

It is a good exercise for you to recalculate the integral, integrating first in the y-variable, to see that the same answer results.

Changing the Order of Integration

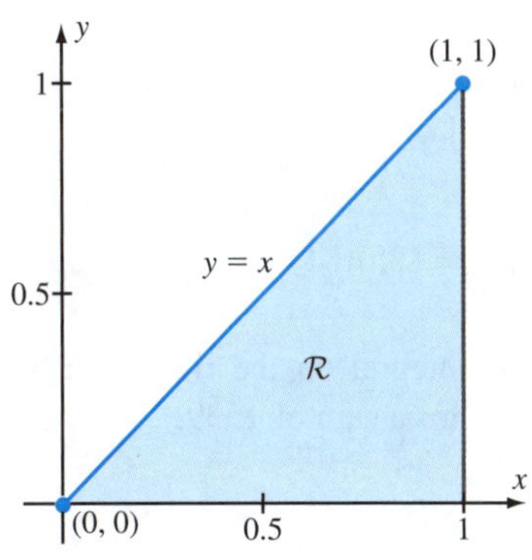

Figure 7
The region $\mathcal{R}$ is x-simple and y-simple.

Sometimes an iterated integral that is difficult to calculate becomes much easier if the order of integration is reversed. Theorems 1 and 2 taken together guarantee that if a region is both x-simple and y-simple, then we get the same answer no matter which order we choose to do the integration.

Example 3 Calculate the integral $\int_0^1 \int_y^1 6ye^{x^3}\, dx\, dy$.

Solution The inner integral with respect to x is not one that we can do with the techniques we have learned. Let us instead change the order of integration in an attempt to reformulate the problem. A glance at Figure 7 shows that the region of integration $\mathcal{R}$ is not only x-simple but is also y-simple:

$$\mathcal{R} = \{(x, y) : 0 \le x \le 1, 0 \le y \le x\}.$$

Therefore, the given iterated integral may be rewritten as

$$\int_0^1 \left(\int_0^x 6ye^{x^3}\, dy \right) dx,$$

an integral that we easily evaluate as follows:

$$\int_0^1 \left(\int_0^x 6ye^{x^3}\, dy \right) dx = \int_0^1 \left(3y^2 e^{x^3} \Big|_{y=0}^{y=x} \right) dx = \int_0^1 3x^2 e^{x^3}\, dx = e^{x^3} \Big|_{x=0}^{x=1} = e - 1.$$

inSIGHT

When you are changing the order of integration, there is no substitute for drawing a good picture. It tells you immediately the new limits of integration and the order in which they should appear.

Part of our work in switching the order of integration in Example 3 was to reexpress the boundary curves in terms of the x-variable instead of the y-variable. Sometimes the domain of a double integral must be broken up if it is to be evaluated in a certain order. However, it may be possible to avoid breaking up the integral if we evaluate in the opposite order. Here is an example.

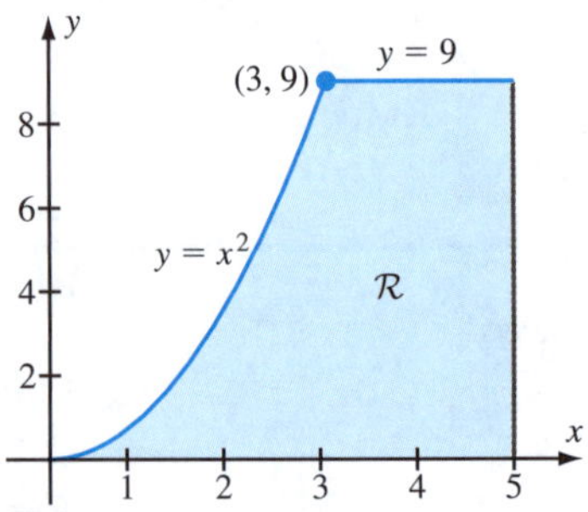

Figure 8
The region $\mathcal{R}$ is x-simple but not y-simple.

Example 4 Evaluate the integral $\iint_{\mathcal{R}} 4xy\, dA$ where $\mathcal{R}$ is the region shown in Figure 8.

Solution Notice that the region of integration is the *union* of two y-simple regions. If we choose to apply Theorem 2 and make y the inside variable, then we must note that for points (x, y) in $\mathcal{R}$ with $0 \le x \le 3$, the ordinate y satisfies $0 \le y \le x^2$, whereas when $3 \le x \le 5$, the ordinate y satisfies $0 \le y \le 9$. Because of the different formulas for y at the upper limits of integration ($y = x^2$ over the interval $[0, 3]$ and $y = 9$ over the interval $[3, 5]$), we must write the given double integral as a sum of two iterated integrals:

$$\iint_{\mathcal{R}} 4xy\, dA = \int_0^3 \left(\int_0^{x^2} 4xy\, dy \right) dx + \int_3^5 \left(\int_0^9 4xy\, dy \right) dx. \tag{14.5}$$

On the other hand, $\mathcal{R}$ is x-simple. In fact, $\mathcal{R} = \{(x, y) : 0 \le y \le 9, \sqrt{y} \le x \le 5\}$. Therefore, by Theorem 1,

$$\begin{aligned}\iint_{\mathcal{R}} 4xy\, dA &= \int_0^9 \left(\int_{\sqrt{y}}^5 4xy\, dx \right) dy = \int_0^9 \left(2x^2 y \Big|_{x=\sqrt{y}}^{x=5} \right) dy \\ &= \int_0^9 (50y - 2y^2)\, dy = \left(25y^2 - \frac{2}{3}y^3 \right) \Bigg|_0^9 = (25)(81) - \frac{2}{3}9^3 = 1539.\end{aligned}$$

You may verify this answer by showing that the first integral on the right side of equation (14.5) evaluates to 243 and the second to 1296, for a sum of 1539.

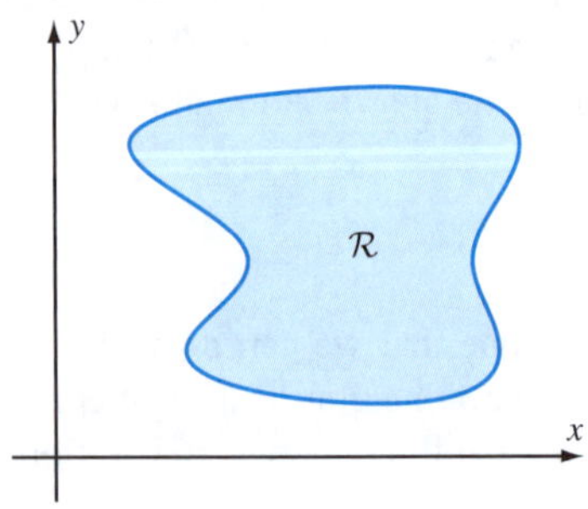

Figure 9
A bounded region in the plane.

The Area of a Planar Region

Suppose that $\mathcal{R}$ is a bounded region in the plane, as in Figure 9. Fix a rectangle Q that contains $\mathcal{R}$, and partition Q into N^2 congruent subrectangles $\{Q_{i,j}\}$. Next, let f be the function that is identically 1 on $\mathcal{R}$; that is, define f by the formula $f(x, y) = 1$ for every point (x, y) in $\mathcal{R}$. Then, whatever choice of points $\{\xi_{i,j}\}$ is made, the Riemann sum for f over $\mathcal{R}$ is equal to

$$\sum_{Q_{i,j}\subseteq\mathcal{R}} 1\,\Delta A. \tag{14.6}$$

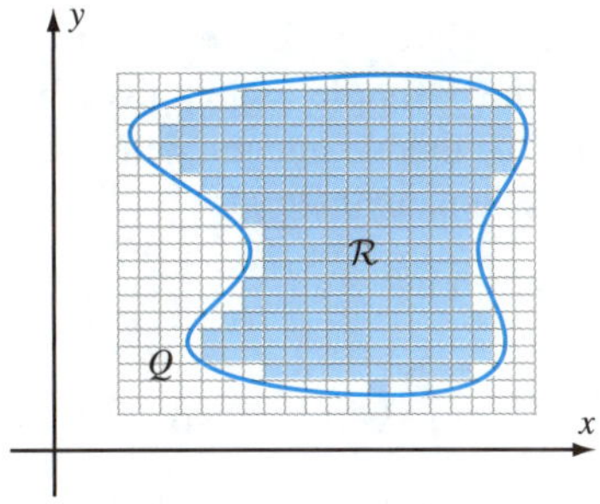

Figure 10
The area of the shaded region is the Riemann sum $\sum_{Q_{i,j}\subseteq R} 1\Delta A$.

This quantity is just the sum of the areas of the small rectangles $Q_{i,j}$ that are contained in $\mathcal{R}$. Refer to Figure 10. When the number N^2 of subrectangles of the partition is large, formula (14.6) gives a good approximation to the area of $\mathcal{R}$. It can be shown that if the boundary of $\mathcal{R}$ is a piecewise smooth curve, then the Riemann sums in line (14.6) tend to a limit as N tends to infinity. We denote this limit by $\iint_{\mathcal{R}} 1\,dA$ and make the following definition.

Definition If $\mathcal{R}$ is a region in the plane bounded by finitely many continuously differentiable curves, then the area of $\mathcal{R}$ is defined to be

$$\iint_{\mathcal{R}} 1\,dA.$$

Example 5

Calculate the area $\mathcal{A}$ inside the ellipse $x^2/a^2 + y^2/b^2 = 1$.

Solution The given ellipse is a y-simple region:

$$\mathcal{R} = \left\{(x, y) : -a \le x \le a,\ -b\sqrt{1-\frac{x^2}{a^2}} \le y \le b\sqrt{1-\frac{x^2}{a^2}}\right\}$$

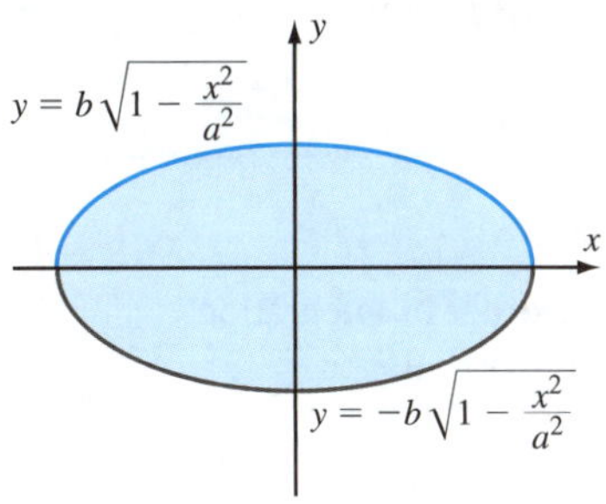

Figure 11
The ellipse $\frac{x^2}{a^2} + \frac{y^2}{b^2} = 1$ is a y-simple region.

(see Figure 11). We have

$$\begin{aligned}\mathcal{A} &= \iint_{\mathcal{R}} 1\,dA = \int_{-a}^{a}\left(\int_{-b\sqrt{1-x^2/a^2}}^{b\sqrt{1-x^2/a^2}} 1\,dy\right)dx \\ &= \int_{-a}^{a}\left(b\sqrt{1-\frac{x^2}{a^2}} - \left(-b\sqrt{1-\frac{x^2}{a^2}}\right)\right)dx = 2b\int_{-a}^{a}\sqrt{1-\frac{x^2}{a^2}}\,dx. \end{aligned} \tag{14.7}$$

In Section 7.4, we learned that this type of integral can be handled by the change of variables $x = a\sin(\theta)$, $dx = a\cos(\theta)\,d\theta$, which leads to

$$\mathcal{A} = 2ab\int_{-\pi/2}^{\pi/2}\sqrt{1-\sin^2(\theta)}\cos(\theta)\,d\theta = 2ab\int_{-\pi/2}^{\pi/2}\cos^2(\theta)\,d\theta \overset{(7.18)}{=} 2ab\left(\frac{\theta}{2} + \frac{\sin(2\theta)}{4}\right)\Bigg|_{-\pi/2}^{\pi/2} = \pi ab. \quad ■$$

inSIGHT

It sometimes happens that an integrand suggests several substitutions. If the substitution $x = au$, $dx = a\,du$ is made in the last integral of line (14.7), then the expression $2ab\int_{-1}^{1}\sqrt{1-u^2}\,du$ results. This leads to the same answer, πab, that we obtained in Example 5, because the integral in this expression represents the area, $\pi \cdot 1^2/2$, of the upper half of the unit circle.

Example 6 Suppose that $\mathcal{R}_1$ is the x-simple region given by $\mathcal{R}_1 = \{(x, y) : c \leq y \leq d, \alpha_1(y) \leq x \leq \alpha_2(y)\}$. Suppose that $\mathcal{R}_2$ is the y-simple region given by $\mathcal{R}_2 = \{(x, y) : a \leq x \leq b, \beta_1(x) \leq y \leq \beta_2(x)\}$. Using the double integral definition of planar area, express the areas of $\mathcal{R}_1$ and $\mathcal{R}_2$ as single Riemann integrals.

Solution According to Theorem 1, the area of $\mathcal{R}_1$ is given by

$$\iint_{\mathcal{R}} 1\,dA = \int_c^d \left(\int_{\alpha_1(y)}^{\alpha_2(y)} 1\,dx\right) dy = \int_c^d \left(x\Big|_{x=\alpha_1(y)}^{x=\alpha_2(y)}\right) dy = \int_c^d (\alpha_2(y) - \alpha_1(y))\,dy.$$

Similarly, using Theorem 2, the area of $\mathcal{R}_2$ is given by

$$\iint_{\mathcal{R}} 1\,dA = \int_a^b \left(\int_{\beta_1(x)}^{\beta_2(x)} 1\,dy\right) dx = \int_a^b \left(y\Big|_{y=\beta_1(x)}^{y=\beta_2(x)}\right) dx = \int_a^b (\beta_2(x) - \beta_1(x))\,dx.$$

A LOOK BACK

In Section 5.6, we learned to express the area between two curves using a single integral. Example 6 shows that, in such a situation, the definition of area by means of a double integral leads to the very same formula that our earlier method prescribed. Of course, it is essential that the two methods be consistent when they are applied to the same region. The advantage of the double integral definition is that it has wider applicability than our earlier definition.

quickquiz

1. What are x-simple and y-simple regions?
2. Why is it sometimes advantageous to switch the order of integration?
3. Evaluate the integral $\iint_{\mathcal{R}} 2x^3y\,dA$ where $\mathcal{R} = \{(x, y) : 0 \leq x \leq 1, 0 \leq y \leq x^2\}$.
4. Describe how the area of a region in the plane is defined.

EXERCISES

Problems for Practice

In Exercises 1–6, draw a sketch of the region $\mathcal{R}$ that is described. Decide whether $\mathcal{R}$ is x-simple, y-simple, or both x-simple and y-simple. If $\mathcal{R}$ is x-simple (or, respectively, y-simple), state the functions $y \mapsto \alpha_1(y)$ and $y \mapsto \alpha_2(y)$ (respectively, $x \mapsto \beta_1(x)$ and $x \mapsto \beta_2(x)$) whose graphs form part of its boundary.

1. The region between the curves $x = -y^2 + 2$ and $x = -2y - 1$
2. The region between the curves $y = x$ and $y = x^2$
3. The region between the curves $x = y^2$ and $x = -2y^2 - 3y + 18$
4. The region between the curves $y = x^2$ and $y = 8x^{1/2}$
5. The region above the curve $y = x^{3/2}$, below the curve $y = 6x - 1$, and between $x = 1$ and $x = 3$
6. The region below $y = 2$, above $y = -1$, and between the curves $y = x^3$ and $y = x + 8$

In Exercises 7–14, evaluate the iterated integral.

7. $\int_{-1}^{1} \int_{(x+1)^2}^{8-(x+1)^2} 3x\,dy\,dx$
8. $\int_{1}^{64} \int_{x^{1/2}}^{x^{1/3}} (y/x)\,dy\,dx$
9. $\int_{0}^{2} \int_{x/4}^{4x} (2x - 1)y^{1/2}\,dy\,dx$

10. $\int_0^1 \int_y^{\sqrt{y}} 2xe^y \, dx \, dy$

11. $\int_{\pi/2}^{\pi} \int_0^{\sin(x)} 4x \, dy \, dx$

12. $\int_1^4 \int_y^{y+1} 4x \, dx \, dy$

13. $\int_{\pi/4}^{\pi/2} \int_0^{\sin(y)} 5 \, dx \, dy$

14. $\int_1^3 \int_{x-1}^{x+1} y^{1/2}(x+1) \, dy \, dx$

In Exercises 15–22, calculate the integral of the function f over the region $\mathcal{R}$.

15. $f(x, y) = 6y - 4x$; $\mathcal{R}$ is the region between the curves $y = 2x$ and $y = x^2$.
16. $f(x, y) = \cos(x)$; $\mathcal{R}$ is the region between the curves $y = x$ and $y = 2x$, $0 \leq x \leq \pi$.
17. $f(x, y) = xe^y$; $\mathcal{R}$ is the region above $y = 0$, below $y = 2x - 6$, and between $x = 4$ and $x = 7$.
18. $f(x, y) = 9\sqrt{y}/4$; $\mathcal{R}$ is the region below $y = 4$, above $y = 1$, and between $x = y^3$ and $x = y^2 + 48$.
19. $f(x, y) = y\sqrt{x}$; $\mathcal{R}$ is the region between $x = 3$ and $x = 4$, above $y = x^2 - 3x$, and below $y = x^2 + 8$.
20. $f(x, y) = y^2/x^2$; $\mathcal{R}$ is the region above $y = -5$, below $y = -3$, and between $x = y^2$ and $x = 5$.
21. $f(x, y) = e^x$; $\mathcal{R}$ is the region between $x = y$ and $x = 6y$, $0 \leq x \leq 1$.
22. $f(x, y) = y\sin(x)$; $\mathcal{R}$ is the region bounded by $y = \cos(x)$ and $y = \sin(x)$, $0 \leq x \leq \pi/4$.

In Exercises 23–26, switch the order of integration to make the integral manageable, and then calculate the iterated integral.

23. $\int_0^{\pi/2} \int_x^{\pi/2} \sin(y)/y \, dy \, dx$

24. $\int_1^2 \int_1^{x^2} x/y \, dy \, dx$

25. $\int_0^2 \int_{x^2}^4 e^y/\sqrt{y} \, dy \, dx$

26. $\int_0^1 \int_y^1 \exp(-x^2) \, dx \, dy$

Further Theory and Practice

In Exercises 27–34, calculate the integral.

27. $\int_0^{\sqrt{\pi}} \int_x^{\sqrt{\pi}} \sin(y^2) \, dy \, dx$
28. $\int_0^1 \int_0^{\arccos(y)} \cos^4(x) \, dx \, dy$
29. $\int_0^1 \int_{y^{1/2}}^1 y(1-x^2)^2/x^3 \, dx \, dy$
30. $\int_0^2 \int_0^{\sqrt{4-x^2}} (4-y^2)^{3/2} \, dy \, dx$
31. $\iint_{\mathcal{R}} (x-y) \, dA$ where $\mathcal{R}$ is the region bounded by $y = \cos(x)$ and $y = \sin(x)$, $0 \leq x \leq \pi/4$
32. $\iint_{\mathcal{R}} (y + \sec^2(x)) \, dA$ where $\mathcal{R}$ is the region bounded by $y = \tan(x)$, $0 \leq x \leq \pi/4$; $y = -x$, $-1 \leq x \leq 0$; and $y = 1$, $-1 \leq x \leq \pi/4$
33. $\int_0^1 \left(\int_y^1 x^2 \sin(xy) \, dx \right) dy$
34. $\int_0^3 \left(\int_0^{\sqrt{9-y^2}} \sqrt{9-x^2} \, dx \right) dy$

In Exercises 35–44, sketch the set $\mathcal{R}$ that is the union of the two regions of integration. Reverse the order of integrations, and write the given sum as one iterated integral over $\mathcal{R}$. Calculate this iterated integral.

35. $\int_{-1}^0 \int_{-x}^1 y \, dy \, dx + \int_0^1 \int_{x^2}^1 y \, dy \, dx$
36. $\int_0^1 \int_0^y xy \, dx \, dy + \int_1^2 \int_0^{2-y} xy \, dx \, dy$
37. $\int_{-1}^0 \int_{-y}^1 xy \, dx \, dy + \int_0^1 \int_{\sqrt{y}}^1 xy \, dx \, dy$
38. $\int_0^1 \int_0^x xy \, dy \, dx + \int_1^3 \int_0^1 xy \, dy \, dx$
39. $\int_0^1 \int_{-\sqrt{y}}^{\sqrt{y}} x \, dx \, dy + \int_1^4 \int_{y-2}^{\sqrt{y}} x \, dx \, dy$
40. $\int_0^1 \int_1^2 1 \, dy \, dx + \int_1^2 \int_1^{x^2-4x+5} 1 \, dy \, dx$
41. $\int_{-\sqrt{2}}^{-1} \int_{-\sqrt{2-y^2}}^{\sqrt{2-y^2}} x \, dx \, dy + \int_{-1}^1 \int_y^1 x \, dx \, dy$
42. $\int_0^1 \int_{-\sqrt{y}}^{\sqrt{y}} y \, dx \, dy + \int_1^2 \int_{-\sqrt{2-y}}^{\sqrt{2-y}} y \, dx \, dy$
43. $\int_0^{1/\sqrt{2}} \int_0^{\arcsin(y)} y \, dx \, dy + \int_{1/\sqrt{2}}^1 \int_0^{\arccos(y)} y \, dx \, dy$
44. $\int_0^1 \int_{-1}^{-y} |x| \, dx \, dy + \int_0^1 \int_y^1 |x| \, dx \, dy$

In Exercises 45–55, calculate $\iint_{\mathcal{R}} f(x, y) \, dA$ for the function f and region $\mathcal{R}$. If the integral is improper, do the iterated integration in the correct order to see that it converges.

45. $f(x, y) = ax + by$; $\mathcal{R}$ is the triangle with vertices $(0, 0)$, $(1, 3)$, $(10, 4)$.
46. $f(x, y) = x^2 + y$; $\mathcal{R}$ is the parallelogram with vertices $(-1, 0)$, $(0, 0)$, $(1, 1)$, $(0, 1)$.
47. $f(x, y) = x + y$; $\mathcal{R}$ is the region in the first quadrant that is underneath $y = 1 + x^2$ and to the left of $y = 5(x-1)$.
48. $f(x, y) = x^2 y$; $\mathcal{R}$ is the region bounded above by $y = 4 + 2x - x^2$ and below by $y = x^2$.
49. $f(x, y) = x + y$; $\mathcal{R}$ is that part of the unit circle that lies in the first quadrant.
50. $f(x, y) = x^2 + y^2$; $\mathcal{R}$ is that part of the unit circle that lies in the first quadrant.
51. $f(x, y) = y\sin(x)$; $\mathcal{R}$ is the region in the first quadrant that is bounded above by $y = 1 - \cos(x)$ and on the right by $x = \pi/2$.
52. $f(x, y) = 1/x^3$; $\mathcal{R} = \{(x, y) : 1 \leq x \leq 2,\ x\sin(1/x) \leq y \leq 2\}$
53. $f(x, y) = |y - x|$; $\mathcal{R}$ is the region in the first quadrant that is bounded above by $y = 2x$ and on the right by $x = 1$.
54. $f(x, y) = 2y/(2-x)$; $\mathcal{R}$ is the region bounded by $y = \sqrt{4-x^2}$ and $y = 0$.
55. Show that the error function

$$\operatorname{erf}(x) = \frac{2}{\sqrt{\pi}} \int_0^x \exp(-y^2) \, dy$$

satisfies

$$\int_0^z \operatorname{erf}(x)\,dx = z\operatorname{erf}(z) - \frac{1}{\sqrt{\pi}}(1 - \exp(-z^2)).$$

(*Hint:* Write the left side as an iterated integral and change the order of integration.) Deduce that

$$\int_0^z (1-\operatorname{erf}(x))\,dx = z(1-\operatorname{erf}(z))+\frac{1}{\sqrt{\pi}}(1-\exp(-z^2)).$$

Given that $\lim_{z\to\infty}\operatorname{erf}(z) = 1$, use l'Hôpital's Rule to evaluate the convergent improper integral $\int_0^\infty (1 - \operatorname{erf}(x))\,dx$.

Calculator/Computer Exercises

In Exercises 56–59, a function f and a region $\mathcal{R}$ are given. Calculate $\iint_{\mathcal{R}} f(x, y)\,dA$ by writing the double integral as an iterated integral and finding the outer limits of integration.

56. $f(x, y) = 1 + x + xy$; $\mathcal{R}$ is the region bounded above by $y = 1 - x^4$ and below by $y = x^3$.

57. $f(x, y) = xy$; $\mathcal{R}$ is the region bounded above by $y = 1 + 3x$ and below by $y = \exp(x)$.

58. $f(x, y) = x$; $\mathcal{R}$ is the region in the first quadrant that is bounded above by $y = 4 - x^{1/8}$ and below by $y = 1 + \sqrt{x}$.

59. $f(x, y) = x$; $\mathcal{R}$ is the region in the first quadrant that is bounded above by $y = 1/(1 + x^2)$ and below by $y = x^{10}$.

In Exercises 60–63, a function f and a region $\mathcal{R}$ are given. Approximate $\iint_{\mathcal{R}} f(x, y)\,dA$ to two decimal places by writing the double integral as an iterated integral for which the inner integration can be evaluated exactly. Use Simpson's Rule for the outer integration.

60. $f(x, y) = \sqrt{1 - x^4}$; $\mathcal{R}$ is that part of the unit circle that lies in the first quadrant.

61. $f(x, y) = \sqrt{x^2 + y}$; $\mathcal{R}$ is that part of the unit circle that lies in the first quadrant.

62. $f(x, y) = \sqrt{x + y}$; $\mathcal{R}$ is the region in the first quadrant that is bounded above by $y = x^3$ and on the right by $x = 1$.

63. $f(x, y) = \exp(-y^2)$; $\mathcal{R}$ is the region in the first quadrant that is bounded on the left by $y = \tan(x)$, $0 \le x \le \pi/4$, and on the right by $y = \sqrt{2}\cos(x)$, $\pi/4 \le x \le \pi/2$.

The iterated integral $I = \int_a^b (\int_{\beta_1(x)}^{\beta_2(x)} f(x, y)\,dy)\,dx$ may be approximated by using an appropriate Trapezoidal Rule. Let $\Delta x = (b - a)/N$, $x_i = a + i\Delta x$ for $0 \le i \le N$, $(\Delta y)_i = (\beta_2(x_i) - \beta_1(x_i))/N$, and $y_{i,j} = \beta_1(x_i) + j\,(\Delta y)_i$ for $0 \le j \le N$. Then

$$I \approx \frac{\Delta x}{2}\sum_{i=0}^{N}\frac{(\Delta y)_i}{2}\sum_{j=0}^{N} c_{i,j} f(x_i, y_{i,j})$$

where $c_{i,j} = 1$ if $\{i, j\} \subset \{0, N\}$, $c_{i,j} = 4$ if $\{i, j\} \subset \{1, 2, \ldots, N - 1\}$, and $c_{i,j} = 2$ otherwise (that is, if exactly one of i and j is 0 or N). In Exercises 64–67, approximate the given integral using $N = 4$. Calculate the integral exactly. What is the approximation error?

64. $\int_0^1 \int_0^x (x - y)\,dy\,dx$

65. $\int_{-1}^1 \int_0^{1-x^2} y\,dy\,dx$

66. $\int_{\pi/6}^{\pi/4} \int_0^{\tan(x)} \cot(x)\,dy\,dx$

67. $\int_0^1 \int_{x^2}^x \sqrt{x}\,dy\,dx$

14.3 Calculation of Volumes of Solids

In this section, we use double integrals to calculate volumes of solids in three-dimensional space. To be specific, given a nonnegative continuous function f defined on a region $\mathcal{R}$ of the xy-plane, we are interested in determining the volume of the solid that is above the region $\mathcal{R}$ and below the graph of f. See Figure 1. Recall that we motivated integration of functions of two variables by volume considerations. We now use the integral to define precisely what we mean by the volume of a solid.

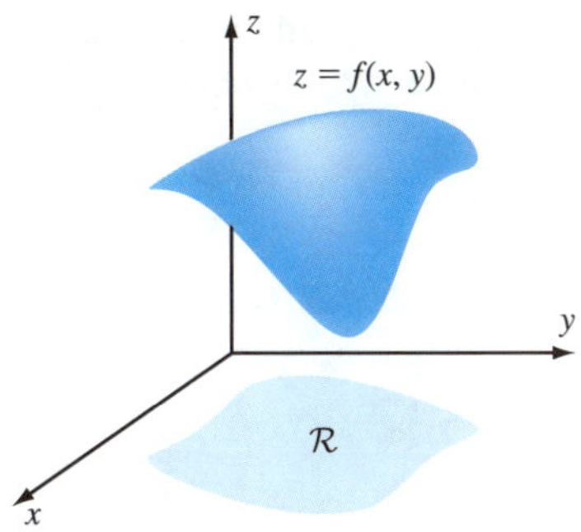

Figure 1a
The graph of f over a planar region $\mathcal{R}$

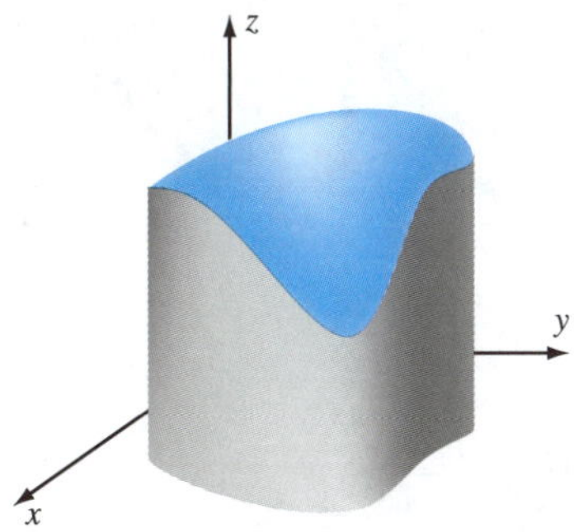

Figure 1b
The solid $\mathcal{U}$ that lies under the graph of f and over the planar region $\mathcal{R}$

Definition Suppose f is a nonnegative, continuous function defined on a region $\mathcal{R}$ of the xy-plane. Let $\mathcal{U}$ be the three-dimensional solid that lies above $\mathcal{R}$ and below the graph of f. The volume of $\mathcal{U}$ is defined to be

$$\iint_{\mathcal{R}} f(x, y)\, dA.$$

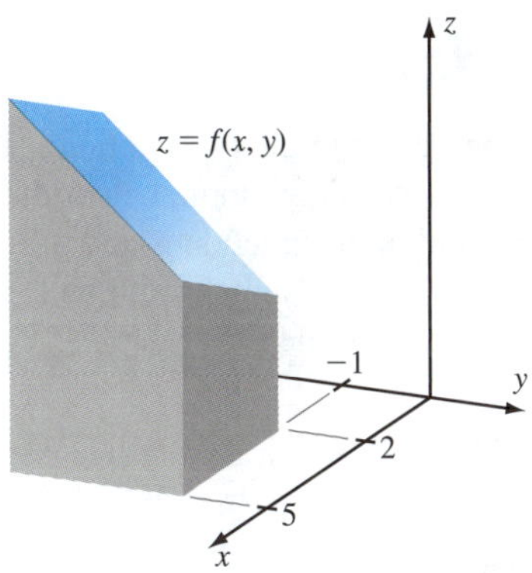

Figure 2
The solid that lies under the graph of f and over a planar rectangle

Example 1 Figure 2 illustrates the three-dimensional solid $\mathcal{U}$ that lies below the plane $f(x, y) = 2x - 6y$ and above the planar region

$$\mathcal{R} = \{(x, y) : 2 \le x \le 5, -3 \le y \le -1\}.$$

Find the volume of $\mathcal{U}$.

Solution Because $x > 0$ and $y < 0$ on $\mathcal{R}$, we have $f(x, y) > 0$ on $\mathcal{R}$. The volume of $\mathcal{U}$ is therefore

$$\iint_{\mathcal{R}} f(x, y)\, dA = \int_{-3}^{-1} \left(\int_{2}^{5} (2x - 6y)\, dx \right) dy = \int_{-3}^{-1} (x^2 - 6xy)\Big|_{x=2}^{x=5} dy$$

$$= \int_{-3}^{-1} (21 - 18y)\, dy = (21y - 9y^2)\Big|_{-3}^{-1} = 114.$$

Sometimes, as in the next example, a little geometric analysis is required before we can calculate a volume.

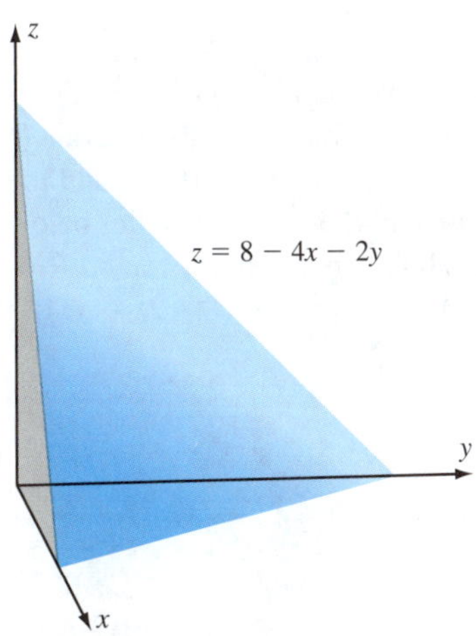

Figure 3a
The solid in the first octant that lies below the plane $z = 8 - 4x - 2y$

Example 2 Calculate the volume below the plane $8x + 4y + 2z = 16$, above the xy-plane, and in the first octant (see Figure 3a).

Solution Solving for z, we think of the plane as the graph of the function

$$z = f(x, y) = 8 - 4x - 2y.$$

We notice that $f(x, y) = 0$ when (x, y) lies on the line $4x + 2y = 8$ in the xy-plane. Therefore, our problem is to find the volume below the graph of f and above the region

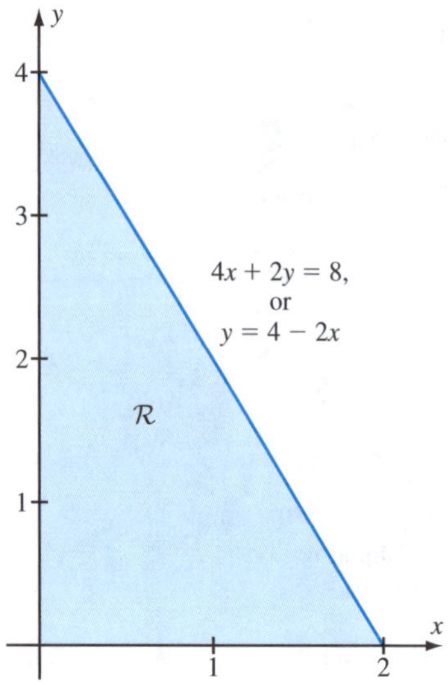

Figure 3b

$\mathcal{R}$ in the xy-plane that is bounded by the x-axis, the y-axis, and the line $4x + 2y = 8$ (see Figure 3b). The region $\mathcal{R}$ is y-simple:

$$\mathcal{R} = \{(x, y) : 0 \le x \le 2,\ 0 \le y \le 4 - 2x\}.$$

The required volume $\iint_{\mathcal{R}} f(x, y)\, dA$ is therefore

$$\begin{aligned}\int_0^2 \left(\int_0^{4-2x} (8 - 4x - 2y)\, dy\right) dx &= \int_0^2 \left((8y - 4xy - y^2)\Big|_{y=0}^{y=4-2x}\right) dx\\ &= \int_0^2 (16 - 16x + 4x^2)\, dx\\ &= \left(16x - 8x^2 + \frac{4}{3}x^3\right)\Bigg|_0^2\\ &= \frac{32}{3}.\end{aligned}$$

■

The Volume between Two Surfaces

So far we have discussed the volume of a region lying below the graph of a continuous function and above a region in the xy-plane. Now let us treat the volume below the graph of a continuous function f, above the graph of a continuous function g, and over a region $\mathcal{R}$ in the xy-plane. In this instance, the relevant Riemann sums would be for the function $f - g$, and the integral representing the desired volume is

$$\iint_{\mathcal{R}} (f(x, y) - g(x, y))\, dA.$$

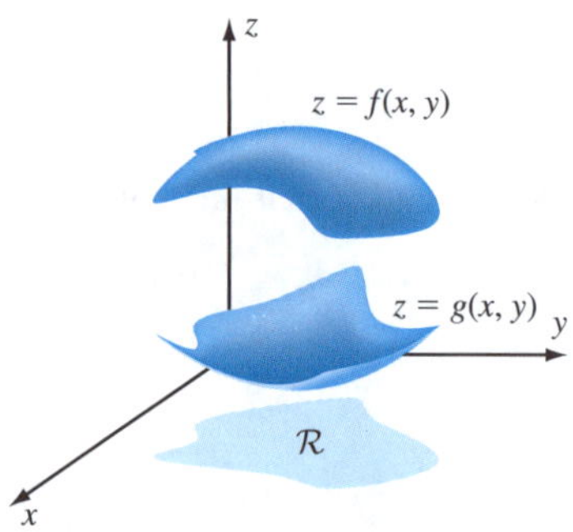

Figure 4a
The graphs of f and g over a planar region $\mathcal{R}$

The geometry is illustrated in Figure 4. Following is an example.

Example 3 Set up (but do not calculate) the integral for finding the volume of the solid bounded below by the paraboloid that is the graph of $g(x, y) = 8 + x^2 + y^2$ and above by the plane that is the graph of $f(x, y) = 2x + 2y + 15$.

Solution The two surfaces intersect when $8 + x^2 + y^2 = 2x + 2y + 15$, or when $(x^2 - 2x) + (y^2 - 2y) = 7$. After completing the squares, we obtain $(x-1)^2 + (y-1)^2 = 3^2$. Thus, the solid we wish to study "casts a shadow in the xy-plane" on the interior $\mathcal{R}$ of the circle with center $(1, 1)$ and radius 3. Figure 5 shows the two surfaces and the region $\mathcal{R}$ of integration. We need to determine the volume of the solid $\mathcal{U}$ that lies below the graph of f, above the graph of g, and over disk $\mathcal{R}$. The boundary of the region $\mathcal{R}$ of integration may be written as $y - 1 = \pm(9 - (x-1)^2)^{1/2}$, or $y = 1 \pm (8 + 2x - x^2)^{1/2}$. We write $\mathcal{R}$ as the y-simple region given by

$$\mathcal{R} = \{(x, y) : -2 \le x \le 4,\ 1 - \sqrt{8 + 2x - x^2} \le y \le 1 + \sqrt{8 + 2x - x^2}\}.$$

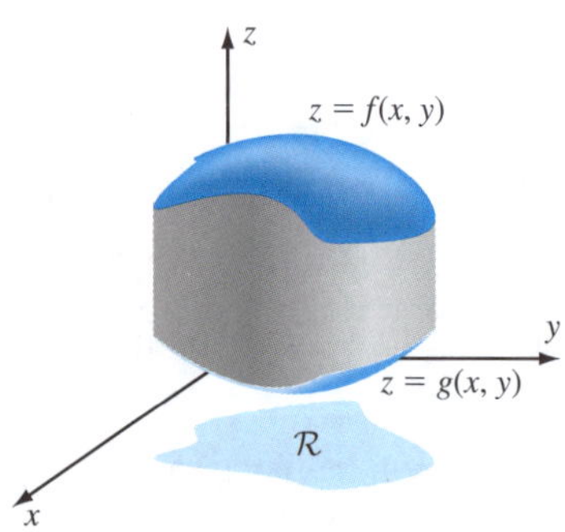

Figure 4b
The solid $\mathcal{U}$ that lies under the graph of f, above the graph of g, and over the planar region $\mathcal{R}$

The desired volume is therefore

$$\iint_{\mathcal{R}} (f(x, y) - g(x, y))\, dA = \int_{-2}^{4} \int_{1-\sqrt{8+2x-x^2}}^{1+\sqrt{8+2x-x^2}} \left((2x + 2y + 15) - (8 + x^2 + y^2)\right) dy\, dx.$$

■

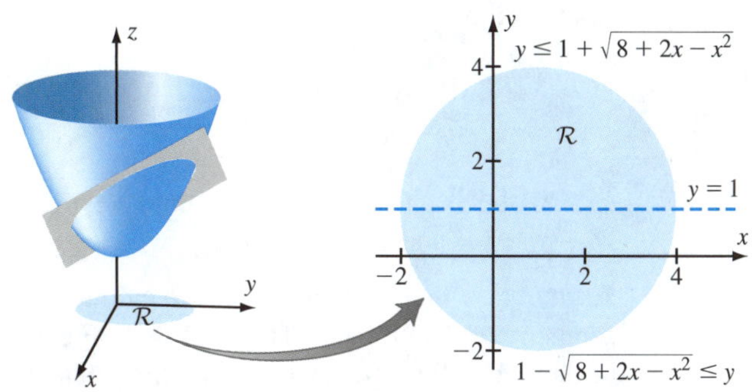

Figure 5
The paraboloid $z = 8 + x^2 + y^2$ and the plane $z = 2x + 2y + 15$

Example 4 Find the volume V of the solid $\mathcal{U}$ above the plane $z = x + y$, below the plane $z = 2x + 5y$, and bounded on the sides by the parabolic cylinder $x = (y-4)^2+3$ and the plane $x + 2y = 11$.

Solution Look at the graph of the solid $\mathcal{U}$ in Figure 6. Clearly, the region $\mathcal{R}$ of integration in the xy-plane is determined by the vertical sides of the solid—the parabolic cylinder and the plane. These intersect the xy-plane in the parabola $x = (y-4)^2+3$ and the line $x + 2y = 11$, respectively. These two curves, and their points of intersection $(3, 4)$ and $(7, 2)$, are shown in Figure 6. We see that the region $\mathcal{R}$ is x-simple: $\mathcal{R} = \{(x, y) : 2 \le y \le 4,\ (y-4)^2+3 \le x \le 11-2y\}$. Since the plane $z = 2x+5y$ forms the upper boundary of the solid and the plane $z = x + y$ forms the lower boundary, the volume of $\mathcal{U}$ is given by

$$V = \iint_{\mathcal{R}} ((2x+5y)-(x+y))\,dA = \int_2^4 \int_{(y-4)^2+3}^{11-2y} (x+4y)\,dx\,dy = \int_2^4 \left(\frac{x^2}{2} + 4xy\right)\Bigg|_{x=(y-4)^2+3}^{x=11-2y} dy.$$

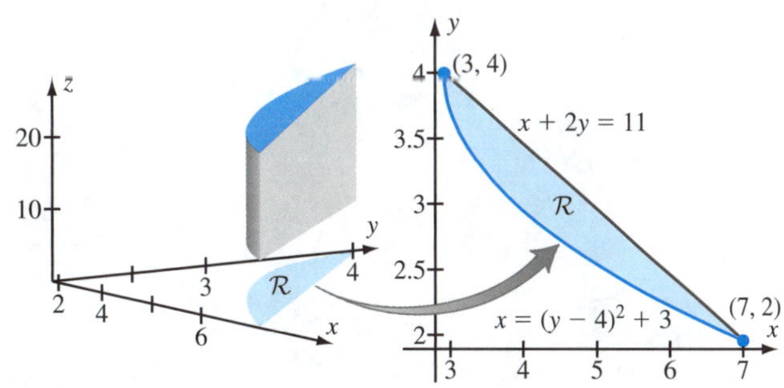

Figure 6

On evaluating, expanding, and simplifying this integrand, we obtain

$$\left(\frac{x^2}{2} + 4xy\right)\Bigg|_{x=(y-4)^2+3}^{x=11-2y} = \left(\frac{(11-2y)^2}{2} + 4(11-2y)y\right) - \left(\frac{((y-4)^2+3)^2}{2} + 4((y-4)^2+3)y\right)$$
$$= -\frac{1}{2}y^4 + 4y^3 - 25y^2 + 98y - 120$$

and therefore

$$V = \int_2^4 \left(-\frac{1}{2}y^4 + 4y^3 - 25y^2 + 98y - 120\right) dy = \left(-\frac{1}{10}y^5 + y^4 - \frac{25}{3}y^3 + 49y^2 - 120y\right)\Bigg|_2^4 = \frac{332}{15}.$$

Example 5 Calculate the volume V of the solid $\mathcal{U}$ enclosed by the cylinders $x^2 + y^2 = a^2$ and $x^2 + z^2 = a^2$.

Solution Figure 7a shows the two cylinders. We can use symmetry to simplify the geometry of the problem. The left side of Figure 7b shows the part of the intersection of the two cylinders that lies in the first octant. We will compute the volume V_0 of this smaller solid $\mathcal{U}_0$ and multiply by 8 to obtain V. Because $\mathcal{U}_0$ lies over the quarter disk shown in the right side of Figure 7b, the region of integration is $\mathcal{R} = \{(x, y) : 0 \le x \le a,\ 0 \le y \le \sqrt{a^2 - x^2}\}$, which is y-simple. The cylinder $x^2 + z^2 = a^2$ is the upper surface of $\mathcal{U}_0$. By solving for z, we realize this surface as the graph of $f(x, y) = \sqrt{a^2 - x^2}$. The lower surface of $\mathcal{U}_0$ is the xy-plane, which we realize as $g(x, y) = 0$. Therefore,

$$V = 8V_0 = 8\int_0^a \int_0^{\sqrt{a^2-x^2}} (f(x, y) - g(x, y))\, dy\, dx = 8\int_0^a \left(\int_0^{\sqrt{a^2-x^2}} \sqrt{a^2 - x^2}\, dy\right) dx.$$

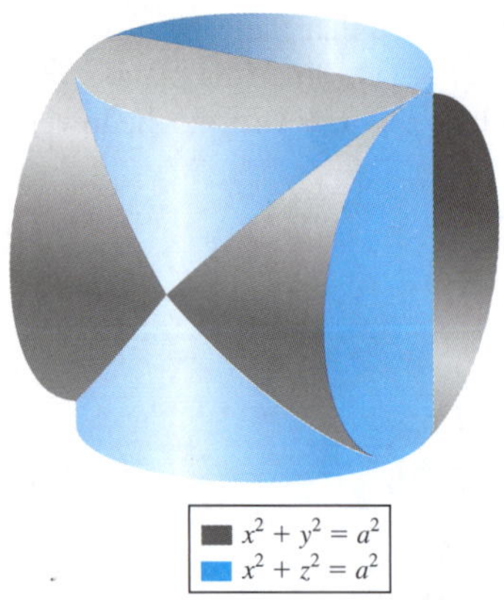

Figure 7a

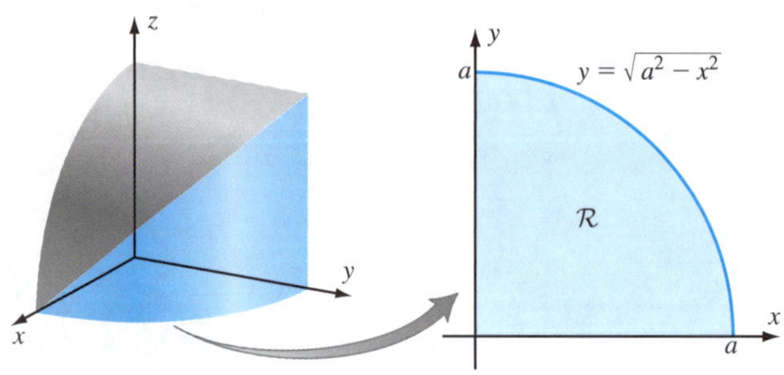

Figure 7b

Finally, we obtain

$$V = 8\int_0^a \left(\left(\sqrt{a^2 - x^2}\right)\cdot y\right)\Big|_{y=0}^{y=\sqrt{a^2-x^2}} dx = 8\int_0^a (a^2 - x^2)\, dx = 8\left(a^2x - \frac{x^3}{3}\right)\Bigg|_0^a = \frac{16}{3}a^3.$$

quickquiz

1. How do we calculate the volume of the solid lying above a region in the plane and below the graph of a nonnegative function f?
2. How do we calculate the volume between the graphs of two functions f and g with $g(x, y) \le f(x, y)$?

3. Calculate the volume of the solid that lies below the graph of $f(x, y) = x + y^2$ and over the rectangle $[0, 1] \times [0, 1]$ in the xy-plane.
4. Calculate the volume of the solid that lies below the graph of $f(x, y) = 1 + x$, above the graph of $g(x, y) = 2x$, and over the rectangle $[0, 1] \times [0, 1]$ in the xy-plane.

EXERCISES

Problems for Practice

In Exercises 1–8, calculate the volume of the solid that lies below the graph of $f(x, y) = 5 + x$ and over the given region in the xy-plane.

1. The region bounded by the curves $y = x^2 + x$ and $y = 2x^2 + 2x - 12$
2. The region bounded by the curves $x = y^2 + y + 2$ and $x = 2y^2 + 4y + 4$
3. The region bounded by the curves $y = x^2 - 2$ and $y = x$
4. The region bounded by the curves $x = y^4$ and $x = 16$
5. The region bounded by the curves $y = 5x^2 - 7x + 3$ and $y = x^3$
6. The region bounded by the curves $y = |x|$ and $y = 2$
7. The region bounded by the curves $x + y = 2$ and $y = x^2 - 4$
8. The region bounded by the curves $y = x^2$ and $y = 5 - (x - 1)^2$

In Exercises 9–23, calculate the volume of the solid $\mathcal{U}$ that lies below the graph of $f(x, y)$ and over the region $\mathcal{R}$ in the xy-plane.

9. $f(x, y) = x - 2y$,
$\mathcal{R} = \{(x, y) : 8 \le x \le 10, 2 \le y \le 4\}$
10. $f(x, y) = x^2 + y^2$,
$\mathcal{R} = \{(x, y) : 2 \le x \le 4, 1 \le y \le 6\}$
11. $f(x, y) = x^2y^2$,
$\mathcal{R} = \{(x, y) : -4 \le x \le -2, -5 \le y \le -2\}$
12. $f(x, y) = \sin(x)$,
$\mathcal{R} = \{(x, y) : 0 \le x \le \pi, 1 \le y \le 3\}$
13. $f(x, y) = 10x - 3y + 40$,
$\mathcal{R} = \{(x, y) : 3 \le x \le 6, 3x \le y \le 2x + 8\}$
14. $f(x, y) = x^{-3/2}y^{1/2}$,
$\mathcal{R} = \{(x, y) : 1 \le y \le 4, y^2 \le x \le y^4\}$
15. $f(x, y) = x - y$,
$\mathcal{R} = \{(x, y) : 3 \le x \le 5, -6x \le y \le -2x\}$
16. $f(x, y) = x/y^2$,
$\mathcal{R} = \{(x, y) : -3 \le y \le -2, y^2 \le x \le y^4\}$
17. $f(x, y) = x + 2$;
$\mathcal{R}$ is the region between the curves $y = x^2 + x$ and $y = 3x + 8$.
18. $f(x, y) = 12 + x + y$;
$\mathcal{R}$ is the region between the curves $y = -x^2 + 4$ and $y = x^2 + 2x - 8$.
19. $f(x, y) = \exp(x)$;
$\mathcal{R}$ is the region bounded by the curves $y = 2x + 3$, $y = x$, and $x = 2$.
20. $f(x, y) = x/y$;
$\mathcal{R}$ is the region below $y = 1$ and above $y = \exp(x^2 - x)$.
21. $f(x, y) = x + 2y$;
$\mathcal{R}$ is the region bounded by the curves $y = x^{1/2}$, $x = 0$, $y = x + 2$, and $x = 4$.
22. $f(x, y) = 8 - 2x + y$;
$\mathcal{R}$ is the region between the curves $y = x^3$ and $y = x^3 + x^2 - 2x - 3$.
23. $f(x, y) = \sin(y)$;
$\mathcal{R}$ is the region bounded by the curves $y = x + 1$, $y = -x + 1$, and $x = 1$.

In Exercises 24–30, calculate the volume of the solid $\mathcal{U}$.

24. $\mathcal{U}$ lies below the graph of $x + 2y + 3z = 6$ and in the first octant.
25. $\mathcal{U}$ lies below the graph of $z = x + 2y^2 + 3$, above the graph of $z = x + y$, and over the region $\mathcal{U}$ in the xy-plane that lies between $x = y^2$ and $x = -y^2 + 2$.
26. $\mathcal{U}$ lies below the graph of $z = 3x + y + 7$, above the graph of $z = -x - 2y - 4$, and over the region $\mathcal{U}$ in the xy-plane that is cut off by the line $x + 3y = 12$ in the first quadrant.
27. $\mathcal{U}$ is the solid bounded above by the cylinder $y^2 + z^2 = 49$, below by the xy-plane, and laterally by the cylinder $x^2 + y^2 = 49$.
28. $\mathcal{U}$ is the solid bounded by the two paraboloids $z = x^2 + y^2 + 1$ and $z = -x^2 - y^2 + 19$.

29. $\mathcal{U}$ is the solid above the graph of $z = x - y + 1$, below the graph of $z = 3x + 5y + 7$, and bounded on the sides by the vertical planes $x = y$, $x = -y$, and $y = -4x + 4$.

30. $\mathcal{U}$ is the region bounded above by the plane $3x - 2y - z = -8$, below by the plane $z = -2x - 3y - 6$, and with vertical sides given by the parabolic cylinders $x = y^2$ and $x = -y^2 - 8y + 10$.

Further Theory and Practice

31. Let $\mathcal{R}$ be the smaller region inside the circle $x^2 + y^2 - 4y + 8x = 6$ that is cut off by the line $y = x$. Calculate the volume of the solid that lies above $\mathcal{R}$ and below the plane $z = 2x + 8$.

32. Let $\mathcal{R}$ be the bounded region between the line $y = x + 12$ and the left branch of the hyperbola $x^2 - 4y^2 = 84$. Calculate the volume of the solid that lies over $\mathcal{R}$ and under the plane $z = -x$.

33. Find the volume of the solid that lies above the planar disk $(x + 4)^2 + y^2 \leq 1$ and below the plane $z = x + 4$.

34. Find the volume of the solid that lies below the graph of $z = 1/(x^2 + y^2)$ and that is bounded laterally by the cylinder set $y = |x|$ and the planes $y = 2$ and $y = 8$.

35. Find the volume of the solid bounded by the two paraboloids $z = x^2 + y^2 + 3x + 6y - 2$ and $z = -x^2 - y^2 - 5x + 2y + 6$.

36. Find the volume in the first octant that is below both the plane $z = 2x + y$ and the plane $z = 2 - x - y$.

37. Find the volume in the first octant that is inside the cylinder $x^2 + y^2 = 1$ and below the surface $z = xy$.

38. Find the volume of the solid bounded by the planes $z = 0$, $y = 0$, and $z = 1 - x + y$ and by the parabolic cylinder $y = 1 - x^2$.

39. Find the volume of the solid bounded above by the parabolic cylinder $z = 1 - y^2$ and below by the parabolic cylinder $z = x^2$.

40. Find the volume of the infinite solid bounded that lies below the surface $z = \exp(-x^2)$ and above the planar region $\{(x, y, 0) : x \geq |y|\}$.

41. Find the volume of the solid that is bounded above by the surface $z = 1/(x + y)$ and that lies over the region in the xy-plane between $y = 1/x$ and $y = 1$ for $1 \leq x \leq 2$.

42. Find the volume of the solid bounded by the paraboloid $z = 1 + x^2 + y^2$ and by the planes $x = 0$, $y = 0$, $z = 0$, and $x + 2y = 2$.

43. Find the volume of the solid region bounded above by $z = xy$, below by $z = 0$, and laterally by the plane $y = x$ and by the cylinder set $y = x^3$.

44. Suppose that $a < b$. Let $\mathcal{R}$ be the triangle with vertices (a, a), (b, a), and (b, b). Let u and v be continuously differentiable functions of one variable on $[a, b]$. Calculate $I = \iint_{\mathcal{R}} u'(x)v'(y)\,dA$ by using iterated integrals. Next, reverse the order of integration, and calculate I again. Equate the two expressions for I, and solve for $\int_a^b u(y)v'(y)\,dy$. What formula have you found?

Calculator/Computer Exercises

45. Let $\mathcal{R}$ be the region in the first quadrant of the xy-plane that lies between the curves $y = 10x - 1 - 6x^2$ and $y = x^3 - 3x^2 + 3x$. Find the volume of the solid that lies above $\mathcal{R}$ and below the plane $z = x + y$.

46. Let $\mathcal{R}$ be the region in the first quadrant of the xy-plane that lies between the curves $y = e^x$ and $y = 1 + 3x$. Find the volume of the solid that lies above $\mathcal{R}$ and below the plane $z = 10 - y$.

47. Let $\mathcal{R}$ be the region in the xy-plane that lies between the curves $y = 2 - x^3$ and $y = 1 + x^4$. Find the volume of the solid that lies above $\mathcal{R}$ and below the plane $z = y$.

48. Let $\mathcal{R}$ be the region in the xy-plane that is bounded by the lines $x = 0$, $y = 0$, and $y = 1 - x$. Use the Trapezoidal Rule approximation with $N = 4$ to estimate the volume of the solid region that lies above $\mathcal{R}$ and below $z = 1/(1 + xy)$. (The Trapezoidal Rule approximation for double integrals is discussed in the instructions for Exercises 65–68 of Section 14.2.)

14.4 Polar Coordinates

You are familiar with the idea of locating a point in the plane with two real numbers: In the Cartesian setup, one of these numbers corresponds to the x-displacement and the other to the y-displacement. These numbers represent signed distances to perpendicular axes. There are certain situations, however, in which it is natural to express the location of a point *relative to a fixed point,* which is taken to be the origin. For example, if a central force field (such as the gravitational field that we studied in Section 12.5) acts on a particle at point P, then we may find it more convenient to work with the distance $r = \|\overrightarrow{OP}\|$ of the particle to the center O than with the distances to the x- and y-axes. Of course, the distance r alone does not determine the location of the point P. For that, we will need information that tells us the direction of $\overrightarrow{OP}$. The angle that $\overrightarrow{OP}$ makes with a fixed half line emanating from O makes a convenient choice.

The Polar Coordinate System

Definition Let the *polar axis* be the half line emanating from the origin and pointing to the right. (The polar axis coincides with the positive x-axis of Cartesian coordinates.) Refer to Figure 1. Suppose that $r > 0$ and that $\theta \in \mathbb{R}$. The point in the plane with *polar coordinates* (r, θ) is that point $P = P(r, \theta)$ of distance r from the origin such that the ray $\overrightarrow{OP}$ subtends an angle of θ with the polar axis. The angle is measured counterclockwise when θ is positive and clockwise when θ is negative.

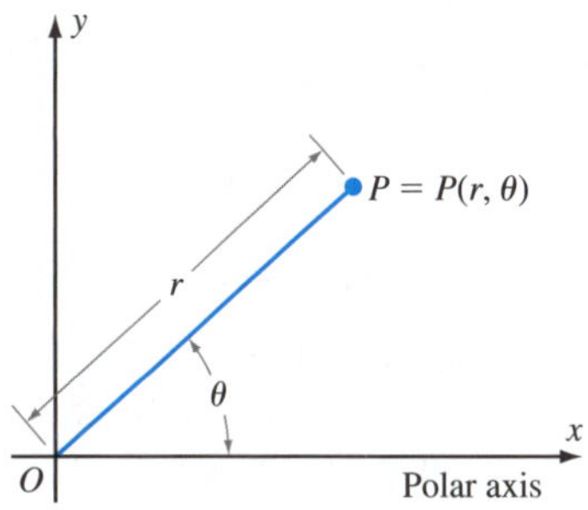

Figure 1

Once you have specified the radial coordinate r of a point, then you have already restricted the point to lie on a certain circle (see Figure 2). The angular coordinate θ specifies which point on the circle it is. Alternatively, once you have specified the angular coordinate θ, you have restricted the point to lie on a certain ray. Then the radial coordinate r tells you how far out on that ray the point lies. These ideas are also shown in Figure 2.

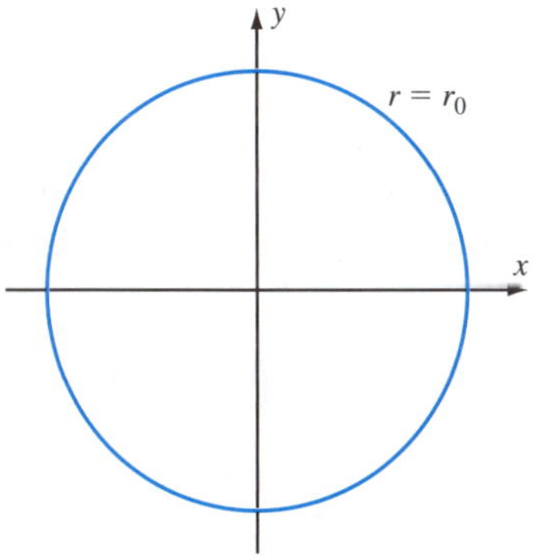

Figure 2a
The circle $r = r_0$

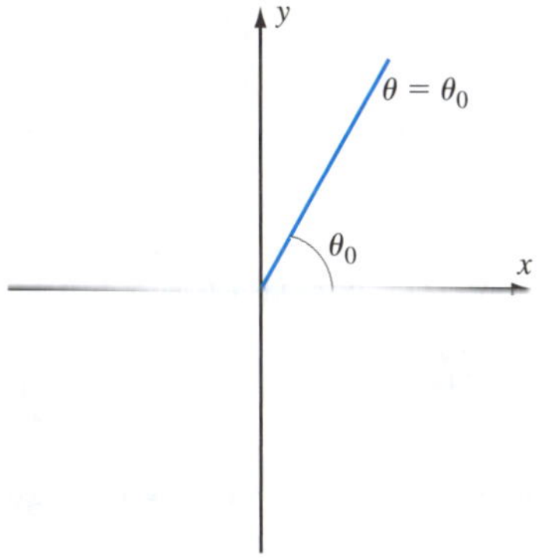

Figure 2b
The ray $\theta = \theta_0$

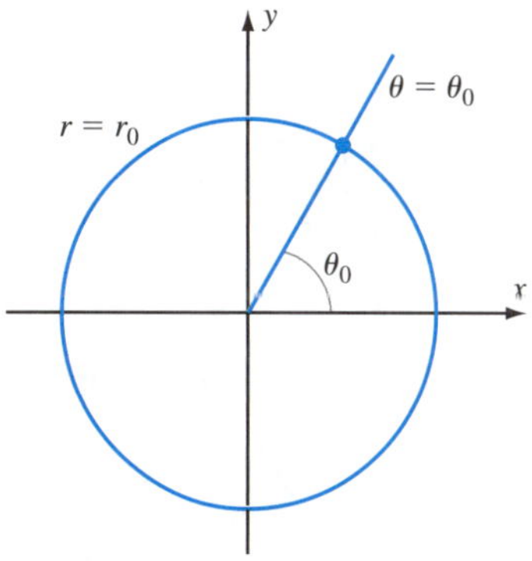

Figure 2c
The point whose polar coordinates are given by $r = r_0$ and $\theta = \theta_0$

Example 1 Use polar coordinates to plot the points $P = (2, \pi/4)$, $Q = (4, 3\pi/2)$, $R = (1/2, -\pi)$, $S = (5, 3\pi/4)$, $T = (1, 4\pi/3)$, $U = (3, -7\pi/6)$, and $V = (5/2, -7\pi/4)$.

Solution The points are exhibited in Figure 3, next page. ■

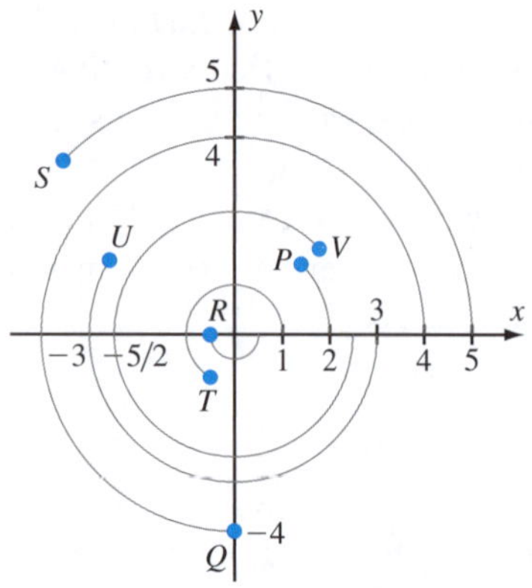

Figure 3

$P = (2, \pi/4)$ $\quad Q = (4, 3\pi/2)$
$R = (1/2, -\pi)$ $\quad S = (5, 3\pi/4)$
$T = (1, 4\pi/3)$ $\quad U = (3, -7\pi/6)$
$V = (5/2, -7\pi/4)$

inSIGHT

In Example 1, compare the first point P and the last point V. They lie on the same ray emanating from the origin. That is because their angular coordinates differ by an integer multiple of 2π: $\pi/4 = -7\pi/4 + 2\pi$. As you can imagine from Figure 3, if the radial coordinates of points P and V had been equal, then the points would have coincided, despite their different angular coordinates. The polar coordinates of a point in the plane are *not* unique (unlike its rectangular coordinates). Indeed, the polar coordinates $(r_0, \theta_0 + 2n\pi)$ represent the same point in the plane for all integer values of n. Thus, every point in the plane can be specified in infinitely many different ways in the polar coordinate system. The polar coordinates of the origin are especially ambiguous: The radial coordinate r must be 0, but the angular coordinate θ may have any value.

Negative Values of the Radial Variable

It is often convenient to allow the radial coordinate r to be negative. If $r < 0$, then the polar coordinates (r, θ) refer to the point P in the plane that is a distance $|r|$ from the origin and that lies on the ray opposite the ray that forms an angle θ with the polar axis. In other words, the ordered pairs (r, θ) and $(-r, \theta)$ are polar coordinates for points that are reflections of each other through the origin. The ordered pairs (r, θ) and $(-r, \theta + \pi)$ are polar coordinates for the same point. See Figure 4.

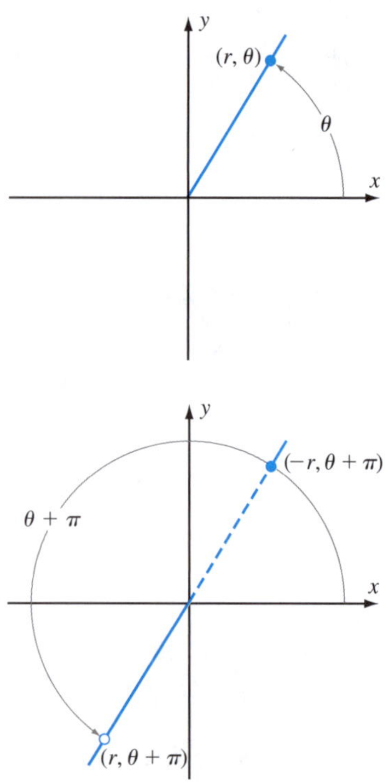

Figure 4

Example 2 Give all possible polar coordinates of the point P whose distance to the origin is 4 such that $\overrightarrow{OP}$ subtends an angle of $5\pi/6$ with the positive x-axis.

Solution In effect, we have been told that $(4, 5\pi/6)$ is a pair of polar coordinates for P. However, we may add any multiple of 2π to the angle and obtain another valid set of polar coordinates for this point. Therefore,

$$\ldots \left(4, -\frac{19\pi}{6}\right), \left(4, -\frac{7\pi}{6}\right), \left(4, \frac{5\pi}{6}\right), \left(4, \frac{17\pi}{6}\right), \left(4, \frac{29\pi}{6}\right) \ldots$$

are all polar coordinates for P. We summarize this by saying

$$\left(4, \frac{5\pi}{6} + 2n\pi\right), \qquad n \in \mathbb{Z}$$

are polar coordinates for P. Is this a complete list of all possible polar coordinates for P? No, because we can allow r to be negative. Thus, P may be thought of as the point $(-4, -\pi/6)$. See Figure 5. Again, we can add any multiple of 2π to the angle. Therefore, the pairs

$$\left(-4, -\frac{\pi}{6} + 2n\pi\right), \qquad n \in \mathbb{Z}$$

are also valid polar coordinates for P. Since P must have radial coordinate ± 4 and must have angular coordinate either $5\pi/6 + 2n\pi$ or $-\pi/6 + 2n\pi$, we have accounted for all possible polar coordinate representations of P. ■

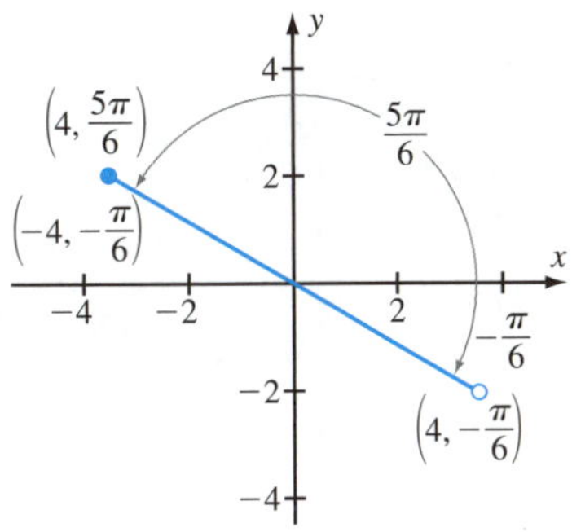

Figure 5

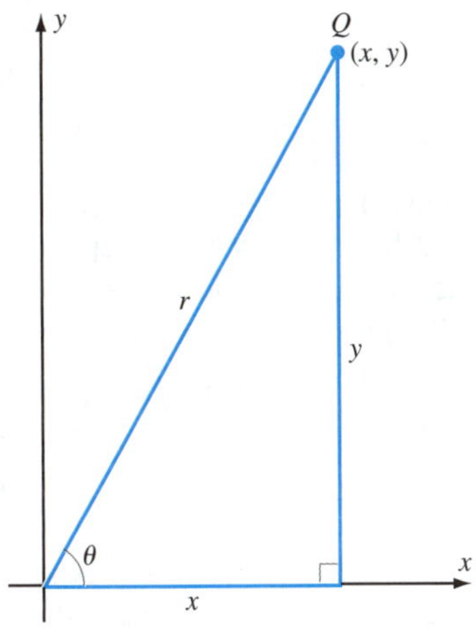

Figure 6

Relating Polar Coordinates to Rectangular Coordinates

Now we discuss the connection between polar coordinates and Cartesian or rectangular coordinates. Look at Figure 6. The point Q has polar coordinates (r, θ) and rectangular coordinates (x, y). By examining the right triangle, we see that

$$x = r\cos(\theta) \quad \text{and} \quad y = r\sin(\theta). \tag{14.8}$$

As expected, the rectangular coordinates x and y of a point are uniquely determined by its polar coordinates r and θ. We also have equations that tell us the *possible* values of the polar coordinates r and θ of a point, given its rectangular coordinates x and y:

$$r^2 = x^2 + y^2 \quad \text{and, if } x \neq 0, \quad \tan(\theta) = \frac{y}{x}. \tag{14.9}$$

As expected from our earlier discussion, these formulas do not determine r and θ uniquely. The first formula in line (14.9) tells us the magnitude—$|r| = \sqrt{x^2 + y^2}$—of the radial coordinate r but not its sign. The second formula in line (14.9) is satisfied by infinitely many values of the angular coordinate θ.

Example 3 Compute the rectangular coordinates of the point P whose polar coordinates are $(4, 5\pi/6)$.

Solution Observe that $r = 4$ and $\theta = 5\pi/6$, so $x = 4\cos(5\pi/6) = -2\sqrt{3}$ and $y = 4\sin(5\pi/6) = 2$. Thus, P has the ordered pair $(-2\sqrt{3}, 2)$ for its rectangular coordinates. ■

inSIGHT

Notice that P is the same point in Examples 2 and 3. In Example 2, we determined infinitely many ways to express P in polar coordinates. Whichever of these representations we choose to work with, the conversion to rectangular coordinates will be the ordered pair $(-2\sqrt{3}, 2)$ that we obtained in Example 3. For instance, the polar coordinates $(-4, -\pi/6)$ lead to $x = (-4)\cos(-\pi/6) = -2\sqrt{3}$ and $y = (-4)\sin(-\pi/6) = 2$.

Example 4 Calculate all possible polar coordinates for the point Q with rectangular coordinates $(-5, 5)$.

Solution We know that $x = -5$, and $y = 5$. Therefore, $r^2 = x^2 + y^2 = 25 + 25 = 50$. If we take $r = +\sqrt{50} = 5\sqrt{2}$, then the identities $x = r\cos(\theta)$ and $y = r\sin(\theta)$

yield

$$\cos(\theta) = -\frac{1}{\sqrt{2}} \quad \text{and} \quad \sin(\theta) = \frac{1}{\sqrt{2}}.$$

It follows that $\theta = 3\pi/4 + 2n\pi, n \in \mathbb{Z}$. So we have found the polar coordinates $(5\sqrt{2}, 3\pi/4 + 2n\pi)$, $n \in \mathbb{Z}$. However, r could also have the value $-\sqrt{50} = -5\sqrt{2}$. In this case, the equations $x = r\cos(\theta)$ and $y = r\sin(\theta)$ yield

$$\cos(\theta) = \frac{1}{\sqrt{2}} \quad \text{and} \quad \sin(\theta) = -\frac{1}{\sqrt{2}}.$$

It follows that $\theta = 7\pi/4 + 2n\pi$, $n \in \mathbb{Z}$. So we have found the polar coordinates $(-5\sqrt{2}, 7\pi/4 + 2n\pi)$, $n \in \mathbb{Z}$. This accounts for all possible polar coordinates for Q. ■

inSIGHT

If we encounter the ordered pair $(-5\sqrt{2}, 7\pi/4)$, how can we tell whether it represents polar coordinates or rectangular coordinates? The answer is that we must tell from *context* whether polar or rectangular coordinates are intended.

Graphing in Polar Coordinates

Graphing in polar coordinates is different from graphing in rectangular coordinates. We mention two new features in particular. The first is that the pictorial notions we have developed for graphing in rectangular coordinates—slope, x- and y-intercepts, and so on—no longer apply. The second is that the ambiguity in the polar coordinates of a point means that we have to be careful not to forget part of the graph. Because of these twists, we begin with the most elementary concepts.

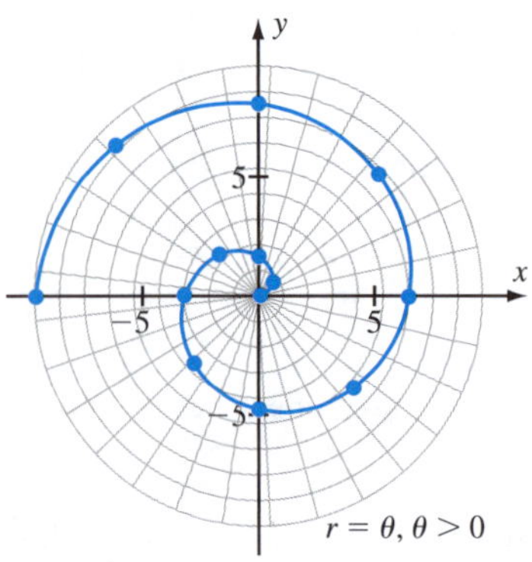

Figure 7

Example 5 Sketch the graph of $r = \theta$.

Solution The graph of $y = x$ in *rectangular* coordinates is a line, but do not expect to see anything like that now. In Figure 7, we have plotted a few points corresponding to $\theta = 0, \pi/4, \pi/2, 3\pi/4, \pi$, and so on. We notice that as θ increases from $\theta = 0$, so does r increase. Therefore, it is logical to connect the points with a smooth curve. Notice that the resulting curve is a spiral that grows with rotation in the counterclockwise direction. The graph is not complete, because we have not considered negative values of θ. We plot some points corresponding to $\theta = -\pi/4, -\pi/2, -3\pi/4, -\pi, \ldots$ and connect them as we did for $\theta > 0$. Thus, the complete graph consists of two spirals, one expanding with clockwise rotation and the other expanding with counterclockwise rotation (Figure 8). The curve we have graphed is called a *Spiral of Archimedes* (as is any curve whose equation in polar coordinates is $r = a\theta$ where a is a constant). ■

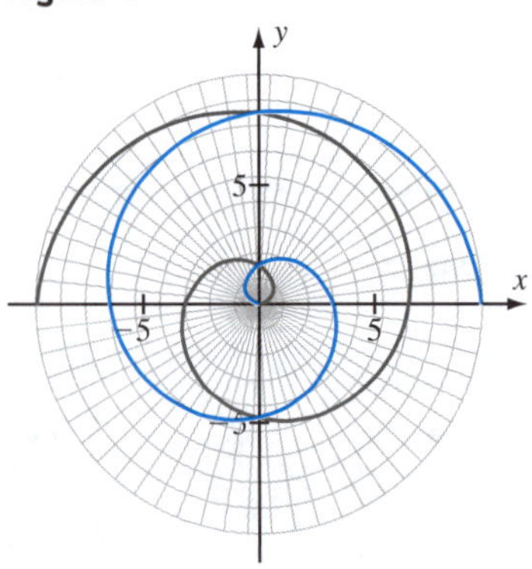

Figure 8
Spiral of Archimedes
$r = \theta$

Example 6 Describe the graphs of $r\cos(\theta) = 2$ and $r\sin(\theta) = -1$.

Solution The easiest thing to do is convert these equations to rectangular coordinates. They become $x = 2$ and $y = -1$. The graphs are lines parallel to the rectangular coordinate axes. ■

Example 7 Sketch the graph of $r = 2\sin(\theta)$ by two different methods.

Solution For the first method, we create a table of values.

$\boldsymbol{\theta}$	0	$\pi/6$	$\pi/4$	$\pi/3$	$\pi/2$	$2\pi/3$	$3\pi/4$	$5\pi/6$	π	$7\pi/6$	$5\pi/4$	$4\pi/3$	$\pi/2$	$5\pi/3$	$7\pi/4$	$11\pi/6$	2π
$\boldsymbol{r}$	0	1	$\sqrt{2}$	$\sqrt{3}$	2	$\sqrt{3}$	$\sqrt{2}$	1	0	-1	$-\sqrt{2}$	$-\sqrt{3}$	-2	$-\sqrt{3}$	$-\sqrt{2}$	-1	0

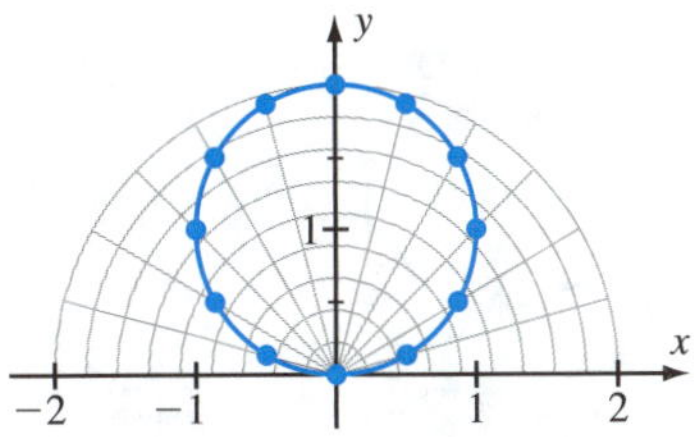

Figure 9a

The corresponding points are plotted in Figure 9a. These points are connected, *in increasing order of* θ, by a smooth curve. We see that a circular loop beginning and ending at the origin is traced when θ increases from 0 to π. This loop is retraced when θ increases from π to 2π.

An alternative method for sketching this graph is to multiply both sides of the equation by r to obtain $r^2 = 2r\sin(\theta)$. Then it is easy to convert to rectangular coordinates: $x^2 + y^2 = 2y$. By completing the square, we obtain the equation $x^2 + (y-1)^2 = 1$ of the circle with center (0, 1) and radius 1. ■

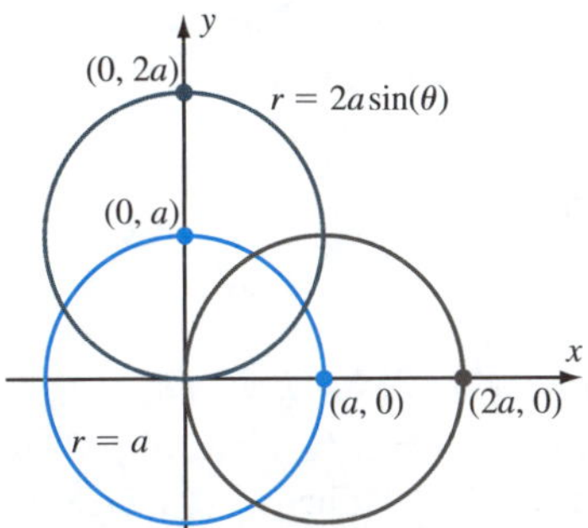

Figure 9b

inSIGHT

> The graph of $r = a$ is a circle of radius a. However, $r = a$ is not the only polar equation that has a circle of radius a as its graph. When generalized, the calculations of Example 7 show that the graph of $r = 2a\sin(\theta)$ is a circle of radius a. In a similar manner, we may show that the graph of $r = 2a\cos(\theta)$ is also a circle of radius a. Figure 9b shows these three circles in one viewing window.

Example 8 Sketch $r = 3(1 + \cos(\theta))$.

Solution Notice that as θ increases from 0 to $\pi/2$, $\cos(\theta)$ decreases from 1 to 0; hence, r decreases from 6 to 3. Likewise, as θ increases from $\pi/2$ to π, we see that r decreases from 3 to 0. This information is indicated in Figure 10. We now have the upper half of the curve. We could continue this type of analysis to get the lower half of the curve. Instead, we notice that when θ is replaced by $-\theta$, the equation does not change. So, for every point (r, θ) on the graph, there is a corresponding point $(r, -\theta)$. This means the graph of $r = 3(1 + \cos(\theta))$ is symmetric about the horizontal axis. The complete sketch appears in Figure 11. A curve whose polar equation has the form $r = a(1 + \cos(\theta))$ or $r = a(1 - \cos(\theta))$ or $r = a(1 + \sin(\theta))$ or $r = a(1 - \sin(\theta))$ is called a *cardioid*. ■

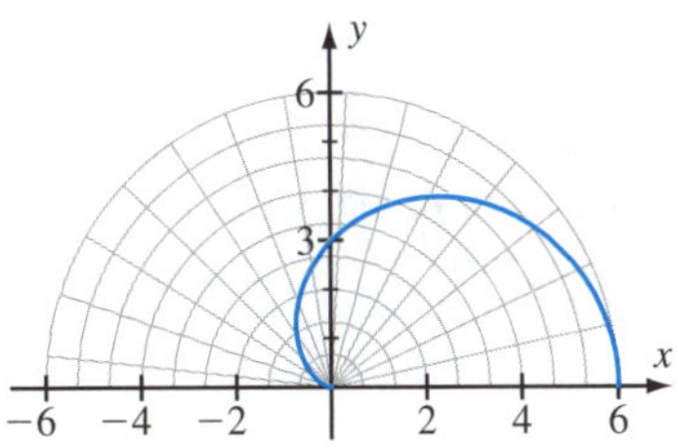

$r = 3(1 + \cos(\theta)),\ 0 \le \theta \le \pi$

Figure 10

Example 9 Sketch the graph of $r = 4\sin(2\theta)$.

Solution When θ increases from 0 to $\pi/4$, then r increases from 0 to 4. When θ then passes from $\pi/4$ to $\pi/2$, the value of r decreases monotonically back to 0. The corresponding sketch is in Figure 12, next page. Figure 13 shows the added information obtained from considering $\pi/2 \le \theta \le \pi$: The value of r decreases from 0 to -4 and then increases again back to 0. Notice that although the values of θ correspond to the second quadrant, this portion of the plot appears in the diagonally *opposite* quadrant, the fourth quadrant, because r is negative.

Now let us see what we can learn from symmetry. Notice that

$$\sin(2(\theta + \pi)) = \sin(2\theta).$$

So, to each point (r, θ) on the curve, there is a corresponding point $(r, \theta + \pi)$. This tells us that the graph is symmetric with respect to reflection through the origin. Glancing at Figure 13 and taking this symmetry into account, we complete our sketch as in Figure 14. This curve is called a *four-leafed rose*. ■

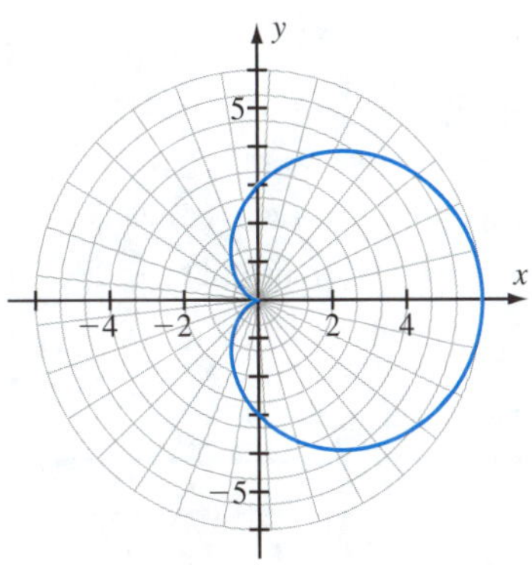

$r = 3(1 + \cos(\theta))$

Figure 11

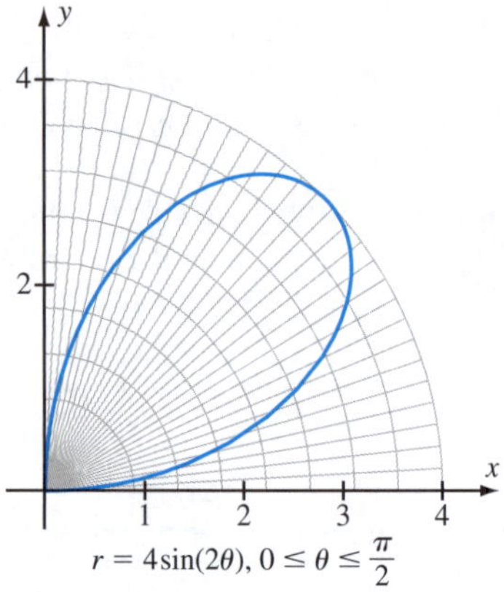

Figure 12

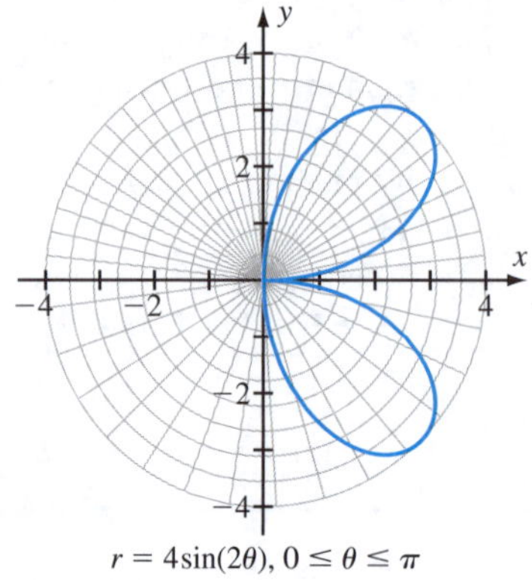

Figure 13

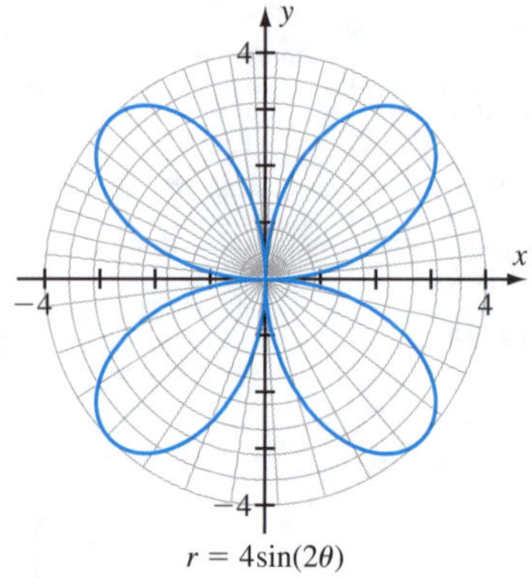

Figure 14

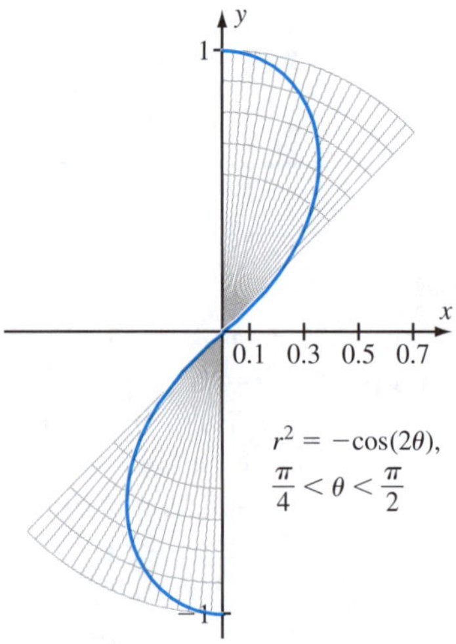

Figure 15

Example 10

Sketch the curve $r^2 = -\cos(2\theta)$.

Solution We notice that r is undefined when $\cos(2\theta)$ is positive, so certain values of θ must be ruled out. For instance, r is undefined for $0 < \theta < \pi/4$. However, when θ increases from $\pi/4$ to $\pi/2$, then $\cos(2\theta)$ decreases from 0 to -1; hence, r^2 increases from 0 to 1. This will give rise to *two branches* of our curve: one corresponding to r increasing from 0 to 1 and the other corresponding to r decreasing from 0 to -1. These are illustrated in Figure 15. Once again, notice the phenomenon of plotting that part of a curve with $r < 0$: It appears in the reflection through the origin of the sector defined by the range of θ.

Next we observe that $\cos(2(\pi - \theta)) = \cos(2\theta)$; hence, the equation $r^2 = -\cos(2\theta)$ is unchanged under the map $\theta \mapsto \pi - \theta$. It follows that for each point (r, θ) on the curve, there is a corresponding point $(r, \pi - \theta)$ that is also on the curve. This property tells us that the graph is symmetric with repect to the y-axis. See Figure 16. The graph of $r^2 = -\cos(2\theta)$ is called a *lemniscate*. ■

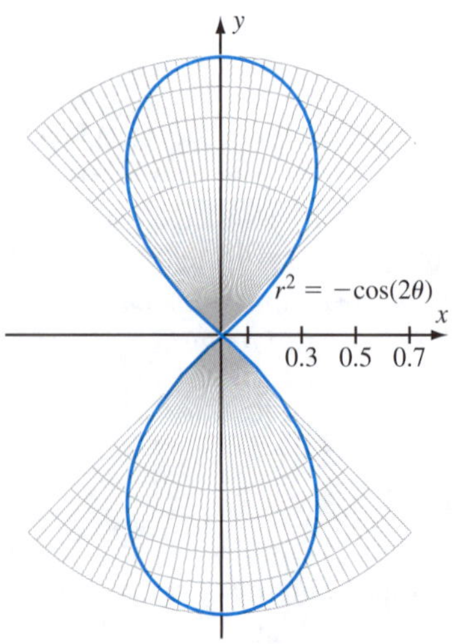

Figure 16

inSIGHT

To be sure that we have not forgotten any part of the graph in Example 10, we can perform a check. We have already considered $0 < \theta < \pi/2$. For $\pi/2 \le \theta \le 3\pi/4$, the quantity $-\cos(2\theta)$ decreases from 1 to 0, which is indicated in Figure 16. For $3\pi/4 < \theta < \pi$, we see that $-\cos(2\theta)$ is negative, so that r is undefined. That is also indicated by our figure. We need not check any other values of θ, since $\cos(2\theta)$ has period π. The picture is just repeated as θ ranges from π to 2π. Another way of making this last observation is to notice that the equation is unchanged if θ is replaced by $-\theta$; hence, to every point (r, θ) on the graph, there corresponds a point $(r, -\theta)$. This corresponds to symmetry in the polar axis.

Symmetry Principles in Graphing

We have seen that symmetry can be a powerful tool in graphing polar equations. We summarize here the basic tests for symmetry.

Symmetry in	Test
x-axis	$\theta \mapsto -\theta$
y-axis	$\theta \mapsto \pi - \theta$
origin	$r \mapsto -r$ or $\theta \mapsto \theta + \pi$

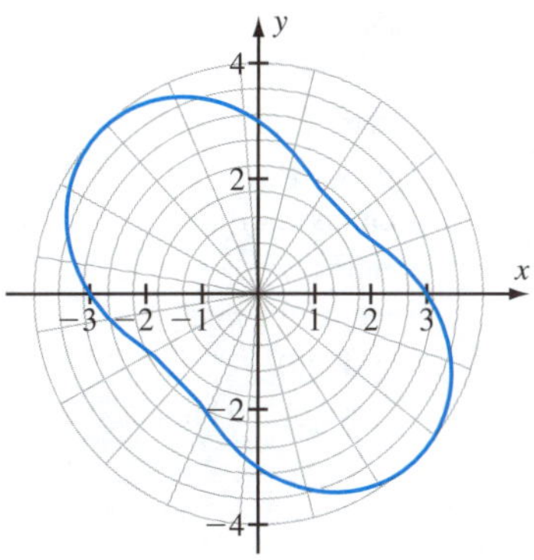

$r = 3 - \sin(2\theta)$

Figure 17

Example 11 Test the equation $r = 3 - \sin(2\theta)$ for symmetry.

Solution Since

$$\sin 2(\pi - \theta) = \sin(2\pi - 2\theta) = \sin(-2\theta) = -\sin(2\theta),$$

the test for symmetry in the y-axis fails. Since $\sin(2(-\theta)) \neq \sin(2\theta)$, the test for symmetry in the x-axis fails. Finally, because $\sin(2(\theta + \pi)) = \sin(2\theta + 2\pi) = \sin(2\theta)$, the equation is unaltered if θ is replaced by $\theta + \pi$. We conclude that there *is* symmetry in the origin. The curve is sketched in Figure 17. ■

inSIGHT

Notice that there are two tests for symmetry in the origin. The one that we used in Example 11 is to replace θ by $\theta + \pi$. The other test is to replace r by $-r$. The curve in Example 11 *fails* this second test. Again, the fact that a single point has several different polar representations is causing confusion. Always remember that our symmetry tests are *sufficient conditions* for symmetry, not necessary conditions. This means that if a curve passes the test, then it is certainly symmetric; if it does not pass the test, it may or may not be symmetric. In practice, you should just apply the tests that you know and use to advantage whatever information they give you.

quickquiz

1. What are the polar coordinates of the point with rectangular coordinates $(1, \sqrt{3})$?
2. What are the rectangular coordinates of the point with polar coordinates $(-2, \pi/2)$?
3. Describe the curve that has polar equation $r = 1 + 2\theta, \theta > 0$.
4. If f is an odd function, then what symmetry does the curve defined by $r = f(\theta)$ have? What if f is even?

EXERCISES

Problems for Practice

In Exercises 1–8, plot the four points given as polar coordinates.

1. $(3, \pi/4), (-3, \pi/4), (3, -\pi/4), (-3, -\pi/4)$
2. $(2, 5\pi/6), (-2, 5\pi/6), (2, -5\pi/6), (-2, -5\pi/6)$
3. $(0, \pi/2), (0, \pi), (0, -\pi), (0, -\pi/2)$
4. $(6, 9\pi/4), (-6, 9\pi/4), (6, -9\pi/4), (-6, -9\pi/4)$
5. $(4, -7\pi/2), (-4, -7\pi/2), (4, 7\pi/2), (-4, 7\pi/2)$
6. $(1, \pi/3), (-1, \pi/3), (1, -\pi/3), (-1, -\pi/3)$
7. $(2, 7\pi/3), (-2, 7\pi/3), (2, -7\pi/3), (-2, -7\pi/3)$
8. $(10, 0), (-10, 0), (6, 0), (0, 0)$

In Exercises 9–16, the polar coordinates of a point are given. What are its rectangular coordinates?

9. $(4, \pi/4)$
10. $(7, 9\pi/2)$
11. $(0, \pi)$
12. $(-8, -3\pi/4)$
13. $(-4, -17\pi/6)$
14. $(-3, 5\pi/6)$
15. $(-6, \pi/3)$
16. $(3, 0)$

In Exercises 17–22, give all possible polar coordinates for the point that is given in rectangular coordinates.

17. $(4, -4)$
18. $\left(-4\sqrt{2}, -4\sqrt{2}\right)$
19. $(0, -9)$
20. $(-5, -5)$
21. $(-1, 0)$
22. $\left(-8\sqrt{3}, 8\right)$

In Exercises 23–32, test the equation for symmetry in the x-axis, the y-axis, and the origin.

23. $r^2 = 3\sin(\theta)$
24. $r = 4 - 2\tan(2\theta)$
25. $r = 3$
26. $\theta = \pi/2$
27. $r = 1 + \cos^2(\theta)$
28. $r\sin(\theta) = \cos(\theta)$
29. $r = 6 - \cos(3\theta)$
30. $r = 4 + \sin(3\theta)$
31. $r = \sin(\theta) - \cos(\theta)$
32. $r^2 = 1 - 2\sin(\theta)$

In Exercises 33–44, sketch the curve that is described by the equation given in polar coordinates. (The name of each curve has been provided in parentheses.)

33. $r = 1 - \cos(\theta)$ (cardioid)
34. $r = 3 + 3\sin(\theta)$ (cardioid)
35. $r = 3\cos(2\theta)$ (four-leafed rose)
36. $r^2 = 9\cos(2\theta)$ (lemniscate)
37. $r = 4 - 2\sin(\theta)$ (limaçon)
38. $r = 2\sin(3\theta)$ (three-leafed rose)
39. $r = -1 + 3\cos(\theta)$ (limaçon)
40. $r^2 = -3\sin(2\theta)$ (lemniscate)
41. $r = 3\sin(\theta) - 2\cos(\theta)$ (circle)
42. $r = -3\sin(5\theta)$ (five-leafed rose)
43. $r = 3 - 3\sin(\theta)$ (cardioid)
44. $r = -2\theta$ (Spiral of Archimedes)

Further Theory and Practice

45. Give all possible polar coordinates for the following points given in rectangular coordinates.
    a. $(3, \pi/4)$ **b.** $(-6, 7\pi/2)$
    c. $(\pi/4, -8\pi/3)$
46. Give rectangular coordinates for the following points given in polar coordinates.
    a. $(2, -6)$ **b.** $(-4, 7)$
    c. $(\pi/3, 9)$ **d.** $(7\pi/3, -6)$
47. Give a formula in polar coordinates for the distance between two points in the plane with polar coordinates (r_1, θ_1) and (r_2, θ_2). What is the polar equation of a circle of radius 1 with center given by $(3, \pi/4)$ in polar coordinates?
48. A regular N-gon (a polygon with N equal sides) is inscribed in a circle of radius r that is centered at the origin. If one of the vertices is on the positive x-axis, what are the polar coordinates of the vertices?

In Exercises 49–54, find all points of intersection of the two polar curves.

49. $r = \cos(\theta)$ and $r = 1 - \cos(\theta)$
50. $r = 1 + \sin(\theta)$ and $r = 3(1 - \sin(\theta))$
51. $r = 1$ and $r = \tan(\theta)$
52. $r = 3\sec(\theta)$ and $r = 4\cos(\theta)$
53. $r = 1 + \cos(\theta)$ and $r^2 = (1/2)\cos(\theta)$

54. $r = 4 - 5\sin(\theta)$ and $r = 3\sin(\theta)$
55. Find each ordered pair (a, b) that represents the same point in the plane in both polar and rectangular coordinates.
56. Suppose f is a continuously differentiable scalar-valued function of one variable. If $\mathcal{C}$ is a curve given in polar coordinates by the equation $r = f(\theta)$ for $\alpha \le \theta \le \beta$, then the arc length of vector **C** is

$$\int_{\alpha}^{\beta} \sqrt{f(\theta)^2 + f'(\theta)^2}\, d\theta.$$

Prove this formula. (*Hint:* Describe the curve in terms of the position vector $\mathbf{r}(\theta) = \langle x(\theta), y(\theta)\rangle$, and use the formula for arc length that was developed in Chapter 12.)

Refer to Exercise 56 for the arc length formula in polar coordinates. In Exercises 57–60, calculate the arc length of the given curve.

57. $r = e^{\theta}, -\pi/2 \le \theta \le \pi/2$
58. $r = a(1 - \cos(\theta))$
59. $r = \sin^3(\theta/3), 0 \le \theta \le \pi/2$
60. $r = \sec(\theta), 0 \le \theta \le \pi/4$
61. Sketch the graph of $r = \exp(\theta/5), -3\pi \le \theta \le 2\pi$.
62. Sketch the graph of $r = 2\sin^2(\theta/2)$.
63. Sketch the graph of $r^2 = -3\sin(4\theta)$.
64. Sketch the graphs of $r^2 = \sin^2(\theta)$ and $r = \sin(\theta)$.

Explain how and why they differ.

65. Sketch the graphs of $r = \theta^2$ and $\theta = r^2$.
66. Sketch the graph of $r^3 = \sin^2(\theta)$.
67. Sketch the graph of $r^{1/2} = \sin(\theta)$.
68. Sketch the graph of $r^{1/3} = \sin(\theta)$.
69. In a θr-plane, plot the Cartesian equation $r = \theta + \cos(\theta), 0 \le \theta \le 2\pi$. Then plot this equation by interpreting it as a polar equation in the xy-plane.
70. In a θr-plane, plot the Cartesian equation $r = (2 + \theta^2)/(1 + \theta^2), 0 \le \theta \le 4\pi$. Then plot this equation by interpreting it as a polar equation in the xy-plane.

Calculator/Computer Exercises

71. Plot the polar curves $r = \sin(j\theta)$ for $j = 1, 2, 3, 4, 5$. Describe the graphs of $r = \sin(50 \cdot \theta)$ and $r = \sin(51\theta)$. Suppose j is a positive integer. For what positive integer values of k does the graph of $r = \sin(k \cdot \theta)$ have more petals than the graph of $r = \sin(j\theta)$?
72. Plot the polar curves $r = j + \sin(\theta)$ for $j = 0, 1, 2, 3, 5, 10, 20, 30$. How does changing the additive constant affect the graph?
73. Plot the polar curves $r = \sin(\theta + j)$ for $j = 0, 1, 2, 3, 5, 10, 20, 30$. How does changing the additive constant affect the graph?
74. Plot Fay's Butterfly: $r = \exp(\cos(\theta)) - 2\cos(4\theta) + \sin^5(\theta/12)$ for $0 \le \theta \le 2\pi$. Repeat, using the range $0 \le \theta \le 4\pi$ and again for $0 \le \theta \le 6\pi$. Continue. For what range $0 \le \theta \le 2N\pi$ do you obtain the complete plot without any repetition?

14.5 Integrating in Polar Coordinates

The theory of area in Cartesian coordinates is predicated on the basic notion that any area can be thought of as a sum of areas of small rectangles, up to a relatively small error. This is sensible because the Cartesian coordinate system is a rectangular one—the axes are perpendicular, and the notion of area *fits* the notion of coordinates. Now that we are dealing with polar coordinates, there is no longer a standard pair of perpendicular directions. In fact, we have a radial direction and an angular direction, which are qualitatively quite different from the information coming from rectangular coordinates. As a result, the building blocks we use for area are not rectangles. Instead, they are sectors.

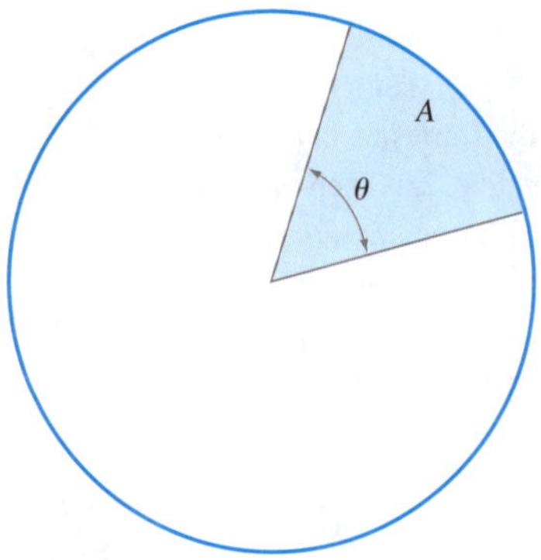

Figure 1
The area of the sector is $A = \frac{1}{2}r^2\theta$.

A *sector* is a portion of a circle cut off by two rays (see Figure 1). The area A of the sector depends on the radius r of the circle and the angle θ between the two rays. By comparing the sector with the full circle and using proportionality, we have $A/(\pi r^2) = \theta/(2\pi)$, or

$$A = \frac{1}{2}r^2\theta. \qquad \textbf{(14.10)}$$

Our method for calculating area in polar coordinates will be determined by this formula.

Areas of More General Regions

We now consider a region contained inside a curve $r = \varphi(\theta)$. More precisely, we consider a region of the form

$$\mathcal{R} = \{(r, \theta) : 0 \le r \le \varphi(\theta), \alpha \le \theta \le \beta\}$$

for a given function $\varphi(\theta) \ge 0$ and limiting values α and β for θ. To avoid ambiguities, we usually restrict α and β to differ by not more than 2π. The region is shown in Figure 2. We estimate the area of the region $\mathcal{R}$ by partitioning the interval $[\alpha, \beta]$, which is the domain of the independent variable:

$$\alpha = \theta_0 < \theta_1 < \theta_2 < \cdots < \theta_{N-1} < \theta_N = \beta.$$

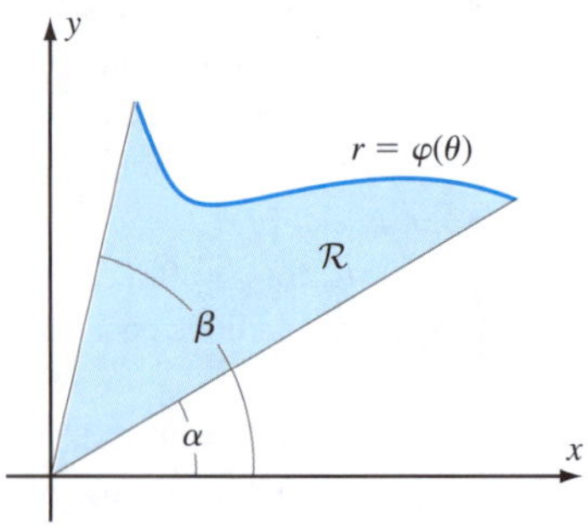

Figure 2

For simplicity, we will use a uniform partition

$$\Delta\theta = \theta_j - \theta_{j-1} = \frac{\beta - \alpha}{N} \quad \text{for all} \quad 1 \le j \le N.$$

Figure 3 shows how the arc corresponding to $\alpha \le \theta \le \beta$ is broken into subarcs by the partition. Now each subarc determines (not a rectangle but) a sector. The sum of the areas of the sectors,

$$\mathcal{S} = \sum_{j=1}^{N} S_j$$

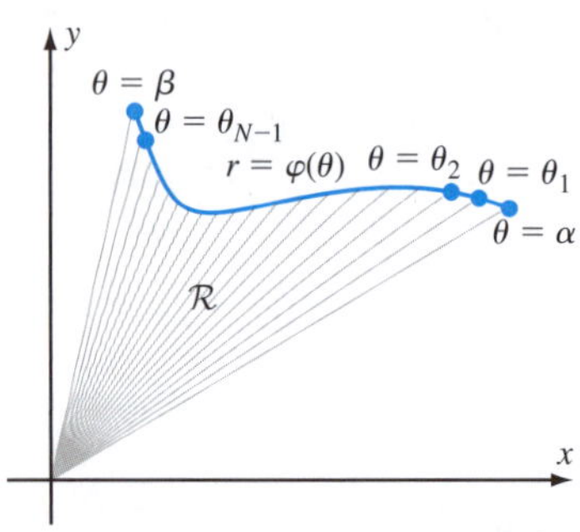

Figure 3

(where S_j is the area of the jth sector), approximates the area of the region $\mathcal{R}$.

To calculate this sum, we need to know the area of the jth sector. The angle the sector makes at the origin is $\Delta\theta = \theta_j - \theta_{j-1}$. The radius is approximately given by $r_j = \varphi(\xi_j)$ where ξ_j is an element chosen from the interval $[\theta_{j-1}, \theta_j]$. Refer to Figure 4. Thus, according to formula (14.10), the area of the jth sector is approximately

$$S_j \approx \frac{1}{2}\varphi(\xi_j)^2\Delta\theta.$$

The area of the region $\mathcal{R}$ is therefore approximately

$$\mathcal{S} \approx \sum_{j=1}^{N} \frac{1}{2}\varphi(\xi_j)^2\Delta\theta.$$

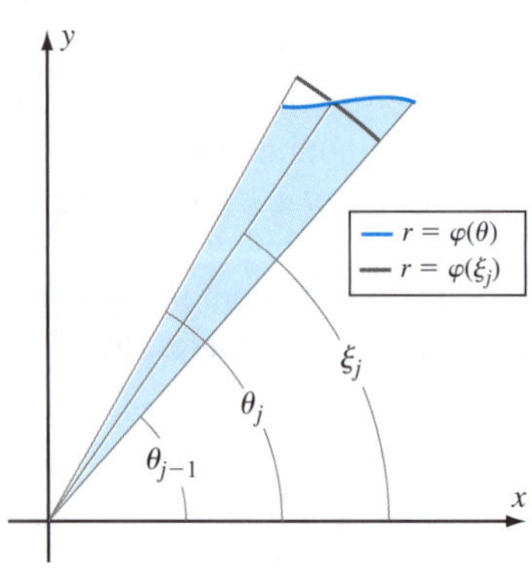

Figure 4

By recognizing that this sum is a Riemann sum for the integral

$$\int_{\alpha}^{\beta} \frac{1}{2}\,\varphi(\theta)^2\,d\theta,$$

we obtain the following theorem.

Theorem 1 If φ is a nonnegative function on the interval $[\alpha, \beta]$ and if $\beta - \alpha \leq 2\pi$, then the area of the region $\mathcal{R} = \{(r, \theta) : 0 \leq r \leq \varphi(\theta),\ \alpha \leq \theta \leq \beta\}$ is given by the integral

$$\frac{1}{2}\int_{\alpha}^{\beta} \varphi(\theta)^2\,d\theta. \qquad \textbf{(14.11)}$$

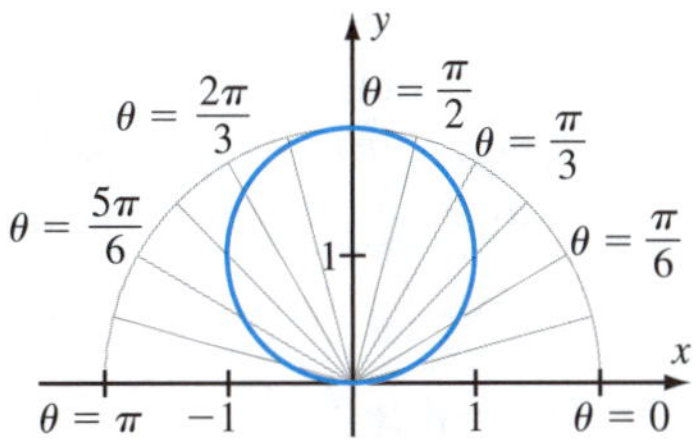

Figure 5
The curve described by the equation $r = 2\sin(\theta)$ in polar coordinates

Example 1 Find the area inside the curve $r = 2\sin(\theta)$.

Solution In Section 14.4, we learned that the curve $r = 2\sin(\theta)$, shown in Figure 5, is a circle of radius 1 that is traced out once as θ varies from 0 to π. Therefore, we expect to obtain $\pi = \pi \cdot 1^2$ as our answer. That is exactly the answer we obtain when we apply Theorem 1 using the correct interval of integration, $[0, \pi]$. Thus,

$$\text{area} = \frac{1}{2}\int_0^{\pi} (2\sin(\theta))^2\,d\theta = 2\int_0^{\pi} \sin^2(\theta)\,d\theta \overset{(7.21)}{=} 2 \cdot \frac{\pi}{2} = \pi.$$

inSIGHT

Example 1 illustrates an important point. Because of the ambiguity built into polar coordinates, the circle in Example 1 is generated twice as θ ranges from 0 to 2π. If we attempt to calculate the area inside it by integrating from 0 to 2π, we count the same area twice and obtain the incorrect value 2π as the answer. As a result of this observation, we must begin the solution to each area problem with a brief analysis to be sure we are avoiding duplication of area. A sketch will usually play a vital role in this analysis.

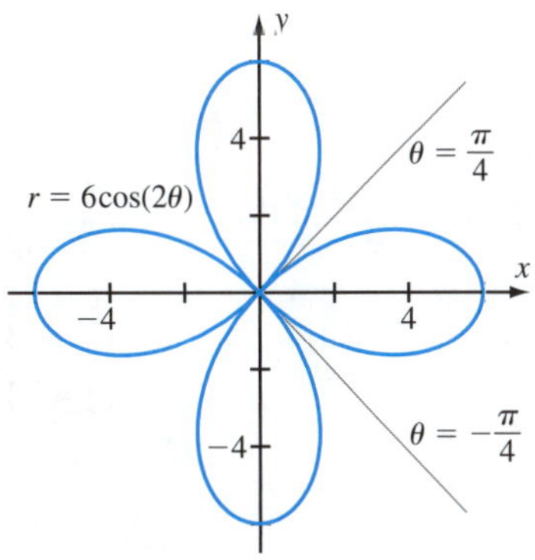

Figure 6

Example 2 Find the area inside one leaf of the rose $r = 6\cos(2\theta)$.

Solution By the periodicity of the cosine function, each of the leaves has the same area, so we choose for convenience the leaf on the right (look at Figure 6). We must integrate from $\theta = -\pi/4$ to $\theta = \pi/4$. By Theorem 1, the area of the leaf is

$$A = \frac{1}{2}\int_{-\pi/4}^{\pi/4} (6\cos(2\theta))^2\,d\theta = 18\int_{-\pi/4}^{\pi/4} \cos^2(2\theta)\,d\theta.$$

As we learned in Section 7.3, we can integrate $\cos^2(k\theta)$ by using the identity $\cos^2(k\theta) = (1+\cos(2k\theta))/2$. Using this identity with $k = 2$, we obtain

$$A = 18\int_{-\pi/4}^{\pi/4} \frac{(1+\cos(4\theta))}{2}\,d\theta = 9\left(\theta + \frac{\sin(4\theta)}{4}\bigg|_{-\pi/4}^{\pi/4}\right)$$
$$= 9\left(\left(\frac{\pi}{4} - \left(-\frac{\pi}{4}\right)\right) - (\sin(\pi) - \sin(-\pi))\right) = \frac{9\pi}{2}.$$

The area of one leaf of the rose is $9\pi/2$. ■

Example 3 Find the area of the region inside the curve $r = 4+\cos(\theta)$ and outside the curve $r = 2+\cos(\theta)$.

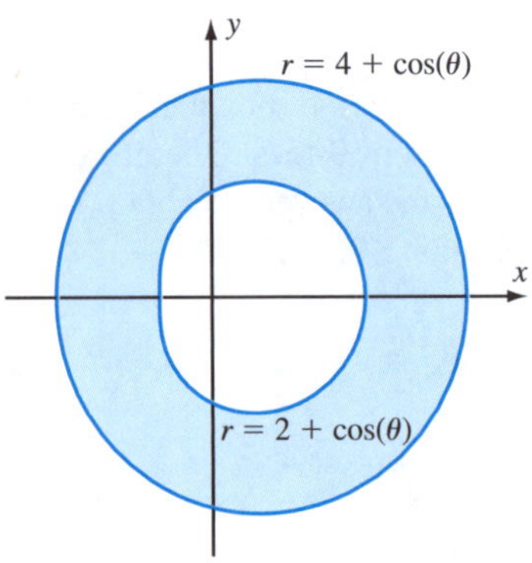

Figure 7

Solution These curves are called *limaçons*. Refer to Figure 7. We find the area of this region by calculating the area inside the limaçon $r = 4+\cos(\theta)$ and subtracting from it the area inside the limaçon $r = 2+\cos(\theta)$. Notice that the limaçon $r = a+\cos(\theta)$, $a > 1$ is generated once, and only once, as θ passes from 0 to 2π. Therefore, the area inside the limaçon is

$$A = \frac{1}{2}\int_0^{2\pi} (a+\cos(\theta))^2\,d\theta = \frac{1}{2}\int_0^{2\pi} (a^2 + 2a\cos(\theta) + \cos^2(\theta))\,d\theta.$$

The first two summands are easy to integrate, and the third summand is treated either by using the Half-Angle Formula for the cosine, as in Example 2, or by appealing directly to formula (7.21). The result of the integration is

$$A = \frac{1}{2}(2\pi a^2 + 0 + \pi) = \frac{\pi}{2} + \pi a^2.$$

Taking $a = 4$, we obtain $\pi/2 + 16\pi$ for the area inside the larger limaçon, and taking $a = 2$, we obtain $\pi/2 + 4\pi$ as the area inside the smaller limaçon. The area of the region between the two curves is therefore

$$A = \left(\frac{\pi}{2} + 16\pi\right) - \left(\frac{\pi}{2} + 4\pi\right) = 12\pi.$$ ■

inSIGHT

If $\varphi_0(\theta) \le \varphi_1(\theta)$ for $\alpha \le \theta \le \beta$, then the formula

$$\frac{1}{2}\int_\alpha^\beta \varphi_1(\theta)^2\,d\theta - \frac{1}{2}\int_\alpha^\beta \varphi_0(\theta)^2\,d\theta \qquad \textbf{(14.12)}$$

gives the area between the two curves $r = \varphi_0(\theta)$ and $r = \varphi_1(\theta)$. This difference of integrals is equivalent to

$$\frac{1}{2}\int_\alpha^\beta (\varphi_1(\theta)^2 - \varphi_0(\theta)^2)\,d\theta.$$

Notice the error we avoided in Example 3. It would be a mistake to subtract first and then square—the integral $(1/2)\int_\alpha^\beta(\varphi_1(\theta) - \varphi_0(\theta))^2\,d\theta$ has nothing to do with the area between the two curves.

Using Iterated Integrals to Calculate Area in Polar Coordinates

Suppose that $\mathcal{R}$ is a planar region given in polar coordinates by

$$\mathcal{R} = \{(r, \theta) : \alpha \le \theta \le \beta,\ 0 \le r \le \varphi(\theta)\}.$$

To avoid ambiguities or counting a part of the area twice, we assume that $\beta - \alpha \le 2\pi$. Although a single integral suffices to calculate the area of $\mathcal{R}$, it is instructive to derive an area formula that uses iterated integrals. We begin with equation (14.11): area of $\mathcal{R} = \int_\alpha^\beta \varphi^2(\theta)/2\, d\theta$. Next, we write the integrand as $\varphi^2(\theta)/2 = \int_0^{\varphi(\theta)} r\, dr$. Substituting this identity into the formula for area gives

$$\text{area of } \mathcal{R} = \int_\alpha^\beta \int_0^{\varphi(\theta)} r\, dr\, d\theta.$$

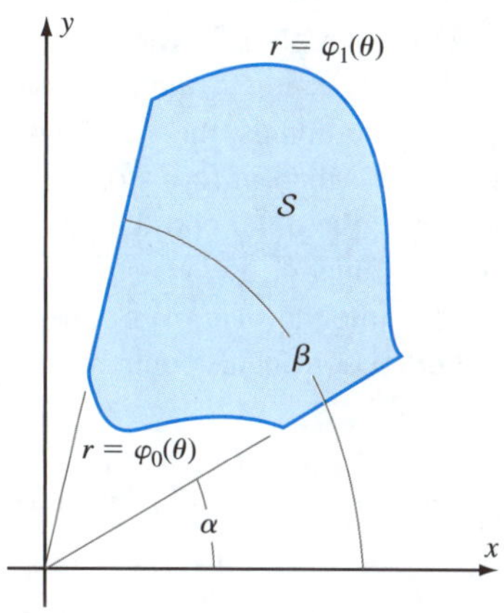

Figure 8

It is just as easy to calculate the area of a region $\mathcal{S}$ bounded by two polar curves. Suppose that $\mathcal{S} = \{(r, \theta) : \alpha \le \theta \le \beta,\ \varphi_0(\theta) \le r \le \varphi_1(\theta)\}$, as in Figure 8. Then the area of $\mathcal{S}$ is clearly

$$\int_\alpha^\beta \int_0^{\varphi_1(\theta)} r\, dr\, d\theta - \int_\alpha^\beta \int_0^{\varphi_0(\theta)} r\, dr\, d\theta = \int_\alpha^\beta \left(\int_0^{\varphi_1(\theta)} r\, dr - \int_0^{\varphi_0(\theta)} r\, dr \right) d\theta$$
$$= \int_\alpha^\beta \int_{\varphi_0(\theta)}^{\varphi_1(\theta)} r\, dr\, d\theta.$$

We state our observations as a theorem.

Theorem 2 Suppose that $0 \le \beta - \alpha \le 2\pi$ and $0 \le \varphi_0(\theta) \le \varphi_1(\theta)$ for all θ. Then the area of $\mathcal{S} = \{\, (r, \theta) : \alpha \le \theta \le \beta,\ \varphi_0(\theta) \le r \le \varphi_1(\theta)\,\}$ is given by

$$\int_\alpha^\beta \int_{\varphi_0(\theta)}^{\varphi_1(\theta)} r\, dr\, d\theta.$$

Example 4 Calculate the area of the planar region lying between the curves $r = \cos(\theta)$ and $r = 1 + \sin(\theta)$ for θ ranging between 0 and $\pi/4$.

Solution Notice that for this range of θ, we have $\cos(\theta) \le 1 + \sin(\theta)$. Refer to Figure 9. According to Theorem 2, this area is

$$A = \int_0^{\pi/4} \int_{\cos(\theta)}^{1+\sin(\theta)} r\, dr\, d\theta = \int_0^{\pi/4} \left(\frac{1}{2} r^2 \Big|_{r=\cos(\theta)}^{r=1+\sin(\theta)} \right) d\theta$$
$$= \frac{1}{2} \int_0^{\pi/4} (1 + 2\sin(\theta) + \sin^2(\theta) - \cos^2(\theta))\, d\theta.$$

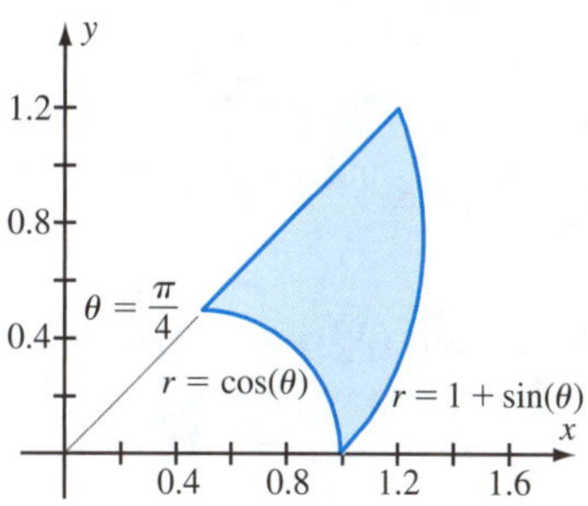

Figure 9

We could proceed as in previous examples in this section by using the Half-Angle Formulas for sine and cosine. Here, however, it is simpler to exploit the combination of

$\cos^2(\theta)$ and $\sin^2(\theta)$ that has arisen. We use the Double Angle Formula for the cosine: $\cos(2\theta) = \cos^2(\theta) - \sin^2(\theta)$. We have

$$A = \frac{1}{2}\int_0^{\pi/4} (1 + 2\sin(\theta) - \cos(2\theta))\, d\theta = \left(\frac{\theta}{2} - \cos(\theta) - \frac{\sin(2\theta)}{4}\right)\Bigg|_0^{\pi/4}$$
$$= \left(\frac{\pi}{8} - \frac{\sqrt{2}}{2} - \frac{1}{4}\right) - (0 - 1 - 0) = \frac{3}{4} + \frac{\pi}{8} - \frac{\sqrt{2}}{2}.$$

Integrating Functions in Polar Coordinates

Because $\iint_{\mathcal{R}} 1\, dA$ and $\iint_{\mathcal{R}} 1\, dx\, dy$ represent the area of a planar region $\mathcal{R}$ when we are working in rectangular coordinates, we sometimes use the expression *element of area* to describe dA and $dx\, dy$. Theorem 2 tells us that in polar coordinates, the *element of area* is $r\, dr\, d\theta$. Suppose that $\mathcal{S}$ is a planar region that is described in terms of polar coordinates. Let $(r, \theta) \mapsto g(r, \theta)$ be a continuous function of the polar coordinates of the points that lie in $\mathcal{S}$. Just as we define the integral of a function $(x, y) \mapsto f(x, y)$ over a planar region $\mathcal{R}$ to be $\iint_{\mathcal{R}} f(x, y)\, dA$, we define the integral of g over $\mathcal{S}$ to be $\iint_{\mathcal{S}} g(r, \theta)\, r\, dr\, d\theta$. Observe that the element of area in polar coordinates contributes a *nontrivial* factor—namely, r— to the integrand.

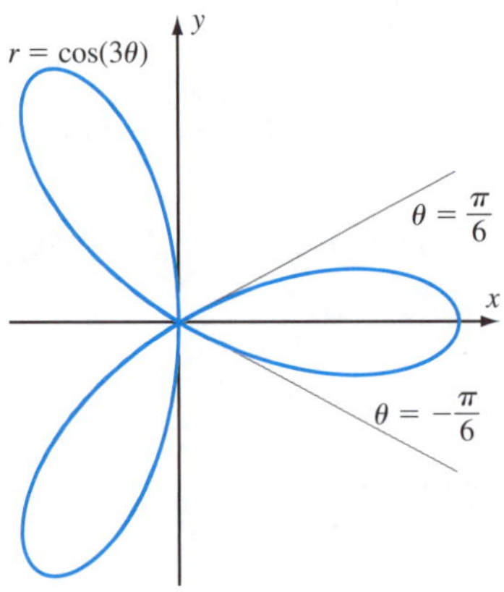

Figure 10

Example 5 Integrate the function $g(r, \theta) = r\sin^2(3\theta)$ over the rightmost petal of the three-leafed rose $r = \cos(3\theta)$.

Solution The plot of $r = \cos(3\theta)$ appears in Figure 10. We need to find the range of θ for the rightmost loop. Since the origin is the "initial" point and "endpoint" of the loop, we must find the values of θ in the first and fourth quadrants for which $\cos(3\theta) = 0$. These are $\theta = \pi/6$ and $\theta = -\pi/6$, respectively. As θ increases from $-\pi/6$ to 0, the value of $r = \cos(3\theta)$ increases from 0 to 1, generating the bottom half of the petal. As θ increases from 0 to $\pi/6$, the value of $r = \cos(3\theta)$ decreases from 1 to 0, generating the top half of the petal. The region of integration is therefore $\mathcal{S} = \{(r, \theta) : -\pi/6 \le \theta \le \pi/6,\ 0 \le r \le \cos(3\theta)\}$. According to Theorem 3, the value of the integral is

$$\int_{-\pi/6}^{\pi/6}\int_0^{\cos(3\theta)} r\sin^2(3\theta)\cdot r\, dr\, d\theta = \int_{-\pi/6}^{\pi/6}\left(\sin^2(3\theta)\int_0^{\cos(3\theta)} r^2\, dr\right) d\theta = \frac{1}{3}\int_{-\pi/6}^{\pi/6} \sin^2(3\theta)\cos^3(3\theta)\, d\theta.$$

We learned the technique for evaluating such integrals in Chapter 7. We write $\cos^3(3\theta) = \cos^2(3\theta)\cos(3\theta) = (1 - \sin^2(3\theta))\cos(3\theta)$ and make the substitution $u = \sin(3\theta)$, $du = 3\cos(3\theta)\, d\theta$. The integral becomes

$$\frac{1}{3}\int_{-\pi/6}^{\pi/6} \sin^2(3\theta)\cos^3(3\theta)\, d\theta = \frac{1}{3}\int_{-\pi/6}^{\pi/6} \sin^2(3\theta)(1 - \sin^2(3\theta))\cos(3\theta)\, d\theta$$
$$= \frac{1}{9}\int_{-1}^{1} u^2(1 - u^2)\, du$$
$$= \frac{1}{9}\int_{-1}^{1} (u^2 - u^4)\, du$$
$$= \frac{4}{135}.$$

It sometimes happens that an integration problem originally presented in rectangular coordinates can be tackled more easily in polar coordinates. However, we cannot just write $x = r\cos(\theta)$ and $y = r\sin(\theta)$ and use $f(r\cos(\theta), r\sin(\theta))$ instead of $f(x, y)$. We must take into account the elements of area, which are different in the two coordinate systems. In other words, we must remember that the element of area $dx\,dy$ in rectangular coordinates becomes $r\,dr\,d\theta$ in polar coordinates. It follows that when we convert an integration problem from rectangular to polar coordinates, the iterated integral

$$\iint f(x, y)\,dx\,dy \quad \text{becomes} \quad \iint f(r\cos(\theta), r\sin(\theta))r\,dr\,d\theta.$$

We will give a more formal treatment of the change from rectangular coordinates to polar coordinates at the end of this section. For now, we state the resulting theorem and turn to some examples.

Theorem 3 Let $(x, y) \mapsto f(x, y)$ be a continuous function that is defined on a planar region $\mathcal{S}$. Suppose that $\mathcal{S}$ can be described in polar coordinates by $\mathcal{S} = \{(r, \theta) : \alpha \le \theta \le \beta,\ \varphi_0(\theta) \le r \le \varphi_1(\theta)\}$ where $0 \le \beta - \alpha \le 2\pi$ and $0 \le \varphi_0(\theta) \le \varphi_1(\theta)$ for all θ. Then the integral of f over $\mathcal{S}$ is given by

$$\iint_{\mathcal{S}} f(x, y)\,dA = \int_\alpha^\beta \int_{\varphi_0(\theta)}^{\varphi_1(\theta)} f(r\cos(\theta), r\sin(\theta))\,r\,dr\,d\theta.$$

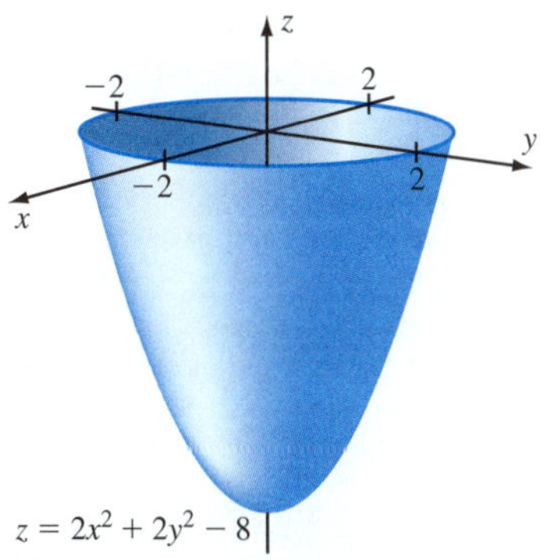

Figure 11

Example 6 Calculate the volume V of the solid bounded by the paraboloid $z = 2x^2 + 2y^2 - 8$ and the xy-plane.

Solution Look at Figure 11. The paraboloid intersects the xy-plane in the circle whose equation is $2x^2 + 2y^2 - 8 = 0$, or $x^2 + y^2 = 4$. Thus, we calculate the volume by integrating the height function $f(x, y) = 0 - (2x^2 + 2y^2 - 8) = 8 - 2(x^2 + y^2)$ over the disk $\mathcal{S}$ that is bounded by the circle with radius 2 centered at the origin of the xy-plane. We can accomplish this by integrating the function $f(r\cos(\theta), r\sin(\theta)) = 8 - 2(r^2\cos^2(\theta) + r^2\sin^2(\theta)) = 8 - 2r^2$ over the region $\mathcal{S} = \{(r, \theta) : 0 \le \theta \le 2\pi, 0 \le r \le 2\}$ using the polar element of area. We obtain

$$V = \int_0^{2\pi}\int_0^2 (8 - 2r^2)\,r\,dr\,d\theta = \int_0^{2\pi}\int_0^2 (8r - 2r^3)\,dr\,d\theta$$
$$= \int_0^{2\pi} \left(4r^2 - \frac{2}{4}r^4\right)\Bigg|_{r=0}^{r=2} d\theta = \int_0^{2\pi} 8\,d\theta = 16\pi.$$

inSIGHT

Rectangular coordinates can also be used to calculate the volume of the solid in Example 6. The required integral is $V = \int_{-2}^{2}\int_{-\sqrt{4-x^2}}^{\sqrt{4-x^2}} (8 - 2(x^2 + y^2))\,dy\,dx$, which, with some effort, does evaluate to 16π. The calculation is decidedly simpler in polar coordinates because of the symmetry about the z-axis. Problems involving circular symmetry are often best solved using polar coordinates.

Example 7 Set up, but do not evaluate, the integral for calculating the volume V of the solid $\mathcal{U}$ that consists of the points inside the lower branch of the cone $(z-3)^2 = x^2 + y^2$, above the xy-plane, and inside the circular cylinder $x^2 + (y-1)^2 = 1$.

Solution Look at Figure 12. Notice that the solid $\mathcal{U}$ lies entirely inside the cylinder; its upper boundary is determined by the lower branch of the cone and its lower boundary by the xy-plane. The equation for the lower half of the cone is $z = 3 - (x^2 + y^2)^{1/2}$. Therefore, we must integrate the function $f(x, y) = 3-(x^2+y^2)^{1/2}$ over the disk that is bounded by the circle $x^2+(y-1)^2 = 1$. This circle is the graph of the polar coordinate equation $r = 2\sin(\theta)$ for $0 \leq \theta \leq \pi$, as we saw in Example 7 from Section 14.4. Therefore, we can calculate V by using polar coordinates and integrating the function $f(r\cos(\theta), r\sin(\theta)) = 3 - r$ over the region $\mathcal{S} = \{(r, \theta) : 0 \leq r \leq \pi, 0 \leq r \leq 2\sin(\theta)\}$. The volume of $\mathcal{U}$ is

$$V = \int_0^{\pi} \int_0^{2\sin(\theta)} (3-r)r\,dr\,d\theta = \int_0^{\pi} \int_0^{2\sin(\theta)} (3r - r^2)\,dr\,d\theta.$$

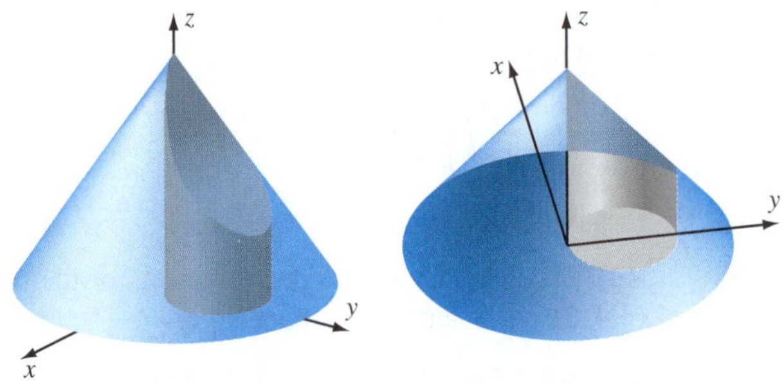

Figure 12a

Figure 12b
Viewed from below

Having set up the iterated integral, we note that the inner integration presents no difficulty:

$$V = \int_0^{\pi} \left(\frac{3}{2}r^2 - \frac{1}{3}r^3\right)\Bigg|_{r=0}^{r=2\sin(\theta)} d\theta = \int_0^{\pi} \left(6\sin^2(\theta) - \frac{8}{3}\sin^3(\theta)\right) d\theta.$$

The interested reader may verify that the value of V is $3\pi - 32/9$. (Use equation (7.21) for the integration of $\sin^2(\theta)$ and refer to Section 7.3 for the integration of $\sin^3(\theta)$.) ■

Change of Variable and the Jacobian

Our discussion of the element of area in polar coordinates suggests an important question: How does a change of variable affect a double integral? That is the problem we now investigate. Suppose that $\mathcal{S}$ is a planar region on which two differentiable functions ϕ and ψ are defined. These are the functions we use to effect the change of variables on $\mathcal{S}$. Let $\mathcal{R}$ be the region defined by $\mathcal{R} = \{(x, y) : x = \phi(u, v)$ and $y = \psi(u, v)$ for some point $(u, v) \in \mathcal{S}\}$. Figure 13 shows that we can imagine ϕ and ψ transforming the region $\mathcal{S}$ into the region $\mathcal{R}$. In other words, we define a function

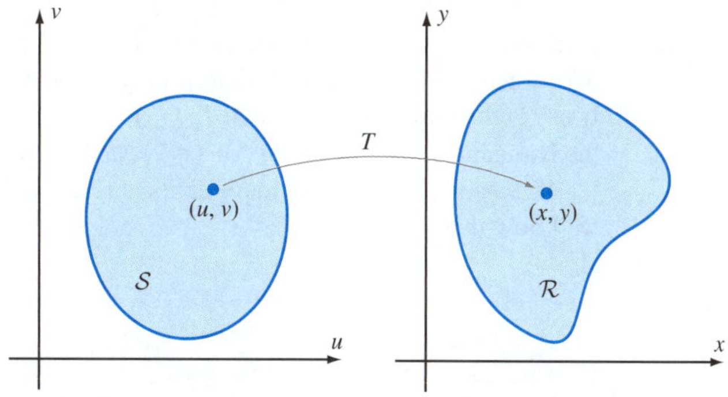

Figure 13
The equation $(x, y) = (\phi(u, v), \psi(u, v))$ transforms the region $\mathcal{S}$ into the region $\mathcal{R}$.

$T : \mathcal{S} \to \mathbb{R}^2$ by the equation $T(u, v) = (\phi(u, v), \psi(u, v))$ and define $\mathcal{R}$ to be the image of $\mathcal{S}$ under T. We assume that T is one-to-one, so that given ξ in $\mathcal{R}$, there is a unique point η in $\mathcal{S}$ for which $T(\eta) = \xi$. Theorem 1 from Section 3.6 asserts that T is an invertible function from $\mathcal{S}$ onto $\mathcal{R}$.

If f is a continuous function on $\mathcal{R}$ and if J is a continuous function on $\mathcal{S}$, then we may consider the integrals $\iint_{\mathcal{R}} f(x, y)\,dx\,dy$ and $\iint_{\mathcal{S}} f(\phi(u, v), \psi(u, v))J(u, v)\,du\,dv$. The second of these integrals may be written a bit more compactly as $\iint_{\mathcal{S}} f(T(u, v))J(u, v)\,du\,dv$. Our task is to determine a factor $J(u, v)$ so that

$$\iint_{\mathcal{R}} f(x, y)\,dx\,dy = \iint_{\mathcal{S}} f(T(u, v))J(u, v)\,du\,dv.$$

Before continuing, you may find it helpful to study the schematic diagram of the functions T, f, $f \circ T$, and J given in Figure 14.

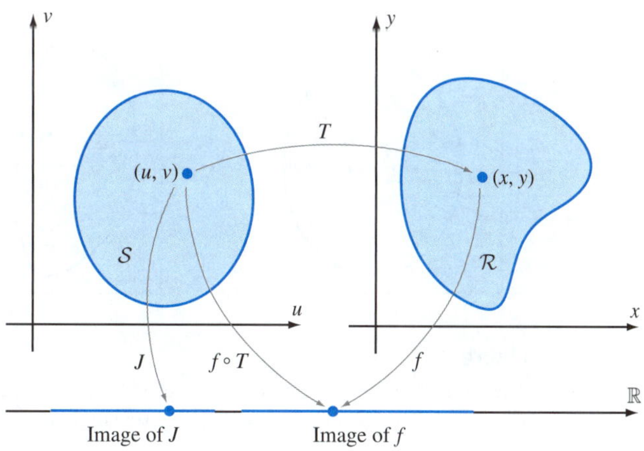

Figure 14

Consider a partition of $\mathcal{R}$. Let the lengths and widths of the rectangles in this partition be Δx and Δy, respectively. Let $\xi_{i,j}$ be a point in the i, jth rectangle in the partition of $\mathcal{R}$. These choices determine a Riemann sum $\sum f(\xi_{i,j})\Delta x \Delta y$ for f over $\mathcal{R}$. If $\eta_{i,j}$ is the point of $\mathcal{S}$ such that $T(\eta_{i,j}) = \xi_{i,j}$, our job is to find a function J such that the Riemann sum $\sum f(T(\eta_{i,j}))J(\eta_{i,j})\Delta u \Delta y$ over $\mathcal{S}$ corresponds to $\sum f(\xi_{i,j})\Delta x \Delta y$.

If Δu and Δv are small increments in u and v, then from Theorem 3 of Section 13.7, we know that

$$\Delta x \approx \phi_u \Delta u + \phi_v \Delta v \quad \text{and} \quad \Delta y \approx \psi_u \Delta u + \psi_v \Delta v.$$

Interpreted geometrically, this means that a small horizontal displacement in the xy-coordinates, which we may represent by the vector $\langle \Delta x, 0\rangle$, corresponds to a displacement in the uv-coordinates given by the vector $\langle \phi_u \Delta u, \phi_v \Delta v\rangle$. Similarly, a small vertical displacement in the xy-coordinates, which we may represent by the vector $\langle 0, \Delta y\rangle$, corresponds to a displacement in the uv-coordinates given by the vector $\langle \psi_u \Delta u, \psi_v \Delta v\rangle$.

Refer to Figure 15. We see that the rectangle with sides Δx and Δy in the xy-plane corresponds to a parallelogram that is determined by the vectors $\langle \phi_u \Delta u, \phi_v \Delta v\rangle$ and $\langle \psi_u \Delta u, \psi_v \Delta v\rangle$ in the uv-plane. Therefore, the element of area $\Delta x \Delta y$ in the xy-plane corresponds to the area ΔA of the parallelogram that is determined by the vectors $\langle \phi_u \Delta u, \phi_v \Delta v\rangle$ and $\langle \psi_u \Delta u, \psi_v \Delta v\rangle$ in the uv-plane. To calculate ΔA, we think of the uv-plane as a coordinate plane in three-dimensional space. Then, as we learned in Section 11.4, ΔA is equal to the length of the cross product of the two vectors $\langle \phi_u \Delta u, \phi_v \Delta v, 0\rangle$ and $\langle \psi_u \Delta u, \psi_v \Delta v, 0\rangle$. As a result, we obtain

$$\Delta A = \left\| \det\left(\begin{bmatrix} \mathbf{i} & \mathbf{j} & \mathbf{k} \\ \phi_u \Delta u & \phi_v \Delta v & 0 \\ \psi_u \Delta u & \psi_v \Delta v & 0 \end{bmatrix}\right) \right\| = \|\,(\phi_u \psi_v - \psi_u \phi_v)\,\Delta u \Delta v\,\mathbf{k}\,\| = |\phi_u \psi_v - \psi_u \phi_v|\,\Delta u \Delta v.$$

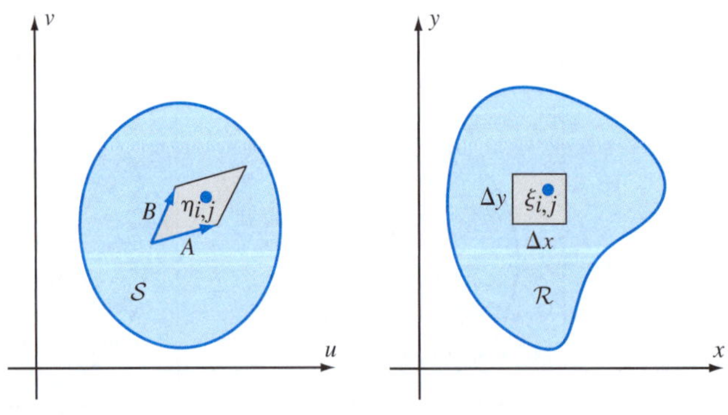

Figure 15
The parallelogram determined by vectors $\vec{A} = \langle \phi_u \Delta u, \phi_v \Delta v\rangle$ and $\vec{B} = \langle \psi_u \Delta u, \psi_v \Delta v\rangle$ corresponds to the rectangle in $\mathcal{R}$ with sides Δx and Δy.

From this computation, we see that the Riemann sum $\sum f(\xi_{i,j})\Delta x \Delta y$ for f over $\mathcal{R}$ corresponds to the Riemann sum in uv-coordinates given by

$$\sum f(T(\eta_{i,j}))|(\phi_u \psi_v - \psi_u \phi_v)(\eta_{i,j})|\Delta u \Delta v.$$

(Remember that in the algebra of functions, $(\phi_u \psi_v - \psi_u \phi_v)(\eta)$ means $\phi_u(\eta)\psi_v(\eta) - \psi_u(\eta)\phi_v(\eta)$.) Passing to the limit, we obtain the following theorem.

Theorem 4 Let $\mathcal{S}$ be a region in the uv-plane bounded by finitely many continuously differentiable curves. Suppose $T(u, v) = (\phi(u, v), \psi(u, v))$ is a transformation of $\mathcal{S}$ to a region $\mathcal{R}$ in the xy-plane that is also bounded by finitely many continuously differentiable curves. We assume that T is one-to-one and onto and that the component functions ϕ, ψ are continuously differentiable. If f is a continuous function on the region $\mathcal{R}$, then

$$\iint_{\mathcal{R}} f(x, y)\, dx\, dy = \iint_{\mathcal{S}} f(\phi(u, v), \psi(u, v))\, |J_T(u, v)|\, du\, dv \tag{14.13}$$

where

$$J_T(u, v) = \det\left(\begin{bmatrix} \phi_u(u, v) & \phi_v(u, v) \\ \psi_u(u, v) & \psi_v(u, v) \end{bmatrix}\right) = (\phi_u \psi_v - \psi_u \phi_v)(u, v). \tag{14.14}$$

We call $J_T(u, v)$ the *Jacobian* (or, more precisely, the *Jacobian determinant*) of the transformation T.

Example 8 Let $\mathcal{R}$ be the rectangle $\{(x, y) : 0 \le x \le 2, 0 \le y \le 5\}$ in the xy-plane. Let $\mathcal{S}$ be the square $\{(x, y) : 0 \le x \le 1, 0 \le y \le 1\}$ in the uv-plane. Let $T(u, v) = (2u, 5v)$ define a one-to-one transformation of $\mathcal{S}$ onto $\mathcal{R}$. Use Theorem 4 to calculate the area of $\mathcal{R}$ by means of a double integral over $\mathcal{S}$.

Solution Of course, we do not need calculus to know that the area of a rectangle with side lengths 2 and 5 is 10. Nevertheless, it is instructive to see how we can obtain this value by integrating over $\mathcal{S}$, a square of side length 1. Observe that T transforms $\mathcal{S}$ into $\mathcal{R}$ by expanding one edge of $\mathcal{S}$ by a factor 2 and another edge by a factor 5. If we are to obtain the area of $\mathcal{R}$ by integrating over $\mathcal{S}$, then our integrand must contain a term that accounts for this expansion. Theorem 4 tells us that the Jacobian determinant of T is the factor we need. The entries of T are $\phi(u, v) = 2u$ and $\psi(u, v) = 5v$. We calculate

$$J_T(u, v) = \det\left(\begin{bmatrix} \phi_u(u, v) & \phi_v(u, v) \\ \psi_u(u, v) & \psi_v(u, v) \end{bmatrix}\right) = \det\left(\begin{bmatrix} 2 & 0 \\ 0 & 5 \end{bmatrix}\right) = 10.$$

Therefore, according to Theorem 4 with $f(x, y) = 1$, the area of $\mathcal{R}$ is

$$\iint_{\mathcal{R}} 1\, dx\, dy = \iint_{\mathcal{S}} 1 \cdot |J_T(u, v)|\, du\, dv = \int_0^1 \int_0^1 10\, du\, dv = 10. \quad \blacksquare$$

Example 9 Apply Theorem 4 to justify the formula for the polar element of area that is used in Theorems 2 and 3.

Solution Instead of the variables u and v, we use the more familiar variables r and θ. Then the transformation T from polar to rectangular coordinates is given by $x = \phi(r, \theta) = r\cos(\theta)$, $y = \psi(r, \theta) = r\sin(\theta)$. We calculate that

$$J_T(r, \theta) = (\phi_r\psi_\theta - \psi_r\phi_\theta)(r, \theta) = \cos(\theta)(r\cos(\theta)) - \sin(\theta)(-r\sin(\theta)) = r(\cos^2(\theta) + \sin^2(\theta)) = r.$$

In the setting of Theorems 2 and 3, the coordinate r is nonnegative. Therefore, $|J_T(r, \theta)| = r$ and, according to Theorem 4,

$$\iint_{\mathcal{R}} f(x, y)\,dx\,dy = \iint_{\mathcal{S}} f(\phi(r, \theta), \psi(r, \theta))\,|J_T(r, \theta)|dr\,d\theta = \iint_{\mathcal{S}} f(r\cos(\theta), r\sin(\theta))\,r\,dr\,d\theta.$$

This computation shows that the polar element of area is $r\,dr\,d\theta$, which is exactly what we learned in Theorems 2 and 3. ■

Example 10 Evaluate the integral

$$\int_0^3 \int_{x/3}^{(x/3)+2} \left(\frac{x+y}{2}\right) dy\,dx$$

using the transformation

$$\frac{x}{3} = u, \qquad \frac{x+y}{2} = v.$$

Solution In this problem, the region $\mathcal{R}$ is given by $\{(x, y) : x/3 \le y \le x/3+2, 0 \le x \le 3\}$. We begin by rewriting the transformation in a form that fits the theorem. The first equation is equivalent to $x = 3u$. The second equation is equivalent to $y = -x+2v$, or $y = -3u+2v$. Our transformation T is therefore $x = \phi(u, v) = 3u$, $y = \psi(u, v) = -3u+2v$. To find the region $\mathcal{S}$ that corresponds to $\mathcal{R}$, notice that $0 \le x \le 3$ corresponds to $0 \le u \le 1$ and $(x/3) \le y \le (x/3) + 2$ corresponds to $u \le 2v - 3u \le u + 2$. This last chain of inequalities simplifies, by adding $3u$ to all parts, to $4u \le 2v \le 4u + 2$, or $2u \le v \le 2u + 1$. Thus, $\mathcal{S} = \{u, v) : 2u \le v \le 2u + 1, 0 \le u \le 1\}$. See Figure 16. Since the Jacobian of the transformation is

$$J_T(u, v) = (\phi_u\psi_v - \psi_u\phi_v)(u, v) = (3)(2) - (-3)(0) = 6,$$

the integral becomes

$$\int_0^1 \int_{2u}^{2u+1} v|6|\,dv\,du = 6\int_0^1 \left(\frac{v^2}{2}\bigg|_{v=2u}^{v=2u+1}\right) du = 3\int_0^1 (4u+1)\,du = 3\left((2u^2+u)\Big|_0^1\right) = 9.$$

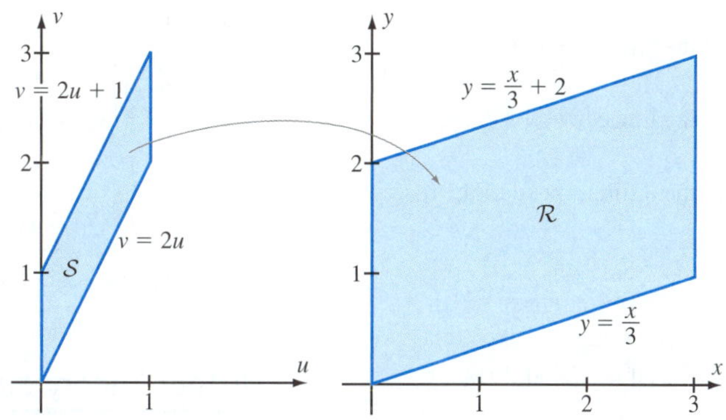

Figure 16
The transformation $x = \phi(u, v) = 3u$, $y = \psi(u, v) = -3u + 2v$

quickquiz

1. A curve in polar coordinates is described by the equation $r = f(\theta)$, with f a positive continuous function and $\alpha \le \theta \le \beta$. Write the single integral that equals the area of the region determined by this curve and between the limits α and β.
2. What is the "element of area" in polar coordinates?
3. Calculate the integral of $f(x, y) = x^2 + y^2$ over the region that lies outside the unit circle and inside the circle $x^2 + y^2 = 4$.
4. What is the formula for change of coordinates in a double integral?

EXERCISES

Problems for Practice

In Exercises 1–6, use formula (14.11) to evaluate the area enclosed by the curve.

1. $r = 1 - \sin(\theta)$
2. $r = 7 - 3\cos(\theta)$
3. $r = 4\sin(3\theta)$
4. $r = 5\sin(2\theta)$
5. $r^2 = -\sin(\theta)$
6. $r^2 = \cos(\theta)$

In Exercises 7–12, use formula (14.12) to find the area of the region described.

7. The region between the graphs of $r = 6 - \sin(\theta)$ and $r = 1 + \cos(\theta)$
8. The region in the first quadrant between the graphs of $r = 3\sin(2\theta)$ and $r = 4 - \cos(\theta)$
9. The region enclosed by the curves $r = \sin(\theta)$ and $r = \cos(\theta)$
10. The "lune" outside the cardioid $r = 1 + \sin(\theta)$ and inside the circle $r = 3\sin(\theta)$
11. The region in the first quadrant that is inside the circle $r = \sqrt{3}\cos(\theta)$ and outside the circle $r = \sin(\theta)$
12. The region inside the circle $r = \sqrt{3}\cos(\theta)$ and outside the cardioid $r = 1 + \sin(\theta)$

In Exercises 13–18, calculate the area of region $\mathcal{R}$ using a double integral in polar coordinates.

13. $\mathcal{R} = \{(r, \theta) : -\pi/4 \le \theta \le \pi/4, 2 \le r \le 4\cos(\theta)\}$
14. $\mathcal{R}$ is one leaf of the rose $r = 3\sin(4\theta)$.
15. $\mathcal{R}$ is the region inside the circle with center the origin and radius 5 and outside the limaçon $r = 3 + \cos(\theta)$.
16. $\mathcal{R} = \{(r, \theta) : \pi/6 \le \theta \le \pi/4, \sin(\theta) \le r \le \cos(\theta)\}$
17. $\mathcal{R}$ is the region outside the cardioid $r = 1 + \cos(\theta)$ and inside the limaçon $r = 4 - \sin(\theta)$.
18. $\mathcal{R}$ is the region inside the circle $r = 4\cos(\theta)$ and outside the rightmost leaf of the rose $r = \cos(2\theta)$.

In Exercises 19–26, integrate the function f over the planar region $\mathcal{R}$, using a double integral in polar coordinates.

19. $f(r, \theta) = \cos(\theta)$;
$\mathcal{R}$ is the region inside the circle $r = \cos(\theta)$.

20. $f(r, \theta) = \sin^2(\theta)$;
$\mathcal{R}$ is the region inside the limaçon $r = 3 - \cos(\theta)$.

21. $f(r, \theta) = \cos(\theta)$;
$\mathcal{R}$ is the region inside the leftmost petal of the rose $r = \cos(2\theta)$.

22. $f(r, \theta) = \theta$,
$\mathcal{R} = \{(r, \theta) : 0 \le \theta \le \pi/2, 0 \le r \le \cos^{1/2}(\theta)\}$

23. $f(r, \theta) = \sin(\theta)$;
$\mathcal{R}$ is the region inside the circle $r = \sin(\theta)$.

24. $f(r, \theta) = r$,
$\mathcal{R} = \{(r, \theta) : 0 \le \theta \le \pi/2, 0 \le r \le \sin(\theta)\}$

25. $f(r, \theta) = 3r - 10$;
$\mathcal{R}$ is the region outside the circle $r = 3$ and inside the limaçon $r = 4 - \cos(\theta)$.

26. $f(r, \theta) = \sin(\theta)$;
$\mathcal{R}$ is the region in the first quadrant that is inside the circle $r = 1$ and outside the cardioid $r = 1 - \cos(\theta)$.

In Exercises 27–38, use a double integral in polar coordinates to determine the volume of the solid.

27. The solid bounded by the surface $z = -x^2 - y^2 + 4$ and the xy-plane

28. The solid bounded by cylinder $(x - 2)^2 + y^2 = 4$, the paraboloid $z = x^2 + y^2 + 6$, and the xy-plane

29. The solid bounded by the paraboloids $z = 3x^2 + 3y^2 - 7$ and $z = -x^2 - y^2 + 9$

30. The solid bounded by the paraboloids $z = 2x^2 + 2y^2 + 4$ and $z = -4x^2 - 4y^2 - 4$ and by the cylinder $x^2 + y^2 = 16$

31. The solid bounded by the upper nappe of the cone $(z + 3)^2 = 4x^2 + 4y^2$ and by the paraboloid $z = 5 - x^2 - y^2$

32. The solid bounded by the paraboloids $z = 4x^2 + 4y^2 + 2$ and $z = 7 - x^2 - y^2$

33. The solid bounded by the hyperboloid $z^2 - x^2 - y^2 = 1$ and by the plane $z = 4$

34. The solid bounded by the upper nappe of the cone $z^2 = 4x^2 + 4y^2$ and by the plane $z = 12$

35. The solid bounded by the cylinder $x^2 + y^2 = 4$ and by the hyperboloid $z^2 - x^2 - y^2 = 4$

36. The solid bounded by the the sphere $x^2 + y^2 + z^2 = 20$ and by the paraboloid $z = x^2 + y^2$

37. The solid outside the cone $z^2 = x^2 + y^2$ and inside the cylinder $x^2 + y^2 = 4$

38. The solid bounded by the cylinder $x^2 + (y - 1)^2 = 1$ and by the sphere $x^2 + y^2 + z^2 = 4$

In Exercises 39–44, perform the indicated change of variable in the double integral, and then evaluate the integral.

39. $\int_0^2 \int_{y^2/4}^{y/2} (x - y)^2\, dx\, dy$; $u = (x - y)/2$, $v = y/2$

40. $\int_0^4 \int_0^{\sqrt{16-x^2}} xy\, dy\, dx$; $x = r\cos(2\theta)$, $y = r\sin(2\theta)$

41. $\int_1^5 \int_3^9 y^2/x^2\, dx\, dy$, $u = 1/x$, $v = y$

42. $\int_0^4 \int_{y/2}^{(y/2)+1} (2x - y)/2\, dx\, dy$; $u = (2x - y)/2$, $v = y/2$

43. $\int_0^1 \int_3^4 \sqrt{4 - xy}\, dy\, dx$; $u = xy$, $v = y$

44. $\int_0^{16} \int_0^{\sqrt{x}} \sqrt{y/x}\, dy\, dx$; $u^2 = x$, $v^2 = y$

Further Theory and Practice

In Exercises 45–52, find the area of the region described.

45. The region between the graphs of $r = 3\sin(2\theta)$ and $r = 4 - \cos(\theta)$

46. The region enclosed by *both* $r = 1 + \sin(\theta)$ and $r = 3\sin(\theta)$

47. The region inside both the cardioid $r = 1 - \cos(\theta)$ and the circle $r = \cos(\theta)$

48. The region inside both the circle $r = \sqrt{3}\cos(\theta)$ and the cardioid $r = 1 + \sin(\theta)$

49. The region inside both the rose $r = \cos(2\theta)$ and the circle $r = 1/2$

50. The region between the two loops of $r = \sqrt{3} - 2\cos(\theta)$

51. The region inside the smaller loop of $r = 1 - 2\sin(\theta)$

52. The region in the first and second quadrants that is bounded by the x-axis, the spiral arc $r = \theta$ with $0 \le \theta \le \pi$, and the spiral arc $r = \theta$ with $2\pi \le \theta \le 3\pi$

Perform the integrations in Exercises 53–56 by converting them to polar coordinates.

53. $\int_{-2}^2 \int_{-(4-x^2)^{1/2}}^{(4-x^2)^{1/2}} (x^2 + y^2 + 1)^{-1/2}\, dy\, dx$

54. $\int_0^1 \int_0^{(1-x^2)^{1/2}} \cos(x^2 + y^2)\, dy\, dx$

55. $\int_{-3}^0 \int_{-(9-y^2)^{1/2}}^0 (x^2 + y^2)^{1/3}\, dx\, dy$

56. $\int_{-3}^3 \int_0^{(9-y^2)^{1/2}} (1 + x^2 + y^2)^{3/2}\, dx\, dy$

57. Integrate the function $f(r, \theta) = \theta$ over the region $\mathcal{R} = \{(r, \theta) : 0 \le \theta \le \pi/2, 0 \le r \le \cos(\theta)\}$.

58. Integrate the function $f(r, \theta) = \cos(\theta)$ over the region $\mathcal{R} = \{(r, \theta) : 0 \le \theta \le \pi, 0 \le r \le \theta\}$.

59. Calculate the area in the first quadrant bounded by the lines $y = x$ and $y = 2x$ and by the curves $xy = 2$ and $xy = 4$. After setting up the integral, simplify matters by using the change of variables $u = xy$, $v = y/x$.

60. The convergent improper integral

$$I = \int_{-\infty}^{\infty} e^{(-x^2)}\, dx$$

arises in probability, statistics, error analysis, and many other topics. It cannot be evaluated by the techniques of Chapter 7. Instead, we write

$$I^2 = \int_{-\infty}^{\infty} e^{(-x^2)}\,dx \int_{-\infty}^{\infty} e^{(-y^2)}\,dy$$
$$= \int_{-\infty}^{\infty}\int_{-\infty}^{\infty} e^{-(x^2+y^2)}\,dx\,dy.$$

Introduce polar coordinates in this integral, and you will be able to evaluate it. Then solve for I.

61. Use polar coordinates to determine for which real values of α the improper integral

$$\int_{-\infty}^{\infty}\int_{\infty}^{\infty} \frac{1}{(1+x^2+y^2)^\alpha}\,dx\,dy$$

converges. Here you should understand this integral to mean

$$\lim_{R\to\infty}\int_{-R}^{R}\int_{-R}^{R} \frac{1}{(1+x^2+y^2)^\alpha}\,dx\,dy.$$

62. Prove that for *any* real value of α, the improper integral

$$\int_{-\infty}^{\infty}\int_{-\infty}^{\infty} \frac{e^{-(x^2+y2)}}{(1+x^2+y^2)^\alpha}\,dx\,dy$$

converges.

Calculator/Computer Exercises

63. Find the value θ_0 of θ in $[0, \pi/2]$ for which $\theta = 3\cos^2(\theta)$. Calculate the area of the region in the first quadrant that is outside the curve $r = \theta$, $0 \le \theta \le \theta_0$ and inside the curve $r = 3\cos^2(\theta)$, $0 \le \theta \le \theta_0$.

64. Find the value θ_0 of θ in $[0, \pi/2]$ for which $2+\theta = \exp(\theta)$. Calculate the area of the region in the first quadrant that is between the curves $r = 2+\theta$, $0 \le \theta \le \theta_0$ and $r = \exp(\theta)$, $0 \le \theta \le \theta_0$.

If an iterated integral $\int_\alpha^\beta \int_{\varphi_0(\theta)}^{\varphi_1(\theta)} f(r\cos(\theta), r\sin(\theta))\,r\,dr\,d\theta$ arises from integration in polar coordinates, then we may estimate its value by applying any approximation technique for double integrals. We do so by using the product $f(r\cos(\theta), r\sin(\theta))\,r$ as the integrand in the approximation formula. In Exercises 65 and 66, use the Trapezoidal Rule with $N = 4$ to approximate the given integral. (Refer to the instructions for Exercise 65–68 from Section 14.2 for a discussion of the Trapezoidal Rule for double integrals.)

65. The integral of $f(r,\theta) = 1/\left(1+\theta+r^3\right)$ over the region bounded by the y-axis and the spiral $r = \theta$, $0 \le \theta \le \pi/2$

66. The integral of $f(r,\theta) = \sqrt{\pi+\theta+\sqrt{r}}$ over the half-disk $0 \le r \le 1$, $-\pi/2 \le \theta \le \pi/2$

14.6 Triple Integrals

In this section, we discuss the theory of the triple integral. In fact, integration can be done in any number of dimensions. We concentrate here on three dimensions because that is sufficient to study basic properties of bodies in space.

The Concept of the Triple Integral

Let $\mathcal{U}$ be a solid in space and f a continuous function defined on $\mathcal{U}$. We want to develop a concept of integrating f over $\mathcal{U}$. This will parallel the development in Section 14.2, and we shall be brief.

Suppose that $\mathcal{U}$ is contained in a box $Q = \{(x, y, z) : a \le x \le b, c \le y \le d,\ e \le z \le f\}$. We partition each interval $[a, b]$, $[c, d]$, and $[e, f]$ into N equal subintervals, giving rise to N^3 small boxes $Q_{j,k,\ell}$. We select a point $\xi_{j,k,\ell}$ in each $Q_{j,k,\ell}$ and form the Riemann sum

$$\sum_{Q_{j,k,\ell}\subset\mathcal{U}} f(\xi_{j,k,\ell})\Delta x\Delta y\Delta z = \sum_{Q_{j,k,\ell}\subset\mathcal{U}} f(\xi_{j,k,\ell})\Delta V$$

where $\Delta V = (\Delta x)(\Delta y)(\Delta z)$ is the increment of volume. As in Section 14.2, we sum only those boxes that are contained in $\mathcal{U}$. If the geometry of $\mathcal{U}$ is relatively simple (a concept that is made precise in this section's first theorem), then it is known that the Riemann sums will have a limiting value as N tends to infinity. The limiting value of these Riemann sums is called the *triple Riemann integral* or the *Riemann integral in three variables*. This integral is denoted by the symbol

$$\iiint_{\mathcal{U}} f\,dV.$$

The most important fact for us about triple integrals is that we have a means of evaluating them when the solid $\mathcal{U}$ has a regular form. The next theorem tells us how to calculate triple integrals in terms of iterated integrals.

Theorem 1 Suppose that α_1, α_2, γ_1, and γ_2 are continuously differentiable functions of one variable. Suppose that β_1 and β_2 are continuously differentiable functions of two variables. Assume that $\mathcal{U}_1$ and $\mathcal{U}_2$ are solids in space given by

$$\mathcal{U}_1 = \{(x, y, z) : a \le x \le b, \alpha_1(x) \le y \le \alpha_2(x), \beta_1(x, y) \le z \le \beta_2(x, y)\}$$

and

$$\mathcal{U}_2 = \{(x, y, z) : c \le y \le d, \gamma_1(y) \le x \le \gamma_2(y), \beta_1(x, y) \le z \le \beta_2(x, y)\}.$$

If f is a continuous function defined on $\mathcal{U}_1$, then

$$\iiint_{\mathcal{U}_1} f(x, y, z)\,dV = \int_a^b \int_{\alpha_1(x)}^{\alpha_2(x)} \int_{\beta_1(x,y)}^{\beta_2(x,y)} f(x, y, z)\,dz\,dy\,dx.$$

If f is a continuous function defined on $\mathcal{U}_2$, then

$$\iiint_{\mathcal{U}_2} f(x, y, z)\,dV = \int_c^d \int_{\gamma_1(y)}^{\gamma_2(y)} \int_{\beta_1(x,y)}^{\beta_2(x,y)} f(x, y, z)\,dz\,dy\,dx.$$

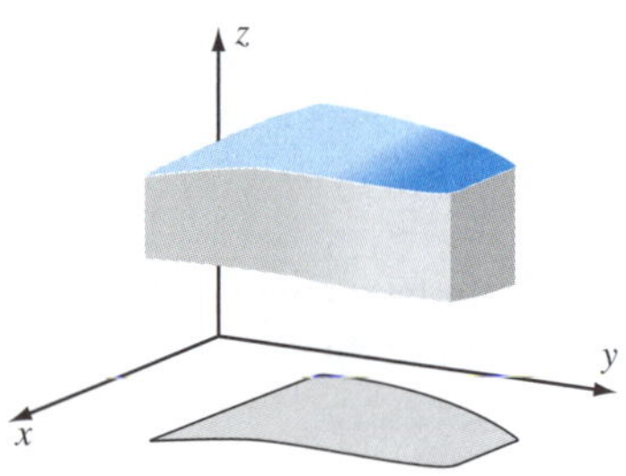

Figure 1

Solids of the type described in Theorem 1 are called *z-simple*. Figure 1 shows a z-simple solid of the form $\mathcal{U} = \{(x, y, z) : c \le y \le d, \gamma_1(y) \le x \le \gamma_2(y), \beta_1(x, y) \le z \le \beta_2(x, y)\}$. The cutaway image of $\mathcal{U}$ that is given in Figure 2 reveals a typical box of volume ΔV that arises in the definition of $\iiint_{\mathcal{U}} f(x, y, z)\,dV$. In the iterated integral $\int_c^d \int_{\gamma_1(y)}^{\gamma_2(y)} \int_{\beta_1(x,y)}^{\beta_2(x,y)} f(x, y, z)\,dz\,dx\,dy$, the inner integral—namely, $\int_{\beta_1(x,y)}^{\beta_2(x,y)} f(x, y, z)\,dz$—integrates up the vertical column that is shown. This result is then integrated, by means of $\int_{\gamma_1(y)}^{\gamma_2(y)} \ldots dx$, along the strip depicted in Figure 2. The result of this calculation is then integrated over all such strips from $y = c$ to $y = d$.

An analogue of Theorem 1 is true for x-simple and y-simple solids, which are defined similarly. You will see all three of these types of solids in the examples that follow.

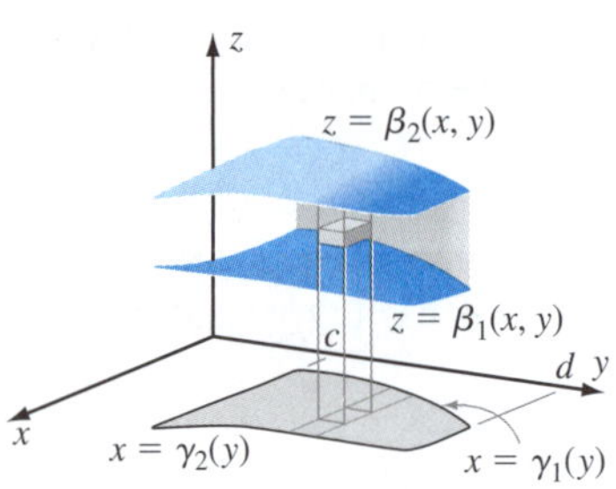

Figure 2

Example 1 Integrate the function $f(x, y, z) = 4x - 12z$ over the solid $\mathcal{U} = \{(x, y, z) : 1 \le x \le 2, x \le y \le 2x, y - x \le z \le y\}$.

Solution By Theorem 1, we have

$$\iiint_{\mathcal{U}} f(x, y, z)\, dV = \int_1^2 \int_x^{2x} \int_{y-x}^{y} (4x - 12z)\, dz\, dy\, dx.$$

We begin by evaluating the inside integral, and then we work our way outward. The last line equals

$$\begin{aligned}
&\int_1^2 \int_x^{2x} (4xz - 6z^2)\Big|_{z=y-x}^{z=y} dy\, dx \\
&= \int_1^2 \int_x^{2x} \Big((4xy - 6y^2) - \big(4x(y-x) - 6(y-x)^2\big) \Big) dy\, dx \\
&= \int_1^2 \int_x^{2x} (10x^2 - 12xy)\, dy\, dx \\
&= \int_1^2 (10x^2 y - 6xy^2)\Big|_{y=x}^{y=2x} dx \\
&= \int_1^2 (-8x^3)\, dx \\
&= -2x^4\Big|_1^2 \\
&= -30.
\end{aligned}$$

■

If we take the function f that is being integrated to be the constant function $f(x, y, z) = 1$, then the Riemann sums

$$\sum_{Q_{j,k,\ell} \subset \mathcal{R}} f(\xi_{j,k,\ell}) \Delta V = \sum_{Q_{j,k,\ell} \subset \mathcal{R}} \Delta V$$

form a good approximation to the volume of $\mathcal{U}$. This observation motivates the following definition.

Definition The volume of a solid $\mathcal{U}$ is defined to be $\iiint_{\mathcal{U}} 1\, dV$ when the integral exists.

Here is an example.

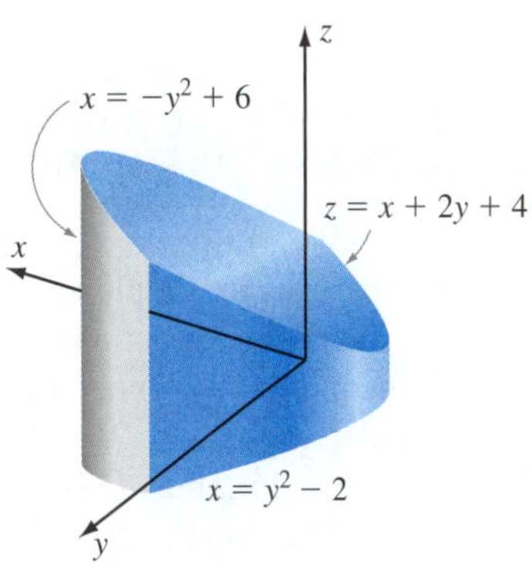

Figure 3

Example 2 Compute the volume V of the three-dimensional solid $\mathcal{U}$ that is bounded laterally by the parabolic cylinders $x = y^2 - 2$ and $x = -y^2 + 6$, above by the plane $z = x + 2y + 4$, and below by the plane $z = -x - 4$.

Solution The solid $\mathcal{U}$ is depicted in Figure 3. Observe that the graphs of $x = y^2 - 2$ and $x = -y^2 + 6$ are vertical sides of $\mathcal{U}$. The projection of $\mathcal{U}$ into the xy-plane is the region bounded by the parabolas $x = y^2 - 2$ and $x = -y^2 + 6$. These parabolas, shown in Figure 4 (next page), intersect at the points $(2, 2)$ and $(2, -2)$. From Figure 4, we see that the y-variable ranges between -2 and 2. So our outer integral will be

$$\int_{-2}^{2} \ldots\, dy.$$

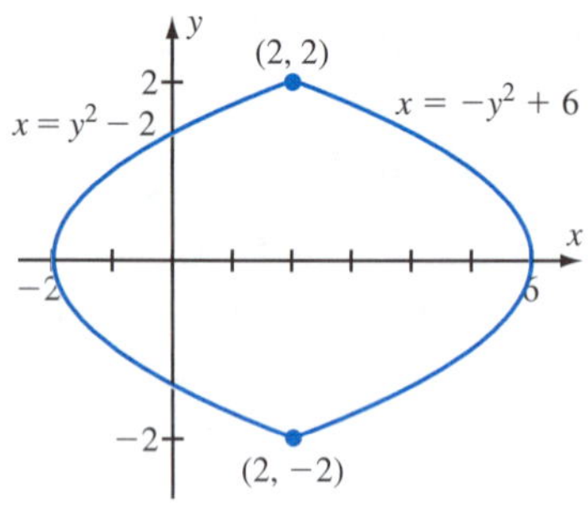

Figure 4

For each given value of y, Figure 4 shows that x ranges from $y^2 - 2$ to $-y^2 + 6$. Thus, the next integral is the integral in the y-variable, and we have

$$\int_{-2}^{2} \int_{y^2-2}^{-y^2+6} \ldots \, dx\, dy.$$

Finally, for each choice of x and y, the least value of z is $-x - 4$, (because $\mathcal{U}$ is bounded below by $z = -x - 4$) and the greatest value of z is $x + 2y + 4$ (because $\mathcal{U}$ is bounded above by $z = x + 2y + 4$). Our formula for the volume finally takes the form

$$V = \int_{-2}^{2} \int_{y^2-2}^{-y^2+6} \int_{-x-4}^{x+2y+4} 1 \, dz\, dx\, dy.$$

Working from the inside out, it is now straightforward to evaluate this integral:

$$\begin{aligned} V &= \int_{-2}^{2} \int_{y^2-2}^{-y^2+6} ((x + 2y + 4) - (-x - 4))\, dx\, dy \\ &= \int_{-2}^{2} \int_{y^2-2}^{-y^2+6} (2x + 2y + 8)\, dx\, dy \\ &= \int_{-2}^{2} (x^2 + 2(y + 4)x)\Big|_{x=y^2-2}^{x=-y^2+6} dy. \end{aligned}$$

After some simplification, we obtain

$$V = \int_{-2}^{2} (96 + 16y - 24y^2 - 4y^3)\, dy = (96y + 8y^2 - 8y^3 - y^4)\Big|_{-2}^{2} = 256. \quad \blacksquare$$

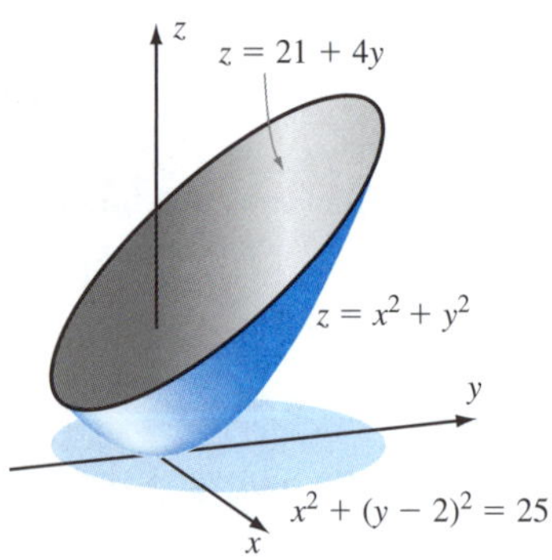

Figure 5

inSIGHT

Take careful note of the analysis that we used in Example 2 to set up the triple integral. This is the key to success in triple integral problems: Work from the outside in to set up the integral, but work from the inside out to evaluate the integral.

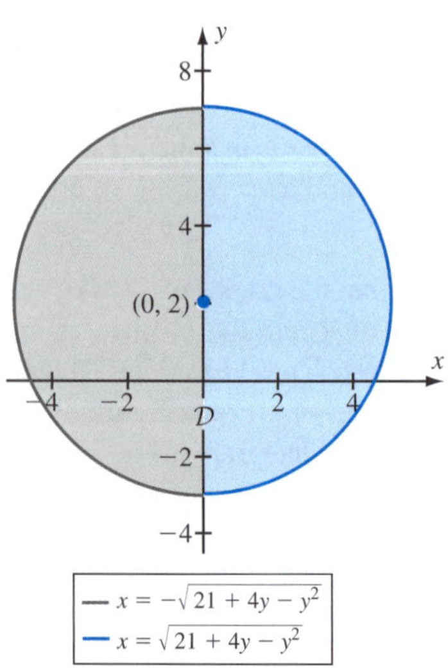

Figure 6

Example 3 Set up, but do not evaluate, the integral of $(x, y, z) \mapsto f(x, y, z)$ over the solid $\mathcal{U}$ bounded by the paraboloid $z = x^2 + y^2$ and the plane $z = 21 + 4y$. Treat $\mathcal{U}$ as a z-simple solid.

Solution The graph of $\mathcal{U}$ appears in Figure 5. When we consider $\mathcal{U}$ as a z-simple solid, we think of x and y as the outside variables of integration and z as the inside variable of integration. Let $\mathcal{D}$ be the projection of $\mathcal{U}$ to the xy-plane. If the point $(x, y, 0)$ belongs to $\mathcal{D}$, then we integrate z from $x^2 + y^2$ to $21 + 4y$. In particular, $x^2 + y^2 \le 21 + 4y$ and, on the boundary of $\mathcal{D}$, $x^2 + y^2 = 21 + 4y$, or $x^2 + (y - 2)^2 = 25$. Thus, $\mathcal{D}$ is the disk in the xy-plane that is bounded by the circle with center $(0, 2)$ and radius 5. Refer to Figure 6. We see that y ranges from -3 to 7. We conclude that the integral will have the form

$$\int_{-3}^{7} \ldots \, dy.$$

For each choice of y in this interval of integration, x ranges from $-\sqrt{25-(y-2)^2}$, or $-\sqrt{21+4y-y^2}$, to $\sqrt{21+4y-y^2}$. Therefore, our integral will have the form

$$\int_{-3}^{7}\int_{-\sqrt{21+4y-y^2}}^{\sqrt{21+4y-y^2}} \ldots\, dx\, dy.$$

Once y and x are chosen, z ranges from x^2+y^2 to $21+4y$. Our integral takes the form

$$\int_{-3}^{7}\int_{-\sqrt{21+4y-y^2}}^{\sqrt{21+4y-y^2}}\int_{x^2+y^2}^{21+4y} f(x, y, z)\, dz\, dx\, dy.$$

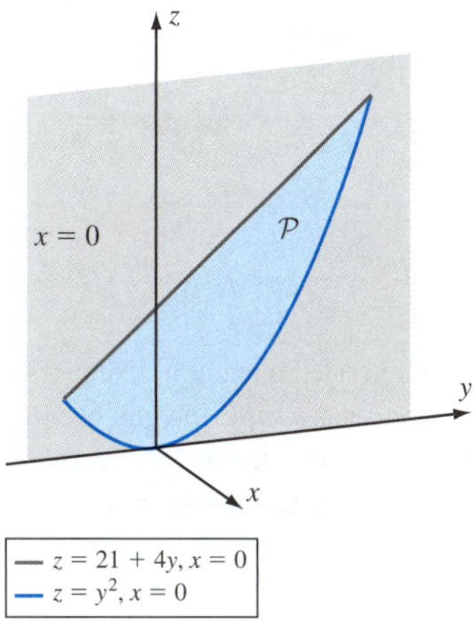

Figure 7

Example 4 Set up the integral from Example 3, treating $\mathcal{U}$ as an x-simple solid.

Solution Refer to Figure 5 again, and imagine the projection $\mathcal{P}$ of $\mathcal{U}$ into the yz-plane. In this case, the projection $\mathcal{P}$ is actually the intersection of $\mathcal{U}$ with the yz-plane. Setting $x=0$, we see that $\mathcal{P}$ is the region of the yz-plane that is bounded below by the parabola $z=y^2$ and above by the line $z=21+4y$. See Figure 7. Our analysis proceeds as follows: First, the variable y ranges from $y=-3$ to $y=7$. We conclude that the integral will have the form

$$\int_{-3}^{7} \ldots\, dy.$$

Next, for each choice of y in the interval of integration, the z-variable ranges from $z=y^2$ to $z=21+4y$. Therefore, our integral will have the form

$$\int_{-3}^{7}\int_{y^2}^{21+4y} \ldots\, dz\, dy.$$

Finally, for each choice of y and z, the variable x ranges from the side of the paraboloid $z=x^2+y^2$ on which $x=-\sqrt{z-y^2}$ to the side of the paraboloid on which $x=\sqrt{z-y^2}$. Our integral therefore takes the form

$$\int_{-3}^{7}\int_{y^2}^{21+4y}\int_{-\sqrt{z-y^2}}^{\sqrt{z-y^2}} f(x, y, z)\, dx\, dz\, dy.$$

quickquiz

1. What is a triple integral, and how does it relate to an iterated integral?
2. Express by means of a triple integral the volume inside a sphere of radius r.
3. Express by means of a triple integral the volume inside a cone of height 8 and with base having radius 5.
4. Let $\mathcal{U}$ be the solid that lies below the paraboloid $z=9-x^2-y^2$ and above the plane $z=5$. Set up, but do not evaluate, the triple integral $\iiint_{\mathcal{U}} f\, dV$ as an iterated integral.

EXERCISES

Problems for Practice

Evaluate the iterated integral in Exercises 1–10.

1. $\int_2^4 \int_x^{x+1} \int_x^{x+y} (6x - 3z)\, dz\, dy\, dx$

2. $\int_{-2}^2 \int_0^{2+y} \int_{y-x}^{y} 3x\, dz\, dx\, dy$

3. $\int_1^4 \int_{-z}^{z} \int_{y-z}^{y+z} (x + 3)\, dx\, dy\, dz$

4. $\int_3^5 \int_2^z \int_0^{(z^2-x^2)^{1/2}} 6y\, dy\, dx\, dz$

5. $\int_e^{e^2} \int_{1/(2y)}^{1/y} \int_1^y (4/z)\, dz\, dx\, dy$

6. $\int_0^3 \int_y^{2y} \int_z^{z+y} 8xyz\, dx\, dz\, dy$

7. $\int_0^{\pi/2} \int_0^x \int_x^z 2\cos(y)\, dy\, dz\, dx$

8. $\int_{-3}^{-2} \int_0^{2y} \int_0^z 16ye^{-(z^2)}\, dx\, dz\, dy$

9. $\int_0^\pi \int_{-y}^{\sqrt{\pi}} \int_{-x}^0 4y\cos(x^2)\, dz\, dx\, dy$

10. $\int_0^{\pi/2} \int_{-5}^3 \int_0^{\cos(y)} 3\sqrt{1 + \sin(y)}\, dx\, dz\, dy$

In Exercises 11–18, use a triple integral to express the volume of the solid that is bounded by the given surfaces in space. Evaluate the volume by means of iterated integrals.

11. $z = x^2 + y^2,\ z = 8,\ z = 2$

12. $z = x^2 + y^2,\ z = 8 - x^2 - y^2$

13. $y = x^2,\ z = 1 - y,\ z = -1 + y$

14. $x + 2y + 4z = 8,\ y = x/2,\ x = 0,\ z = 0$

15. $x^2 + y^2 = 4,\ z = x + y,\ z = -6 - x$

16. $y = x^2,\ z = -y + 4,\ z = 0$

17. $z = -x^2 - y^2 + 9,\ z = 2x + 4y - 2$

18. $y = x^2 + z^2 - 3,\ y = -x^2 - z^2 + 5$

In Exercises 19–24, evaluate the iterated integral. Describe the solid $\mathcal{U}$ over which the integration is performed.

19. $\int_0^2 \int_0^{3-3x/2} \int_0^{6-3x-2y} x\, dz\, dy\, dx$

20. $\int_0^6 \int_0^6 \int_0^{xy/6} z\, dz\, dx\, dy$

21. $\int_0^1 \int_0^{\sqrt{1-x^2}} \int_0^{\sqrt{1-x^2-y^2}} xz\, dz\, dy\, dx$

22. $\int_0^1 \int_0^{\sqrt{1-x^2}} \int_0^{1-y^2} x\, dz\, dy\, dx$

23. $\int_0^1 \int_0^x \int_0^{\sqrt{1-x^2}} z\, dz\, dy\, dx$

24. $\int_0^1 \int_0^{\sqrt{1-y^2}} \int_0^{\sqrt{x^2+y^2}} x\,(1 + y)\, dz\, dx\, dy$

Further Theory and Practice

25. The iterated integral in Exercise 19 represents a triple integral $\iiint_{\mathcal{U}} x\, dV$ over a solid $\mathcal{U}$ that is z-simple, as we can see from the inner variable of integration. However, the solid $\mathcal{U}$ is also y-simple. Reflecting this property, $\iiint_{\mathcal{U}} x\, dV$ can also be expressed as an iterated integral with y the inner variable of integration. Both

$$\int_0^2 \int_0^{6-3x} \int_0^{3-3x/2-z/2} x\, dy\, dz\, dx$$

and

$$\int_0^6 \int_0^{2-z/3} \int_0^{3-3x/2-z/2} x\, dy\, dx\, dz$$

do the job. Examine each iterated integral in Exercises 20–24. If the integral represents a triple integral over a solid that is y-simple, then express the integral as an iterated integral with y the inner variable of integration.

Draw a sketch of the domain of integration of each iterated integral in Exercises 26–31.

26. $\int_0^2 \int_x^{2x+1} \int_x^y f(x, y, z)\, dz\, dy\, dx$

27. $\int_{-1}^1 \int_y^{y+3} \int_{x+2}^{x+4} f(x, y, z)\, dz\, dx\, dy$

28. $\int_{-1}^0 \int_z^4 \int_0^{2x+4} f(x, y, z)\, dy\, dx\, dz$

29. $\int_{-3}^3 \int_{-(9-y^2)^{1/2}}^{(9-y^2)^{1/2}} \int_{-x-y-2}^{x+y+2} f(x, y, z)\, dz\, dx\, dy$

30. $\int_{-1}^1 \int_{-(4-4x^2)^{1/2}}^{(4-4x^2)^{1/2}} \int_{-4-x-z}^{2+x+z} f(x, y, z)\, dy\, dz\, dx$

31. $\int_1^4 \int_y^6 \int_y^{x^2} f(x, y, z)\, dz\, dx\, dy$

32. Suppose that a, b, and c are positive constants. Use a triple integral to evaluate the volume inside an ellipsoid of the form $x^2/a^2 + y^2/b^2 + z^2/c^2 = 1$.

33. Use a triple integral to give a formula for the volume of the solid bounded by the paraboloid $z = x^2 + y^2$ and the plane $z = ax + by + c$, $a > 0$, $b > 0$, $c > 0$.

34. Let a, b, and c be positive. Use a triple integral to find a formula for the volume of the solid in the first octant cut off by the plane $x/a + y/b + z/c = 1$.

35. Integrate $f(x, y, z) = x^2 + y^2 + z^2$ over the solid that is bounded above by the cone $z = m\sqrt{x^2 + y^2}$ and below by the paraboloid $z = x^2 + y^2$.

36. Use a triple integral to find the volume of the solid bounded by the cone $z^2 = x^2 + y^2$ and the planes $z = a$ and $z = b$, $0 < a < b$.
37. Calculate $\int_0^1 \int_x^1 \int_0^x z \sin(y^4)\, dz\, dy\, dx$

In Exercises 38 and 39, a solid $\mathcal{U}$ is described. Evaluate $\iiint_{\mathcal{U}} (xy + z)\, dV$.

38. $\mathcal{U}$ is the solid in the first octant that is bounded by the three coordinate planes, by the plane $x = 1$, and by the plane $3y + 2z = 6$.
39. $\mathcal{U}$ is the solid in the first octant that is bounded by the three coordinate planes, by the plane $z = 2 + 7x$, and by the parabolic cylinder $y = 4 - x^2$.

In Exercises 40 and 41, a solid $\mathcal{U}$ is described. Write $\iiint_{\mathcal{U}} f(x, y, z)\, dV$ in each of the six possible orders of integration.

40. $\mathcal{U}$ is the solid in the first octant that is bounded by the three coordinate planes and by the plane $5x + 3y + 2z = 30$.
41. $\mathcal{U}$ is the solid in the first octant that is bounded by the planes $x = 0$, $x = y$, $y = 1$, $z = 0$, and $z = 1$.
42. Write the iterated integral

$$\int_{-4}^{4} \int_{9}^{25-x^2} \int_{-\sqrt{25-x^2-y}}^{\sqrt{25-x^2-y}} (x^2 - yz)\, dz\, dy\, dx.$$

in all possible orders.
43. Write the iterated integral

$$\int_{-1}^{0} \int_{2}^{37} \int_{x+1}^{\sqrt{x+1}} y\sqrt{1-x}\, dy\, dz\, dx.$$

in all possible orders.
44. Suppose that a, b, and c are positive constants with $b > a$. Integrate $x^3\sqrt{c^2 - y^2}$ over the solid that is bounded by the parabolic cylinder $y = x^2$ and by the planes $x = 0$, $y = c$, $z = by$, and $z = ay$.
45. Suppose that a, b, c are positive constants. Let $\mathcal{U}$ be the solid that is bounded below by the xy-plane, above by the graph of $\sqrt{x/a} + \sqrt{y/b} + \sqrt{z/c} = 1$, and on the side by the xz- and yz-planes. Show that the volume of $\mathcal{U}$ is $abc/90$. Show that

$$\iiint_{\mathcal{U}} xyz\, dV = \frac{a^2b^2c^2}{277200}.$$

Computer/Calculator Exercises

46. Let $\mathcal{R}$ be the region of the xy-plane that is bounded above by $y = 3 - x^2$ and below by $y = 1 + (x+1)^3$. Let $\mathcal{U}$ be the solid that is bounded below by $\mathcal{R}$, above by the plane $z = y$, and laterally by the cylinder sets $y = 3 - x^2$ and $y = 1 + (x+1)^3$. Calculate $\iiint_{\mathcal{U}} x^2yz\, dV$.
47. Let $\mathcal{R}$ be the region in the first quadrant of the xy-plane that is bounded above by $y = \sin(x)$ and below by $y = 1/(1+x^2)$. Let $\mathcal{U}$ be the solid in the first octant that is bounded below by $\mathcal{R}$, above by the plane $z = x$, and laterally by the cylinder sets $y = \sin(x)$ and $y = 1/(1+x^2)$. Calculate $\iiint_{\mathcal{U}} z\, dV$.
48. Let $\mathcal{U}$ be the solid in the first octant that is bounded below by the plane $z = x$, above by the plane $z = 2$, and laterally by the cylinder sets $y = \exp(x)$ and $y = 6x - x^2 - 1$. Calculate $\iiint_{\mathcal{U}} z\, dV$.

14.7 Physical Applications

Multiple integrals are useful for calculating a number of physical quantities. In this section, we learn how to calculate the mass of a body, the center of mass of a body, the first moment of a body about an axis, and the moment of inertia of a body about an axis.

Mass

Imagine a thin metal plate (sometimes called a *lamina*) made of material that varies in composition from point to point. We think of the plate as occupying a region $\mathcal{R}$ in the xy-plane and assume that it has mass density given by a continuous function $\delta(x, y)$ at the point (x, y). In this context, "mass density" at a point (x, y) means mass per unit area *at the point* (x, y). This physical concept is defined like a partial derivative. To be specific, let $M(x, y, h)$ be the mass of a square of side length h that is centered at the point (x, y). To obtain the density at (x, y), we divide this mass by the area of the square and take the limit as h tends to 0:

$$\delta(x, y) = \lim_{h \to 0^+} \frac{M(x, y, h)}{h^2}. \qquad \textbf{(14.15)}$$

We now answer the following question: If the mass density of a plate is known at each point, then how can we calculate the mass of the entire plate?

As usual, we think of the plate as lying inside a rectangle $Q = \{(x, y) : a \leq x \leq b, c \leq y \leq d\}$. Partitioning the intervals $[a, b]$ and $[c, d]$ into N subintervals each, we obtain rectangles $Q_{i,j}$. We consider only the rectangles that lie entirely inside $\mathcal{R}$. If $\xi_{i,j} = (x_i, y_j)$ is a point in $Q_{i,j}$, then the mass $M_{i,j}$ of the square $Q_{i,j}$ is approximately

$$M_{i,j} = \delta(\xi_{i,j})\Delta x \Delta y = \delta(\xi_{i,j})\Delta A.$$

Notice that we estimate the mass by multiplying $\delta(\xi_{i,j})$, which represents mass per unit of area, by the area of the small rectangle. Limit formula (14.15) justifies this approximation. The total mass of the plate is about

$$\sum_{Q_{i,j} \subset \mathcal{R}} M_{i,j} = \sum_{Q_{i,j} \subset \mathcal{R}} \delta(\xi_{i,j})\Delta A.$$

Letting the number of subrectangles of the partition tend to infinity leads us to define the *mass of the plate* to be

$$M = \iint_{\mathcal{R}} \delta(x, y)\, dA. \qquad \textbf{(14.16)}$$

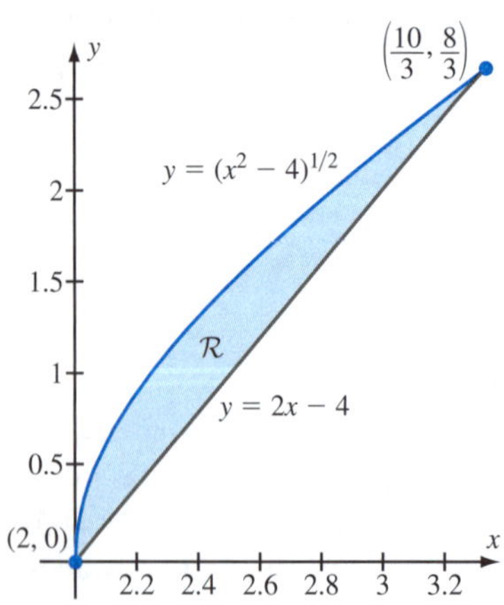

Figure 1

Example 1 Suppose that a thin plate $\mathcal{R}$ is in the shape shown in Figure 1. The bounding curves are $y = (x^2 - 4)^{1/2}$ and $y = 2x - 4$. Assume also that the density of $\mathcal{R}$ at the point (x, y) is $\delta(x, y) = 4x$. Find the mass of $\mathcal{R}$.

Solution The curves intersect when $x = 2$ and $x = 10/3$. We may write $\mathcal{R}$ as a y-simple region as follows:

$$\mathcal{R} = \left\{(x, y) : 2 \leq x \leq \frac{10}{3}, 2x - 4 \leq y \leq (x^2 - 4)^{1/2}\right\}.$$

Then the mass of $\mathcal{R}$ is

$$\begin{aligned} M &= \iint_{\mathcal{R}} \delta(x, y)\, dA \\ &= \int_2^{10/3} \int_{2x-4}^{(x^2-4)^{1/2}} 4x\, dy\, dx \\ &= \int_2^{10/3} 4x((x^2-4)^{1/2} - (2x-4))\, dx \\ &= \frac{4}{3}(x^2-4)^{3/2} - \frac{8}{3}x^3 + 8x^2 \Big|_2^{10/3} \\ &= \frac{384}{81}. \end{aligned}$$

The mass of the plate is 384/81, or 128/27. ■

First Moments

Next we consider the first moments of a lamina about each axis. For a point mass, the *first moment* about an axis is the product of the mass times the distance to the axis. For each i and j, the first moment of the small rectangle $Q_{i,j}$ about the x-axis is approximately $y_j \cdot \delta(\xi_{i,j})\Delta A$ where $\xi_{i,j}$ is a point arbitrarily chosen from $Q_{i,j}$. Look at Figure 2. There are two approximations in estimating this first moment. The different points that make up $Q_{i,j}$ do not have constant distance to the x-axis; however, if $Q_{i,j}$ is small, then the distances of the constituent points do not vary greatly, and y_j can be used as an approximation. Also, we approximate the mass of $Q_{i,j}$ by $\delta(\xi_{i,j})\Delta A$ as usual. Summing over i and j, we obtain

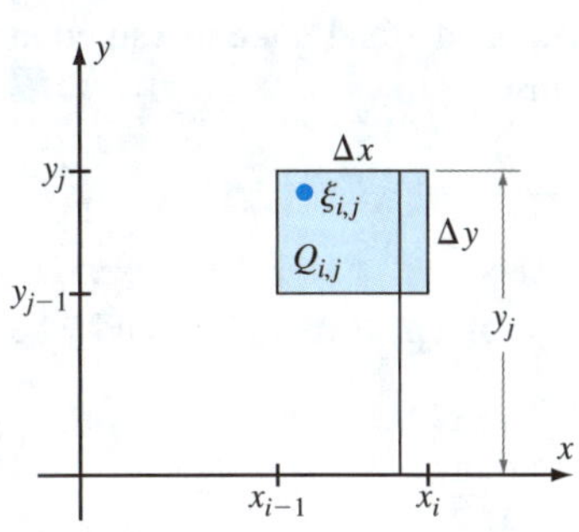

Figure 2

$$\sum_{Q_{i,j} \subseteq \mathcal{R}} y_j \cdot \delta(\xi_{i,j})\Delta A$$

as an approximation to the first moment of the plate about the x-axis. Letting the number of subrectangles tend to infinity leads us to define the *first moment* of the plate with respect to the x-axis, which we write as $y = 0$, to be

$$M_{y=0} = \iint_{\mathcal{R}} y\delta(x, y)\, dA. \tag{14.17}$$

Similar computations lead to the definition that the *first moment* of the plate with respect to the y-axis, which we write as $x = 0$, is

$$M_{x=0} = \iint_{\mathcal{R}} x\delta(x, y)\, dA. \tag{14.18}$$

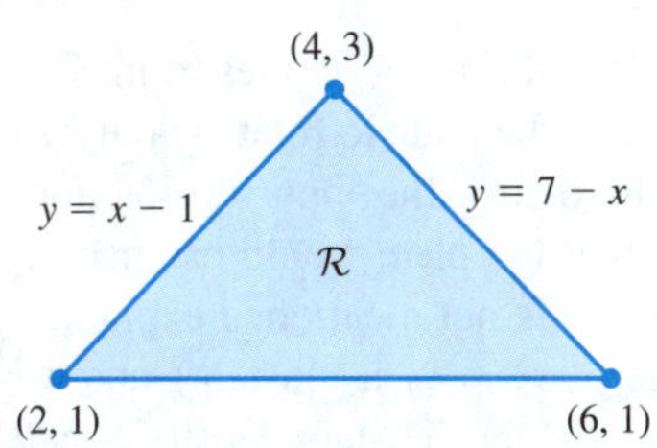

Figure 3

Example 2 A metal plate $\mathcal{R}$ sitting in the xy-plane is shown in Figure 3. Its density at a point (x, y) is $\delta(x, y) = 6x + 9y$. Calculate its first moment with respect to the x-axis.

Solution First, write the region $\mathcal{R}$ in x-simple form:

$$\mathcal{R} = \{(x, y) : 1 \leq y \leq 3,\ y + 1 \leq x \leq -y + 7\}.$$

We have

$$\begin{aligned}
M_{y=0} &= \iint_{\mathcal{R}} y\delta(x, y)\, dA \\
&= \int_1^3 \int_{y+1}^{-y+7} y(6x + 9y)\, dx\, dy \\
&= \int_1^3 (3yx^2 + 9y^2 x)\Big|_{x=y+1}^{x=-y+7} dy \\
&= \int_1^3 (-18y^3 + 6y^2 + 144y)\, dy \\
&= \left(-\frac{9}{2}y^4 + 2y^3 + 72y^2\right)\Big|_1^3 \\
&= 268.
\end{aligned}$$

We conclude that the first moment of $\mathcal{R}$ with respect to the x-axis is 268 units. For example, if δ is measured in kilograms per square meter and x and y are measured in meters, then our answer is measured in meter-kilograms. ■

Center of Mass

If M, $M_{x=0}$, and $M_{y=0}$ denote the mass and first moments of a plate, then let $\bar{x}$ and $\bar{y}$ be defined by

$$\bar{x} = \frac{M_{x=0}}{M} \quad \text{and} \quad \bar{y} = \frac{M_{y=0}}{M}.$$

Referring to the definition of $M_{x=0}$, we see that $\bar{x}$ is the average value of the x-coordinate function on the plate, weighted according to the varying density function $\delta(x, y)$. Similarly, $\bar{y}$ is the average y-value on the plate, weighted according to the mass density. The *center of mass* (or *center of gravity*) of the plate is now defined to be the point

$$\bar{c} = (\bar{x}, \bar{y}).$$

It is not difficult to see where the definition of "center of gravity" comes from. The point $\bar{c}$ is determined by the property that the tendency of the plate to rotate about $\bar{c}$ in any given direction is balanced exactly by its tendency to rotate in the opposite direction. The physical significance of the center of gravity is that the plate would balance on the point of a pin placed at $\bar{c}$. When the density function is not mentioned explicitly, we assume that mass is uniformly distributed (that is, $\delta(x, y) = \delta_0$, a constant), and we speak of the "centroid" of a region. In this case, notice that the constant density δ_0 can be brought outside each of the integrals that define M, $M_{x=0}$, and $M_{y=0}$. The constant δ_0 then cancels in the ratios $\bar{x} = M_{x=0}/M$ and $\bar{y} = M_{y=0}/M$, which is why the value of a uniform density is often not specified.

Example 3 Consider a plate $\mathcal{R}$ in the xy-plane bounded by the curve $y = -3x^2 + 12$ and the x-axis. Assume that the plate has constant density $\delta(x, y) = 1$. Calculate the plate's center of mass.

Solution The curve crosses the x-axis at $x = \pm 2$. The mass is therefore

$$M = \iint_{\mathcal{R}} 1\, dA = \int_{-2}^{2} \int_{0}^{12-3x^2} 1\, dy\, dx = \int_{-2}^{2} (12 - 3x^2)\, dx = 32.$$

Further, we know that the first moment with respect to the x-axis is

$$M_{y=0} = \iint_{\mathcal{R}} y \cdot 1\, dA = \int_{-2}^{2} \int_{0}^{12-3x^2} y\, dy\, dx = \int_{-2}^{2} \left. \frac{y^2}{2} \right|_{y=0}^{y=12-3x^2} dx$$

$$= \int_{-2}^{2} \left(\frac{9}{2}x^4 - 36x^2 + 72 \right) dx = \frac{768}{5}.$$

Similarly, the first moment with respect to the y-axis is

$$M_{x=0} = \iint_{\mathcal{R}} x \cdot 1\, dA = \int_{-2}^{2} \int_{0}^{12-3x^2} x\, dy\, dx = \int_{-2}^{2} (12 - 3x^2)x\, dx = 0.$$

In physical terms, the first moment with respect to the y-axis is equal to 0 because the plate is symmetric with respect to the y-axis. We conclude that

$$\bar{x} = \frac{M_{x=0}}{M} = \frac{0}{32} = 0 \quad \text{and} \quad \bar{y} = \frac{M_{y=0}}{M} = \frac{768/5}{32} = \frac{24}{5}.$$

Therefore, the center of gravity is $\bar{c} = (\bar{x}, \bar{y}) = (0, 24/5)$. ■

Moment of Inertia

The second moments, or moments of inertia, of a plate $\mathcal{R}$ about the axes are defined as follows:

$$I_{y=0} = \iint_{\mathcal{R}} y^2 \delta(x, y)\, dA$$

$$I_{x=0} = \iint_{\mathcal{R}} x^2 \delta(x, y)\, dA.$$

Here, $I_{y=0}$ is the moment of inertia about the x-axis, and $I_{x=0}$ is the moment of inertia about the y-axis. Notice that the moment of inertia differs from the first moment in that in the latter we use the *square* of the distance to the axis. The second moment can be used, for instance, to measure the kinetic energy of rotation of a spinning body. Moment of inertia is the rotational analogue of mass. These matters are studied in more detail in a physics text.

Example 4 Calculate the second moment $I_{y=0}$ of the lamina with constant density $\delta = 3$ that occupies the region bounded by the curves $y = -x^2 + 4$ and $y = 0$.

Solution The x limits of integration are -2 and 2. Thus, according to the definition,

$$\begin{aligned}
I_{y=0} &= \iint_{\mathcal{R}} y^2 \cdot 3\,dA \\
&= \int_{-2}^{2}\int_{0}^{-x^2+4} 3y^2\,dy\,dx \\
&= \int_{-2}^{2} y^3\Big|_0^{-x^2+4}\,dx \\
&= \int_{-2}^{2} (-x^2+4)^3\,dx \\
&= \int_{-2}^{2} (-x^6+12x^4-48x^2+64)\,dx \\
&= \left(-\frac{x^7}{7}+\frac{12x^5}{5}-16x^3+64x\right)\Bigg|_{-2}^{2} \\
&= \frac{4096}{35}.
\end{aligned}$$

Mass, First Moment, Moment of Inertia, and Center of Mass in Three Dimensions

The reasoning that we used to derive double integral formulas for mass, first moment, moment of inertia, and center of mass in two dimensions can also be used to derive triple integral formulas for these quantities in three dimensions. If $\mathcal{U}$ is a solid in space and if the density of the solid at the point (x, y, z) is $\delta(x, y, z)$, then an argument similar to that in the first part of this section leads us to define the *mass* of $\mathcal{U}$ to be

$$M = \iiint_{\mathcal{U}} \delta(x, y, z)\,dV.$$

Similarly, the first moments with respect to the xy-plane, yz-plane, and xz-plane are given by

$$M_{z=0} = \iiint_{\mathcal{U}} z\delta(x, y, z)\,dV, \qquad M_{x=0} = \iiint_{\mathcal{U}} x\delta(x, y, z)\,dV, \quad \text{and}$$

$$M_{y=0} = \iiint_{\mathcal{U}} y\delta(x, y, z)\,dV.$$

The coordinates of the center of mass of the body are then given by

$$\bar{x} = \frac{M_{x=0}}{M},$$

$$\bar{y} = \frac{M_{y=0}}{M}, \quad \text{and}$$

$$\bar{z} = \frac{M_{z=0}}{M}.$$

The center of mass is the point

$$\bar{c} = (\bar{x}, \bar{y}, \bar{z}).$$

As in two dimensions, when the density function is not mentioned explicitly, we assume that mass is uniformly distributed, and we speak of the "centroid" of a region.

Finally, the second moments, or moments of inertia, of a solid $\mathcal{U}$ with density δ are given by

$$I_x = \iiint\limits_{\mathcal{U}} (y^2 + z^2)\delta(x, y, z)\, dV,$$

$$I_y = \iiint\limits_{\mathcal{U}} (x^2 + z^2)\delta(x, y, z)\, dV,$$

and

$$I_z = \iiint\limits_{\mathcal{U}} (x^2 + y^2)\delta(x, y, z)\, dV.$$

Example 5 Assuming that the body

$$\mathcal{U} = \{(x, y, z) : -2 \le y \le 0, y + 3 \le x \le y + 5, 2x + y + 4 \le z \le 2y + 14\}$$

has a uniform mass distribution $\delta(x, y, z) = 3$, calculate its first moment with respect to the xz-plane.

Solution According to the definition, the first moment of $\mathcal{U}$ with respect to the xz-plane is

$$\begin{aligned}
M_{y=0} &= \iint\limits_{\mathcal{R}} 3y\, dV \\
&= \int_{-2}^{0} \int_{y+3}^{y+5} \int_{2x+y+4}^{2y+14} 3y\, dz\, dx\, dy \\
&= \int_{-2}^{0} \int_{y+3}^{y+5} 3yz\Big|_{z=2x+y+4}^{z=2y+14} dx\, dy \\
&= \int_{-2}^{0} \int_{y+3}^{y+5} (3y^2 - 6xy + 30y)\, dx\, dy \\
&= \int_{-2}^{0} -6y^2 + 12y\, dy \\
&= -40.
\end{aligned}$$

■

Example 6 Assume that the solid body

$$\mathcal{U} = \{(x, y, z) : -1 \le z \le 1, 1 \le y \le 2, 0 \le x \le z + y\}$$

has density function $\delta(x, y, z) = 2x + y$. Compute the center of mass of $\mathcal{U}$.

Solution First we have

$$\begin{aligned} M &= \iiint_{\mathcal{R}} \delta(x, y, z)\, dV \\ &= \int_{-1}^{1} \int_{1}^{2} \int_{0}^{z+y} (2x + y)\, dx\, dy\, dz \\ &= \int_{-1}^{1} \int_{1}^{2} (x^2 + yx)\Big|_{0}^{z+y} dy\, dz \\ &= \int_{-1}^{1} \int_{1}^{2} (z^2 + 3zy + 2y^2)\, dy\, dz \\ &= \int_{-1}^{1} \left(z^2 y + \frac{3}{2} z y^2 + \frac{2}{3} y^3\right)\Big|_{y=1}^{y=2} dz \\ &= \int_{-1}^{1} z^2 + \frac{9}{2} z + \frac{14}{3}\, dz \\ &= 10. \end{aligned}$$

Similar calculations show that

$$\begin{aligned} M_{x=0} &= \int_{-1}^{1} \int_{0}^{1} \int_{0}^{z+y} x(2x + y)\, dx\, dy\, dz = \frac{45}{4}, \\ M_{y=0} &= \int_{-1}^{1} \int_{0}^{1} \int_{0}^{z+y} y(2x + y)\, dx\, dy\, dz = 16, \\ M_{z=0} &= \int_{-1}^{1} \int_{0}^{1} \int_{0}^{z+y} z(2x + y)\, dx\, dy\, dz = 3. \end{aligned}$$

Therefore,

$$\begin{aligned} \bar{x} &= \frac{M_{x=0}}{M} = \frac{45/4}{10} = \frac{9}{8}, \\ \bar{y} &= \frac{M_{y=0}}{M} = \frac{16}{10} = \frac{8}{5}, \\ \bar{z} &= \frac{M_{z=0}}{M} = \frac{3}{10}. \end{aligned}$$

We conclude that the center of mass of $\mathcal{U}$ is the point $(9/8, 8/5, 3/10)$. ■

quickquiz

1. How do we define the mass of a region in the plane? Of a solid in space?
2. How do we define the first moments of a region in the plane? Of a solid in space?
3. How do we define the moments of inertia of a region in the plane? Of a solid in space?
4. How do we define the center of mass of a region in the plane? Of a solid in space?

EXERCISES

Problems for Practice

In Exercises 1 and 2, use a double integral to find the mass of the planar region $\mathcal{R}$ with density function $\delta(x, y)$.

1. $\mathcal{R} = \{(x, y) : 0 \leq x \leq 2, x + 1 \leq y \leq 2x + 3\}$, $\delta(x, y) = 3x + 4y$

2. $\mathcal{R} = \{(x, y) : -2 \leq x \leq 2, 0 \leq y \leq 4 - x^2\}$, $\delta(x, y) = x + 3$

In Exercises 3–6, find the first moment of the region $\mathcal{R}$ with density function $\delta(x, y)$ about the indicated axis.

3. $\mathcal{R} = \{(x, y) : 1 \leq y \leq 4, \sqrt{y} \leq x \leq y^2\}$, $\delta(x, y) = 10$, x-axis

4. $\mathcal{R} = \{(x, y) : 0 \leq x \leq 1, 0 \leq y \leq x^2\}$, $\delta(x, y) = \sqrt{x}$, y-axis

5. $\mathcal{R} = \{(x, y) : -3 \leq x \leq -1, 2 \leq y \leq 4\}$, $\delta(x, y) = -x/y$, y-axis

6. $\mathcal{R} = \{(x, y) : -1 \leq y \leq 0, 0 \leq x \leq y + 1\}$, $\delta(x, y) = \sqrt{|y|}$, y-axis

In Exercises 7–10, find the center of mass of the planar region $\mathcal{R}$ with density function $\delta(x, y)$.

7. $\mathcal{R} = \{(x, y) : 0 \leq x \leq 1, x \leq y \leq \sqrt{x}\}$, $\delta(x, y) = x + 6$

8. $\mathcal{R}$ is the region between the parabolas $y = -x^2 + 4$ and $y = x^2 - 4$; $\delta(x, y) = 3 - x$.

9. $\mathcal{R} = \{(x, y) : -1 \leq y \leq 1, 2 \leq x \leq y + 3\}$, $\delta(x, y) = x(y + 1)$

10. $\mathcal{R} = \{(x, y) : 3 \leq x \leq 4, -2 \leq y \leq 1\}$, $\delta(x, y) = x + y + 2$

In Exercises 11–14, calculate the mass of the solid $\mathcal{U}$ with density function $\delta(x, y, z)$.

11. $\mathcal{U} = \{(x, y, z) : 0 \leq x \leq 1, 0 \leq y \leq 2, 0 \leq z \leq xy\}$, $\delta(x, y, z) = x + 1$

12. $\mathcal{U} = \{(x, y, z) : 0 \leq y \leq 1, y \leq x \leq 4, y \leq z \leq x\}$, $\delta(x, y, z) = \sqrt{x} + \sqrt{y}$

13. $\mathcal{U} = \{(x, y, z) : 0 \leq z \leq 2, z \leq y \leq 3, y \leq x \leq y + 1\}$, $\delta(x, y, z) = x + z$

14. $\mathcal{U} = \{(x, y, z) : 0 \leq x \leq 1, 0 \leq y \leq 1, 0 \leq z \leq x + y\}$, $\delta(x, y, z) = x + y + z$

In Exercises 15–18, calculate the three first moments and the center of mass of the given solid in space.

15. $\mathcal{U} = \{(x, y, z) : -1 \leq x \leq 0, x \leq y \leq 1, x + y \leq z \leq 2\}$, $\delta(x, y, z) = x + 3$

16. $\mathcal{U} = \{(x, y, z) : 1 \leq z \leq 2, z^2 \leq y \leq 4, 2 \leq x \leq 2z\}$, $\delta(x, y, z) = \sqrt{y}$

17. $\mathcal{U} = \{(x, y, z) : 1 \leq z \leq 2, 0 \leq x \leq z, x \leq y \leq z\}$ $\delta(x, y, z) = y - x$

18. $\mathcal{U} = \{(x, y, z) : x^2 + y^2 \leq 4, 0 \leq z \leq 4, \}$ $\delta(x, y, z) = x^2 + y^2 + z^2$

In Exercises 19–22, calculate the moments of inertia with respect to all axes.

19. $\mathcal{U} = \{(x, y, z) : 3 \leq x \leq 5, 1 \leq y \leq 4, 1 \leq z \leq 2\}$, $\delta(x, y, z) = xy$

20. $\mathcal{R} = \{(x, y) : -1 \leq y \leq 2, 0 \leq x \leq 2 - y, 0 \leq z \leq x\}$, $\delta(x, y, z) = y + 2$

21. $\mathcal{U} = \{(x, y, z) : 0 \leq x \leq 3, 1 \leq y \leq 2, x + y \leq z \leq x + y + 1\}$, $\delta(x, y, z) = 4$

22. $\mathcal{U} = \{(x, y, z) : 1 \leq z \leq 2, 1 \leq x \leq z, 1 \leq y \leq x\}$, $\delta(x, y, z) = x$

Further Theory and Practice

23. Give an example of a planar region $\mathcal{R}$ such that the center of gravity $\mathcal{R}$ is not an element of $\mathcal{R}$.

24. Suppose that a region in the plane consists of all points that lie in one or more of the following sets:

$$\mathcal{R} = \{(x, y) : x^2 + y^2 \leq 1\},$$
$$\mathcal{R} = \{(x, y) : (x - 4)^2 + (y - 3)^2 \leq 1\},$$
$$\mathcal{R} = \{(x, y) : x^2 + (y - 12)^2 \leq 4\}.$$

Assume that the mass distribution is uniform in this region. Find the center of mass.

25. A solid $\mathcal{U}$ rotates about the x-axis with constant angular speed ω. Express the solid's kinetic energy $KE_{\mathcal{U}}$ in terms of ω and the moment of inertia I_x of $\mathcal{U}$.

26. Pappus's Theorem says that if a region in the right half of the xy-plane (that is, in the set $\{(x, y, 0) : x > 0\}$) is rotated about the y-axis, then the volume of the resulting solid is the area of the region times the circumference of the circle traversed by the centroid of the region. Prove Pappus's Theorem.

27. Apply Pappus's Theorem (as stated in Exercise 26) to calculate the volume obtained when a disk of radius r and center $(R, 0)$ with $r < R$ is rotated about the y-axis.

Calculator/Computer Exercises

28. Let $\mathcal{R}$ be the region of the xy-plane that is bounded above by $y = 3 - x^2$ and below by $y = 1 + (x + 1)^3$. Calculate the center of mass of $\mathcal{R}$ given that its mass density is $\delta(x, y) = x + 3y$.

29. Let $\mathcal{R}$ be the region in the first quadrant of the xy-plane that is bounded above by $y = 1 + x + x^2$ and below by $y = \exp(x)$. Calculate the center of mass of $\mathcal{R}$ given that its mass density is $\delta(x, y) = 1 + x$.

30. Let $\mathcal{R}$ be the region in the first quadrant of the xy-plane that is bounded above by $y = 1 + x + 2x^2$ and below by $y = 1 + x^3$. The mass density of $\mathcal{R}$ is $\delta(x, y) = 1 + x\sqrt{y}$. Calculate the second moments of $\mathcal{R}$ about the x- and y-axes.

31. Let $\mathcal{R}$ be the square in the xy-plane with vertices $(0, 0)$, $(0, 1)$, $(1, 1)$, and $(1, 0)$. The mass density of $\mathcal{R}$ is $\delta(x, y) = 2 + \sin(x^2) + \exp(-y^4)$. Approximate the center of mass of $\mathcal{R}$.

14.8 Other Coordinate Systems

Earlier in this chapter, we learned that polar coordinates make it much easier to study the equations of planar curves with circular symmetry. A similar situation occurs when we are doing calculus in space. In this section, we learn about two special coordinate systems for handling surfaces and solids with special symmetries.

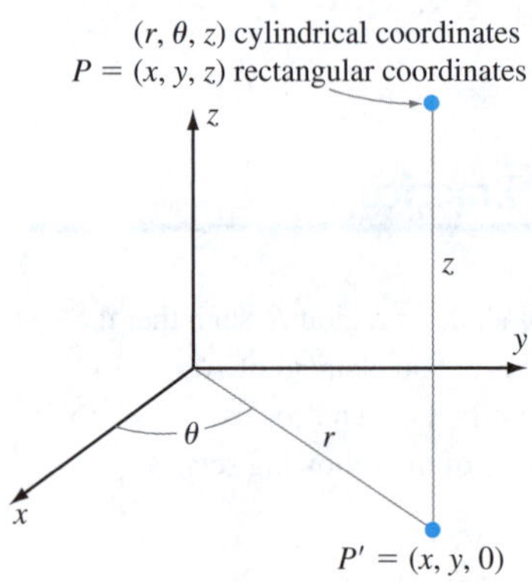

Figure 1

Cylindrical Coordinates

The cylindrical coordinate system is created by using polar coordinates in the xy-plane and the standard Euclidean z-coordinate in the vertical direction. The idea is illustrated in Figure 1. Notice that we locate the point P in space by first looking at the projection P' of P into the xy-plane. The distance of P' from the origin is denoted by r, and the angle that $\overrightarrow{OP'}$ subtends with the positive x-axis is called θ. The third coordinate is simply z, the signed distance of P from P'. We write the cylindrical coordinates for P as (r, θ, z). The formulas that we learned in Section 14.4 for converting from rectangular to polar coordinates and back are valid in the context of cylindrical coordinates:

$$x = r\cos(\theta), \quad y = r\sin(\theta), \quad x^2 + y^2 = r^2, \quad \text{and, if } x \neq 0, \quad \tan(\theta) = \frac{y}{x}.$$

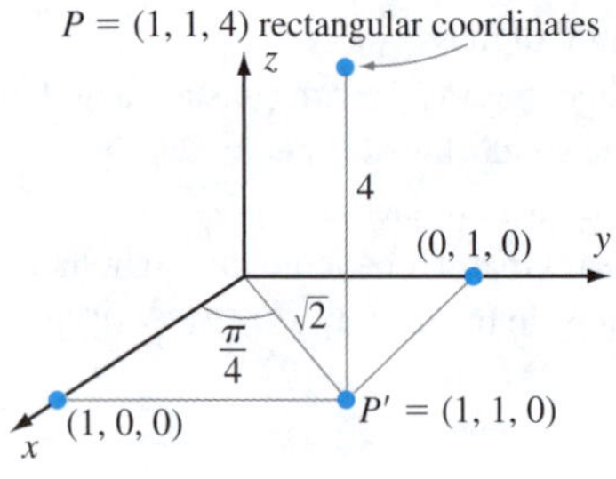

Figure 2

Example 1 The point P has rectangular coordinates $(1, 1, 4)$. What are its cylindrical coordinates?

Solution The projection of P into the xy-plane is $P' = (1, 1, 0)$, as shown in Figure 2. Then we see that $r = \sqrt{2}$ and $\theta = \pi/4$. Also, $z = 4$. Therefore, the cylindrical coordinates for P are $(\sqrt{2}, \pi/4, 4)$. ∎

We can calculate many integrals easily by using cylindrical coordinates. To do so, we replace the rectangular element of area $dx\,dy$ with the polar element of area

$r\,dr\,d\theta$. So our element of volume, which is $dV = dx\,dy\,dz$ in rectangular coordinates, becomes

$$dV = r\,dr\,d\theta\,dz$$

in cylindrical coordinates.

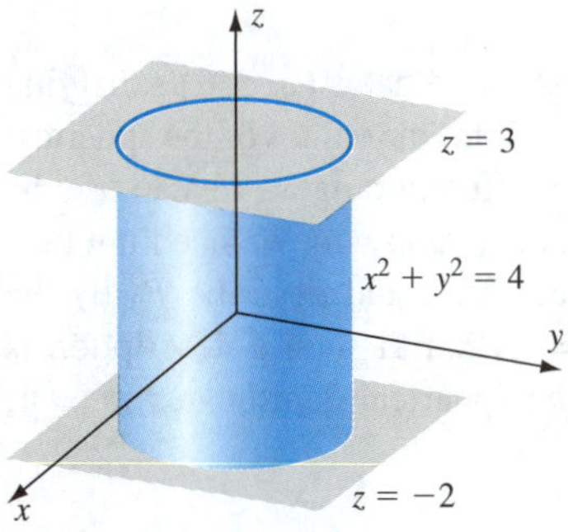

Figure 3

Example 2 Integrate the function $f(x, y, z) = x^2 + y^2$ over the solid $\mathcal{U}$ that is inside the cylinder $x^2 + y^2 = 4$, below the plane $z = 3$, and above the plane $z = -2$.

Solution The solid $\mathcal{U}$ is drawn in Figure 3. It is convenient to convert the problem to cylindrical coordinates. Then $f(r\cos(\theta), r\sin(\theta), z) = r^2$, and the solid over which we wish to integrate becomes

$$\mathcal{U} = \{(r, \theta, z) : 0 \le r \le 2,\ 0 \le \theta \le 2\pi,\ -2 \le z \le 3\}.$$

Bearing in mind that the volume element is $r\,dr d\theta dz$, we find that our integral is

$$\iiint\limits_{\mathcal{U}} f\,dV = \int_{-2}^{3}\int_{0}^{2\pi}\int_{0}^{2} r^2 \cdot r\,dr\,d\theta\,dz = \int_{-2}^{3}\int_{0}^{2\pi} \frac{r^4}{4}\bigg|_{r=0}^{r=2} d\theta\,dz$$

$$= \frac{16}{4} \cdot 2\pi \cdot (3 - (-2)) = 40\pi.$$

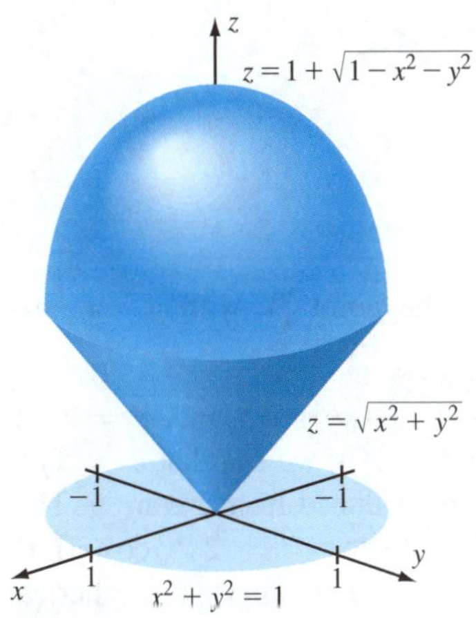

Figure 4

Example 3 Calculate the volume of the solid $\mathcal{U}$ that lies above the cone $z = \sqrt{x^2 + y^2}$ and below the hemisphere $z = 1 + \sqrt{1 - x^2 - y^2}$.

Solution The solid $\mathcal{U}$ appears in Figure 4. Its projection in the xy-plane is the disk $\mathcal{D}$ that is described by the inequality $x^2 + y^2 \le 1$ in rectangular coordinates. In terms of cylindrical coordinates, $\mathcal{D}$ is described by the inequalities $0 \le \theta \le 2\pi$, $0 \le r \le 1$. We observe that the cone may be written as $z = \sqrt{x^2 + y^2} = r$ and the hemisphere may be written as $z = 1 + \sqrt{1 - r^2}$. A point with cylindrical coordinates (r, θ, z) lies in $\mathcal{U}$ if and only if $r \le z \le 1 + \sqrt{1 - r^2}$. Thus, the volume of $\mathcal{U}$ is

$$\iiint\limits_{\mathcal{U}} 1\,dV = \int_{0}^{2\pi}\int_{0}^{1}\int_{r}^{1+\sqrt{1-r^2}} 1 \cdot r\,dz\,dr\,d\theta = \int_{0}^{2\pi}\int_{0}^{1} \left(1 + \sqrt{1 - r^2} - r\right) \cdot r\,dr\,d\theta.$$

Hence, we have

$$\iiint\limits_{\mathcal{U}} 1\,dV = \int_{0}^{2\pi} \left(\frac{r^2}{2} - \frac{(1 - r^2)^{3/2}}{3} - \frac{r^3}{3}\right)\bigg|_{r=0}^{r=1} d\theta$$

$$= 2\pi\left(\left(\frac{1}{2} - 0 - \frac{1}{3}\right) - \left(0 - \frac{1}{3} - 0\right)\right) = \pi.$$

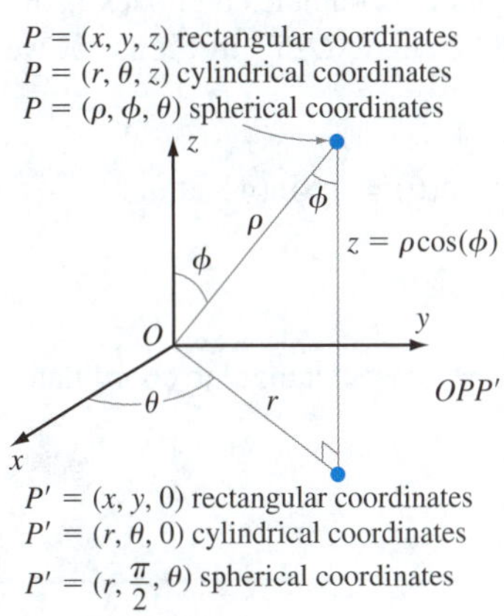

Figure 5

Spherical Coordinates

For calculations involving objects in three-dimensional space that are symmetrical in the origin, spherical coordinates are most convenient. Referring to Figure 5, we see that in this new coordinate system, we locate a point P according to its distance ρ from the origin, the angle ϕ that its ray makes with the positive z-axis, and the angle θ that its projected ray makes in the xy-plane with the positive x-axis. We write the spherical coordinates for P as (ρ, ϕ, θ).

Every point in space can be located with spherical coordinates (ρ, ϕ, θ) satisfying $0 \leq \rho, 0 \leq \phi \leq \pi$, and $0 \leq \theta \leq 2\pi$. By convention, we limit the values of the spherical coordinates to these ranges. Notice that θ and ϕ have different ranges. We can use the point P_0 with rectangular coordinates $(0, -1, 0)$ to understand why we need not ever take ϕ to be greater than π. If convention permitted, we could describe P_0 by the spherical coordinates $\rho = 1$, $\phi = 3\pi/2$, and $\theta = \pi/2$. But such a description is unnecessary because we can also describe P_0 by the spherical coordinates $\rho = 1$, $\phi = \pi/2$, and $\theta = 3\pi/2$. See Figure 6.

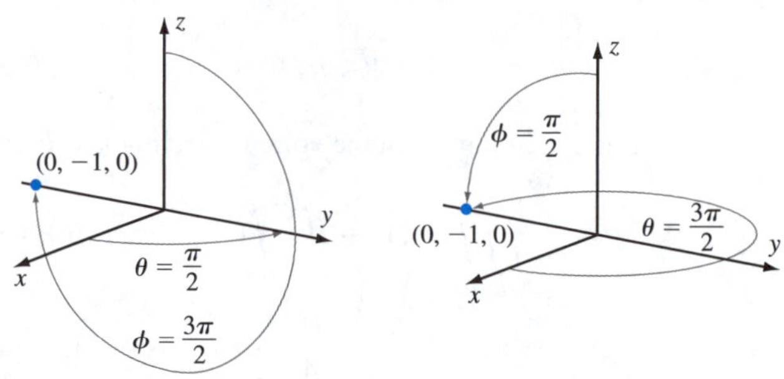

Figure 6

Example 4 What are spherical coordinates for the point P_0 with rectangular coordinates $(1, 1, \sqrt{6})$?

Solution The distance of P_0 from 0 is $\rho = (1^2 + 1^2 + (\sqrt{6})^2)^{1/2} = \sqrt{8} = 2\sqrt{2}$. Since the projection of P_0 into the xy-plane is $P_0' = (1, 1, 0)$, we see that $\theta = \pi/4$. Next, we use the equation $z = \rho\cos(\phi)$, which can be deduced from Figure 5. This equation tells us that the spherical coordinate ϕ of P_0 satisfies $\sqrt{6} = 2\sqrt{2}\cos(\phi)$, or $\cos(\phi) = \sqrt{6}/(2\sqrt{2}) = \sqrt{3}/2$. It follows that $\phi = \pi/6$. In summary, the spherical coordinates of P_0 are $(2\sqrt{2}, \pi/6, \pi/4)$. ■

Converting from spherical coordinates to rectangular coordinates and back again requires some calculation. As can be seen from right triangle OPP' in Figure 5, we have $z = \rho\cos(\phi)$ and $r = \rho\sin(\phi)$. Therefore,

$$x = r\cos(\theta) = \rho\sin(\phi)\cos(\theta) \quad \text{and} \quad y = r\sin(\theta) = \rho\sin(\phi)\sin(\theta).$$

We summarize in the following theorem.

Theorem 1 The point in space with spherical coordinates (ρ, ϕ, θ) has rectangular coordinates (x, y, z) given by

$$\begin{aligned} x &= \rho\cos(\theta)\sin(\phi) \\ y &= \rho\sin(\theta)\sin(\phi) \\ z &= \rho\cos(\phi). \end{aligned}$$

Observe that, according to the formulas in Theorem 1,

$$\begin{aligned} x^2 + y^2 + z^2 &= \rho^2 \cos^2(\theta)\sin^2(\phi) + \rho^2 \sin^2(\theta)\sin^2(\phi) + \rho^2\cos^2(\phi) \\ &= \rho^2 \sin^2(\phi)(\cos^2(\theta) + \sin^2(\theta)) + \rho^2 \cos^2(\phi) \\ &= \rho^2. \end{aligned}$$

Thus, we have confirmed that ρ is the radial coordinate in space, analogous to the coordinate r in polar coordinates in the plane.

Example 5 A point P_0 in space has spherical coordinates given by $(2, 2\pi/3, 5\pi/6)$. Calculate its rectangular and cylindrical coordinates.

Solution From Theorem 1, we see that

$$x = 2\cos\left(\frac{5\pi}{6}\right)\sin\left(\frac{2\pi}{3}\right) = 2\left(-\frac{\sqrt{3}}{2}\right)\left(\frac{\sqrt{3}}{2}\right) = -\frac{3}{2},$$

$$y = 2\sin\left(\frac{5\pi}{6}\right)\sin\left(\frac{2\pi}{3}\right) = 2\left(\frac{1}{2}\right)\left(\frac{\sqrt{3}}{2}\right) = \frac{\sqrt{3}}{2},$$

$$z = 2\cos\left(\frac{2\pi}{3}\right) = 2\left(-\frac{1}{2}\right) = -1.$$

The rectangular coordinates of P_0 are $(-3/2, \sqrt{3}/2, -1)$. It is now easy to convert to cylindrical coordinates: $r = (x^2 + y^2)^{1/2} = \sqrt{3}$, θ has the same value $5\pi/6$ that it has in spherical coordinates, and z can be read off from the rectangular coordinates as -1. So, the cylindrical coordinates of P_0 are $(\sqrt{3}, 5\pi/6, -1)$. ■

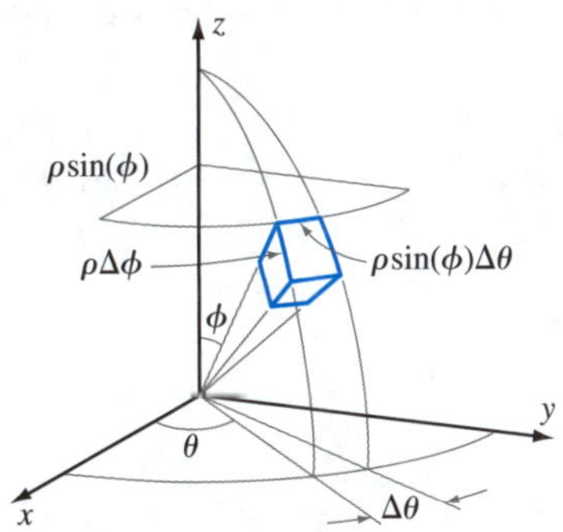

Figure 7
A small "box" in spherical coordinates

Naturally, we are interested in performing integrations in spherical coordinates. We need to determine the volume element in this new coordinate system. Figure 7 exhibits an increment of volume swept out by increments $\Delta\rho$, $\Delta\phi$, and $\Delta\theta$ of the coordinates. This solid is approximately a box, with sides that are determined by arcs of circles and with a radial height $\Delta\rho$. One circular edge is an arc of a circle that is parallel to the xy-plane and that has radius $\rho\sin(\phi)$. Because this edge subtends an angle $\Delta\theta$, it has arc length $\rho\sin(\phi)\cdot\Delta\theta$. Another circular edge of the box is an arc of a circle of radius ρ. This edge subtends an angle $\Delta\phi$ and therefore has arc length $\rho\Delta\phi$. The height of the box in the radial direction is $\Delta\rho$. Thus, the product

$$(\rho\sin(\phi)\,\Delta\theta)\cdot(\rho\,\Delta\phi)\cdot(\Delta\rho) = \rho^2\sin(\phi)\Delta\theta\,\Delta\phi\,\Delta\rho$$

approximates the volume of the small box. By using familiar limiting arguments, we conclude that the volume element for spherical coordinates is

$$dV = \rho^2 \sin(\phi)\, d\rho\, d\phi\, d\theta.$$

Example 6 Use spherical coordinates to calculate the volume V of the solid $\mathcal{U}$ that lies above the cone $z = \sqrt{x^2 + y^2}$ and below the hemisphere $z = 1 + \sqrt{1 - x^2 - y^2}$.

Solution This solid is pictured in Figure 4. In Example 3, we use cylindrical coordinates to calculate its volume to be π.

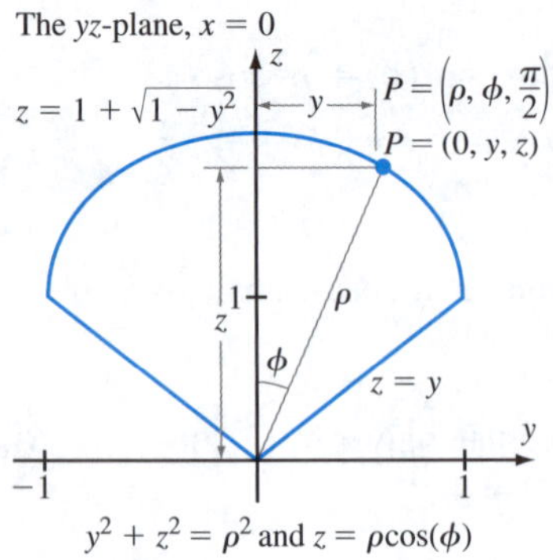

Figure 8

The intersection of $\mathcal{U}$ with the yz-plane is shown in Figure 8. Since $x = 0$ in this plane, the equation of the hemisphere is $z = 1 + \sqrt{1 - y^2}$, or $(z-1)^2 = 1 - y^2$. After expanding the left side of this equation, we get $y^2 + z^2 = 2z$. But, as Figure 8 shows, in the yz-plane the relationships between y, z, ρ, and ϕ are $y^2 + z^2 = \rho^2$ and $z = \rho\cos(\phi)$. Therefore, the equation of the intersection of the hemisphere with the yz-plane is $\rho^2 = 2\rho\cos(\phi)$, or $\rho = 2\cos(\phi)$. It follows that the solid $\mathcal{U}$ is described in spherical coordinates by the inequalities $0 \le \theta \le 2\pi, 0 \le \phi \le \pi/4, 0 \le \rho \le 2\cos(\phi)$. According to Theorem 1,

$$V = \int_0^{2\pi}\int_0^{\pi/4}\int_0^{2\cos(\phi)} 1 \cdot \rho^2 \sin(\phi)\, d\rho\, d\phi\, d\theta.$$

We evaluate this integral as follows:

$$V = \frac{1}{3}\int_0^{2\pi}\int_0^{\pi/4} 8\cos^3(\phi)\sin(\phi)\, d\rho\, d\phi\, d\theta = \frac{8}{3}\int_0^{2\pi}\left(-\frac{\cos^4(\phi)}{4}\Bigg|_{\phi=0}^{\phi=\pi/4}\right) d\theta$$

$$= 2\pi \cdot \frac{8}{3}\left(\frac{1}{4} - \frac{(1/\sqrt{2})^4}{4}\right) = \pi.$$

Example 7 Calculate the center of gravity of the solid in Example 6, assuming a uniform mass distribution $\delta(x, y, z) = 1$. (Refer to Section 14.7 for terminology and notation.)

Solution It follows from symmetry considerations that $\bar{x} = \bar{y} = 0$. So, we must calculate $\bar{z}$. Now

$$M_{z=0} = \iiint_{\mathcal{U}} z\delta(x, y, z)\, dV = \int_0^{2\pi}\int_0^{\pi/4}\int_0^{2\cos(\phi)} \rho\cos(\phi) \cdot 1 \cdot \rho^2\sin(\phi)\, d\rho\, d\phi\, d\theta$$

$$= \int_0^{2\pi}\int_0^{\pi/4}\int_0^{2\cos(\phi)} \rho^3\cos(\phi)\sin(\phi)\, d\rho\, d\phi\, d\theta.$$

Therefore,

$$M_{z=0} = \int_0^{2\pi}\left(\int_0^{\pi/4} \frac{1}{4}\cos(\phi)\sin(\phi)\rho^4\Bigg|_{\rho=0}^{\rho=2\cos(\phi)} d\phi\right) d\theta = 4\int_0^{2\pi}\left(\int_0^{\pi/4}\cos(\phi)\sin(\phi)\cos^4(\phi)\, d\phi\right) d\theta.$$

Since the inner integral does not depend on θ, we have

$$M_{z=0} = 2\pi \cdot 4\int_0^{\pi/4}\cos^5(\phi)\sin(\phi)\, d\phi.$$

We make the substitution $u = \cos(\phi)$, $du = -\sin(\phi)\, d\phi$, obtaining

$$M_{z=0} = 8\pi\int_1^{1/\sqrt{2}}(-u^5)\, du = -\frac{8\pi}{6}u^6\Bigg|_1^{1/\sqrt{2}} = -\frac{8\pi}{6}\left(\frac{1}{8} - 1\right) = \frac{7}{6}\pi.$$

From Examples 3 and 6, we know that the volume of the solid is π. The mass is therefore $M = \iiint_{\mathcal{U}} \delta(x, y, z)\, dV = \iiint_{\mathcal{U}} 1\, dV = \pi$. We conclude that

$$\bar{z} = \frac{M_{z=0}}{M} = \frac{7\pi/6}{\pi} = \frac{7}{6}.$$

Therefore, the center of gravity is $(0, 0, 7/6)$.

quickquiz

1. How are cylindrical coordinates specified?
2. How are spherical coordinates specified?
3. What is the element of volume in each of our new coordinate systems?
4. How do we pass from rectangular to spherical coordinates?

EXERCISES

Problems for Practice

In Exercises 1–8, a point P is given in rectangular coordinates. Give cylindrical coordinates for P.

1. $P = \left(2\sqrt{3}, 2, 5\right)$
2. $P = (4, 0, 6)$
3. $P = (-2, 2, -1)$
4. $P = \left(\sqrt{2}, -\sqrt{2}, 2\right)$
5. $P = (0, -3, -2)$
6. $P = (5, 5, 0)$
7. $P = \left(1, -\sqrt{3}, -2\right)$
8. $P = (0, 0, -3)$

In Exercises 9–16, rectangular coordinates are given for a point P. Give spherical coordinates for P.

9. $P = \left(1, 1, \sqrt{2}\right)$
10. $P = \left(-1, \sqrt{3}, 2\right)$
11. $P = (4, 0, 4)$
12. $P = (0, 0, 3)$
13. $P = \left(1, 1, \sqrt{6}\right)$
14. $P = \left(2, 0, -2\sqrt{3}\right)$
15. $P = \left(-4, 4, -4\sqrt{6}\right)$
16. $P = \left(\sqrt{3}, -3, \sqrt{12}\right)$

In Exercises 17–24, rectangular or spherical or cylindrical coordinates are given for the point P. Calculate coordinates in the other two coordinate systems.

17. $P = (3, \pi/3, 2\pi/3)$ spherical
18. $P = (1, 5\pi/6, 7\pi/4)$ spherical
19. $P = \left(2, -2, -2\sqrt{2}\right)$ rectangular
20. $P = \left(0, \sqrt{3}, -1\right)$ rectangular
21. $P = (2, -3\pi/4, -2)$ cylindrical
22. $P = (6, -7\pi/4, -6)$ cylindrical
23. $P = (3, 0, -3)$ rectangular
24. $P = (5, 3\pi/4, 2\pi/3)$ spherical

In Exercises 25–30, integrate the function f over the solid $\mathcal{U}$ using cylindrical coordinates.

25. $f(x, y, z) = x^2 + y^2$; $\mathcal{U}$ is the region below the paraboloid $z = -x^2 - y^2$ and above the plane $z = -4$.
26. $f(x, y, z) = x + y$; $\mathcal{U}$ is the solid in the first octant bounded by $z = x^2 + y^2$ and $z = (x^2 + y^2)^{1/2}$.
27. $f(x, y, z) = (x^2 + y^2)^{1/2}$; $\mathcal{U}$ is the solid bounded by the cylinder $x^2 + y^2 = 4$ and by $z = x^2 + y^2 + 1$, $z = -3$.
28. $f(x, y, z) = 1/(1 + x^2 + y^2)$; $\mathcal{U}$ is the solid bounded by the paraboloids $z = x^2 + y^2 + 2$ and $z = -x^2 - y^2 + 10$.
29. $f(x, y, z) = 1/(x^2 + y^2)$; $\mathcal{U}$ is the solid bounded by the cylinders $x^2 + y^2 = 2$ and $x^2 + y^2 = 4$ and by the planes $z = 8$ and $z = -2$.
30. $f(x, y, z) = x^2$; $\mathcal{U}$ is the solid inside the cone $z^2 = x^2 + y^2$, outside the cylinder $x^2 + y^2 = 9$, and between $z = -5$ and $z = 5$.

In Exercises 31–38, integrate the function f over the solid $\mathcal{U}$ using spherical coordinates.

31. $f(x, y, z) = 5$; $\mathcal{U}$ is the solid inside the sphere $x^2 + y^2 + z^2 = 9$.
32. $f(x, y, z) = x^2 + y^2 + z^2$; $\mathcal{U}$ is the solid outside the cone $z^2 = x^2 + y^2$ and inside the sphere $x^2 + y^2 + z^2 = 8$.
33. $f(x, y, z) = (x^2 + y^2 + z^2)^{1/2}$; $\mathcal{U}$ is the solid inside the sphere $x^2 + y^2 + z^2 = 9$ and outside the sphere $x^2 + y^2 + z^2 = 4$.
34. $f(x, y, z) = \cos((x^2 + y^2 + z^2)^{3/2})$; $\mathcal{U}$ is the solid inside the sphere $x^2 + y^2 + z^2 = 9$ and above the xy-plane.
35. $f(x, y, z) = z$; $\mathcal{U}$ is the solid below the surface $z = (x^2 + y^2)^{1/2}$ and inside the sphere $x^2 + y^2 + z^2 = 4$.
36. $f(x, y, z) = x^2 + y^2$; $\mathcal{U}$ is the region outside the cone $z^2 = x^2 + y^2$ and inside the sphere $x^2 + y^2 + z^2 = 4$.
37. $f(x, y, z) = (x^2 + y^2 + z^2)^{-1/2}$; $\mathcal{U}$ is the solid between the spheres $x^2 + y^2 + z^2 = 2$ and $x^2 + y^2 + z^2 = 3$.
38. $f(x, y, z) = (1 + x^2 + y^2 + z^2)^{-2}$; $\mathcal{U}$ is the solid above the cone $z = \sqrt{x^2 + y^2}$ and below the sphere $x^2 + y^2 + z^2 = 3$.

Further Theory and Practice

39. Use cylindrical coordinates to calculate the volume outside the cone $z^2 = x^2 + y^2$ and inside the cylinder $x^2 + y^2 = 4$.

40. A point P has rectangular coordinates $(3, 5\pi/6, 2\pi/3)$. What are the cylindrical coordinates of P? What are the spherical coordinates of P?

41. A point P has spherical coordinates $(3, 2, 5)$. What are the rectangular coordinates of P? What are the cylindrical coordinates of P?

42. Determine all points that have spherical coordinates equal to their rectangular coordinates.

43. Determine all points that have cylindrical coordinates equal to their rectangular coordinates.

44. Find formulas for converting directly from cylindrical to spherical coordinates and vice versa.

45. Sketch the graph of the surface $\rho = \cos(\phi)$ in spherical coordinates.

46. Sketch the graph of $\phi = \pi/3$ in spherical coordinates.

47. Sketch the graph of $z = 2 + r^2$ in cylindrical coordinates.

48. Sketch the graph of $\phi = \theta$ in spherical coordinates.

Computer/Calculator Exercises

49. Let $\mathcal{R}$ be the planar region in the first quadrant that lies outside the curve $r = \sin(\theta)^3$ and inside the curve $r = 3 - 2\cos(\theta)$. Let $\mathcal{U}$ be the solid that lies above $\mathcal{R}$ and below $z = y$. Calculate

$$\iiint_{\mathcal{U}} (x^2 + y^2)^{1/3}\, dV.$$

50. Let $V(\alpha)$ be the volume of that part of the unit ball that consists of points with spherical coordinates $0 \le \theta \le \alpha$ and $0 \le \phi \le \alpha$. For what value of α is $V(\alpha)$ equal to $1/4$ the volume of the unit ball?

Summary of Key Topics

Double Integrals over Rectangular Regions (Section 14.1)

The double integral of a continuous function f over a rectangular region

$$\mathcal{R} = \{(x, y) : a \le x \le b,\ c \le y \le d\}$$

is defined to be the limit of a Riemann sum over the product of partitions of $[a, b]$ and $[c, d]$. The integral is written

$$\iint_{\mathcal{R}} f(x, y)\, dA.$$

This double integral is evaluated by computing either of the following iterated integrals:

$$\int_a^b \left(\int_c^d f(x, y)\, dy \right) dx \quad \text{or} \quad \int_c^d \left(\int_a^b f(x, y)\, dx \right) dy.$$

Integration over More General Planar Regions (Section 14.2)

If $\mathcal{R}$ is a planar region bounded by finitely many smooth curves and contained in a bounded rectangle $Q = \{(x, y) : a \le x \le b, c \le y \le d\}$, then for each positive integer N, we partition Q into N^2 subrectangles by dividing each interval $[a, b]$ and $[c, d]$ into N equal subintervals. To integrate a continuous function f over $\mathcal{R}$, we form Riemann sums using only subrectangles of the partition that lie completely in $\mathcal{R}$. The limit of the Riemann sums, as the number of subrectangles tends to infinity, gives the double integral; we denote the integral by

$$\iint_{\mathcal{R}} f \, dA.$$

Simple Regions (Section 14.2)

A region $\mathcal{R}$ is x-simple if it has the form

$$\mathcal{R} = \{(x, y) : c \leq y \leq d, \ \alpha_1(y) \leq x \leq \alpha_2(y)\}.$$

It is y-simple if it has the form

$$\mathcal{R} = \{(x, y) : a \leq x \leq b, \beta_1(x) \leq y \leq \beta_2(x)\}.$$

The integral of a continuous f over an x-simple region $\mathcal{R}$ is calculated with the iterated integral

$$\int_c^d \left(\int_{\alpha_1(y)}^{\alpha_2(y)} f(x, y) \, dx \right) dy.$$

The integral of a continuous f over a y-simple region $\mathcal{R}$ is calculated with the iterated integral

$$\int_a^b \left(\int_{\beta_1(x)}^{\beta_2(x)} f(x, y) \, dy \right) dx.$$

If a region is both x-simple and y-simple, then the integral may be evaluated in either order; this observation is often useful in simplifying integration problems.

Area and Volume (Section 14.3)

The area of a bounded region $\mathcal{R}$ whose boundary consists of finitely many smooth curves is given by

$$\iint_{\mathcal{R}} 1 \, dA.$$

The volume under the graph of a continuous function $z = f(x, y)$ and above a region $\mathcal{R}$ in the xy-plane is given by

$$\iint_{\mathcal{R}} f(x, y) \, dA.$$

The volume under the graph of a function $z = f(x, y)$ and above the graph of $z = g(x, y)$, both defined on the region $\mathcal{R}$ in the xy-plane, is given by

$$\iint_{\mathcal{R}} (f - g) \, dA.$$

Polar Coordinates (Section 14.5)

A planar point P is located by polar coordinates (r, θ) where r is the distance of P from the origin and θ is the signed angle subtended by the ray from O to P with the positive x-axis. If $r < 0$, then (r, θ) denotes the point that is the reflection of $(|r|, \theta)$ in the origin.

Polar coordinates (r, θ) are related to Cartesian coordinates of a point P by the formulas

$$x = r\cos(\theta),\ \ y = r\sin(\theta),\ \ r^2 = x^2 + y^2.$$

Graphing in Polar Coordinates (Section 14.4)

To graph a curve $r = \varphi(\theta)$, plotting points is a vital tool. We also take note of intervals of increase of $\varphi(\theta)$ and intervals of decrease of $\varphi(\theta)$. We test for symmetry as follows:

Symmetry in the x-axis if $\varphi(\theta) = \varphi(-\theta)$
Symmetry in the y-axis if $\varphi(\theta) = \varphi(\pi - \theta)$
Symmetry in the origin if $\varphi(\theta) = \varphi(\theta + \pi)$

If the curve is defined implicitly by an equation of the form $\Psi(r, \theta) = 0$, then the curve is symmetric about the origin if $\Psi(-r, \theta) = \Psi(r, \theta)$.

Area in Polar Coordinates (Section 14.5)

If $\varphi(\theta) \geq 0$ for $\theta \in [\alpha, \beta]$ and if $\beta - \alpha \leq 2\pi$, then the area of the region

$$\mathcal{R} = \{(r, \theta) : 0 \leq r \leq \varphi(\theta), \alpha \leq \theta \leq \beta\}$$

is given by

$$\frac{1}{2}\int_{\alpha}^{\beta} \varphi^2(\theta)\, d\theta.$$

If $0 \leq \varphi_0(\theta) \leq \varphi_1(\theta)$ for $\theta \in [\alpha, \beta]$, then the area of the region

$$\mathcal{R} = \{(r, \theta) : \varphi_0(\theta) \leq r \leq \varphi_1(\theta)\}$$

is given by

$$\frac{1}{2}\int_{a}^{b} \left(\varphi_1^2(\theta) - \varphi_0^2(\theta)\right) d\theta.$$

Double Integrals in Polar Coordinates (Section 14.5)

If, in polar coordinates, a planar region has the form

$$\mathcal{R} = \{(r, \theta) : \alpha \leq \theta \leq \beta,\ \varphi_0(\theta) \leq r \leq \varphi_1(\theta)\}$$

with $\beta - \alpha \leq 2\pi$ and $0 \leq \varphi_0(\theta) \leq \varphi_1(\theta)$ for all θ, then the area of $\mathcal{R}$ is given by

$$\int_{\alpha}^{\beta} \int_{\varphi_0(\theta)}^{\varphi_1(\theta)} r \, dr \, d\theta.$$

If $f(r, \theta)$ is a continuous function on $\mathcal{R}$, then the integral of f over $\mathcal{R}$ is given by

$$\int_{\alpha}^{\beta} \int_{\varphi_0(\theta)}^{\varphi_1(\theta)} f(r, \theta) r \, dr \, d\theta.$$

Integrals in polar coordinates can also be used to calculate volumes.

Change of Variable in Double Integrals (Section 14.5)

Let $\mathcal{S}$ be a region in the uv-plane bounded by finitely many continuously differentiable curves. Suppose that $T(u, v) = (\phi(u, v), \psi(u, v))$ is a transformation of $\mathcal{S}$ to a region $\mathcal{R}$ in the xy-plane that is also bounded by finitely many continuously differentiable curves. We assume that T is one-to-one and onto and that the component functions ϕ, ψ are continuously differentiable.

If $\mathcal{S}$ and $\mathcal{R}$ and the transformation T are as above and if f is a continuous function on the region $\mathcal{R}$, then

$$\iint_{\mathcal{R}} f(x, y) \, dx \, dy = \iint_{\mathcal{S}} f(\phi(u, v), \psi(u, v)) |(\phi_u \psi_v - \psi_u \phi_v)(u, v)| \, du \, dv.$$

The expression $(\phi_u \psi_v - \psi_u \phi_v)(u, v)$ is called the Jacobian of the transformation T and is denoted by $J_T(u, v)$. The Jacobian may also be expressed as a determinant.

Triple Integrals (Section 14.6)

For a solid $\mathcal{U}$ in space contained in a cube

$$Q = \{(x, y, z) : a \leq x \leq b, c \leq y \leq d, \ e \leq z \leq f\}$$

and a continuous function $f(x, y, z)$ on $\mathcal{U}$, the Riemann sums for f over $\mathcal{U}$ are defined using a product of partitions of the intervals $[a, b]$, $[c, d]$, $[e, f]$. The triple integral is a limit of the Riemann sums whenever the limit exists. The integral is denoted

$$\iiint_{\mathcal{U}} f(x, y, z) \, dV.$$

When $\mathcal{U}$ is a solid of the form

$$\mathcal{U} = \{(x, y, z) : a \leq x \leq b, \alpha_1(x) \leq y \leq \alpha_2(x), \ \beta_1(x, y) \leq z \leq \beta_2(x, y)\},$$

the triple integral may be evaluated as the iterated integral

$$\int_a^b \int_{\alpha_1(x)}^{\alpha_2(x)} \int_{\beta_1(x,y)}^{\beta_2(x,y)} f(x, y, z) \, dz \, dy \, dx.$$

The roles of x, y, and z may be permuted in this last formula when they are similarly permuted in the definition of $\mathcal{R}$.

Mass, Moment of Inertia, and Center of Mass (Section 14.7)

If a planar plate (or lamina) $\mathcal{R}$ has density $\delta(x, y)$ at the point (x, y), then the mass of the plate is given by

$$M = \iint_{\mathcal{R}} \delta \, dA.$$

The first moment of the plate with respect to the x-axis is

$$M_{y=0} = \iint_{\mathcal{R}} y\delta(x, y)\, dA,$$

and the first moment with respect to the y-axis is

$$M_{x=0} = \iint_{\mathcal{R}} x\delta(x, y)\, dA.$$

The respective moments of inertia are

$$I_{y=0} = \iint_{\mathcal{R}} y^2\delta(x, y)\, dA$$

and

$$I_{x=0} = \iint_{\mathcal{R}} x^2\delta(x, y)\, dA.$$

The center of mass (or gravity) of the plate is the point $\bar{c} = (\bar{x}, \bar{y})$ where

$$\bar{x} = \frac{M_{x=0}}{M}$$

and

$$\bar{y} = \frac{M_{y=0}}{M}.$$

Mass, moments, and centers of mass are defined similarly for three-dimensional solids using triple integrals.

Cylindrical Coordinates (Section 14.8)

For a point P in three-dimensional space, cylindrical coordinates have the form (r, θ, z) where r and θ are the polar coordinates of the projection P' of P into the xy-plane and z is the vertical signed distance of the point from the xy-plane. Cylindrical coordinates are related to rectangular coordinates by the formulas

$$x = r\cos(\theta), \quad y = r\sin(\theta), \quad x^2 + y^2 = r^2.$$

The volume element in cylindrical coordinates is $r\, dz\, dr\, d\theta$.

Spherical Coordinates (Section 14.8)

Suppose that P is a point in three-dimensional space. If P has rectangular coordinates (x, y, z), then the projection P_0 of P in the xy-plane has rectangular coordinates $(x, y, 0)$. The spherical coordinates of P are (ρ, ϕ, θ) where ρ is the distance of P from the origin O, ϕ is the angle between $\overrightarrow{OP}$ and the positive z-axis, and θ is the angle between $\overrightarrow{OP_0}$ and the positive x-axis. Spherical coordinates are related to rectangular coordinates by the formulas

$$x = \rho \cos(\theta) \sin(\phi), \quad y = \rho \sin(\theta) \sin(\phi), \quad z = \rho \cos(\phi).$$

When symmetry about the z-axis is present, it is often easiest to integrate using cylindrical coordinates. The volume element in cylindrical coordinates is $r\, dr\, d\theta\, dz$. When symmetry about the origin is present, it is often easiest to integrate using spherical coordinates. The volume element in spherical coordinates is $\rho^2 \sin(\phi)\, d\rho\, d\phi\, d\theta$.

genesis & DEVELOPMENT

The double integral was introduced by Euler in 1769. Four years later, Lagrange used triple integrals to solve a problem of mathematical physics. Both of these mathematicians discovered the method for changing variables in an integral. In particular, Lagrange obtained the formula for integration in spherical coordinates and used it to study gravitational attraction. The general theory of changing variables in a multiple integral was developed by the Ukrainian mathematician Mikhail Vasilevich Ostrogradskii (1801–1862) in 1836. However, Ostrogradskii's work did not receive wide dissemination in the West. Thus, when Carl Gustav Jacobi covered much the same ground in 1841, he received credit for the "Jacobian," a determinant that already had appeared in the neglected papers of Ostrogradskii.

Exotic Functions

When 19th-century mathematicians enlarged the concept of function to include any correspondence of real numbers, they opened the door to some exceedingly complicated functions. For some time, it was believed that the assumption of continuity would be enough to exclude the functions that are beyond the realm of differential calculus. Indeed, it is difficult to imagine a continuous function that has a graph without *any* smoothness. However, many efforts by notable mathematicians, such as André Marie Ampère (1775–1836), failed to demonstrate that a continuous function must be differentiable on a "substantial" set. Eventually, some mathematicians began to suspect that continuity does not necessarily entail any differentiability.

The first "everywhere-continuous, nowhere-differentiable" function was conceived by Bernhard Bolzano (1781–1848) in 1830. Because communication beyond the established mathematical centers was then very spotty, Bolzano's work remained unknown until 1922. Around 1860, the Swiss mathematician Charles Cellérier (1818–1890) also constructed an everywhere-continuous nowhere-differentiable function:

$$C(x) = \sum_{n=1}^{\infty} \frac{\sin(2^n x)}{2^n}.$$

Had this example been published in a timely fashion, Cellérier would have earned a secure position in the history of mathematical analysis. Instead, Cellérier's function was posthumously published after a delay of 30 years, and by that time, his method of construction had become old hat.

Indeed, at more or less the same time that Cellérier was concocting his novel function, Bernhard Riemann (1826–1866) was thinking along the same lines. Some time before 1861, Riemann defined and studied the everywhere-continuous function

$$R(x) = \sum_{n=1}^{\infty} \frac{\sin(n^2 x)}{n^2}.$$

Although Riemann's function is differentiable at some points, it is nondifferentiable at infinitely many points of any open interval. The graph of R over the interval [0, 1] is shown in Figure 1. The plotted curve does not seem too irregular in some places, such as the small rectangle drawn in Figure 1. However, when we zoom in, as in Figure 2, we find that the apparent smoothness is an illusion.

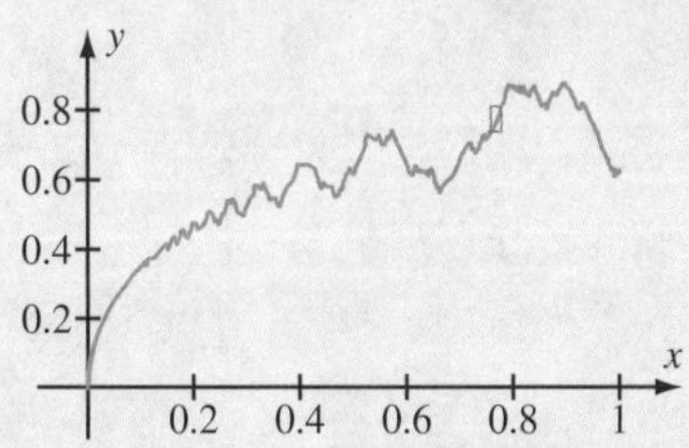

Figure 1
The graph of Riemann's function $R_1(x) = \sum_{n=1}^{\infty} \frac{\sin(n^2 x)}{n^2}$

There is no record that any mathematician other than Karl Weierstrass (1815–1897) was aware of Riemann's example. Weierstrass took Riemann's construction a step further and, in 1872, announced that

$$W(x) = \sum_{n=0}^{\infty} \frac{\cos(3^n \pi x)}{2^n}$$

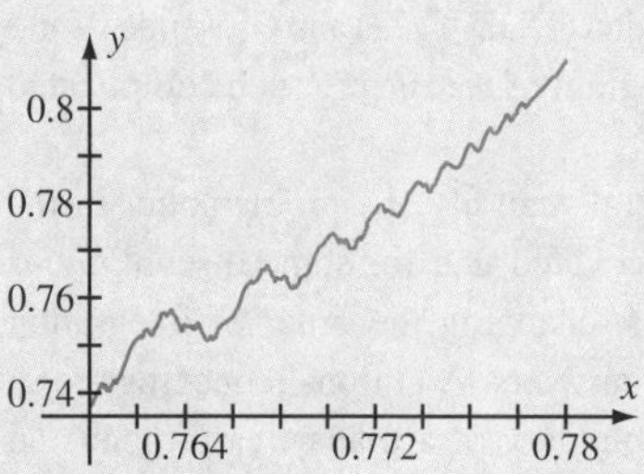

Figure 2
A detail of the graph of Riemann's function R_1, obtained by zooming in on the small viewing window shown in Figure 1

is an everywhere-continuous nowhere-differentiable function. Weierstrass's function is plotted over the interval [0,1] in Figure 3. A detail appears in Figure 4.

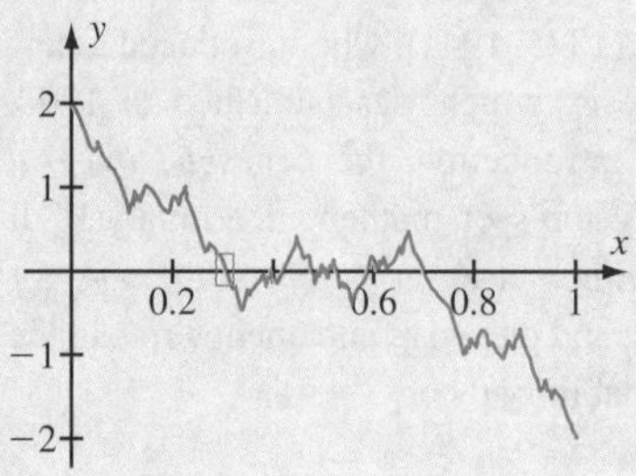

Figure 3
The graph of Weierstrass's everywhere-continuous nowhere-differentiable function on the interval [0, 1]

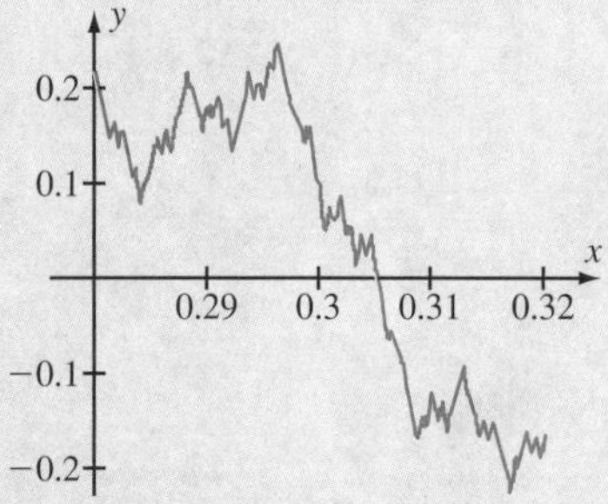

Figure 4
A detail of the graph of Weierstrass's function, obtained by zooming in on the small rectangle drawn in Figure 3.

In 1873, Gaston Darboux (1842–1917) defined yet another nondifferentiable continuous function,

$$G(x) = \sum_{n=1}^{\infty} \frac{\sin((n+1)! \cdot x)}{n!},$$

and it was far from the last. By the end of the 19th century, such pathological functions had become so commonplace that Charles Hermite wrote, "I turn away with fright and horror from this lamentable plague of functions that do not have derivatives." Such reactions notwithstanding, the very existence of such intractable functions prompted Ulisse Dini (1845–1918) and others to undertake a deeper study of the theory of differentiation.

The Lebesgue Integral

As badly behaved as they are for differential calculus, the continuous functions that we have been discussing pose no difficulties for the Riemann integral. However, as mathematicians became familiar with ever more troublesome functions, they discovered some undesirable limitations of the Riemann integral. Above all, the Riemann integral does not "respect" some of the most fundamental processes of calculus.

Let us consider one situation in which the Riemann integral comes up short. Set $S_1 = \{0, 1\}$, $S_2 = \{0, 1/2, 1\}$, $S_3 = \{0, 1/3, 1/2, 2/3, 1\}$, and, in general,

$$S_n = S_{n-1} \cup \left\{ \frac{1}{n}, \frac{2}{n}, \frac{3}{n}, \ldots, \frac{n-1}{n} \right\}.$$

We define f_n on the interval [0, 1] by the formula

$$f_n(x) = \begin{cases} 0 & \text{if } x \in S_n \\ 1 & \text{if } x \notin S_n \end{cases}.$$

The graphs of f_4 and f_5 are shown in Figure 5. In general, the graph of f_n is obtained from the graph of f_{n-1} by moving, at most,

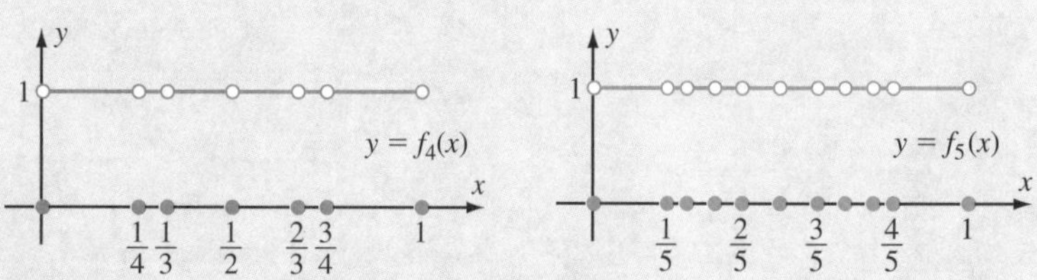

Figure 5

$n - 1$ points from the line $y = 1$ to the line $y = 0$. Next, define the function f on $[0, 1]$ by

$$f(x) = \begin{cases} 0 & \text{if } x \text{ is a rational number} \\ 1 & \text{if } x \text{ is an irrational number} \end{cases}.$$

It is not hard to deduce that $f_n(x) \to f(x)$ for every x. Moreover, this convergence is as nice as can be: The values $\{f_n(x)\}$ decrease to $f(x)$, and the graphs of all the functions are contained in the square $[0, 1] \times [0, 1]$. Under these conditions, it is highly desirable that the limit function f have an integral and that we can interchange limit and integral:

$$\lim_{n\to\infty} \int_0^1 f_n(x)\,dx = \int_0^1 f(x)\,dx.$$

In the case of the sequence we have just constructed, however, the limit function f does not have a Riemann integral. No matter how many subintervals of $[0, 1]$ we take, some of the Riemann sums of f will be 1 and some will be 0.

Double integrals highlight another weakness of the Riemann integral. In a first course in calculus, we limit our attention to *simple* planar regions. These regions are enclosed by a finite number of well-behaved curves. By the 1880s, serious attempts were under way to understand the integral $\iint_{\mathcal{R}} f(x, y)\,dA$ when $\mathcal{R}$ is a very irregular set of points in the plane. The most simple intuitive concepts of area are inadequate for such planar point sets. Therefore, Camille Jordan (1838–1922) and Giuseppe Peano (1858–1932) initiated a program of *measuring* such complicated point sets.

Iterated Riemann integrals can also be problematic. In the 1870s, mathematicians discovered that for some discontinuous integrands of two variables, changing the order of integration does not result in the same answers. Additional concerns arose. As we have learned, the boundary of a planar region plays an important role when we calculate a double integral by means of an iterated integral. For a simple planar region, the boundary is "infinitely thin" and therefore has zero area. Yet, in 1890, Peano constructed a continuous curve that passes through every point of the square $[0, 1] \times [0, 1]$. Such *space-filling curves* added to the urgency of coming to a better understanding of multiple integrals.

Jordan's theory of measure did not prove to be the right instrument for resolving all the sore points that beset the theory of integration. However, in the 1890s, Emile Borel (1871–1956) published refinements to Jordan's theory that guided the way to a successful resolution. This successful resolution was accomplished by Henri Lebesgue (1875–1941), who introduced a new integral in his doctoral thesis, which was published in 1902. Within a few years of its introduction, the *Lebesgue integral* became one of the primary tools of mathematical analysis. It revitalized the study of trigonometric series, paved the way for the study of function spaces, and played a fundamental role in the axiomatic treatment of probablility theory.

15

Vector Calculus

PREVIEW

Imagine a fluid flowing through a planar region. Since velocity has both magnitude and direction, the velocity of the flow at various points can be indicated by vectors. In this chapter, we will learn to relate the flow through the boundary of the region to the flow in the interior of the region. We will see that these physical considerations are related to a remarkable multidimensional version of the Fundamental Theorem of Calculus.

This chapter begins by developing the mathematical tools that are the basis of vector calculus. It is plain that a number of new ideas are needed. After we have learned the necessary tools in the first few sections, we will come to the theorems—Green's Theorem, Stokes's Theorem, and the Divergence Theorem—which give precise formulations of the relationship between the behavior of a vector field on the boundary and its behavior in the interior.

First, we need a type of function that is appropriate for describing flows. Such a function would have to assign a direction and a magnitude to each point in the plane or in space; in other words, it would have to be a vector-valued function on a planar or spatial region. Next, we will need to develop certain integrals that involve vector-valued functions. They will enable us to integrate over curves and surfaces.

Also, if we are going to use information on the boundary curve to obtain information inside the curve, then we need a mathematical way to tell the inside from the outside: This distinction will be closely connected to the idea of "orienting" a surface.

Finally, we will need vector operations that help measure the flow properties of a vector-valued function. Together with the gradient (which we studied in Chapter 13), we will use two new operations, called *divergence* and *curl*.

15.1 Vector Fields

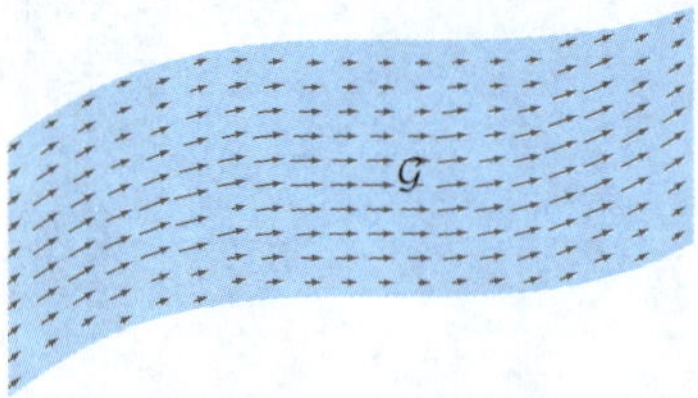

Figure 1

Imagine a fluid flowing through a region $\mathcal{G}$ in the plane or in space. At each point P, we may use a vector $\mathbf{v}(P)$ to represent the magnitude and direction of the velocity of the flow. Several of these vectors are sketched in Figure 1. By associating a vector $\mathbf{v}(P)$ with each point P in $\mathcal{G}$, we are in effect creating a vector-valued function with domain $\mathcal{G}$. Such a function is said to be a *vector field* on the region $\mathcal{G}$. Let us make these concepts precise.

Definition

We say that a planar set $\mathcal{G}$ is *open* if each of its points is the center of a disk (having positive radius) that is contained in $\mathcal{G}$. A set $\mathcal{G}$ in $\mathbb{R}^3$ is said to be *open* if each of its points is the center of a ball (with positive radius) that is contained in $\mathcal{G}$. A set $\mathcal{G}$ in the plane or in space is said to be *path-connected* if every pair of points in $\mathcal{G}$ can be connected by a continuous curve that lies entirely in $\mathcal{G}$. A set that is both open and path-connected is called a *region*.

Figure 2 shows a set that is both open and path-connected: It is a region. Figure 3 shows an open set $\mathcal{G}$ that comprises two components. This set is not a region because it is not path-connected: A point in one component cannot be connected to a point in the other component by means of a continuous curve that lies entirely within $\mathcal{G}$. In Figure 4, a new set $\mathcal{S}$ is formed by connecting the two components of $\mathcal{G}$ with a line segment $\mathcal{L}$. Although $\mathcal{S}$ is path-connected, it is not a region because it is not an open set: Each point of $\mathcal{L}$ is in $\mathcal{S}$ but is not the center of any non-trivial disk that is contained in $\mathcal{S}$.

Figure 2
A set that is open and path-connected: It is a region.

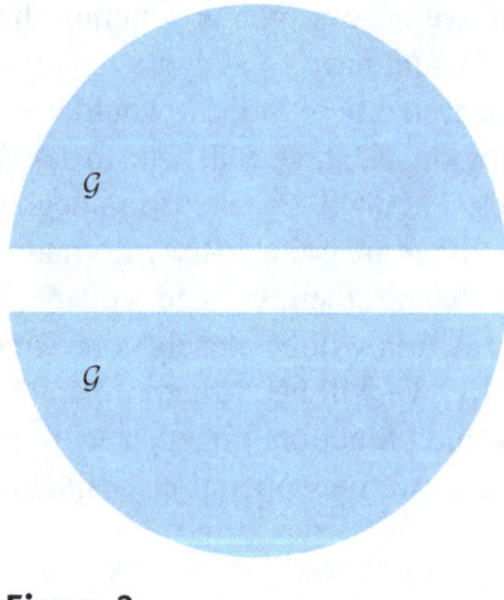

Figure 3
The set $\mathcal{G}$ comprises two components. $\mathcal{G}$ is open but not path-connected; it is *not* a region.

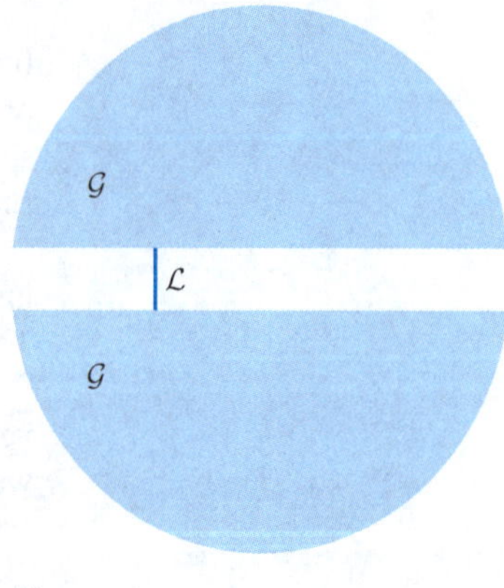

Figure 4
The set $\mathcal{S} = \mathcal{G} \cup \mathcal{L}$ is path-connected but not open; it is *not* a region.

Definition

A *vector field* $\mathbf{v}$ on a region $\mathcal{G}$ is a function that assigns to each point of $\mathcal{G}$ a vector. If $\mathcal{G}$ is a planar region, then the values of $\mathbf{v}$ are planar vectors. In other words, $\mathbf{v}$ has the form $\mathbf{v}(x, y) = v_1(x, y)\mathbf{i} + v_2(x, y)\mathbf{j}$ for each point (x, y) in $\mathcal{G}$. If $\mathcal{G}$ is a region in space, then the values of $\mathbf{v}$ are vectors in space, and $\mathbf{v}$ has the form $\mathbf{v}(x, y, z) = v_1(x, y, z)\mathbf{i} + v_2(x, y, z)\mathbf{j} + v_3(x, y, z)\mathbf{k}$ for each point (x, y, z) in $\mathcal{G}$.

> **inSIGHT**
>
> In Chapter 12, we used vector-valued functions $t \mapsto \mathbf{r}(t)$ of a scalar variable t to parameterize curves. A vector field $P \mapsto \mathbf{v}(P)$ is also a vector-valued function. It differs from a curve in that its domain consists not of scalars but of points P that have two or three coordinates.

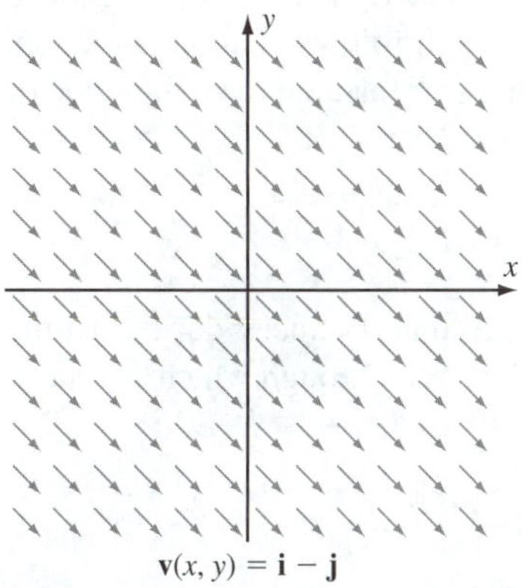

Figure 5

Example 1 Let $\mathcal{G}$ be the entire plane, and define $\mathbf{v}(x, y) = \mathbf{i} - \mathbf{j}$ and $\mathbf{w}(x, y) = x\mathbf{i} - (y/4)\mathbf{j}$. Sketch these vector fields.

Solution The vector field $\mathbf{v}$ assigns to each point (x, y) in the plane the vector $\mathbf{i} - \mathbf{j}$. It is a *constant* vector field. A sketch is indicated in Figure 5. Note that we make a sketch by drawing *some* of the vectors (after all, if we drew them all, the picture would be solid ink). Now consider the vector field $\mathbf{w}$. This is a vector field that assigns to the point (x, y) in the plane the vector $x\mathbf{i} - (y/4)\mathbf{j}$. Some of the values of this vector field are indicated in Table 1.

(x, y)	$(1, 0)$	$(0, 1)$	$(-1, 0)$	$(0, -1)$	$(1, 1)$	$(-1, 1)$	$(1, -1)$	$(-1, -1)$	$(2, 2)$	$(-2, 2)$
$\mathbf{w}(x, y)$	$\mathbf{i}$	$-\frac{1}{4}\mathbf{j}$	$-\mathbf{i}$	$\frac{1}{4}\mathbf{j}$	$\mathbf{i} - \frac{1}{4}\mathbf{j}$	$-\mathbf{i} - \frac{1}{4}\mathbf{j}$	$\mathbf{i} + \frac{1}{4}\mathbf{j}$	$-\mathbf{i} + \frac{1}{4}\mathbf{j}$	$2\mathbf{i} - \frac{1}{2}\mathbf{j}$	$-2\mathbf{i} - \frac{1}{2}\mathbf{j}$

Table 1

These values of the vector field $\mathbf{w}$, and a few others, are shown in Figure 6. ■

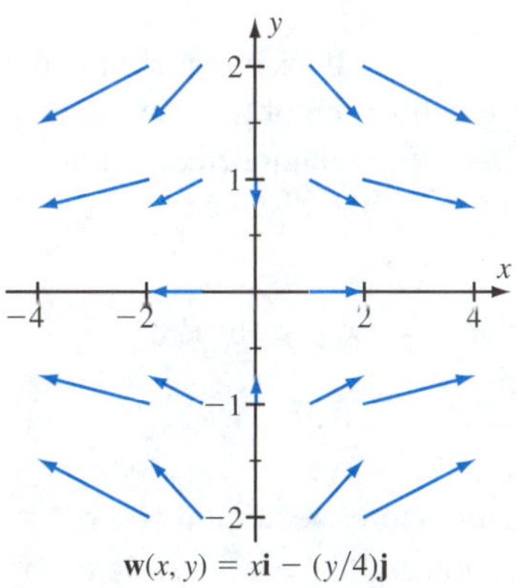

Figure 6

Vector Fields in Physics

Vector fields arise naturally in many physical contexts. For example, gravitational, magnetic, and electrostatic forces generate vector fields that are called force fields. Another example from physics that will be important for us is the velocity vector field.

Example 2 Let $\mathbf{v}$ denote the velocity of a wind that blows through a wind tunnel parallel to the sides of the tunnel. Suppose that, at each point $P = (x, y, z)$, the magnitude of $\mathbf{v}(P)$ is proportional to the height of P above the ground. Assume that the ground is at height $z = 0$ and that the vector $\mathbf{i}$ points down the tunnel in the direction that the wind is blowing. See Figure 7. Describe the velocity as a vector field.

Solution We let c be the constant of proportionality. At $P = (x, y, z)$, the magnitude of $\mathbf{v}(P)$ is cz, and the direction of $\mathbf{v}(P)$ is $\mathbf{i}$. The vector field we seek is therefore $\mathbf{v}(x, y, z) = cz\mathbf{i}$. ■

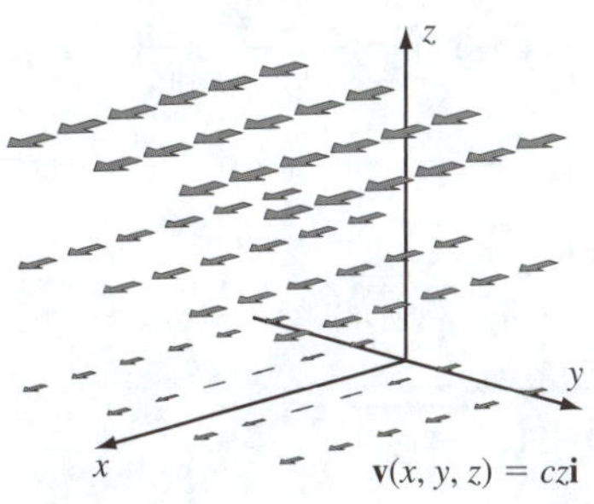

Figure 7

Example 3 Let ρ be a positive constant. Suppose that a solid ball $\mathcal{B} = \{(x, y, z) : x^2 + y^2 + z^2 \leq \rho^2\}$ is rotating about the z-axis with constant angular speed ω. Looking down at the xy-plane from the point $(0, 0, 2\rho)$ on the positive z-axis, the rotation of $\mathcal{B}$ is counterclockwise. Show that there is a vector $\mathbf{\Omega}$ such that the velocity vector field $P \mapsto \mathbf{v}(P)$ is given by $\mathbf{v}(P) = \mathbf{\Omega} \times \overrightarrow{OP}$ for each point $P = (x, y, z)$ in $\mathcal{B}$. (The vector $\mathbf{\Omega}$ is called the *angular velocity*.)

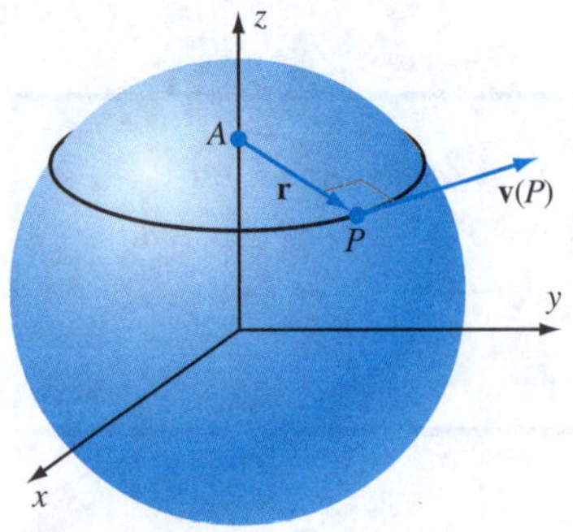

Figure 8
The velocity vector $\mathbf{v}(P)$ at $P = (x, y, z)$ is perpendicular to vector $\mathbf{r} = \langle x, y, 0\rangle$.

Solution Each point $P = (x, y, z)$ in $\mathcal{B}$ traces out a circle $\mathcal{C}_P$ in a horizontal plane. The center of the circle is the point $A = (0, 0, z)$ on the z-axis. The radius vector from A to P is $\mathbf{r} = \overrightarrow{AP} = \langle x - 0, y - 0, z - z\rangle = \langle x, y, 0\rangle$. Because P traces out $\mathcal{C}_P$, the velocity vector $\mathbf{v}(P)$ at $P = (x, y, z)$ is tangent to $\mathcal{C}_P$. Since $\mathcal{C}_P$ is a circle, we deduce that $\mathbf{v}(P)$ is perpendicular to $\mathbf{r} = \langle x, y, 0\rangle$. See Figure 8. It follows that the direction, $\mathbf{dir}(\mathbf{v}(P))$, of $\mathbf{v}(P)$ is equal to $\|\mathbf{r}\|^{-1}\langle -y, x, 0\rangle$. Letting $v(P)$ denote the magnitude of $\mathbf{v}(P)$, we obtain

$$\mathbf{v}(P) = \|\mathbf{v}(P)\|\,\mathbf{dir}(\mathbf{v}(P)) = v(P)\|\mathbf{r}\|^{-1}\langle -y, x, 0\rangle.$$

To determine $v(P)$, notice that the arc length Δs that P traces during a time interval Δt is equal to the product of the radius and the angle through which P rotates in time Δt; that is, $\Delta s = (\omega\Delta t)\,\|\mathbf{r}\|$. This equation tells us that

$$v(P) = \frac{ds}{dt} = \lim_{\Delta t\to 0}\frac{\Delta s}{\Delta t} = \lim_{\Delta t\to 0}\frac{(\omega\Delta t)\,\|\mathbf{r}\|}{\Delta t} = \omega\|\mathbf{r}\|.$$

Substituting this value for $v(P)$ into the formula for $\mathbf{v}(P)$, we obtain

$$\mathbf{v}(P) = (\omega\|\mathbf{r}\|)\,\|\mathbf{r}\|^{-1}\langle -y, x, 0\rangle = \omega\langle -y, x, 0\rangle = -\omega y\mathbf{i} + \omega x\mathbf{j} + 0\mathbf{k}.$$

We may now verify that $\boldsymbol{\Omega} = \omega\mathbf{k}$ is a vector that satisfies the requirements:

$$\boldsymbol{\Omega}\times\overrightarrow{OP} = \det\left(\begin{bmatrix}\mathbf{i} & \mathbf{j} & \mathbf{k}\\ 0 & 0 & \omega\\ x & y & z\end{bmatrix}\right) = -\omega y\mathbf{i} + \omega x\mathbf{j} + 0\mathbf{k} = \mathbf{v}(P).$$ ■

Example 4 A mass is located at the point $P_0 = (x_0, y_0, z_0)$. It exerts a gravitational attraction $\mathbf{F}(P)$ at each point $P \neq P_0$ in space. At P, the direction of the gravitational force is from P toward P_0 and, according to Newton's Law of Gravitation, the magnitude of the force is proportional to the reciprocal of the square of the distance of P to P_0. Using α as the proportionality constant, express $\mathbf{F}(P)$ as a vector field.

Solution Let $\mathbf{r} = \overrightarrow{OP}$ and $\mathbf{r}_0 = \overrightarrow{OP_0}$. The distance of $P = (x, y, z)$ to P_0 is

$$\|\mathbf{r} - \mathbf{r}_0\| = ((x - x_0)^2 + (y - y_0)^2 + (z - z_0)^2)^{1/2}.$$

Newton's law tells us that the magnitude of the gravitational force vector is $\alpha\|\mathbf{r}-\mathbf{r}_0\|^{-2}$. The gravitational force vector points in the same direction as $\overrightarrow{PP_0} = \mathbf{r}_0 - \mathbf{r}$. Since the direction vector of $\overrightarrow{PP_0}$ is $(-1/\|\mathbf{r} - \mathbf{r}_0\|)\,(\mathbf{r} - \mathbf{r}_0)$, we conclude that

$$\mathbf{F}(P) = \|\mathbf{F}(P)\|\,\mathbf{dir}(\mathbf{F}(P)) = \alpha\|\mathbf{r} - \mathbf{r}_0\|^{-2}\left(\frac{-1}{\|\mathbf{r} - \mathbf{r}_0\|}\right)(\mathbf{r} - \mathbf{r}_0) = -\frac{\alpha}{\|\mathbf{r} - \mathbf{r}_0\|^3}(\mathbf{r}-\mathbf{r}_0).$$

In component form, the force is given by

$$\begin{aligned}\mathbf{F}(x, y, z) = &-\frac{\alpha(x - x_0)}{((x - x_0)^2 + (y - y_0)^2 + (z - z_0)^2)^{3/2}}\mathbf{i}\\ &-\frac{\alpha(y - y_0)}{((x - x_0)^2 + (y - y_0)^2 + (z - z_0)^2)^{3/2}}\mathbf{j}\\ &-\frac{\alpha(z - z_0)}{((x - x_0)^2 + (y - y_0)^2 + (z - z_0)^2)^{3/2}}\mathbf{k}.\end{aligned}$$ ■

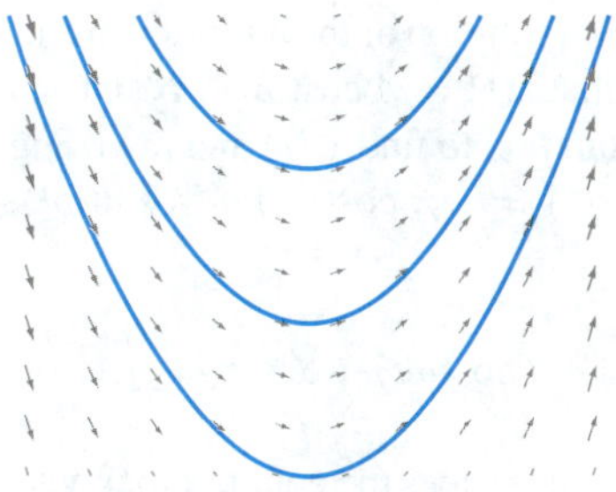

Figure 9
A vector field and some of its integral curves

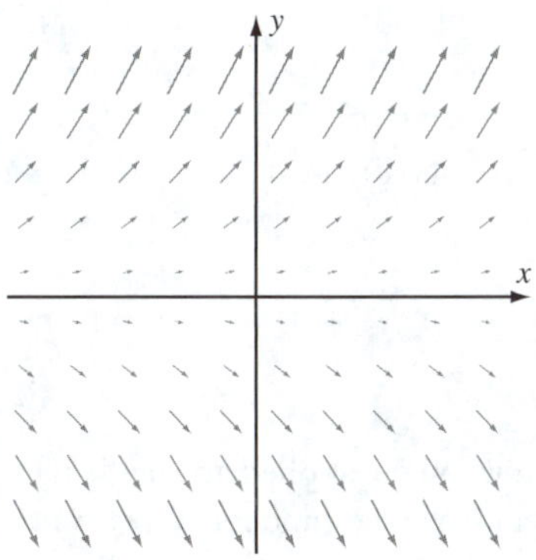

$\mathbf{F}(x, y) = \mathbf{i} + y\mathbf{j}$

Figure 10

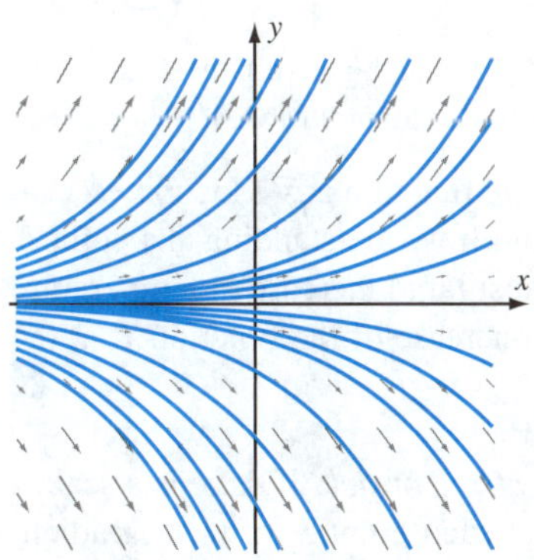

Figure 11
Field lines of $\mathbf{F}(x, y) = \mathbf{i} + y\mathbf{j}$

Integral Curves (Streamlines)

If $\mathbf{F}$ is a vector field on a region $\mathcal{G}$ and $t \mapsto \mathbf{r}(t)$ is a vector-valued function whose values lie in $\mathcal{G}$, then we say that the curve $\mathcal{C}$ described by $\mathbf{r}$ is an *integral curve* of $\mathbf{F}$ if

$$\mathbf{r}'(t) = \mathbf{F}(\mathbf{r}(t)). \tag{15.1}$$

Equation (15.1) tells us that for each point $P = \mathbf{r}(t)$ of the integral curve $\mathcal{C}$, the vector $\mathbf{F}(P)$ is tangent to $\mathcal{C}$ at P. This behavior can be seen in Figure 9, which shows a vector field and three of its integral curves. Integral curves are often called by other names, such as *field lines*. (A curve that passes through a vector field is often called a "line," whether or not it is straight.) If we think of $\mathbf{r}$ as the position vector of a point mass moving along $\mathcal{C}$, then $\mathbf{F}$ may be interpreted as a velocity field. In this case, the integral curves are called *flow lines,* or *streamlines*. As equation (15.1) suggests, determining the integral curves of a vector field usually amounts to solving a differential equation or a system of differential equations.

Example 5 Determine the field lines of the vector field $\mathbf{F}(x, y) = \mathbf{i} + y\mathbf{j}$ that is shown in Figure 10.

Solution We seek $\mathbf{r}(t) = x(t)\mathbf{i} + y(t)\mathbf{j}$ such that

$$x'(t)\mathbf{i} + y'(t)\mathbf{j} = \mathbf{r}'(t) = \mathbf{F}(\mathbf{r}(t)) = \mathbf{i} + y(t)\mathbf{j}.$$

By separating components, we obtain two differential equations: $x'(t) = 1$ and $y'(t) = y(t)$. The first of these differential equations can be integrated to give $x = t + c$ for some constant c. We know from Chapter 6 that all solutions of the second differential equation have the form $y = Ae^t$ for some constant A. Having shown that $t = x - c$, we conclude that $y = Ae^{x-c} = (Ae^{-c})e^x = Ce^x$ for a constant C. Thus, the field lines of $\mathbf{F}(x, y) = \mathbf{i} + y\mathbf{j}$ are the graphs of the equations $y = Ce^x$ where C is a constant. Several of these field lines are shown in Figure 11. ■

Example 6 Let ω be a positive constant. The differential equation

$$\frac{d^2}{dt^2}u(t) + \omega^2 u(t) = 0 \tag{15.2}$$

is called the *harmonic oscillator equation*. It is known that $u(t)$ is a solution of this equation if and only if $u(t) = A\cos(\omega t) + B\sin(\omega t)$ for some constants A and B. Use this fact to determine the integral curves of the vector field $\mathbf{F}(x, y) = -\omega y\mathbf{i} + \omega x\mathbf{j}$.

Solution We seek $\mathbf{r}(t) = x(t)\mathbf{i} + y(t)\mathbf{j}$ such that

$$x'(t)\mathbf{i} + y'(t)\mathbf{j} = \mathbf{r}'(t) = \mathbf{F}(\mathbf{r}(t)) = -\omega y(t)\mathbf{i} + \omega x(t)\mathbf{j}.$$

By separating components, we obtain the following system of two differential equations:

$$x'(t) = -\omega y(t) \tag{15.3}$$
$$y'(t) = \omega x(t). \tag{15.4}$$

By differentiating equation (15.3) with respect to t and then using equation (15.4) to substitute for $y'(t)$, we obtain $x''(t) = -\omega y'(t) = (-\omega)(\omega x(t)) = -\omega^2 x(t)$, which may be

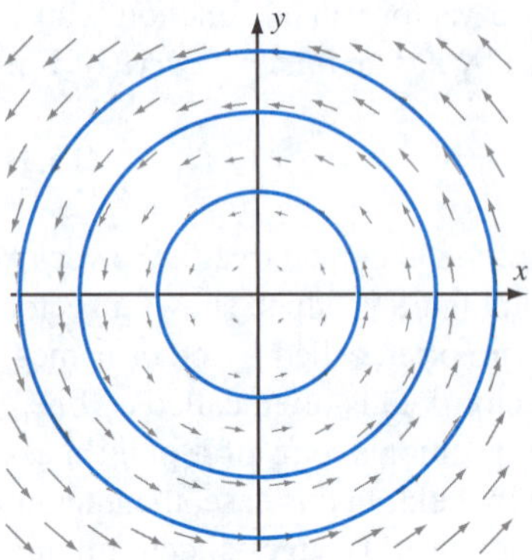

Figure 12
The vector field $\mathbf{F}(x, y) = -\omega\, y\mathbf{i} + \omega\, x\mathbf{j}$ with three of its integral curves

rewritten as $x''(t) + \omega^2 x(t) = 0$. In other words, $x(t)$ is a solution of the harmonic oscillator equation. From the stated information, it follows that $x(t) = A\cos(\omega t) + B\sin(\omega t)$ for some constants A and B. By differentiating this equation to find $x'(t)$ and rewriting equation (15.3) as $y(t) = (-1/\omega)x'(t)$, we obtain $y(t) = -B\cos(\omega t) + A\sin(\omega t)$. Our integral curves therefore have the form

$$\mathbf{r}(t) = \langle x(t),\ y(t)\rangle = \langle A\cos(\omega t) + B\sin(\omega t),\ -B\cos(\omega t) + A\sin(\omega t)\rangle.$$

If B is 0, then it is clear that $\mathbf{r}(t) = A\langle\cos(\omega t), \sin(\omega t)\rangle$ describes the counterclockwise traversal of a circle of radius $|A|$. In general, the calculation

$$\begin{aligned}\|\mathbf{r}(t)\|^2 &= (A\cos(\omega t) + B\sin(\omega t))^2 + (-B\cos(\omega t) + A\sin(\omega t))^2 \\ &= A^2(\cos^2(\omega t) + \sin^2(\omega t)) + B^2(\cos^2(\omega t) + \sin^2(\omega t)) \\ &\quad + 2AB\cos(\omega t)\sin(\omega t) - 2AB\cos(\omega t)\sin(\omega t) \\ &= A^2 + B^2\end{aligned}$$

shows that $\mathbf{r}$ parameterizes a circle of radius $\sqrt{A^2 + B^2}$. See Figure 12. ■

Gradient Vector Fields and Potential Functions

In Section 13.6, we introduced the gradient operator. When applied to the scalar-valued function $u(x, y)$, the gradient results in the vector-valued function $\nabla u(x, y) = u_x(x, y)\mathbf{i} + u_y(x, y)\mathbf{j}$, and when applied to the scalar-valued function $u(x, y, z)$, the gradient results in the vector-valued function $\nabla u(x, y, z) = u_x(x, y, z)\mathbf{i} + u_y(x, y, z)\mathbf{j} + u_z(x, y, z)\mathbf{k}$. Notice that $(x, y) \mapsto \nabla u(x, y)$ is a planar vector field and $(x, y, z) \mapsto \nabla u(x, y, z)$ is a spatial vector field.

Example 7 Let the temperature at a point in the plane be given by $T(x, y) = x^2 + 6y^2$. Calculate the gradient of T, and give a physical interpretation.

Solution The gradient vector field of the temperature function is $\nabla T(x, y) = 2x\mathbf{i} + 12y\mathbf{j}$. Physically, this vector field tells us the following: If we are standing at a point P in the plane, then $\nabla T(P)$ points in the direction of most rapid increase of temperature at P. The magnitude of $\nabla T(P)$ indicates the rate of increase of the function T at the point P in the direction of $\nabla T(P)$. ■

Definition We say that a vector field $\mathbf{F}$ on a region $\mathcal{G}$ is a *gradient vector field* if there is a scalar-valued function u on $\mathcal{G}$ for which $\mathbf{F}(P) = \nabla u(P)$ at each point P of $\mathcal{G}$. A gradient vector field is also called a *conservative vector field*. If V is a scalar-valued function on $\mathcal{G}$ such that $\mathbf{F}(P) = -\nabla V(P)$ for each point P in $\mathcal{G}$, then we say that V is a *potential function* for $\mathbf{F}$.

If u is a scalar-valued function and $V = -u$, then $\mathbf{F}(P) = \nabla u(P)$ if and only if $\mathbf{F}(P) = -\nabla V(P)$. In other words, a vector field is conservative if and only if it has a potential function. Although the equation $\mathbf{F}(P) = \nabla u(P)$ suffices for mathematical applications, the equivalent equation $\mathbf{F}(P) = -\nabla V(P)$ has an important physical significance, as the next example shows.

Example 8 Suppose that a force field $\mathbf{F}$ has a potential function V in a region $\mathcal{G}$. The force acts on a particle of mass m that moves with a continuously differentiable position vector $\mathbf{r}(t)$, $a \le t \le b$. Show that

$$\frac{1}{2}m\,\|\mathbf{r}'(t)\|^2 + V(\mathbf{r}(t)) = C \qquad \textbf{(15.5)}$$

for some scalar constant C.

Solution According to Newton's law, force equals mass times acceleration. We therefore have $\mathbf{F}(\mathbf{r}(t)) = m\mathbf{r}''(t)$. Because $\mathbf{F}$ has a potential function V, we may rewrite this equation as $-\nabla V(\mathbf{r}(t)) = m\mathbf{r}''(t)$, or $m\mathbf{r}''(t) + \nabla V(\mathbf{r}(t)) = \vec{\mathbf{0}}$. If we form the dot product of each side of this equation with $\mathbf{r}'(t)$, then we obtain

$$m\mathbf{r}'(t)\cdot\mathbf{r}''(t) + \nabla V(\mathbf{r}(t))\cdot\mathbf{r}'(t) = 0. \qquad \textbf{(15.6)}$$

Now

$$\frac{d}{dt}(\mathbf{r}'(t)\cdot\mathbf{r}'(t)) = 2\mathbf{r}'(t)\cdot\mathbf{r}''(t),$$

and, by equation (13.11),

$$\frac{d}{dt}(V(\mathbf{r}(t))) = \nabla V(\mathbf{r}(t))\cdot\mathbf{r}'(t).$$

We may use these last two equations to rewrite equation (15.6) as

$$\frac{d}{dt}\left(\frac{1}{2}m\mathbf{r}'(t)\cdot\mathbf{r}'(t) + V(\mathbf{r}(t))\right) = 0.$$

It follows that

$$\frac{1}{2}m\mathbf{r}'(t)\cdot\mathbf{r}'(t) + V(\mathbf{r}(t)) = C$$

for some constant C. We obtain equation (15.5) by noticing that $\mathbf{r}'(t)\cdot\mathbf{r}'(t) = \|\mathbf{r}'(t)\|^2$. ∎

Suppose that, as in Example 8, a conservative force field $\mathbf{F}$ acts on a particle that has mass m and position vector $\mathbf{r}(t)$. If V_0 is a potential function for $\mathbf{F}$ and if k is any scalar constant, then since $\nabla(V_0 + k) = \nabla V_0 + \nabla k = -\mathbf{F} + \vec{\mathbf{0}} = -\mathbf{F}$, we see that $V = V_0 + k$ is also a potential function for $\mathbf{F}$. Once a choice of potential function V has been fixed, the quantity $V(P)$ is said to be the *potential energy* of the mass at point $P = \mathbf{r}(t)$. The quantity $m\|\mathbf{r}'(t)\|^2/2$ is defined to be the *kinetic energy* of the mass at $P = \mathbf{r}(t)$. Equation (15.5) states that the total mechanical energy of the particle—kinetic plus potential—is constant. For this reason, equation (15.5) is called the *Law of Conservation of Mechanical Energy* for the conservative force $\mathbf{F}$.

In Section 15.3, we will learn how to recognize a conservative vector field. The next two examples show how to find a potential function for a vector field that is known to be conservative.

Example 9 The vector field $\mathbf{F}(x, y) = (y - 3x^2)\mathbf{i} + (x + \sin(y))\mathbf{j}$ is known to be conservative. Find a scalar-valued function u for which $\mathbf{F} = \nabla u$.

Solution We want u to satisfy the vector equation

$$(y - 3x^2)\mathbf{i} + (x + \sin(y))\mathbf{j} = \mathbf{F}(x, y) = \nabla u(x, y) = u_x(x, y)\mathbf{i} + u_y(x, y)\mathbf{j}.$$

This vector equation is equivalent to the two simultaneous scalar equations

$$u_x(x, y) = y - 3x^2 \tag{15.7}$$

and

$$u_y(x, y) = x + \sin(y). \tag{15.8}$$

To remove the partial derivative from u in equation (15.7), we antidifferentiate each side of $u_x(x, y) = y - 3x^2$ with respect to the x-variable. The new feature here is that the constant of integration we obtain is constant only with respect to x. It *can* depend on y. The result of our antidifferentiation is

$$u(x, y) = xy - x^3 + \phi(y) \tag{15.9}$$

where the constant of integration is some unknown function $\phi(y)$. To verify this assertion, calculate the partial derivative with respect to x of each side of equation (15.9) to see that you recover equation (15.7). Our goal now is to determine the expression $\phi(y)$ that has arisen in the integration. To do so, we differentiate each side of equation (15.9) with respect to y, obtaining

$$u_y(x, y) = x + \phi'(y). \tag{15.10}$$

Notice that the expression $\phi'(y)$ is an ordinary derivative, since $\phi(y)$ depends only on y. Comparing equation (15.10) with equation (15.8), we see that $\phi'(y) = \sin(y)$. After antidifferentiating with respect to y, we obtain $\phi(y) = -\cos(y) + C$. Observe that in this last antidifferentiation, which we performed to remove an *ordinary* derivative, the constant of integration C is a numerical constant. By substituting our formula for $\phi(y)$ into equation (15.9), we obtain

$$u(x, y) = xy - x^3 - \cos(y) + C.$$

Example 10 The vector field $\mathbf{F}(x, y, z) = y^2\mathbf{i} + (2xy + z)\mathbf{j} + (y + 3)\mathbf{k}$ is known to be conservative. Find a scalar-valued function u for which $\mathbf{F} = \nabla u$.

Solution We wish to find a function u whose partial derivatives satisfy the equations

$$u_x(x, y, z) = y^2, \tag{15.11}$$

$$u_y(x, y, z) = 2xy + z, \tag{15.12}$$

$$u_z(x, y, z) = y + 3. \tag{15.13}$$

We antidifferentiate equation (15.11) with respect to the x-variable. The result is $u(x, y, z) = xy^2 + \phi(y, z)$. Notice that the constant of integration $\phi(y, z)$ is constant only *with respect to* x. It can depend on y and z.

Next, we determine the expression $\phi(y, z)$. To do so, we differentiate each side of the equation $u(x, y, z) = xy^2 + \phi(y, z)$ with respect to y. This differentiation yields the equation $u_y(x, y, z) = 2xy + \phi_y(y, z)$. Comparing this last equation with equation (15.12), we see that $\phi_y(y, z) = z$. We antidifferentiate this with respect to y, noting that the constant of integration can depend on z. The result is $\phi(y, z) = yz + \psi(z)$. We substitute this formula for $\phi(y, z)$ into our equation for u, obtaining

$$u(x, y, z) = xy^2 + yz + \psi(z). \tag{15.14}$$

Our next task is to identify the unknown expression $\psi(z)$. To do so, we take the partial derivative of each side of equation (15.14) with respect to z. The result is $u_z(x, y, z) = y + \psi'(z)$. Comparing the right side of this last equation with the right side of equation (15.13), we see that $\psi'(z) = 3$. It follows that $\psi(z) = 3z + C$ where C is a numerical constant. We conclude our computation by substituting our expression for $\psi(z)$ into equation (15.14), obtaining

$$u(x, y, z) = xy^2 + yz + 3z + C.$$

■

Continuously Differentiable Vector Fields

In previous chapters, we learned that it is convenient to deal with functions that are continuously differentiable. These functions have much nicer properties than arbitrary functions. The same is true for vector fields. As a result, we will deal primarily with *continuously differentiable vector fields*. We call a planar vector field $\mathbf{v}(x, y) = v_1(x, y)\mathbf{i} + v_2(x, y)\mathbf{j}$ or a spatial vector field $\mathbf{w}(x, y, z) = w_1(x, y, z)\mathbf{i} + w_2(x, y, z)\mathbf{j} + w_3(x, y, z)\mathbf{k}$ *continuously differentiable* if each of the component functions—v_1, v_2 or w_1, w_2, w_3—is continuously differentiable.

Example 11 Discuss the differentiability properties of the vector fields $\mathbf{v}(x, y) = x^2y\mathbf{i} - y^5\mathbf{j}$ and $\mathbf{w}(x, y, z) = xy\mathbf{i} - |x|\mathbf{j} + y^3\mathbf{k}$.

Solution The vector field $\mathbf{v}(x, y) = x^2y\mathbf{i} - y^5\mathbf{j}$ is continuously differentiable at all points (x, y) because each of the component functions—x^2y and $-y^5$—is continuously differentiable. However, the vector field $\mathbf{w}(x, y, z) = xy\mathbf{i} - |x|\mathbf{j} + y^3\mathbf{k}$ is *not* continuously differentiable at any point of the form $(0, y, z)$ because the coefficient of $\mathbf{j}$, namely, $-|x|$, is not continuously differentiable (indeed, it is not even differentiable) at $x = 0$. However, $\mathbf{w}$ *is* continuously differentiable at all other points. ■

quickquiz

1. What is a vector field?
2. What does it mean for a vector field to be continuously differentiable?
3. What calculus operation on functions gives rise to vector fields in a natural way?
4. What does it mean for a vector field to be *conservative*?

EXERCISES

Problems for Practice

In Exercises 1–8, sketch the given vector field in the plane by drawing the vectors that are assigned to several points.

1. $\mathbf{F}(x, y) = 3\mathbf{i} + \mathbf{j}$
2. $\mathbf{F}(x, y) = x\mathbf{i} + y\mathbf{j}$
3. $\mathbf{F}(x, y) = 2y\mathbf{i} - 3\mathbf{j}$
4. $\mathbf{F}(x, y) = 3x\mathbf{i} + x\mathbf{j}$
5. $\mathbf{F}(x, y) = y\mathbf{i} - 5x\mathbf{j}$
6. $\mathbf{F}(x, y) = x^2\mathbf{i} + y^2\mathbf{j}$
7. $\mathbf{F}(x, y) = x^2\mathbf{i} - y^2\mathbf{j}$
8. $\mathbf{F}(x, y) = \mathbf{i} + \sin(y)\mathbf{j}$

In Exercises 9–12, determine an explicit formula for the spatial vector field that is described.

9. A wind blows down a 10 ft high hall with vertical walls and width 8 ft. The speed of the wind at a point is proportional to the distance of the point from a line running down the middle of the floor. The line down the middle of the floor is the y-axis, with the positive direction being the direction in which the wind blows. Write the velocity vector field.
10. A cylindrical tank is 5 m high with a radius of 4 m. Water circulates about the tank's central axis, the line segment [0,5] on the z-axis, in such a way that the flow is clockwise as you look down on it and the speed at any point is proportional to the distance of the point from the center of the tank. Write the velocity vector field.
11. In the tank described in Exercise 10, water circulates about the tank's central axis in such a way that the flow is clockwise as you look down on it and the speed at any point is proportional to the distance of the point from the central axis of the tank. Write the velocity vector field.
12. Particles circulate in space in such a way that rotation takes place about the y-axis. Each particle rotates in a plane perpendicular to the y-axis in such a fashion that its oriented axis of rotation, determined by the Right Hand Rule, is the vector $\mathbf{j}$. The speed of any particle is proportional to the square of its distance from the y-axis. Write the velocity vector field.

In Exercises 13–20, write the gradient vector field associated with the given scalar-valued function.

13. $u(x, y) = xy - y^2 + 3x$
14. $u(x, y) = \sin(xy^2)$
15. $u(x, y) = \ln(x^3 - y^2)$
16. $u(x, y) = \exp(2y - 5x)$
17. $u(x, y, z) = x^2y^3z$
18. $u(x, y, z) = \sin(xy)/\cos(z)$
19. $u(x, y, z) = \ln(x - y)/\ln(y - z)$
20. $u(x, y, z) = e^{\cos(x)}e^{\sin(y)}e^z$

In Exercises 21–26, the given vector field $\mathbf{F}(x, y) = M(x, y)\mathbf{i} + N(x, y)\mathbf{j}$ is conservative. Find an expression $u(x, y)$ such that $\mathbf{F} = \nabla u$.

21. $\mathbf{i} + \mathbf{j}$
22. $x\mathbf{i} + y\mathbf{j}$
23. $y\mathbf{i} + x\mathbf{j}$
24. $y\mathbf{i} + (x + 2)\mathbf{j}$
25. $(y^3 + 2xy)\mathbf{i} + (3xy^2 + x^2)\mathbf{j}$
26. $2\exp(2x - y)\mathbf{i} - \exp(2x - y)\mathbf{j}$

In Exercises 27–32, the given vector field $\mathbf{F}(x, y, z) = M(x, y, z)\mathbf{i} + N(x, y, z)\mathbf{j} + R(x, y, z)\mathbf{k}$ is conservative. Find an expression $u(x, y, z)$ such that $\mathbf{F} = \nabla u$.

27. $2x\mathbf{i} + \mathbf{j} + 0\mathbf{k}$
28. $\mathbf{i} + \mathbf{j} + 2z\mathbf{k}$
29. $2xyz^3\mathbf{i} + (x^2z^3 + 5)\mathbf{j} + 3x^2yz^2\mathbf{k}$
30. $y\mathbf{i} + (x + 2z)\mathbf{j} + 2y\mathbf{k}$
31. $(z^3 - 2xy^2)\mathbf{i} + (z^2 - 2yx^2)\mathbf{j} + (3xz^2 + 2yz)\mathbf{k}$
32. $(\sin(y) - z\sin(x))\mathbf{i} + x\cos(y)\mathbf{j} + \cos(x)\mathbf{k}$

In Exercises 33–38, find the integral curves $t \mapsto \mathbf{r}(t)$ of the given vector field $\mathbf{F}(x, y) = M(x, y)\mathbf{i} + N(x, y)\mathbf{j}$.

33. $\mathbf{F}(x, y) = \mathbf{i} + \mathbf{j}$
34. $\mathbf{F}(x, y) = \mathbf{i} + 2\mathbf{j}$
35. $\mathbf{F}(x, y) = x\mathbf{i} + 2\mathbf{j}$
36. $\mathbf{F}(x, y) = x\mathbf{i} - y\mathbf{j}$
37. $\mathbf{F}(x, y) = 2y\mathbf{i} + \mathbf{j}$
38. $\mathbf{F}(x, y) = \mathbf{i} - x\mathbf{j}$

Further Theory and Practice

In Exercises 39–42, the given vector field $\mathbf{F}(x, y) = M(x, y)\mathbf{i} + N(x, y)\mathbf{j}$ is conservative. Find an expression $u(x, y)$ such that $\mathbf{F} = \nabla u$.

39. $(\exp(2x) + 2x\exp(2x))\mathbf{i} + \mathbf{j}$
40. $(y/(1 + x^2y^2))\mathbf{i} + (x/(1 + x^2y^2))\mathbf{j}$
41. $y^2\cos(xy)\mathbf{i} + (\sin(xy) + xy\cos(xy))\mathbf{j}$
42. $(\ln(x + y) + x/(x + y))\mathbf{i} + (x/(x + y))\mathbf{j}$

In Exercises 43–46, the given vector field $\mathbf{F}(x, y, z) = M(x, y, z)\mathbf{i} + N(x, y, z)\mathbf{j} + R(x, y, z)\mathbf{k}$ is conservative. Find an expression $u(x, y, z)$ such that $\mathbf{F} = \nabla u$.

43. $z(y + z)^{-1}\mathbf{i} - xz(y + z)^{-2}\mathbf{j} + xy(y + z)^{-2}\mathbf{k}$
44. $(y^2 + z^2)^{-1}\mathbf{i} - 2xy(y^2 + z^2)^{-2}\mathbf{j} - 2xz(y^2 + z^2)^{-2}\mathbf{k}$

45. $x^{y-1}y\mathbf{i} + x^y \ln(x)\mathbf{j} + \ln(z)\mathbf{k}$
46. $\ln(y)^z\mathbf{i} + (xz/y)\ln(y)^{z-1}\mathbf{j} + x\ln(y)^z \ln(\ln(y))\mathbf{k}$
47. Show that $\mathbf{F}(x, y) = y\mathbf{i} + 2x\mathbf{j}$ is not the gradient vector field of any continuously differentiable function.
48. A planar vector field satisfying a certain nondegeneracy condition and vanishing at the origin exhibits one of eight qualitatively different modes of behavior near the origin. The vector fields given in this exercise illustrate these different behaviors. On separate sets of axes, sketch each given **F** in a neighborhood of (0, 0).
   a. $\mathbf{F}(x, y) = x\mathbf{i} + y\mathbf{j}$ **b.** $\mathbf{F}(x, y) = -x\mathbf{i} - y\mathbf{j}$
   c. $\mathbf{F}(x, y) = -x\mathbf{i} + y\mathbf{j}$ **d.** $\mathbf{F}(x, y) = x\mathbf{i} - y\mathbf{j}$
   e. $\mathbf{F}(x, y) = y\mathbf{i} + x\mathbf{j}$ **f.** $\mathbf{F}(x, y) = -y\mathbf{i} + x\mathbf{j}$
   g. $\mathbf{F}(x, y) = y\mathbf{i} - x\mathbf{j}$ **h.** $\mathbf{F}(x, y) = -y\mathbf{i} - x\mathbf{j}$
49. Let $r(x, y, z) = \|x\mathbf{i} + y\mathbf{j} + z\mathbf{k}\|$. Calculate $\nabla(1/r)$ at all points where the derivative exists.
50. Coulomb's law says that the force **F** acting on a charge q at position **r** because of a charge Q at the origin is

$$\mathbf{F} = \frac{\epsilon Q q}{\|\mathbf{r}\|^3}\mathbf{r}$$

where ϵ is a positive constant that depends on the physical units that are used. Show that $V = \epsilon Qq/\|\mathbf{r}\|$ is a potential function for **F**. If $Qq > 0$, then **F** is said to be *repulsive;* whereas if $Qq < 0$, then **F** is said to be *attractive*. Explain why this terminology is appropriate.
51. At the point (x, y), the elevation of the terrain is $h(x, y) = -x^2 + 3xy + 2y^2 + 100$ ft. Sketch the vector field that shows the direction of quickest descent at each point.
52. At the point (x, y) on a sheet of aluminum, the temperature is $T(x, y) = 150 - xy + 4y^3$. Sketch the vector field that shows at each point the direction of greatest increase in temperature.

If r_1 and r_2 are distinct real numbers, then all solutions of the differential equation

$$\frac{d^2}{dt^2}u(t) - (r_1 + r_2)\frac{d}{dx}u(t) + r_1 r_2 u(t) = 0$$

have the form $u(t) = A\exp(r_1 t) + B\exp(r_2 t)$ for some constants A and B. In Exercises 53–56, use this fact to determine the integral curves $t \mapsto \mathbf{r}(t)$ of the given vector field $\mathbf{F}(x, y) = M(x, y)\mathbf{i} + N(x, y)\mathbf{j}$.

53. $\mathbf{F}(x, y) = (3x + y)\mathbf{i} + 2y\mathbf{j}$
54. $\mathbf{F}(x, y) = 2y\mathbf{i} + (3y - x)\mathbf{j}$
55. $\mathbf{F}(x, y) = (3x - 4y)\mathbf{i} - x\mathbf{j}$
56. $\mathbf{F}(x, y) = (3x + 2y)\mathbf{i} + (12x - 7y)\mathbf{j}$
57. Find a continuous vector field $\mathbf{F}(x, y) = M(x, y)\mathbf{i} + N(x, y)\mathbf{j}$ such that at each point (x, y) on the circle $\mathcal{C} = \{(x, y) : x^2 + y^2 = 1\}$, the vector $\mathbf{F}(x, y)$ satisfies the following:
   a. $\|\mathbf{F}(x, y)\| = 1$
   b. $\mathbf{F}(x, y)$ is tangent to $\mathcal{C}$.
   According to the Combing the Hairy Sphere Theorem, there is no such continuous vector field on the unit sphere in three-dimensional space.

Calculator/Computer Exercises

58. Draw the given planar vector fields in the viewing window $[-1, 1] \times [-1, 1]$.
   a. $(x + y)\mathbf{i} + (1 + xy)\mathbf{j}$
   b. $xy\mathbf{i} + (1 + y - x)\mathbf{j}$
   c. $(x + y)\mathbf{i} + (x - y^2)\mathbf{j}$
   d. $(x^2 - y^2)\mathbf{i} + (x^2 + y^2)\mathbf{j}$

15.2 Line Integrals

In Chapter 12, we learned to integrate a vector-valued function componentwise. The result of such an operation is a vector. In certain applications, however, it is appropriate to integrate a vector-valued function with the aim of obtaining a *scalar* as the result. For instance, if we apply a force at each point of a curve, then we need to integrate the force (a vector field) along the curve in order to calculate the work (a scalar) that is performed. Similarly, if a thin wire takes the shape of a curve, we might wish to integrate a mass density over the curve (which we describe by a vector-valued function) in order to obtain the total mass (a scalar) of the wire. In this section, we study the mathematical ideas that are the basis of these and other applications.

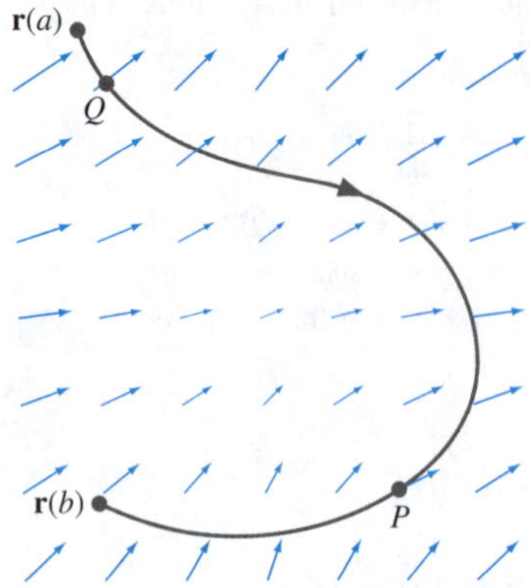

Figure 1

Work along a Curved Path

In Section 8.5, we developed a method for computing the *work* performed in moving a mass along a line segment. Now we would like to calculate work along a curved path. To be specific, we suppose that $\mathbf{F}$ is a continuous force field in a region of the plane or in space. If a particle moves through this force field with position vector $\mathbf{r}(t)$ for $a \le t \le b$, then how much work is performed? See Figure 1. The interesting feature here is that sometimes the force might be directly opposed to the path (point P in Figure 1), whereas at other times it will be skew to the path (point Q in Figure 1). In other words, we do not assume that there is any relationship between the direction of the force, $\mathbf{dir}(\mathbf{F}(P))$, and the direction $\mathbf{T}(t) = \text{dir}(\mathbf{r}'(t))$ of the curve at a point $P = \mathbf{r}(t)$ of the curve.

We estimate the work W by partitioning the domain $[a, b]$ of $\mathbf{r}$ into N subintervals $I_j = [t_{j-1}, t_j]$ with equal lengths $\Delta t = (b-a)/N$. For each j, we let W_j denote the work performed as t passes from t_{j-1} to t_j. If ξ_j is an arbitrary point in I_j, then we have the approximation

$$\begin{aligned} W_j \approx\ &(\text{component of force in direction of motion at } \mathbf{r}(\xi_j)) \\ &\cdot (\text{distance between } \mathbf{r}(t_{j-1}) \text{ and } \mathbf{r}(t_j)). \end{aligned}$$

Using formula (11.19) to evaluate the component of $\mathbf{F}$ along the tangent vector to the curve at $\mathbf{r}(\xi_j)$, we obtain

$$\begin{aligned} W_j &\approx (\mathbf{F}(\mathbf{r}(\xi_j)) \cdot \mathbf{T}(\xi_j))\|\mathbf{r}(t_j) - \mathbf{r}(t_{j-1})\| \\ &= \left(\mathbf{F}(\mathbf{r}(\xi_j)) \cdot \frac{1}{\|\mathbf{r}'(\xi_j)\|}\mathbf{r}'(\xi_j)\right) \|\mathbf{r}(t_j) - \mathbf{r}(t_{j-1})\|. \end{aligned}$$

Since we can approximate the velocity vector $\mathbf{r}'(\xi_j)$ by the difference quotient

$$\frac{1}{\Delta t}(\mathbf{r}(t_j) - \mathbf{r}(t_{j-1})),$$

it follows that $\|\mathbf{r}(t_j) - \mathbf{r}(t_{j-1})\| \approx \|\mathbf{r}'(\xi_j)\|\Delta t$. When we substitute this last expression into our approximation for W_j, we obtain $W_j \approx (\mathbf{F}(\mathbf{r}(\xi_j)) \cdot \mathbf{r}'(\xi_j))\Delta t$. Therefore, the work W performed in traversing the entire curve from $\mathbf{r}(a)$ to $\mathbf{r}(b)$ is approximately

$$\sum_{j=1}^{N} \mathbf{F}(\mathbf{r}(\xi_j)) \cdot \mathbf{r}'(\xi_j)\Delta t.$$

We notice that this expression is a Riemann sum. As N increases, the approximation becomes more accurate, and the sum tends to the integral of $\mathbf{F}(\mathbf{r}(t)) \cdot \mathbf{r}'(t)$ over the interval $[a, b]$. This observation leads to the definition of work.

Definition The *work* done by a force $\mathbf{F}$ along a parameterized curve $t \mapsto \mathbf{r}(t)$, $a \le t \le b$, is

$$\int_a^b \mathbf{F}(\mathbf{r}(t)) \cdot \mathbf{r}'(t)\, dt. \tag{15.15}$$

Example 1 A particle with position vector $\mathbf{r}(t) = \langle t^3, t^2 \rangle$, $0 \le t \le 1$, moves through a planar force field $\mathbf{F}(x, y) = \langle y, 1 + x \rangle$. What is the work that is done by $\mathbf{F}$? If the particle retraces its path with trajectory $\tilde{\mathbf{r}}(t) = \langle (2-t)^3, (2-t)^2 \rangle$, $1 \le t \le 2$, what is the work done by $\mathbf{F}$ during this "return trip"?

Solution We calculate $\mathbf{F}(\mathbf{r}(t)) = \mathbf{F}(t^3, t^2) = \langle t^2, 1 + t^3 \rangle$ and $\mathbf{r}'(t) = \langle 3t^2, 2t \rangle$. Thus, we have $\mathbf{F}(\mathbf{r}(t)) - \mathbf{r}'(t) = (t^2)(3t^2) + (1 + t^3)(2t) = 5t^4 + 2t$. According to formula (15.15), the work performed is equal to

$$\int_0^1 \mathbf{F}(\mathbf{r}(t)) \cdot \mathbf{r}'(t)\, dt = \int_0^1 (5t^4 + 2t)\, dt = (t^5 + t^2)\Big|_0^1 = 2.$$

Similarly, we have $\mathbf{F}(\tilde{\mathbf{r}}(t)) = \langle (2-t)^2, 1+(2-t)^3 \rangle$ and $\tilde{\mathbf{r}}(t) = \langle -3(2-t)^2, -2(2-t) \rangle$. Thus,

$$\mathbf{F}(\tilde{\mathbf{r}}(t)) \cdot \tilde{\mathbf{r}}'(t) = (2-t)^2(-3(2-t)^2) + (1 + (2-t)^3)(-2(2-t)) = -5(2-t)^4 - 2(2-t).$$

The work performed along the return path is therefore

$$\int_1^2 \mathbf{F}(\tilde{\mathbf{r}}(t)) \cdot \tilde{\mathbf{r}}'(t)\, dt = -\int_1^2 \left(5(2-t)^4 + 2(2-t)\right) dt = ((2-t)^5 + (2-t)^2)\Big|_1^2 = -2. \quad \blacksquare$$

inSIGHT

Work is a scalar quantity that can be positive, negative, or zero. In Example 1, the values of work associated with the paths traced by $\mathbf{r}$ and $\tilde{\mathbf{r}}$ are opposite in sign. The *physical* meaning of "negative work" will be explored in Section 15.3. However, we can understand right away why the work done by $\mathbf{F}$ over $\tilde{\mathbf{r}}$ is negative. Observe that the functions $\mathbf{r}$ and $\tilde{\mathbf{r}}$ both trace the graph of $y = x^{2/3}$, $0 \le x \le 1$, but in *opposite* directions (see Figure 2). We can see from the plot that the angles between the vectors $\mathbf{F}(\mathbf{r}(t))$ and $\mathbf{r}'(t)$ are acute. The integrand $\mathbf{F}(\mathbf{r}(t)) \cdot \mathbf{r}'(t)$ is therefore nonnegative, and $\int_0^1 \mathbf{F}(\mathbf{r}(t)) \cdot \mathbf{r}'(t)\, dt > 0$. On the other hand, $\tilde{\mathbf{r}}'$ points in the opposite direction of $\mathbf{r}'$, the angles between $\mathbf{F}(\tilde{\mathbf{r}}(t))$ and $\tilde{\mathbf{r}}'(t)$ are obtuse, and the integrand $\mathbf{F}(\tilde{\mathbf{r}}(t)) \cdot \tilde{\mathbf{r}}'(t)$ is negative. It follows that $\int_0^1 \mathbf{F}(\tilde{\mathbf{r}}(t)) \cdot \tilde{\mathbf{r}}'(t)\, dt < 0$.

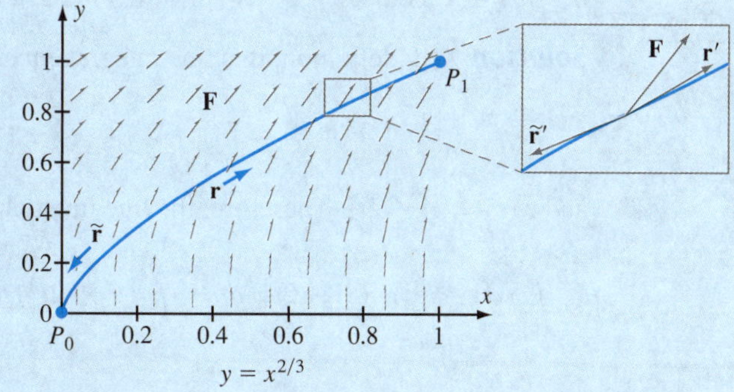

Figure 2

Line Integrals

Now we wish to define the integral of a vector field $\mathbf{F}$ over a curve. Let us recall that a *smooth parametric curve* is a continuously differentiable, vector-valued function $t \mapsto \mathbf{r}(t),\ a \le t \le b$, such that $\mathbf{r}'(t) \neq \vec{\mathbf{0}}$. In general we will identify such a curve with its image $\mathcal{C} = \{\mathbf{r}(t) : a \le t \le b\}$. We refer to $\mathcal{C}$ as a *directed* or *oriented curve* because $\mathbf{r}$ imparts to $\mathcal{C}$ an initial point $\mathbf{r}(a)$, a terminal point $\mathbf{r}(b)$, and a direction for traversing $\mathcal{C}$ from $\mathbf{r}(a)$ to $\mathbf{r}(b)$. We wish to define the integral of a vector field $\mathbf{F}$ over a directed curve $\mathcal{C}$. Our calculation of work motivates the definition of the integral of $\mathbf{F}$ over $\mathcal{C}$.

Definition Let $\mathcal{C}$ be a directed curve (in the plane or in space) with parameterization $t \mapsto \mathbf{r}(t)$, $a \le t \le b$. Suppose that $\mathbf{F}$ is a continuous vector field defined on $\mathcal{C}$. Then the *line integral* (or *path integral*) of $\mathbf{F}$ over $\mathcal{C}$ is denoted by the symbol $\int_{\mathcal{C}} \mathbf{F} \cdot d\mathbf{r}$ and is defined to be

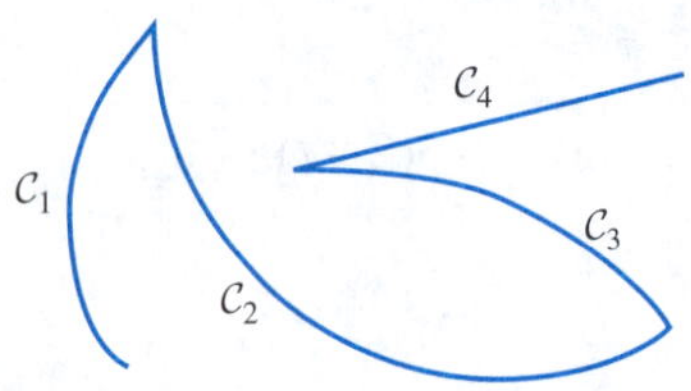

Figure 3

$$\int_{\mathcal{C}} \mathbf{F} \cdot d\mathbf{r} = \int_a^b \mathbf{F}(\mathbf{r}(t)) \cdot \mathbf{r}'(t)\, dt. \tag{15.16}$$

If $\mathcal{C}$ is a piecewise-smooth directed curve that is the union of smooth segments $\mathcal{C}_1$, $\mathcal{C}_2, \ldots, \mathcal{C}_n$ (as in Figure 3), then $\int_{\mathcal{C}} \mathbf{F} \cdot d\mathbf{r}$ is defined by

$$\int_{\mathcal{C}} \mathbf{F} \cdot d\mathbf{r} = \int_{\mathcal{C}_1} \mathbf{F} \cdot d\mathbf{r} + \int_{\mathcal{C}_2} \mathbf{F} \cdot d\mathbf{r} + \cdots + \int_{\mathcal{C}_n} \mathbf{F} \cdot d\mathbf{r}$$

where each summand on the right side of the equation is defined by formula (15.16). (In each of the expressions $\int_{\mathcal{C}_i} \mathbf{F} \cdot d\mathbf{r}$, it is implicit that $\mathbf{r}$ is a parameterization of the particular curve $\mathcal{C}_i$ that appears in the expression. In other words, the symbol $d\mathbf{r}$ signifies different parameterizations in the different summands.)

If $\mathbf{F}$ is a force field, then equation (15.15) shows that the line integral $\int_{\mathcal{C}} \mathbf{F} \cdot d\mathbf{r}$ has a physical interpretation as the work performed in traversing $\mathcal{C}$ from initial point to terminal point. However, we may form the line integral $\int_{\mathcal{C}} \mathbf{F} \cdot d\mathbf{r}$ of *any* continuous vector field $\mathbf{F}$ defined on the directed curve $\mathcal{C}$. The vector field $\mathbf{F}$ does not have to be a force field, and the line integral $\int_{\mathcal{C}} \mathbf{F} \cdot d\mathbf{r}$ does not have to represent work. Notice that the value of a line integral is a scalar: The dot product notation $\mathbf{F} \cdot d\mathbf{r}$ serves to remind us of this fact.

Example 2 Let $\mathcal{C}$ be the planar directed curve parameterized by $\mathbf{r}(t) = \langle t, -t^2 \rangle$, $0 \le t \le 1$. Calculate $\int_{\mathcal{C}} \mathbf{F} \cdot d\mathbf{r}$ for $\mathbf{F}(x, y) = e^x\mathbf{i} + xy\mathbf{j}$.

Solution By definition, this line integral equals $\int_0^1 \mathbf{F}(\mathbf{r}(t)) \cdot \mathbf{r}'(t)\, dt$. Now

$$\mathbf{F}(\mathbf{r}(t)) = \mathbf{F}(t, -t^2) = e^t\mathbf{i} + t(-t^2)\mathbf{j} = e^t\mathbf{i} - t^3\mathbf{j},$$

and $\mathbf{r}'(t) = 1\mathbf{i} - 2t\mathbf{j}$. Therefore, the line integral equals

$$\int_0^1 (e^t\mathbf{i} - t^3\mathbf{j}) \cdot (1\mathbf{i} - 2t\mathbf{j})\, dt = \int_0^1 (e^t + 2t^4)\, dt = \left(e^t + \frac{2}{5}t^5\right)\Bigg|_{t=0}^{t=1} = e - \frac{3}{5}.$$ ■

As we know from Chapter 12, a directed curve $\mathcal{C}$ may be parameterized in several different ways. Formula (15.16) seems to imply that the line integral $\int_{\mathcal{C}} \mathbf{F} \cdot d\mathbf{r}$ of $\mathbf{F}$ over $\mathcal{C}$ depends on the parameterization $\mathbf{r}$ of $\mathcal{C}$ that is used in the calculation. The next theorem demonstrates that the value of the line integral $\int_{\mathcal{C}} \mathbf{F} \cdot d\mathbf{r}$ does *not* actually depend on the particular parameterization of $\mathcal{C}$ that is chosen (provided that the parameterization respects the direction of $\mathcal{C}$).

Theorem 1 Let $t \mapsto \mathbf{r}(t)$, $a \leq t \leq b$, be a continuously differentiable parameterization for a directed curve $\mathcal{C}$ in the plane or in space. Suppose that $\mathbf{F}$ is a continuous vector field whose domain contains $\mathcal{C}$. Let $s \mapsto \mathbf{p}(s)$, $0 \leq s \leq L$, be the arc length parameterization of $\mathcal{C}$. Let $\mathbf{T}(s) = \mathbf{p}'(s)$ denote the unit tangent vector to $\mathcal{C}$ at the point $\mathbf{p}(s)$. Then the line integral of $\mathbf{F}$ over $\mathcal{C}$ is given by the equation

$$\int_{\mathcal{C}} \mathbf{F} \cdot d\mathbf{r} = \int_0^L \mathbf{F}(\mathbf{p}(s)) \cdot \mathbf{T}(s)\, ds. \qquad \mathbf{(15.17)}$$

In particular, since the right side of equation (15.17) does not involve the parameterization $\mathbf{r}$, the line integral of $\mathbf{F}$ over $\mathcal{C}$ does not depend on the chosen parameterization of $\mathcal{C}$.

Proof The derivation of equation (15.17) is very similar to the proof of Theorem 1, Section 12.3. If $s = \int_a^t \|\mathbf{r}'(\tau)\|\, d\tau$, then $\mathbf{p}(s) = \mathbf{r}(t)$, $\mathbf{T}(s) = \frac{1}{\|\mathbf{r}'(t)\|}\mathbf{r}'(t)$, and $\frac{ds}{dt} = \|\mathbf{r}'(t)\|$. Making the substitution $s = \int_a^t \|\mathbf{r}'(\tau)\|\, d\tau$, $ds = \|\mathbf{r}'(t)\|\, dt$ in the integral on the right side of equation (15.17), we obtain

$$\begin{aligned}\int_0^L \mathbf{F}(\mathbf{p}(s)) \cdot \mathbf{T}(s)\, ds &= \int_a^b \mathbf{F}(\mathbf{r}(t)) \cdot \left(\frac{1}{\|\mathbf{r}'(t)\|}\mathbf{r}'(t)\right) \|\mathbf{r}'(t)\| dt \\ &= \int_a^b \mathbf{F}(\mathbf{r}(t)) \cdot \mathbf{r}'(t)\, dt = \int_{\mathcal{C}} \mathbf{F} \cdot d\mathbf{r}. \quad \blacksquare\end{aligned}$$

Example 3

Let $\mathcal{C}$ be the quarter circle in the xy-plane that is parameterized by $\mathbf{r}(t) = t\mathbf{i} + \sqrt{1 - t^2}\mathbf{j}$, $0 \leq t \leq 1$. Verify Theorem 1 for the vector field $\mathbf{F}(x, y) = x\mathbf{i} + xy\mathbf{j}$.

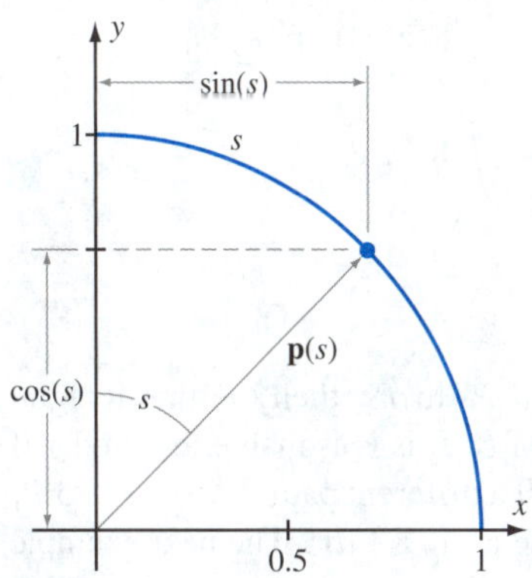

Figure 4

Solution Since $\mathbf{F}(\mathbf{r}(t)) = \mathbf{F}\big(t, \sqrt{1 - t^2}\big) = t\mathbf{i} + t\sqrt{1 - t^2}\mathbf{j}$ and $\mathbf{r}'(t) = \mathbf{i} - (t/\sqrt{1 - t^2})\mathbf{j}$, we have $\mathbf{F}(\mathbf{r}(t)) \cdot \mathbf{r}'(t) = t - t^2$. Therefore,

$$\int_{\mathcal{C}} \mathbf{F} \cdot d\mathbf{r} = \int_0^1 \mathbf{F}(\mathbf{r}(t)) \cdot \mathbf{r}'(t)\, dt = \int_0^1 (t - t^2)\, dt = \left(\frac{t^2}{2} - \frac{t^3}{3}\right)\bigg|_0^1 = \frac{1}{6}.$$

In this example, the arc length parameterization of $\mathcal{C}$ can be established from geometric principles: The length of an arc of a circle of radius a is a times the angle at the center that is subtended by the arc. As Figure 4 shows, this implies that $\mathbf{p}(s) = \sin(s)\mathbf{i} + \cos(s)\mathbf{j}$, $0 \leq s \leq \pi/2$, is the arc length parameterization of $\mathcal{C}$. Thus, $L = \pi/2$, $\mathbf{F}(\mathbf{p}(s)) = \sin(s)\mathbf{i} + \sin(s)\cos(s)\mathbf{j}$, $\mathbf{T}(s) = \mathbf{p}'(s) = \cos(s)\mathbf{i} - \sin(s)\mathbf{j}$, and

$\mathbf{F}(\mathbf{p}(s)) \cdot \mathbf{T}(s) = \sin(s)\cos(s) - \sin^2(s)\cos(s)$. We therefore have

$$\int_0^L \mathbf{F}(\mathbf{p}(s)) \cdot \mathbf{T}(s)\,ds = \int_0^{\pi/2} (\sin(s)\cos(s) - \sin^2(s)\cos(s))\,ds$$
$$= \left(\frac{\sin^2(s)}{2} - \frac{\sin^3(s)}{3}\right)\Bigg|_0^{\pi/2} = \frac{1}{6}.$$

For every directed curve $\mathcal{C}$ from P_0 to P_1, there is a directed curve, often denoted by $-\mathcal{C}$, that is *opposite* to $\mathcal{C}$. It consists of the same points as $\mathcal{C}$ but has P_1 as its initial point and P_0 as its terminal point. The line integrals of a field $\mathbf{F}$ over the two directed curves $\mathcal{C}$ and $-\mathcal{C}$ are related by the formula

$$\int_{\mathcal{C}} \mathbf{F} \cdot d\mathbf{r} = -\int_{-\mathcal{C}} \mathbf{F} \cdot d\tilde{\mathbf{r}} \tag{15.18}$$

where $\mathbf{r}$ and $\tilde{\mathbf{r}}$ denote parameterizations of $\mathcal{C}$ and $-\mathcal{C}$, respectively.

Example 4 Let $t \mapsto \mathbf{r}(t)$, $t \in [a, b]$ be a parameterization of a directed curve $\mathcal{C}$. Prove that $t \mapsto \tilde{\mathbf{r}}(t) = \mathbf{r}(a + b - t),\ t \in [a, b]$ is a parameterization of the opposite curve $-\mathcal{C}$. Verify equation (15.18).

Solution As t increases from a to b, the expression $a + b - t$ decreases from b to a. It follows that the values $\tilde{\mathbf{r}}(t) = \mathbf{r}(a + b - t)$, $a \le t \le b$, are the points of $\mathcal{C}$ traversed from $\mathbf{r}(b)$ to $\mathbf{r}(a)$. Since $\tilde{\mathbf{r}}'(t) = -\mathbf{r}'(a + b - t)$, we have

$$\int_{-\mathcal{C}} \mathbf{F} \cdot d\tilde{\mathbf{r}} = \int_a^b \mathbf{F}(\tilde{\mathbf{r}}(t)) \cdot \tilde{\mathbf{r}}'(t)\,dt = -\int_a^b \mathbf{F}(\mathbf{r}(a + b - t)) \cdot \mathbf{r}'(a + b - t)\,dt.$$

By making the change of variable $u = a + b - t$, $du = (-1)\,dt$ in this last integral, we obtain

$$\int_{-\mathcal{C}} \mathbf{F} \cdot d\tilde{\mathbf{r}} = \int_b^a \mathbf{F}(\mathbf{r}(u)) \cdot \mathbf{r}'(u)\frac{du}{dt}\,dt = \int_b^a \mathbf{F}(\mathbf{r}(u)) \cdot \mathbf{r}'(u)\,du$$
$$= -\int_a^b \mathbf{F}(\mathbf{r}(u)) \cdot \mathbf{r}'(u)\,du = -\int_{\mathcal{C}} \mathbf{F} \cdot d\mathbf{r}.$$

inSIGHT

Equation (15.18) is a generalization of the formula $\int_b^a f(t)\,dt = -\int_a^b f(t)\,dt$. Just as on the real axis, reversing the direction of integration in a line integral changes the sign of the integral's value.

Dependence on Path

We have observed that if $\mathcal{C}$ is a directed curve from P_0 to P_1, then the line integral $\int_{\mathcal{C}} \mathbf{F} \cdot d\mathbf{r}$ does not depend on the parameterization of $\mathcal{C}$. It is reasonable to wonder if the line integral depends on the path at all; that is, will a different path $\mathcal{C}_*$ from P_0 to P_1 lead to a line integral $\int_{\mathcal{C}_*} \mathbf{F} \cdot d\mathbf{r}_*$ with the same value as $\int_{\mathcal{C}} \mathbf{F} \cdot d\mathbf{r}$? The next example shows that the answer to this question is generally "*No*!"

Example 5 Let $\mathbf{F}(x, y) = -y\mathbf{i} + x\mathbf{j}$. Set $P_0 = (1, 0)$ and $P_1 = (-1, 0)$. Consider two paths from P_0 to P_1: $\mathcal{C}$ parameterized by

$$\mathbf{r}(t) = \cos(t)\mathbf{i} + \sin(t)\mathbf{j}, \qquad 0 \le t \le \pi$$

and $\mathcal{C}_*$ parameterized by

$$\mathbf{r}_*(t) = \cos(t)\mathbf{i} - \sin(t)\mathbf{j}, \qquad 0 \le t \le \pi.$$

Is the line integral of **F** over $\mathcal{C}$ equal to the line integral of **F** over $\mathcal{C}_*$?

Solution We calculate that

$$\begin{aligned}\int_{\mathcal{C}} \mathbf{F} \cdot d\mathbf{r} &= \int_0^{\pi} \mathbf{F}(\mathbf{r}(t)) \cdot \mathbf{r}'(t)\, dt \\ &= \int_0^{\pi} (-\sin(t)\mathbf{i} + \cos(t)\mathbf{j}) \cdot (-\sin(t)\mathbf{i} + \cos(t)\mathbf{j})\, dt \\ &= \int_0^{\pi} (\sin^2(t) + \cos^2(t))\, dt = \pi.\end{aligned}$$

However,

$$\begin{aligned}\int_{\mathcal{C}_*} \mathbf{F} \cdot d\mathbf{r}_* &= \int_0^{\pi} \mathbf{F}(\mathbf{r}_*(t)) \cdot \mathbf{r}_*'(t)\, dt \\ &= \int_0^{\pi} (\sin(t)\mathbf{i} + \cos(t)\mathbf{j}) \cdot (-\sin(t)\mathbf{i} - \cos(t)\mathbf{j})\, dt \\ &= -\int_0^{\pi} (\sin^2(t) + \cos^2(t))\, dt = -\pi.\end{aligned}$$

The line integrals have different values. ■

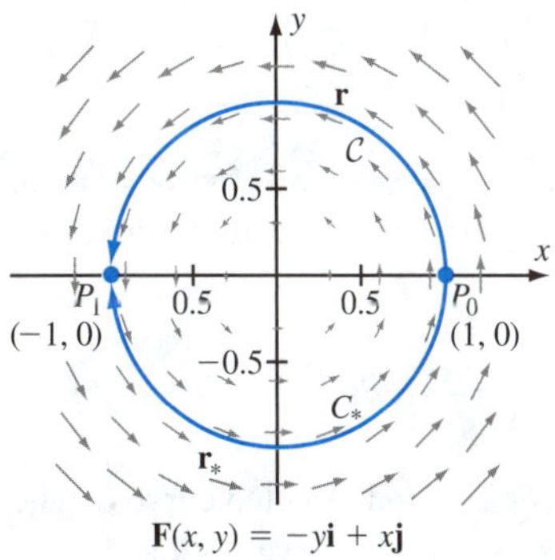

Figure 5

inSIGHT

Figure 5 suggests why it is plausible that the two line integrals of Example 5 should be different. The line integral over $\mathcal{C}$ is counterclockwise along the upper half of the unit circle: It is *in the direction of the vector field* **F**. However, the line integral over $\mathcal{C}_*$ is clockwise along the lower half of the unit circle: It is *against* the direction of the vector field. Thus, we would anticipate the values of the integrals to have opposite signs. It makes sense that the value of a line integral over a path from P_0 to P_1 ought to depend on the path between the points. But it turns out that line integrals of conservative vector fields *are* independent of path. Section 15.3 treats these ideas in depth.

Other Notation for the Line Integral

If $\mathbf{F} = M(x, y, z)\mathbf{i} + N(x, y, z)\mathbf{j} + R(x, y, z)\mathbf{k}$ is a vector field defined on a directed curve $\mathcal{C}$ that is parameterized by $\mathbf{r}(t) = \langle x(t), y(t), z(t)\rangle$, $a \le t \le b$, then the line integral $\int_{\mathcal{C}} \mathbf{F} \cdot d\mathbf{r} = \int_a^b \mathbf{F}(\mathbf{r}(t)) \cdot \mathbf{r}'(t)\, dt$ becomes

$$\int_{\mathcal{C}} \mathbf{F} \cdot d\mathbf{r} = \int_a^b \left(M(x(t), y(t), z(t))\frac{dx}{dt} + N(x(t), y(t), z(t))\frac{dy}{dt} + R(x(t), y(t), z(t))\frac{dz}{dt} \right) dt \qquad \textbf{(15.19)}$$

when expanded. Using the formalism of calculus,

$$\frac{dx}{dt}dt = dx, \qquad \frac{dy}{dt}dt = dy, \qquad \text{and} \qquad \frac{dz}{dt}dt = dz,$$

we write equation (15.19) more compactly as

$$\int_C \mathbf{F} \cdot d\mathbf{r} = \int_C M\,dx + N\,dy + R\,dz. \tag{15.20}$$

Example 6 Let C be the directed curve that is parameterized by $\mathbf{r}(t) = \langle 1 - t, 2 - t^2, 1 \rangle$, $-1 \le t \le 2$. Calculate $\int_C yz\,dx + z\,dy + x\,dz$.

Solution We use formula (15.20) with $x(t) = 1 - t$, $y(t) = 2 - t^2$, and $z(t) = 1$. Then $\frac{dx}{dt} = -1$, $\frac{dy}{dt} = -2t$, and $\frac{dz}{dt} = 0$. As a result, we have

$$\begin{aligned}\int_C xy\,dx + z\,dy + x\,dz &= \int_{-1}^{2} \left((1-t)(2-t^2)\frac{dx}{dt} + 1\frac{dy}{dt} + (1-t)\frac{dz}{dt} \right) dt \\ &= \int_{-1}^{2} ((1-t)(2-t^2) \cdot (-1) + 1 \cdot (-2t) + (1-t) \cdot 0)\,dt.\end{aligned}$$

On expanding the integrand, we obtain

$$\begin{aligned}\int_C xy\,dx + z\,dy + x\,dz &= \int_{-1}^{2} (-2 + t^2 - t^3)\,dt = \left(-2t + \frac{t^3}{3} - \frac{t^4}{4} \right)\Bigg|_{-1}^{2} \\ &= \left(-4 + \frac{8}{3} - \frac{16}{4} \right) - \left(2 - \frac{1}{3} - \frac{1}{4} \right) = -\frac{27}{4}.\end{aligned}$$

■

Closed Curves

If a directed curve C is parameterized by $t \mapsto \mathbf{r}(t)$ for $a \le t \le b$ and if $\mathbf{r}(a) = \mathbf{r}(b)$, then we say that the curve C is *closed*. A closed curve begins and ends at the same point. We often write

$$\oint_C \mathbf{F} \cdot d\mathbf{r} \qquad \text{or} \qquad \oint_C M\,dx + N\,dy + R\,dz$$

for the line integral of a vector field $\mathbf{F} = \langle M, N, R \rangle$ over a closed directed curve C. The small circle in the middle of the integral sign reminds us that the curve is closed.

Example 7 Calculate $\oint_C (x + y)\,dx + z\,dy + x^2\,dz$ where C comprises the line segment from $P_0 = (1, 0, 0)$ to $P_1 = (0, 1, 0)$, the line segment from P_1 to $P_2 = (0, 0, 2)$, and the line segment from P_2 to P_0.

Solution Let C_1 be the line segment from $P_0 = (1, 0, 0)$ to $P_1 = (0, 1, 0)$. We may parameterize it by $\mathbf{r}_1(t) = \overrightarrow{OP_0} + t(\overrightarrow{P_0P_1}) = \langle 1 - t, t, 0 \rangle$, $0 \le t \le 1$. Then $\frac{dx}{dt} = -1$, $\frac{dy}{dt} = 1$, and $\frac{dz}{dt} = 0$. We have

$$\begin{aligned}\int_{C_1} (x + y)\,dx + z\,dy + x^2\,dz &= \int_0^1 \left(((1-t) + t)\frac{dx}{dt} + 0\frac{dy}{dt} + (1-t)^2\frac{dz}{dt} \right) dt \\ &= \int_0^1 (-1 + 0 + 0)\,dt = -1.\end{aligned}$$

Similarly, we parameterize the directed line segment C_2 from $P_1 = (0, 1, 0)$ to $P_2 = (0, 0, 2)$ by $\mathbf{r}_2(t) = \overrightarrow{OP_1} + t(\overrightarrow{P_1P_2}) = \langle 0, 1-t, 2t\rangle, 0 \le t \le 1$. Then $\frac{dx}{dt} = 0$, $\frac{dy}{dt} = -1$, and $\frac{dz}{dt} = 2$, and

$$\begin{aligned}\int_{C_2} (x+y)\,dx + z\,dy + x^2\,dz &= \int_0^1 \left((0+(1-t))\frac{dx}{dt} + 2t\frac{dy}{dt} + 0^2\frac{dz}{dt}\right) dt \\ &= \int_0^1 (0 - 2t + 0)\,dt = -1.\end{aligned}$$

Finally, we parameterize the directed line segment C_3 from $P_2 = (0, 0, 2)$ to $P_0 = (1, 0, 0)$ by $\mathbf{r}_3(t) = \overrightarrow{OP_2} + t(\overrightarrow{P_2P_0}) = \langle 0, 0, 2\rangle + t\langle 1, 0, -2\rangle = \langle t, 0, 2-2t\rangle, 0 \le t \le 1$. Then $\frac{dx}{dt} = 1$, $\frac{dy}{dt} = 0$, and $\frac{dz}{dt} = -2$, and

$$\begin{aligned}\int_{C_3} (x+y)\,dx + z\,dy + x^2\,dz &= \int_0^1 \left((t+0)\frac{dx}{dt} + (2-2t)\frac{dy}{dt} + t^2\frac{dz}{dt}\right) dt \\ &= \int_0^1 (t + 0 - 2t^2)\,dt = -\frac{1}{6}.\end{aligned}$$

It follows that $\oint_C (x+y)\,dx + z\,dy + x^2\,dz = -1 + (-1) + (-1/6) = -13/6$. ■

Other Applications of Line Integrals

Suppose that C is a directed curve parameterized by arc length $s \mapsto \mathbf{p}(s)$, $0 \le s \le L$. Then equation (15.17) tells us that

$$\int_C \mathbf{F}\cdot d\mathbf{r} = \int_0^L f(\mathbf{p}(s))\,ds \tag{15.21}$$

where $f(\mathbf{p}(s)) = \mathbf{F}(\mathbf{p}(s)) \cdot \mathbf{T}(s)$. Now the integral on the right side of equation (15.21) actually makes sense for any scalar-valued function f whose domain contains C. For example, if $f(x, y, z)$ is identically 1, then integral (15.21) is the arc length of C, a topic we studied in Chapter 12.

We obtain a new application of the line integral by supposing that $f(x, y, z)$ is the linear mass density of a thin wire that has the shape C. If we divide C into N arcs of equal length $\Delta s = L/N$ and choose a point $\mathbf{p}(s_j)$ in the jth arc, then the mass of the wire is approximately $\sum_{j=1}^N f(\mathbf{p}(s_j))\Delta s$. The approximation tends to become more accurate as N tends to infinity. At the same time,

$$\lim_{N\to\infty} \sum_{j=1}^N f(\mathbf{p}(s_j))\Delta s = \int_0^L f(\mathbf{p}(s))\,ds.$$

Thus, when f is a linear mass density, line integral (15.21) represents the total mass of the curve. Of course, similar interpretations can be made when f is another type of density.

The right side of equation (15.21) is often written as $\int_C f\,ds$. The proof of Theorem 1 shows that, in general, if $\mathbf{r}$ is any parameterization of C, not necessarily the arc length parameterization, then

$$\int_C f\,ds = \int_a^b f(\mathbf{r}(t))\|\mathbf{r}'(t)\|\,dt. \tag{15.22}$$

Example 8 Suppose that a wire has the shape of the curve $\mathcal{C}$ that is parameterized by $\mathbf{r}(t) = \langle 6t, 8t, 5t^2 \rangle$, $0 \le t \le 2$. The charge density of the wire is given by $q(x, y, z) = x - y + \sqrt{5z}$. What is the total charge held by the wire?

Solution We use formula (15.22). Observe that $\mathbf{r}'(t) = \langle 6, 8, 10t \rangle$ and $\|\mathbf{r}'(t)\| = \sqrt{36 + 64 + 100t^2} = 10\sqrt{1 + t^2}$. The total charge is therefore given by

$$\int_{\mathcal{C}} q\, ds = \int_0^2 q(\mathbf{r}(t))\, \|\mathbf{r}'(t)\|\, dt = \int_0^2 \left(6t - 8t + \sqrt{25t^2}\right) 10\sqrt{1 + t^2}\, dt$$
$$= 10(1 + t^2)^{3/2}\Big|_0^2 = 10\left(5\sqrt{5} - 1\right).$$

quickquiz

1. What is a directed curve?
2. What is the formula for work performed by a force field $\mathbf{F}$ over a directed curve $\mathcal{C}$?
3. How is a parameterization of $\mathcal{C}$ used to calculate $\int_{\mathcal{C}} P\, dx + Q\, dy + R\, dz$?
4. If $\mathbf{r}(t) = \langle t, t^2 \rangle$, $1 \le t \le 3$, and $\mathbf{F}(x, y) = \langle y, x \rangle$, then what is $\int_{\mathcal{C}} \mathbf{F} \cdot d\mathbf{r}$?

EXERCISES

Problems for Practice

In Exercises 1–8, calculate $\int_{\mathcal{C}} \mathbf{F} \cdot d\mathbf{r}$ for the planar vector field $\mathbf{F}$ and the parametric curve $\mathbf{r}$.

1. $\mathbf{F}(x, y) = xy\mathbf{i} + y^2\mathbf{j}$,
   $\mathbf{r}(t) = \cos(t)\mathbf{i} + \sin(t)\mathbf{j}$, $0 \le t \le \pi/3$
2. $\mathbf{F}(x, y) = y^2\mathbf{i} - x^2\mathbf{j}$,
   $\mathbf{r}(t) = \sin(t)\mathbf{i} + \cos(t)\mathbf{j}$, $0 \le t \le \pi$
3. $\mathbf{F}(x, y) = \sin(x)\mathbf{i} - \cos(y)\mathbf{j}$,
   $\mathbf{r}(t) = t^2\mathbf{i} + t^3\mathbf{j}$, $0 \le t \le \sqrt{\pi}$
4. $\mathbf{F}(x, y) = \ln(y)\mathbf{i} - \exp(x)\mathbf{j}$,
   $\mathbf{r}(t) = \ln(t)\mathbf{i} + t^3\mathbf{j}$, $1 \le t \le e$
5. $\mathbf{F}(x, y) = xy\mathbf{i} + x^2y^2\mathbf{j}$,
   $\mathbf{r}(t) = t^{1/2}\mathbf{i} + t^{-3/2}\mathbf{j}$, $1 \le t \le 4$
6. $\mathbf{F}(x, y) = y^{1/2}\mathbf{i} + x^{3/2}\mathbf{j}$,
   $\mathbf{r}(t) = t^2\mathbf{i} + t^4\mathbf{j}$, $-1 \le t \le 2$
7. $\mathbf{F}(x, y) = (x^2 + 1)^{-1}\mathbf{j} - xy\mathbf{i}$,
   $\mathbf{r}(t) = t\mathbf{i} + t^2\mathbf{j}$, $-4 \le t \le -1$
8. $\mathbf{F}(x, y) = y\mathbf{i} - x\mathbf{j}$,
   $\mathbf{r}(t) = \cos(t)\mathbf{i} + \sin(t)\mathbf{j}$, $-\pi/2 \le t \le \pi/2$

In Exercises 9–16, calculate $\int_{\mathcal{C}} \mathbf{F} \cdot d\mathbf{r}$ for the spatial vector field $\mathbf{F}$ and the parametric curve $\mathbf{r}$.

9. $\mathbf{F}(x, y, z) = xz\mathbf{i} - yx\mathbf{j} + xy\mathbf{k}$,
   $\mathbf{r}(t) = t\mathbf{i} + t^2\mathbf{j} + t^3\mathbf{k}$, $-1 \le t \le 1$
10. $\mathbf{F}(x, y, z) = y\mathbf{i} - \sin(4z)\mathbf{j} + x\mathbf{k}$,
    $\mathbf{r}(t) = \cos(t)\mathbf{i} + \sin(t)\mathbf{j} + t/4\mathbf{k}$, $-3\pi \le t \le -\pi/2$
11. $\mathbf{F}(x, y, z) = \cos(y)\mathbf{i} + \sin(z)\mathbf{j} + x\mathbf{k}$,
    $\mathbf{r}(t) = \cos(t)\mathbf{i} + t\mathbf{j} + t\mathbf{k}$, $\pi/4 \le t \le \pi/2$
12. $\mathbf{F}(x, y, z) = (x/y)\mathbf{i} - (y/z)\mathbf{j} + xz\mathbf{k}$,
    $\mathbf{r}(t) = t^3\mathbf{i} + t^2\mathbf{j} + (1/t)\mathbf{k}$, $-2 \le t \le -1$
13. $\mathbf{F}(x, y, z) = z\mathbf{i} + \mathbf{k}$,
    $\mathbf{r}(t) = t\mathbf{i} + \mathbf{j} + \cos^2(t)\mathbf{k}$, $0 \le t \le \pi$
14. $\mathbf{F}(x, y, z) = y\mathbf{i} + (x - z)\mathbf{j} + z\mathbf{k}$,
    $\mathbf{r}(t) = \sin(t)\mathbf{i} + \cos(t)\mathbf{j} + \sin(t)\mathbf{k}$, $0 \le t \le 2\pi$
15. $\mathbf{F}(x, y, z) = x/(x^2 + y^2)\mathbf{k}$,
    $\mathbf{r}(t) = \sin(t)\mathbf{i} + \cos(t)\mathbf{j} + t\mathbf{k}$, $0 \le t \le \pi$
16. $\mathbf{F}(x, y, z) = yz\mathbf{i} + \mathbf{j} + 1/(1 + y^2)\mathbf{k}$,
    $\mathbf{r}(t) = t\mathbf{i} + \tan(t)\mathbf{j} + \tan(t)\mathbf{k}$, $0 \le t \le \pi/4$

In Exercises 17–20, calculate the work performed by a force $\mathbf{F}$ in moving a particle along the parametric curve $\mathbf{r}$.

17. $\mathbf{F}(x, y) = yx\mathbf{i} + x^3\mathbf{j}$,
    $\mathbf{r}(t) = t^{1/2}\mathbf{i} + t^{1/4}\mathbf{j}$, $1 \le t \le 16$
18. $\mathbf{F}(x, y) = 3y^2\mathbf{i} - 2x^3y\mathbf{j}$,
    $\mathbf{r}(t) = \cos(t)\mathbf{i} + \sin(t)\mathbf{j}$, $0 \le t \le \pi$
19. $\mathbf{F}(x, y, z) = x^2\mathbf{i} + y^2\mathbf{j} + z^2\mathbf{k}$,
    $\mathbf{r}(t) = (1 + t^2)\mathbf{i} + t^2\mathbf{j} + (2 + \sin(\pi t))\mathbf{k}$, $0 \le t \le 1$
20. $\mathbf{F}(x, y, z) = (x + y)\mathbf{i} - (y - x)\mathbf{j} + 4z\mathbf{k}$,
    $\mathbf{r}(t) = \sin(3t)\mathbf{i} + \cos(3t)\mathbf{j} + 3t\mathbf{k}$, $0 \le t \le 2\pi$

In Exercises 21–24, calculate the line integral over the directed curve $\mathcal{C}$ that is parameterized by $\mathbf{r}(t) = t^2\mathbf{i} + t^3\mathbf{j} + t^4\mathbf{k}$, $0 \le t \le 1$.

21. $\int_{\mathcal{C}} xyz\,dx$ **22.** $\int_{\mathcal{C}} dx + dy + dz$

23. $\int_{\mathcal{C}} xz\,dx + yz\,dy + xy\,dz$

24. $\int_{\mathcal{C}} xe^z\,dx + yz\,dy$

In Exercises 25–28, calculate the line integral $\oint_{\mathcal{C}} M\,dx + N\,dy$ where $\mathcal{C}$ is the triangle with vertices $P_0 = (0, 1)$, $P_1 = (2, 1)$, and $P_2 = (3, 4)$ with counterclockwise orientation.

25. $\oint_{\mathcal{C}} x\,dy$ **26.** $\oint_{\mathcal{C}} x\,dx + dy$

27. $\oint_{\mathcal{C}} (2x + y)\,dx + y\,dy$

28. $\oint_{\mathcal{C}} y\,dx - x\,dy$

In Exercises 29–32, calculate the line integral $\int_{\mathcal{C}} f\,ds$ for the function f and the parametric curve $\mathbf{r}$.

29. $f(x, y) = x^2 + y^2$, $\mathbf{r}(t) = \sin(t)\mathbf{i} + \cos(t)\mathbf{j}$, $0 \le t \le 2\pi$

30. $f(x, y) = y/x$, $\mathbf{r}(t) = t\mathbf{i} + t^2\mathbf{j}$, $1 \le t \le 2$

31. $f(x, y) = x + \sqrt{y}$, $\mathbf{r}(t) = t^2\mathbf{i} + t^4\mathbf{j}$, $1 \le t \le 2$

32. $f(x, y) = ye^x$, $\mathbf{r}(t) = t\mathbf{i} + e^t\mathbf{j}$, $0 \le t \le 1$

Further Theory and Practice

In Exercises 33–36, a vector field $\mathbf{F}$ is given. Calculate $\int_{\mathcal{C}_i} \mathbf{F} \cdot d\mathbf{r}_i$ for $i = 1, 2, 3$ where $\mathcal{C}_1$ is parameterized by

$$\mathbf{r}_1(t) = \langle \cos(2t), \sin(t), 1 - 2t/\pi \rangle, \quad 0 \le t \le \pi;$$

$\mathcal{C}_2$ is parameterized by

$$\mathbf{r}_2(t) = \langle 1 + \sin(2\pi t), \cos(2\pi t) - 1, \cos(\pi t) \rangle, \quad 0 \le t \le 1;$$

and $\mathcal{C}_3$ is parameterized by

$$\mathbf{r}_3(t) = \langle 1, 0, 1 - 2t \rangle, \quad 0 \le t \le 1.$$

Notice that all three paths begin at $(1, 0, 1)$ and terminate at $(1, 0, -1)$. Verify that for the given $\mathbf{F}$, the values of the line integrals are the same for all three paths. (The vector fields of Exercises 33–36 are path-independent. Section 15.3 is devoted to this type of vector field.)

33. $\mathbf{F}(x, y) = 3\mathbf{i} - 2\mathbf{j}$ **34.** $\mathbf{F}(x, y) = \mathbf{i} + y\mathbf{j}$

35. $\mathbf{F}(x, y) = x\mathbf{i} + y\mathbf{j}$ **36.** $\mathbf{F}(x, y) = y\mathbf{i} + x\mathbf{j}$

37. Let $\mathbf{F}(x, y) = 2y\mathbf{i} + x\mathbf{j}$. Calculate $\int_{\mathcal{C}_i} \mathbf{F} \cdot d\mathbf{r}_i$ for $i = 1, 2, 3$ where $\mathcal{C}_1$, $\mathcal{C}_2$, and $\mathcal{C}_3$ are the parametric curves described in the instructions to Exercises 33–36. The values will not all be the same in this case.

38. Calculate $\int_{\mathcal{C}} xe^z\,dx + yz\,dy + xe^y\,dz$ over the directed curve $\mathcal{C}$ that is parameterized by $\mathbf{r}(t) = t^2\mathbf{i} + t^3\mathbf{j} + t^4\mathbf{k}$, $0 \le t \le 1$.

In Exercises 39–42, a closed curve $\mathcal{C}$ with counterclockwise orientation is specified. Calculate

$$\oint_{\mathcal{C}} \frac{-y\,dx + x\,dy}{x^2 + y^2}.$$

39. $\mathcal{C}$ is the circle $x^2 + y^2 = a^2$.

40. $\mathcal{C}$ is the square with vertices $(\pm 1, \pm 1)$.

41. $\mathcal{C}$ is the arc of the circle $x^2 + y^2 = 2$, from $(1, -1)$ to $(1, 1)$ followed by the line segment from $(1, 1)$ to $(1, -1)$.

42. $\mathcal{C}$ is the arc of the circle $x^2 + y^2 = 2$, from $(-1, -1)$ to $(-1, 1)$ followed by the line segment from $(-1, 1)$ to $(-1, -1)$.

43. Calculate $\oint_{\mathcal{C}} xyz\,dz$ where $\mathcal{C}$ is the closed curve of intersection of the cylinder $x^2 + y^2 = 1$ and the plane $z = 1 + x$, oriented counterclockwise when viewed from above.

44. Calculate $\oint_{\mathcal{C}} x\,dx + x^2\,dy + (y - z)\,dz$ where $\mathcal{C}$ is the arc of the curve of intersection of the cylinders $x^2 + y^2 = 1$ and $x^2 + z^2 = 1$ that lies in the first octant and that begins at the point $(1, 0, 0)$ and terminates at $(0, 1, 1)$.

Calculator/Computer Exercises

45. Let a and b denote the real roots of $4 + 2x - 3x^4$ with $a < b$. Calculate $\oint_{\mathcal{C}} dx + x^2\,dy$ where $\mathcal{C}$ is the piecewise-smooth closed curve that comprises the line segment from $(a, 0)$ to $(b, 0)$ and the arc of the graph of $y = 4 + 2x - 3x^4$ from $(b, 0)$ to $(a, 0)$.

46. Let P denote the point of intersection of the graphs of $y = x^{1/3}$ and $y = 1 - x^2$ in the first quadrant. Calculate $\oint_{\mathcal{C}} y^3\,dx + x\,dy$ where $\mathcal{C}$ is the piecewise-smooth closed curve that comprises the arc of the graph of $y = x^{1/3}$ from $(0, 0)$ to P, the arc of the graph of $y = 1 - x^2$ from P to $(0, 1)$, and the line segment from $(0, 1)$ to $(0, 0)$.

47. Let $\mathcal{C}$ be the planar curve whose equation is $\mathbf{r}(t) = \langle t, t^2 \rangle$, $0 \le t \le 1$. Calculate $\int_{\mathcal{C}} xe^y\,ds$.

48. Let $\mathcal{C}$ be the spatial curve whose equation is $\mathbf{r}(t) = \langle t, t^2, t^3 \rangle$, $0 \le t \le 1$. Calculate $\int_{\mathcal{C}} \sqrt{x + y + z}\,ds$.

49. Let $\mathcal{C}$ be the planar curve whose equation in polar coordinates is $r = \theta$, $0 \le \theta \le 2\pi$. Calculate $\int_{\mathcal{C}} x^2\,ds$.

50. Calculate $\oint_{\mathcal{C}} (x^2 + y^2)\,ds$ where $\mathcal{C}$ is the ellipse $x^2 + 2y^2 = 1$.

15.3 Conservative Vector Fields and Path-Independence

In Example 5 from Section 15.2, we learned that the value of a line integral over a path from P_0 to P_1 can depend on the path that connects the two points. There is, however, an important class of vector fields whose line integrals are *independent of the path*. In this section, we will study and characterize such vector fields.

Definition Suppose that $\mathbf{F}$ is a vector field defined on a region $\mathcal{G}$. We say that $\mathbf{F}$ is *path-independent* on $\mathcal{G}$ if, for any two points P_0 and P_1 in $\mathcal{G}$, the line integral $\int_{\mathcal{C}} \mathbf{F} \cdot d\mathbf{r}$ has the same value for all directed curves $\mathcal{C}$ in $\mathcal{G}$ with initial point P_0 and terminal point P_1.

Our first theorem characterizes path-independent vector fields by means of their line integrals over *closed* curves.

Theorem 1 Let $\mathbf{F}$ be a vector field on a region $\mathcal{G}$. Then $\mathbf{F}$ is path-independent on $\mathcal{G}$ if and only if $\oint_{\mathcal{C}} \mathbf{F} \cdot d\mathbf{r} = 0$ for every closed curve $\mathcal{C}$ in $\mathcal{G}$.

Proof If $\mathcal{C}_1$ and $\mathcal{C}_2$ are two directed curves from P_0 to P_1, then the union of $\mathcal{C}_1$ and $-\mathcal{C}_2$ is a closed curve $\mathcal{C}$. See Figure 1. On the other hand, if $\mathcal{C}$ is a closed curve with initial and terminal point P_0, then we may (arbitrarily) choose another point P_1 on $\mathcal{C}$ and write $\mathcal{C} = -\mathcal{C}_2 \cup \mathcal{C}_1$ where $\mathcal{C}_1$ is the path along $\mathcal{C}$ from P_0 to P_1 and $-\mathcal{C}_2$ is the path along $\mathcal{C}$ from P_1 to P_0. Then

$$\oint_{\mathcal{C}} \mathbf{F} \cdot d\mathbf{r} = \int_{\mathcal{C}_1} \mathbf{F} \cdot d\mathbf{r} + \int_{-\mathcal{C}_2} \mathbf{F} \cdot d\mathbf{r} \overset{(15.18)}{=} \int_{\mathcal{C}_1} \mathbf{F} \cdot d\mathbf{r} - \int_{\mathcal{C}_2} \mathbf{F} \cdot d\mathbf{r}.$$

Thus,

$$\oint_{\mathcal{C}} \mathbf{F} \cdot d\mathbf{r} = 0 \qquad \text{if and only if} \qquad \int_{\mathcal{C}_1} \mathbf{F} \cdot d\mathbf{r} = \int_{\mathcal{C}_2} \mathbf{F} \cdot d\mathbf{r}.$$ ■

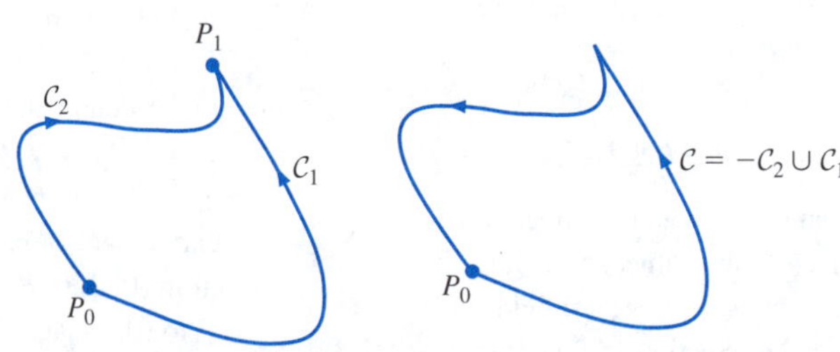

Figure 1

At first glance, it might seem that few vector fields could have the remarkable property of path-independence. The next theorem produces an abundant supply of path-independent vector fields.

Theorem 2 Suppose that u is a continuously differentiable scalar-valued function on a region $\mathcal{G}$ in the plane or in space. Let $\mathbf{F}$ be the gradient vector field defined by $\mathbf{F} = \nabla u$. Let $\mathcal{C}$ be a piecewise-smooth directed curve in $\mathcal{G}$ that has P_0 as its initial point and P_1 as its terminal point. Then

$$\int_{\mathcal{C}} \mathbf{F} \cdot d\mathbf{r} = u(P_1) - u(P_0). \tag{15.23}$$

In particular, the value of the line integral $\int_{\mathcal{C}} \mathbf{F} \cdot d\mathbf{r}$ depends on the initial and terminal points P_0 and P_1 but not on the path $\mathcal{C}$ that joins them. In other words, a gradient vector field is path-independent.

Proof Suppose $\mathcal{C}$ is a smooth curve. Let $t \mapsto \mathbf{r}(t)$, $a \le t \le b$ be any parameterization of $\mathcal{C}$. Then, using the Chain Rule, we see that

$$\begin{aligned}\int_{\mathcal{C}} \mathbf{F} \cdot d\mathbf{r} &= \int_a^b \mathbf{F}(\mathbf{r}(t)) \cdot \mathbf{r}'(t)\,dt = \int_a^b \nabla u(\mathbf{r}(t)) \cdot \mathbf{r}'(t)\,dt \\ &= \int_a^b \frac{d}{dt}(u(\mathbf{r}(t)))\,dt = u(\mathbf{r}(t))\Big|_a^b \\ &= u(\mathbf{r}(b)) - u(\mathbf{r}(a)) = u(P_1) - u(P_0).\end{aligned}$$

The line integral depends on P_0, P_1, and u alone, as asserted. This proves the theorem in the smooth case.

Now suppose $\mathcal{C}$ is a continuous, directed curve comprising two smooth pieces, $\mathcal{C}_1$ from P_0 to P' and $\mathcal{C}_2$ from P' to P_1. According to the smooth case we just proved,

$$\begin{aligned}\int_{\mathcal{C}} \mathbf{F} \cdot d\mathbf{r} &= \int_{\mathcal{C}_1} \mathbf{F} \cdot d\mathbf{r}_1 + \int_{\mathcal{C}_2} \mathbf{F} \cdot d\mathbf{r}_2 = (u(P') - u(P_0)) + (u(P_1) - u(P')) \\ &= u(P_1) - u(P_0).\end{aligned}$$

This proves the theorem in the case that the piecewise-smooth curve is made up of two smooth pieces. The case in which $\mathcal{C}$ consists of an arbitrary number of smooth pieces is proved in the same way. ∎

inSIGHT

It is worth noting the similarity between equation (15.23) and the Fundamental Theorem of Calculus. If $\mathbf{F}(x, y, z) = \nabla u(x, y, z)$, then u can be thought of as an antiderivative of the first component of $\mathbf{F}$ with respect to x, the second component of $\mathbf{F}$ with respect to y, and the third component of $\mathbf{F}$ with respect to z. Theorem 1 tells us that the line integral of $\mathbf{F}$ over $\mathcal{C}$ can be calculated by evaluating the "antiderivative" u at the endpoints of $\mathcal{C}$.

Recall from Section 15.1 that we say that a gradient vector field $\mathbf{F} = \nabla u$ is conservative and that $V = -u$ is a potential function for $\mathbf{F}$. Theorem 2 states that a conservative vector field is path-independent. An equivalent formulation is that a vector field with a potential function is path-independent.

Example 1 Consider the vector field $\mathbf{F}(x, y, z) = yz\mathbf{i} + xz\mathbf{j} + xy\mathbf{k}$. Show that $\mathbf{F}$ is path-independent. Calculate the line integral of $\mathbf{F}$ from $P_0 = (0, -1, 2)$ to $P_1 = (2, 1, 4)$.

Solution Observe that $\mathbf{F} = \nabla u$ where $u(x, y, z) = xyz$. Therefore, $\mathbf{F}$ is a conservative vector field. Equation (15.23) tells us that if $\mathcal{C}$ is any directed curve from P_0 to P_1, then

$$\int_{\mathcal{C}} \mathbf{F} \cdot d\mathbf{r} = u(P_1) - u(P_0) = (2)(1)(4) - (0)(-1)(2) = 8.$$

Notice that we have evaluated $\int_C \mathbf{F} \cdot d\mathbf{r}$ without actually having calculated a line integral. Let us verify our result by calculating $\int_C \mathbf{F} \cdot d\mathbf{r}$ as a line integral. For simplicity, we take C to be a straight line path joining P_0 to P_1. We can then parameterize C by

$$\mathbf{r}(t) = \overrightarrow{OP_0} + t\,\overrightarrow{P_0P_1} = \langle 0, -1, 2\rangle + t\langle 2, 2, 2\rangle = \langle 2t, -1+2t, 2+2t\rangle, \quad 0 \le t \le 1.$$

With these choices, we have

$$\mathbf{F}(\mathbf{r}(t)) = (-1+2t)(2+2t)\mathbf{i} + 2t(2+2t)\mathbf{j} + 2t(-1+2t)\mathbf{k} \quad \text{and} \quad \mathbf{r}'(t) = 2\mathbf{i} + 2\mathbf{j} + 2\mathbf{k}.$$

We calculate $\mathbf{F}(\mathbf{r}(t)) \cdot \mathbf{r}'(t) = (-1+2t)(2+2t)(2) + 2t(2+2t)(2) + 2t(-1+2t)(2) = -4 + 8t + 24t^2$. Therefore, the line integral of $\mathbf{F}$ from P_0 to P_1 is

$$\int_C \mathbf{F} \cdot d\mathbf{r} = \int_0^1 (-4 + 8t + 24t^2)\, dt = 8. \qquad ■$$

A Characterization of Path-Independent Vector Fields

It is a remarkable fact that the *only* continuous path-independent vector fields are the conservative ones. This is the content of the next theorem.

Theorem 3 Suppose that $\mathbf{F}$ is a continuous vector field on a region $\mathcal{G}$ in the plane or in space. If $\mathbf{F}$ is path-independent, then $\mathbf{F} = \nabla u$ for some continuously differentiable function u.

Proof Let $\mathbf{F} = \langle M, N, R\rangle$ be a continuous, path-independent vector field in a spatial region $\mathcal{G}$. (The planar case is proved in the same way.) Fix a point P_0 in $\mathcal{G}$. For each point $P = (x, y, z)$ in $\mathcal{G}$, set

$$u(x, y, z) = \int_{\mathcal{C}_P} \mathbf{F} \cdot d\mathbf{r}$$

where $\mathcal{C}_P$ is *any* curve that connects P_0 to P and that lies entirely in $\mathcal{G}$. Such a curve exists because $\mathcal{G}$ is a region (and therefore path-connected). Moreover, because $\mathbf{F}$ is path-independent, the value of $u(x, y, z)$ will be the same no matter what connecting path $\mathcal{C}_P$ we choose. This tells us that there is no ambiguity in the definition of $u(x, y, z)$. For sufficiently small Δx, we can choose the path $\mathcal{C}_{P'}$ from P_0 to $P' = (x + \Delta x, y, z)$ to be the piecewise-smooth curve that comprises $\mathcal{C}_P$ and the directed line segment $\overrightarrow{PP'}$ (Figure 2). We can parameterize this segment by $\tilde{\mathbf{r}}(t) = \langle x + t\Delta x, y, z\rangle$, $0 \le t \le 1$. Since $\tilde{\mathbf{r}}'(t) = \langle \Delta x, 0, 0\rangle$, it follows that $\mathbf{F}(\tilde{\mathbf{r}}(t)) \cdot \tilde{\mathbf{r}}'(t) = M(x + t\Delta x, y, z)\Delta x$ and

$$\int_{\overrightarrow{PP'}} \mathbf{F} \cdot d\tilde{\mathbf{r}} = \int_0^1 M(x + t\Delta x, y, z)\Delta x\, dt = \Delta x \int_0^1 M(x + t\Delta x, y, z)\, dt.$$

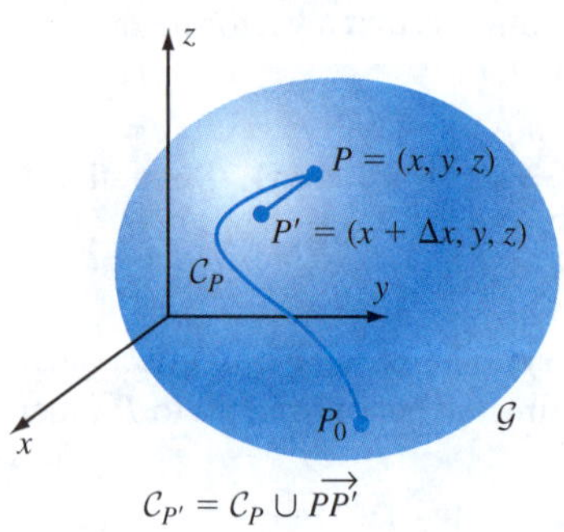

Figure 2

Thus,

$$u(x + \Delta x, y, z) - u(x, y, z) = \left(\int_{\mathcal{C}_P} \mathbf{F} \cdot d\mathbf{r} + \int_{\overrightarrow{PP'}} \mathbf{F} \cdot d\tilde{\mathbf{r}}\right) - \int_{\mathcal{C}_P} \mathbf{F} \cdot d\mathbf{r}$$

$$= \Delta x \int_0^1 M(x + t\Delta x, y, z)\, dt$$

and

$$u_x(x, y, z) = \lim_{\Delta x \to 0} \frac{u(x + \Delta x, y, z) - u(x, y, z)}{\Delta x} = \lim_{\Delta x \to 0} \int_0^1 M(x + t\Delta x, y, z)\, dt.$$

In advanced calculus courses, it is shown that this limit can be taken inside the integral:

$$\lim_{\Delta x \to 0} \int_0^1 M(x + t\Delta x, y, z)\, dt = \int_0^1 \lim_{\Delta x \to 0} M(x + t\Delta x, y, z)\, dt.$$

From this equality, it follows that

$$u_x(x, y, z) = \int_0^1 \lim_{\Delta x \to 0} M(x + t\Delta x, y, z)\, dt = \int_0^1 M(x, y, z)\, dt = M(x, y, z).$$

Similar arguments show that $u_y(x, y, z) = N(x, y, z)$ and $u_z(x, y, z) = R(x, y, z)$. Together these three equations show that $\mathbf{F} = \nabla u$. ■

Example 2 Let $\mathbf{F}$ be a path-independent force field on a region $\mathcal{G}$. If we move a particle through the field from one point P_0 in $\mathcal{G}$ to another point P_1 in $\mathcal{G}$, how does the work we do relate to the change in potential energy of the particle?

Solution To move a particle from P_0 to P_1 along a path $\mathcal{C}$, we must apply a force opposite to $\mathbf{F}$. The work W that we do is therefore given by $W = \int_{\mathcal{C}} (-\mathbf{F}) \cdot d\mathbf{r}$. According to Theorem 2, there is a scalar-valued function u for which $\mathbf{F} = \nabla u$. Theorem 1 tells us that

$$W = -\int_{\mathcal{C}} \mathbf{F} \cdot d\mathbf{r} = -(u(P_1) - u(P_0)) = V(P_1) - V(P_0) \qquad \textbf{(15.24)}$$

where $V = -u$. Since V is a potential function for $\mathbf{F}$, the potential energy at each point P of $\mathcal{G}$ may be defined to be $V(P)$, as discussed in Section 15.1. Observe that the change in potential energy between any two points in $\mathcal{G}$ does not depend on the choice of potential function that is used to define potential energy. Indeed, equation (15.24) tells us that *the work we do is equal to the change in the particle's potential energy*. If the particle has gained potential energy, then we have performed positive work. If the particle has lost potential energy, then we have performed negative work. ■

inSIGHT

If $P_1 = P_0$, then $V(P_1) = V(P_0)$ and $W = V(P_1) - V(P_0) = 0$. Example 2 shows that when a force field is conservative (hence, path-independent), zero work is performed when a particle is moved around a closed curve to the point at which the motion began.

Closed Vector Fields

Because it is not usually feasible to check path-independence directly, it is important to have a practical method for recognizing when a vector field is conservative. Our next theorem is a first step toward this goal: It provides us with an easily verifiable condition that will help us identify planar vector fields that are *not* conservative.

Theorem 4 Let $\mathbf{F}(x, y) = M(x, y)\mathbf{i} + N(x, y)\mathbf{j}$ be a continuously differentiable vector field in a region $\mathcal{G}$. If $\mathbf{F}$ is conservative, then

$$\frac{\partial M}{\partial y} = \frac{\partial N}{\partial x} \qquad \textbf{(15.25)}$$

at each point in $\mathcal{G}$.

Proof Since $\mathbf{F}$ is conservative and continuously differentiable, there is a twice continuously differentiable function u such that

$$M\mathbf{i} + N\mathbf{j} = \mathbf{F} = \nabla u = \frac{\partial u}{\partial x}\mathbf{i} + \frac{\partial u}{\partial y}\mathbf{j}.$$

In other words, $M = \frac{\partial u}{\partial x}$ and $N = \frac{\partial u}{\partial y}$. It follows that

$$\frac{\partial}{\partial y}M = \frac{\partial}{\partial y}\left(\frac{\partial}{\partial x}u\right) = \frac{\partial}{\partial x}\left(\frac{\partial}{\partial y}u\right) = \frac{\partial}{\partial x}N.$$

A vector field $\mathbf{F}(x, y) = M(x, y)\mathbf{i} + N(x, y)\mathbf{j}$ that satisfies equation (15.25) at each point of a region $\mathcal{G}$ is said to be *closed* on $\mathcal{G}$. Theorem 4 says that every conservative vector field is closed. A logically equivalent reformulation is that any vector field that is not closed is not conservative.

Example 3 Is the vector field $\mathbf{F}(x, y) = (x^2 - \sin(y))\mathbf{i} + (y^3 + \cos(x))\mathbf{j}$ conservative?

Solution Notice that

$$\frac{\partial}{\partial y}(x^2 - \sin(y)) = -\cos(y) \qquad \text{and} \qquad \frac{\partial}{\partial x}(y^3 + \cos(x)) = -\sin(x).$$

Since these expressions are unequal, Theorem 4 guarantees that $\mathbf{F}$ could not be a conservative vector field.

As Example 3 shows, Theorem 4 gives us a handy way to tell when a vector field is *not* conservative: If the vector field does not satisfy equation (15.25), then the vector field is not conservative. However, Theorem 4 does *not* say anything about a vector field that does satisfy equation (15.25). In particular, Theorem 4 does *not* assert that a closed vector field is conservative. Indeed, the next example shows that a closed vector field need not be conservative.

Example 4 Show that the vector field

$$\mathbf{F}(x, y) = \left(\frac{-y}{x^2 + y^2}\right)\mathbf{i} + \left(\frac{x}{x^2 + y^2}\right)\mathbf{j}$$

is closed but not conservative on the region $\mathcal{G} = \{(x, y) : (x, y) \neq (0, 0)\}$.

Solution Observe that $\mathbf{F}$ is defined and differentiable on $\mathcal{G} = \{(x, y) : (x, y) \neq (0, 0)\}$. To verify equation (15.25) on $\mathcal{G}$, we set $M(x, y) = -y/(x^2+y^2)$ and $N(x, y) = x/(x^2 + y^2)$, and we calculate

$$\frac{\partial M}{\partial y} = \frac{\partial}{\partial y}\left(\frac{-y}{x^2 + y^2}\right) = -\frac{x^2 - y^2}{(x^2 + y^2)^2} = \frac{\partial}{\partial x}\left(\frac{x}{x^2 + y^2}\right) = \frac{\partial N}{\partial x}.$$

Thus, $\mathbf{F}$ is closed on $\mathcal{G}$. To show that $\mathbf{F}$ is not conservative on $\mathcal{G}$, it suffices to observe that $\mathbf{F}$ is not path-independent. Set $P_0 = (1, 0)$ and $P_1 = (-1, 0)$, and consider two paths from P_0 to P_1: $\mathcal{C}$ parameterized by

$$\mathbf{r}(t) = \cos(t)\mathbf{i} + \sin(t)\mathbf{j}, \qquad 0 \le t \le \pi,$$

and $\tilde{C}$ parameterized by

$$\tilde{\mathbf{r}}(t) = \cos(t)\mathbf{i} - \sin(t)\mathbf{j}, \qquad 0 \le t \le \pi.$$

We calculate that

$$\int_C \mathbf{F} \cdot d\mathbf{r} = \int_0^{\pi} \mathbf{F}(\mathbf{r}(t)) \cdot \mathbf{r}'(t)\, dt = \int_0^{\pi} \left(-\frac{\sin(t)}{1}\mathbf{i} + \frac{\cos(t)}{1}\mathbf{j} \right) \cdot (-\sin(t)\mathbf{i} + \cos(t)\mathbf{j})\, dt$$
$$= \int_0^{\pi} (\sin^2(t) + \cos^2(t))\, dt = \pi.$$

However,

$$\int_{\tilde{C}} \mathbf{F} \cdot d\tilde{\mathbf{r}} = \int_0^{\pi} \mathbf{F}(\tilde{\mathbf{r}}(t)) \cdot \tilde{\mathbf{r}}'(t)\, dt = \int_0^{\pi} \left(\frac{\sin(t)}{1}\mathbf{i} + \frac{\cos(t)}{1}\mathbf{j} \right) \cdot (-\sin(t)\mathbf{i} - \cos(t)\mathbf{j})\, dt$$
$$= -\int_0^{\pi} (\sin^2(t) + \cos^2(t))\, dt = -\pi.$$

Our calculations show that $\mathbf{F}$ is not path-independent. We use Theorem 2 to conclude that $\mathbf{F}$ is not conservative. ■

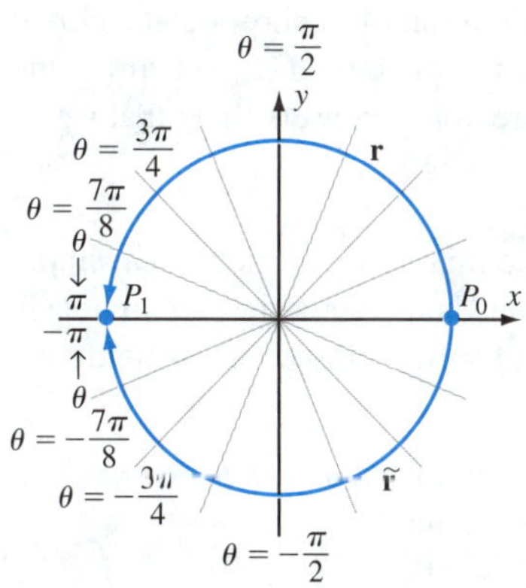

Figure 3

inSIGHT

Let $u(x, y)$ denote the polar coordinate $\theta \in (-\pi, \pi)$ of each point (x, y) that is not on the negative x-axis $\{(x, 0) : x \le 0\}$. For $x > 0$, we have $u(x, y) = \arctan(y/x)$ and $\nabla u(x, y) = \mathbf{F}(x, y)$, the vector field of Example 4. If $\mathbf{F}$ had a potential function on $\mathcal{G}$, then it would have to be of the form $u + C$ for some constant C. But, as we traverse the curve $\mathcal{C}$ from $P_0 = (1, 0)$ to $P_1 = (-1, 0)$, the polar angle θ increases from 0 to π. See Figure 3. On the other hand, as we traverse the curve $\tilde{\mathcal{C}}$ from $P_0 = (1, 0)$ to $P_1 = (-1, 0)$, the polar angle θ decreases from 0 to $-\pi$. Consequently, there is no continuous and unambiguous way to define a potential function for $\mathbf{F}$ on $\mathcal{G} = \{(x, y) : (x, y) \ne (0, 0)\}$.

Definition A planar or spatial region $\mathcal{G}$ is called *simply connected* if any closed curve in it can be continuously deformed to a point in $\mathcal{G}$ while staying within the region.

Roughly speaking, a planar region is simply connected if it has no holes in it. Figure 4 (next page) illustrates some regions that are *not* simply connected: Any closed curve that encircles a hole in one of these regions *cannot* be continuously deformed to a point while staying within the region. By contrast, the regions in Figure 5 (next page) *are* simply connected.

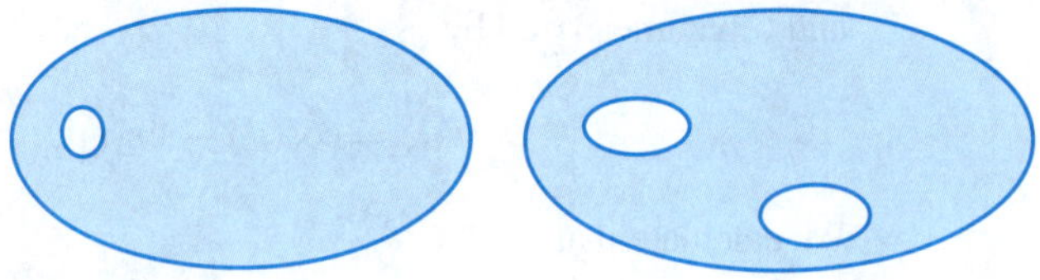

Figure 4
The two shaded regions are *not* simply connected.

Figure 5
The two shaded regions *are* simply connected.

If we remove one point from a simply connected planar region, then the resulting punctured region will not be simply connected. For instance, the region $\mathcal{G} = \{(x, y) : (x, y) \neq (0, 0)\}$ in Example 4 is not simply connected. Example 4 shows that a closed vector field need not be conservative on a region that is punctured. It is a surprising fact that a closed vector field on a simply connected region *is* conservative; that is, we have the following theorem.

Theorem 5 Let $\mathbf{F} = M(x, y)\mathbf{i} + N(x, y)\mathbf{j}$ be a continuously differentiable vector field on a simply connected planar region $\mathcal{G}$. If $\frac{\partial M}{\partial y} = \frac{\partial N}{\partial x}$ on $\mathcal{G}$, then there is a twice continuously differentiable function u on $\mathcal{G}$ such that $\nabla u = \mathbf{F}$. In short, a closed vector field on a simply connected region is conservative.

Proof We will outline the proof for a rectangle $\mathcal{G} = \{(x, y) : a < x < b, c < y < d\}$. For simplicity, we assume that the origin lies in $\mathcal{G}$. We define

$$u(x, y) = \int_0^x M(s, y)\,ds + \int_0^y N(0, t)\,dt.$$

Notice that because our domain $\mathcal{G}$ is a rectangle, we are integrating along paths that lie entirely in $\mathcal{G}$, and these integrals are guaranteed to make sense.

We assert that u satisfies $\nabla u = \mathbf{F}$. Let us see why. First, we compute u_x. The second expression in the formula for u does not even depend on x, so we have

$$u_x(x, y) = \frac{\partial}{\partial x} \int_0^x M(s, y)\,ds.$$

The Fundamental Theorem of Calculus tells us that this last expression is $M(x, y)$, which is just what we want. Next, we calculate u_y. We have

$$u_y(x, y) = \frac{\partial}{\partial y} \int_0^x M(s, y)\,ds + \frac{\partial}{\partial y} \int_0^y N(0, t)\,dt = \frac{\partial}{\partial y} \int_0^x M(s, y)\,ds + N(0, y).$$

It is a deep but plausible result from advanced calculus that we may pass the differentiation under the integral sign to obtain

$$u_y(x, y) = \int_0^x \frac{\partial M}{\partial y}(s, y)\, ds + N(0, y).$$

Now we use the hypothesis that $\frac{\partial M}{\partial y} = \frac{\partial N}{\partial x}$ to obtain

$$u_y(x, y) = \int_0^x \frac{\partial N}{\partial x}(s, y)\, ds + N(0, y).$$

By the Fundamental Theorem of Calculus, we obtain

$$u_y(x, y) = (N(x, y) - N(0, y)) + N(0, y) = N(x, y),$$

as desired. ■

Example 5 Let $M(x, y) = \sin(y) + y\sin(x)$ and $N(x, y) = x\cos(y) - \cos(x) + 1$. Is the vector field $\mathbf{F}(x, y) = M(x, y)\mathbf{i} + N(x, y)\mathbf{j}$ conservative on the rectangle $\mathcal{G} = \{(x, y) : -1 < x < 1, -1 < y < 1\}$? If it is, then find a function u such that $\mathbf{F} = \nabla u$.

Solution We notice that $\mathcal{G}$ is simply connected and

$$\frac{\partial M}{\partial y}(x, y) = \cos(y) + \sin(x) = \frac{\partial N}{\partial x}(x, y).$$

Our vector field $\mathbf{F}$ passes the test: It *is* closed on $\mathcal{G}$. By Theorem 5, $\mathbf{F}$ is conservative.

Let us find a function u on $\mathcal{G}$ whose gradient is $\mathbf{F}$. Proceeding as in Example 9 from Section 15.1, we first study the equation

$$\frac{\partial u}{\partial x} = M(x, y) = \sin(y) + y\sin(x).$$

This equation tells us that u is an antiderivative with respect to x of the right side. Therefore, $u(x, y) = x\sin(y) - y\cos(x) + \phi(y)$ where the "constant of integration" $\phi(y)$ is independent of x but possibly dependent on y. We differentiate this formula for u, with respect to y, obtaining

$$\frac{\partial u}{\partial y} = x\cos(y) - \cos(x) + \phi'(y).$$

Since $\nabla u = \mathbf{F}$, we also know that

$$\frac{\partial u}{\partial y} = N(x, y) = x\cos(y) - \cos(x) + 1.$$

On equating our two expressions for $\frac{\partial u}{\partial y}$, we obtain $\phi'(y) = 1$. It follows that $\phi(y) = y + C$ where C is a numerical constant. Putting this information together yields $u(x, y) = x\sin(y) - y\cos(x) + y + C$. ■

Vector Fields in Space

Theorems 4 and 5 pertain to planar vector fields, but they have analogues that apply to vector fields in space. Our next theorem states a necessary condition for a spatial vector field to be conservative.

Theorem 6 If $\mathbf{F}(x, y, z) = M(x, y, z)\mathbf{i} + N(x, y, z)\mathbf{j} + R(x, y, z)\mathbf{k}$ is a continuously differentiable conservative vector field on a region $\mathcal{G}$, then all three of the following equalities hold on $\mathcal{G}$:

$$\frac{\partial M}{\partial y} = \frac{\partial N}{\partial x}, \qquad \frac{\partial M}{\partial z} = \frac{\partial R}{\partial x}, \qquad \frac{\partial N}{\partial z} = \frac{\partial R}{\partial y}. \tag{15.26}$$

To understand what Theorem 6 says, let us put it in words: If $\mathbf{F}(x, y, z)$ is conservative, then the derivative of the first component of $\mathbf{F}$ with respect to the second variable equals the derivative of the second component of $\mathbf{F}$ with respect to the first variable; the derivative of the first component of $\mathbf{F}$ with respect to the third variable equals the derivative of the third component of $\mathbf{F}$ with respect to the first variable; and the derivative of the second component of $\mathbf{F}$ with respect to the third variable equals the derivative of the third component of $\mathbf{F}$ with respect to the second variable. A vector field $\mathbf{F}(x, y, z) = M(x, y, z)\mathbf{i} + N(x, y, z)\mathbf{j} + R(x, y, z)\mathbf{k}$ that satisfies the three equations of line (15.26) at every point of a region $\mathcal{G}$ is said to be a *closed* vector field on $\mathcal{G}$. Theorem 6 states that every conservative vector field on a spatial region is closed. Equivalently, a vector field that is not closed is not conservative.

inSIGHT

Whereas a planar vector field must satisfy only one equation to be conservative, a spatial vector field must satisfy three. The vector field $\mathbf{F}(x, y, z) = M(x, y, z)\mathbf{i} + N(x, y, z)\mathbf{j} + R(x, y, z)\mathbf{k}$ of Example 6 *does* satisfy two of the three necessary equations to be conservative, namely, $M_y = 0 = N_x$ and $N_z = y = R_y$, but that *does not* suffice.

Example 6 Is the vector field in space given by $\mathbf{F}(x, y, z) = xz\mathbf{i} + yz\mathbf{j} + (y^2/2)\mathbf{k}$ a conservative vector field?

Solution Let us write $\mathbf{F}(x, y, z) = M(x, y, z)\mathbf{i} + N(x, y, z)\mathbf{j} + R(x, y, z)\mathbf{k}$ where $M(x, y, z) = xz$, $N(x, y, z) = yz$, and $R(x, y, z) = y^2/2$. We notice that the partial derivative of the first component of $\mathbf{F}$ with respect to the third variable z does not equal the partial derivative of the third component of $\mathbf{F}$ with respect to the first variable x:

$$\frac{\partial M}{\partial z} = x \neq 0 = \frac{\partial R}{\partial x}.$$

Because one of the equations of line (15.26) is not satisfied, the vector field $\mathbf{F}$ cannot be conservative. ■

Theorem 6 has a converse that will work on geometrically simple regions.

Theorem 7 Suppose that $\mathbf{F}(x, y, z) = M(x, y, z)\mathbf{i} + N(x, y, z)\mathbf{j} + R(x, y, z)\mathbf{k}$ is a continuously differentiable vector field on a simply connected region $\mathcal{G}$. If $\mathbf{F}$ satisfies all three equations of line (15.26), then there is a twice continuously differentiable function u on $\mathcal{G}$ such that $\nabla u = \mathbf{F}$. In short, a closed vector field on a simply connected region is conservative.

Example 7 Is the vector field $\mathbf{F}(x, y, z) = yz^2\mathbf{i} + (xz^2 - z)\mathbf{j} + (2xyz - y)\mathbf{k}$ conservative on the box $\mathcal{G} = \{(x, y, z) : 0 < x < 2, 0 < y < 2, 0 < z < 2\}$? If it is, find a function u for which $\mathbf{F} = \nabla u$.

Solution First we check whether **F** is closed. Since

$$\frac{\partial}{\partial y}(yz^2) = z^2 = \frac{\partial}{\partial x}(xz^2 - z), \qquad \frac{\partial}{\partial z}(yz^2) = 2yz = \frac{\partial}{\partial x}(2xyz - y),$$

and

$$\frac{\partial}{\partial z}(xz^2 - z) = 2xz - 1 = \frac{\partial}{\partial y}(2xyz - y),$$

we conclude that **F** is closed. By Theorem 7, it is conservative. Thus, $\mathbf{F} = \nabla u$ for some function u. Let us find u. First, u is the antiderivative in the x-variable of yz^2. It follows that

$$u(x, y, z) = xyz^2 + \phi(y, z). \qquad \textbf{(15.27)}$$

Here the "constant of integration" does not depend on x, but it can depend on y and z. We now differentiate each side of equation (15.27) with respect to y, treating x and z as constants:

$$\frac{\partial u}{\partial y} = xz^2 + \frac{\partial}{\partial y}\phi(y, z).$$

But the equality of the second components of the vector equation $\nabla u = \mathbf{F}$ also gives us

$$\frac{\partial u}{\partial y} = xz^2 - z.$$

We conclude that

$$xz^2 + \frac{\partial}{\partial y}\phi(y, z) = xz^2 - z,$$

or equivalently,

$$\frac{\partial}{\partial y}\phi(y, z) = -z.$$

It follows that

$$\phi(y, z) = -yz + \psi(z)$$

where our "constant of integration" is now independent of x and y but may depend on z. If we substitute our formula for $\phi(y, z)$ into equation (15.27), we obtain

$$u(x, y, z) = xyz^2 - yz + \psi(z). \qquad \textbf{(15.28)}$$

Finally, when we differentiate each side of equation (15.28) with respect to z, we obtain

$$\frac{\partial u}{\partial z} = 2xyz - y + \psi'(z).$$

But the equality of the third components of the vector equation $\nabla u = \mathbf{F}$ tells us that

$$\frac{\partial u}{\partial z} = 2xyz - y.$$

By equating the two expressions for $\frac{\partial u}{\partial z}$, we see that $\psi'(z) = 0$. We conclude that $\psi(z) = C$ for some numerical constant C. Substituting this into equation (15.28), we arrive at our final formula for u: $u(x, y, z) = xyz^2 - yz + C$. ■

Summary of Principal Ideas

Let us summarize the key ideas that we have learned so far in this section. Let $\mathbf{F}$ be a continuous vector field either on a planar or spatial region $\mathcal{G}$. Then the following three properties are equivalent—if $\mathbf{F}$ possesses any one of them, then $\mathbf{F}$ possesses all three:

1. $\oint_C \mathbf{F} \cdot d\mathbf{r} = 0$ for every closed curve $\mathcal{C}$ in $\mathcal{G}$.
2. $\mathbf{F}$ is conservative: $\mathbf{F} = \nabla u$ for some u.
3. $\mathbf{F}$ is path-independent: $\int_{C_1} \mathbf{F} \cdot d\mathbf{r} = \int_{C_2} \mathbf{F} \cdot d\mathbf{r}$ for any two directed curves in $\mathcal{G}$ that share the same initial point and the same terminal point.

If $\mathbf{F}$ possesses any one (and hence all) of Properties 1, 2, or 3, then $\mathbf{F}$ also possesses the following property:

4. $\mathbf{F}$ is closed: $\mathbf{F}$ satisfies equation (15.25) if planar; it satisfies the three equations of line (15.26) if spatial.

If $\mathcal{G}$ is simply connected and $\mathbf{F}$ possesses Property 4, then $\mathbf{F}$ possesses any one (and hence all) of Properties 1, 2, and 3. (The schematic diagram of implications and equivalences in Figure 6 may help you organize these relationships.)

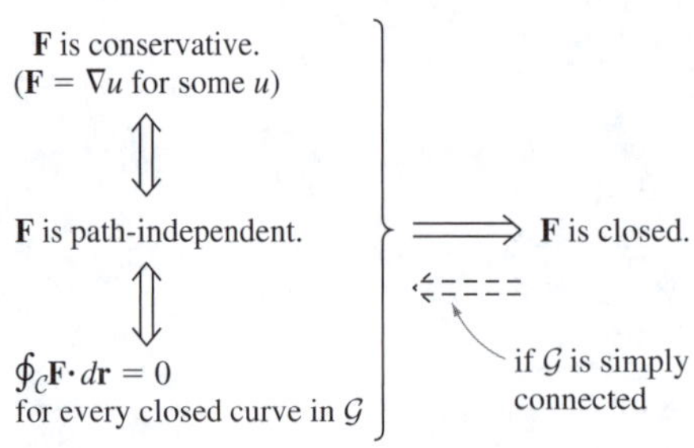

Figure 6

quickquiz

1. What is the definition of a closed vector field in two dimensions?
2. What is the definition of a closed spatial vector field?
3. Is a conservative vector field necessarily closed?
4. What geometrical restriction must be put on a region $\mathcal{G}$ to ensure that closed vector fields on $\mathcal{G}$ are conservative?

EXERCISES

Problems for Practice

Determine which of the vector fields in Exercises 1–12 is closed on the unit square $Q = \{(x, y) : -1 < x < 1, -1 < y < 1\}$. If the vector field is closed, then find a potential function.

1. $\mathbf{F}(x, y) = (x + \pi)\mathbf{i} + \mathbf{j}$
2. $\mathbf{F}(x, y) = (x^2 - y^2)\mathbf{i} + (2xy - y^3)\mathbf{j}$
3. $\mathbf{F}(x, y) = y^3\mathbf{i} + (3y^2x - 7y^6)\mathbf{j}$
4. $\mathbf{F}(x, y) = 2x\ln(y + 2)\mathbf{i} - (x^2/(y + 2))\mathbf{j}$
5. $\mathbf{F}(x, y) = (y^3 - xy^2)\mathbf{i} - (x^2y^2 + x)\mathbf{j}$
6. $\mathbf{F}(x, y) = (y/(x + 1)^2)\mathbf{i} + (x/(x + 1))\mathbf{j}$
7. $\mathbf{F}(x, y) = (x - 2)^{-2}\mathbf{i} + (y - 1/2)^2\mathbf{j}$
8. $\mathbf{F}(x, y) = x\cos(xy)\mathbf{i} + y\sin(xy)\mathbf{j}$
9. $\mathbf{F}(x, y) = 2x\sin(x^2 - y)\mathbf{i} - \sin(x^2 - y)\mathbf{j}$
10. $\mathbf{F}(x, y) = (x/(1 + x^2 + y^2))\mathbf{i} + (y/(1 + x^2 + y^2))\mathbf{j}$
11. $\mathbf{F}(x, y) = (-y/(x^2 + y^2))\mathbf{i} + (x/(x^2 + y^2))\mathbf{j}$
12. $\mathbf{F}(x, y) = (y/(3 + x) - 2\ln(3 + y))\mathbf{i} + (\ln(3 + x) - 2x/(3 + y))\mathbf{j}$

Determine which of the vector fields in Exercises 13–22 is closed on the unit cube $Q = \{(x, y, z) : -1 < x, y, z < 1\}$. If the vector field is closed, then find a potential function.

13. $\mathbf{F}(x, y, z) = (yz + 1)\mathbf{i} + (xz + 2)\mathbf{j} + (xy + 3)\mathbf{k}$
14. $\mathbf{F}(x, y, z) = z^3\mathbf{i} - 2yz\mathbf{j} + (3xz^2 - y^2)\mathbf{k}$
15. $\mathbf{F}(x, y, z) = z/(2 + x + y)(\mathbf{i} + \mathbf{j}) + (y + \ln(2 + x + y))\mathbf{k}$
16. $\mathbf{F}(x, y, z) = 3x^2y^2z^3\mathbf{i} + 2x^3yz^3\mathbf{j} + 3x^3y^2z\mathbf{k}$
17. $\mathbf{F}(x, y, z) = -z^2/(x + y - 4)^2(\mathbf{i} + \mathbf{j}) + 2z/(x + y - 4)\mathbf{k}$
18. $\mathbf{F}(x, y, z) = (3 + y^2)\mathbf{i} + (2xy + z^3)\mathbf{j} + (3yz^2 + 8z)\mathbf{k}$
19. $\mathbf{F}(x, y, z) = x\cos(z)\mathbf{i} - z\sin(y)\mathbf{j} + y\sin(x)\mathbf{k}$
20. $\mathbf{F}(x, y, z) = z\sin(y)\mathbf{i} + xz\cos(y)\mathbf{j} + z^2\sin(y)\mathbf{k}$
21. $\mathbf{F}(x, y, z) = z\sec^2(x + y)\mathbf{i} + z\sec^2(x + y)\mathbf{j} + \tan(x + y)\mathbf{k}$
22. $\mathbf{F}(x, y, z) = (xy + z + 2)^{-1}(y\mathbf{i} + x\mathbf{j} + \mathbf{k})$

In Exercises 23–28, a vector field $\mathbf{F}$ is given. Only the initial point P_0 and terminal point P_1 of a directed curve $\mathcal{C}$ are specified. Verify that $\mathbf{F}$ is closed in a simply connected region that contains P_0 and P_1. Then calculate the line integral $\int_{\mathcal{C}} \mathbf{F} \cdot d\mathbf{r}$.

23. $\mathbf{F}(x, y) = 5/(5 + xy)^2\mathbf{i} - x^2/(5 + xy)^2\mathbf{j}$, $P_0 = (1, -1)$, $P_1 = (2, 2)$
24. $\mathbf{F}(x, y) = (1 + \ln(xy))\mathbf{i} + (1 + x/y)\mathbf{j}$, $P_0 = (1, 1)$, $P_1 = (\sqrt{2}, \sqrt{2})$
25. $\mathbf{F}(x, y) = (y^2 + y\exp(xy))\mathbf{i} + (2xy + x\exp(xy))\mathbf{j}$, $P_0 = (0, 0)$, $P_1 = (-1, 1)$
26. $\mathbf{F}(x, y, z) = (y^2 + z^3)\mathbf{i} + (2xy + 2yz^3)\mathbf{j} + (3y^2z^2 + 3xz^2)\mathbf{k}$, $P_0 = (1, -1, -1)$, $P_1 = (-1, 1, 0)$
27. $\mathbf{F}(x, y, z) = \exp(y + 2z))\mathbf{i} + x\exp(y + 2z)\mathbf{j} + 2x\exp(y + 2z)\mathbf{k}$, $P_0 = (0, 0, 0)$, $P_1 = (-1, 2, 1)$
28. $\mathbf{F}(x, y, z) = (3x^2z + 1/x)\mathbf{i} + (1/z - 1/y)\mathbf{j} + (x^3 - y/z^2)\mathbf{k}$, $P_0 = (1, 1, 1)$, $P_1 = (4, 2, 2)$

In Exercises 29–32, a conservative vector field $\mathbf{F}$ and three points P_0, P_1, and P_2 are given. Verify that

$$\int_{\overrightarrow{P_0P_1}} \mathbf{F} \cdot d\mathbf{r} + \int_{\overrightarrow{P_1P_2}} \mathbf{F} \cdot d\mathbf{r} + \int_{\overrightarrow{P_2P_0}} \mathbf{F} \cdot d\mathbf{r} = 0.$$

29. $\mathbf{F}(x, y) = (y^2 + 2y)\mathbf{i} + (2xy + 2x)\mathbf{j}$, $P_0 = (0, 0)$, $P_1 = (0, 1)$, $P_2 = (1, 1)$
30. $\mathbf{F}(x, y) = (2x - y)\mathbf{i} - x\mathbf{j}$, $P_0 = (1, 1)$, $P_1 = (3, 4)$, $P_2 = (2, 5)$
31. $\mathbf{F}(x, y, z) = 2xy\mathbf{i} + x^2\mathbf{j} + 2z\mathbf{k}$, $P_0 = (0, 0, 1)$, $P_1 = (0, 1, 1)$, $P_2 = (1, 1, 1)$
32. $\mathbf{F}(x, y, z) = yz^2\mathbf{i} + xz^2\mathbf{j} + 2xyz\mathbf{k}$, $P_0 = (0, 0, 0)$, $P_1 = (-1, 2, 1)$, $P_2 = (0, 2, 0)$

Further Theory and Practice

33. If $\mathbf{F}(x, y, z) = yz\mathbf{i} + xz\mathbf{j} + xy\mathbf{k}$ and if $\mathcal{C}$ is any directed curve from $P_0 = (0, -1, 2)$ to $P_1 = (2, 1, 4)$, then $\int_{\mathcal{C}} \mathbf{F} \cdot d\mathbf{r} = 8$, as shown in Example 1. Verify this equation for the parameterized curve $\mathbf{r}(t) = \langle 12t^2 - 10t, 8t^3 - 8t^2 + 2t - 1, 4t^2 - 2t + 2 \rangle)$, $0 \leq t \leq 1$.

The gravitational force that a point mass m at the origin O exerts on a unit point mass at $P = (x, y, z) \neq O$ is $\mathbf{F} = -(Gmr^{-3})\overrightarrow{OP}$ where $r = \|\overrightarrow{OP}\|$. In Exercises 34 and 35, let $P_0 = (x_0, y_0, z_0) \neq O$ be a fixed point, and let V be a fixed choice of potential function for $\mathbf{F}$. Let W_P be the work done in moving a unit mass from P_0 to $P = (x, y, z) \neq O$.

34. Calculate the work W_P. For what points Q is $W_Q = W_P$?
35. Show that $W_\infty = \lim_{P\to\infty} W_P$ exists in the following sense: For any $\epsilon > 0$, there is an R such that $|W_P - W_\infty| < \epsilon$ for all P with $\|\overrightarrow{OP}\| > R$. (Frequently in physics, a potential function is chosen so that W_∞ has a convenient value for a point P_0 of interest.)
36. Consider the square $Q = \{(x, y) : |x| < 1, |y| < 1\}$. Let $L_c = \{(0, y) : y > c\}$, $\mathcal{G} = \{(x, y) \in Q : (x, y) \notin L_c\}$. For what values of c is the region $\mathcal{G}$ simply connected?

37. Consider the square $Q = \{(x, y) : |x| < 1, |y| < 1\}$. Let $L_{c,d} = \{(0, y) : c < y < d\}$, $\mathcal{G} = \{(x, y) \in Q : (x, y) \notin L_{c,d}\}$. Describe the pairs (c, d) for which the region $\mathcal{G}$ is simply connected.

38. Is the solid $\mathcal{U} = \{(x, y, z) : 1 < x^2 + y^2 + z^2 < 4\}$ simply connected? What about the solid $\mathcal{V} = \{(x, y, z) : 1 < x^2 + y^2, x^2 + y^2 + z^2 < 4\}$?

39. Sketch two simply connected spatial regions, the intersection of which is not a simply connected region.

40. Verify that

$$\mathbf{F}(x, y) = \frac{2}{\sqrt{\pi}} \exp(-(x - y)^2)(\mathbf{i} - \mathbf{j})$$

is conservative. Calculate $\int_{\mathcal{C}} \mathbf{F} \cdot d\mathbf{r}$ for a curve that joins $(0, 0)$ to $(1, 1)$.

Calculator/Computer Exercises

41. Verify that $\mathbf{F}(x, y) = \sin(\pi x y^2)(x^{-1}\,\mathbf{i} + 2y^{-1}\mathbf{j})$ is conservative in the first quadrant. Calculate $\int_{\mathcal{C}} \mathbf{F} \cdot d\mathbf{r}$ on a directed curve $\mathcal{C}$ from $(1, 1/2)$ to $(1/2, 1)$.

42. Verify that $\mathbf{F}(x, y) = 2\pi^{-1/2} \exp(-(x - y)^2)(\mathbf{i} - \mathbf{j})$ is conservative, and calculate $\int_{\mathcal{C}} \mathbf{F} \cdot d\mathbf{r}$ on a directed curve $\mathcal{C}$ from $(0, 0)$ to $(1, 2)$.

15.4 Divergence, Gradient, and Curl

In this section, we will learn about the differential operators that we will need to formulate the fundamental theorems of vector analysis. Each of the constructs we will see has an interesting physical interpretation.

Divergence of a Vector Field

Suppose that $\mathbf{F}(x, y) = M(x, y)\mathbf{i} + N(x, y)\mathbf{j}$ is a differentiable vector field defined on a region $\mathcal{G}$ of the plane. The scalar-valued function $(x, y) \mapsto M_x(x, y) + N_y(x, y)$ is said to be the *divergence* of $\mathbf{F}$. We denote this function by $\mathrm{div}(\mathbf{F})$:

$$\mathrm{div}(\mathbf{F}) = \frac{\partial M}{\partial x} + \frac{\partial N}{\partial y}.$$

Similarly, if $\mathbf{F}(x, y, z) = M(x, y, z)\mathbf{i} + N(x, y, z)\mathbf{j} + R(x, y, z)\mathbf{k}$ is a differentiable vector field on a region $\mathcal{G}$ in space, then the *divergence* of $\mathbf{F}$ is defined to be

$$\mathrm{div}(\mathbf{F}) = \frac{\partial M}{\partial x} + \frac{\partial N}{\partial y} + \frac{\partial R}{\partial z}. \tag{15.29}$$

If $\mathrm{div}(\mathbf{F}) = 0$ at every point of $\mathcal{G}$, then we say that $\mathbf{F}$ is *solenoidal* on $\mathcal{G}$.

Example 1 Let $\mathbf{F}(x, y, z) = xy\mathbf{i} + yz\mathbf{j} - 2xz\mathbf{k}$ and $\mathbf{G}(x, y, z) = 2xy\mathbf{i} + (z - y^2)\mathbf{j} + xy\mathbf{k}$. Calculate $\mathrm{div}(\mathbf{F})$ and $\mathrm{div}(\mathbf{G})$. Is either vector field solenoidal on $\mathbb{R}^3$?

Solution We have

$$\mathrm{div}(\mathbf{F})(x, y, z) = \frac{\partial}{\partial x}(xy) + \frac{\partial}{\partial y}(yz) + \frac{\partial}{\partial z}(-2xz) = y + z - 2x$$

and

$$\mathrm{div}(\mathbf{G})(x, y, z) = \frac{\partial}{\partial x}(2xy) + \frac{\partial}{\partial y}(z - y^2) + \frac{\partial}{\partial z}(xy) = 2y - 2y + 0 = 0.$$

Thus, $\mathbf{G}$ is solenoidal on $\mathbb{R}^3$. Although $\operatorname{div}(\mathbf{F})(x, y, z) = 0$ at each point of the plane $y + z - 2x = 0$, these are the only points at which $\operatorname{div}(\mathbf{F})$ vanishes. Consequently, $\mathbf{F}$ is not solenoidal on $\mathbb{R}^3$. ■

What is the significance of the divergence? Imagine that $\mathbf{F}$ represents the velocity of a fluid flow. Then $\operatorname{div}(\mathbf{F})$ sums the rate of change of the $\mathbf{i}$ component of $\mathbf{F}$ in the $\mathbf{i}$ direction, the rate of change of the $\mathbf{j}$ component of $\mathbf{F}$ in the $\mathbf{j}$ direction, and the rate of change of the $\mathbf{k}$ component of $\mathbf{F}$ in the $\mathbf{k}$ direction. If, at a point $P = (x, y, z)$, the divergence is positive, then the fluid is tending to flow outward from P. If $\operatorname{div}(\mathbf{F})$ is negative at P, then the fluid is tending to flow inward at P. Thus, the divergence measures the tendency of a vector field to expand or contract. These ideas are illustrated in the next two examples.

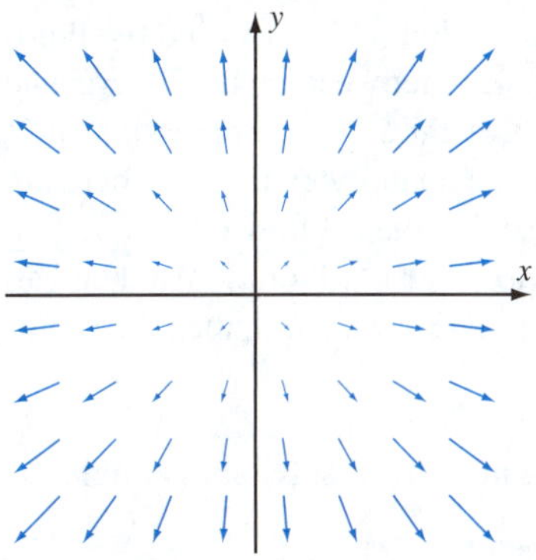

$\mathbf{F}(x, y) = x\mathbf{i} + y\mathbf{j}$

Figure 1

Example 2 Define $\mathbf{F}(x, y) = x\mathbf{i} + y\mathbf{j}$, $\mathbf{G}(x, y) = -x\mathbf{i} - y^3\mathbf{j}$, and $\mathbf{H}(x, y) = x^2\mathbf{i} - y^2\mathbf{j}$. Calculate the divergence of each vector field at the origin, and relate your answer to the physical properties of the fluid flow that the vector field represents.

Solution We begin with $\mathbf{F}$. Look at Figure 1. The flow at the origin seems to be outward from 0. We calculate that $\operatorname{div}(\mathbf{F})(0, 0) = 1 + 1 = 2$, which, being positive, is the indication of an outward flow, as expected. Next, we calculate that $\operatorname{div}(\mathbf{G})(x, y) = -1 - 3y^2$. Therefore, $\operatorname{div}(\mathbf{G})(0, 0) = -1$, which, being negative, is the indication of an inward flow. Indeed, Figure 2 shows that the flow of $\mathbf{G}$ at the origin is inward. Finally, $\operatorname{div}(\mathbf{H})(x, y) = 2x - 2y = 2(x - y)$, which is 0 at the origin. The (net) flow at the origin is neither outward nor inward, as the plot of vector field $\mathbf{H}$ suggests (Figure 3). ■

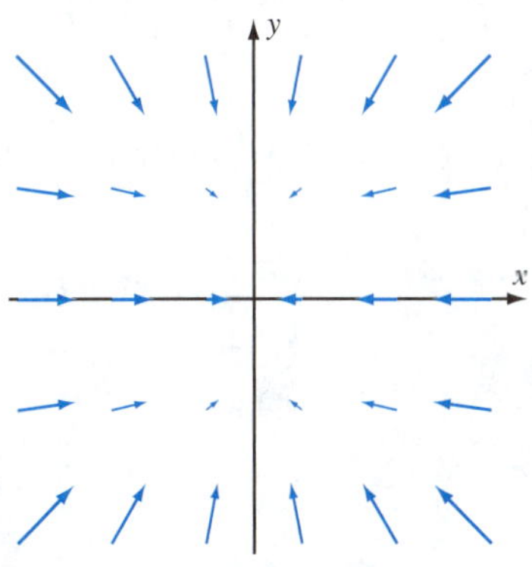

$\mathbf{G}(x, y) = -x\mathbf{i} - y^3\mathbf{j}$

Figure 2

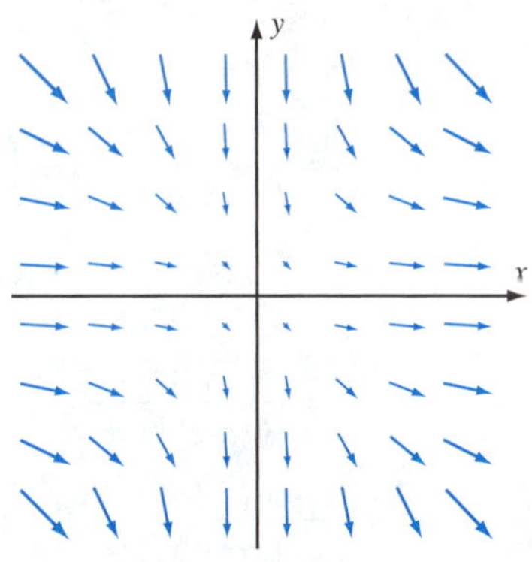

$\mathbf{H}(x, y) = x^2\mathbf{i} - y^2\mathbf{j}$

Figure 3

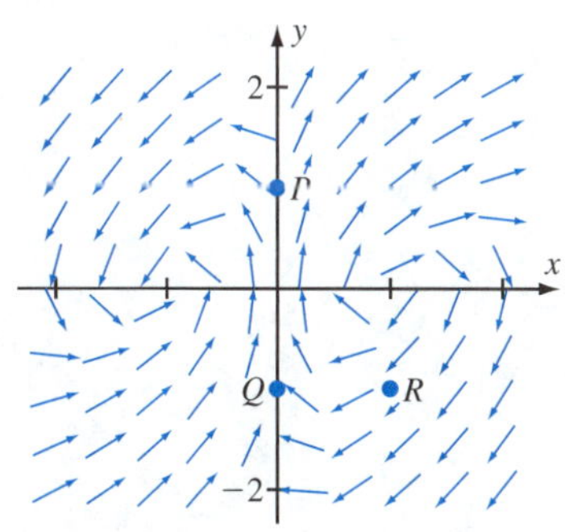

Figure 4

Example 3 Define $\mathbf{F}(x, y) = 2xy\mathbf{i} + (1 + 2xy - x^2)\mathbf{j}$. Figure 4 represents a sketch of the vector field $(x, y) \mapsto (1/2)\,\mathbf{dir}(\mathbf{F}(x, y))$, which shows us the directions of $\mathbf{F}$ scaled to a convenient size. The points $P = (0, 1)$, $Q = (0, -1)$, and $R = (1, -1)$ are also plotted. Use Figure 4 to determine the sign of $\operatorname{div}(\mathbf{F})(P)$ and $\operatorname{div}(\mathbf{F})(Q)$. Confirm your deductions by calculating these values exactly. What can you say about $\operatorname{div}(\mathbf{F})(R)$?

Solution Because the flow at P is outward, we deduce that $\operatorname{div}(\mathbf{F})(P) > 0$. Similarly, the inward flow at Q tells us that $\operatorname{div}(\mathbf{F})(Q) < 0$. However, the figure is less revealing at R (and nearby points). Although the net flow does not appear to be strongly inward or strongly outward at such points, the visual evidence is not sufficiently accurate to determine whether the divergence is exactly 0 or merely a small positive or small negative value. To be sure, we calculate

$$\operatorname{div}(\mathbf{F})(x, y) = \frac{\partial}{\partial x}(2xy) + \frac{\partial}{\partial y}(1 + 2xy - x^2) = 2y + 2x.$$

Thus, $\operatorname{div}(\mathbf{F})(P) = 2 > 0$ and $\operatorname{div}(\mathbf{F})(Q) = -2 < 0$, confirming our deductions. We also discover that $\operatorname{div}(\mathbf{F})(R) = 0$. The points just above R have a small positive divergence, and the points just below have a small negative divergence. ■

Many terms related to divergence have been coined to describe fluid flow. Suppose the vector field $\mathbf{v}$ represents the velocity of a fluid in a region $\mathcal{G}$. Then $\mathbf{v}$ (and the fluid) is called *incompressible* if $\operatorname{div}(\mathbf{v}) = 0$ at each point of $\mathcal{G}$. After we learn the Divergence Theorem in Section 15.8, we will understand that if a fluid is incompressible in a region $\mathcal{G}$, then the quantity of fluid entering $\mathcal{G}$ equals the quantity exiting $\mathcal{G}$. There are circumstances in which the fluid can be compressed so that the net flow into the region exceeds the net flow out. This situation corresponds to $\operatorname{div}(\mathbf{F}) < 0$ on $\mathcal{G}$. We then call the vector field (and the fluid) *compressible*. Finally, if $\operatorname{div}(\mathbf{F}) > 0$ on $\mathcal{G}$, then the vector field and the fluid are called *expanding*.

Example 4 Suppose that a point mass at the origin exerts a gravitational field

$$\mathbf{F}(x, y, z) = \frac{\lambda}{\|\mathbf{r}\|^3}\mathbf{r}, \qquad (x, y, z) \neq (0, 0, 0)$$

where λ is a proportionality constant and $\mathbf{r}$ is the position vector $x\mathbf{i} + y\mathbf{j} + z\mathbf{k}$ of the point (x, y, z). Show that $\operatorname{div}(\mathbf{F}) = 0$ at each point of space other than the origin.

Solution We calculate

$$\begin{aligned}
\frac{\partial}{\partial x}\frac{x}{(x^2+y^2+z^2)^{3/2}} &= \frac{1}{(x^2+y^2+z^2)^3}\left((x^2+y^2+z^2)^{3/2} - x\frac{\partial}{\partial x}(x^2+y^2+z^2)^{3/2}\right) \\
&= \frac{1}{(x^2+y^2+z^2)^3}((x^2+y^2+z^2)^{3/2} - 3x^2(x^2+y^2+z^2)^{1/2}) \\
&= \frac{(x^2+y^2+z^2)^{1/2}}{(x^2+y^2+z^2)^3}((x^2+y^2+z^2) - 3x^2) \\
&= \frac{(y^2+z^2-2x^2)}{(x^2+y^2+z^2)^{5/2}}.
\end{aligned}$$

By symmetry, we deduce that

$$\begin{aligned}
\operatorname{div}(\mathbf{F}) &= (\lambda)\left(\frac{(y^2+z^2-2x^2)}{(x^2+y^2+z^2)^{5/2}} + \frac{(x^2+z^2-2y^2)}{(x^2+y^2+z^2)^{5/2}} + \frac{(x^2+y^2-2z^2)}{(x^2+y^2+z^2)^{5/2}}\right) \\
&= (\lambda)\left(\frac{0}{(x^2+y^2+z^2)^{5/2}}\right) = 0.
\end{aligned}$$

■

The Curl of a Vector Field

If $\mathbf{F}(x, y) = M(x, y)\mathbf{i} + N(x, y)\mathbf{j}$ is a vector field in the plane, then we define its *curl* to be the *spatial* vector field

$$\mathbf{curl}(\mathbf{F})(x, y) = \left(\frac{\partial N}{\partial x} - \frac{\partial M}{\partial y}\right)\mathbf{k}. \tag{15.30}$$

If $\mathbf{F}(x, y, z) = M(x, y, z)\mathbf{i} + N(x, y, z)\mathbf{j} + R(x, y, z)\mathbf{k}$ is a vector field in space, then we define its *curl* to be the vector field

$$\mathbf{curl}(\mathbf{F})(x, y, z) = \left(\frac{\partial R}{\partial y} - \frac{\partial N}{\partial z}\right)\mathbf{i} - \left(\frac{\partial R}{\partial x} - \frac{\partial M}{\partial z}\right)\mathbf{j} + \left(\frac{\partial N}{\partial x} - \frac{\partial M}{\partial y}\right)\mathbf{k}. \tag{15.31}$$

Formula (15.31) can be symbolically written as

$$\mathbf{curl}(\mathbf{F}) = \det\left(\begin{bmatrix} \mathbf{i} & \mathbf{j} & \mathbf{k} \\ \dfrac{\partial}{\partial x} & \dfrac{\partial}{\partial y} & \dfrac{\partial}{\partial z} \\ M & N & R \end{bmatrix}\right),$$

a form that should be regarded as a memory aid.

Example 5 Define $\mathbf{F}(x, y, z) = y^2z\mathbf{i} - x^3\mathbf{j} + xy\mathbf{k}$. Calculate $\mathbf{curl}(\mathbf{F})$.

Solution We have

$$\frac{\partial R}{\partial y} - \frac{\partial N}{\partial z} = x - 0, \qquad \frac{\partial R}{\partial x} - \frac{\partial M}{\partial z} = y - y^2, \qquad \text{and} \qquad \frac{\partial N}{\partial x} - \frac{\partial M}{\partial y} = -3x^2 - 2yz.$$

Therefore, $\mathbf{curl}(\mathbf{F})(x, y, z) = x\mathbf{i} - (y - y^2)\mathbf{j} + (-3x^2 - 2yz)\mathbf{k}$. ■

What is the geometric or physical meaning of the curl of a vector field? The vector field $\mathbf{curl}(\mathbf{F})$ measures the tendency of the vector field to "curl" or "rotate." More specifically, $\mathbf{curl}\ \mathbf{F}$ points along the *axis of rotation* of $\mathbf{F}$ (determined by the Right Hand Rule), and the length of $\mathbf{curl}(\mathbf{F})$ measures the amount of curling. This is best seen by studying a simple example.

Example 6 Sketch the vector field $\mathbf{F}(x, y, z) = -y\mathbf{i} + x\mathbf{j} + 0\mathbf{k}$ and its curl.

Solution Set $M = -y$, $N = x$, and $R = 0$. We calculate $\frac{\partial R}{\partial y} = 0$, $\frac{\partial N}{\partial z} = 0$, $\frac{\partial R}{\partial x} = 0$, $\frac{\partial M}{\partial z} = 0$, $\frac{\partial N}{\partial x} = 1$, and $\frac{\partial M}{\partial y} = -1$. Substituting these values into formula (15.31), we obtain $\mathbf{curl}(\mathbf{F})(x, y, z) = 2\mathbf{k}$. Vector fields $\mathbf{F}$ and $\mathbf{curl}(\mathbf{F})$ are sketched in Figure 5, next page. We see that the vector field $\mathbf{F}$ curls counterclockwise about the origin in the xy-plane. If you curl your right hand in the direction of the curl, your thumb points in the direction $\mathbf{k}$. Thus, $\mathbf{curl}(\mathbf{F})$ coincides with our physical perception of the curl of a flow. If the vector field were rotating in the opposite direction—say, that $\mathbf{H} = y\mathbf{i} - x\mathbf{j}$—then we would expect the curl to have opposite sign. And, indeed, $\mathbf{curl}(\mathbf{H}) = -2\mathbf{k}$. When we study Stokes's Theorem in Section 15.7, we will be able to flesh out this discussion. ■

The notation $\mathbf{rot}(\mathbf{F})$, an abbreviation for "rotation of $\mathbf{F}$," is sometimes found in older texts instead of $\mathbf{curl}(\mathbf{F})$. In physics, a vector field for which $\mathbf{curl}(\mathbf{F}) = \vec{\mathbf{0}}$ at all points is called *irrotational*. Comparing equation (15.25) with equation (15.30) in the

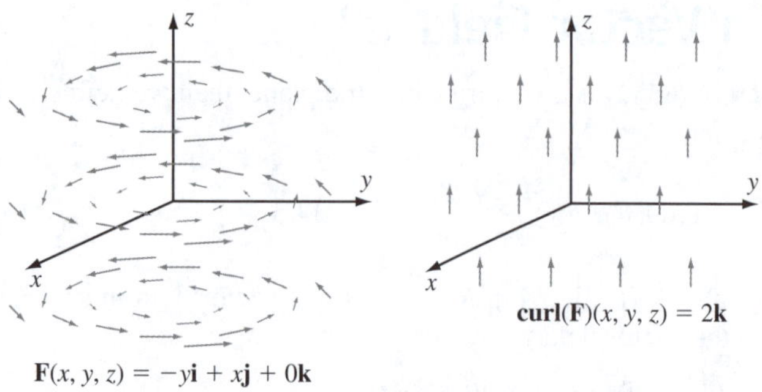

Figure 5

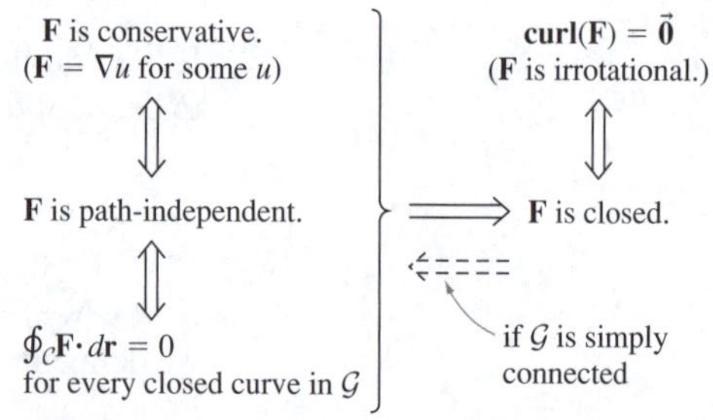

Figure 6

planar case and equation (15.26) with equation (15.31) in the spatial case, we see that $\mathbf{curl}(\mathbf{F}) = \vec{\mathbf{0}}$ if and only if $\mathbf{F}$ is closed.

From our work in Section 15.3, it follows that, on a simply connected region, $\mathbf{curl}(\mathbf{F}) = \vec{\mathbf{0}}$ if and only if $\mathbf{F}$ is conservative. This information, which is recorded in the diagram in Figure 6, will be of use when we study Stokes's Theorem.

Example 7 Verify that the vector field $\mathbf{F}(x, y, z) = 2x\mathbf{i} + (z^2 - 2y)\mathbf{j} + 2yz\mathbf{k}$ is irrotational on all of three-dimensional space.

Solution We must show that $\mathbf{curl}(\mathbf{F})(x, y, z) = \vec{\mathbf{0}}$ for all x, y, z. Setting $M(x, y, z) = 2x$, $N(x, y, z) = z^2 - 2y$, and $R(x, y, z) = 2yz$, we have

$$\frac{\partial R}{\partial y} - \frac{\partial N}{\partial z} = 2z - 2z = 0, \quad \frac{\partial R}{\partial x} - \frac{\partial M}{\partial z} = 0 - 0 = 0, \quad \text{and} \quad \frac{\partial N}{\partial x} - \frac{\partial M}{\partial y} = 0 - 0 = 0.$$ ■

The Del Notation

Let us introduce the formal notation

$$\nabla = \frac{\partial}{\partial x}\mathbf{i} + \frac{\partial}{\partial y}\mathbf{j} + \frac{\partial}{\partial z}\mathbf{k}.$$

This is not a vector, but it is a convenient notation in calculations.

We are already accustomed to writing the gradient of a function u as ∇u. (The gradient of a function u may also be written as $\mathbf{grad}(u)$.) Now we note that the divergence of a vector field $\mathbf{F} = M\mathbf{i} + N\mathbf{j} + R\mathbf{k}$ may be written as

$$\text{div}(\mathbf{F}) = \nabla \cdot \mathbf{F}.$$

This means that if we treat ∇ algebraically, as though it were a vector, then

$$\nabla \cdot \mathbf{F} = \left(\frac{\partial}{\partial x}\mathbf{i} + \frac{\partial}{\partial y}\mathbf{j} + \frac{\partial}{\partial z}\mathbf{k}\right) \cdot (M\mathbf{i} + N\mathbf{j} + R\mathbf{k}) = \frac{\partial M}{\partial x} + \frac{\partial N}{\partial y} + \frac{\partial R}{\partial z} = \text{div}(\mathbf{F}).$$

In a similar vein, we may write the curl of the vector field $\mathbf{F} = M\mathbf{i} + N\mathbf{j} + R\mathbf{k}$ as

$$\mathbf{curl}(\mathbf{F}) = \nabla \times \mathbf{F}.$$

This means that if we treat ∇ algebraically, as though it were a vector, then

$$\nabla\times\mathbf{F} = \det\left(\begin{bmatrix} \mathbf{i} & \mathbf{j} & \mathbf{k} \\ \dfrac{\partial}{\partial x} & \dfrac{\partial}{\partial y} & \dfrac{\partial}{\partial z} \\ M & N & R \end{bmatrix}\right) = \left(\frac{\partial R}{\partial y}-\frac{\partial N}{\partial z}\right)\mathbf{i}-\left(\frac{\partial R}{\partial x}-\frac{\partial M}{\partial z}\right)\mathbf{j}+\left(\frac{\partial N}{\partial x}-\frac{\partial M}{\partial y}\right)\mathbf{k}.$$

In physics books, it is quite common to see divergence and curl expressed using the notation ∇. We call this the del notation. The symbol ∇ is also referred to as "nabla."

Probably the most important partial differential operator of mathematical physics is the *Laplace operator,* which is often denoted by the symbol $\triangle$. Applied to a twice continuously differentiable scalar-valued function u of x and y, this operator is given by the formula

$$\triangle u = \frac{\partial^2 u}{\partial x^2} + \frac{\partial^2 u}{\partial y^2}.$$

For a twice continuously differentiable scalar-valued function v of x, y, and z, the Laplace operator (or "Laplacian") is given by

$$\triangle v = \frac{\partial^2 v}{\partial x^2} + \frac{\partial^2 v}{\partial y^2} + \frac{\partial^2 v}{\partial z^2}.$$

Notice that we must not confuse the Laplacian $\triangle v$ with its look-alike, Δv, which represents a small increment in the value of v. In general, we can infer which meaning is intended from the context. It is often convenient to define the Laplacian of a vector-valued function $\mathbf{F} = \langle M, N, R\rangle$ by mapping $\triangle$ onto the components of $\mathbf{F}$:

$$\triangle\mathbf{F} = \langle \triangle M, \triangle N, \triangle R\rangle.$$

Sometimes we write the Laplacian as $\nabla\cdot\nabla$, since

$$\triangle v = \frac{\partial^2 v}{\partial x^2} + \frac{\partial^2 v}{\partial y^2} + \frac{\partial^2 v}{\partial z^2} = \frac{\partial}{\partial x}\frac{\partial v}{\partial x} + \frac{\partial}{\partial y}\frac{\partial v}{\partial y} + \frac{\partial}{\partial z}\frac{\partial v}{\partial z} = \nabla\cdot(\nabla v). \qquad \textbf{(15.32)}$$

The notation $\nabla\cdot\nabla$ is often contracted to ∇^2.

Many important differential equations in mathematical physics involve the Laplacian. For example, if the edges of a metal plate are heated, then the second law of thermodynamics implies that the steady state heat distribution $h(x, y)$ in the plate satisfies the equation $\triangle h = 0$. A function u that satisfies $\triangle u = 0$ is called *harmonic*. Harmonic functions are important objects of study in mathematical analysis, physics, and engineering.

Example 8 Show that the functions $u(x, y) = x^2 - y^2$ and $v(x, y, z) = e^{(x+y)}\cos(\sqrt{2}z)$ are harmonic.

Solution We have

$$\frac{\partial^2 u}{\partial x^2} + \frac{\partial^2 u}{\partial y^2} = 2 + (-2) = 0$$

and

$$\frac{\partial^2 v}{\partial x^2} + \frac{\partial^2 v}{\partial y^2} + \frac{\partial^2 v}{\partial z^2} = e^{(x+y)}\cos\left(\sqrt{2}z\right) + e^{(x+y)}\cos\left(\sqrt{2}z\right) + \left(-\left(\sqrt{2}\right)^2 e^{(x+y)}\cos\left(\sqrt{2}z\right)\right) = 0. \quad \blacksquare$$

Identities Involving div, curl, grad, and Δ

The next theorem gathers some basic identities that involve the operators we have been studying. Several other identities may be found in the exercises for this section.

Theorem 1 Let u be a twice continuously differentiable scalar-valued function on a region $\mathcal{G}$ in the plane or in space. Let $\mathbf{F}$ be a twice continuously differentiable vector field on $\mathcal{G}$. Then

a. $\operatorname{div}(\mathbf{grad}(u)) = \Delta u$
b. $\operatorname{div}(\mathbf{curl}(\mathbf{F})) = 0$
c. $\mathbf{curl}(\mathbf{grad}(u)) = \vec{\mathbf{0}}$
d. $\mathbf{curl}(\mathbf{curl}(\mathbf{F})) = \mathbf{grad}(\operatorname{div}(\mathbf{F})) - \Delta\mathbf{F}$
e. $\operatorname{div}(u\mathbf{F}) = u\operatorname{div}(\mathbf{F}) + \mathbf{grad}(u)\cdot\mathbf{F}$

Proof The first assertion is simply a restatement of equation (15.32). Part b follows from the equality of the mixed partial derivatives of the components of $\mathbf{F} = \langle M, N, R\rangle$:

$$\operatorname{div}(\mathbf{curl}(\mathbf{F})) = \frac{\partial}{\partial x}\left(\frac{\partial R}{\partial y} - \frac{\partial N}{\partial z}\right) + \frac{\partial}{\partial y}\left(\frac{\partial M}{\partial z} - \frac{\partial R}{\partial x}\right) + \frac{\partial}{\partial z}\left(\frac{\partial N}{\partial x} - \frac{\partial M}{\partial y}\right),$$

or

$$\operatorname{div}(\mathbf{curl}(\mathbf{F})) = \left(\frac{\partial^2 R}{\partial x\partial y} - \frac{\partial^2 R}{\partial y\partial x}\right) + \left(\frac{\partial^2 N}{\partial z\partial x} - \frac{\partial^2 N}{\partial x\partial z}\right) + \left(\frac{\partial^2 M}{\partial y\partial z} - \frac{\partial^2 M}{\partial z\partial y}\right) = 0.$$

Part c asserts that the conservative vector field $\mathbf{grad}(u)$ is closed, a fact we established in Section 15.3. Part d can be proved by a brute force calculation that is similar in spirit to part b. Finally, part e is obtained as follows:

$$\begin{aligned}\operatorname{div}(u\mathbf{F}) &= \frac{\partial(uM)}{\partial x} + \frac{\partial(uN)}{\partial y} + \frac{\partial(uR)}{\partial z}\\ &= \left(u\frac{\partial M}{\partial x} + M\frac{\partial u}{\partial x}\right) + \left(u\frac{\partial N}{\partial y} + N\frac{\partial u}{\partial y}\right) + \left(u\frac{\partial R}{\partial z} + R\frac{\partial u}{\partial z}\right)\\ &= u\left(\frac{\partial M}{\partial x} + \frac{\partial N}{\partial y} + \frac{\partial R}{\partial z}\right) + \left(M\frac{\partial u}{\partial x} + N\frac{\partial u}{\partial y} + R\frac{\partial u}{\partial z}\right)\\ &= u\operatorname{div}(\mathbf{F}) + \mathbf{grad}(u)\cdot\mathbf{F}.\end{aligned}$$

■

Example 9 For the vector field $\mathbf{F}(x, y, z) = \langle y, z^3, \cos(x) + e^z\rangle$, verify the identity $\mathbf{curl}(\mathbf{curl}(\mathbf{F})) = \mathbf{grad}(\operatorname{div}(\mathbf{F})) - \Delta\mathbf{F}$.

Solution We have

$$\begin{aligned}\mathbf{curl}(\mathbf{F}) &= \left(\frac{\partial}{\partial y}(\cos(x) + e^z) - \frac{\partial}{\partial z}z^3\right)\mathbf{i} + \left(\frac{\partial}{\partial z}y - \frac{\partial}{\partial x}(\cos(x) + e^z)\right)\mathbf{j} + \left(\frac{\partial}{\partial x}z^3 - \frac{\partial}{\partial y}y\right)\mathbf{k}\\ &= -3z^2\mathbf{i} + \sin(x)\mathbf{j} + (-1)\mathbf{k},\end{aligned}$$

and therefore

$$\begin{aligned}\mathbf{curl}(\mathbf{curl}(\mathbf{F})) &= \left(\frac{\partial}{\partial y}(-1) - \frac{\partial}{\partial z}\sin(x)\right)\mathbf{i} + \left(\frac{\partial}{\partial z}(-3z^2) - \frac{\partial}{\partial x}(-1)\right)\mathbf{j} + \left(\frac{\partial}{\partial x}\sin(x) - \frac{\partial}{\partial y}(-3z^2)\right)\mathbf{k}\\ &= -6z\mathbf{j} + \cos(x)\mathbf{k}.\end{aligned}$$

On the other hand,

$$\mathbf{grad}(\text{div}(\mathbf{F})) = \text{grad}\left(\frac{\partial}{\partial x}y + \frac{\partial}{\partial y}z^3 + \frac{\partial}{\partial z}(\cos(x) + e^z)\right) = \mathbf{grad}(e^z) = e^z\mathbf{k}$$

and

$$\Delta\mathbf{F} = (\Delta y)\mathbf{i} + (\Delta z^3)\mathbf{j} + (\Delta(\cos(x) + e^z))\mathbf{k} = 6z\mathbf{j} + (-\cos(x) + e^z)\mathbf{k}.$$

Therefore, $\mathbf{grad}(\text{div}(\mathbf{F})) - \Delta\mathbf{F} = e^z\mathbf{k} - (6z\mathbf{j} + (-\cos(x) + e^z)\mathbf{k}) = -6z\mathbf{j} + \cos(x)\mathbf{k}$, which equals $\mathbf{curl}(\mathbf{curl}(\mathbf{F}))$. ■

quickquiz

1. What is curl? To what sort of function do we apply the curl operation?
2. What is divergence? To what sort of function do we apply the divergence operator?
3. What does curl tell us about path-independence?
4. What is the physical significance of curl?

EXERCISES

Problems for Practice

In Exercises 1–6, calculate the divergence of the planar vector field **F**.

1. $\mathbf{F}(x, y) = \cos(x)\mathbf{i} - \sin^2(xy)\mathbf{j}$
2. $\mathbf{F}(x, y) = \ln(x + y)\mathbf{i} + e^{xy}\mathbf{j}$
3. $\mathbf{F}(x, y) = x/(x + y)\mathbf{i} - 2y/(x - y)\mathbf{j}$
4. $\mathbf{F}(x, y) = xe^y\mathbf{i} - ye^x\mathbf{j}$
5. $\mathbf{F}(x, y) = \tan(x/y)\mathbf{j} + \cot(y/x)\mathbf{i}$
6. $\mathbf{F}(x, y) = (x^2 + y^2)e^{x+y}(\mathbf{i} + \mathbf{j})$

In Exercises 7–12, calculate the divergence of the spatial vector field **F**.

7. $\mathbf{F}(x, y, z) = x^2\mathbf{i} + y^2\mathbf{j} + z^2\mathbf{k}$
8. $\mathbf{F}(x, y, z) = x^3y\mathbf{i} - y^2zx\mathbf{j} + xyz\mathbf{k}$
9. $\mathbf{F}(x, y, z) = e^{x-z}\mathbf{i} + e^{z-y}\mathbf{j} - e^{y-x}\mathbf{k}$
10. $\mathbf{F}(x, y, z) = x^2\sin(y)(\mathbf{i} - \mathbf{j} + \mathbf{k})$
11. $\mathbf{F}(x, y, z) = \cos(xy)\mathbf{i} - \sin(yz)\mathbf{j} + \cos(y)\sin(x)\mathbf{k}$
12. $\mathbf{F}(x, y, z) = x^y\mathbf{i} + y^z\mathbf{j} - x^z\mathbf{k}$

In Exercises 13–18, calculate the curl of the vector field **F**.

13. $\mathbf{F}(x, y, z) = zx\mathbf{i} - y^3\mathbf{j} + xyz\mathbf{k}$
14. $\mathbf{F}(x, y, z) = \cos^2(z)\mathbf{i} - \sin(yx)\mathbf{j} + x^4z\mathbf{k}$
15. $\mathbf{F}(x, y, z) = \ln(x + z)\mathbf{i} - e^{yz}\mathbf{j} + xy\mathbf{k}$
16. $\mathbf{F}(x, y, z) = z^2\mathbf{i} - xz\mathbf{j} + x^3\mathbf{k}$
17. $\mathbf{F}(x, y, z) = \tan(xy)\mathbf{i} + \cos(xy)\mathbf{j} - \sin(xy)\mathbf{k}$
18. $\mathbf{F}(x, y, z) = xyz\mathbf{i} + (xyz)^2\mathbf{j} + (xyz)^3\mathbf{k}$

In Exercises 19–24, calculate the curl of the vector field **F**. State whether **F** is closed.

19. $\mathbf{F}(x, y, z) = y^2z\mathbf{i} + 2xyz\mathbf{j} + xy^2\mathbf{k}$
20. $\mathbf{F}(x, y, z) = (y/x)\mathbf{i} + (z/y)\mathbf{j} + (x/z)\mathbf{k}$
21. $\mathbf{F}(x, y, z) = \cos(yz)\mathbf{i} - xz\sin(yz)\mathbf{j} - xy\sin(yz)\mathbf{k}$
22. $\mathbf{F}(x, y, z) = y/(x + z)\mathbf{i} + \ln(x + z)\mathbf{j} + y/(x + z)\mathbf{k}$
23. $\mathbf{F}(x, y, z) = x\cos(y)\mathbf{i} - z\sin(x)\mathbf{j} + (xy)\sin(z)\mathbf{k}$
24. $\mathbf{F}(x, y, z) = e^x\mathbf{i} - e^{x-z}\mathbf{j} + e^{y-x}\mathbf{k}$

In Exercises 25–30, a scalar-valued function u is given. Calculate $\text{div}(\mathbf{grad}(u))$.

25. $u(x, y) = x^2 - y^2$
26. $u(x, y) = e^x\cos(y)$
27. $u(x, y, z) = x/(z - y^2)$
28. $u(x, y, z) = y\sin(x - z)$
29. $u(x, y, z) = x(y + 2z)$
30. $u(x, y, z) = (2x - z)/yz$

Further Theory and Practice

Suppose that g and h are twice continuously differentiable scalar-valued functions and that **F** and **G** are continuously differentiable vector fields. Prove each identity in Exercises 31–36.

31. $\nabla(gh) = g\nabla(h) + h\nabla(g)$
32. $\nabla \cdot (g\mathbf{F}) = (\nabla g) \cdot \mathbf{F} + g(\nabla \cdot \mathbf{F})$
33. $\nabla \times (g\mathbf{F}) = (\nabla g) \times \mathbf{F} + g(\nabla \times \mathbf{F})$
34. $\nabla \cdot (\mathbf{F} \times \mathbf{G}) = (\nabla \times \mathbf{F}) \cdot \mathbf{G} - \mathbf{F} \cdot (\nabla \times \mathbf{G})$
35. $\nabla \times (g\nabla h - h\nabla g) = 2\nabla g \times \nabla h$
36. $\nabla \times (g\nabla h + h\nabla g) = \vec{\mathbf{0}}$
37. Let $\mathbf{r} = \langle x, y, z\rangle$. Define $\mathbf{F}(x, y, z) = \lambda\|\mathbf{r}\|^{-3}\mathbf{r}$, as in Example 4. Verify that $\nabla \times \mathbf{F} = \vec{\mathbf{0}}$ at all points of space except for the origin.

Harmonic functions on the plane have a remarkable mean value property: The average value of a harmonic function on a circle is equal to the value of the function at the center. To be precise, if u is harmonic, then for any $r > 0$,

$$u(x_0, y_0) = \frac{1}{2\pi}\int_0^{2\pi} u(x_0 + r\cos(t), y_0 + r\sin(t))\, dt.$$

In Exercises 38 and 39, verify that the function u is harmonic, and then verify the mean value property at the point (x_0, y_0).

38. $u(x, y) = x^2 - y^2$, $(x_0, y_0) = (3, 2)$
39. $u(x, y) = x^3 - 3xy^2 + 1$, $(x_0, y_0) = (2, 1)$
40. Suppose that u is a twice continuously differentiable scalar-valued function. Set $\mathbf{r}(x, y, z) = x\mathbf{i} + y\mathbf{j} + z\mathbf{k}$. Prove that $\mathbf{r} \times \nabla u$ is solenoidal.
41. Prove that $\nabla \times (\boldsymbol{\omega} \times \mathbf{r}) = 2\boldsymbol{\omega}$ when $\boldsymbol{\omega}$ is a constant vector and $\mathbf{r}(x, y, z) = x\mathbf{i} + y\mathbf{j} + z\mathbf{k}$.
42. Prove that $\mathbf{F} \times \mathbf{G}$ is solenoidal when both spatial vector fields $\mathbf{F}$ and $\mathbf{G}$ are irrotational.
43. Suppose that the curl of a spatial vector field $\mathbf{F}$ has a potential function V. Prove that V is harmonic and $\Delta\mathbf{F} = \mathbf{grad}(\mathrm{div}(\mathbf{F}))$.
44. Theorem 1 tells us that the curl of any twice continuously differentiable vector field on a region $\mathcal{G}$ is solenoidal on $\mathcal{G}$. The converse is also true, provided $\mathcal{G}$ has certain properties. In this exercise, assume that $\mathcal{G}$ is an open ball containing the origin O. Let $\mathbf{F}$ be a differentiable vector field such that $\mathrm{div}(\mathbf{F}) = 0$ on $\mathcal{G}$. For each $P = (x, y, z) \in \mathcal{G}$, let $\mathbf{r}_P$ be the parameterization of $\overrightarrow{OP}$ given by $\mathbf{r}_P(t) = \langle tx, ty, tz\rangle$, $0 \leq t \leq 1$. Define

$$\mathbf{G}(P) = \int_0^1 t\mathbf{F}(\mathbf{r}_P(t)) \times \mathbf{r}'_P(t)\, dt.$$

Show that $\mathbf{F} = \mathbf{curl}(\mathbf{G})$.

Calculator/Computer Exercises

In Exercises 45 and 46, verify the mean value property of the harmonic function u at the point (x_0, y_0). (See the instructions for Exercises 38 and 39). Use $r = 1$.

45. $u(x, y) = e^x\cos(y)$, $(x_0, y_0) = (1, 3)$
46. $u(x, y) = \ln(x^2 + y^2)$, $(x_0, y_0) = (3, 4)$

15.5 Green's Theorem

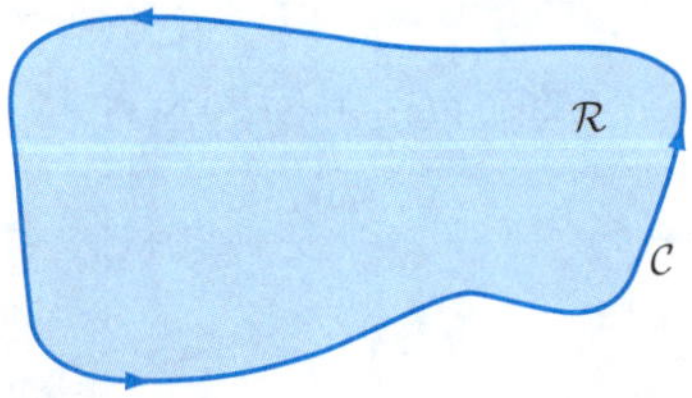

Figure 1

In this section, we learn our first theorem that relates the behavior of a flow at the boundary of a region to the behavior in the interior. There will be several interesting applications. Everything in this section takes place in the two-dimensional plane. Throughout this section (and in the remaining sections), we will write dA for the element of area in the plane.

We begin our study with a simply connected bounded region $\mathcal{R}$ in the plane. Its boundary consists of a single closed curve $\mathcal{C}$ that we parameterize by a vector-valued function $t \mapsto \mathbf{r}(t)$, $a \leq t \leq b$. Refer to Figure 1. We assume four things about $\mathbf{r}$:

1. $\mathbf{r}$ is piecewise smooth.
2. $\mathbf{r}(a) = \mathbf{r}(b)$ ($\mathcal{C}$ is *closed*).
3. $\mathbf{r}(t_1) \neq \mathbf{r}(t_2)$ for $a \leq t_1 < t_2 \leq b$.
4. As $\mathbf{r}(t)$ traverses $\mathcal{C}$ with increasing t, the enclosed region $\mathcal{R}$ lies to the left.

The first condition states that $\mathcal{C}$ is made up of finitely many smooth arcs. The second condition says that $\mathbf{r}$ begins and ends at the same point. The third condition states that

$\mathcal{C}$ does not cross itself. It rules out curves like the one in Figure 2. A closed curve that satisfies the third condition is said to be *simple*. The fourth condition says that $\mathcal{C}$ is oriented as in Figure 3a, *not* as in Figure 3b. This is said to be the *positive* orientation of $\mathcal{C}$. If we regard the xy-plane as a subset of xyz-space, then the Right Hand Rule for cross products tells us that $\mathbf{k} \times \mathbf{r}'$ points *into the* region $\mathcal{R}$ when $\mathcal{C}$ is positively oriented.

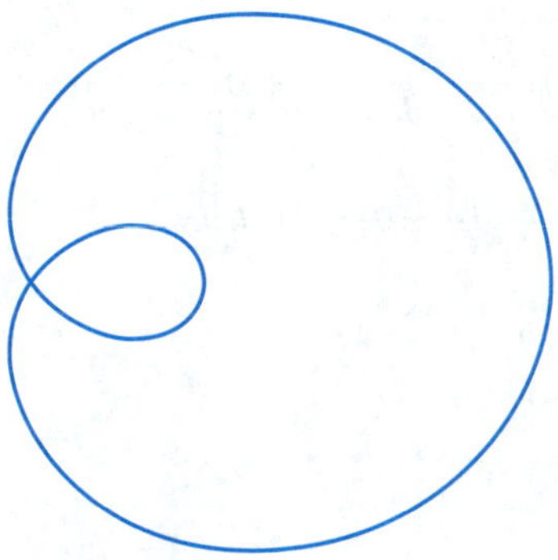

Figure 2
A curve that intersects itself is *not* simple.

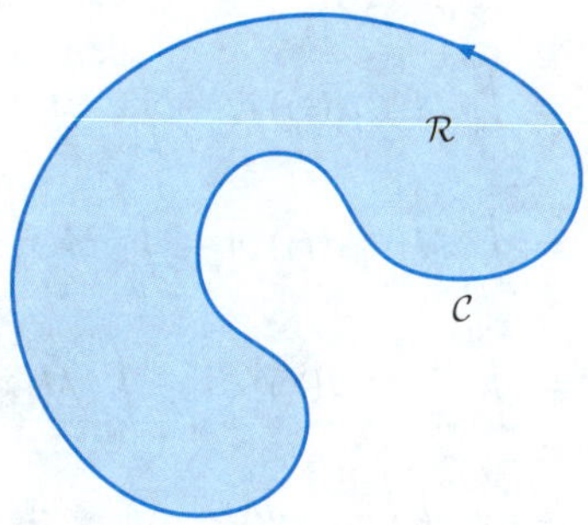

Figure 3a
Positive orientation of boundary

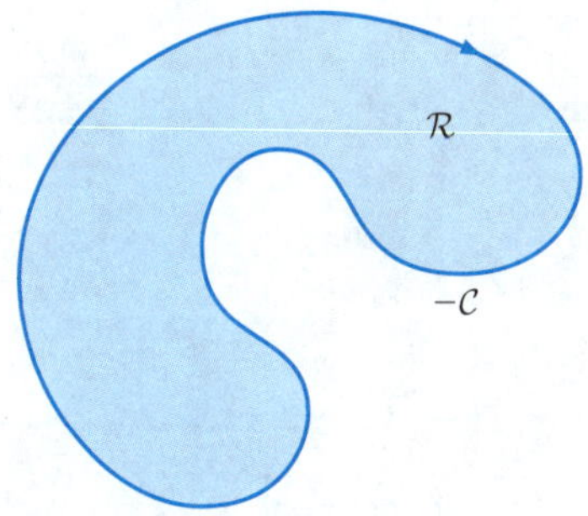

Figure 3b
Negative orientation of boundary

Theorem 1 ***Green's Theorem*** Suppose that $\mathcal{R}$ is a simply connected region with boundary $\mathcal{C}$ that is parameterized by a vector-valued function $\mathbf{r}$ that has the four properties enumerated above. If $\mathbf{F}(x, y) = M(x, y)\mathbf{i} + N(x, y)\mathbf{j}$ is a continuously differentiable vector field on a region containing $\mathcal{R}$ and its boundary, then

$$\oint_{\mathcal{C}} \mathbf{F} \cdot d\mathbf{r} = \iint_{\mathcal{R}} \left(\frac{\partial N}{\partial x} - \frac{\partial M}{\partial y} \right) dA. \tag{15.33}$$

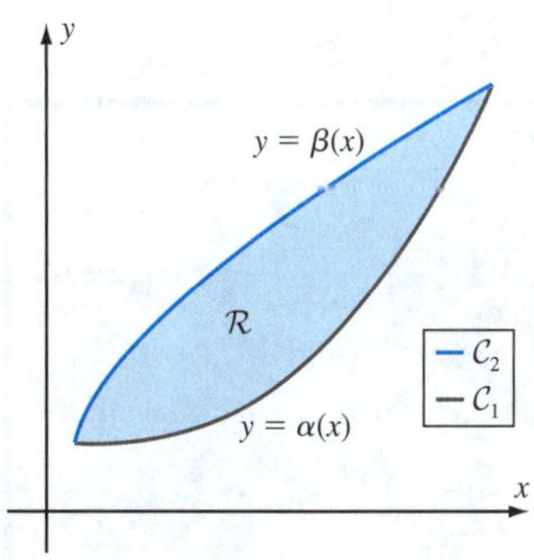

Figure 4
$\mathcal{R}$ is both x-simple and y-simple.

Proof A number of subtleties arise in the proof of Green's Theorem in its full generality, and these subtleties are a proper topic for a course in advanced calculus. Therefore, we will settle for a proof of Green's Theorem in a special case. We assume that our region is both x-simple and y-simple, as in Figure 4, with boundary consisting of the graphs of two functions, $y = \alpha(x)$ and $y = \beta(x)$. We can parameterize the lower curve $\mathcal{C}_1$ by $\mathbf{r}_1(t) = \langle t, \alpha(t) \rangle,\ a \le t \le b$ and the upper curve $\mathcal{C}_2$ by $\mathbf{r}_2(t) = \langle b + a - t, \beta(b + a - t) \rangle,\ a \le t \le b$. Now we begin with the right side of the formula in Green's Theorem:

$$\iint_{\mathcal{R}} \left(\frac{\partial N}{\partial x} - \frac{\partial M}{\partial y} \right) dA = \iint_{\mathcal{R}} \frac{\partial N}{\partial x}\, dA - \iint_{\mathcal{R}} \frac{\partial M}{\partial y}\, dA.$$

We can calculate the second double integral on the right side by expressing it as an iterated integral and using the Fundamental Theorem of Calculus:

$$\int_a^b \int_{\alpha(x)}^{\beta(x)} \frac{\partial M}{\partial y}\, dy\, dx = \int_a^b M(x, y)\Big|_{y=\alpha(x)}^{y=\beta(x)} dx = \int_a^b M(x, \beta(x))\, dx - \int_a^b M(x, \alpha(x))\, dx. \tag{15.34}$$

Now

$$\begin{aligned}
\oint_C M\mathbf{i}\cdot d\mathbf{r} &= \int_{C_1} M\mathbf{i}\cdot d\mathbf{r}_1 + \int_{C_2} M\mathbf{i}\cdot d\mathbf{r}_2 \\
&= \int_a^b \left(M(t,\alpha(t))\frac{dt}{dt} + 0\frac{d\alpha(t)}{dt} \right) dt \\
&\quad + \int_a^b \left(M(b+a-t,\beta(b+a-t))\,\frac{d(b+a-t)}{dt} + 0\frac{d\beta(b+a-t)}{dt} \right) dt \\
&= \int_a^b M(t,\alpha(t))\,dt - \int_a^b M(b+a-t,\beta(b+a-t))\,dt \\
&= \int_a^b M(t,\alpha(t))\,dt + \int_b^a M(x,\beta(x))\,dx \qquad (x=b+a-t,\ dx=-dt) \\
&= \int_a^b M(x,\alpha(x))\,dx - \int_a^b M(x,\beta(x))\,dx \\
&\overset{(15.34)}{=} -\iint_{\mathcal{R}} \frac{\partial M}{\partial y}\,dA
\end{aligned}$$

This is half the job. A similar calculation shows that

$$\int_C N\mathbf{j}\cdot d\mathbf{r} = \iint_{\mathcal{R}} \frac{\partial N}{\partial x}\,dA.$$

Adding the last two equations results in the identity that is Green's Theorem. ■

inSIGHT

Expressed in component form, Green's Theorem says that

$$\oint_C M\,dx + N\,dy = \iint_{\mathcal{R}} \left(\frac{\partial N}{\partial x} - \frac{\partial M}{\partial y} \right) dA. \tag{15.35}$$

It is instructive to compare equation (15.35) with the Fundamental Theorem of Calculus,

$$f(b) - f(a) = \int_a^b f'(t)\,dt,$$

which relates the values of f on the boundary of an interval $[a, b]$—namely, the points a and b—to the integral of the derivative of f on the interval. Observe that the boundary orientation is important in the Fundamental Theorem of Calculus: The term $f(b)$ appears with a $+$ and the term $f(a)$ appears with a $-$. It is plain that Green's Theorem is a two-dimensional version of the Fundamental Theorem of Calculus in that it relates the integral of $\mathbf{F}$ on the oriented boundary of a region $\mathcal{R}$ to the integral of a certain derivative of $\mathbf{F}$ on the region.

Example 1 Verify Green's Theorem for $\mathbf{F}(x, y) = -3y\mathbf{i} + 6x\mathbf{j}$ and $\mathcal{R} = \{(x, y) : x^2 + y^2 < 1\}$.

Solution The boundary of $\mathcal{R}$ is the unit circle $\mathcal{C}$ traversed counterclockwise when positively oriented. We may use the parameterization $\mathbf{r}(t) = \cos(t)\mathbf{i} + \sin(t)\mathbf{j}$, $0 \leq t \leq 2\pi$. We have

$$\begin{aligned}\oint_{\mathcal{C}} \mathbf{F} \cdot d\mathbf{r} &= \int_0^{2\pi} \left(-3\sin(t)\frac{d}{dt}\cos(t) + 6\cos(t)\frac{d}{dt}\sin(t) \right) dt \\ &= \int_0^{2\pi} (3\sin^2(t) + 6\cos^2(t))\, dt.\end{aligned}$$

But $\int_0^{2\pi} \sin^2(t)\, dt = \int_0^{2\pi} \cos^2(t)\, dt = \pi$, by formula (7.21). It follows that $\oint_{\mathcal{C}} \mathbf{F} \cdot d\mathbf{r} = 9\pi$. Thus, the left side of the formula in Green's Theorem equals 9π for this example. Next we compute the right side. We have

$$\begin{aligned}\iint_{\mathcal{R}} \left(\frac{\partial N}{\partial x} - \frac{\partial M}{\partial y} \right) dA &= \iint_{\mathcal{R}} \left(\frac{\partial(6x)}{\partial x} - \frac{\partial(-3y)}{\partial y} \right) dA \\ &= \iint_{\mathcal{R}} 9\, dA = 9 \cdot (\text{area of unit disk}) = 9\pi.\end{aligned}$$

Thus, we have verified Green's Theorem for this particular example. ■

At first, Example 1 may seem pointless. We now have two ways of calculating something instead of one. That's just twice as much work! But this example was only for mechanical practice, to help us understand the *components* of Green's Theorem. It is not the way we usually use Green's Theorem. The next theorem shows one of the really striking applications.

Theorem 2 Suppose that $\mathcal{R}$, $\mathcal{C}$, and $\mathbf{r}$ satisfy the hypotheses of Green's Theorem. Then we have

$$(\text{area of } \mathcal{R}) = \frac{1}{2} \oint_{\mathcal{C}} (-y)\, dx + x\, dy. \qquad \textbf{(15.36)}$$

Proof For the vector field $\mathbf{F} = M\mathbf{i} + N\mathbf{j} = (-y/2)\mathbf{i} + (x/2)\mathbf{j}$, we have $N_x - M_y = (1/2) - (-1/2) = 1$. Green's Theorem tells us that

$$\oint_{\mathcal{C}} -\frac{y}{2}dx + \frac{x}{2}dy = \iint_{\mathcal{R}} (N_x - M_y)\, dA = \iint_{\mathcal{R}} (1)\, dA.$$

This, of course, is just the area of $\mathcal{R}$. ■

Example 2 Use formula (15.36) to calculate the area A inside the ellipse $x^2/a^2 + y^2/b^2 = 1$.

Solution We can parameterize the ellipse $\mathcal{C}$ by $\mathbf{r}(t) = a\cos(t)\mathbf{i} + b\sin(t)\mathbf{j}$, $0 \leq t \leq 2\pi$. Then

$$A = \frac{1}{2} \oint_{\mathcal{C}} (-y)\, dx + x\, dy = \frac{1}{2} \int_0^{2\pi} \left((-b\sin(t))\frac{d}{dt}a\cos(t) + (a\cos(t))\frac{d}{dt}b\sin(t) \right) dt.$$

By simplifying the integrand, we obtain

$$A = \frac{1}{2}\int_0^{2\pi} ab(\cos^2(t) + \sin^2(t))\,dt = \frac{1}{2}\int_0^{2\pi} ab \cdot 1\,dt = \pi ab.$$

Theorem 5 from Section 15.3 asserts that a closed vector field on a simply connected planar region is conservative. This is a deep result whose proof we have outlined for rectangular regions. Now we can use Green's Theorem to obtain a more general proof.

Example 3 Use Green's Theorem to show that a closed vector field $\mathbf{F} = \langle M, N\rangle$ on a simply connected planar region $\mathcal{G}$ is conservative.

Solution If $\mathbf{F}$ is closed on $\mathcal{G}$ and if $\mathcal{C}$ is a simple closed curve in $\mathcal{G}$ that encloses a region $\mathcal{R}$, then

$$\oint_{\mathcal{C}} \mathbf{F}\cdot d\mathbf{r} = \iint_{\mathcal{R}} \left(\frac{\partial N}{\partial x} - \frac{\partial M}{\partial y}\right) dA = \iint_{\mathcal{R}} 0\,dA = 0.$$

By Theorem 1 from Section 15.3, this implies that $\mathbf{F}$ is path-independent. This, in turn, implies that $\mathbf{F}$ is conservative (by Theorem 3, Section 15.3).

A Vector Form of Green's Theorem

Green's Theorem has many variants, each of which reveals new information. In particular, there is a formulation of Green's Theorem for each of the differential operators—div, curl, and Δ—that we studied in Section 15.4. For example, for $\mathbf{F} = M\mathbf{i} + N\mathbf{j}$, we have $\mathbf{curl}(\mathbf{F}) = (N_x - M_y)\mathbf{k}$, and therefore $N_x - M_y = \mathbf{curl}(\mathbf{F})\cdot\mathbf{k}$. We can therefore state Green's Theorem as

$$\oint_{\mathcal{C}} \mathbf{F}\cdot d\mathbf{r} = \iint_{\mathcal{R}} \mathbf{curl}(\mathbf{F})\cdot\mathbf{k}\,dA. \qquad \textbf{(15.37)}$$

Notice that $\mathbf{k}$ is a unit normal to the region $\mathcal{R}$. The direction of this unit normal (as opposed to $-\mathbf{k}$) is determined by the Right Hand Rule: Curl your fingers of your right hand in the positive orientation of $\mathcal{C}$, and your thumb points in the direction of $\mathbf{k}$. Formula (15.37) tells us that we can compute the line integral of $\mathbf{F}$ over a simple closed curve $\mathcal{C}$ by integrating the component of $\mathbf{curl}(\mathbf{F})$ in the direction of the outward unit normal $\mathbf{k}$ to the enclosed region $\mathcal{R}$, the integration being performed over $\mathcal{R}$.

Next we will develop a divergence form of Green's Theorem. Suppose that $\mathcal{R}$, $\mathcal{C}$, $t \mapsto \mathbf{r}(t) = x(t)\mathbf{i} + y(t)\mathbf{j}$, and $\mathbf{F}(x, y) = M(x, y)\mathbf{i} + N(x, y)\mathbf{j}$ all satisfy the hypotheses of Green's Theorem. At each point $\mathbf{r}(t)$ of $\mathcal{C}$, we define the unit vector

$$\mathbf{n}(\mathbf{r}(t)) = \frac{y'(t)}{\sqrt{x'(t)^2 + y'(t)^2}}\mathbf{i} + \frac{(-x'(t))}{\sqrt{x'(t)^2 + y'(t)^2}}\mathbf{j} = \frac{1}{\|\mathbf{r}'(t)\|}(y'(t)\mathbf{i} - x'(t)\mathbf{j}).$$

Notice that $\mathbf{r}'(t)\cdot\mathbf{n}(\mathbf{r}(t)) = (1/\|\mathbf{r}'(t)\|)(x'(t)y'(t) + y'(t)(-x'(t))) = 0$. Since $\mathbf{r}'(t)$ is tangent to $\mathcal{C}$ at the point $\mathbf{r}(t)$, we deduce that $\mathbf{n}(\mathbf{r}(t))$ is normal to $\mathcal{C}$ at $\mathbf{r}(t)$. By observing the signs of $x'(t)$ and $y'(t)$ as $\mathbf{r}(t)$ traverses $\mathcal{C}$ in its positive orientation, we see that $\mathbf{n}(\mathbf{r}(t))$ is the *outward* unit normal to $\mathcal{C}$ at $\mathbf{r}(t)$.

To obtain the divergence form of Green's Theorem, we apply formula (15.33) to the vector field $\widetilde{\mathbf{F}}(x, y) = -N(x, y)\mathbf{i} + M(x, y)\mathbf{j}$. We obtain

$$\oint_{\mathcal{C}} \widetilde{\mathbf{F}} \cdot d\mathbf{r} = \iint_{\mathcal{R}} \left(\frac{\partial M}{\partial x} + \frac{\partial N}{\partial y} \right) dx\,dy = \iint_{\mathcal{R}} \operatorname{div}(\mathbf{F})\, dA. \qquad \textbf{(15.38)}$$

Now let us look at the left side of equation (15.38). We have

$$\begin{aligned}
\oint_{\mathcal{C}} \widetilde{\mathbf{F}} \cdot d\mathbf{r} &= \int_a^b \big(-N(x(t), y(t))x'(t) + M(x(t), y(t))y'(t) \big)\, dt \\
&= \int_a^b \left(M(x(t), y(t)) \frac{y'(t)}{\|\mathbf{r}'(t)\|} + N(x(t), y(t)) \frac{-x'(t)}{\|\mathbf{r}'(t)\|} \right) \|\mathbf{r}'(t)\|\, dt \\
&= \int_a^b (\mathbf{F} \cdot \mathbf{n})(\mathbf{r}(t)) \|\mathbf{r}'(t)\|\, dt.
\end{aligned}$$

But formula (15.22) allows us to identify this last expression as the line integral $\int_{\mathcal{C}} \mathbf{F} \cdot \mathbf{n}\, ds$. Thus, we have

$$\int_{\mathcal{C}} \mathbf{F} \cdot \mathbf{n}\, ds = \iint_{\mathcal{R}} \operatorname{div}(\mathbf{F})\, dA. \qquad \textbf{(15.39)}$$

This form of Green's Theorem is sometimes called the *Divergence Theorem* in two dimensions. The expression on the left side of equation (15.39) is called the *flux* of vector field $\mathbf{F}$ across $\mathcal{C}$. In the case that $\mathbf{F}$ is the velocity field of a fluid, the flux is the sum total of the scalar component of the fluid's velocity in the direction of the outward normal along the boundary curve.

Example 4 Let $\mathbf{F}(x, y) = x\mathbf{i} + y\mathbf{j}$ and $\mathcal{R} = \{(x, y) : x^2 + y^2 < 1\}$. Assume that $\mathbf{F}$ represents the velocity of the flow of a fluid in the region. Explain what Green's Theorem tells us about this fluid flow.

Solution Let $\mathcal{C}$ denote the positively oriented unit circle. The integral

$$\int_{\mathcal{C}} \mathbf{F} \cdot \mathbf{n}\, ds$$

measures the fluid flow across the boundary of $\mathcal{R}$. If the integral is positive, then more fluid is flowing out of the region than into it (indicating that there is a *source* in the region); if the integral is negative, then more fluid is flowing into the region than out of it (indicating that there is a *sink* in the region). According to equation (15.39), we have

$$\int_{\mathcal{C}} \mathbf{F} \cdot \mathbf{n}\, ds = \iint_{\mathcal{R}} \operatorname{div}(\mathbf{F})\, dA = \iint_{\mathcal{R}} (1 + 1)\, dA = 2\pi.$$

Since this number is positive, we see that the overall fluid flow is out of the region. A glance at the plot of $\mathbf{F}$ in Figure 5 confirms our conclusion. ■

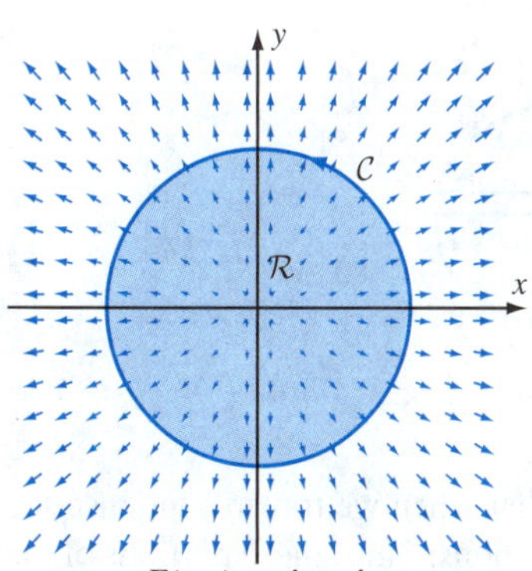

$\mathbf{F}(x, y) = x\mathbf{i} + y\mathbf{j}$

Figure 5

Green's Theorem for More General Regions

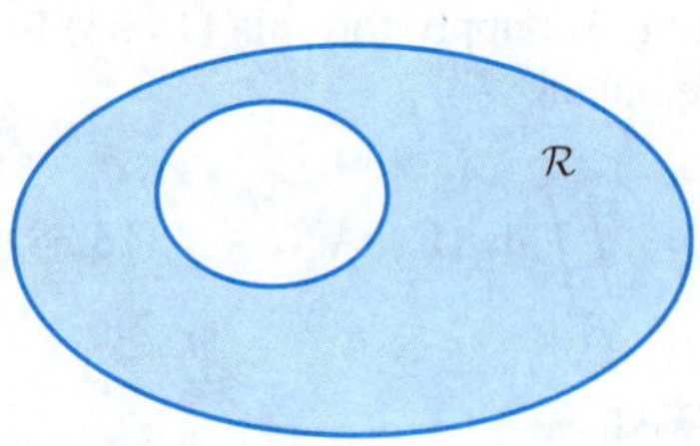

Figure 6

Now we want to discuss Green's Theorem on some regions different from those specified in Theorem 1. For example, the boundary of the region in Figure 6 consists of two smooth, simple closed curves. This new feature poses no problem. We orient *each* curve so that the region $\mathcal{R}$ is to its left as it is traversed (as in Figure 7). Then we calculate the line integral by integrating over each curve, one at a time, and adding up the results.

Theorem 3 Let $\mathcal{R}$ be a region in the plane bounded by one or more continuously differentiable curves $\mathcal{C}_1, \ldots, \mathcal{C}_N$ with parameterizations $\boldsymbol{r}_1, \boldsymbol{r}_2, \ldots, \boldsymbol{r}_N$. Assume that these curves are oriented so that the region $\mathcal{R}$ is on the left as each curve is traversed. If $\mathbf{F}(x, y) = M(x, y)\mathbf{i} + N(x, y)\mathbf{j}$ is a continuously differentiable vector field on a neighborhood of $\mathcal{R}$ and its boundary, then

$$\sum_{j=1}^{N} \int_{\mathcal{C}_j} \mathbf{F} \cdot d\mathbf{r}_j = \iint_{\mathcal{R}} \left(\frac{\partial N}{\partial x} - \frac{\partial M}{\partial y} \right) dA. \tag{15.40}$$

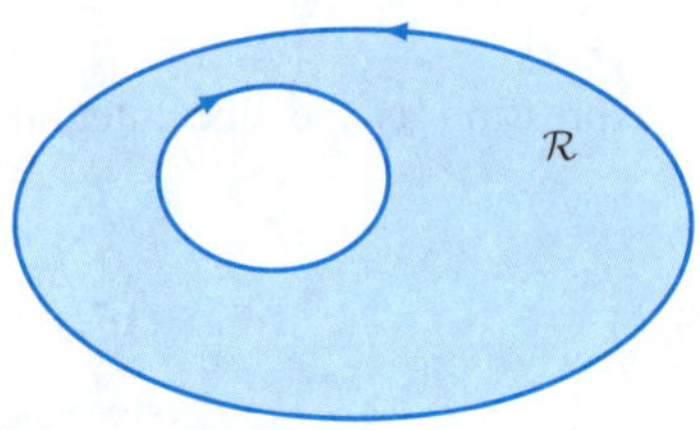

Figure 7

Proof Rather than give a complete proof, we will simply indicate how to prove Green's Theorem for a region that is not simply connected. Look again at the region $\mathcal{R}$ shown in Figure 6. Connect the two curves by line segments, as in Figure 8. Doing so creates two simply connected subregions. By applying Green's Theorem to each subregion, we obtain

$$\begin{aligned}
\iint_{\mathcal{R}} \left(\frac{\partial N}{\partial x} - \frac{\partial M}{\partial y} \right) dA &= \iint_{\mathcal{R}_1} \left(\frac{\partial N}{\partial x} - \frac{\partial M}{\partial y} \right) dA + \iint_{\mathcal{R}_2} \left(\frac{\partial N}{\partial x} - \frac{\partial M}{\partial y} \right) \\
&= \int_{\mathcal{C}_1} \mathbf{F} \cdot d\mathbf{r}_1 + \int_{\mathcal{C}_2} \mathbf{F} \cdot d\mathbf{r}_2 \\
&= \int_{\mathcal{C}} \mathbf{F} \cdot d\mathbf{r}.
\end{aligned}$$

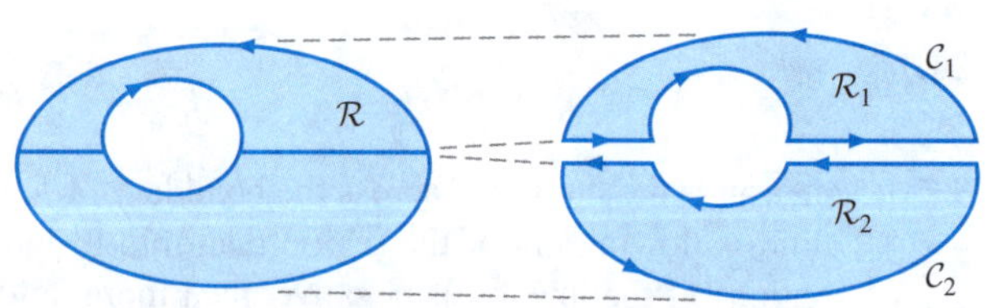

Figure 8

To obtain the last equality in this chain, we notice that when we traverse the inserted line segments in $\mathcal{C}_1$ and $\mathcal{C}_2$, we travel in opposite directions. Thus, these portions of the line integrals cancel when we add $\int_{\mathcal{C}_1} \mathbf{F} \cdot d\mathbf{r}_1$ and $\int_{\mathcal{C}_2} \mathbf{F} \cdot d\mathbf{r}_2$. ■

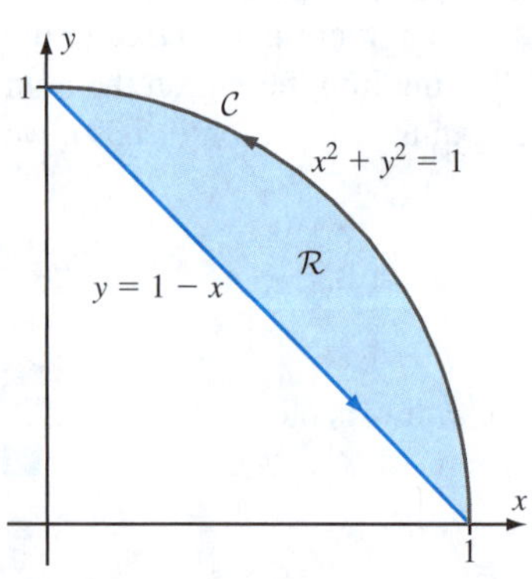

Figure 9

Example 5 Consider the region $\mathcal{R}$ (with piecewise-smooth boundary) that is shown in Figure 9. Verify equation (15.40) for $\mathbf{F} = \mathbf{i} + x^2\mathbf{j}$.

Solution Here $M = 1$, $N = x^2$, and $N_x - M_y = 2x$. We have

$$\iint_{\mathcal{R}} \left(\frac{\partial N}{\partial x} - \frac{\partial M}{\partial y}\right) dA = \int_0^1 \int_{1-x}^{\sqrt{1-x^2}} 2x\, dy\, dx = \int_0^1 2xy\Big|_{y=1-x}^{y=\sqrt{1-x^2}} dx$$

$$= \int_0^1 \left(2x\sqrt{1-x^2} - 2x(1-x)\right) dx.$$

Making the substitution $u = 1 - x^2$, $du = -2x\, dx$, we calculate

$$\int_0^1 (2x\sqrt{1-x^2})\, dx = -\int_1^0 \sqrt{u}\, du = \frac{2}{3}.$$

Since

$$\int_0^1 (-2x(1-x))\, dx = \int_0^1 (2x^2 - 2x)\, dx = \frac{2}{3}x^3 - x^2\Big|_{x=0}^{x=1} = -\frac{1}{3},$$

we conclude that

$$\iint_{\mathcal{R}} \left(\frac{\partial N}{\partial x} - \frac{\partial M}{\partial y}\right) dA = \frac{2}{3} + \left(-\frac{1}{3}\right) = \frac{1}{3}.$$

We can parameterize the boundary of $\mathcal{R}$ by using $\mathbf{r}_1(t) = \cos(t)\mathbf{i} + \sin(t)\mathbf{j},\ 0 \le t \le \pi/2$ for the circular arc $\mathcal{C}_1$ and $\mathbf{r}_2(t) = t\mathbf{i} + (1-t)\mathbf{j},\ 0 \le t \le 1$ for the line segment $\mathcal{C}_2$. Then

$$\int_{\mathcal{C}_1} \mathbf{F} \cdot d\mathbf{r} = \int_0^{\pi/2} \left(1\frac{d}{dt}\cos(t) + \cos^2(t)\frac{d}{dt}\sin(t)\right) dt = \int_0^{\pi/2} (-\sin(t) + \cos^3(t))\, dt.$$

An antiderivative of $\cos^3(t)$, namely, $\sin(t) - \sin^3(t)/3$, is calculated in Example 5 from Section 7.3. It follows that

$$\int_{\mathcal{C}_1} \mathbf{F} \cdot d\mathbf{r} = \cos(t) + \left(\sin(t) - \frac{\sin^3(t)}{3}\right)\Big|_{t=0}^{t=\pi/2} = \left(1 - \frac{1}{3}\right) - 1 = -\frac{1}{3}.$$

Finally, we have

$$\int_{\mathcal{C}_2} \mathbf{F} \cdot d\mathbf{r} = \int_0^1 \left(1\frac{d}{dt}t + t^2\frac{d}{dt}(1-t)\right) dt = \int_0^1 (1 - t^2)\, dt = \left(t - \frac{t^3}{3}\right)\Big|_{t=0}^{t=1} = \frac{2}{3}.$$

Thus,

$$\int_{\mathcal{C}} \mathbf{F} \cdot d\mathbf{r} = \int_{\mathcal{C}_1} \mathbf{F} \cdot d\mathbf{r} + \int_{\mathcal{C}_2} \mathbf{F} \cdot d\mathbf{r} = \left(-\frac{1}{3}\right) + \frac{2}{3} = \frac{1}{3} = \iint_{\mathcal{R}} \left(\frac{\partial N}{\partial x} - \frac{\partial M}{\partial y}\right) dA. \quad \blacksquare$$

Example 6 Let $\mathcal{R}$ be the region bounded by the circles $x^2 + y^2 = 1$ and $x^2 + y^2 = 4$. Verify Green's Theorem for the vector field $\mathbf{F} = -y^3\mathbf{i} + 2\mathbf{j}$ defined on $\mathcal{R}$.

Solution Here $M = -y^3$, $N = 2$, and $N_x - M_y = 3y^2$. Using polar coordinates, we have

$$\iint_{\mathcal{R}} \left(\frac{\partial N}{\partial x} - \frac{\partial M}{\partial y}\right) dA = \iint_{\mathcal{R}} 3y^2\, dA = \int_0^{2\pi} \int_1^2 3r^2 \sin^2(\theta) r\, dr\, d\theta$$

$$= 3\int_0^{2\pi} \sin^2(\theta) \left.\frac{r^4}{4}\right|_{r=1}^{r=2} d\theta = \frac{45}{4}\int_0^{2\pi} \sin^2(\theta)\, d\theta.$$

In Example 1, we observed that the value of the last integral is π. Thus, we have

$$\iint_{\mathcal{R}} \left(\frac{\partial N}{\partial x} - \frac{\partial M}{\partial y}\right) dA = \frac{45}{4}\pi.$$

Next we integrate over the oriented boundary $\mathcal{C}$ of $\mathcal{R}$. If we let $\mathcal{C}_a$ denote the circle of radius a that is centered at the origin and that has counterclockwise orientation, then we observe that $\mathcal{C} = \mathcal{C}_2 \cup (-\mathcal{C}_1)$ and

$$\int_{\mathcal{C}} \mathbf{F} \cdot d\mathbf{r} = \int_{\mathcal{C}_2} \mathbf{F} \cdot d\mathbf{r}_2 - \int_{\mathcal{C}_1} \mathbf{F} \cdot d\mathbf{r}_1. \qquad \textbf{(15.41)}$$

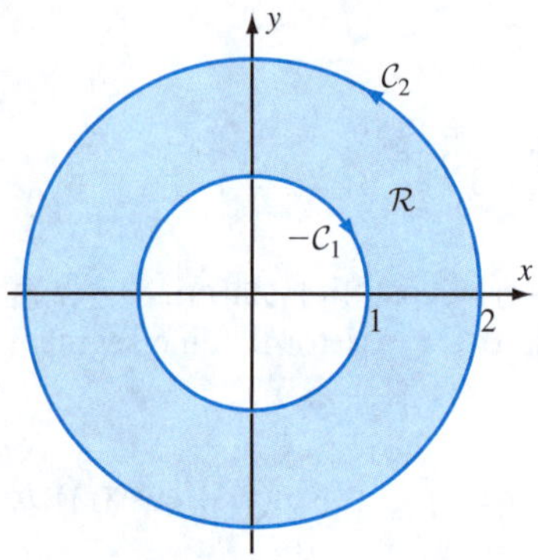

Figure 10

See Figure 10. We parameterize $\mathcal{C}_a$ by $\mathbf{r}(t) = a\cos(t)\mathbf{i} + a\sin(t)\mathbf{j}$, $0 \le t \le 2\pi$ and calculate

$$\int_{\mathcal{C}_a} \mathbf{F} \cdot d\mathbf{r} = \int_0^{2\pi} \left(-a^3 \sin^3(t)\frac{d}{dt}a\cos(t) + 2\frac{d}{dt}a\sin(t)\right) dt$$

$$= \int_0^{2\pi} (a^4 \sin^4(t) + 2a\cos(t))\, dt,$$

or

$$\int_{\mathcal{C}_a} \mathbf{F} \cdot d\mathbf{r} = a^4 \int_0^{2\pi} \sin^4(t)\, dt + 2a \underbrace{\int_0^{2\pi} \cos(t)\, dt}_{0}.$$

We can calculate the first integral on the right using formula (7.22) with $n = 4$. We obtain

$$\int_{\mathcal{C}_a} \mathbf{F} \cdot d\mathbf{r} = a^4 \int_0^{2\pi} \sin^4(t)\, dt = a^4 \left(-\frac{1}{4}\sin^3(t)\cos(t)\Big|_{t=0}^{t=2\pi} + \frac{3}{4}\int_0^{2\pi} \sin^2(t)\, dt\right)$$

$$= \frac{3}{4}\pi a^4. \qquad \textbf{(15.42)}$$

We conclude by using formula (15.42) with $a = 1$ and $a = 2$ to evaluate the two summands of the right side of (15.41). We obtain

$$\int_{\mathcal{C}} \mathbf{F} \cdot d\mathbf{r} = \int_{\mathcal{C}_2} \mathbf{F} \cdot d\mathbf{r}_2 - \int_{\mathcal{C}_1} \mathbf{F} \cdot d\mathbf{r}_1 = \frac{3}{4}\pi \cdot 2^4 - \frac{3}{4}\pi \cdot 1^4 = \frac{45}{4}\pi,$$

as required. ■

quickquiz

1. Explain how we positively orient the boundary of a region.
2. State Green's Theorem.
3. State the forms of Green's Theorem that involve div and curl.
4. Explain how Green's Theorem can be used to calculate area.

EXERCISES

Problems for Practice

In Exercises 1–6, calculate both sides of the formula in Green's Theorem and verify that they are equal.

1. $\mathbf{F}(x, y) = y^2\mathbf{i} - 3x\mathbf{j}$, $\mathbf{r}(t) = \sin(t)\mathbf{i} - \cos(t)\mathbf{j}$, $\pi \le t \le 3\pi$
2. $\mathbf{F}(x, y) = xy\mathbf{i} - x^2\mathbf{j}$, $\mathbf{r}(t) = \cos(t)\mathbf{i} + 3\sin(t)\mathbf{j}$, $0 \le t \le 2\pi$
3. $\mathbf{F}(x, y) = (y - 2x)\mathbf{i} + (3x - 4y)\mathbf{j}$, $\mathbf{r}(t) = \cos(2t)\mathbf{i} + \sin(2t)\mathbf{j}$, $0 \le t \le \pi$
4. $\mathbf{F}(x, y) = xy^2\mathbf{i} - yx\mathbf{j}$, $\mathbf{r}(t) = \sin(2t)\mathbf{i} - \cos(2t)\mathbf{j}$, $\pi/2 \le t \le 3\pi/2$
5. $\mathbf{F}(x, y) = y\mathbf{i} + 2x\mathbf{j}$, $\mathbf{r}(t) = \sin(t)\mathbf{i} - \cos(t)\mathbf{j}$, $-\pi \le t \le \pi$
6. $\mathbf{F}(x, y) = y\mathbf{i} + y^2\mathbf{j}$, $\mathbf{r}(t) = \cos(4t)\mathbf{i} + \sin(4t)\mathbf{j}$, $-\pi/2 \le t \le 0$

In Exercises 7–10, determine the region bounded by the oriented curve $\mathcal{C}$. Then use Green's Theorem to calculate $\int_{\mathcal{C}} \mathbf{F} \cdot d\mathbf{r}$.

7. $\mathbf{F}(x, y) = yx\mathbf{i} - x^2\mathbf{j}$, $\mathbf{r}(t) = \cos(t)\mathbf{i} + \sin(t)\mathbf{j}$, $0 \le t \le 2\pi$
8. $\mathbf{F}(x, y) = 3y\mathbf{i} - 4x\mathbf{j}$, $\mathbf{r}(t) = 3\sin(t)\mathbf{i} - 4\cos(t)\mathbf{j}$, $\pi \le t \le 3\pi$
9. $\mathbf{F}(x, y) = -y^2\mathbf{i} + x^3\mathbf{j}$, $\mathbf{r}(t) = \sin(t)\mathbf{i} - \cos(t)\mathbf{j}$, $-\pi \le t \le \pi$
10. $\mathbf{F}(x, y) = (y - x)\mathbf{i} + (x - 3y)\mathbf{j}$, $\mathbf{r}(t) = 4\cos(t)\mathbf{i} + \sin(t)\mathbf{j}$, $0 \le t \le 2\pi$

In Exercises 11–16, the given region has a piecewise-smooth boundary. Apply Green's Theorem to evaluate the line integral of **F** around the positively oriented boundary curve.

11. $\{(x, y) : -1 \le x \le 1, -2 \le y \le 2\}$; $\mathbf{F}(x, y) = xy^2\mathbf{i} - y^2x^3\mathbf{j}$
12. the triangle with vertices $(\pi/4, 0)$, $(0, \pi/4)$, and $(0, 0)$; $\mathbf{F}(x, y) = \cos(y)\mathbf{i} + \sin(x)\mathbf{j}$
13. the square with vertices $(1, 0)$, $(0, 1)$, $(-1, 0)$, and $(0, -1)$; $\mathbf{F}(x, y) = \ln(3 + y)\mathbf{i} - xy\mathbf{j}$
14. the rectangle with vertices $(-2, -1)$, $(2, -1)$, $(2, 1)$, and $(-2, 1)$; $\mathbf{F}(x, y) = \sin^3(y)\mathbf{i} - \cos^3(x)\mathbf{j}$
15. the region bounded by $y = x^2$ and $y = 4x + 5$; $\mathbf{F}(x, y) = x^2y\mathbf{i} + xy\mathbf{j}$
16. $\{(x, y) : 0 \le y \le \sqrt{4 - x^2}, -2 \le x \le 2\}$; $\mathbf{F}(x, y) = e^x\mathbf{i} - xy\mathbf{j}$

In Exercises 17–22, use formula (15.36) to calculate the area of the region.

17. the triangle with vertices $(2, 5)$, $(3, -6)$, and $(4, 1)$
18. the trapezoid with vertices $(4, 1)$, $(-2, 1)$, $(3, 3)$, and $(-1, 3)$
19. the region bounded by the parabola $y = -2x^2 + 6$ and the line $y = 4x$
20. $\{(x, y) : 1/3 \le y \le 1/x, 1/2 \le x \le 3\}$
21. the area inside the circle $x^2 + y^2 = 1$ and lying above the line $y = (x + 1)/\sqrt{3}$
22. the region obtained by removing the disk with boundary $x^2 + y^2 = 4$ from the triangle with vertices $(-14, 16)$, $(8, 12)$, and $(1, -20)$

In Exercises 23–25, compute the flux of **F** across the boundary of the region $\mathcal{R}$.

23. $\mathbf{F}(x, y) = xy^2\mathbf{i} - x\mathbf{j}$, $\mathcal{R} = \{(x, y) : |x| < y < 1\}$
24. $\mathbf{F}(x, y) = y/(x^2 + 1)\mathbf{i} - 3x/(y^2 + 1)\mathbf{j}$, $\mathcal{R} = \{(x, y) : 0 < x < 2, -1 < y < 2\}$
25. $\mathbf{F}(x, y) = xy^{7/2}\mathbf{i} - yx^{5/2}\mathbf{j}$, $\mathcal{R} = \{(x, y) : 0 < x < y < 1\}$

Further Theory and Practice

26. Use formula (15.36) to calculate the area of the region bounded by $\mathbf{r}(t) = \langle \cos(t) - 2\sin(t), \cos(t) \rangle$, $0 \le t \le 2\pi$.
27. Use formula (15.36) to calculate the area inside the cardioid that is described by the polar equation $r = 2 + 2\cos(\theta)$.

28. Use formula (15.36) to calculate the area inside the limaçon with polar equation $r = 5 - 3\sin(\theta)$.

29. Calculate $\oint_{\mathcal{C}} (y + \arcsin(x/2))\,dx + (2xy + \ln(1 + y^4))\,dy$ where $\mathcal{C}$ is the closed counterclockwise path consisting of the graphs $y = x^2, 0 \leq x \leq 1$ and $y = \sqrt{x}, 0 \leq x \leq 1$.

30. Calculate $\oint_{\mathcal{C}} (x - y\sec^2(x))\,dx + \arctan(y^2)\,dy$ where $\mathcal{C}$ is the closed counterclockwise path consisting of the graphs $y = \tan(x)^2, 0 \leq x \leq \pi/4$ and $y = \tan(x)^4, 0 \leq x \leq \pi/4$.

31. Sketch the curve $\mathcal{C}$ parameterized by $\mathbf{r}(t) = \langle \sin(t), \sin(2t) \rangle, 0 \leq t \leq \pi$. Use formula (15.36) to calculate the area of the region enclosed by $\mathcal{C}$.

32. Sketch the curve $\mathcal{C}$ parameterized by $\mathbf{r}(t) = \langle t - t^2, t - t^3 \rangle, 0 \leq t \leq 1$. Use formula (15.36) to calculate the area of the region enclosed by $\mathcal{C}$.

33. The Folium of Descartes defined by $x^3 + y^3 = 3xy$ is shown in Figure 11. The loop in the folium is a simple closed curve that can be parameterized by $\mathbf{r}(t) = 3t/(1 + t^3)\mathbf{i} + 3t^2/(1 + t^3)\mathbf{j}, 0 \leq t < \infty$. Although the interval of the parameter is unbounded, formula (15.36) does yield the correct area of the region enclosed by the loop. Calculate the area.

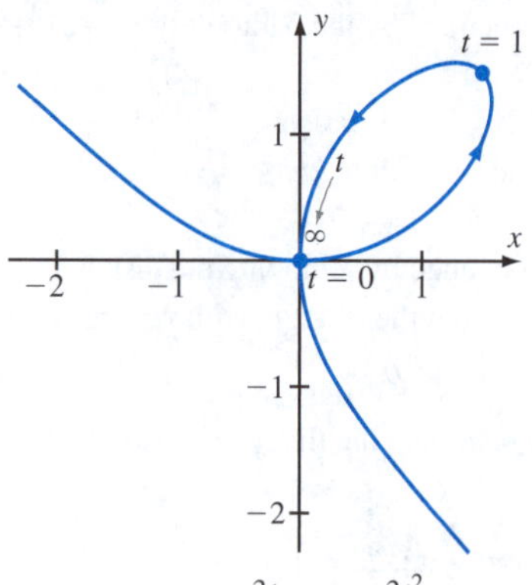

Figure 11

34. Suppose that u and v are twice continuously differentiable functions in a neighborhood of a region $\mathcal{R}$ and its boundary curve $\mathcal{C}$. Suppose that $\mathcal{R}$ and $\mathcal{C}$ are as in the statement of Green's Theorem. Prove the following identities:

a. Green's First Identity

$$\iint_{\mathcal{R}} (u \,\Delta\, v + \nabla u \cdot \nabla v)\,dA = \int_{\mathcal{C}} (uv_x\,dy - uv_y\,dx)$$

b. Green's Second Identity

$$\iint_{\mathcal{R}} (u \,\Delta\, v - v \,\Delta\, u)\,dA$$

$$= \int_{\mathcal{C}} (uv_x - vu_x)\,dy + (vu_y - uv_y)\,dx$$

(*Hint:* For part b, use part a by reversing the roles of u and v.)

35. Use Green's Second Identity (from Exercise 34) to show that

$$\int_{\mathcal{C}} u_y\,dx - u_x\,dy = 0$$

if u is harmonic.

Calculator/Computer Exercises

36. Plot the simple closed curve that is parameterized by $\mathbf{r}(t) = \sin(2t)\mathbf{i} - \sin^2(3t)\mathbf{j}, \pi/12 < t < 5\pi/12$. Calculate the area that is enclosed.

In Exercises 37–40, verify Green's Theorem for the vector field $\mathbf{F}$ and the region $\mathcal{R}$.

37. $\mathbf{F}(x, y) = \sin(x + y^2)\mathbf{i}$, $\mathcal{R} = \{(x, y) : x^2 < y < x, 0 < x < 1\}$

38. $\mathbf{F}(x, y) = \exp(yx^2)\mathbf{i} + x\exp(xy^2)\mathbf{j}$, $\mathcal{R} = \{(x, y) : 0 < y < 1 - x^2, -1 < x < 1\}$

39. $\mathbf{F}(x, y) = x^2/(1 + y^4)\mathbf{i} + \ln(1 + x^4)\mathbf{j}$, $\mathcal{R} = \{(x, y) : 0 < y < \sin(x), 0 < x < \pi\}$

40. $\mathbf{F}(x, y) = (1 + x^2 + y^2)^{-1/2}\mathbf{i} + \sqrt{1 + x^2 + y^2}\mathbf{j}$, $\mathcal{R} = \{(x, y) : (x-1)^2 < y < 1-x+x^2-x^5, 0 < x < 1\}$

15.6 Surface Integrals

In this section, we develop the final concept that we will need for the principal theorems of vector analysis—the concept of the surface integral, which enables us to integrate a function defined on a surface. We begin with the problem of calculating the area of a surface that is the graph of a function.

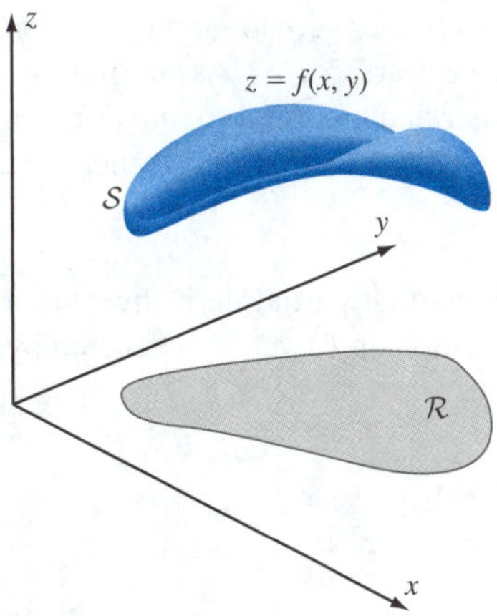

Figure 1

Suppose that a surface $\mathcal{S}$ is given as the graph of a function $z = f(x, y)$ over a region $\mathcal{R}$ in the xy-plane, as in Figure 1. How do we calculate the surface area of $\mathcal{S}$? As you might suspect, we can develop a method for calculating surface area by using a double integral of a certain expression over the region $\mathcal{R}$, which is what we now do.

The Integral for Surface Area

Suppose that a surface $\mathcal{S}$ is the graph of a function f over a bounded region $\mathcal{R}$ in the xy-plane, as in Figure 1. Let $Q = \{(x, y) : a < x < b, c < y < d\}$ be a rectangle that contains $\mathcal{R}$. We partition the x-interval $[a, b]$ as

$$a = x_0 < x_1 < x_2 < \cdots < x_{N-1} < x_N = b \quad \text{where } x_i = a + i\Delta x, \quad \Delta x = \frac{b-a}{N},$$

and we partition the y-interval $[c, d]$ as

$$c = y_0 < y_1 < y_2 < \cdots < y_{N-1} < y_N = d \quad \text{where } y_j = c + j\Delta y, \quad \Delta y = \frac{d-c}{N}.$$

Corresponding to each pair x_{i-1}, x_i in the partition of the x-variable and each pair y_{j-1}, y_j in the partition of the y-variable, there is a small subrectangle $Q_{i,j}$ in the xy-plane. The only relevant rectangles $Q_{i,j}$ for our calculation are the ones that lie entirely in $\mathcal{R}$. See Figure 2. We wish to approximate the area $A_{i,j}$ of the small piece of surface lying above $Q_{i,j}$. Examine Figure 3. We estimate $A_{i,j}$ by the area of a parallelogram that closely approximates the graph of f over $Q_{i,j}$. To do this, we select a point $\xi_{i,j}$ inside $Q_{i,j}$ and use the parallelogram that is tangent to $\mathcal{S}$ at the point over $\xi_{i,j}$ and that has the vectors $\mathbf{v} = \langle \Delta x, 0, f_x(\xi_{i,j})\Delta x\rangle$ and $\mathbf{w} = \langle 0, \Delta y, f_y(\xi_{i,j})\Delta y\rangle$ for its sides. Refer to Figure 4. We know from Section 11.4 that the area of this parallelogram is

$$\begin{aligned}\|\mathbf{v}\times\mathbf{w}\| = \left\|\det\left(\begin{bmatrix}\mathbf{i} & \mathbf{j} & \mathbf{k}\\ \Delta x & 0 & f_x(\xi_{i,j})\Delta x\\ 0 & \Delta y & f_y(\xi_{i,j})\Delta y\end{bmatrix}\right)\right\| &= \|(-f_x(\xi_{i,j})\Delta x\Delta y)\mathbf{i} + (-f_y(\xi_{i,j})\Delta x\Delta y)\mathbf{j} + (\Delta x\Delta y)\mathbf{k}\|\\ &= \sqrt{1 + f_x(\xi_{i,j})^2 + f_y(\xi_{i,j})^2}\,\Delta x\Delta y.\end{aligned}$$

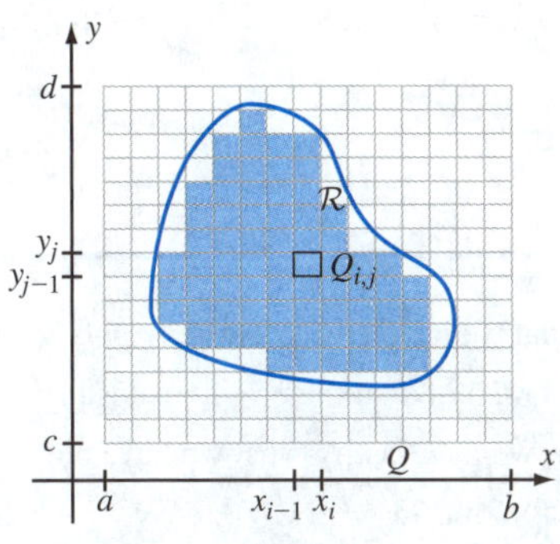

Figure 2

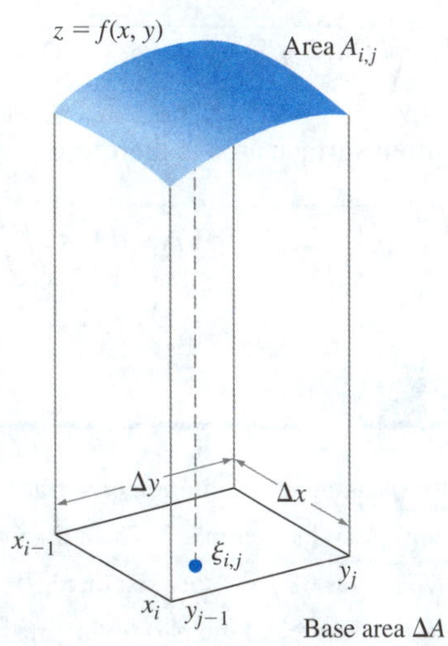

Figure 3

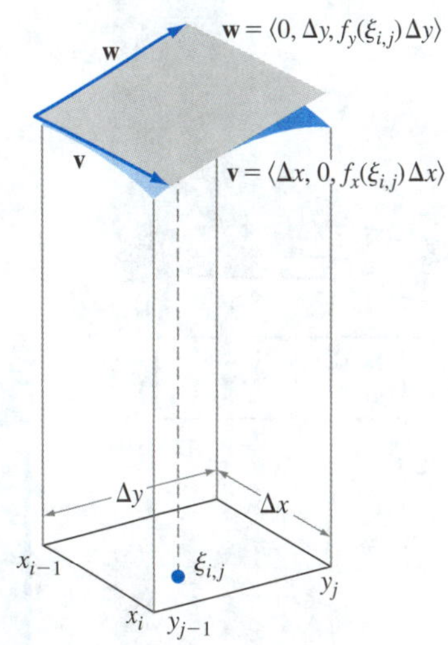

Figure 4

If we sum over rectangles $Q_{i,j}$ that lie in $\mathcal{R}$, then we get an approximation $\sum_{Q_{i,j}\subset\mathcal{R}} \sqrt{1+f_x(\xi_{i,j})^2+f_y(\xi_{i,j})^2}\,\Delta x\Delta y$ to the surface area of $\mathcal{S}$. As the number N increases, our approximation improves. On the other hand, our approximation is a Riemann sum for the double integral $\iint_{\mathcal{R}} \sqrt{1+f_x(x,y)^2+f_y(x,y)^2}\,dA$. We therefore make the following definition.

Definition Let $\mathcal{S}$ be the graph of a function f that has continuous first partial derivatives on a region $\mathcal{R}$ in the xy-plane. Then the surface area of the graph of f over $\mathcal{R}$ is denoted by $\iint_{\mathcal{S}} dS$ and is given by

$$\iint_{\mathcal{R}} \sqrt{1+(f_x)^2+(f_y)^2}\,dA.$$

We refer to dS as the *element of surface area* on $\mathcal{S}$.

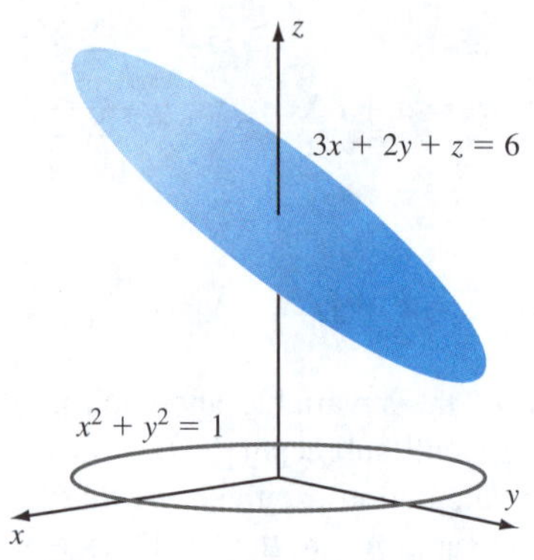

Figure 5

Example 1 Find the area of the portion of the plane $3x+2y+z=6$ that lies over the interior of the circle $x^2+y^2=1$ in the xy-plane.

Solution The surface whose area we are computing is *not itself a disk.* It is the interior of an ellipse. See Figure 5. But we can do the area calculation without determining the exact nature of this ellipse. We need only notice that the plane is the graph of the function $z=f(x,y)=6-3x-2y$. We need to calculate the surface area of the graph over the region $\mathcal{R}=\{(x,y): x^2+y^2<1\}$. According to the definition, the required surface area is

$$\iint_{\mathcal{R}} \sqrt{1+(f_x)^2+(f_y)^2}\,dA = \iint_{\mathcal{R}} \sqrt{1+(-3)^2+(-2)^2}\,dA = \sqrt{14}\iint_{\mathcal{R}} dA = \sqrt{14}\cdot\pi\cdot 1^2 = \sqrt{14}\pi. \quad \blacksquare$$

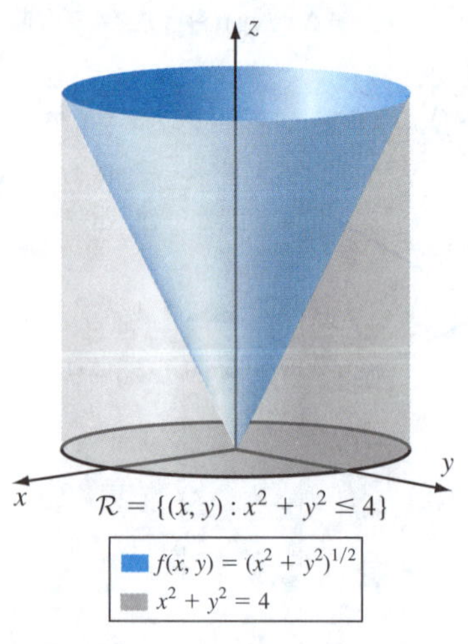

Figure 6

Example 2 Calculate the surface area of that portion of the graph of $z=\sqrt{x^2+y^2}$ that lies inside the cylinder $x^2+y^2=4$.

Solution See Figure 6. To determine the surface area of the graph of $z=\sqrt{x^2+y^2}$, we let $f(x,y)=(x^2+y^2)^{1/2}$ and calculate

$$\sqrt{1+f_x(x,y)^2+f_y(x,y)^2} = \sqrt{1+\frac{x^2}{x^2+y^2}+\frac{y^2}{x^2+y^2}} = \sqrt{2}.$$

The required surface area is therefore

$$\iint_{\mathcal{R}} \sqrt{1+(f_x)^2+(f_y)^2}\,dA = \iint_{\mathcal{R}} \sqrt{2}\,dA = \sqrt{2}\cdot(\text{area of }\mathcal{R}) = 4\pi\sqrt{2}. \quad \blacksquare$$

inSIGHT

Strictly speaking, the definition of surface area that we have given does not apply to the conical surface in Example 2. The corner on the graph of f at the origin means that the first derivatives of f do not exist there. However, we may justify what we have done by calculating the area of the part of the graph over the ring $\mathcal{R}_\epsilon=\{(x,y):\epsilon^2<x^2+y^2<4\}$ and then letting ϵ tend to 0. See Figure 7.

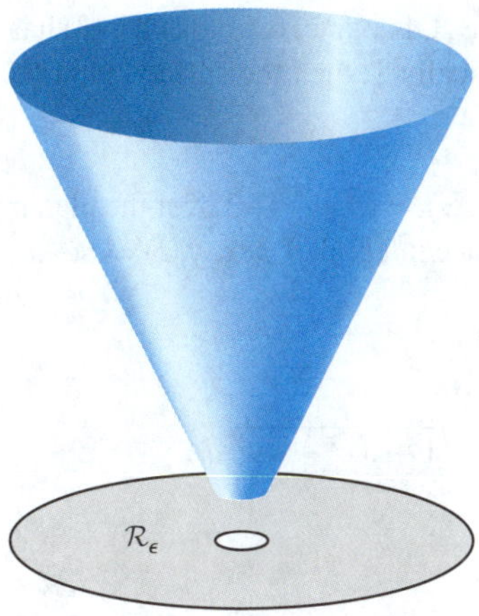

$\mathcal{R}_\epsilon = \{(x, y) : \epsilon^2 < x^2 + y^2 < 4\}$

Figure 7

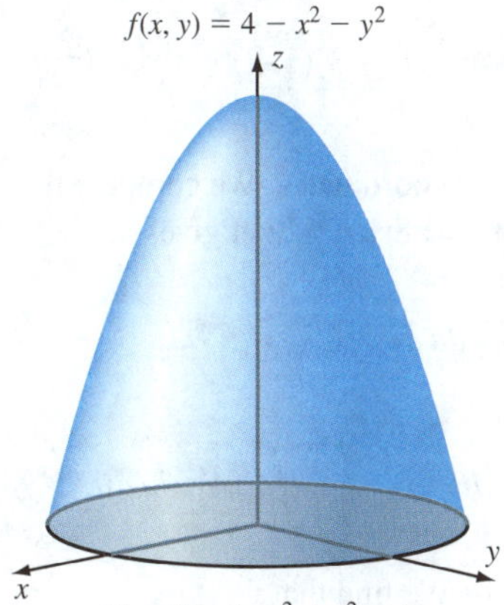

$\mathcal{R} = \{(x, y) : x^2 + y^2 \le 4\}$

Figure 8

Example 3 Calculate the surface area of that part of the graph of $f(x, y) = 4 - x^2 - y^2$ that lies above the xy-plane.

Solution Look at Figure 8. We are required to calculate the surface area of the graph of f over the region $\mathcal{R} = \{(x, y) : x^2 + y^2 \le 4\}$:

$$\iint_{\mathcal{R}} \sqrt{1 + (f_x)^2 + (f_y)^2}\, dA = \iint_{\mathcal{R}} \sqrt{1 + 4x^2 + 4y^2}\, dA.$$

This integration is easiest if we use polar coordinates. The region of integration is given by $\mathcal{R} = \{(r, \theta) : 0 \le \theta \le 2\pi, 0 \le r \le 2\}$ in polar coordinates. The integral becomes

$$\int_0^{2\pi} \int_0^2 \sqrt{1 + 4r^2} r\, dr\, d\theta = \int_0^{2\pi} \frac{1}{12}(1 + 4r^2)^{3/2} \Big|_{r=0}^{r=2} d\theta$$

$$= \int_0^{2\pi} \frac{17^{3/2} - 1}{12}\, d\theta = \frac{17^{3/2} - 1}{6}\pi.$$ ■

Integrating a Function over a Surface

Once again, consider the surface $\mathcal{S}$ shown in Figure 1. Suppose that at each point (x, y, z) of $\mathcal{S}$, we know the weight density $\varphi(x, y, z)$ of the surface. To calculate the total weight of the surface, we repeat the procedure we used to find surface area; that is, we enclose the region $\mathcal{R}$ in a rectangle Q, we subdivide each side of Q into N equal-length subintervals (Figure 2), and we choose a point $\xi_{i,j}$ in each of the resulting small rectangles $Q_{i,j}$ that are entirely contained within $\mathcal{R}$. The weight of that part of $\mathcal{S}$ that lies over $Q_{i,j}$ (as shown in Figure 3) is approximated by

$$\underbrace{\varphi(\xi_{i,j}, f(\xi_{i,j}))}_{\text{weight per unit area}} \cdot \underbrace{\sqrt{1 + f_x(\xi_{i,j})^2 + f_y(\xi_{i,j})^2} \cdot \Delta x \cdot \Delta y}_{\text{approximate surface area of patch over } Q_{i,j}}.$$

Summing over all the subrectangles $Q_{i,j}$ in $\mathcal{R}$, we obtain the Riemann sum

$$\sum_{i,j} \varphi(\xi_{i,j}, f(\xi_{i,j})) \sqrt{1 + f_x(\xi_{i,j})^2 + f_y(\xi_{i,j})^2} \Delta x \Delta y$$

as an approximation of the weight of $\mathcal{S}$. This approximation tends to improve as the number N of subdivisions increases. Since our approximation is a Riemann sum for the double integral $\iint_{\mathcal{R}} \phi(x, y, f(x, y)) \sqrt{1 + f_x(x, y)^2 + f_y(x, y)^2}\, dA$, we make the following definition.

Definition Let $\mathcal{S}$ be the graph of a function f that has continuous first partial derivatives on a region $\mathcal{R}$ in the xy-plane. If φ is a continuous function on $\mathcal{S}$, then the *surface integral* of φ over $\mathcal{S}$, denoted by $\iint_{\mathcal{S}} \varphi\, dS$, is given by

$$\iint_{\mathcal{S}} \varphi\, dS = \iint_{\mathcal{R}} \phi(x, y, f(x, y)) \sqrt{1 + f_x(x, y)^2 + f_y(x, y)^2}\, dA. \qquad \textbf{(15.43)}$$

Formula (15.43) is often used when φ is a density function (such as weight density, mass density, charge density, and so on). The surface integral then represents the total value of the quantity whose density is represented by φ. Of course, formula (15.43) has meaning for *any* continuous function φ. If φ is identically 1, then the surface integral reduces to the integral for surface area.

Example 4 Let the surface $\mathcal{S}$ be the graph of $f(x, y) = x^2 + y^2 + 3$ over the planar region $\mathcal{R} = \{(x, y) : x^2 + y^2 < 2\}$. Calculate the surface integral of $\varphi(x, y, z) = z - x^2$ over the surface $\mathcal{S}$.

Solution The element of surface area is

$$dS = \sqrt{1 + f_x(x, y)^2 + f_y(x, y)^2}\, dA = \sqrt{1 + 4x^2 + 4y^2}\, dA$$

and

$$\varphi(x, y, f(x, y)) = f(x, y) - x^2 = (x^2 + y^2 + 3) - x^2 = y^2 + 3.$$

The surface integral therefore equals

$$\iint_{\mathcal{S}} \varphi\, dS = \iint_{\mathcal{R}} (y^2+3)\sqrt{1 + 4x^2 + 4y^2}\, dA = \int_0^{2\pi} \int_0^{\sqrt{2}} (r^2 \sin^2(\theta) + 3)(1 + 4r^2)^{1/2}\, r\, dr\, d\theta$$

where the last integral is obtained by converting to polar coordinates. We compute this last integral by making the substitution $u = 1 + 4r^2$, $du = 8r\,dr$, which gives us

$$\iint_{\mathcal{S}} \varphi\, dS = \int_0^{2\pi} \int_1^9 \left(\frac{u-1}{4} \sin^2(\theta) + 3 \right) u^{1/2} \frac{1}{8}\, du\, d\theta$$

$$= \frac{1}{32} \int_0^{2\pi} \sin^2(\theta) \int_1^9 (u^{3/2} - u^{1/2})\, du\, d\theta + \frac{3}{8} \int_0^{2\pi} \int_1^9 u^{1/2}\, du\, d\theta.$$

The integrals involving u may be routinely determined: We find that

$$\int_1^9 (u^{3/2} - u^{1/2})\, du = \left(\frac{2}{5} u^{5/2} - \frac{2}{3} u^{3/2} \right) \Big|_1^9 = \frac{1192}{15} \quad \text{and} \quad \int_1^9 u^{1/2}\, du = \left(\frac{2}{3} u^{3/2} \right) \Big|_1^9 = \frac{52}{3}.$$

Using these integrations and formula (7.21), we obtain

$$\iint_{\mathcal{S}} \varphi\, dS = \frac{1}{32} \cdot \frac{1192}{15} \int_0^{2\pi} \sin^2(\theta)\, d\theta + \frac{3}{8} \cdot \frac{52}{3} \int_0^{2\pi} d\theta = \frac{929}{60} \pi.$$ ■

Example 5 Integrate the function $\varphi(x, y, z) = xy - 3z$ over the surface $\mathcal{S}$ given by the graph of $f(x, y) = xy$ over the region $\mathcal{R} = \{(x, y) : 0 < x < 1, 0 < y < 2\}$.

Solution The element of surface area is

$$dS = \sqrt{1 + f_x(x, y)^2 + f_y(x, y)^2}\, dA = (1 + y^2 + x^2)^{1/2}\, dx\, dy$$

and

$$\varphi(x, y, f(x, y)) = xy - 3f(x, y) = xy - 3xy = -2xy.$$

Thus, our required surface integral $\iint_S \varphi \, dS$ is given by

$$\int_0^2 \int_0^1 -2xy(1+y^2+x^2)^{1/2}\,dx\,dy = -2\int_0^2 y \int_0^1 x(1+y^2+x^2)^{1/2} dx\,dy$$
$$= -2\int_0^2 y\left(\frac{1}{3}(1+y^2+x^2)^{3/2}\Big|_{x=0}^{x=1}\right)dy.$$

It follows that

$$\iint_{\mathcal{S}} \varphi\, dS = -\frac{2}{3}\int_0^2 y((2+y^2)^{3/2}-(1+y^2)^{3/2})\,dy$$
$$= -\frac{2}{15}\left((2+y^2)^{5/2}-(1+y^2)^{5/2}\right)\Big|_{y=0}^{y=2} = \frac{2}{15}(2^{5/2}+5^{5/2}-6^{5/2}-1).$$ ■

An Application

We now give an application of the concept of surface integral to the determination of a center of mass.

Example 6 Assuming that it has uniform mass distribution 1, determine the center of mass of the upper half of the sphere $x^2+y^2+z^2=a^2$.

Solution Refer to Section 14.7 for the concept of center of mass. The surface to be studied is $\mathcal{S}=\{(x,y,z): x^2+y^2+z^2=a^2, z>0\}$. By symmetry considerations, it is clear that $\bar{x}=0$ and $\bar{y}=0$. Thus, we need only calculate $\bar{z}$. We think of the surface as the graph of the function $f(x,y)=\sqrt{a^2-x^2-y^2}$ over the region $\mathcal{R}=\{(x,y): x^2+y^2<a^2\}$. We see that

$$M_{z=0} = \iint_{\mathcal{R}} 1\cdot z\cdot\sqrt{1+f_x(x,y)^2+f_y(x,y)^2}\,dA.$$

Now

$$f_x(x,y)=\frac{-x}{\sqrt{a^2-x^2-y^2}} \quad\text{and}\quad f_y(x,y)=\frac{-y}{\sqrt{a^2-x^2-y^2}}.$$

It follows that

$$\sqrt{1+f_x(x,y)^2+f_y(x,y)^2}=\sqrt{1+\left(\frac{x^2}{a^2-x^2-y^2}\right)+\left(\frac{y^2}{a^2-x^2-y^2}\right)}$$
$$=\frac{a}{\sqrt{a^2-x^2-y^2}}. \qquad \textbf{(15.44)}$$

Therefore,

$$M_{z=0}=\iint_{\mathcal{R}} z\cdot\frac{a}{\sqrt{a^2-x^2-y^2}}\,dA = a\iint_{\mathcal{R}}\frac{\sqrt{a^2-x^2-y^2}}{\sqrt{a^2-x^2-y^2}}\,dA = a\iint_{\mathcal{R}} 1\,dA.$$

This double integral is just the area of $\mathcal{R}$. Therefore, $M_{z=0}=\pi a^3$.

We also need to calculate the mass of $\mathcal{S}$:

$$M = \iint_{\mathcal{R}} 1 \cdot \sqrt{1 + f_x(x,y)^2 + f_y(x,y)^2}\, dA \overset{(15.44)}{=} \iint_{\mathcal{R}} \frac{a}{\sqrt{a^2 - x^2 - y^2}}\, dA.$$

This double integral is best evaluated using polar coordinates:

$$M = a \int_0^{2\pi} \int_0^a \frac{1}{\sqrt{a^2 - r^2}}\, r\, dr\, d\theta = a \int_0^{2\pi} \int_0^1 r(a^2 - r^2)^{-1/2}\, dr\, d\theta.$$

The change of variable $u = a^2 - r^2$, $du = -2r\, dr$ yields

$$M = -\frac{a}{2} \int_0^{2\pi} \int_{a^2}^0 u^{-1/2}\, du\, d\theta = -a \int_0^{2\pi} u^{1/2} \Big|_{u=a^2}^{u=0} d\theta = a^2 \int_0^{2\pi} 1\, d\theta = 2\pi a^2.$$

In conclusion, $\bar{z} = M_{z=0}/M = \pi a^3/2\pi a^2 = a/2$. Therefore, the center of mass is $(0, 0, a/2)$. ∎

The Element of Area for a Surface That Is Given Parametrically

In many applications, it is most convenient to describe a surface parametrically. We now briefly describe how to calculate the element of area on a surface $\mathcal{S}$ that is parameterized by a vector-valued function $\mathbf{r}(u, v) = \langle x(u,v), y(u,v), z(u,v)\rangle$ where u and v range over a parameter set $\mathcal{R}$ in the plane (Figure 9). Fix values u_0 and v_0. Figure 10 shows a small rectangle Q with vertex at (u_0, v_0), width Δu, and height Δv. The patch on $\mathcal{S}$ that is the image $\mathbf{r}(Q)$ of Q is also shown. In analogy with our preceding investigation, we wish to find the area of a parallelogram that approximates $\mathbf{r}(Q)$. The functions

$$u \mapsto \langle x(u, v_0), y(u, v_0), z(u, v_0)\rangle \qquad \text{and} \qquad v \mapsto \langle x(u_0, v), y(u_0, v), z(u_0, v)\rangle$$

describe curves on the surface $\mathcal{S}$ that pass through the point $P_0 = \mathbf{r}(u_0, v_0)$. The velocity vectors $\mathbf{r}_u(u_0, v_0)$ and $\mathbf{r}_v(u_0, v_0)$ are tangent to $\mathcal{S}$ at P_0, as are their scalar multiples $(\Delta u)\mathbf{r}_u(u_0, v_0)$ and $(\Delta v)\mathbf{r}_v(u_0, v_0)$. (Here the subscripts u and v denote partial differentiation with respect to u and v, respectively.) By reasoning similar to that

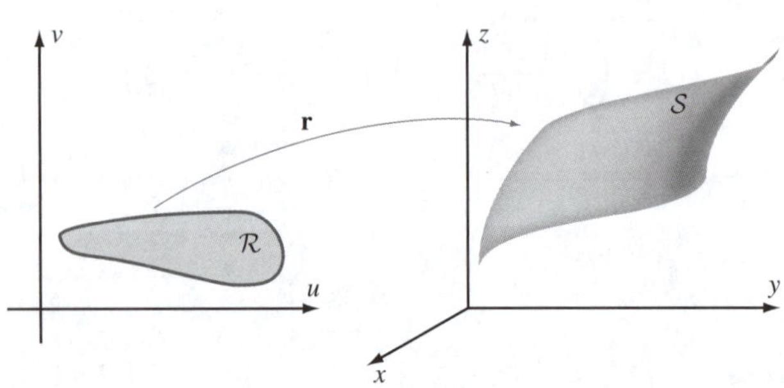

Figure 9

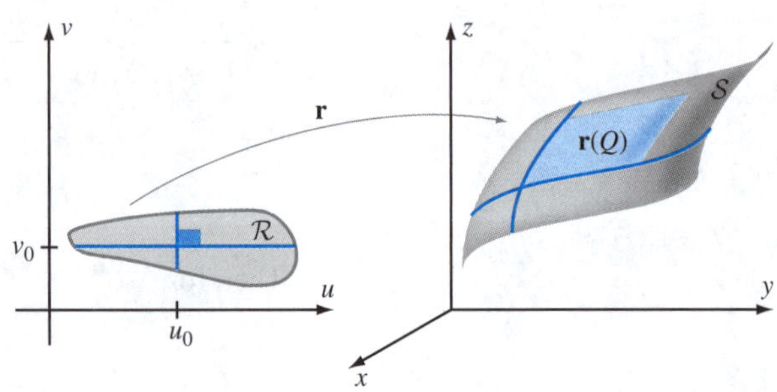

Figure 10
Rectangle Q of width Δu and height Δv

used at the beginning of this section, we find that $\mathbf{r}(Q)$ is closely approximated by the area of the parallelogram determined by the vectors $(\Delta u)\mathbf{r}_u(u_0, v_0)$ and $(\Delta v)\mathbf{r}_v(u_0, v_0)$. Refer to Figure 11. That area is equal to $\|(\Delta u)\mathbf{r}_u(u_0, v_0) \times (\Delta v)\mathbf{r}_v(u_0, v_0)\|$, or $\|(\mathbf{r}_u \times \mathbf{r}_v)(u_0, v_0)\|\Delta u \Delta v$. Adding these expressions, we obtain a Riemann sum for the integral $\iint_{\mathcal{R}} \|(\mathbf{r}_u \times \mathbf{r}_v)(u, v)\|\, du\, dv$. We are therefore led to the following definition of surface area.

Definition Let $\mathbf{r}(u, v) = \langle x(u, v), y(u, v), z(u, v)\rangle$ be a continuously differentiable vector-valued function with domain $\mathcal{R}$ in the plane. Let $\mathcal{S}$ be the surface that is the image of $\mathcal{R}$ under $\mathbf{r}$. Then the surface area of $\mathcal{S}$ is given by the formula

$$\iint_{\mathcal{R}} \|(\mathbf{r}_u \times \mathbf{r}_v)(u, v)\|\, du\, dv.$$

We refer to the expression $\|(\mathbf{r}_u \times \mathbf{r}_v)(u_0, v_0)\|\, du\, dv$ as the *element of area for the surface* $\mathcal{S}$.

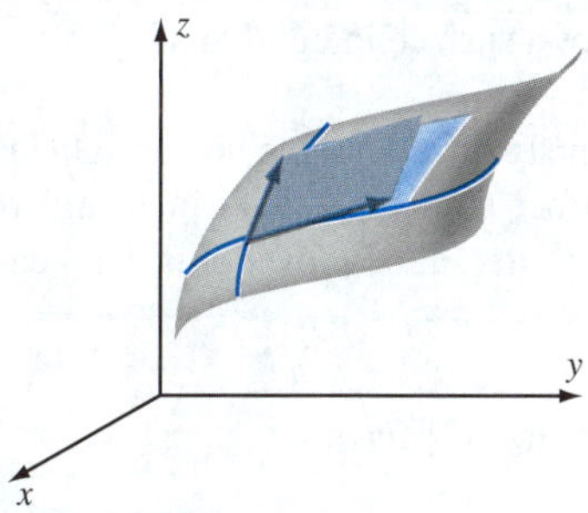

Figure 11
The vectors $\Delta u(\mathbf{r}_u(u_0, v_0))$, $\Delta v(\mathbf{r}_v(u_0, v_0))$, and the parallelogram that they determine

Example 7 Use the parameterization

$$\mathbf{r}(\theta, \phi) = \langle a\cos(\theta)\sin(\phi), a\sin(\theta)\sin(\phi), a\cos(\phi)\rangle, \quad 0 \le \theta \le 2\pi, \quad 0 \le \phi \le \pi$$

to determine the surface area of a sphere of radius $a > 0$.

Solution We have

$$\mathbf{r}_\theta(\theta, \phi) = a\langle -\sin(\theta)\sin(\phi), \cos(\theta)\sin(\phi), 0\rangle$$

and

$$\mathbf{r}_\phi(\theta, \phi) = a\langle \cos(\theta)\cos(\phi), \sin(\theta)\cos(\phi), -\sin(\phi)\rangle.$$

Thus,

$$(\mathbf{r}_\theta \times \mathbf{r}_\phi)(\theta, \phi) = a^2 \det\left(\begin{bmatrix} \mathbf{i} & \mathbf{j} & \mathbf{k} \\ -\sin(\theta)\sin(\phi) & \cos(\theta)\sin(\phi) & 0 \\ \cos(\theta)\cos(\phi) & \sin(\theta)\cos(\phi) & -\sin(\phi) \end{bmatrix}\right),$$

or

$$(\mathbf{r}_\theta \times \mathbf{r}_\phi)(\theta, \phi) = -a^2(\cos(\theta)\sin^2(\phi)\mathbf{i} + \sin(\theta)\sin^2(\phi)\mathbf{j} + (\sin^2(\theta)\sin(\phi)\cos(\phi) + \cos^2(\theta)\cos(\phi)\sin(\phi))\mathbf{k}).$$

On simplification, we have

$$(\mathbf{r}_\theta \times \mathbf{r}_\phi)(\theta, \phi) = -a^2\sin(\phi)(\cos(\theta)\sin(\phi)\mathbf{i} + \sin(\theta)\sin(\phi)\mathbf{j} + \cos(\phi)\mathbf{k})$$

and so

$$\|(\mathbf{r}_\theta \times \mathbf{r}_\phi)(\theta, \phi)\| = a^2\sin(\phi)\sqrt{\cos^2(\theta)\sin^2(\phi) + \sin^2(\theta)\sin^2(\phi) + \cos^2(\phi)} = a^2\sin(\phi).$$

It follows that the surface area of a sphere of radius a is

$$\int_0^{2\pi}\int_0^{\pi} a^2 \sin(\phi)\, d\phi\, d\theta = a^2 \int_0^{2\pi} (-\cos(\phi))\Big|_{\phi=0}^{\phi=\pi} d\theta = a^2 \int_0^{2\pi} (-(-1) - (-1))\, d\theta = 4\pi a^2. \quad \blacksquare$$

Surface Integrals over Parameterized Surfaces

Having already determined the element of surface area for a parametrically defined surface $\mathcal{S}$, we may proceed directly to the definition of a surface integral on $\mathcal{S}$.

Definition Let $\mathbf{r}(u, v) = \langle x(u, v), y(u, v), z(u, v)\rangle$ be a continuously differentiable vector-valued function with domain $\mathcal{R}$ in the plane. Let $\mathcal{S}$ be the surface that is the image of $\mathcal{R}$ under $\mathbf{r}$. If φ is a continuous function on $\mathcal{S}$, then the surface integral of φ over $\mathcal{S}$ is denoted by $\iint_{\mathcal{S}} \varphi\, dS$ and is given by the formula

$$\iint_{\mathcal{S}} \varphi\, dS = \iint_{\mathcal{R}} \varphi(\mathbf{r}(u, v))\, \|(\mathbf{r}_u \times \mathbf{r}_v)(u, v)\|\, du\, dv.$$

Example 8 Integrate the function $\varphi(x, y, z) = z - xy$ over the surface $\mathcal{S}$ that is parameterized by

$$\mathbf{r}(u, v) = \langle u + v, u - v, 3u - 2v\rangle, \qquad 0 \le u \le 1, 0 \le v \le 2.$$

Solution We calculate that $\mathbf{r}_u(u, v) = \langle 1, 1, 3\rangle$ and $\mathbf{r}_v(u, v) = \langle 1, -1, -2\rangle$. Then

$$\|(\mathbf{r}_u \times \mathbf{r}_v)(u, v)\| = \|\mathbf{i} + 5\mathbf{j} - 2\mathbf{k}\| = \sqrt{30}.$$

The integral we wish to evaluate is

$$\iint_{\mathcal{S}} \varphi\, dS = \int_0^2 \int_0^1 \varphi(u + v, u - v, 3u - 2v)\sqrt{30}\, du\, dv = \sqrt{30}\int_0^2\int_0^1 ((3u - 2v) - (u + v)(u - v))\, du\, dv.$$

Performing the necessary algebra, we find that

$$\begin{aligned}\iint_{\mathcal{S}} \varphi\, dS &= \sqrt{30}\int_0^2 \left(\frac{3}{2}u^2 - 2uv - \frac{1}{3}u^3 + uv^2\right)\Big|_{u=0}^{u=1} dv \\ &= \sqrt{30}\int_0^2 \left(\frac{7}{6} - 2v + v^2\right) dv \\ &= \sqrt{30}\left(-v^2 + \frac{7}{6}v + \frac{1}{3}v^3\right)\Big|_0^2 \\ &= \sqrt{30}.\end{aligned}$$

$\blacksquare$

quickquiz

1. What is the element of area for a surface that is given as the graph of a function $f(x, y)$?
2. Write the integral representing the area of the portion of the paraboloid $z = x^2 + y^2$ that lies over the planar region $\{(x, y) : x^2 + y^2 < 8\}$.

3. Write the integral representing the area of the portion of the surface $z = 4 - 2x^2 - 6y^2$ that lies above the xy-plane.
4. What is the element of surface area for a surface given in parametric form?

EXERCISES

Problems for Practice

In Exercises 1–12, calculate the surface area of the graph of the function f over the region $\mathcal{R}$ in the xy-plane.

1. $f(x, y) = 3x - 4y + 8$, $\mathcal{R} = \{(x, y) : 2 \le x \le 5, -3 \le y \le -2\}$
2. $f(x, y) = x^2 + y^2 + 6$, $\mathcal{R} = \{(x, y) : x^2 + y^2 < 16\}$
3. $f(x, y) = (x^2 + y^2)^{1/2}$, $\mathcal{R} = \{(x, y) : 1 < x^2 + y^2 < 9\}$
4. $f(x, y) = 2\sqrt{2}x + (3/2)y^2$, $\mathcal{R} = \{(x, y) : 0 \le y \le 3, 0 \le x \le 2y\}$
5. $f(x, y) = x^2 - y^2 + 3$, $\mathcal{R} = \{(x, y) : x^2 + y^2 < 4\}$
6. $f(x, y) = xy$, $\mathcal{R} = \{(x, y) : x^2 + y^2 < 9\}$
7. $f(x, y) = x^2/2$, $\mathcal{R} = \{(x, y) : 2 \le x \le 6, 0 \le y \le x\}$
8. $f(x, y) = 2y^2 - 2x^2$, $\mathcal{R} = \{(x, y) : 16 < x^2 + y^2 < 25\}$
9. $f(x, y) = x + 6y$, $\mathcal{R} = \{(x, y) : 1 \le x \le 4; 2 \le y \le 2x\}$
10. $f(x, y) = x^2 - \sqrt{2}y$, $\mathcal{R} = \{(x, y) : 3 \le x \le 7, 0 \le y \le x\}$
11. $f(x, y) = 2x^2 - 2y^2 + 2$, $\mathcal{R} = \{(x, y) : 4 < x^2 + y^2 < 9\}$
12. $f(x, y) = \sqrt{1 - x^2}$, $\mathcal{R} = \{(x, y) : -1 < x < 1, 0 < y < 1\}$

In Exercises 13–20, calculate the area of the given surface.

13. the cone with base of radius 5 and height 7 (Do not count the area of the base.)
14. the portion of the sphere $x^2 + y^2 + z^2 = 9$ that lies above the plane $z = 2$
15. the portion of the cone $z^2 = x^2 + y^2$ that lies between $z = 4$ and $z = 10$
16. the portion of the surface $z = xy$ that lies within the cylinder $x^2 + y^2 = 16$
17. the portion of the surface $z = (2/3)(x^{3/2} + y^{3/2})$ that lies over the region $\mathcal{R} = \{(x, y) : 1 \le x \le 3, 2 \le y \le 5\}$
18. the portion of the surface $z = (2/3)x^{3/2} + y + 2$ that lies over the region $\mathcal{R} = \{(x, y) : 1 \le x \le 4, 1 \le y \le 4\}$
19. the portion of the surface $3x - 3y + z = 12$ that lies over the interior of the ellipse $x^2 + 4y^2 = 4$
20. the portion of the surface $2x + 4y + z = 11$ that lies inside the paraboloid $z = x^2 + y^2$

In Exercises 21–28, integrate the function $\phi(x, y, z)$ over the surface given by the graph of f over the region $\mathcal{R}$.

21. $\phi(x, y, z) = x - 2y + 3z$, $f(x, y) = x^2 + y^2 + 8$, $\mathcal{R} = \{(x, y) : x^2 + y^2 < 6\}$
22. $\phi(x, y, z) = 3x^2 - 4y^2 + 7z$, $f(x, y) = x - 3y$, $\mathcal{R} = \{(x, y) : |x| < 2, |y| < 3\}$
23. $\phi(x, y, z) = (3x^2 + 3y^2 + z + 1)^{1/2}$, $f(x, y) = x^2 + y^2$, $\mathcal{R} = \{(x, y) : 1 < x^2 + y^2 < 4\}$
24. $\phi(x, y, z) = x^2 - y^2 + z^2$, $f(x, y) = 2x - 2y$, $\mathcal{R} = \{(x, y) : x^2 + y^2 < 1\}$
25. $\phi(x, y, z) = y^2 + z$, $f(x, y) = 8 - y^2$, $\mathcal{R} = \{(x, y) : 0 < x < y < \sqrt{2}\}$
26. $\phi(x, y, z) = \sin(x) + \sin(z)$, $f(x, y) = 9x - 4y$, $\mathcal{R} = \{(x, y) : 0 < x < \pi, 0 < y < \pi\}$
27. $\phi(x, y, z) = x - xy + z$, $f(x, y) = 3x - 7y$, $\mathcal{R} = \{(x, y) : 1 < 2y < x < 4\}$
28. $\phi(x, y, z) = z$, $f(x, y) = \sqrt{x^2 + y^2}$, $\mathcal{R} = \{(x, y) : 0 < x^2 + y^2 < 1\}$

In Exercises 29–32, write the integral that gives the surface area of the parametrically defined surface. Do not evaluate the integral.

29. $x = u^2 - v$, $y = u + v^2$, $z = v$, $0 \le u \le 1, 0 \le v \le 4$
30. $x = \cos(u)$, $y = \sin(v)$, $z = u - v$, $\pi/6 \le u \le \pi/3, \pi/4 \le v \le 3\pi/4$
31. $x = e^{u+v}$, $y = e^{u-v}$, $z = u$, $0 \le u \le 1, -1 \le v \le 1$
32. $x = u/(u + v)$, $y = v/(u + v)$, $z = u + v$, $1 \le u \le 2, 1 \le v \le 2$

In Exercises 33–36, integrate the function ϕ over the parametrically defined surface.

33. $x = 2u - v,\ y = v + 2u,\ z = v - u,$
$0 \le u \le 2, 0 \le v \le 3,\ \phi(x, y, z) = x + y + z$

34. $x = u + 3v,\ y = 4u,\ z = -3v,$
$-2 \le u \le 1, -1 \le v \le 0,$
$\phi(x, y, z) = \sin(\pi(x - y + z)/3)$

35. $x = v,\ y = u,\ z = u + v,$
$3 \le u \le 5, 2 \le v \le 3,\ \phi(x, y, z) = \sqrt{(z - x)y}$

36. $x = u^2,\ y = v,\ z = u + 3v,$
$\sqrt{2} \le u \le \sqrt{3}, 1 \le v \le 3,\ \phi(x, y, z) = z - 3y$

Further Theory and Practice

37. Let f be a positive, continuously differentiable function of one variable. Let $\mathcal{S}$ be the surface of revolution that results when the graph of f over $[a, b]$, realized as the space curve $\{(x, f(x), 0) : a \le x \le b\}$, is rotated about the x-axis. Parameterize $\mathcal{S}$ and obtain a formula for its surface area.

38. Suppose that $b > a > 0$. Let $\mathcal{S}$ be the surface that results when the circle $\{(0, y, z) : (y - b)^2 + z^2 = a^2\}$ is rotated about the z-axis. ($\mathcal{S}$ is called a *torus*.) Calculate the surface area of $\mathcal{S}$.

39. Let $\mathcal{S}$ be the graph of a function $f(x, y)$ over a planar region $\mathcal{R}$. For each $(x, y) \in \mathcal{R}$, let $\mathbf{n}(x, y)$ be the unit upward normal to $\mathcal{S}$ at the point $(x, y, f(x, y))$. Let $\gamma(x, y)$ be the angle between $\mathbf{k}$ and $\mathbf{n}$. Give an argument to demonstrate that the surface area of $\mathcal{S}$ is given by

$$\iint_{\mathcal{R}} \sec(\gamma)\, dA$$

and the integral of a continuous function $\phi(x, y, z)$ over M is given by

$$\iint_{\mathcal{R}} \phi(x, y, f(x, y)) \sec(\gamma)\, dA.$$

40. Define a concept of average temperature over a surface. If the temperature of any point in space is given by

$$T(x, y, z) = 3x - 8y + z,$$

then find the average temperature over the upper hemisphere of the unit sphere centered at the origin.

41. Use an improper integral to integrate the function

$$\phi(x, y, z) = \frac{1}{(x^2 + y^2 + 1)^3}$$

over the surface that is the graph of $f(x, y) = x^2 + y^2$.

42. Use an improper integral to integrate the function $\phi(x, y) = ze^{-(x^2+y^2)}$ over the surface that is the graph of the function $f(x, y) = x^2 - y^2$.

43. Assuming that it has uniform mass distribution, find the center of gravity of the cone that is the graph of $z = 2\sqrt{x^2 + y^2}, 0 \le x^2 + y^2 \le 4$.

44. Assuming that it has uniform mass distribution, find the center of gravity of the portion of the sphere $x^2 + y^2 + z^2 = 4$ that lies above the plane $z = -1$.

45. Assuming that it has uniform mass distribution, find the center of gravity of the portion of the paraboloid $z = 4 - x^2 - y^2$ that lies above the xy-plane.

46. Let F be a continuously differentiable function of three variables. Let $\mathcal{R}$ be a region in the xy-plane. Suppose that for each point $(x, y) \in \mathcal{R}$, there is exactly one value of z such that $F(x, y, z) = 0$. Further assume that $\frac{\partial F}{\partial z}$ is not 0 at each such point. Prove that the area $\mathcal{A}$ of the portion of the graph of $\{(x, y, z) : F(x, y, z) = 0\}$ lying over $\mathcal{R}$ is given by

$$\mathcal{A} = \iint_{\mathcal{R}} \frac{\|\nabla F\|}{|\nabla F \cdot \mathbf{k}|}\, dA.$$

47. Use the formula developed in Exercise 46 to calculate the surface area of a hemisphere of radius a.

48. Use the formula developed in Exercise 46 to calculate the surface area of a cone with base of radius r and height h.

49. A *prolate spheroid* is an ellipsoid with equal shorter axes. Calculate the surface area of the prolate spheroid $x^2/a^2 + y^2/a^2 + z^2/c^2 = 1$ where $0 < a < c$.

50. An *oblate spheroid* is an ellipsoid with equal longer axes. Calculate the surface area of the oblate spheroid $x^2/a^2 + y^2/c^2 + z^2/c^2 = 1$ where $0 < a < c$.

Calculator/Computer Exercises

51. Calculate the surface area of that portion of the graph of $f(x, y) = \exp(x^2)$ that lies over the interior of the square $\{(x, y) : -1 < x < 1, -1 < y < 1\}$.

52. Calculate the surface area of that portion of the graph of $f(x, y) = \cos(\sqrt{y})$ that lies over the interior of the rectangle $\{(x, y) : 0 < x < 1, 0 < y < \pi^2\}$.

53. Calculate the surface area of that portion of the graph of $f(x, y) = 2 - x^2y^4$ that lies over the interior of the square $\{(x, y) : -1 < x < 1, -1 < y < 1\}$.

54. Calculate the surface area of that portion of the graph of $f(x, y) = x^2 + y^2$ that lies over the interior of the ellipse $\{(x, y) : x^2 + 4y^2 < 4\}$.

15.7 Stokes's Theorem

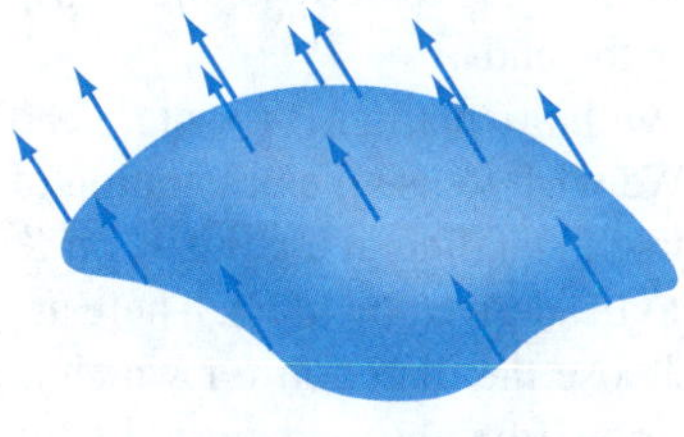

Figure 1

When we studied Green's Theorem in Section 15.5, we used the notion of planar flows to gain a physical understanding of the theorem. Now imagine a flow through a surface (see Figure 1). We may ask how the flow through the interior of the surface is related to the flow at the boundary of the surface. This is the subject of Stokes's Theorem, which generalizes Green's Theorem from planar regions to surfaces in space.

Orientable Surfaces and Their Boundaries

A crucial idea in formulating and using Stokes's Theorem is that of orientation. Intuitively, we want to say two things about the orientation of a surface and its boundary. The first of our requirements is that we may continuously assign to each point of $\mathcal{S}$ a "preferred" unit normal vector.

Definition A continuous vector field $P \mapsto \mathbf{n}(P)$ defined on a smooth surface $\mathcal{S}$ is said to be an *orientation* of $\mathcal{S}$ if at each point P of $\mathcal{S}$, the vector $\mathbf{n}(P)$ is a unit normal to $\mathcal{S}$ at P. We say that a surface $\mathcal{S}$ in space is *orientable* if it possesses an orientation. An orientable surface together with a fixed choice of orientation is called an *oriented surface*.

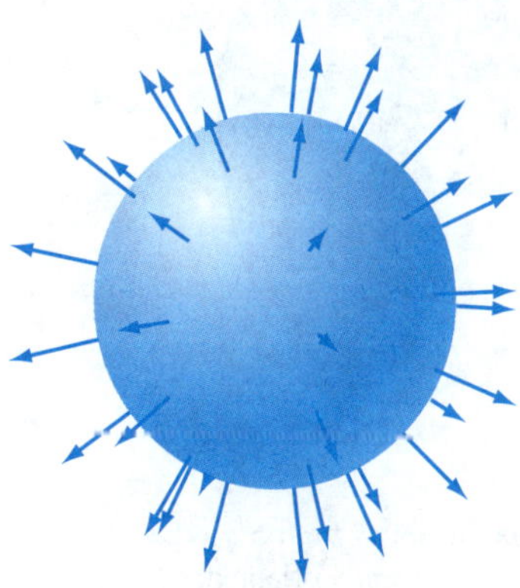

Figure 2
The outward orientation of the sphere

inSIGHT

If $P \mapsto \mathbf{n}(P)$ is an orientation of $\mathcal{S}$, then so is the opposite vector field $P \mapsto -\mathbf{n}(P)$. We therefore say that an orientable surface is *two-sided*. If $\mathcal{S}$ is oriented, then we say that the side out of which the orientation $\mathbf{n}$ points is the *positive* side and the side out of which $-\mathbf{n}$ points is the negative side.

Example 1 Suppose that $a > 0$. What are the two orientations of the sphere $x^2 + y^2 + z^2 = a^2$?

Solution By elementary geometry, we know that the position vector $\langle x, y, z\rangle$ is perpendicular to the sphere at the point (x, y, z). Since $\|\langle x, y, z\rangle\| = \sqrt{x^2 + y^2 + z^2} = a$, it follows that the function $(x, y, z) \mapsto \langle x/a, y/a, z/a\rangle$ is a continuous unit normal on the sphere. It is the *outward* unit normal. See Figure 2. The opposite orientation of the sphere is given by the *inward* unit normal $(x, y, z) \mapsto -\langle x/a, y/a, z/a\rangle$. ■

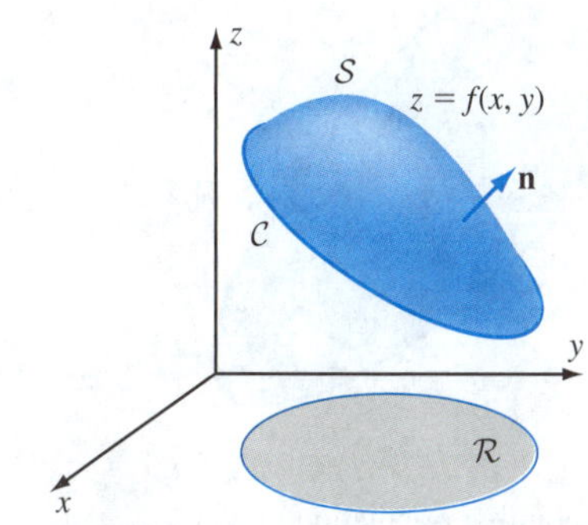

Figure 3

inSIGHT

A surface with no boundary is said to be *closed*. The sphere is an example. With such a surface, we can speak of an *inward normal* or an *outward normal* vector field. When a surface is the graph of a function, we may speak of an *upward* normal vector field (as shown in Figure 3) or a *downward* normal vector field.

Although familiar surfaces, such as the sphere, are orientable, nonorientable surfaces do exist. For example, August Ferdinand Möbius (1790–1868) published a "one-sided" surface, now called the *Möbius strip,* that is not orientable. See Figure 4. Notice that when you move a unit normal continuously around a Möbius strip, it winds up opposite to its initial position. You can make your own Möbius strip by taking a strip of paper, giving one end a 180° twist, and then joining the ends.

Figure 4
A Möbius strip

Suppose now that $\mathcal{S}$ is a smooth bounded surface with orientation $P \mapsto \mathbf{n}(P)$ and with closed curve $\mathcal{C}$ as its boundary. See Figure 5a. We wish to use the orientation of $\mathcal{S}$ to induce a particular direction on $\mathcal{C}$. Imagine yourself standing at a point P on $\mathcal{C}$, perpendicular to $\mathcal{S}$ with the direction from your feet to head given by $\mathbf{n}(P)$. There are two directions in which you may walk around $\mathcal{C}$. Choose the direction for which $\mathcal{S}$ lies to your *left*. See Figure 5b. We say that this direction around $\mathcal{C}$ is *induced* by the oriented surface $\mathcal{S}$. In the case that $\mathcal{C}$ is a parameterized curve, we say that $\mathcal{C}$ is *oriented consistently* with $\mathcal{S}$ if the orientation of the parameterization is the same as the one induced by $\mathcal{S}$.

These notions can be formulated more precisely by observing that if $\mathbf{r}$ is a parameterization of $\mathcal{C}$, then the vector $\mathbf{n}(P) \times \mathbf{r}'(P)$ is perpendicular to $\mathcal{C}$ and tangent to $\mathcal{S}$. It points either away from $\mathcal{S}$ or into $\mathcal{S}$. See Figure 6. When $\mathcal{C}$ is oriented consistently with $\mathcal{S}$, the vector $\mathbf{n}(P) \times \mathbf{r}'(P)$ points *into* $\mathcal{S}$. Figure 7 shows two possible orientations of

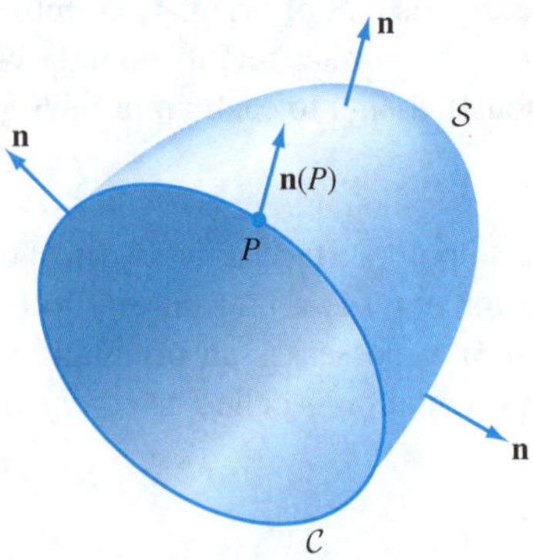

Figure 5a

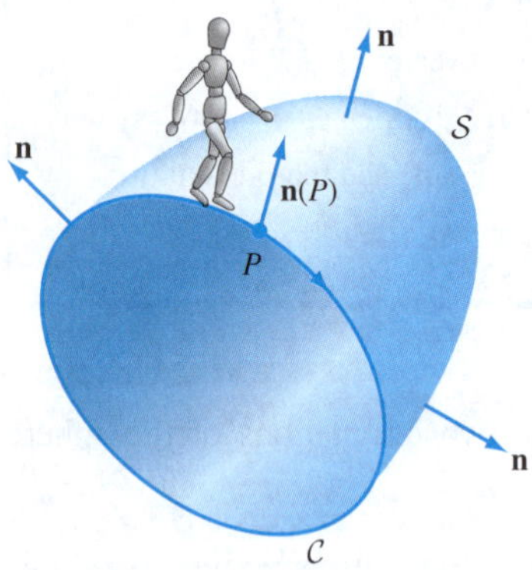

Figure 5b
$\mathcal{C}$ is oriented so that $\mathcal{S}$ lies to the left when $\mathcal{C}$ is traversed in the direction of the orientation.

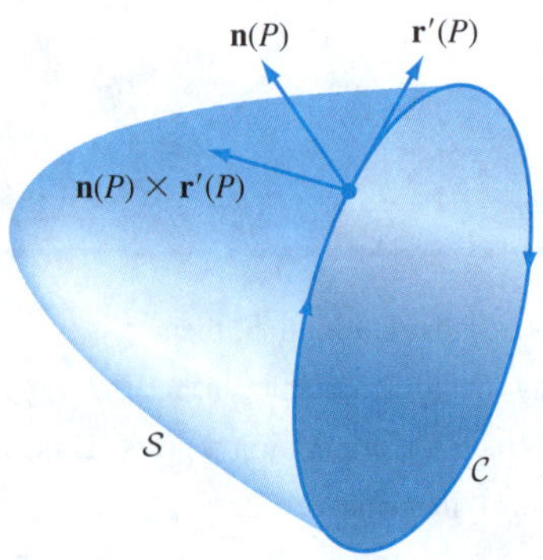

Figure 6a
$\mathbf{n}(P) \times \mathbf{r}'(P)$ points *into* $\mathcal{S}$: $\mathcal{S}$ and $\mathcal{C}$ are consistently oriented.

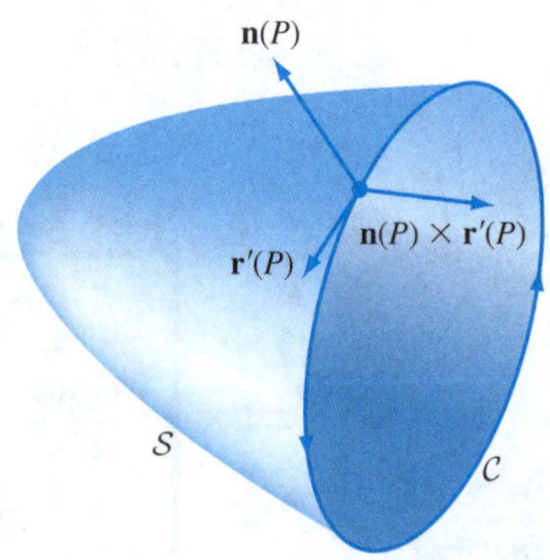

Figure 6b
$\mathbf{n}(P) \times \mathbf{r}'(P)$ points *away from* $\mathcal{S}$: $\mathcal{S}$ and $\mathcal{C}$ are *not* consistently oriented.

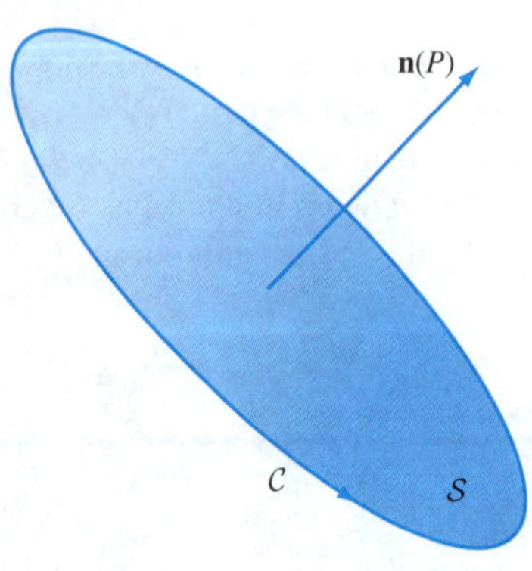

Figure 7a
Upward-pointing orientation of $\mathcal{S}$ with consistent boundary orientation

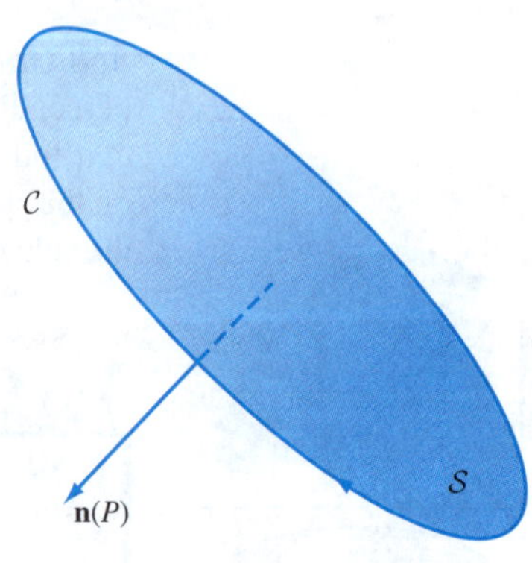

Figure 7b
Downward-pointing orientation of $\mathcal{S}$ with consistent boundary orientation

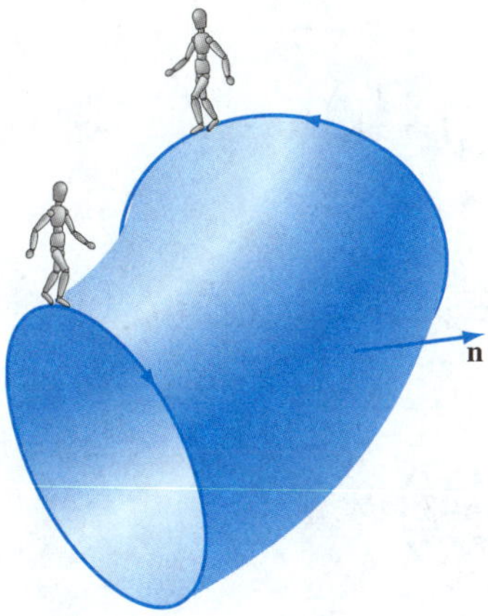

Figure 8

a disk in space with consistently oriented boundary curve. Figure 8 shows another oriented surface with a consistently oriented boundary that comprises two disconnected pieces. We shall discuss such surfaces at the end of this section. For now, just note that the definition of "consistent orientation" can be applied to each component of a disconnected boundary.

The Component of Curl in the Normal Direction

Recall that the curl operator, when applied to a vector field **F**, can be interpreted as a measure of the vorticity of the vector field about a given point. In our calculations, we saw that **curl(F)** points in the direction of the axis of rotation. For example, if $\mathbf{F}(x, y) = M(x, y)\mathbf{i} + N(x, y)\mathbf{j}$ is a planar vector field, then equation (15.30) defines **curl(F)** to be a multiple of **k**. In particular, the curl points out of the plane in which the circulation is taking place. These observations suggest that if we want to measure the magnitude of the curl of a vector field as it relates to a surface $\mathcal{S}$, then we should consider the component of the curl in the direction normal to the surface. We therefore use $\mathbf{curl}(\mathbf{F}(P)) \cdot \mathbf{n}(P)$ to measure how much the vector field is curling in the surface at P. The integral

$$\iint_{\mathcal{S}} \mathbf{curl}(\mathbf{F}) \cdot \mathbf{n}\, dS \tag{15.45}$$

represents the "total curl" of the vector field **F** on $\mathcal{S}$. Bear in mind that, in general, the vector field **F** curls clockwise with respect to $\mathbf{n}(P)$ at some points (the points P at which $\mathbf{curl}(\mathbf{F}(P)) \cdot \mathbf{n}(P) < 0$) and counterclockwise at other points (the points at which $\mathbf{curl}(\mathbf{F}(P)) \cdot \mathbf{n}(P) > 0$). So the surface integral (15.45) takes into account certain cancellations.

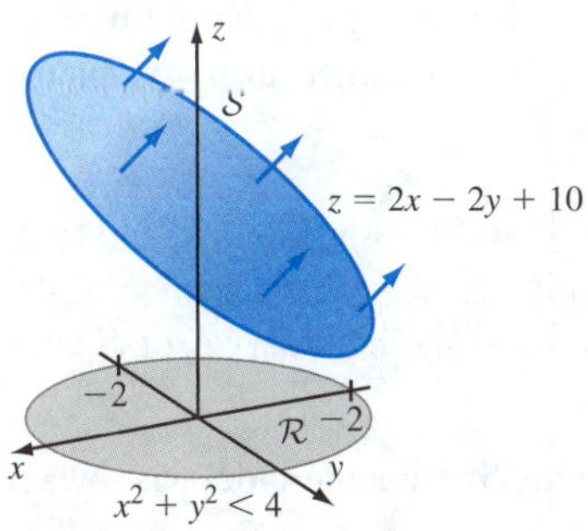

Figure 9

Example 2 Let $\mathcal{S}$ be the graph of $z = 2x - 2y + 10$ over the circular region $\mathcal{R} = \{(x, y) : x^2 + y^2 < 4\}$. Orient $\mathcal{S}$ with the upward-pointing unit normal **n**. Refer to Figure 9. Define $\mathbf{F}(x, y, z) = (y - z)\mathbf{i} - (x + y)\mathbf{j} + 2x\mathbf{k}$. Calculate $\iint_{\mathcal{S}} \mathbf{curl}(\mathbf{F}) \cdot \mathbf{n}\, dS$.

Solution Our surface is just that portion of the plane $2x - 2y - z = -10$ that lies above the interior of the circle centered at the origin and with radius 2. The surface $\mathcal{S}$ is therefore elliptic. We use the coefficients that appear in the equation of the plane, $2x - 2y - z = -10$, to obtain a vector $2\mathbf{i} - 2\mathbf{j} - 1\mathbf{k}$ that is normal to $\mathcal{S}$. An upward normal at any point is then given by $-2\mathbf{i} + 2\mathbf{j} + 1\mathbf{k}$ (we know this is upward because the **k** component is positive). Dividing by the length of this vector, we obtain $\mathbf{n}(x, y, z) = (-2/3)\mathbf{i} + (2/3)\mathbf{j} + (1/3)\mathbf{k}$. Next, we calculate

$$\begin{aligned}\mathbf{curl}(\mathbf{F}) &= \left(\frac{\partial}{\partial y}(2x) - \frac{\partial}{\partial z}(-(x + y))\right)\mathbf{i} - \left(\frac{\partial}{\partial x}(2x) - \frac{\partial}{\partial z}(y - z)\right)\mathbf{j} \\ &\quad + \left(\frac{\partial}{\partial x}(-(x + y)) - \frac{\partial}{\partial y}(y - z)\right)\mathbf{k} = 0\mathbf{i} - 3\mathbf{j} - 2\mathbf{k}\end{aligned}$$

and

$$\mathbf{curl}(\mathbf{F}) \cdot \mathbf{n} = (0\mathbf{i} - 3\mathbf{j} - 2\mathbf{k}) \cdot \left(-\frac{2}{3}\mathbf{i} + \frac{2}{3}\mathbf{j} + \frac{1}{3}\mathbf{k}\right) = -\frac{8}{3}.$$

Now the element of area dS on $\mathcal{S}$ is

$$dS = \sqrt{1 + f_x(x, y)^2 + f_y(x, y)^2}\, dA = \sqrt{1 + 2^2 + (-2)^2}\, dA = 3\, dA.$$

Therefore, we have

$$\iint_{\mathcal{S}} \mathbf{curl}(\mathbf{F}) \cdot \mathbf{n}\, dS = \iint_{\mathcal{R}} \left(-\frac{8}{3} \cdot 3\right) dA.$$

Because $\mathcal{R}$ is a disk of radius 2, we have

$$\iint_{\mathcal{S}} \mathbf{curl}(\mathbf{F}) \cdot \mathbf{n}\, dS = (-8)(\text{area of } \mathcal{R}) = (-8)(\pi \cdot 2^2) = -32\pi.$$

Stokes's Theorem

Since the line integral $\oint_{\mathcal{C}} \mathbf{F} \cdot d\mathbf{r}$ sums the tangential component of $\mathbf{F}$, namely, $\mathbf{F} \cdot \mathbf{r}'$, over $\mathcal{C}$, we can regard the integral as a measure of the tendency of $\mathbf{F}$ to flow about the boundary of $\mathcal{S}$. It therefore makes good physical sense for this line integral over the boundary of $\mathcal{S}$ to be related to the surface integral $\iint_{\mathcal{S}} \mathbf{curl}(\mathbf{F}) \cdot \mathbf{n}\, dS$. Indeed, Green's Theorem, in the form of equation (15.37), states that

$$\oint_{\mathcal{C}} \mathbf{F} \cdot d\mathbf{r} = \iint_{\mathcal{S}} \mathbf{curl}(\mathbf{F}) \cdot \mathbf{n}\, dS \tag{15.46}$$

when $\mathcal{S}$ is a planar region with boundary $\mathcal{C}$ and upward-pointing orientation $\mathbf{n} = \mathbf{k}$. The next example, which pertains to a surface that is not contained in the xy-plane, provides further evidence of this relationship.

Example 3 As in Example 2, let $\mathbf{F}(x, y, z) = (y - z)\mathbf{i} - (x + y)\mathbf{j} + 2x\mathbf{k}$, and let $\mathcal{S}$ be the graph of $z = 2x - 2y + 10$ over the circular region $\mathcal{R} = \{(x, y) : x^2 + y^2 < 4\}$. Let $\mathbf{n}$ be the upward-pointing unit normal on $\mathcal{S}$, and orient the boundary $\mathcal{C}$ of $\mathcal{S}$ consistently. Verify equation (15.46).

Solution The orientation of $\mathcal{C}$ that is consistent with $\mathcal{S}$ is counterclockwise when viewed from above. The vector-valued function

$$\mathbf{r}(t) = 2\cos(t)\mathbf{i} + 2\sin(t)\mathbf{j} + (2(2\cos(t)) - 2(2\sin(t)) + 10)\mathbf{k}, \quad 0 \le t \le 2\pi,$$

parameterizes $\mathcal{C}$ with the proper orientation. We calculate

$$\begin{aligned}\mathbf{F}(\mathbf{r}(t)) &= (2\sin(t) - (2(2\cos(t)) - 2(2\sin(t)) + 10))\mathbf{i} \\ &\qquad - ((2\cos(t)) + (2\sin(t)))\mathbf{j} + 2(2\cos(t))\mathbf{k} \\ &= (6\sin(t) - 4\cos(t) - 10)\mathbf{i} - (2\cos(t) + 2\sin(t))\mathbf{j} + 4\cos(t)\mathbf{k},\end{aligned}$$

and

$$\mathbf{r}'(t) = -2\sin(t)\mathbf{i} + 2\cos(t)\mathbf{j} + (-4\sin(t) - 4\cos(t))\mathbf{k}.$$

After some simplification, we obtain

$$\begin{aligned}\mathbf{F}(\mathbf{r}(t)) \cdot \mathbf{r}'(t) &= -12\sin(t)\cos(t) + 20\sin(t) - 8\cos^2(t) - 12 \\ &= -6\sin(2t) + 20\sin(t) - 8\cos^2(t) - 12.\end{aligned}$$

Therefore, using formula (7.21) to evaluate the definite integral of $\cos^2(t)$, we have

$$\begin{aligned}\oint_{\mathcal{C}} \mathbf{F} \cdot d\mathbf{r} &= \int_0^{2\pi} (-6\sin(2t) + 20\sin(t) - 8\cos^2(t) - 12)\, dt \\ &= 0 + 0 - 8 \cdot \pi - 12 \cdot (2\pi) = -32\pi.\end{aligned}$$

This equation, together with the result of Example 2, establishes the validity of equation (15.46) for $\mathbf{F}$ and $\mathcal{S}$. ■

Our next theorem tells us that equation (15.46) holds quite generally.

Theorem 1 ***Stokes's Theorem*** Suppose that $(x, y, z) \mapsto \mathbf{F}(x, y, z) = M(x, y, z)\mathbf{i} + N(x, y, z)\mathbf{j} + R(x, y, z)\mathbf{k}$ is a continuously differentiable vector field defined on an oriented surface $\mathcal{S}$ and its boundary $\mathcal{C}$. Let $P \mapsto \mathbf{n}(P)$ denote the orientation of $\mathcal{S}$, and let $\mathbf{r}$ be a parameterization of $\mathcal{C}$ that orients $\mathcal{C}$ consistently with $\mathcal{S}$. Then

$$\oint_{\mathcal{C}} \mathbf{F} \cdot d\mathbf{r} = \iint_{\mathcal{S}} \mathbf{curl}(\mathbf{F}) \cdot \mathbf{n}\, dS.$$

Proof Our scheme is to reduce Stokes's Theorem to a calculation involving Green's Theorem. For ease of calculation, we will assume that our surface $\mathcal{S}$ is the graph of a continuously differentiable function f over a bounded planar region $\mathcal{R}$. We suppose that the boundary of $\mathcal{R}$ is a continuously differentiable curve $\mathcal{C}_0$ with a parameterization $t \mapsto \mathbf{r}_0(t) = \xi(t)\mathbf{i} + \eta(t)\mathbf{j}$, $a \le t \le b$, that gives $\mathcal{C}_0$ a counterclockwise orientation when viewed from above. The vector-valued function $\mathbf{r}(t) = \langle \mathbf{r}_0(t), f(\mathbf{r}_0(t))\rangle$, $a \le t \le b$ is then a parameterization of $\mathcal{C}$ that orients $\mathcal{C}$ consistently with the upward-pointing orientation $P \mapsto \mathbf{n}(P)$ of $\mathcal{S}$. Figure 10 illustrates the geometry.

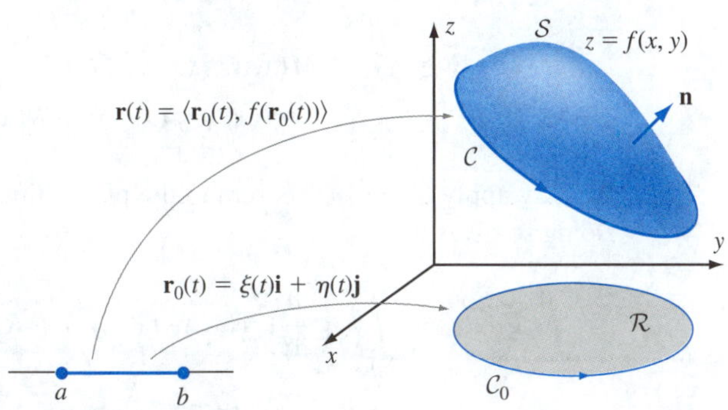

Figure 10

Let us begin by rewriting the surface integral. The element of area on $\mathcal{S}$ is

$$dS = \sqrt{1 + f_x(x, y)^2 + f_y(x, y)^2}\, dA.$$

Since $\mathcal{S}$ is the graph of f, a normal to $\mathcal{S}$ at each point $P = (x, y, f(x, y))$ is given by $f_x(x, y)\mathbf{i} + f_y(x, y)\mathbf{j} - 1\mathbf{k}$. Of course, an upward normal is then $-f_x(x, y)\mathbf{i} - f_y(x, y)\mathbf{j} + 1\mathbf{k}$ (since the $\mathbf{k}$ component is then positive). Thus,

$$\mathbf{n}(P) = \frac{1}{\sqrt{1 + f_x(x, y)^2 + f_y(x, y)^2}}(-f_x(x, y)\mathbf{i} - f_y(x, y)\mathbf{j} + 1\mathbf{k}).$$

Using these expressions for dS and $\mathbf{n}(P)$, we obtain

$$\begin{aligned}\iint_{\mathcal{S}} \mathbf{curl}(\mathbf{F}) \cdot \mathbf{n}\, dS &= \iint_{\mathcal{R}} \mathbf{curl}(\mathbf{F}) \cdot (-f_x(x, y)\mathbf{i} - f_y(x, y)\mathbf{j} + 1\mathbf{k})\, dA \\ &= \iint_{\mathcal{R}} (-(R_y - N_z) f_x - (M_z - R_x) f_y + (N_x - M_y))\, dA. \qquad \mathbf{(15.47)}\end{aligned}$$

In the last integration, each partial derivative of M, N, and R is evaluated at the points $(x, y, f(x, y))$ of $\mathcal{S}$. This is an explicit expression for the right side of the formula in Stokes's Theorem.

Now we look at the line integral portion of Stokes's Theorem. We have

$$\oint_{\mathcal{C}} \mathbf{F} \cdot d\mathbf{r} = \int_a^b \mathbf{F}(\mathbf{r}(t)) \cdot \mathbf{r}'(t)\, dt = \int_a^b M(\mathbf{r}(t))\xi'(t) + N(\mathbf{r}(t))\eta'(t) + R(\mathbf{r}(t)) \left(\frac{d}{dt} f(\xi(t), \eta(t))\right) dt.$$

Using the Chain Rule, we obtain

$$\oint_{\mathcal{C}} \mathbf{F} \cdot d\mathbf{r} = \int_a^b M(\mathbf{r}(t))\xi'(t) + N(\mathbf{r}(t))\eta'(t) + R(\mathbf{r}(t)) f_x(\mathbf{r}_0(t))\xi'(t) + R(\mathbf{r}(t)) f_y(\mathbf{r}_0(t))\eta'(t)\, dt,$$

or

$$\begin{aligned}\oint_{\mathcal{C}} \mathbf{F} \cdot d\mathbf{r} = \int_a^b & (M(\mathbf{r}(t)) + R(\mathbf{r}(t)) f_x(\mathbf{r}_0(t)))\xi'(t) \\ & + (N(\mathbf{r}(t)) + R(\mathbf{r}(t)) f_y(\mathbf{r}_0(t)))\eta'(t)\, dt = \oint_{\mathcal{C}_0} \widetilde{\mathbf{F}} \cdot d\mathbf{r}_0,\end{aligned}$$

where

$$\begin{aligned}\widetilde{\mathbf{F}}(x, y) = &\, (M(x, y, f(x, y)) + R(x, y, f(x, y)) f_x(x, y))\mathbf{i} \\ & + (N(x, y, f(x, y)) + R(x, y, f(x, y)) f_y(x, y))\mathbf{j}.\end{aligned}$$

We may apply Green's Theorem to the planar line integral $\oint_{\mathcal{C}_0} \widetilde{\mathbf{F}} \cdot d\mathbf{r}_0$ that has arisen. Doing so gives us

$$\begin{aligned}\oint_{\mathcal{C}_0} \widetilde{\mathbf{F}} \cdot d\mathbf{r}_0 = \iint_{\mathcal{R}} & \left(\frac{\partial}{\partial x}(N(x, y, f(x, y)) + R(x, y, f(x, y)) f_y(x, y))\right. \\ & \left. - \frac{\partial}{\partial y}(M(x, y, f(x, y)) + R(x, y, f(x, y)) f_x(x, y))\right) dA. \qquad \mathbf{(15.48)}\end{aligned}$$

If these differentiations are written out, using the Product Rule and the Chain Rule, then

$$\frac{\partial}{\partial x}(N + R \cdot f_y) = N_x + N_z f_x + R_x f_y + R_z f_x f_y + R f_{yx}$$

and

$$\frac{\partial}{\partial y}(M + R \cdot f_x) = M_y + M_z f_y + R_y f_x + R_z f_y f_x + R f_{xy}.$$

If we substitute these expressions into the right side of equation (15.48), then the integral reduces to the right side of equation (15.47), which establishes Stokes's Theorem. ■

Although this proof is limited, it should be stressed that Stokes's Theorem is true for a region in any smooth surface—whether or not that surface is the graph of a function. In practice, it is often convenient to treat a general surface by breaking it into pieces, each of which can be realized as the graph of a function. In addition to its theoretical importance, Stokes's Theorem can be used to advantage when the computation of one side of equation (15.46) is substantially easier than the computation of the other side.

Example 4 Let $\mathcal{S}$ be the graph of $f(x, y) = 4x - 8y + 5$ over $\mathcal{R} = \{(x, y) : (x-1)^2 + 9(y-3)^2 < 36\}$. Let $\mathbf{n}(P)$ be the upward unit normal at each point P of $\mathcal{S}$. Consider the vector field $\mathbf{F}(x, y, z) = -3z\mathbf{i} + (x+y)\mathbf{j} + y\mathbf{k}$. Let $\mathcal{C}$ be the boundary of $\mathcal{S}$ with consistent orientation. Evaluate $\oint_{\mathcal{C}} \mathbf{F} \cdot d\mathbf{r}$.

Solution To calculate the given line integral directly would involve an awkward parameterization of the boundary of $\mathcal{S}$. With Stokes's Theorem, matters become much simpler. We calculate

$$dS = \sqrt{1 + f_x(x, y)^2 + f_y(x, y)^2}\, dA = \sqrt{1 + 4^2 + (-8)^2}\, dA = 9\, dA$$

and

$$\mathbf{curl}(\mathbf{F}) = \det\left(\begin{bmatrix} \mathbf{i} & \mathbf{j} & \mathbf{k} \\ \dfrac{\partial}{\partial x} & \dfrac{\partial}{\partial y} & \dfrac{\partial}{\partial z} \\ -3z & x+y & y \end{bmatrix}\right) = \mathbf{i} - 3\mathbf{j} + \mathbf{k}.$$

Finally, since our surface is contained in the plane $4x - 8y - z = -5$, the vector $-4\mathbf{i} + 8\mathbf{j} + 1\mathbf{k}$ is an upward normal (the coefficient of $\mathbf{k}$ is positive) and $\mathbf{n}(x, y, z) = (-4/9)\mathbf{i} + (8/9)\mathbf{j} + (1/9)\mathbf{k}$. Thus,

$$\begin{aligned}\oint_{\mathcal{C}} \mathbf{F} \cdot d\mathbf{r} &= \iint_{\mathcal{S}} \mathbf{curl}(\mathbf{F}) \cdot \mathbf{n}\, dS = \iint_{\mathcal{R}} (\mathbf{i} - 3\mathbf{j} + \mathbf{k}) \cdot ((-4/9)\mathbf{i} + (8/9)\mathbf{j} + (1/9)\mathbf{k})9\, dA \\ &= \iint_{\mathcal{R}} (-27)\, dA = -27 \cdot (\text{area of } \mathcal{R}).\end{aligned}$$

But $\mathcal{R}$ is an ellipse with semimajor axis 6 and semiminor axis 2. Therefore, it has area $6 \cdot 2 \cdot \pi = 12\pi$. We conclude that $\oint_{\mathcal{C}} \mathbf{F} \cdot d\mathbf{r} = -27 \cdot 12\pi = -324\pi$. ■

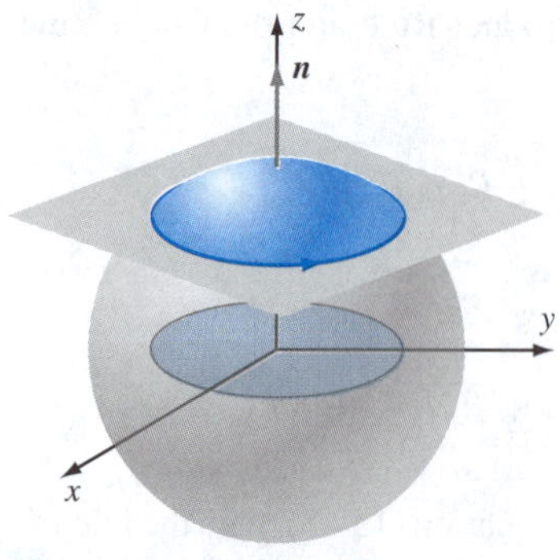

Figure 11

Example 5 Consider the surface $\mathcal{S}$ consisting of the part of the sphere of radius 3 centered at the origin that lies above the plane $z = \sqrt{5}$. Let $P \mapsto \mathbf{n}(P)$ denote the upward-pointing normal on $\mathcal{S}$. Define $\mathbf{F}(x, y, z) = -y\mathbf{i} + xz\mathbf{j} + y^2\mathbf{k}$. Calculate $\iint_{\mathcal{S}} \mathbf{curl}(\mathbf{F}) \cdot \mathbf{n}\, dS$.

Solution The intersection of the level plane $z = \sqrt{5}$ and the sphere $x^2 + y^2 + z^2 = 9$ is the circle $\{(x, y, z) : x^2 + y^2 = 4,\ z = \sqrt{5}\}$. See Figure 11. The surface $\mathcal{S}$ is therefore the graph of the function $f(x, y) = (9 - x^2 - y^2)^{1/2}$ over the region $\mathcal{R} = \{(x, y) : x^2 + y^2 < 4\}$. The counterclockwise-oriented boundary curve $\mathcal{C}$ can be parameterized by

$$\mathbf{r}(t) = 2\cos(t)\mathbf{i} + 2\sin(t)\mathbf{j} + \sqrt{5}\mathbf{k}, \qquad 0 \le t \le 2\pi.$$

We have

$$\mathbf{F}(\mathbf{r}(t)) = -2\sin(t)\mathbf{i} + 2\sqrt{5}\cos(t)\mathbf{j} + 4\sin^2(t)\mathbf{k} \quad \text{and} \quad \mathbf{r}'(t) = -2\sin(t)\mathbf{i} + 2\cos(t)\mathbf{j} + 0\mathbf{k}.$$

Therefore, $\mathbf{F}(\mathbf{r}(t)) \cdot \mathbf{r}'(t) = 4\sin^2(t) + 4\sqrt{5}\cos^2(t)$ and, by Stokes's Theorem,

$$\iint_{\mathcal{S}} \mathbf{curl}(\mathbf{F}) \cdot \mathbf{n}\, dS = \oint_{\mathcal{C}} \mathbf{F} \cdot d\mathbf{r} = \int_0^{2\pi} \mathbf{F}(\mathbf{r}(t)) \cdot \mathbf{r}'(t)\, dt$$

$$= \int_0^{2\pi} 4\sin^2(t)\, dt + \int_0^{2\pi} 4\sqrt{5}\cos^2(t)\, dt \overset{(7.21)}{=} \left(4 + 4\sqrt{5}\right)\pi. \quad \blacksquare$$

Stokes's Theorem on a Region with Piecewise-Smooth Boundary

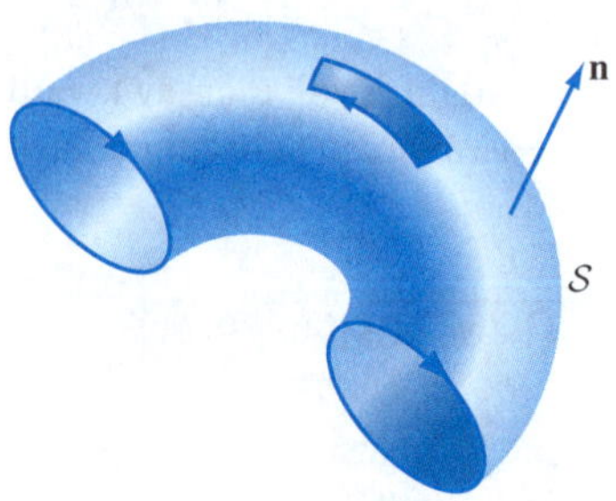

Figure 12
A surface whose boundary consists of six smooth curves that form three closed curves (one of which is only piecewise smooth)

In many applications, the boundary of the surface $\mathcal{S}$ is not given by a single curve but rather by finitely many continuously differentiable curves. Stokes's Theorem is also valid in that setting.

The setup is this: The oriented surface $\mathcal{S}$ is the image of a continuously differentiable vector-valued function $(u, v) \mapsto \mathbf{r}(u, v)$ defined on a bounded region $\mathcal{R}$ of the uv-plane. The boundary of $\mathcal{S}$ consists of finitely many continuously differentiable curves $\mathcal{C}_1, \ldots, \mathcal{C}_N$, having parameterizations $\mathbf{r}_1, \ldots, \mathbf{r}_N$, each of which is consistent with the orientation $P \mapsto \mathbf{n}(P)$ of $\mathcal{S}$. Refer to Figure 12. Now the statement of Stokes's Theorem is that if $\mathbf{F}$ is a continuously differentiable vector field on a region of space that contains $\mathcal{S}$ and its boundary, then

$$\iint_{\mathcal{S}} \mathbf{curl}(\mathbf{F}) \cdot \mathbf{n}\, dS = \sum_{j=1}^{N} \int_{\mathcal{C}_j} \mathbf{F} \cdot d\mathbf{r}_j. \tag{15.49}$$

(Because the curve $\mathcal{C}_j$ might not be closed, we use the symbol $\int_{\mathcal{C}_j}$ instead of $\oint_{\mathcal{C}}$. Even if the curve $\mathcal{C}_j$ is closed, there is no harm in using the more general notation.)

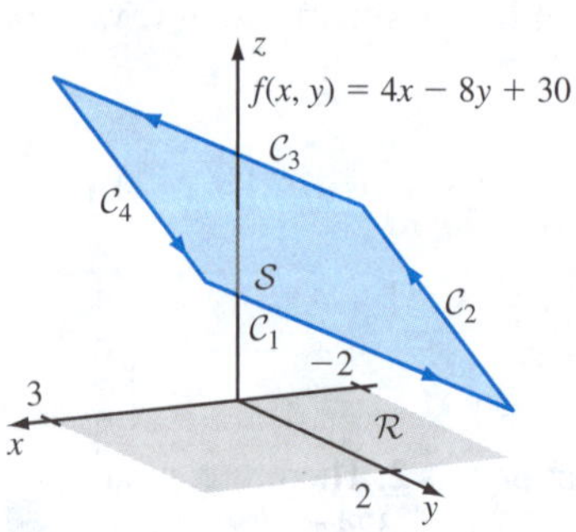

Figure 13

Example 6 Let $\mathcal{S}$ be the graph of $f(x, y) = 4x - 8y + 30$ over the rectangle $\mathcal{R} = \{(x, y) : -2 < x < 3,\ 0 < y < 2\}$ (as shown in Figure 13). Consider the vector field $\mathbf{F}(x, y, z) = -x^2\mathbf{i} + xz\mathbf{j} + yx\mathbf{k}$. Use Stokes's Theorem to evaluate $\oint_{\mathcal{C}} \mathbf{F} \cdot d\mathbf{r}$ where $\mathcal{C}$ is the boundary of $\mathcal{S}$ oriented in the counterclockwise direction when viewed from above.

Solution We could certainly calculate the required line integral directly, but to do so would involve parameterizing four separate line segments and evaluating four separate integrals. Using Stokes's Theorem, the task becomes much simpler. We have

$$dS = \sqrt{1 + f_x(x,y)^2 + f_y(x,y)^2}\,dA = \sqrt{1 + 4^2 + (-8)^2} = 9\,dA.$$

Also,

$$\mathbf{curl}(\mathbf{F})(x,y,z) = \det\left(\begin{bmatrix} \mathbf{i} & \mathbf{j} & \mathbf{k} \\ \dfrac{\partial}{\partial x} & \dfrac{\partial}{\partial y} & \dfrac{\partial}{\partial z} \\ -x^2 & xz & yx \end{bmatrix}\right) = 0\mathbf{i} - y\mathbf{j} + z\mathbf{k}.$$

Finally, since our surface is $4x - 8y - z = -30$, the vector $-4\mathbf{i} + 8\mathbf{j} + 1\mathbf{k}$ is an upward normal and $\mathbf{n}(P) = (-4/9)\mathbf{i} + (8/9)\mathbf{j} + (1/9)\mathbf{k}$. Thus, on $\mathcal{S}$, we have

$$\begin{aligned}\mathbf{curl}(\mathbf{F}) \cdot \mathbf{n} &= (0\mathbf{i} - y\mathbf{j} + z\mathbf{k}) \cdot \left(-\frac{4}{9}\mathbf{i} + \frac{8}{9}\mathbf{j} + \frac{1}{9}\mathbf{k}\right) \\ &= \frac{1}{9}(z - 8y) = \frac{1}{9}((4x - 8y + 30) - 8y) = \frac{1}{9}(4x - 16y + 30).\end{aligned}$$

We conclude that

$$\begin{aligned}\oint_{\mathcal{C}} \mathbf{F} \cdot d\mathbf{r} &= \sum_{j=1}^{4} \int_{\mathcal{C}_j} \mathbf{F} \cdot d\mathbf{r}_j = \iint_{\mathcal{S}} \mathbf{curl}(\mathbf{F}) \cdot \mathbf{n}\,dS = \frac{1}{9} \iint_{\mathcal{R}} (4x - 16y + 30)\,9\,dA \\ &= \int_0^2 \int_{-2}^3 (4x - 16y + 30)\,dx\,dy.\end{aligned}$$

For the inner integration, we have

$$\int_2^3 (4x - 16y + 30)\,dx = (2x^2 - 16xy + 30x)\Big|_{x=-2}^{x=3} = -80y + 160.$$

Finally,

$$\oint_{\mathcal{C}} \mathbf{F} \cdot d\mathbf{r} = \int_0^2 (-80y + 160)\,dy = (-40y^2 + 160y)\Big|_{y=0}^{y=2} = -160 + 320 = 160. \quad \blacksquare$$

Example 6 involved a line integral over a piecewise-smooth boundary comprising several smooth curves. The next example concerns a surface that has a boundary consisting of disconnected curves.

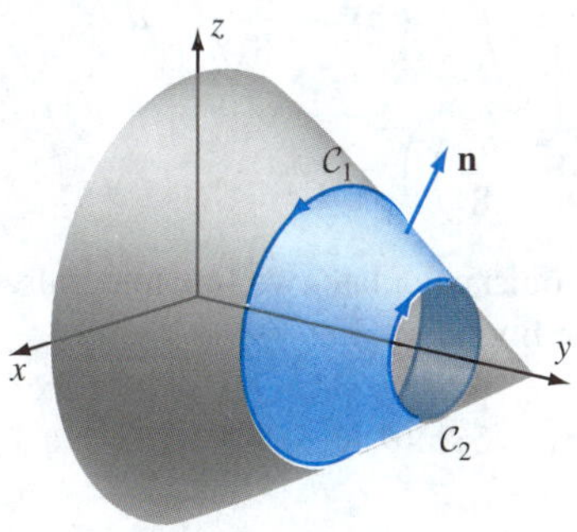

$y = 20 - 2\sqrt{x^2 + z^2}$

Figure 14

Example 7 Let $\mathcal{S}$ be the frustum of the cone $y = 20 - 2\sqrt{x^2 + z^2}$ that lies between the planes $y = 8$ and $y = 14$. Suppose that $\mathcal{S}$ and its boundary curves are consistently oriented, as shown in Figure 14. Consider the vector field $\mathbf{F}(x,y,z) = z\mathbf{i} + 1\mathbf{j} + z^2\mathbf{k}$. Verify Stokes's Theorem for $\mathbf{F}$ and $\mathcal{S}$.

Solution Notice that $\mathcal{S}$ is *not* the graph of a function. We may use polar coordinates in the xz-plane as an aid in parameterizing $\mathcal{S}$. If $x = r\cos(\theta)$ and $z = r\sin(\theta)$, then on $\mathcal{S}$ we have $y = 20 - 2\sqrt{x^2 + z^2} = 20 - 2r$, or $r = 10 - y/2$. Therefore, the vector-valued function

$$\mathbf{r}(\theta, y) = (10 - y/2)\cos(\theta)\mathbf{i} + y\mathbf{j} + (10 - y/2)\sin(\theta)\mathbf{k}, \qquad 0 \le \theta \le 2\pi, \quad 8 \le y \le 14,$$

parameterizes $\mathcal{S}$. We calculate the partial derivatives

$$\mathbf{r}_\theta(\theta, y) = -\left(10 - \frac{y}{2}\right)\sin(\theta)\mathbf{i} + 0\mathbf{j} + \left(10 - \frac{y}{2}\right)\cos(\theta)\mathbf{k},$$
$$\mathbf{r}_y(\theta, y) = -\frac{1}{2}\cos(\theta)\mathbf{i} + 1\mathbf{j} - \frac{1}{2}\sin(\theta)\mathbf{k}.$$

A normal vector at a point $P = (x, y, z)$ on $\mathcal{S}$ is then given by

$$\mathbf{N} = \mathbf{r}_\theta(\theta, y) \times \mathbf{r}_y(\theta, y) = \det\left(\begin{bmatrix} \mathbf{i} & \mathbf{j} & \mathbf{k} \\ -\left(10 - \frac{y}{2}\right)\sin(\theta) & 0 & \left(10 - \frac{y}{2}\right)\cos(\theta) \\ -\left(\frac{1}{2}\right)\cos(\theta) & 1 & -\left(\frac{1}{2}\right)\sin(\theta) \end{bmatrix}\right)$$
$$= \left(-10 + \frac{y}{2}\right)\cos(\theta)\mathbf{i} + \left(-5 + \frac{y}{4}\right)\mathbf{j} + \left(-10 + \frac{y}{2}\right)\sin(\theta)\mathbf{k}.$$

If P is in the upper half of the yz-plane, then $x = r\cos(\theta) = 0$; so $\theta = \pi/2$, and the $\mathbf{k}$ component of $\mathbf{N}$ is $(-10 + y/2)\sin(\pi/2) = y/2 - 10 < 0$. Therefore, $\mathbf{n} = -(1/\|\mathbf{N}\|)\mathbf{N}$ gives the correct orientation of $\mathcal{S}$. We calculate

$$\mathbf{curl}(\mathbf{F}) = \det\left(\begin{bmatrix} \mathbf{i} & \mathbf{j} & \mathbf{k} \\ \frac{\partial}{\partial x} & \frac{\partial}{\partial y} & \frac{\partial}{\partial z} \\ z & 1 & z^2 \end{bmatrix}\right) = 0\mathbf{i} + 1\mathbf{j} + 0\mathbf{k} = \mathbf{j}.$$

Therefore,

$$\mathbf{curl}(\mathbf{F}) \cdot \mathbf{n}\, dS = \mathbf{j} \cdot \left(-\left(\frac{1}{\|\mathbf{N}\|}\mathbf{N}\right)\right) \|\mathbf{N}\|\, dA = \left(5 - \frac{y}{4}\right) dy\, d\theta.$$

Thus,

$$\iint_{\mathcal{S}} \mathbf{curl}(\mathbf{F}) \cdot \mathbf{n}\, dS = \int_0^{2\pi}\int_8^{14} \left(5 - \frac{y}{4}\right) dy\, d\theta = \int_0^{2\pi} \left(5y - \frac{y^2}{8}\right)\Bigg|_8^{14} d\theta$$
$$= 2\pi\left(\left(70 - \frac{14^2}{8}\right) - \left(40 - \frac{8^2}{8}\right)\right) = 27\pi.$$

To compute the right side of equation (15.49), we must calculate two line integrals. Figure 14 shows the two components $\mathcal{C}_1$ and $\mathcal{C}_2$ of the boundary. Let

$$\mathbf{r}_1(\theta) = \mathbf{r}(-\theta, 8) = 6\cos(\theta)\mathbf{i} + 8\mathbf{j} - 6\sin(\theta)\mathbf{k}$$

and

$$\mathbf{r}_2(\theta) = \mathbf{r}(\theta, 14) = 3\cos(\theta)\mathbf{i} + 14\mathbf{j} + 3\sin(\theta)\mathbf{k}$$

be parameterizations of $\mathcal{C}_1$ and $\mathcal{C}_2$, respectively. (Verify that these parameterizations orient $\mathcal{C}_1$ and $\mathcal{C}_2$ consistently with $\mathcal{S}$.) We obtain

$$\begin{aligned}\mathbf{F}(\mathbf{r}_1(\theta)) \cdot \mathbf{r}_1'(\theta) &= (-6\sin(\theta)\mathbf{i} + 1\mathbf{j} + 36\sin^2(\theta)\mathbf{k}) \cdot (-6\sin(\theta)\mathbf{i} + 0\mathbf{j} - 6\cos(\theta)\mathbf{k}) \\ &= 36\sin^2(\theta) - 216\sin^2(\theta)\cos(\theta).\end{aligned}$$

Thus,

$$\int_{\mathcal{C}_1} \mathbf{F} \cdot d\mathbf{r}_1 = \int_0^{2\pi} (36\sin^2(\theta) - 216\sin^2(\theta)\cos(\theta))\, d\theta \overset{(7.21)}{=} 36\pi - 0 = 36\pi.$$

Similarly, we have

$$\begin{aligned}\mathbf{F}(\mathbf{r}_2(\theta)) \cdot \mathbf{r}_2'(\theta) &= (3\sin(\theta)\mathbf{i} + 1\mathbf{j} + 9\sin^2(\theta)\mathbf{k}) \cdot (-3\sin(\theta)\mathbf{i} + 0\mathbf{j} + 3\cos(\theta)\mathbf{k}) \\ &= -9\sin^2(\theta) + 27\sin^2(\theta)\cos(\theta)\end{aligned}$$

and

$$\int_{\mathcal{C}_2} \mathbf{F} \cdot d\mathbf{r}_2 = \int_0^{2\pi} (-9\sin^2(\theta) + 27\sin^2(\theta)\cos(\theta))\, d\theta \overset{(7.21)}{=} -9\pi + 0 = -9\pi.$$

Thus,

$$\int_{\mathcal{C}_1} \mathbf{F} \cdot d\mathbf{r}_1 + \int_{\mathcal{C}_2} \mathbf{F} \cdot d\mathbf{r}_2 = 36\pi - 9\pi = 27\pi = \iint_{\mathcal{S}} \mathbf{curl}(\mathbf{F}) \cdot \mathbf{n}\, dS,$$

which is the assertion of Stokes's Theorem. ■

inSIGHT

Notice that because of cancellation, we did not need to calculate $\|\mathbf{N}\|$ in Example 7. In general, when a surface $\mathcal{S}$ is parameterized by $(u, v) \mapsto \mathbf{r}(u, v)$, $(u, v) \in \mathcal{R}$, we have

$$\mathbf{n} = \pm\frac{1}{\|\mathbf{r}_u \times \mathbf{r}_v\|}\mathbf{r}_u \times \mathbf{r}_v \qquad \text{and} \qquad dS = \|\mathbf{r}_u \times \mathbf{r}_v\|\; du\, dv.$$

Therefore, the scalar $\|\mathbf{r}_u \times \mathbf{r}_v\|$ cancels when $\mathbf{curl}(\mathbf{F}) \cdot \mathbf{n}\, dS$ is calculated, resulting in the formula

$$\iint_{\mathcal{S}} \mathbf{curl}(\mathbf{F}) \cdot \mathbf{n}\, dS = \iint_{\mathcal{R}} \mathbf{curl}(\mathbf{F}) \cdot (\mathbf{r}_u \times \mathbf{r}_v)\, du\, dv. \qquad \textbf{(15.50)}$$

Theorem 2 Let $\mathcal{S}$ be an oriented surface with piecewise-smooth boundary, such as we have been discussing. Let the boundary curves be $\mathcal{C}_1, \ldots, \mathcal{C}_N$ with consistent parameterizations $\mathbf{r}_1, \ldots, \mathbf{r}_N$. Suppose that $\mathbf{F}$ is a continuously differentiable irrotational vector field on a neighborhood of $\mathcal{S}$ and its boundary. Then

$$\sum_{j=1}^{N} \int_{\mathcal{C}_j} \mathbf{F} \cdot d\mathbf{r}_j = 0. \tag{15.51}$$

Proof As discussed in Section 15.4, saying that **F** is irrotational means that $\mathbf{curl}(\mathbf{F}) = 0$ at each point of $\mathcal{S}$. We apply Stokes's Theorem to obtain

$$\sum_{j=1}^{N} \int_{\mathcal{C}_j} \mathbf{F} \cdot d\mathbf{r}_j = \iint_{\mathcal{S}} \mathbf{curl}(\mathbf{F}) \cdot \mathbf{n}\, dS = \iint_{\mathcal{S}} 0\, dS = 0.$$

If the irrotational vector field **F** represents a flow, then we see that the flow around the boundary integrates to 0. Though this is to be expected, there can be some subtleties involved. Remember that if $\mathcal{S}$ is not simply connected, then the property "**F** is irrotational" is weaker than the property "**F** is conservative." If **F** is irrotational but not conservative, then a summand $\int_{\mathcal{C}_j} \mathbf{F} \cdot d\mathbf{r}_j$ in equation (15.51) can be nonzero, even if the path $\mathcal{C}_j$ is closed. The next example illustrates this point.

Example 8 Verify equation (15.51) for the irrotational vector field

$$\mathbf{F}(x, y, z) = \left(\frac{-y}{x^2 + y^2}\right)\mathbf{i} + \left(\frac{x}{x^2 + y^2}\right)\mathbf{j} + 0\mathbf{k}$$

on the planar surface $\mathcal{S} = \{(x, y, z) : z = 0, 1 \le x^2 + y^2 \le 4\}$ oriented by an upward-pointing normal.

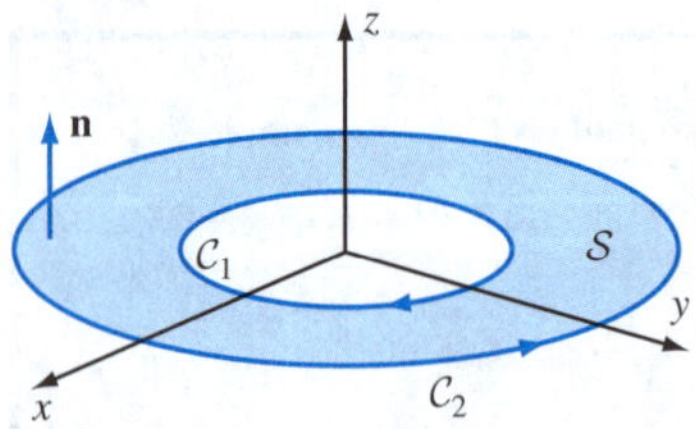

$\{(x, y, z) : z = 0, 1 \le x^2 + y^2 \le 4\}$

Figure 15

Solution It is a routine matter to verify that **F** is irrotational. Let a be any positive constant. The vector-valued function $\mathbf{r}(t) = a\cos(t)\mathbf{i} + a\sin(t)\mathbf{j} + 0\mathbf{k}$, $0 \le t \le 2\pi$, parameterizes a counterclockwise circle $\mathcal{C}$ in the xy-plane of radius a and center $(0, 0, 0)$. We have $\mathbf{F}(\mathbf{r}(t)) = -(1/a)\sin(t)\mathbf{i} + (1/a)\cos(t)\mathbf{j} + 0\mathbf{k}$ and $\mathbf{r}'(t) = -a\sin(t)\mathbf{i} + a\cos(t)\mathbf{j} + 0\mathbf{k}$. Therefore, $\mathbf{F}(\mathbf{r}(t)) \cdot \mathbf{r}'(t) = \sin^2(t) + \cos^2(t)$ and

$$\int_{\mathcal{C}} \mathbf{F} \cdot d\mathbf{r} = \int_0^{2\pi} (\sin^2(t) + \cos^2(t))\, dt = \int_0^{2\pi} 1\, dt = 2\pi.$$

Notice that the value of this line integral does not depend on the radius a. If $\mathcal{C}_1$ and $\mathcal{C}_2$ are the components of the boundary indicated in Figure 15, then $\mathcal{C}_1$ has clockwise orientation, $\mathcal{C}_2$ has counterclockwise orientation, and

$$\int_{\mathcal{C}_1} \mathbf{F} \cdot d\mathbf{r}_1 + \int_{\mathcal{C}_2} \mathbf{F} \cdot d\mathbf{r}_2 = -2\pi + 2\pi = 0.$$

An Application

The background material on curl presented in Section 15.4 provides ample evidence that Stokes's Theorem is important in physics. Here we provide but one example.

Example 9 Imagine that **B** is a magnetic field in space. Let the surface $\mathcal{S}$, together with its continuously differentiable boundary curve $\mathcal{C}$, be the graph of a continuously differentiable function $z = f(x, y)$. Suppose that $\mathcal{S}$ is given the upward unit normal **n** and that $\mathcal{C}$ is parameterized by **r** with consistent orientation. What does Stokes's Theorem tell us about the flow of **B** around $\mathcal{C}$?

Solution By Stokes's Theorem,

$$\oint_C \mathbf{B} \cdot d\mathbf{r} = \iint_S \mathbf{curl(B)} \cdot \mathbf{n}\, dS.$$

Let us normalize this equation so that we can introduce some standard physical terminology. Let c denote the speed of light (about 186,000 mi/s). The vector

$$\mathbf{J} = \frac{c}{4\pi}\mathbf{curl(B)},$$

called the *current density* for the field, measures the current per unit area of the surface. Then the total current I for the field on the surface $\mathcal{S}$ is

$$I = \iint_S \mathbf{J} \cdot \mathbf{n}\, dS.$$

In summary, we see that

$$\oint_C \mathbf{B} \cdot d\mathbf{r} = \frac{4\pi}{c} I. \qquad \textbf{(15.52)}$$

Thus, the flow of **B** around the boundary of $\mathcal{S}$ is a physical constant times the total current I. Equation (15.52) is known as Ampère's law, after its discoverer André-Marie Ampère (1775–1836). It is geometrically obvious that a closed curve $\mathcal{C}$ bounds infinitely many different surfaces $\mathcal{S}$. It is remarkable that Ampère's law does not depend on which such surface $\mathcal{S}$ we are considering. ■

quickquiz

1. Why is the component of curl in the normal direction important?
2. State Stokes's Theorem.
3. How is Stokes's Theorem related to Green's Theorem?
4. Give a physical interpretation of Stokes's Theorem.

EXERCISES

Problems for Practice

In Exercises 1–12, a planar region $\mathcal{R}$, a vector field **F**, and a scalar-valued function f are given. Let $\mathcal{S}$ denote the graph of f over the region $\mathcal{R}$. Calculate the surface integral of $\mathbf{curl(F)} \cdot \mathbf{n}$ over $\mathcal{S}$ in two ways:

a. Do the integration directly as a surface integral.

b. Use Stokes's Theorem to reduce the problem to a line integral.

1. $\mathbf{F}(x, y, z) = z^3\mathbf{i} - xy\mathbf{j} + xz\mathbf{k}$, $f(x, y) = 2x - y$, $\mathcal{R} = \{(x, y) : |x| < 2, |y| < 1\}$

2. $\mathbf{F}(x, y, z) = (y - z)\mathbf{i} - (x + z)\mathbf{j} + (x + y)\mathbf{k}$, $f(x, y) = -5x + 2y$, $\mathcal{R} = \{(x, y) : 1 < x < 2, 0 < y < 2\}$

3. $\mathbf{F}(x, y, z) = e^z\mathbf{i} + 2\mathbf{j} - y\mathbf{k}$, $f(x, y) = x - 2y$, $\mathcal{R} = \{(x, y) : \ln(2) < x < \ln(4), 0 < y < 1\}$

4. $\mathbf{F}(x, y, z) = z\mathbf{i} - z^2\mathbf{j} + x^2\mathbf{k}$, $f(x, y) = 4x - y$, $\mathcal{R} =$ (the triangle with vertices $(1, 0)$, $(0, 1)$, and $(-1, 0)$)

5. $\mathbf{F}(x, y, z) = xz\mathbf{i} - x\mathbf{j} + yz\mathbf{k}$, $f(x, y) = (1 - x^2 - y^2)^{1/2}$, $\mathcal{R} = \{(x, y) : x^2 + y^2 < 1\}$

6. $\mathbf{F}(x, y, z) = \sin(z)\mathbf{i} - \cos(x)\mathbf{j} + \sin(y)\mathbf{k}$, $f(x, y) = 2x - 3y$, $\mathcal{R} = \{(x, y) : |x| < \pi, 0 < y < \pi/2\}$

7. $\mathbf{F}(x, y, z) = 7z\mathbf{i} - 5x\mathbf{j} + 3y\mathbf{k}$, $f(x, y) = x^2$, $\mathcal{R} = \{(x, y) : 0 < x < 3, 1 < y < 2\}$
8. $\mathbf{F}(x, y, z) = 2y\mathbf{i} + (x + y)\mathbf{j} + yz\mathbf{k}$, $f(x, y) = (1 - x^2 - y^2)^2$, $\mathcal{R} = \{(x, y) : x^2 + y^2 < 1\}$
9. $\mathbf{F}(x, y, z) = x\mathbf{i} + (y - z^2)\mathbf{j} + \mathbf{k}$, $f(x, y) = x + y + 3$, $\mathcal{R} = \{(x, y) : |x| < y < 1\}$
10. $\mathbf{F}(x, y, z) = y\mathbf{i} - x\mathbf{j} + \mathbf{k}$, $f(x, y) = 2x - 3y + 1$, $\mathcal{R} = \{(x, y) : |y| < x < 1\}$
11. $\mathbf{F}(x, y, z) = z\mathbf{i} - x\mathbf{k}$, $f(x, y) = x + 3y$, $\mathcal{R} = \{(x, y) : |x| < 1, |y| < 1\}$
12. $\mathbf{F}(x, y, z) = xy\mathbf{i} + zx\mathbf{j} - xy\mathbf{k}$, $f(x, y) = x^2 - y^2$, $\mathcal{R} = \{(x, y) : 0 < y < 1 - x^2\}$

In Exercises 13–18, use Stokes's Theorem to evaluate $\iint_{\mathcal{S}} \mathbf{curl}(\mathbf{F}) \cdot \mathbf{n}\, dS$ for the vector field $\mathbf{F}$ and the oriented surface $\mathcal{S}$.

13. $\mathbf{F}(x, y, z) = y\mathbf{i} + z^2\mathbf{j} + \sqrt{1 + z^4}\mathbf{k}$; $\mathcal{S} = \{(x, y, z) : x^2 + y^2 + (z - 4)^2 = 25, 0 < z\}$, oriented by $P \mapsto \mathbf{n}(P)$ with $\mathbf{n}(0, 0, 9) = \mathbf{k}$
14. $\mathbf{F}(x, y, z) = x\mathbf{i} + x\mathbf{j} + \exp(xyz)\mathbf{k}$; $\mathcal{S} = \left\{(x, y, z) : z = \sqrt{x^2 + y^2}, z < 1\right\}$, oriented by the upward-pointing unit normal vector field
15. $\mathbf{F}(x, y, z) = zx\mathbf{j}$; $\mathcal{S} = \{(x, y, z) : x^2 + y^2 = 9, -2 < z < 5\}$, oriented by $P \mapsto \mathbf{n}(P)$ with $\mathbf{n}(3, 0, 0) = -\mathbf{i}$
16. $\mathbf{F}(x, y, z) = 2y\mathbf{i} + (z + 1)y\mathbf{j} + \mathbf{k}$; $\mathcal{S}$ = the union of the cylinder $\{(x, y, z) : x^2 + y^2 = 4, 0 < z < 1\}$, the disk $\{(x, y, 0) : x^2 + y^2 \leq 4\}$, and the annulus $\{(x, y, 1) : 1 < x^2 + y^2 \leq 4\}$, oriented by $P \mapsto \mathbf{n}(P)$ with $\mathbf{n}(0, 0, 0) = -\mathbf{k}$
17. $\mathbf{F}(x, y, z) = y^2\mathbf{i} + x\mathbf{j} + \sin(z^2)\mathbf{k}$; $\mathcal{S} = \Delta PQT \cup \Delta QRT \cup \Delta RST \cup \Delta SPT$ where $P = (0, 0, 0)$, $Q = (2, 0, 0)$, $R = (2, 2, 0)$, $S = (0, 2, 0)$, $T = (1, 1, 2)$, oriented by the upward-pointing unit normal vector field
18. $\mathbf{F}(x, y, z) = (3x + y)\mathbf{i} - z\mathbf{j} + y^2\mathbf{k}$; $\mathcal{S}$ = the union of the cylinder $\{(x, y, z) : x^2 + y^2 = 4, 2 < z < 4\}$ and the hemisphere $\{(x, y, z) : x^2 + y^2 + (z - 2)^2 = 4, z \leq 2\}$, oriented by $P \mapsto \mathbf{n}(P)$ with $\mathbf{n}(0, 0, 0) = -\mathbf{k}$

Further Theory and Practice

In Exercises 19–30, an oriented surface $\mathcal{S}$ and a vector field $\mathbf{F}$ are given. Orient the boundary of $\mathcal{S}$ consistently and verify Stokes's Theorem.

19. $\mathbf{F}(x, y, z) = y^3\mathbf{i} + z^3\mathbf{j} + x^3\mathbf{k}$; $\mathcal{S}$ = the hemisphere $\{(x, y, z) : x^2 + y^2 + z^2 = 1, y > 0\}$, oriented by $P \mapsto \mathbf{n}(P)$ with $\mathbf{n}(0, 1, 0) = \mathbf{j}$
20. $\mathbf{F}(x, y, z) = z\mathbf{i} + x\mathbf{j} + y\mathbf{k}$; $\mathcal{S} = \{(x, y, z) : y^2 + z^2 = 1, 0 < x < 1, 0 < z\}$, oriented by $P \mapsto \mathbf{n}(P)$ with $\mathbf{n}(1/2, 0, 1) = \mathbf{k}$
21. $\mathbf{F}(x, y, z) = y\mathbf{i} - x\mathbf{j} + 3\mathbf{k}$; $\mathcal{S} = \{(x, y, z) : z = \sqrt{x^2 + y^2}, 0 < z < 2\}$, oriented by $P \mapsto \mathbf{n}(P)$ with $\mathbf{n}(P) \cdot \mathbf{k} < 0$
22. $\mathbf{F}(x, y, z) = -y\mathbf{i} + x\mathbf{j} + z\mathbf{k}$; $\mathcal{S} = \{(x, y, z) : x^2 + y^2 + z^2 = 4, 0 < x, 0 < y, 0 < z\}$, oriented by $P \mapsto \mathbf{n}(P)$ with $\mathbf{n}(P) \cdot \mathbf{k} > 0$
23. $\mathbf{F}(x, y, z) = z\mathbf{i} + y\mathbf{j} + z\mathbf{k}$; $\mathcal{S} = \{(x, y, z) : z = x^2 + y^2, z < x + y\}$, oriented by $P \mapsto \mathbf{n}(P)$ with $\mathbf{n}(P) \cdot \mathbf{k} < 0$
24. $\mathbf{F}(x, y, z) = -xz\mathbf{j}$; $\mathcal{S} = \{(x, y, z) : x^2 + y^2 + z^2 = 2, -1 < z\}$, oriented by $P \mapsto \mathbf{n}(P)$ with $\mathbf{n}(0, 0, \sqrt{2}) = \mathbf{k}$
25. $\mathbf{F}(x, y, z) = xy^2\mathbf{j} - xz\mathbf{k}$; $\mathcal{S} = \{(x, y, z) : x^2 + y^2 + z^2 = 25, 3 < z < 4\}$, oriented by $P \mapsto \mathbf{n}(P)$ with $\mathbf{n}(P) \cdot \mathbf{k} > 0$
26. $\mathbf{F}(x, y, z) = yz\mathbf{i}$; $\mathcal{S} = \{(x, y, z) : x^2 + y^2 + z^2 = 25, 0 < z < 4\}$, oriented by $P \mapsto \mathbf{n}(P)$ with $\mathbf{n}(P) \cdot \mathbf{k} > 0$
27. $\mathbf{F}(x, y, z) = x^3\mathbf{i} - z\mathbf{j} + y\mathbf{k}$; $\mathcal{S} = \{(x, y, z) : x^2 = y^2 + z^2, 1 < x < 4\}$, oriented by $P \mapsto \mathbf{n}(P)$ with $\mathbf{n}(P) \cdot \mathbf{i} < 0$
28. $\mathbf{F}(x, y, z) = x\mathbf{i} - xyz^3\mathbf{j} + z^2\mathbf{k}$; $\mathcal{S} = \{(x, y, z) : x^2 + y^2 = 1, 0 < z < 1\}$, oriented by $P \mapsto \mathbf{n}(P)$ with $\mathbf{n}(1, 0, 1/2) \cdot \mathbf{i} < 0$
29. $\mathbf{F}(x, y, z) = y^2\mathbf{i} - xz\mathbf{j} + x\mathbf{k}$; $\mathcal{S} = \{(x, y, z) : x^2 + y^2 = 4, -2 < z < 3\}$, oriented by $P \mapsto \mathbf{n}(P)$ with $\mathbf{n}(2, 0, 0) = \mathbf{i}$
30. $\mathbf{F}(x, y, z) = yz\mathbf{i} - xyz\mathbf{j} + x^2\mathbf{k}$; $\mathcal{S} = \{(x, y, z) : x^2 + y^2 = 4, -1 < z < 1\}$, oriented by $P \mapsto \mathbf{n}(P)$ with $\mathbf{n}(2, 0, 0) = \mathbf{i}$
31. Let $\mathcal{S} = \{(x, y, z) : x^2 + y^2 + (z - 1)^2 = 2, 0 < z\}$, oriented by $P \mapsto \mathbf{n}(P)$ with $\mathbf{n}\left(0, 0, 1 + \sqrt{2}\right) = \mathbf{k}$. Let $\mathbf{F}(x, y, z) = y\cos(xz)\mathbf{i} - xy\sin(z^2)\mathbf{j} + \exp(x^2)\mathbf{k}$. Calculate $\iint_{\mathcal{S}} \mathbf{curl}(\mathbf{F}) \cdot \mathbf{n}\, dS$ by using Stokes's Theorem twice. Convert the required surface integral to a line integral, and then convert that to a surface integral over which integration is easy.
32. Let $\mathcal{S} = \{(x, y, z) : z = 1 - x^2/4 - y^2 = 2, 0 < z\}$, oriented by $P \mapsto \mathbf{n}(P)$ with $\mathbf{n}(0, 0, 1) = \mathbf{k}$. Let $\mathbf{F}(x, y, z) = y\exp(x\sin(z))\mathbf{i} - x\ln(1 + z^2)\mathbf{j} + \sin(x^2)\mathbf{k}$. Calculate $\iint_{\mathcal{S}} \mathbf{curl}(\mathbf{F}) \cdot \mathbf{n}\, dS$ by using Stokes's Theorem twice. Convert the required surface integral to a line integral, and then convert that to a surface integral over which integration is easy.

33. Consider the surface $\mathcal{S}$, which is the unit sphere in space centered at the origin. Let **n** denote the outward unit normal on $\mathcal{S}$. Use Stokes's Theorem to show that if **F** is a continuously differentiable vector field on a neighborhood of $\mathcal{S}$, then

$$\iint_{\mathcal{S}} \mathbf{curl}(\mathbf{F}) \cdot \mathbf{n}\, dS = 0.$$

Calculator/Computer Exercises

In Exercises 34–37, an oriented surface $\mathcal{S}$ and a vector field **F** are given. Orient the boundary of $\mathcal{S}$ consistently, and verify Stokes's Theorem.

34. $\mathbf{F}(x, y, z) = z^3\mathbf{i} + 3y\mathbf{j}$;
$\mathcal{S} = \{(x, y, z) : z = x^2 + 2y, (x-2)^2 < y < \ln(x)\}$, oriented by $P \mapsto \mathbf{n}(P)$ with $\mathbf{n}(2, 1/2, 5) = \mathbf{k}$

35. $\mathbf{F}(x, y, z) = x\exp(-y)\mathbf{j}$;
$\mathcal{S} = \{(x, y, z) : z = \exp(x^2 + y^2), x^2 + y^2 < 1\}$, oriented by $P \mapsto \mathbf{n}(P)$ with $\mathbf{n}(0, 0, 1) = -\mathbf{k}$

36. $\mathbf{F}(x, y, z) = \ln(2 + y)\mathbf{i}$;
$\mathcal{S} = \{(x, y, z) : z = 1 - x^2/4 - y^2, 0 < z\}$, oriented by $P \mapsto \mathbf{n}(P)$ with $\mathbf{n}(0, 0, 1) = \mathbf{k}$

37. $\mathbf{F}(x, y, z) = \exp(z^2)\mathbf{j}$;
$\mathcal{S} = \{(x, y, z) : x + y + z = 1, 0 < x, 0 < y, 0 < z\}$, oriented by $P \mapsto \mathbf{n}(P)$ with $\mathbf{n} \cdot \mathbf{k} > 0$

15.8 Flux and the Divergence Theorem

The Fundamental Theorem of Calculus relates the boundary data of a function at the endpoints of an interval with information about the derivative of the function on the interval:

$$F(b) - F(a) = \int_a^b F'(t)\, dt.$$

This is a *one-dimensional* theorem.

Green's Theorem relates the boundary data of a vector field on a planar region $\mathcal{R}$ with information about certain derivatives of the vector field $M\mathbf{i} + N\mathbf{j}$ in the interior:

$$\oint_C M\, dx + N\, dy = \iint_{\mathcal{R}} \left(\frac{\partial N}{\partial x} - \frac{\partial M}{\partial y} \right) dA.$$

This is a *two-dimensional* theorem. Stokes's Theorem is a similar two-dimensional theorem for surfaces. Now we will learn a three-dimensional theorem in the same vein—the Divergence Theorem. As with Stokes's Theorem, we must begin by considering the question of orientation.

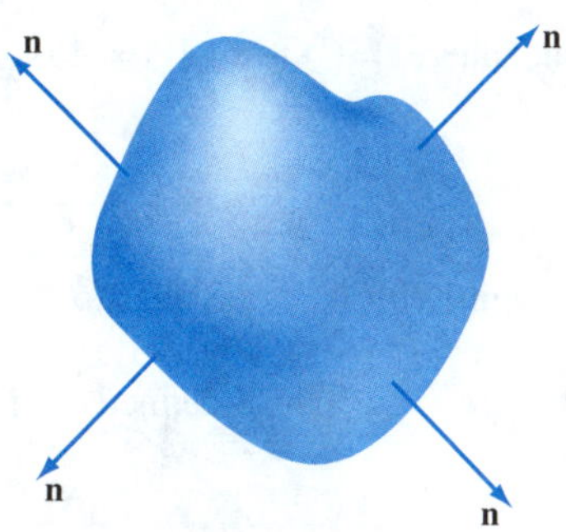

Figure 1

Suppose that $\mathcal{U}$ is a bounded three-dimensional solid in space whose boundary $\mathcal{S}$ consists of one or more continuously differentiable surfaces. To each point (x, y, z) on $\mathcal{S}$, associate the unit normal vector that points *out* of the solid. Refer to Figure 1. Call this vector field **n**. Suppose that **F** is a continuously differentiable vector field on a neighborhood of $\mathcal{U}$ and its boundary. At a point $P = (x, y, z)$ in $\mathcal{S}$, the quantity $\mathbf{F}(x, y, z) \cdot \mathbf{n}(x, y, z)$ is the normal component of **F**. If **F** represents a flow and $\mathbf{F}(P) \cdot \mathbf{n} > 0$, then flow at the point P is outward; if $\mathbf{F}(P) \cdot \mathbf{n} < 0$, then flow at P is inward. The surface integral

$$\iint_{\mathcal{S}} \mathbf{F} \cdot \mathbf{n}\, dS$$

represents the overall normal component of flow at the boundary. We call this quantity the *flux* of the vector field $\mathbf{F}$ across the boundary. If the flux is positive, then, overall, the flow is outward; if the flux is negative, then, overall, the flow is inward.

The Divergence Theorem

It makes good physical sense that the flux of a vector field $\mathbf{F}$ can also be obtained by summing its divergence over the interior. This is, in fact, what the Divergence Theorem tells us.

Theorem 1 ***The Divergence Theorem*** With $\mathcal{U}$, $\mathcal{S}$, $\mathbf{n}$, and $\mathbf{F}$ as described in the introduction to this section, we have

$$\iint_{\mathcal{S}} \mathbf{F} \cdot \mathbf{n}\, dS = \iiint_{\mathcal{U}} \operatorname{div}(\mathbf{F})\, dV. \tag{15.53}$$

A proof of the Divergence Theorem in its full generality is far beyond the scope of this book. We therefore supply only a proof of a relatively simple case. Because even that simplified derivation is quite lengthy, we defer it to the end of the section. For now, we concentrate on understanding the Divergence Theorem in particular examples.

Example 1 Let $\mathcal{U}$ be the unit ball in space. The boundary surface $\mathcal{S}$ is the unit sphere $x^2+y^2+z^2 = 1$. Verify the Divergence Theorem for the vector field $\mathbf{F}(x, y, z) = 3x\mathbf{i} + 3y\mathbf{j} + 3z\mathbf{k}$ defined on $\mathcal{U}$ and its boundary.

Solution First, $\operatorname{div}(\mathbf{F}) = 3 + 3 + 3 = 9$. Therefore,

$$\iiint_{\mathcal{U}} \operatorname{div}(\mathbf{F})\, dV = 9 \cdot (\text{volume of ball of radius 1}) = 9 \cdot \left(\frac{4}{3}\pi \cdot 1^3\right) = 12\pi.$$

Turning our attention to the left side of equation (15.53), we observe that $\mathbf{n}(x, y, z) = x\mathbf{i} + y\mathbf{j} + z\mathbf{k}$ is the unit outward-pointing normal at the point (x, y, z) of $\mathcal{S}$. Therefore,

$$\mathbf{F} \cdot \mathbf{n} = (3x\mathbf{i} + 3y\mathbf{j} + 3z\mathbf{k}) \cdot (x\mathbf{i} + y\mathbf{j} + z\mathbf{k}) = 3(x^2 + y^2 + z^2) \overset{\text{On } \mathcal{S}}{=} 3.$$

It follows that

$$\iint_{\mathcal{S}} \mathbf{F} \cdot \mathbf{n}\, dS = \iint_{\mathcal{S}} 3\, dS = 3 \cdot (\text{surface area of the unit sphere}) = 3 \cdot (4\pi \cdot 1^2) = 12\pi,$$

as required. ■

Example 2 Let $\mathcal{U}$ be the cube centered at the origin with faces parallel to the coordinate planes and side length 2: $\mathcal{U} = \{(x, y, z) : -1 < x < 1, -1 < y < 1, -1 < z < 1\}$. Verify the Divergence Theorem for $\mathbf{F}(x, y, z) = 2x^2\mathbf{i} - y^2\mathbf{j} + z\mathbf{k}$ defined on $\mathcal{U}$ and its boundary.

Solution We have

$$\iiint\limits_{\mathcal{U}} \operatorname{div}(\mathbf{F})\,dV = \iiint\limits_{\mathcal{U}} (4x - 2y + 1)\,dV = \int_{-1}^{1}\int_{-1}^{1}\int_{-1}^{1} (4x - 2y + 1)\,dz\,dy\,dx$$

$$= 2\int_{-1}^{1}\int_{-1}^{1} (4x - 2y + 1)\,dy\,dx.$$

Continuing,

$$\begin{aligned} 2\int_{-1}^{1}\int_{-1}^{1} (4x - 2y + 1)\,dy\,dx &= 2\int_{-1}^{1} (4xy - y^2 + y)\Big|_{y=-1}^{y=1} dx \\ &= 2\int_{-1}^{1} ((4x - 1 + 1) - (-4x - 1 - 1))\,dx \\ &= 2\int_{-1}^{1} (8x + 2)\,dx \\ &= 2\,(4x^2 + 2x)\Big|_{x=-1}^{x=1} \\ &= 8. \end{aligned}$$

Let $\mathcal{S}$ be the boundary of $\mathcal{U}$. It consists of the six faces of the cube: top and bottom, front and back, left and right. On the top, the unit normal field is $\mathbf{n}(x, y, z) = \mathbf{k}$, the element of area is $dx\,dy$, and $z = 1$. Thus, the surface integral of $\mathbf{F} \cdot \mathbf{n}$ over the top of the cube is

$$\int_{-1}^{1}\int_{-1}^{1} (2x^2\mathbf{i} - y^2\mathbf{j} + 1\mathbf{k}) \cdot \mathbf{k}\;dx\,dy = \int_{-1}^{1}\int_{-1}^{1} 1\,dx\,dy = 4.$$

Similarly, we see that the normal to the bottom is $-\mathbf{k}$, the element of area is $dx\,dy$, and $z = -1$. Therefore, the surface integral of $\mathbf{F} \cdot \mathbf{n}$ over the bottom of the cube is

$$\int_{-1}^{1}\int_{-1}^{1} (2x^2\mathbf{i} - y^2\mathbf{j} - 1\mathbf{k}) \cdot (-\mathbf{k})\,dx\,dy = \int_{-1}^{1}\int_{-1}^{1} 1\,dx\,dy = 4.$$

In the same fashion, it can be calculated that the integral over the front is 8, the integral over the back is -8, the integral over the left side is 4, and the integral over the right side is -4.

Summing these values, we see that

$$\iint\limits_{\mathcal{S}} \mathbf{F} \cdot \mathbf{n}\,dS = 4 + 4 + 8 - 8 + 4 - 4 = 8,$$

which verifies the Divergence Theorem. From the positive value of the flux, we conclude that the net flow is *out* of the cube.

Of course, there is no point in calculating the same quantity twice. Examples 1 and 2 are just for practice. The next example is a more typical application of the Divergence Theorem.

Example 3 Use the Divergence Theorem to calculate the flux of $\mathbf{F}(x, y, z) = -x^3\mathbf{i} - y^3\mathbf{j} + 3z^2\mathbf{k}$ across the boundary $\mathcal{S}$ of

$$\mathcal{U} = \{(x, y, z) : x^2 + y^2 < 16, 0 < z < 5\}.$$

Solution Observe that $\mathcal{U}$ is a cylinder with radius 4 and height 5. The flux of $\mathbf{F}$ across $\mathcal{S}$ is defined to be $\iint_{\mathcal{S}} \mathbf{F} \cdot \mathbf{n}\, dS$. To evaluate this integral, we would have to break up the calculation into an integral over the top, an integral over the bottom, and an integral over the cylindrical side. It is therefore considerably easier to use the Divergence Theorem to calculate the flux. Thus,

$$\iint_{\mathcal{S}} \mathbf{F} \cdot \mathbf{n}\, dS = \iiint_{\mathcal{U}} \operatorname{div}(\mathbf{F})\, dV = \iiint_{\mathcal{U}} \left(\frac{\partial}{\partial x}(-x^3) + \frac{\partial}{\partial y}(-y^3) + \frac{\partial}{\partial z}3z^2\right) dV$$
$$= \iiint_{\mathcal{U}} (-3x^2 - 3y^2 + 6z)\, dV.$$

Introducing cylindrical coordinates, we may rewrite this integral as

$$\int_0^5 \int_0^{2\pi} \int_0^4 (-3r^2 + 6z) r\, dr\, d\theta\, dz,$$

the evaluation of which is straightforward. We have

$$\iint_{\mathcal{S}} \mathbf{F} \cdot \mathbf{n}\, dS = \int_0^5 \int_0^{2\pi} \left(-\frac{3}{4}r^4 + 3zr^2\right)\Bigg|_{r=0}^{r=4} d\theta\, dz = \int_0^5 \int_0^{2\pi} (-192 + 48z)\, d\theta\, dz$$
$$= 2\pi \int_0^5 (-192 + 48z)\, dz = -720\pi.$$

The flux is negative, and we conclude that, overall, the flow is *into* the cylinder. ■

Some Applications

The Divergence Theorem has many important uses in several branches of physics. We turn to a few representative applications.

Example 4 Let $\mathcal{S}$ be the boundary of a solid that contains the origin in its interior. Denote the outward unit normal of $\mathcal{S}$ by $\mathbf{n}$. Suppose that a point charge at the origin exerts a gravitational field

$$\mathbf{F}(x, y, z) = \frac{k}{\|\mathbf{r}\|^3}\mathbf{r}, \quad (x, y, z) \neq (0, 0, 0)$$

where k is a proportionality constant and $\mathbf{r}$ is the position vector $x\mathbf{i} + y\mathbf{j} + z\mathbf{k}$ of the point (x, y, z). Show that

$$\iint_{\mathcal{S}} \mathbf{F} \cdot \mathbf{n}\, dS = 4\pi k. \tag{15.54}$$

Solution Let us first calculate $\iint_S \mathbf{F} \cdot \mathbf{n}\, dS$ when $\mathcal{S}$ is the sphere Σ_a of radius a centered at the origin, oriented with outward-pointing unit normal. (The symbol Σ is often used to denote a sphere, particularly when it cannot be confused with its familiar role as summation symbol.) At any point (x, y, z) of Σ_a, the outward unit normal is $(1/a)\mathbf{r}$. Therefore,

$$\mathbf{F} \cdot \mathbf{n} = \frac{k}{a^3}\mathbf{r} \cdot \frac{1}{a}\mathbf{r} = \frac{k}{a^4}\,\|\mathbf{r}\|^2 = \frac{k}{a^2}$$

and

$$\iint_{\Sigma_a} \mathbf{F} \cdot \mathbf{n}\, dS = \iint_{\Sigma_a} \frac{k}{a^2} dS = \frac{k}{a^2} \cdot (\text{surface area of sphere of radius } a) = \frac{k}{a^2} \cdot (4\pi a^2) = 4\pi k.$$

We will use this special case, together with the Divergence Theorem, to derive equation (15.54) in general. Of course, we cannot apply the Divergence Theorem to $\mathbf{F}$ on $\mathcal{U}$ because $\mathbf{F}$ is not differentiable at the origin. However, $\mathbf{F}$ is differentiable on the solid $\mathcal{U}_a$, which is obtained by removing a small ball of radius a from $\mathcal{U}$. Let $\mathcal{S}_a$ denote the boundary of $\mathcal{U}_a$ with outward normal $\mathbf{n}$. Since $\text{div}(\mathbf{F}) = 0$ at each point other than the origin (by Example 3 of Section 15.4), we have

$$\iint_{\mathcal{S}_a} \mathbf{F} \cdot \mathbf{n}\, dS = \iiint_{\mathcal{U}_a} \text{div}(\mathbf{F})\, dV = \iiint_{\mathcal{U}_a} 0\, dV = 0. \qquad \textbf{(15.55)}$$

But $\mathcal{S}_a$ comprises $\mathcal{S}$ and the sphere $-\Sigma_a$ of radius a centered at the origin—the minus sign signifies that the sphere is oriented with the inward-pointing normal. Taking these observations into account in equation (15.55), we obtain

$$0 = \iint_{\mathcal{S}_a} \mathbf{F} \cdot \mathbf{n}\, dS = \iint_{\mathcal{S}} \mathbf{F} \cdot \mathbf{n}\, dS + \iint_{-\Sigma_a} \mathbf{F} \cdot \mathbf{n}\, dS = \iint_{\mathcal{S}} \mathbf{F} \cdot \mathbf{n}\, dS - \iint_{\Sigma_a} \mathbf{F} \cdot \mathbf{n}\, dS = \iint_{\mathcal{S}} \mathbf{F} \cdot \mathbf{n}\, dS - 4\pi k,$$

from which equation (15.54) immediately follows. ■

inSIGHT

Equation (15.54) is known as *Gauss's law*. Since the force between two charged particles has the same mathematical form as Newton's Law of Gravitation, Gauss's law is also applicable to the theory of electrostatics.

There are several variants of the Divergence Theorem. In the next example, we examine one of them that is often useful in applications.

Example 5 Suppose that $\mathcal{U}$, $\mathcal{S}$, and $\mathbf{n}$ are as in the statement of the Divergence Theorem. Suppose that f is a continuously differentiable scalar-valued function on an open set containing $\mathcal{U}$. Prove that

$$\iint_{\mathcal{S}} f\mathbf{n}\, dS = \iiint_{\mathcal{U}} \nabla f\, dV. \qquad \textbf{(15.56)}$$

Solution First of all, observe that the required equation is a *vector* equation. The integrations are performed componentwise. Let $\mathbf{c}$ be a constant vector. Define the vector field $\mathbf{G} = f(x, y, z)\mathbf{c}$. Theorem 1e, Section 15.4, tells us that $\text{div}(\mathbf{G}) = f(x, y, z)\text{div}(\mathbf{c}) + (\nabla f) \cdot \mathbf{c}$. But $\text{div}(\mathbf{c}) = 0$, because $\mathbf{c}$ is constant. Therefore, $\text{div}(\mathbf{G}) = (\nabla f) \cdot \mathbf{c}$. The Divergence Theorem tells us that

$$\iiint_{\mathcal{U}} (\nabla f) \cdot \mathbf{c}\, dV = \iiint_{\mathcal{U}} \text{div}(\mathbf{G})\, dV = \iint_{\mathcal{S}} \mathbf{G} \cdot \mathbf{n}\, dS = \iint_{\mathcal{S}} (f\mathbf{c}) \cdot \mathbf{n}\, dS.$$

Therefore,

$$\left(\iiint\limits_{\mathcal{U}} \nabla f \, dV - \iint\limits_{\mathcal{S}} f \mathbf{n} \, dS \right) \cdot \mathbf{c} = 0$$

for every constant vector $\mathbf{c}$. Since the zero vector is the only vector perpendicular to every constant vector, we conclude that

$$\iiint\limits_{\mathcal{U}} \nabla f \, dV - \iint\limits_{\mathcal{S}} f \mathbf{n} \, dS = \vec{\mathbf{0}},$$

from which the required identity follows. ■

Next we use the Divergence Theorem to derive a famous law of physics, *Archimedes's Law of Buoyancy*.

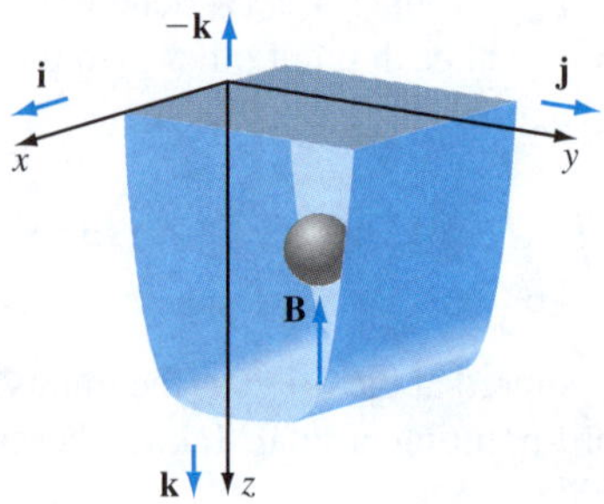

Figure 2

Example 6 Let $\mathcal{U}$ be a solid body with piecewise-smooth boundary $\mathcal{S}$. The body is immersed in water having constant mass density μ. For convenience, we use a coordinate system such that the positive z-axis points down into the water. The surface of the water corresponds to the plane $z = 0$. Refer to Figure 2. The pressure exerted by the water at depth z is $p = \mu g z$ where g is the gravitational acceleration (we arranged for the positive z-axis to be the downward direction so that this formula for pressure would have no minus sign in it). Use the Divergence Theorem to show that the buoyant force exerted on the body in the direction $-\mathbf{k}$ has magnitude equal to the weight of the fluid displaced by the body. This statement is known as *Archimedes's Law of Buoyancy*.

Solution Let $\mathbf{n}$ denote the outward unit normal to $\mathcal{S}$. Let $\mathbf{B}$ be the buoyant force. By Newton's Third Law,

$$\mathbf{B} = -\iint\limits_{\mathcal{S}} p\mathbf{n} \, dS = -\iiint\limits_{\mathcal{U}} \nabla p \, dV,$$

the second equality resulting from equation (15.56). But since $p = \mu g z$, we have $\nabla p = \mu g \mathbf{k}$. This makes sense because the direction of greatest change for the pressure function is down, and that is in the $\mathbf{k}$ direction in the coordinate system that we have chosen. Therefore,

$$\mathbf{B} = -\iiint\limits_{\mathcal{U}} \mu g \mathbf{k} \, dV = -\left(g \iiint\limits_{\mathcal{U}} \mu \, dV \right) \mathbf{k}.$$

Because of the minus sign, we see that the vector $\mathbf{B}$ is directed upward. By definition of density, the value of the triple integral is just the total mass of water displaced. The quantity in the parentheses is therefore the weight of the displaced water. Eureka! ■

Proof of the Divergence Theorem

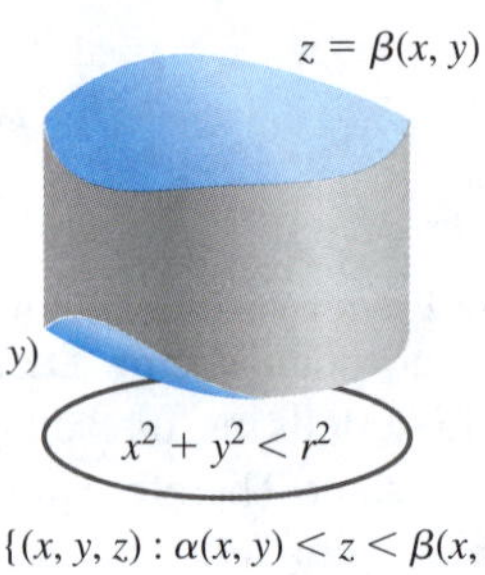

$\mathcal{U} = \{(x, y, z) : \alpha(x, y) < z < \beta(x, y),$
$x^2 + y^2 < r^2\}$

Figure 3

Here we prove the Divergence Theorem in the case that $\mathcal{U}$ is the z-simple solid between the graphs of two functions $\alpha(x, y)$ and $\beta(x, y)$ over a disk $\mathcal{D} = \{(x, y) : x^2 + y^2 < r^2\}$. That is, $\mathcal{U} = \{(x, y, z) : \alpha(x, y) < z < \beta(x, y), \ \ x^2 + y^2 < r^2\}$. Look at Figure 3. The boundary $\mathcal{S}$ of $\mathcal{U}$ consists of three pieces. The upper boundary is the surface

$z = \beta(x, y), x^2 + y^2 < r^2$; the lower boundary is $z = \alpha(x, y), x^2 + y^2 < r^2$; and the boundary on the side is the cylindrical surface $x^2 + y^2 = r^2, \alpha(x, y) < z < \beta(x, y)$.

Let us write our continuously differentiable vector field $\mathbf{F}$ in component form: $\mathbf{F} = M\mathbf{i} + N\mathbf{j} + R\mathbf{k}$. Our job is to prove equation (15.45), which we may write as

$$\iint_{\mathcal{S}} (M\mathbf{i} + N\mathbf{j} + R\mathbf{k}) \cdot \mathbf{n}\, dS = \iiint_{\mathcal{U}} (M_x + N_y + R_z)\, dV, \quad \text{or}$$

$$\iint_{\mathcal{S}} M\mathbf{i} \cdot \mathbf{n}\, dS + \iint_{\mathcal{S}} N\mathbf{j} \cdot \mathbf{n}\, dS + \iint_{\mathcal{S}} R\mathbf{k} \cdot \mathbf{n}\, dS = \iiint_{\mathcal{U}} M_x\, dV + \iiint_{\mathcal{U}} N_y\, dV + \iiint_{\mathcal{U}} R_z\, dV. \quad \textbf{(15.57)}$$

We will show that the third summand on the left equals the third summand on the right. The equalities of the first summands and of the second summands can be obtained by similar calculations.

Let us begin with the surface integral $\iint_{\mathcal{S}} R\mathbf{k} \cdot \mathbf{n}\, dS$. It consists of an integration over the top, an integration over the bottom, and an integration over the side. For the top, $\mathbf{n}$ points upward and

$$\mathbf{n}(x, y, z) = \frac{1}{(1 + \beta_x(x, y, z)^2 + \beta_y(x, y, z)^2)^{1/2}}(-\beta_x(x, y, z)\mathbf{i} - \beta_y(x, y, z)\mathbf{j} + 1\mathbf{k}).$$

Notice that $\mathbf{k} \cdot \mathbf{n} = (1 + \beta_x(x, y, z)^2 + \beta_y(x, y, z)^2)^{-1/2}$ as a result. Also, we have

$$dS = (1 + \beta_x(x, y, z)^2 + \beta_y(x, y, z)^2)^{1/2}\, dA,$$

and so $\mathbf{k} \cdot \mathbf{n}\, dS = dA$. The upshot of these observations is that

$$\iint_{\text{top of } \mathcal{S}} R\mathbf{k} \cdot \mathbf{n}\, dS = \iint_{\mathcal{D}} R(x, y, \beta(x, y))\, dA. \quad \textbf{(15.58)}$$

For the bottom, the analysis is the same except that the normal points downward (and, of course, we use $z = \alpha(x, y)$ instead of $z = \beta(x, y)$). The result is

$$\iint_{\text{bottom of } \mathcal{S}} R\mathbf{k} \cdot \mathbf{n}\, dS = -\iint_{\mathcal{D}} R(x, y, \alpha(x, y))\, dA. \quad \textbf{(15.59)}$$

For the side, $\mathbf{n}$ is horizontal. Therefore, $\mathbf{k} \cdot \mathbf{n} = 0$. We conclude that

$$\iint_{\text{side of } \mathcal{S}} R\mathbf{k} \cdot \mathbf{n}\, dS = 0. \quad \textbf{(15.60)}$$

Summing equations (15.58), (15.59), and (15.60), we have

$$\iint_{\mathcal{S}} R\mathbf{k} \cdot \mathbf{n}\, dS = \iint_{\mathcal{D}} (R(x, y, \beta(x, y)) - R(x, y, \alpha(x, y)))\, dA. \quad \textbf{(15.61)}$$

We turn our attention now to the last summand on the right side of equation (15.57), applying the Fundamental Theorem of Calculus to the integrand *in the z-variable* to obtain

$$\iiint_{\mathcal{U}} R_z\,dV = \iint_{\mathcal{D}} \left(\int_{\alpha(x,y)}^{\beta(x,y)} R_z(x, y, z)\,dz\right) dA$$
$$= \iint_{\mathcal{D}} R(x, y, z)\Big|_{z=\alpha(x,y)}^{z=\beta(x,y)} dA$$
$$= \iint_{\mathcal{D}} (R(x, y, \beta(x, y)) - R(x, y, \alpha(x, y)))\,dA.$$

Since the double integral that terminates this chain of equalities is the right side of equation (15.61), we have proved the desired equality.

quickquiz

1. State the Divergence Theorem.
2. Explain in words what the Divergence Theorem means.
3. What is the physical significance of divergence and the Divergence Theorem?
4. Calculate $\iint_{\mathcal{S}} \mathbf{F}\cdot\mathbf{n}\,dS$ where $\mathbf{F}(x, y, z) = \cos(y)\mathbf{i} + \sin(x)\mathbf{j} + z\mathbf{k}$ and $\mathbf{n}$ is the outward unit normal of $\mathcal{S} = \{(x, y, z) : x^2 + y^2 + z^2 = 1\}$.

EXERCISES

Problems for Practice

In Exercises 1–20, calculate the flux of the given vector field $\mathbf{F}(x, y, z)$ across the boundary $\mathcal{S}$ of the given solid $\mathcal{U}$ in space by doing the following:

a. calculating the surface integral directly
b. using the Divergence Theorem

1. $\mathbf{F}(x, y, z) = 5\mathbf{k}$, $\mathcal{U} = \{(x, y, z) : x^2 + y^2 < 9, -1 < z < 2\}$
2. $\mathbf{F}(x, y, z) = z\mathbf{i}$, $\mathcal{U} = \{(x, y, z) : (x-1)^2 + y^2 < 1, 0 < z < 3\}$
3. $\mathbf{F}(x, y, z) = z\mathbf{k}$, $\mathcal{U} = \{(x, y, z) : x^2 + y^2 < 1, 0 < z < 1\}$
4. $\mathbf{F}(x, y, z) = x\mathbf{i} + y\mathbf{j} + z\mathbf{k}$, $\mathcal{U} = \{(x, y, z) : 0 < z < \sqrt{1 - x^2 - y^2}\}$
5. $\mathbf{F}(x, y, z) = x\mathbf{i}$, $\mathcal{U} = \{(x, y, z) : x^2 + y^2 < 1, 0 < z < 1\}$
6. $\mathbf{F}(x, y, z) = 3xz\mathbf{i} - 2yz\mathbf{j}$, $\mathcal{U} = \{(x, y, z) : x^2 + z^2 < 25, 3 < z, 0 < y < 2\}$
7. $\mathbf{F}(x, y, z) = 7x\mathbf{i} + y\mathbf{j} - 2z\mathbf{k}$, $\mathcal{U} = \{(x, y, z) : x + y + z < 1, 0 < x, 0 < y, 0 < z\}$
8. $\mathbf{F}(x, y, z) = \mathbf{i} + 2\mathbf{j} + 3z\mathbf{k}$, $\mathcal{U} = \{(x, y, z) : \sqrt{x^2 + y^2} < z < 1\}$
9. $\mathbf{F}(x, y, z) = y^2\mathbf{i} - yz\mathbf{j} + xz\mathbf{k}$, $\mathcal{U} = \{(x, y, z) : |x| < 2, |y| < 2, 0 < z < 1\}$
10. $\mathbf{F}(x, y, z) = x/(z+2)\mathbf{i} + y/(x+2)\mathbf{j} + z/(y+2)\mathbf{k}$, $\mathcal{U} = \{(x, y, z) : |x| < 1, |y| < 1, |z| < 1\}$
11. $\mathbf{F}(x, y, z) = xz\mathbf{i} - yz\mathbf{j} + xz\mathbf{k}$, $\mathcal{U} = \{(x, y, z) : 0 < z < x^2 - y^2, 0 < x < 1\}$
12. $\mathbf{F}(x, y, z) = (x - z)\mathbf{i} + (y + z)\mathbf{j} + z\mathbf{k}$, $\mathcal{U} = \{(x, y, z) : 0 < z < 6 - x^2 - y^2, x^2 + y^2 < 4\}$
13. $\mathbf{F}(x, y, z) = (x - z)\mathbf{i} + y^2\mathbf{j} + (x + z)\mathbf{k}$, $\mathcal{U} = \{(x, y, z) : x^2 + z^2 < 9, 0 < x, 0 < z, 1 < y < 2\}$
14. $\mathbf{F}(x, y, z) = x^2\mathbf{i} - 4xy\mathbf{j} - 6xz\mathbf{k}$, $\mathcal{U} = \{(x, y, z) : 0 < x < 1, 0 < y < 1, 0 < z < -x - 4y + 8\}$
15. $\mathbf{F}(x, y, z) = (y - 2z)\mathbf{i} + (y - 5x)\mathbf{j} - 9y\mathbf{k}$, $\mathcal{U} = \{(x, y, z) : 4 < x < 5, -2 < z < 3, x + 2z < y < 8 + 2x + 3z\}$
16. $\mathbf{F}(x, y, z) = z\mathbf{i} + 3y^2\mathbf{j} + \mathbf{k}$, $\mathcal{U}$ = the solid bounded by the planes $x + y = 1$, $-x + y = 1$, $y = -2$, $z = -3$, and $z = 2x + y + 9$
17. $\mathbf{F}(x, y, z) = xy\mathbf{i} + y^2\mathbf{j} + z\mathbf{k}$, $\mathcal{U} = \{(x, y, z) : x^2 + y^2 < 1, x^2 + y^2 + z^2 < 4\}$
18. $\mathbf{F}(x, y, z) = xz\mathbf{i} - yz\mathbf{j} + x^2z\mathbf{k}$, $\mathcal{U} = \{(x, y, z) : x^2 + y^2 < z^2 < 9\}$

19. $\mathbf{F}(x, y, z) = (x^2 - 2xz)\mathbf{i} - (3 - y^2)\mathbf{j} + z^2\mathbf{k}$, $\mathcal{U} = \{(x, y, z) : x^2 + y^2 < z < 4, 0 < y\}$
20. $\mathbf{F} = x^2\mathbf{j} - y^2\mathbf{k}$, $\mathcal{U} = \{(x, y, z) : y^2 + 4z^2 < x < 4, 0 < z\}$

Further Theory and Practice

21. Let $R = \sqrt{x^2 + y^2 + z^2}$, and set $\mathbf{F}(x, y, z) = R^{-3}(x\mathbf{i} + y\mathbf{j} + z\mathbf{k})$. For $0 < \alpha < \beta$, calculate the flux of $\mathbf{F}$ across the boundary of the solid $\mathcal{U} = \{(x, y, z) : \alpha^2 \leq x^2 + y^2 + z^2 \leq \beta^2\}$.
22. Use the result of Exercise 21 to show that the flux of the vector field $\mathbf{F}$ across any sphere $\Sigma_r = \{(x, y, z) : x^2 + y^2 + z^2 = r\}$ is independent of r. (This statement is less precise than Gauss's law.)
23. Use the results of Exercises 21 and 22, together with the Divergence Theorem, to see that if $\mathcal{U}$ is *any* solid with no holes that contains the origin, then the flux of $\mathbf{F}$ across the boundary of $\mathcal{U}$ is the same as the flux across any Σ_r. (This statement is again less precise than Gauss's law.)
24. Suppose that $\mathbf{F}(x, y, z) = (x - yz)\mathbf{i} + (y - yx)\mathbf{j} + (xz - z)\mathbf{k}$. Let

$$\mathcal{U} = \{(x, y, z) : x^2 + 4y^2 < 8, -2 < z < \phi(x, y)\}$$

for some continuously differentiable function $\phi(x, y)$. If the flux of $\mathbf{F}$ through the side of this solid is known to be 9, then what is the total flux through the top and bottom?
25. Suppose that $\mathbf{F}$ is a twice continuously differentiable vector field and that $\mathbf{G} = \mathbf{curl}(\mathbf{F})$. Prove that the total flux of $\mathbf{G}$ across the boundary of a solid, such as the one in the statement of the Divergence Theorem, is 0.
26. Let $\mathbf{F}$ be a constant vector field and $\mathcal{U}$ a solid in space that is bounded by finitely many continuously differentiable surfaces. What is the total flux of $\mathbf{F}$ across the boundary of $\mathcal{U}$?
27. Let $\mathcal{U}$ be the solid in space between the two spheres $\{(x, y, z) : x^2 + y^2 + z^2 = 1\}$ and $\{(x, y, z) : x^2 + y^2 + z^2 = 4\}$. Apply the Divergence Theorem to evaluate the flux of the vector field $\mathbf{F} = x\mathbf{i} - y^2\mathbf{j} + xz\mathbf{k}$ across the boundary of $\mathcal{U}$.
28. Suppose that $\mathbf{F}$ is a continuously differentiable vector field defined on all of space, and assume that $\text{div}(\mathbf{F}) = 0$ at all points. If $\mathcal{S}$ and $\mathcal{S}'$ are two surfaces with a common boundary, then prove that

$$\iint_{\mathcal{S}} \mathbf{F} \cdot \mathbf{n}\, dS = \iint_{\mathcal{S}'} \mathbf{F} \cdot \mathbf{n}\, dS.$$

29. Let $\mathbf{F}(x, y, z) = (1/3)(x\mathbf{i} + y\mathbf{j} + z\mathbf{k})$, and let $\mathcal{U}$ be a solid in space bounded by finitely many continuously differentiable surfaces. Prove that the flux of $\mathbf{F}$ across the boundary of $\mathcal{U}$ equals the volume of $\mathcal{U}$.
30. Let $\mathcal{U}$ be a solid in space bounded by finitely many continuously differentiable surfaces. Let u be a twice continuously differentiable function whose domain contains a neighborhood of $\mathcal{U}$ and its boundary $\mathcal{S}$. Assume that u is harmonic, that is, $u_{xx} + u_{yy} + u_{zz} = 0$. Prove that $\iint_{\mathcal{S}} \nabla u \cdot \mathbf{n}\, dS = 0$.
31. With $\mathcal{U}$, $\mathcal{S}$, $\mathbf{n}$, and u as in Exercise 30, prove that

$$\iint_{\mathcal{S}} (\nabla u \cdot \mathbf{n}) u\, dS = \iiint_{\mathcal{U}} |\nabla u|^2\, dV.$$

32. Let $\mathcal{U}$ be a solid in space. Assume that the boundary $\mathcal{S}$ of $\mathcal{U}$ is piecewise continuously differentiable. Let $\mathbf{F}$ be a continuously differentiable vector field with domain that contains both $\mathcal{U}$ and $\mathcal{S}$. At any smooth point of $\mathcal{S}$, let $\mathbf{n}$ be the outward unit normal. Prove that

$$\iint_{\mathcal{S}} \mathbf{n} \times \mathbf{F}\, dS = \iiint_{\mathcal{U}} \nabla \times \mathbf{F}\, dV.$$

33. We now present a bogus proof that all magnetic fields are zero. (This nugget is taken from G. Arfken, *American Journal of Physics* 27 (1959), 526.) The mistake is purely a mathematical one. Your job is to fill in the details of the argument and then determine where the error lies. One of Maxwell's laws says that if $\mathbf{B}$ is a magnetic field, then $\text{div}(\mathbf{B}) = 0$. (Accept this fact—it is not the problem!) The Divergence Theorem then says that

$$\iint_{\mathcal{S}} \mathbf{B} \cdot \mathbf{n}\, dS = 0.$$

Also, since $\text{div}(\mathbf{B}) = 0$, there is a vector field $\mathbf{F}$ such that $\mathbf{B} = \mathbf{curl}(\mathbf{F})$. Combining all this information yields the equation

$$\iint_{\mathcal{S}} \mathbf{curl}(\mathbf{F}) \cdot \mathbf{n}\, dS = 0.$$

Now we apply Stokes's Theorem to see that

$$\oint_{\mathcal{C}} \mathbf{F} \cdot d\mathbf{r} = \iint_{\mathcal{S}} \mathbf{curl}(\mathbf{F}) \cdot \mathbf{n}\, dS = 0.$$

Thus, $\mathbf{F}$ is path-independent. We conclude that $\mathbf{F} = \mathbf{grad}\,(u)$ for some function u. But then

$$\mathbf{B} = \mathbf{curl}(\mathbf{grad}(u)) = \vec{0}$$

by Theorem 1c from Section 15.4. Of course, this is absurd. Where is the error?

Calculator/Computer Exercises

In Exercises 34–37, use the Divergence Theorem to calculate the flux of the vector field $\mathbf{F}(x, y, z)$ across the boundary $\mathcal{S}$ of the solid $\mathcal{U}$.

34. $\mathbf{F}(x, y, z) = x(1 + z)\sin(y^2)\mathbf{i}$,
$\mathcal{U} = \{(x, y, z) : |x| < 2, |y| < 2, |z| < 1\}$

35. $\mathbf{F}(x, y, z) = \ln(y^2 + z^2)\mathbf{i} + \ln(x^2 + z^2)\mathbf{j} + z^2\ln(2 + y)\mathbf{k}$,
$\mathcal{U} = \{(x, y, z) : 1 < z < 2 - x^2 - y^2\}$

36. $\mathbf{F}(x, y, z) = x\exp(-y^2)\mathbf{i} + y^2\exp(-x^2)\mathbf{j}$,
$\mathcal{U} = \{(x, y, z) : 0 < x, y, z < 1\}$

37. $\mathbf{F}(x, y, z) = (1 + z)^y\mathbf{i} + (1 + x)^z\mathbf{j} + \exp(z)\arctan(1 + x)\mathbf{k}$,
$\mathcal{U} = \{(x, y, z) : 0 < z < 1 - x^2 - y^2\}$

Summary of Key Topics

A vector field in the plane is a function $\mathbf{F}(x, y) = M(x, y)\mathbf{i} + N(x, y)\mathbf{j}$ of two real variables with vector values in the plane. A vector field in space has the form $\mathbf{F}(x, y, z) = M(x, y, z)\mathbf{i}+N(x, y, z)\mathbf{j}+R(x, y, z)\mathbf{k}$. We usually assume that the coefficient functions are continuously differentiable. Vector fields are useful for describing force fields and the velocity of a flow. They also arise when the gradient operator is applied to a scalar-valued function.

Line Integrals (Section 15.2)

If $\mathbf{F}$ is a vector field and if $t \mapsto \mathbf{r}(t)$, $a \le t \le b$ parameterizes a directed curve $\mathcal{C}$ over the interval $[a, b]$, then the line integral of $\mathbf{F}$ over $\mathcal{C}$ is

$$\int_{\mathcal{C}} \mathbf{F} \cdot d\mathbf{r} = \int_a^b \mathbf{F}(\mathbf{r}(t)) \cdot \mathbf{r}'(t)\, dt.$$

If $\mathbf{F}$ is a force field, this integral represents work performed in traversing the curve. The line integral does not depend on the choice of the parameterization (as long as the parameterization traverses $\mathcal{C}$ in its direction).

Conservative and Path-Independent Vector Fields (Section 15.3)

A vector field is conservative if it is the gradient of a scalar-valued function. If $\mathbf{F}$ is conservative, then there is an integration procedure for finding a function u such that $\nabla u = \mathbf{F}$. We then say that $V = -u$ is a potential function for $\mathbf{F}$. (If $\mathbf{F}$ is a conservative force field, then a potential function can be used to define potential energy.) A vector field $\mathbf{F}$ is conservative in a region $\mathcal{R}$ if and only if

$$\int_{\mathcal{C}} \mathbf{F} \cdot d\mathbf{r} = \int_{\tilde{\mathcal{C}}} \mathbf{F} \cdot d\tilde{\mathbf{r}}$$

for every two directed curves $\mathcal{C}$ and $\tilde{\mathcal{C}}$ in $\mathcal{R}$ that begin at the same point and end at the same point. A vector field with this property is called *path-independent*.

A curve $\mathcal{C}$ parameterized by $t \mapsto \mathbf{r}(t)$, $t \in [a, b]$ is said to be closed if $\mathbf{r}(a) = \mathbf{r}(b)$. When we write the path integral over a closed curve, we often use the special integral sign

$$\oint_{\mathcal{C}} \mathbf{F} \cdot d\mathbf{r}.$$

Closed Vector Fields (Section 15.3)

A planar vector field $\mathbf{F} = M\mathbf{i} + N\mathbf{j}$ is closed if

$$\frac{\partial M}{\partial y} = \frac{\partial N}{\partial x}.$$

Conservative vector fields are closed. If the domain of a closed vector field is simply connected, then the vector field is conservative. On regions with holes, this last assertion can be false.

In space, all these statements are still correct, with the definition of a closed vector field $\mathbf{F} = M\mathbf{i} + N\mathbf{j} + R\mathbf{k}$ being

$$\frac{\partial M}{\partial y} = \frac{\partial N}{\partial x}, \qquad \frac{\partial M}{\partial z} = \frac{\partial R}{\partial x}, \qquad \frac{\partial N}{\partial z} = \frac{\partial R}{\partial y}$$

(all three conditions must be satisfied).

Divergence (Section 15.4)

If $\mathbf{F} = M\mathbf{i} + N\mathbf{j}$ is a planar vector field, then its divergence is defined to be

$$\text{div}(\mathbf{F}) = \frac{\partial M}{\partial x} + \frac{\partial N}{\partial y}.$$

For a spatial vector field $\mathbf{F} = M\mathbf{i} + N\mathbf{j} + R\mathbf{k}$, the divergence is

$$\text{div}(\mathbf{F}) = \frac{\partial M}{\partial x} + \frac{\partial N}{\partial y} + \frac{\partial R}{\partial z}.$$

Curl (Section 15.4)

If $\mathbf{F}(x, y) = M(x, y)\mathbf{i} + N(x, y)\mathbf{j}$ is a planar vector field, then the curl of $\mathbf{F}$ is defined to be

$$\mathbf{curl}(\mathbf{F})(x, y) = \left(\frac{\partial}{\partial x}N(x, y) - \frac{\partial}{\partial y}M(x, y)\right)\mathbf{k}.$$

If $\mathbf{F} = M\mathbf{i} + N\mathbf{j} + R\mathbf{k}$ is a spatial vector field, then the curl of $\mathbf{F}$ is defined to be

$$\mathbf{curl}(\mathbf{F}) = \left(\frac{\partial R}{\partial y} - \frac{\partial N}{\partial z}\right)\mathbf{i} - \left(\frac{\partial R}{\partial x} - \frac{\partial M}{\partial z}\right)\mathbf{j} + \left(\frac{\partial N}{\partial x} - \frac{\partial M}{\partial y}\right)\mathbf{k}.$$

The Del Notation (Section 15.4)

The del operator is

$$\nabla = \frac{\partial}{\partial x}\mathbf{i} + \frac{\partial}{\partial y}\mathbf{j} + \frac{\partial}{\partial z}\mathbf{k}.$$

We have used the nabla symbol ∇ for the gradient: $\mathbf{grad}(u) = \nabla u$. Using the del notation, we may write div($\mathbf{F}$) as $\nabla \cdot \mathbf{F}$ and $\mathbf{curl}(\mathbf{F})$ as $\nabla \times \mathbf{F}$.

Operations with Div and Curl (Section 15.4)

A spatial vector field $\mathbf{F} = M\mathbf{i} + N\mathbf{j} + R\mathbf{k}$ is closed if and only if $\mathbf{curl}(\mathbf{F}) = \vec{\mathbf{0}}$. In physics, such a vector field is called *irrotational*. A closely related fact is that $\operatorname{div}(\mathbf{curl}(\mathbf{F})) = 0$ for any twice continuously differentiable vector field $\mathbf{F}$.

Green's Theorem (Section 15.5)

Green's Theorem says that if $\mathbf{F}(x, y) = M(x, y)\mathbf{i} + N(x, y)\mathbf{j}$ is a continuously differentiable vector field defined on a neighborhood of a region $\mathcal{R}$ and its piecewise continuously differentiable boundary curve $\mathcal{C}$, then

$$\oint_{\mathcal{C}} \mathbf{F} \cdot d\mathbf{r} = \iint_{\mathcal{R}} \left(\frac{\partial N}{\partial x} - \frac{\partial M}{\partial y} \right) dA.$$

This result relates the flux of a flow across the boundary to the flow across the interior. The parameterization $\mathbf{r}$ of $\mathcal{C}$ should have the property that the vector obtained by rotating $\mathbf{r}'$ 90° *counterclockwise* points *into* $\mathcal{R}$. If $\mathbf{n}$ is the unit *outward*-pointing normal vector field on the boundary $\mathcal{C}$, then Green's Theorem may be rewritten as

$$\int_{\mathcal{C}} \mathbf{F} \cdot \mathbf{n}\, ds = \iint_{\mathcal{R}} \operatorname{div}(\mathbf{F})\, dA.$$

Surface Area and Surface Integrals (Section 15.6)

If $\mathcal{R}$ is a region in the plane and if $f(x, y)$ is a continuously differentiable function on $\mathcal{R}$, then the element of surface area on $\mathcal{S} = \{(x, y, f(x, y)) : (x, y) \in \mathcal{R}\}$ is

$$dS = \sqrt{1 + f_x(x, y)^2 + f_y(x, y)^2}\, dA$$

where dA is the element of area in the xy-plane. Thus, the area of $\mathcal{S}$ is given by

$$A = \iint_{\mathcal{R}} \sqrt{1 + f_x(x, y)^2 + f_y(x, y)^2}\, dA.$$

If $\varphi(x, y, z)$ is a continuous function defined on the surface $\mathcal{S}$, then the surface integral of φ over $\mathcal{S}$ is denoted by $\iint_{\mathcal{S}} \varphi\, dS$ and defined by

$$\iint_{\mathcal{R}} \varphi(x, y, f(x, y)) \sqrt{1 + f_x(x, y)^2 + f_y(x, y)^2}\, dA.$$

If a surface $\mathcal{S}$ in space is given parametrically by the vector-valued function $(u, v) \mapsto \mathbf{r}(u, v)$, $(u, v) \in \mathcal{R}$, then the element of surface area is $\|\mathbf{r}_u \times \mathbf{r}_v\|\, dA$ where dA is the element of area in the uv-plane. Thus, the area of the surface is

$$A = \iint_{\mathcal{R}} \|\mathbf{r}_u \times \mathbf{r}_v\|\, dA.$$

If φ is a continuous function defined on the surface, then the integral of φ over the surface is

$$\iint_{\mathcal{S}} \varphi\, dS = \iint_{\mathcal{R}} \varphi\,(\mathbf{r}(u, v))\, \|(\mathbf{r}_u \times \mathbf{r}_v)(u, v)\|\, du\, dv.$$

Stokes's Theorem (Section 15.7)

Let $\mathcal{S}$ be a surface in space with boundary curve $\mathcal{C}$. We say that $\mathcal{S}$ is *oriented* if it is possible to make a continuous assignment of a unit normal vector $\mathbf{n}(P)$ to each point P of $\mathcal{S}$. We say that the boundary curve $\mathcal{C}$ is *oriented consistently* with $\mathcal{S}$ if it is parameterized by $\mathbf{r}$ so that at each boundary point, $\mathbf{n} \times \mathbf{r}'$ points *into* the region S. If $\mathcal{S}$ and $\mathcal{C}$ are consistently oriented and $\mathbf{F}$ is a continuously differentiable vector field defined on $\mathcal{S}$ and its boundary, then Stokes's Theorem states that

$$\oint_{\mathcal{C}} \mathbf{F} \cdot d\mathbf{r} = \iint_{\mathcal{S}} \mathbf{curl}(\mathbf{F}) \cdot \mathbf{n}\, dS.$$

The Divergence Theorem (Section 15.8)

If $\mathcal{U}$ is a solid in space bounded by finitely many continuously differentiable surfaces, if the boundary of $\mathcal{U}$ is $\mathcal{S}$, if $\mathbf{n}$ is the unit outward-pointing normal vector field on $\mathcal{S}$, and if $\mathbf{F}$ is a continuously differentiable vector field on an open set that contains $\mathcal{U}$ and its boundary $\mathcal{S}$, then the Divergence Theorem says that

$$\iint_{\mathcal{S}} \mathbf{F} \cdot \mathbf{n}\, dS = \iiint_{\mathcal{U}} \operatorname{div}(\mathbf{F})\, dV.$$

genesis & DEVELOPMENT

Chapter 15 is devoted to multivariable analogues—Green's Theorem, Stokes's Theorem, and the Divergence Theorem—of the Fundamental Theorem of Calculus. In fact, each theorem is a special case of a single very general formula that can be written in a remarkably compact form:

$$\int_M df = \int_{\partial M} f. \qquad \textbf{(15.62)}$$

This *Genesis & Development* will discuss the background and history of this equation.

The Beginning of Potential Theory

In 1752, while investigating a question in fluid flow, Euler introduced the partial differential equation

$$\triangle V = \frac{\partial^2 V}{\partial x^2} + \frac{\partial^2 V}{\partial y^2} + \frac{\partial^2 V}{\partial z^2} \equiv 0. \qquad \textbf{(15.63)}$$

Nine years later, Lagrange encountered this equation in his own research on fluid dynamics. Lagrange's work on celestial mechanics in 1773 provided him the occasion for a deeper study of equation (15.63). He noticed that the force

$$\mathbf{F}(P) = -MG \frac{1}{\|\overrightarrow{P_0P}\|^3} \overrightarrow{P_0P}$$

that a mass M at point P_0 exerts on a unit mass at point P can be expressed (using modern vector notation) as $\mathbf{F}(P) = -\nabla V(P)$ where $V = -MG\|\overrightarrow{P_0P}\|^{-1}$. Lagrange also observed that the potential function V satisfies equation (15.63). In a series of papers that followed, Lagrange laid the foundations for a theory of the potential.

In 1782, Laplace, then only 24 years old, used the theory of potential functions to solve a problem about planetary motion that had baffled Newton, Euler, and Lagrange: the "great inequality of Jupiter and Saturn." According to all observations available at that time, the average angular velocity of Jupiter was increasing, whereas that of Saturn was decreasing. The implication was that Jupiter would eventually be pulled into the sun and that Saturn would fly off from the solar system. Not knowing how to reconcile these observations with his theory of gravitation, Newton believed that divine intervention was occasionally necessary to keep the planets in their place. Laplace was able to show that the planets could not have "secular" (or constant direction) accelerations due to their mutual attractions. Laplace deduced that the inequality of Jupiter and Saturn was not secular but periodic of very long period. Observation has born out the truth of his deductions. His use of equation (15.63) became so well known that it was named *Laplace's equation*.

Laplace ranks as one of the most important mathematicians of his era. As a professor at the École Militaire, Laplace had the opportunity to be the academic examiner of the young Napoléon Bonaparte. After the French revolution, Napoléon appointed Laplace to France's new cabinet (Minister of the Interior) but dismissed him after only six weeks. Napoléon is alleged to have commented, "Laplace saw no question from its true point of view; he sought subtleties everywhere, had only doubtful ideas, and finally carried the spirit of the infinitely small into administration." Regardless, Laplace's reputation as a scientist continued to grow with the publication of his *Traité de Mécanique Céleste,* which appeared in five huge volumes from 1798 to 1825. In addition to the scientific accolades that Laplace received, he was made a Marquis. Laplace's remains are buried in Paris's famous Père Lachaise cemetery (not very close to Jim Morrison's grave).

George Green

George Green (1793–1841) was the self-taught son of a baker and miller. He used the top floor of his father's mill as his study. Inspired by Laplace's mathematical success in celestial mechanics, Green attempted to establish a mathematical foundation for electricity and magnetism. The results of his researches appeared in a paper that was published by subscription in 1828. Green announced his viewpoint with these words:

> *[B]y confining the attention solely on that peculiar function on whose differentials they all depend, I was induced to try whether it would be possible to discover any general relations, existing between this function and the quantities of electricity in the bodies producing it.*

Green coined the name *potential function* for the "peculiar function." Because there were only 52 subscribers to Green's publication, his work had no influence at first. Five years after its publication, in 1833, the 39-year-old Green became an undergraduate at Cambridge. He obtained the B.A. degree and stayed on at Cambridge for a few years until poor health forced his return to Green's Mill in 1840. Green died in obscurity one year later. It was not until the republication of his paper in 1846 that Green's research received the attention it deserved. His work reestablished the significance of the potential function when progress in the theory created by Lagrange and Laplace was in danger of coming to a halt. Green also discovered formulas that are equivalent to what we now call Green's Theorem. In addition, he introduced a construct, now called *Green's function,* that is fundamental to potential theory.

Today Green's Mill and Centre is a tourist attraction. What was once a windmill is now very much part of the 21st century. Included in the exhibits is a weather satellite receiving system, which displays data live from orbiting and geostationary satellites. Among the memorabilia that may be purchased there is a pencil with Green's Theorem inscribed on the side.

William Thomson and George Stokes

Only a few weeks before Green's death, William Thomson (1824–1907) entered Cambridge. Thomson's mathematical interests had already been formed by his reading of Fourier's *Théorie Analytique de la Chaleur* and Laplace's *Mécanique Céleste,* which he had acquired at the age of 15. As an undergraduate, Thomson encountered a reference to Green's work on potential theory but was not able to locate a copy until after his graduation in 1845. Thomson recognized at once the importance of this work and caused it to be republished the next year. He furthered the application of potential theory to electromagnetic theory and developed his own integral formulas. Indeed, Thomson discovered Stokes's Theorem.

George Gabriel Stokes (1819–1903) was Lucasian professor at Cambridge and a leading expert in fluid dynamics. Stokes's Theorem, in the form

$$\int (\alpha\, dx + \beta\, dy + \gamma\, dz) = \pm \iint \left\{ \ell \left(\frac{d\beta}{dz} - \frac{d\gamma}{dy} \right) + m \left(\frac{d\gamma}{dx} - \frac{d\alpha}{dz} \right) + n \left(\frac{d\alpha}{dy} - \frac{d\beta}{dx} \right) \right\} dS,$$

first appeared in a postscript to an 1850 letter that Thomson sent to Stokes. Stokes liked Thomson's equation enough to include it on several examinations at Cambridge, the first of which was given in 1854. Thomson's theorem eventually came to be associated with, and named after, Stokes.

Although Thomson published many mathematical papers, he is better known as a physicist thanks to his fundamental contributions to thermodynamics, electricity, and magnetism. Thomson also made his mark as a teacher. At the University of Glasgow, he established the first teaching laboratory in Great Britain. His *Treatise on Natural Philosophy,* written with P. G. Tait, first appeared in 1867 and remained the leading physics text for decades.

In the mid-1850s, Thomson served on the board of directors of a consortium of British industrialists who were planning to lay an underwater telegraph cable from Ireland to Newfoundland. Over Thomson's objections, the first two attempts were based on unsound electrical practices and failed. The third attempt, which followed Thomson's design, proved successful. Thomson's role in saving an enormous capital investment made him a hero in British financial circles. As knighthood had long been the appropriate reward for such heroism, Thomson became Sir William. He later acquired the title Baron Kelvin of Largs, and you will usually find him referred to as Lord Kelvin in physics texts.

As the 20th century approached, the aging Lord Kelvin rejected many of the scientific theories that were beginning to take hold. In biology, he was a staunch opponent of Darwin; in physics, he rejected the atomic theory of matter. His scientific conservatism manifested itself in mathematics as well. Although we now state Kelvin's best-known mathematical theorem in the language of vector analysis, Kelvin's own opinion was that " 'vector' is a useless survival, or offshoot, from quaternions, and has never been of the slightest use to any creature."

The Divergence Theorem was first published by Mikhail Vasilievich Ostrogradsky (1801–1862) in 1831. Ostrogradsky's ambition had been to become an officer in the military. Instead, because the life of an officer was too expensive for someone of modest means, he entered university in preparation for a career in civil service. After graduating from the University of Kharkov, Ostrogradsky traveled to Paris, where he learned Fourier's theory of heat conduction. Research on the partial differential equation describing heat conduction in three dimensions led Ostrogradsky to the Divergence Theorem. A few years later, in 1839, Gauss discovered the form of the Divergence Theorem that is named for him.

The Fundamental Theorem of Calculus: An Advanced Point of View

Long after the separate discoveries of Green's Theorem, Stokes's Theorem, and the Divergence Theorem, mathematicians came to understand that these formulas are all special cases of one general theorem, a theorem that also contains the Fundamental Theorem of Calculus of one variable. To understand what is involved, we must first recast the Fundamental Theorem of Calculus into an appropriate form.

The equation

$$\int_a^b f'(x)\,dx = f(b) - f(a)$$

contains two basic objects, the function f and the interval $M = [a, b]$. There are also two operations at work on them. One is the operation d of differentiation on the function. We have already used the notation df for $f'(x)\,dx$. The other operation is a geometric one: To the oriented interval M, we associate its boundary $\partial M = \{a, b\}$ with orientation induced from M. When the interval is given its standard left-to-right orientation, ∂M is oriented by assigning a "+" at b and a "−" at a.

Recall that an integral over a region is the limit of Riemann sums associated with decompositions of the region. When that region contains only two points, there is no need to take limits. In view of the preceding discussion, we may write $f(b) - f(a)$ as $\int_{\partial M} f$ and reformulate the Fundamental Theorem of Calculus as

$$\int_M df = \int_{\partial M} f.$$

A theory of integration has been worked out over very general oriented regions. The typical "differentials" ω that are integrated in three-dimensional space vary in form according to the dimension of the geometric object M over which they are being integrated. They may be written in the following forms:

Differental Form ω	Dimension of M	Description of M
$f_1dx + f_2dy + f_3dz$	1	Curve
$g_1dx \wedge dy + g_2dy \wedge dz + g_3dz \wedge dx$	2	Surface
$h\,dx \wedge dy \wedge dz$	3	Solid

Here, $f_1, f_2, f_3, g_1, g_2, g_3$, and h are all real-valued functions of x, y, and z. The basic differentials dx, dy, and dz are provided with an associative algebra structure in which multiplication is denoted by the symbol $\wedge$, and the following rules apply: $dx \wedge dx = dy \wedge dy = dz \wedge dz = 0$, $dy \wedge dx = -dx \wedge dy$, $dz \wedge dy = -dy \wedge dz$, and $dx \wedge dz = -dz \wedge dx$. The integration of $\mathbf{F} \cdot \mathbf{n}$ on an oriented surface S with normal $\mathbf{n}$ is written as

$$\iint_S F_1dy \wedge dz + F_2dz \wedge dx + F_3dx \wedge dy$$

in this notation, instead of $\iint_S \mathbf{F} \cdot \mathbf{n}\,dS$.

Notice that in a general differential form, the coefficients of the basic differentials are functions. We apply d to any function f to obtain the total differential $df = \frac{\partial f}{\partial x}dx + \frac{\partial f}{\partial y}dy + \frac{\partial f}{\partial z}dz$. The differential operator d is applied to a differential form by applying d to each coefficient and then using the calculus of differentials just described. For example,

$$\begin{aligned}
&d(xy^2z\,dx \wedge dy + \sin(x)\,dy \wedge dz) \\
&= (y^2z\,dx + 2xyz\,dy + xy^2\,dz) \wedge dx \wedge dy + \cos(x)\,dx \wedge dy \wedge dz \\
&= (xy^2 + \cos(x))\,dx \wedge dy \wedge dz.
\end{aligned}$$

It is astonishing that the Fundamental Theorem of Calculus, Green's Theorem, Stokes's Theorem, and the Divergence Theorem can all be written as

$$\int_{\partial M} \omega = \int_M d\omega.$$

A formal verification of the Divergence Theorem,

$$\iiint_M \operatorname{div}(\mathbf{F})\,dx\,dy\,dz = \iint_{\partial M} \mathbf{F} \cdot \mathbf{n}\,dS,$$

runs as follows: Since $\mathbf{F} \cdot \mathbf{n}\,dS$ equals $F_1dy \wedge dz + F_2dz \wedge dx + F_3dx \wedge dy$, the calculus of differentials gives

$$\begin{aligned}
d(\mathbf{F} \cdot \mathbf{n}\,dS) &= \left(\frac{\partial F_1}{\partial x}dx + \frac{\partial F_1}{\partial y}dy + \frac{\partial F_1}{\partial z}dz\right) \wedge dy \wedge dz \\
&\quad + \left(\frac{\partial F_2}{\partial x}dx + \frac{\partial F_2}{\partial y}dy + \frac{\partial F_2}{\partial z}dz\right) \wedge dz \wedge dx \\
&\quad + \left(\frac{\partial F_3}{\partial x}dx + \frac{\partial F_3}{\partial y}dy + \frac{\partial F_3}{\partial z}dz\right) \wedge dx \wedge dy \\
&= \left(\frac{\partial F_1}{\partial x} + \frac{\partial F_2}{\partial y} + \frac{\partial F_3}{\partial z}\right) dx\,dy\,dz \\
&= \operatorname{div}(\mathbf{F}).
\end{aligned}$$

A complete understanding of the calculus of differential forms requires concepts from linear algebra, mathematical analysis, and differential geometry. These subjects, like the theory of differential equations, make natural follow-ups to your study of calculus.

APPENDIX: ANSWERS TO SELECTED EXERCISES

Section 11.1

1. $\langle -1, -4\rangle$ **3.** $\langle -5, 1\rangle$ **5.** $\langle 5, -11\rangle$ **7.** $\langle 6, -6\rangle$

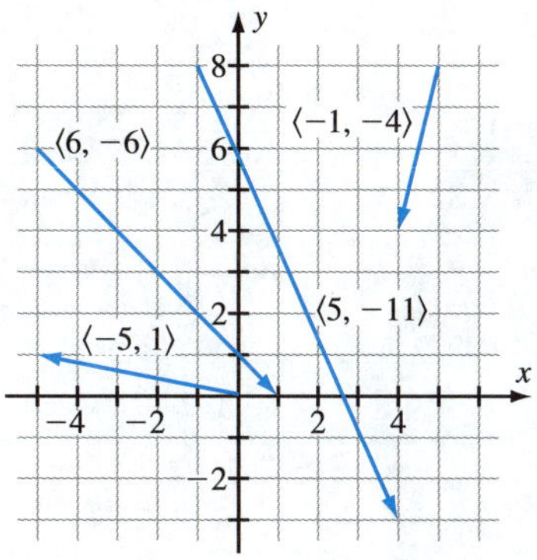

Figure for 1, 3, 5, and 7

9. $\|\overrightarrow{PQ}\| = \sqrt{5}$; $\|\overrightarrow{RS}\| = 3\sqrt{5}$; parallel

11. $\|\overrightarrow{PQ}\| = 3\sqrt{10}$; $\|\overrightarrow{RS}\| = \sqrt{53}$; not parallel

13. $\langle 9, 1\rangle$ **15.** $\langle 2, 8\rangle$

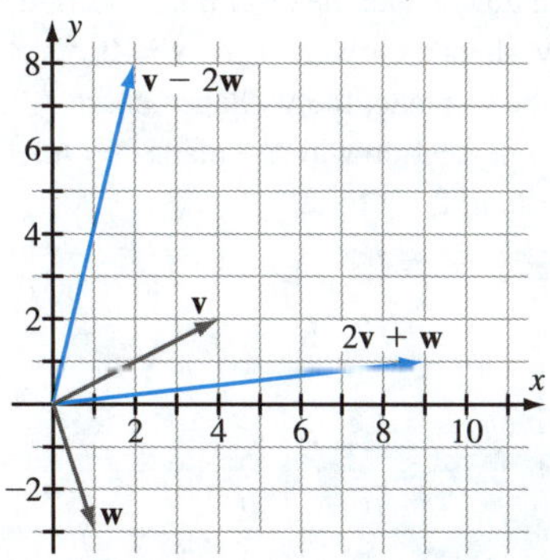

Figure for 13 and 15

17. **a.** $\langle -5, 9\rangle$ **b.** $\sqrt{106}$ **c.** $\langle 5, -9\rangle$
d. $\langle -5/\sqrt{106}, 9/\sqrt{106}\rangle$ **e.** $\langle 5/\sqrt{106}, -9/\sqrt{106}\rangle$

19. **a.** $\langle 4, -16\rangle$ **b.** $4\sqrt{17}$ **c.** $\langle -4, 16\rangle$
d. $\langle 1/\sqrt{17}, -4/\sqrt{17}\rangle$ **e.** $\langle -1/\sqrt{17}, 4/\sqrt{17}\rangle$

21. $\mathbf{v} = -7\mathbf{i} + 2\mathbf{j}$; $\mathbf{w} = -2\mathbf{i} + 9\mathbf{j}$; $-4\mathbf{v} = 28\mathbf{i} - 8\mathbf{j}$;
$3\mathbf{v} - 2\mathbf{w} = -17\mathbf{i} - 12\mathbf{j}$; $4\mathbf{v} + 7\mathbf{j} = -28\mathbf{i} + 15\mathbf{j}$

23. $\mathbf{v} = 6\mathbf{i} - 2\mathbf{j}$; $\mathbf{w} = 9\mathbf{i} - 3\mathbf{j}$; $-4\mathbf{v} = -24\mathbf{i} + 8\mathbf{j}$;
$3\mathbf{v} - 2\mathbf{w} = 0\mathbf{i} + 0\mathbf{j}$; $4\mathbf{v} + 7\mathbf{j} = 24\mathbf{i} - \mathbf{j}$

25. $\mathbf{F} = \langle 3, 4\rangle$; magnitude: 5

27. $\mathbf{F} = 3\mathbf{i} - \mathbf{j}$; magnitude: $\sqrt{10}$

29. $\langle \sqrt{3}/2, 1/2\rangle$ **31.** $\langle -1, 0\rangle$

33. magnitude of Mrs. Woodman's force: $50\sqrt{2}$; magnitude of resultant force: $50(1 + \sqrt{3})$

35. 200 m

37. $(-80\sqrt{3} - 200 - 140\sqrt{2})\mathbf{i} + (80 - 140\sqrt{2})\mathbf{j}$ (each component in N)

49. $\langle 1.85325, 1.44312\rangle$ **51.** 1.89838

53. left x-intercept: -0.65421 at $t = -0.98132$; right x-intercept: 1 at $t = 0$; y-intercept: 0.51447 at $t = -0.63598$

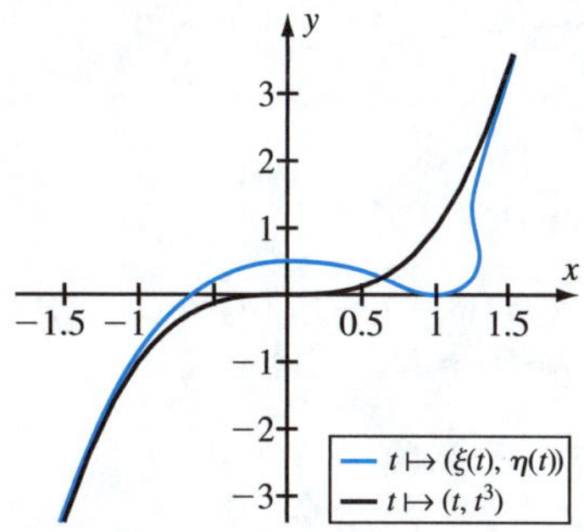

Section 11.2

1.

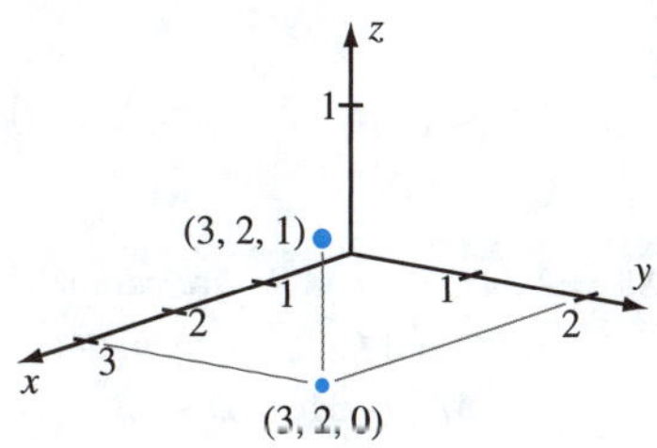

3.

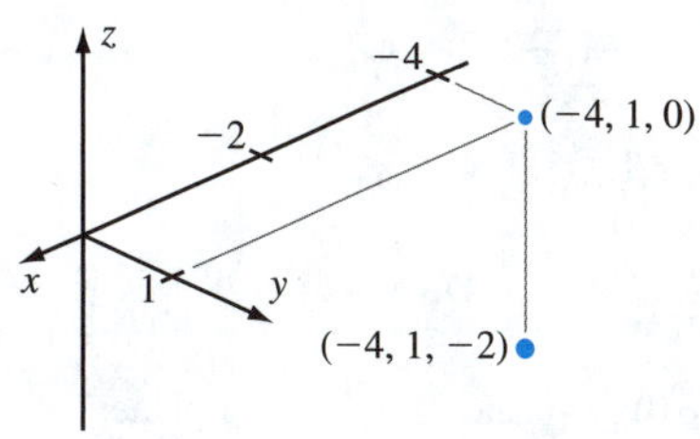

5. $3\sqrt{14}$ **7.** $3\sqrt{3}$

9. center: $(-1, -2, -3)$; radius: $\sqrt{22}$

11. center: $(-1/2, -1/2, -1/2)$; radius: $\sqrt{5}/2$

13. center: $(1/2, 0, 0)$; radius: $1/2$

15. The coefficients of x^2, y^2, and z^2 are not all equal.

17. The equation has the term $4xy$.

19. The equation has the term $-2xy$.

21. $\{(x, y, z) : (x-3)^2 + (y+2)^2 + (z-6)^2 < 16\}$

23. $\{(x, y, z) : (x - \pi)^2 + (y - \pi)^2 + (z + \pi)^2 = \pi^2\}$

25. $\{(x, y, z) : (x - 2)^2 + (y - 1)^2 + z^2 > 4\}$

27. $\overrightarrow{OP} = \langle 1, 4, -2\rangle$; $\overrightarrow{PQ} = \langle 1, -3, 8\rangle$; $\overrightarrow{OQ} = \langle 2, 1, 6\rangle$; $\langle 1, 4, -2\rangle + \langle 1, -3, 8\rangle = \langle 2, 1, 6\rangle$

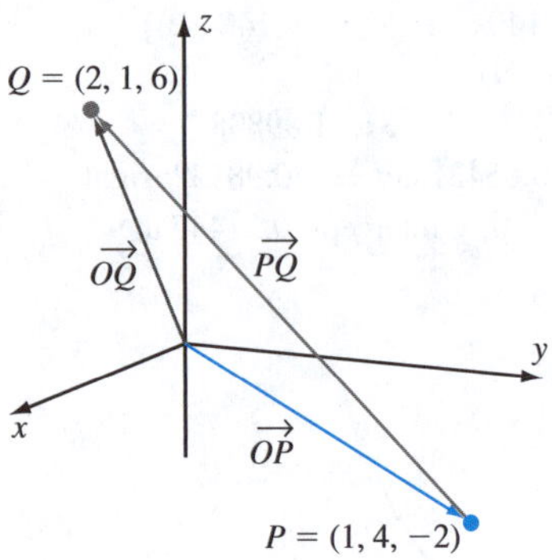

29. $\overrightarrow{OP} = \langle -1, 6, 1\rangle$; $\overrightarrow{PQ} = \langle 3, -7, -1\rangle$; $\overrightarrow{OQ} = \langle 2, -1, 0\rangle$; $\langle -1, 6, 1\rangle + \langle 3, -7, -1\rangle = \langle 2, -1, 0\rangle$

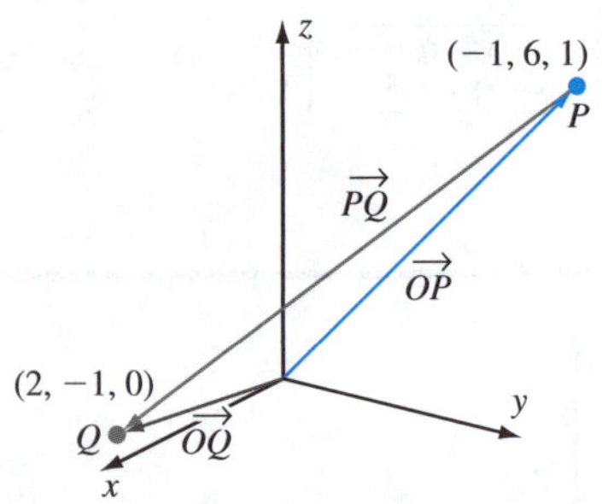

31. $\|\overrightarrow{PQ}\| = \sqrt{41}$; $\|\overrightarrow{RS}\| = 2\sqrt{41}$; $\overrightarrow{PQ}$ and $\overrightarrow{RS}$ *are* parallel.

33. $\|\overrightarrow{PQ}\| = 3$; $\|\overrightarrow{RS}\| = 3\sqrt{5}$; $\overrightarrow{PQ}$ and $\overrightarrow{RS}$ are *not* parallel.

35. $\langle -24, 6, 42\rangle$ **37.** $\langle -31, -6, -22\rangle$

39. $\langle 7, 1, 6\rangle$

41. **a.** $\langle 1, 2, -2\rangle$ **b.** $\langle -1, -2, 2\rangle$ **c.** 3 **d.** $\langle 1/3, 2/3, -2/3\rangle$

43. **a.** $\langle -8, -1, -4\rangle$ **b.** $\langle 8, 1, 4\rangle$ **c.** 9 **d.** $\langle -8/9, -1/9, -4/9\rangle$

45. $-15\mathbf{i} + 20\mathbf{j} - 5\mathbf{k}$ **47.** $\mathbf{i} - 6\mathbf{j} + 2\mathbf{k}$

49. $-4\mathbf{j} + \mathbf{k}$

51. *xy*-plane: $(x_0, y_0, 0)$; *yz*-plane $(0, y_0, z_0)$; *xz*-plane: $(x_0, 0, z_0)$

53. $\{(x, y, z) : x = 0\}$

55. $\{(x, y, z) : y > 0\}$

57. $\{(x, y, z) : |z| > 5\}$

59. The vectors $a\mathbf{i} + b\mathbf{k}$ are parallel to the *xz*-plane. If the vectors $a\mathbf{i} + b\mathbf{k}$ are represented by directed line segments with initial point at the origin, then the endpoints constitute the *xz*-plane.

61. If $\mathbf{m} = \langle m_1, m_2, m_3\rangle$, then $a = (-1/21)m_1 + (10/21)m_2 - (2/21)m_3$, $b = (-11/21)m_1 + (5/21)m_2 - (1/21)m_3$, and $c = (2/3)m_1 - (2/3)m_2 + (1/3)m_3$.

63. $\sqrt{5}$ and 3

65. $(x + 1)^2 + (y - 2)^2 + (z - 3)^2 = 36$

69. the ellipse $x^2 + y^2/9 = 1$, $z = 0$; the ellipse $x^2 + z^2/16 = 1$, $y = 0$; and the ellipse $y^2/9 + z^2/16 = 1$, $x = 0$

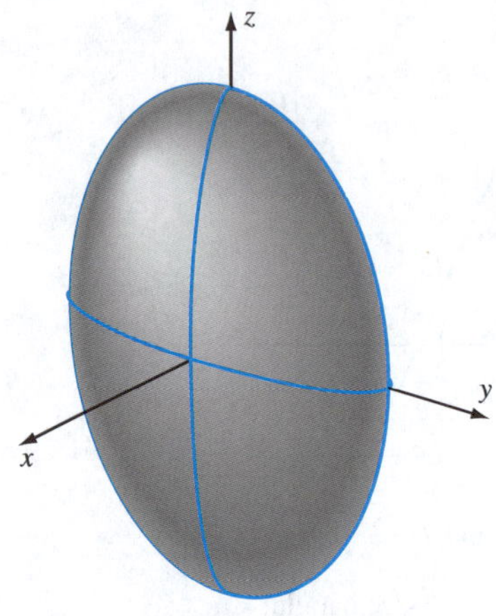

$$x^2 + \frac{y^2}{9} + \frac{z^2}{16} = 1$$

71. intersection with the *xy*-plane: the origin; intersection with the *xz*-plane: the V-shaped curve $z = |x|$, $y = 0$; intersection with the *yz*-plane: the V-shaped curve $z = |y|/2$, $x = 0$; intersection with the plane $z = h$ where h is a positive constant: the ellipse $x^2/h^2 + y^2/(2h)^2 = 1$, $z = h$

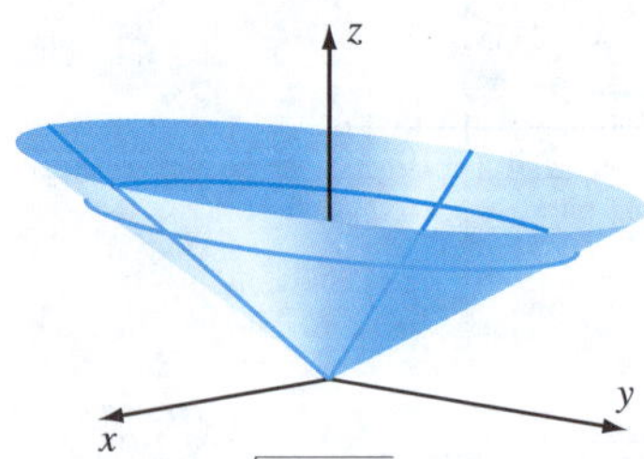

$$z = \sqrt{x^2 + \frac{y^2}{4}}$$

73. ellipses

75. intersection with the *xy*-plane: an ellipse; intersection with the *xz*-plane: a hyperbola; intersection with the *yz*-plane: a hyperbola; intersections with the planes $z = h$: ellipses

Section 11.3

1. 28 **3.** 4 **5.** -31 **7.** $\pi/3$

9. $3\pi/4$ **11.** $\pi/2$

13. Not perpendicular; dot product is 1, not 0.
15. Perpendicular; dot product is 0.
17. $\mathbf{P_v(w)} = \langle -51/62, 119/62, 17/31 \rangle$;
$\mathbf{P_w(v)} = \langle 17/33, 68/33, -68/33 \rangle$; $17/\sqrt{62}$; $17/\sqrt{33}$
19. $\mathbf{P_v(w)} = \langle 4, 8, 12 \rangle$; $\mathbf{P_w(v)} = \langle 2, 4, 6 \rangle$; $4\sqrt{14}$; $2\sqrt{14}$
21. $\mathbf{P_v(w)} = \langle 0, 0, 0 \rangle$; $\mathbf{P_w(v)} = \langle 0, 0, 0 \rangle$; 0; 0
23. direction: $\langle \sqrt{3}/2, -1/2, 0 \rangle$; $\alpha = \pi/6$; $\beta = 2\pi/3$; $\gamma = \pi/2$
25. direction: $\langle 0, 1, 0 \rangle$; $\alpha = \pi/2$; $\beta = 0$; $\gamma = \pi/2$
27. direction: $\langle -1/\sqrt{2}, 0, 1/\sqrt{2} \rangle$; $\alpha = 3\pi/4$; $\beta = \pi/2$;
$\gamma = \pi/4$
29. dot product: -16; lengths 6 and 3; $|-16| \leq 6 \cdot 3$ is true.
31. dot product: 9; lengths $\sqrt{5}$ and $2\sqrt{5}$; $|9| \leq \sqrt{5} \cdot 2\sqrt{5}$ is true.
33. $\mathbf{P_u(v)} = \langle 4, 4, -2 \rangle$; $\mathbf{v} - \mathbf{P_u(v)} = \langle 0, -3, -6 \rangle$, the dot product of which with $\mathbf{P_u(v)}$ is 0.
35. $\mathbf{P_u(v)} = \langle -2/\sqrt{3}, -2/\sqrt{3}, 2/\sqrt{3} \rangle$;
$\mathbf{v} - \mathbf{P_u(v)} = \langle 8/\sqrt{3}, 2/\sqrt{3}, 10/\sqrt{3} \rangle$, the dot product of which with $\mathbf{P_u(v)}$ is 0.
37. -2
39. $-4, 4$
45. $100000\sqrt{3}$ ft-lb
49. **a.** 6/7 **b.** $-11/5$ **c.** $-1/5$ **d.** $-9/25$
51. $\overrightarrow{OP} \cdot \mathbf{j} = (\sqrt{3}/2)\|\overrightarrow{OP}\|$
63. 0.158548, 0.700449, 2.282596; $3.141593 \approx \pi$
65. 0.203888

Section 11.4

1. $\mathbf{v} \times \mathbf{w} = \langle 28, 0, 14 \rangle$; $\mathbf{w} \times \mathbf{v} = \langle -28, 0, -14 \rangle$;
$\mathbf{v} \cdot (\mathbf{v} \times \mathbf{w}) = \mathbf{w} \cdot (\mathbf{v} \times \mathbf{w}) = 0$
3. $\mathbf{v} \times \mathbf{w} = \langle -8, 16, 14 \rangle$; $\mathbf{w} \times \mathbf{v} = \langle 8, -16, -14 \rangle$;
$\mathbf{v} \cdot (\mathbf{v} \times \mathbf{w}) = \mathbf{w} \cdot (\mathbf{v} \times \mathbf{w}) = 0$
5. $\mathbf{v} \times \mathbf{w} = \langle 0, 0, 0 \rangle$; $\mathbf{w} \times \mathbf{v} = \langle 0, 0, 0 \rangle$;
$\mathbf{v} \cdot (\mathbf{v} \times \mathbf{w}) = \mathbf{w} \cdot (\mathbf{v} \times \mathbf{w}) = 0$
7. $\mathbf{v} \times \mathbf{w} = \langle 32, 0, -4 \rangle$; $\mathbf{w} \times \mathbf{v} = \langle -32, 0, 4 \rangle$;
$\mathbf{v} \cdot (\mathbf{v} \times \mathbf{w}) = \mathbf{w} \cdot (\mathbf{v} \times \mathbf{w}) = 0$
9. $\langle 6/\sqrt{118}, -9/\sqrt{118}, -1/\sqrt{118} \rangle$
11. $\langle -1/3, -2/3, 2/3 \rangle$
13. $\langle -2/\sqrt{17}, 3/\sqrt{17}, -2/\sqrt{17} \rangle$
15. $\sqrt{74}/2$
17. $\sqrt{1145}/2$
19. $\sqrt{1073}$
21. $2\sqrt{10}$
23. $\mathbf{u} \times (\mathbf{v} \times \mathbf{w}) = \langle -27, -108, 198 \rangle$;
$(\mathbf{u} \times \mathbf{v}) \times \mathbf{w} = \langle -51, -15, 111 \rangle$
25. $\mathbf{u} \times (\mathbf{v} \times \mathbf{w}) = \langle 11, -33, 11 \rangle$; $(\mathbf{u} \times \mathbf{v}) \times \mathbf{w} = \langle -2, 3, 6 \rangle$
27. $\mathbf{u} \times \mathbf{v} = \langle 1, 11, -3 \rangle$; $\mathbf{v} \times \mathbf{w} = \langle -11, -1, -7 \rangle$;
$(\mathbf{u} \times \mathbf{v}) \cdot \mathbf{w} = \mathbf{u} \cdot (\mathbf{v} \times \mathbf{w}) = -40$
29. $\mathbf{u} \times \mathbf{v} = \langle 0, -2, -1 \rangle$; $\mathbf{v} \times \mathbf{w} = \langle 7, -12, 1 \rangle$;
$(\mathbf{u} \times \mathbf{v}) \cdot \mathbf{w} = \mathbf{u} \cdot (\mathbf{v} \times \mathbf{w}) = -7$
31. 5
33. -16
35. $\mathbf{u} \cdot \mathbf{v} \times \mathbf{w} = 0$
37. $\mathbf{u} \cdot \mathbf{v} \times \mathbf{w} = 0$
39. $s = 19, t = -6$
41. $s = 5, t = -1$
49. 30 ft-lb
51. $d(R, \ell) = \|\overrightarrow{PQ} \times \overrightarrow{RS}\| / \|\overrightarrow{PQ}\|$
55. $7x + 22y + 9z = 78$, the graph of which is a plane that passes through the point $(1, 2, 3)$ and that is perpendicular to $\mathbf{v} \times \mathbf{w}$
57. -7.161655

Section 11.5

1. normals: $\lambda\langle 1, -3, 4 \rangle$, $\lambda \in \mathbb{R}$; $\mathbf{N} = \langle -1, 3, -4 \rangle$
3. normals: $\lambda\langle 3, 0, -4 \rangle$, $\lambda \in \mathbb{R}$; $\mathbf{N} = \langle -12, 0, 16 \rangle$
5. $\lambda\langle -5, 7, -4 \rangle$, $\lambda \in \mathbb{R}$
7. $\lambda\langle 1, 4, -2 \rangle$, $\lambda \in \mathbb{R}$
9. $-3x + 7y + 9z = 79$
11. $2x + 2y + 5z = 18$
13. $-30x - y + 40z = 119$
15. $96x + 80y - 28z = -512$
17. $x = 2 - 3t$, $y = 1 + t$, $z = 9 + 7t$
19. $x = 2t$, $y = 1 - t$, $z = t$
21. $(x - 2)/(-5) = (y - 1)/8 = (z - 5)/2$
23. $x = y - 1 = z/9$
25. $x = 7 + t$, $y = -4 - 3t$, $z = 6 + 7t$
27. $x = -1 + t$, $y = -1 - t$, $z = -8 + t/3$
29. $x - 1 = y = 1 - z$
31. $(x - 1)/2 = 2 - y = 1 - z$
33. $x = 2 - 39t$, $y = 7t$, $z = 12t$
35. $x = 6t$, $y = 1/6 - 8t$, $z = -24t$
37. $(x - 4)/2 = y - 2 = -z$
39. $x = (y + 1)/2 = (z + 3)/4$
41. $(4, -2, -2)$
43. $(2, 2, 7)$
45. $(2, 1, 4)$
47. $(3, 1, 0)$
49. $2/15$
51. $2\sqrt{6}/5$
53. $\sqrt{6}$
55. $9\sqrt{3}/5$
57. on line
59. not on line
61. $(-2, -9, 2)$
63. no intersection
65. $10\sqrt{3}$
67. $57\sqrt{6}/2$
69. $\mathbf{u} \times (\mathbf{v} \times \mathbf{w}) = \langle -8, -6, -1 \rangle = (\mathbf{u} \cdot \mathbf{w})\mathbf{v} - (\mathbf{u} \cdot \mathbf{v})\mathbf{w}$
71. $\mathbf{u} \times (\mathbf{v} \times \mathbf{w}) = \langle 4, 15, -19 \rangle = (\mathbf{u} \cdot \mathbf{w})\mathbf{v} - (\mathbf{u} \cdot \mathbf{v})\mathbf{w}$
73. $3x + 4y + z = 14$
75. $(x - 1) + 2(y - 2) + 3(z - 3) = 0$

77.

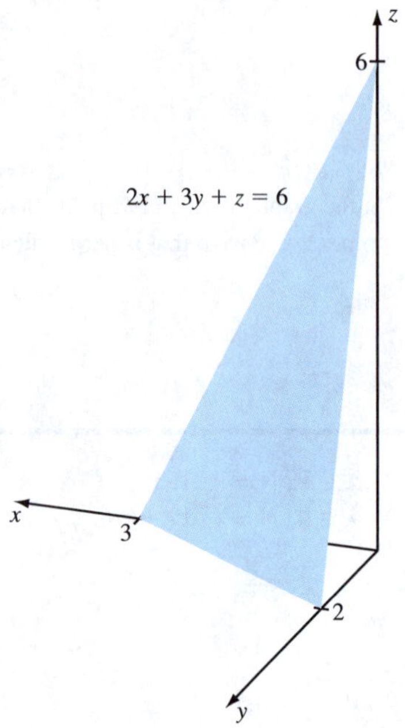

79.

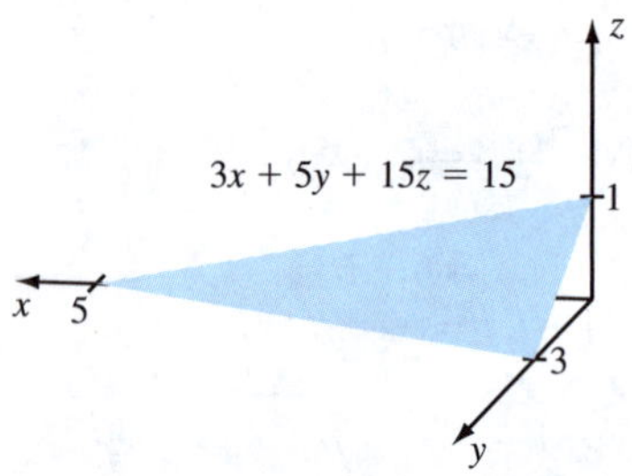

81. $x/a + y/b + z/c = 1$ is the plane that intersects the x-, y-, and z-axes in the nonzero numbers a, b, and c, respectively.

83. **a.** parallel **b.** parallel **c.** not parallel **d.** parallel **e.** parallel

85. **a.** no intersection **b.** The line lies in the plane. **c.** (8, 8, 3) is the point of intersection. **d.** no intersection

91.

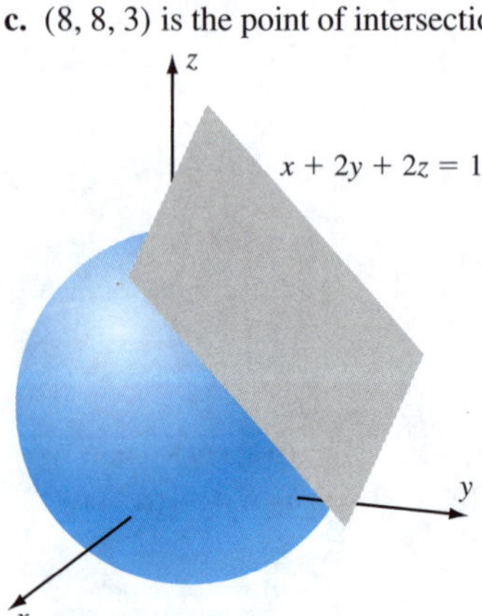

The sphere $(x-2)^2 + (y-1)^2 + (z-2)^2 = 9$ and its tangent plane at (3, 3, 4)

93. 0.191173, 1.186758

Section 12.1

1. The parabola $y = x^2$

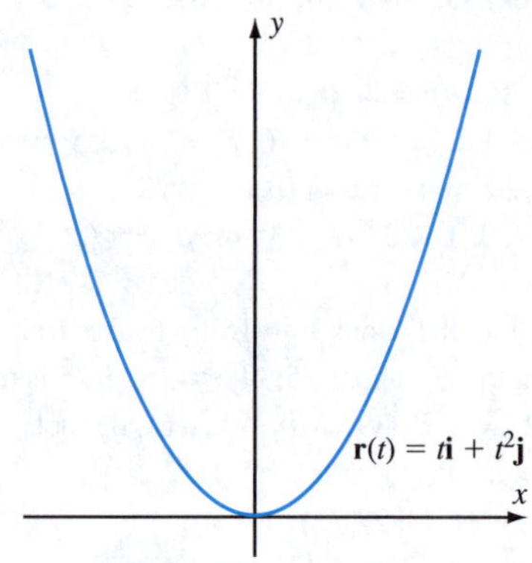

3. The parabola $y = x^2$, $z = 2$

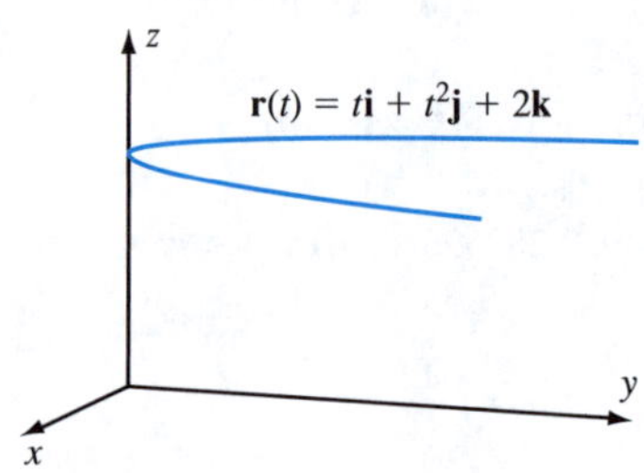

5. The circle $x = 1$, $y^2 + z^2 = 1$

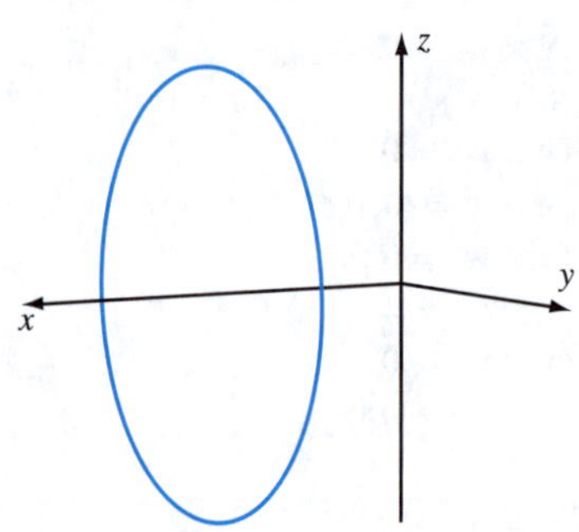

$\mathbf{r}(t) = \mathbf{i} + \cos(t)\mathbf{j} + \sin(t)\mathbf{k}$

7. A straight line segment from $P = (1, 1, 1)$ to $Q = (0, 0, 1)$, followed by the line segment from Q to $R = (4, 4, 1)$

9. One cycle of the graph of $z = \sin(x)$, rotated 45° so that the curve lies in the $x = y$ plane

11. $-\mathbf{i} + (1/2)\mathbf{j}$

13. $\langle 4, 4, 1 \rangle$

15. $(2t^{1/2})^{-1}\mathbf{i} + 2t\mathbf{k}$

17. $\sec^2(t)\mathbf{i} + \csc^2(t)\mathbf{j} + \sec(t)\tan(t)\mathbf{k}$

19. $\mathbf{i} - (t+5)^{-2}\mathbf{j} - 2(t-7)^{-3}\mathbf{k}$

21. $2\mathbf{i} - 3/(4t^{1/2})\mathbf{j} + (15/4)(t+1)^{1/2}\mathbf{k}$

23. $e^{-t}\mathbf{i} - 2(2t^4 + t^2 + 1)/(t(1+t^2))^2\mathbf{j}$

25. $4/(9t^{7/3})\mathbf{i} - ((2t^2-1)/(1+t^2)^{5/2})\mathbf{j} - 3/(16t^{7/4})\mathbf{k}$
27. $(t^3/3 + C_1)\mathbf{i} - (2\sqrt{t} + C_2)\mathbf{j} + (\sin(2t)/2 + C_3)\mathbf{k}$
29. $(t\ln(t) - t + C_1)\mathbf{i} - (\cos(t)\sin(t)/2 + t/2 + C_2)\mathbf{j} + (\ln(|\sec(t)|) + C_3)\mathbf{k}$
31. $(t^2+1)\mathbf{i} + (6-t^3)\mathbf{j} + (t^4-15)\mathbf{k}$
33. $(3+\sin(t))\mathbf{i} + (1+(1/2)\cos(2t))\mathbf{j} + (\sec(t)-3)\mathbf{k}$
35. $3e^{3t}\mathbf{i} + \sin(3t)\mathbf{j} + (6t/(1+9t^2))\mathbf{k}$
37. $15(1+5t)^2\mathbf{i} - 5\mathbf{k}$
39. $(3e^t+5)\mathbf{i} + (3\sin(t)+5)\mathbf{j} + (6t/(1+t^2)+5)\mathbf{k}$
41. $-9t^{-10}\mathbf{i} + 3t^{-4}\mathbf{k}$
43. $6t^5 + 2t$
45. $5(3t^2e^t + t^3e^t - \ln(1+t^2) - 2t^2/(1+t^2))$
47. $(9t^2+10t+1)\mathbf{j}$
49. $(1+t)e^t\mathbf{j}$
51. 0
53. $\mathbf{i}+\mathbf{j}+\mathbf{k}$
55. $-\mathbf{i} - 2\ln(\pi)\mathbf{j} + \mathbf{k}$
57. Continuous on $\{t \in \mathbb{R}: t \neq 5\}$
59. Continuous on $\mathbb{R}$
65. $\mathbf{r}(t) = \langle t, |t|, 0\rangle$, $c = (0,0,0)$

Section 12.2

1. $\mathbf{v}(t) = \mathbf{i} + 2t\mathbf{j} + 3t^2\mathbf{k}$; $v(t) = \sqrt{1+4t^2+9t^4}$; $\mathbf{a}(t) = 2\mathbf{j} + 6t\mathbf{k}$
3. $\mathbf{v}(t) = -2t\,\mathbf{i} + 2\mathbf{j} + (-2t/(1+t^2)^2)\mathbf{k}$; $v(t) = 2\sqrt{1+t^2+t^2/(1+t^2)^4}$; $\mathbf{a}(t) = -2\mathbf{i} + (2(3t^2-1)/(1+t^2)^3)\mathbf{k}$
5. $\mathbf{v}(t) = \mathbf{i} + e^t\mathbf{k}$; $v(t) = \sqrt{1+e^{2t}}$; $\mathbf{a}(t) = e^t\mathbf{k}$
7. $\mathbf{v}(t) = 2t\sin(t^2)\mathbf{i} + 2t\cos(t^2)\mathbf{j} + 3t^2\mathbf{k}$; $v(t) = |t|\sqrt{4+9t^2}$; $\mathbf{a}(t) = (4t^2\cos(t^2) + 2\sin(t^2))\mathbf{i} + (-4t^2\sin(t^2) + 2\cos(t^2))\mathbf{j} + 6t\mathbf{k}$
9. $(2+3t)\mathbf{i} - 2t\mathbf{j} + (-5+t-16t^2)\mathbf{k}$
11. $(3+t)\mathbf{i} + (-2+t)\mathbf{j} + (-1+t-16t^2)\mathbf{k}$
13. $x = 1+s$, $y = 1+2s$, $z = 1+3s$
15. $x = 1+s$, $y = 1-s$, $z = -1+2s$
17. $x = 2+s$, $y = -1$, $z = e + (e/2)s$
19. $x = 5+s$, $y = -5$, $z = 1-(1/5)s$
21. $x = 1-(3/5)s$, $y = 4+4s$, $z = \pi s$
23. $x+2 = (y+4)/4 = (z+8)/12$
25. $x-5 = 5(1-z)$, $y = -5$
27. $2(1-x) = (y-4)/2$, $z = 1$
29. $x = 1$, $y = 2\pi - 2\sqrt{\pi z}$
31. $(x-1) = 2(y+4) = 5(z-1)$
33. Trajectory: $\mathbf{r}(t) = 50\sqrt{3}t\mathbf{i} + (4+50t-16t^2)\mathbf{j}$; time until the arrow hits the ground: $(25+\sqrt{689})/16$ s; horizontal distance: $25\sqrt{3}(25+\sqrt{689})/8$ ft; greatest height: $689/16$ ft
35. $-80\mathbf{i} + 44\mathbf{j} + 50\mathbf{k}$
41. Velocity when the arrow hits the ground: $40\mathbf{i} - 20\sqrt{107}\mathbf{j}$ ft/s; horizontal distance: $25\sqrt{107} - 75\sqrt{3} \approx 128.7$ ft
43. $2\sqrt{r/g}$
51. $\mathbf{r}'(\pi/3) = -\sqrt{3}\mathbf{i} + (1/2)\mathbf{j}$

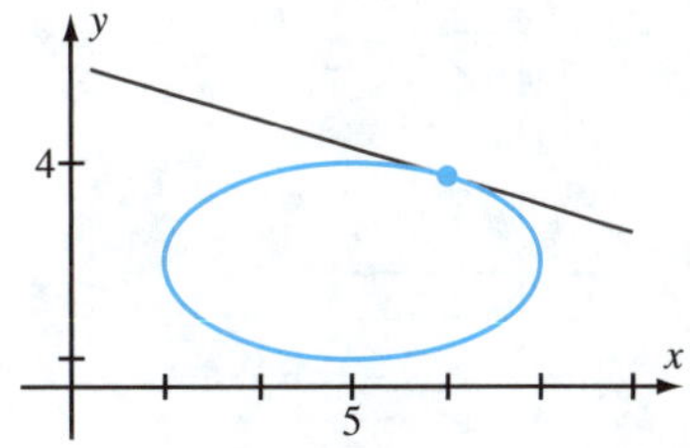

53. $\mathbf{r}'(\pi/4) = 2\mathbf{i} + \sqrt{2}\mathbf{j}$

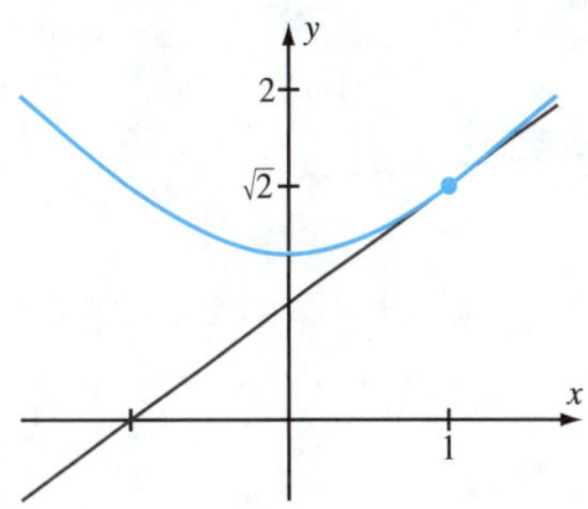

55. $\mathbf{r}'(2) = (e^2 - e^{-2})\mathbf{i} + (e^2 + e^{-2})\mathbf{j}$

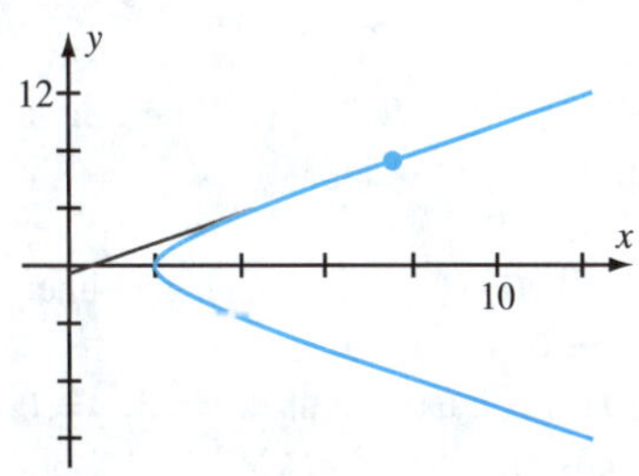

57. $\mathbf{r}'(1/2) = (24/25)\mathbf{i} - (32/25)\mathbf{j}$

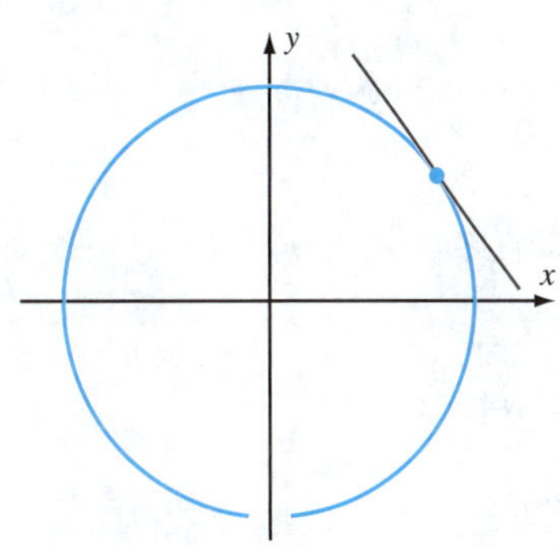

59. Vertical tangent: at $R = (2^{2/3}, 2^{1/3})$; horizontal tangents: at $P = (0, 0)$ and $Q = (2^{1/3}, 2^{2/3})$; tangents parallel to $y = x$: at $S = (1.2067, 0.52540)$ and $T = (0.52540, 1.2067)$

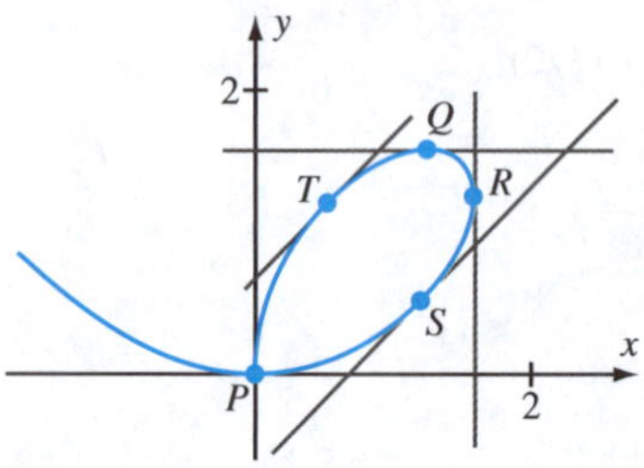

61.

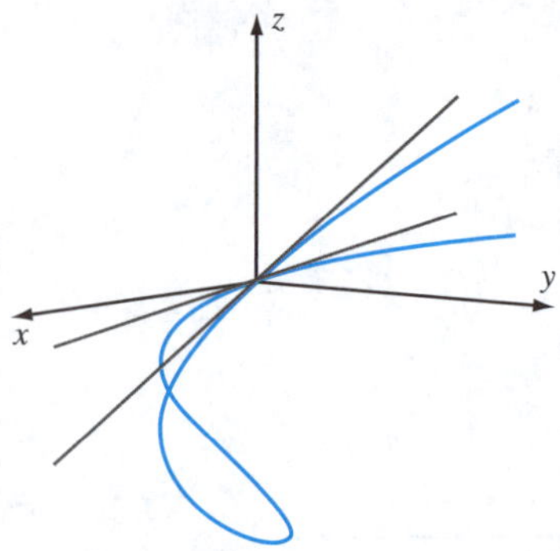

Section 12.3

1. $\mathbf{r}'(t) = \mathbf{i} - 2t\mathbf{j} + 3t^2\mathbf{k}$; $\mathbf{T}(t) = (1 + 4t^2 + 9t^4)^{-1/2}(\mathbf{i} - 2t\mathbf{j} + 3t^2\mathbf{k})$; tangent line: $(x - 1) = (y + 1)/(-2) = (z - 1)/3$

3. $\mathbf{r}'(t) = \mathbf{i} - 1/(2t^{1/2})\mathbf{k}$; $\mathbf{T}(t) = (1 + 1/(4t))^{-1/2}(\mathbf{i} - (t^{-1/2}/2)\mathbf{k})$; tangent line: $1 - x/4 = z + 2, y = 0$

5. $\mathbf{r}'(t) = (e^t + 2t)\mathbf{i}$; $\mathbf{T}(t) = \mathbf{i}$; tangent line: $y = 3, z = 0$

7. $\mathbf{r}'(t) = -\sin(t)\mathbf{i} + \cos(t)\mathbf{j} + \mathbf{k}$; $\mathbf{T}(t) = (1/\sqrt{2})(-\sin(t)\mathbf{i} + \cos(t)\mathbf{j} + \mathbf{k})$; tangent line: $x = \pi/2 - z, y = 1$

9. $\sqrt{2}\pi$

11. 6

13. $3\sqrt{3} - 2\sqrt{2}$

15. Is not

17. Is

19. e^2

21. $e - e^{-1}$

23. $(1 - x)/11 = (y + 1)/4 = (z - 1)/9$

25. $x = 0, z = \pi/2$

27. $x - 1 = y - 1 = -z/2$

29. $(x - 2)/5 = (y - 1)/4 = z - 2$

31. $x - 2y + 3z = 6$

33. $x - y/2 + 4z = 35/2$

35. $x - y = 0$

37. $x - y + 32z = 33$

39.
$$\mathbf{p}(s) = \left(\frac{(27s + 16\sqrt{2})^{2/3} - 8}{9}\right)(\mathbf{i} + \mathbf{j}) + \left(\frac{(27s + 16\sqrt{2})^{2/3} - 8}{9}\right)^{3/2}\mathbf{k}$$

41.
$$\mathbf{p}(s) = \left(\frac{s + \sqrt{s^2 + 4}}{2}\right)\mathbf{i} + \left(\frac{2}{s + \sqrt{s^2 + 4}}\right)\mathbf{j} + \sqrt{2}\ln\left(\frac{s + \sqrt{s^2 + 4}}{2}\right)\mathbf{k}$$

43. Tangent line: $-(x - 6) = \sqrt{3}(2y - 6 - \sqrt{3})$; normal line: $(x - 6) = \sqrt{3}(2y - 6 - \sqrt{3})/12$

45. Tangent line: $y = x/\sqrt{2} + 1/\sqrt{2}$; normal line: $y = -\sqrt{2}(x - 1) + \sqrt{2}$

47. Tangent line: $x = 2$; normal line: $y = 0$

49. Tangent line: $y = -4x/3 + 5/3$; normal line: $y = 3x/4$

51. $\mathbf{T}(\pi/4) = \langle -1/\sqrt{3}, 0, \sqrt{2/3}\rangle$; $\mathbf{N}(\pi/4) = \langle -\sqrt{6}/21, -12\sqrt{3}/21, -\sqrt{3}/21\rangle$; $\mathbf{B}(\pi/4) = \langle 4\sqrt{2}/7, -1/7, 4/7\rangle$

53. $\mathbf{T}(\pi/4) = \langle 2/\sqrt{7}, \sqrt{2}/\sqrt{7}, 1/\sqrt{7}\rangle$; $\mathbf{N}(\pi/4) = \langle 0, 1/\sqrt{3}, -2/\sqrt{6}\rangle$; $\mathbf{B}(\pi/4) = \langle -3/\sqrt{21}, 2\sqrt{2}/\sqrt{21}, 2/\sqrt{21}\rangle$

55. $\mathbf{T}(\ln(2)) = \langle 4/\sqrt{26}, -1/\sqrt{26}, 3/\sqrt{26}\rangle$; $\mathbf{N}(\ln(2)) = \langle -2/\sqrt{78}, 7/\sqrt{78}, 5/\sqrt{78}\rangle$; $\mathbf{B}(\ln(2)) = \langle -1/\sqrt{3}, -1/\sqrt{3}, 1/\sqrt{3}\rangle$

59. Simpson's Rule approximation: 255.707

61. Simpson's Rule approximation: 4.69858

Section 12.4

1. a. $\mathbf{T}(s) = \langle\cos(s), \sin(s), 0\rangle$ **b.** $\mathbf{N}(s) = \langle -\sin(s), \cos(s), 0\rangle$ **c.** $\kappa(s) = 1$

3. a. $\mathbf{T}(s) = (1/\sqrt{2})\langle -\sin(s/\sqrt{2}), -\cos(s/\sqrt{2}), 1\rangle$ **b.** $\mathbf{N}(s) = \langle -\cos(s/\sqrt{2}), \sin(s/\sqrt{2}), 0\rangle$ **c.** $\kappa(s) = 1/2$

5. a. $\mathbf{T}(s) = \langle (3/5)\cos(s), -\sin(s), (4/5)\cos(s)\rangle$ **b.** $\mathbf{N}(s) = \langle -(3/5)\sin(s), -\cos(s), -(4/5)\sin(s)\rangle$ **c.** $\kappa(s) = 1$

7. $\rho(s) = 2$; center: $(-\cos(s/\sqrt{2}), \sin(s/\sqrt{2}), s/\sqrt{2})$

9. $\rho(s) = 1$; center: $(0, 0, 0)$

11. $2/(1 + 4e^{2t})^{3/2}$

13. $t^2/(t^2 + 1)^{3/2}$

15. $|\cos^3(t)|/(2 - \cos^2(t))^{3/2}$

17. $2/(4x + 1)^{3/2}$

19. $|\cos(x)|$

21. $8|x|^3/(x^4 + 1)^2$

23. Radius: $\sqrt{2}(2t^2 + 1)^{3/2}$; center: $(3t^2 + 1, 3t^2 + 1, 1 - 4t^3)$

25. Radius: $(2e^{2t} + 1)^{3/2}/(\sqrt{2}e^t)$; center: $(2e^t + e^{-t}/2, t - 2e^{2t} - 1, 2e^t + e^{2t}/2)$

27. Radius: $\sqrt{2}(12t^2 + 4t + 1)^{3/2}/4$; center: $(3/4 - 6t^3 - 6t^2 - 3t/2, 5/4 + 6t^3 + 6t^2 + 3t/2, -12t^3 - 3t^2)$

29. $4x - 2y - z = 3$

31. $x + y - 2z = 0$
33. $(18t^4 + t^2 + 2, 8t^3 + 2t/3)$
35. $(t - 1 - e^{2t}, 2e^t + e^{-t})$
37. $(t + \sin(t), -1 + \cos(t))$
39. $(t - \tan(t), \ln(\sec(t)) + 1)$
41. $(0, 0)$
43. $(-\sqrt{2}, \ln(4))$ and $(\sqrt{2}, \ln(4))$
45. $2\sqrt{2}|ab|/(a^2 + b^2)^{3/2}$
47. $4\sqrt{3}/9$
57.

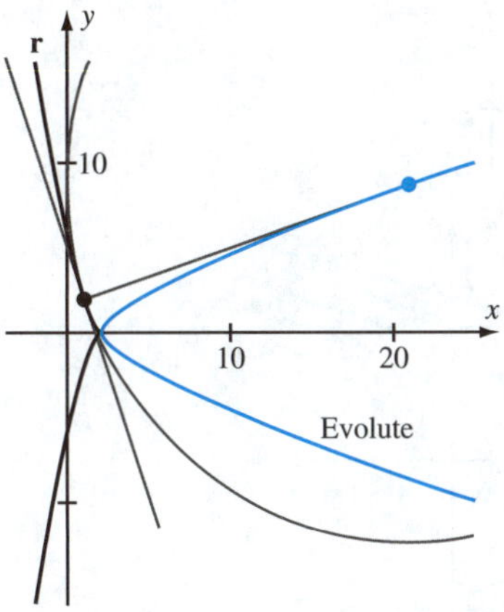

59.

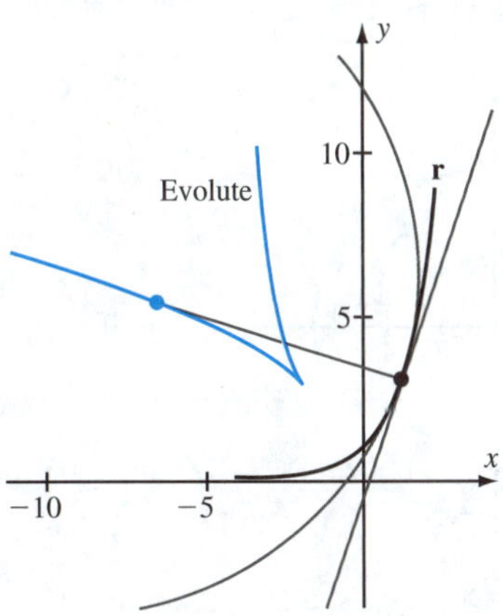

61.

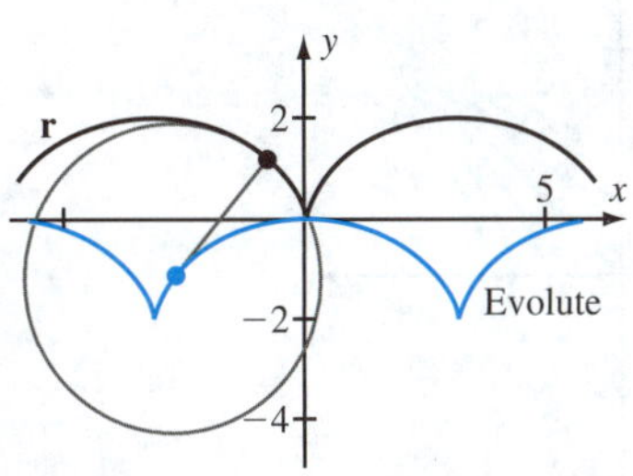

Section 12.5

1. $\mathbf{r}'(t) = \langle 1, -2t \rangle$; $\mathbf{r}''(t) = \langle 0, -2 \rangle$;
$\mathbf{T}(t) = (1 + 4t^2)^{-1/2}\langle 1, -2t \rangle$;
$\mathbf{N}(t) = (1 + 4t^2)^{-1/2}\langle -2t, -1 \rangle$; $\kappa_{\mathbf{r}}(t) = 2/(1 + 4t^2)^{3/2}$;
$a_T = 4t/\sqrt{1 + 4t^2}$; $a_N = 2/\sqrt{1 + 4t^2}$

3. $\mathbf{r}'(t) = \langle 1 - 2t, 1 + 2t \rangle$; $\mathbf{r}''(t) = \langle -2, 2 \rangle$;
$\mathbf{T}(t) = (2 + 8t^2)^{-1/2}\langle 1 - 2t, 1 + 2t \rangle$;
$\mathbf{N}(t) = (2 + 8t^2)^{-1/2}\langle -(1 + 2t), 1$; $\kappa_{\mathbf{r}}(t) = 4/(2 + 8t^2)^{3/2}$;
$a_T = 8t/\sqrt{2 + 8t^2}$; $a_N = 4/\sqrt{2 + 8t^2}$

5. $\mathbf{r}'(t) = \langle 1 - \cos(t), \sin(t) \rangle$; $\mathbf{r}''(t) = \langle \sin(t), \cos(t) \rangle$;
$\mathbf{T}(t) = (2 - 2\cos(t))^{-1/2}\langle 1 - \cos(t), \sin(t) \rangle$;
$\mathbf{N}(t) = (2 - 2\cos(t))^{-1/2}\langle \sin(t), -(1 - \cos(t)) \rangle$;
$\kappa_{\mathbf{r}}(t) = 1/(2\sqrt{2 - 2\cos(t)})$; $a_T = \sin(t)/\sqrt{2 - 2\cos(t)}$;
$a_N = (1 - \cos(t))/\sqrt{2 - 2\cos(t)}$

7. $\mathbf{r}'(t) = \langle 1/t, 1/(2\sqrt{t}) \rangle$; $\mathbf{r}''(t) = \langle -1/t^2, -1/(4t^{3/2}) \rangle$;
$\mathbf{T}(t) = (4 + t)^{-1/2}\langle 2, \sqrt{t} \rangle$; $\mathbf{N}(t) = (4 + t)^{-1/2}\langle -\sqrt{t}, 2 \rangle$;
$\kappa_{\mathbf{r}}(t) = 2\sqrt{t}/(4 + t)^{3/2}$; $a_T = -(8 + t)/(4t^2\sqrt{4 + t})$;
$a_N = 1/(2t^{3/2}\sqrt{4 + t})$

9. $\mathbf{r}'(t) = \langle 1, 2t, 3t^2 \rangle$; $\mathbf{r}''(t) = \langle 0, 2, 6t \rangle$;
$\mathbf{T}(t) = (1 + 4t^2 + 9t^4)^{-1/2}\langle 1, 2t, 3t^2 \rangle$;
$\mathbf{N}(t) = (1 + 4t^2 + 9t^4)^{-1/2}(1 + 9t^2 + 9t^4)^{-1/2}\langle -t(2 + 9t^2), 1 - 9t^4, 3t(2t^2 + 1) \rangle$;
$\kappa_{\mathbf{r}}(t) = 2\sqrt{1 + 9t^2 + 9t^4}/(1 + 4t^2 + 9t^4)^{3/2}$;
$a_T = 2t(2 + 9t^2)/\sqrt{1 + 4t^2 + 9t^4}$;
$a_N = 2\sqrt{1 + 9t^2 + 9t^4}/\sqrt{1 + 4t^2 + 9t^4}$

11. $\mathbf{r}'(t) = \langle (3/2)\sqrt{1 + t}, -(3/2)\sqrt{1 - t}, 1 \rangle$;
$\mathbf{r}''(t) = (3/4)\langle (1 + t)^{-1/2}, (1 - t)^{-1/2}, 0 \rangle$;
$\mathbf{T}(t) = 22^{-1/2}\langle 3\sqrt{1 + t}, -3\sqrt{1 - t}, 2 \rangle$;
$\mathbf{N}(t) = \langle \sqrt{(1 - t)/2}, \sqrt{(1 + t)/2}, 0 \rangle$;
$\kappa_{\mathbf{r}}(t) = (3\sqrt{2})/(22\sqrt{1 - t^2})$; $a_T = 0$;
$a_N = 3\sqrt{2}/(4\sqrt{1 - t^2})$

13. $\mathbf{r}'(t) = \langle e^t, -e^{-t}, \sqrt{2} \rangle$; $\mathbf{r}''(t) = \langle e^t, e^{-t}, 0 \rangle$;
$\mathbf{T}(t) = (e^{2t} + 1)^{-1}\langle e^{2t}, -1, \sqrt{2}e^t \rangle$;
$\mathbf{N}(t) = (e^{2t} + 1)^{-1}\langle \sqrt{2}e^t, \sqrt{2}e^t, 1 - e^{2t} \rangle$;
$\kappa_{\mathbf{r}}(t) = \sqrt{2}/(e^t + e^{-t})^2$; $a_T = e^t - e^{-t}$; $a_N = \sqrt{2}$

15. $\mathbf{r}'(t) = \langle -\sin(t), \cos(t), 2t \rangle$;
$\mathbf{r}''(t) = \langle -\cos(t), -\sin(t), 2 \rangle$;
$\mathbf{T}(t) = (1 + 4t^2)^{-1/2}\langle -\sin(t), \cos(t), 2t \rangle$;
$\mathbf{N}(t) = (1 + 4t^2)^{-1/2}(5 + 4t^2)^{-1/2}\langle -\cos(t) + 4t\sin(t) - 4t^2\cos(t), -\sin(t) - 4t\cos(t) - 4t^2\sin(t), 2 \rangle$;
$\kappa_{\mathbf{r}}(t) = \sqrt{5 + 4t^2}/(1 + 4t^2)^{3/2}$; $a_T = 4t/\sqrt{1 + 4t^2}$;
$a_N = \sqrt{5 + 4t^2}/\sqrt{1 + 4t^2}$

17. $a_T = 4t/\sqrt{5 + 4t^2}$; $a_N = 2\sqrt{5}/\sqrt{5 + 4t^2}$

19. $a_T = (\exp(2t) - \exp(-2t))/\sqrt{1 + \exp(2t) + \exp(-2t)}$;
$a_N = \sqrt{4 + \exp(2t) + \exp(-2t)}/\sqrt{1 + \exp(2t) + \exp(-2t)}$

21. $(x/2)^2 + (y/3)^2 = 1$

23. $(x/5)^2 + (y/3)^2 = 1$

25. $(x - 1)^2/5^2 + (y - 2)^2/13^2 = 1$

27. $(x - 1)^2/3^2 + (y - 2)^2/5^2 = 1$

29. $(x - 2/3)^2/(8/3)^2 + y^2/(4/\sqrt{3})^2 = 1$

35. 3.225×10^{-3} days

37. 6.166×10^9 cm

39. $969750\sqrt{3}$ mi^2/h

41. 4.6×10^7 km

43. 11.87 Earth yr

53. The graphs of $t \mapsto a_N(t)$ and $t \mapsto \|\mathbf{a}(t)\|$ touch when $a_T(t) = 0$.

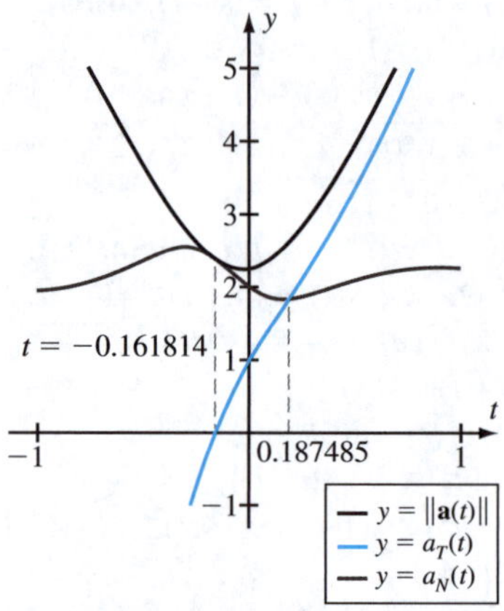

55. The particle slows on the interval (0.608, 0.967). Both components of acceleration decrease on (0.354, 0.6549).

57. The absolute maximum of a_N occurs at 1.63782. The absolute maximum of $\|\mathbf{a}\|$ occurs at 0 and 2π.

Section 13.1

1. $V(h, r) = \pi r^2 h$

3. $V(h, r) = \pi r^2 h/3$

5. $A(a, b) = (b^3 - a^3)/3$

7. $m(M, \tau, T) = M \exp(-(T/\tau)\ln(2))$

9. $2e^3 + 17$

11. $5/2$

13. $6/5$

15. 6

17. $12\sqrt{17}/5$

19.

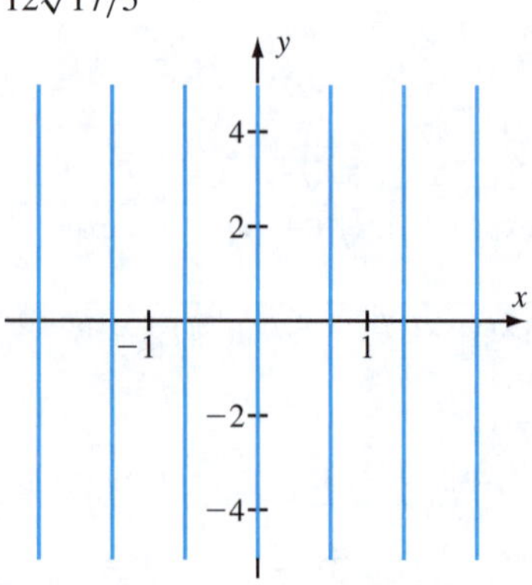

21.

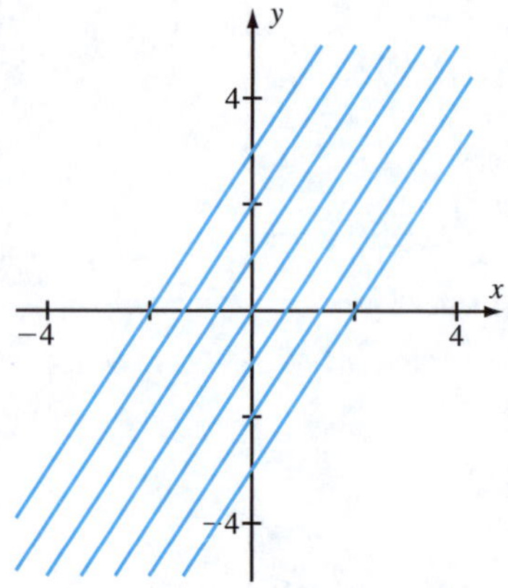

23.

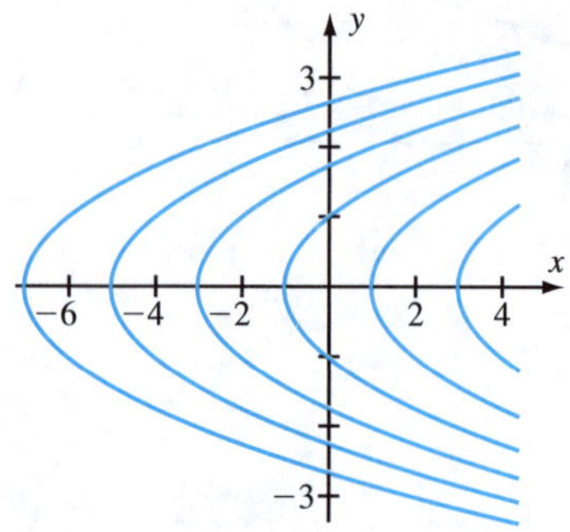

25.

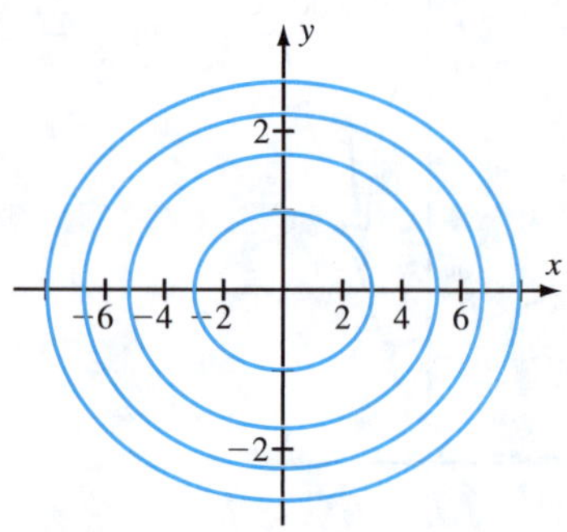

27.

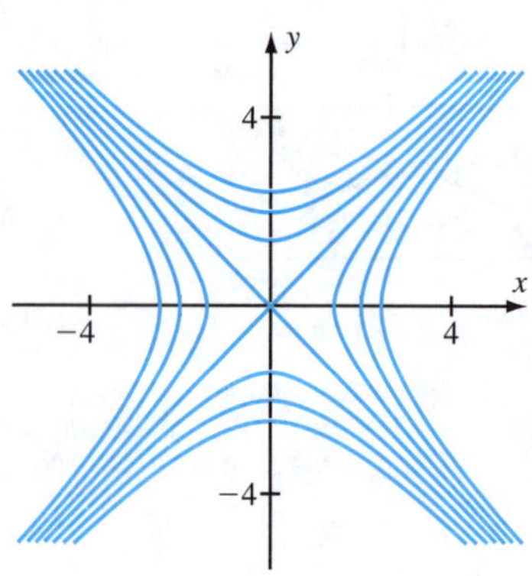

29.

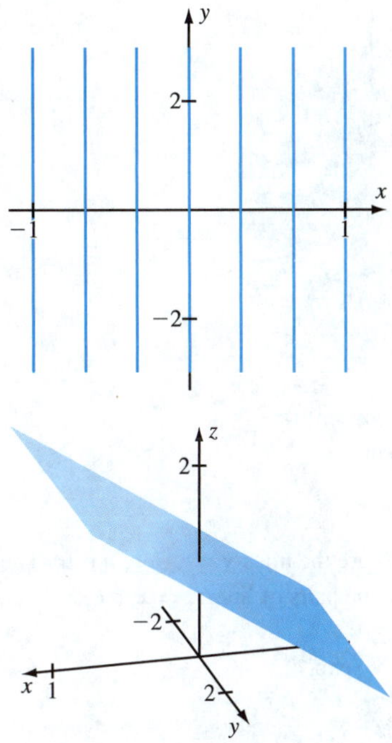
y
2
x
−1
1
−2
z
2
−2
x 1
2
y

31.

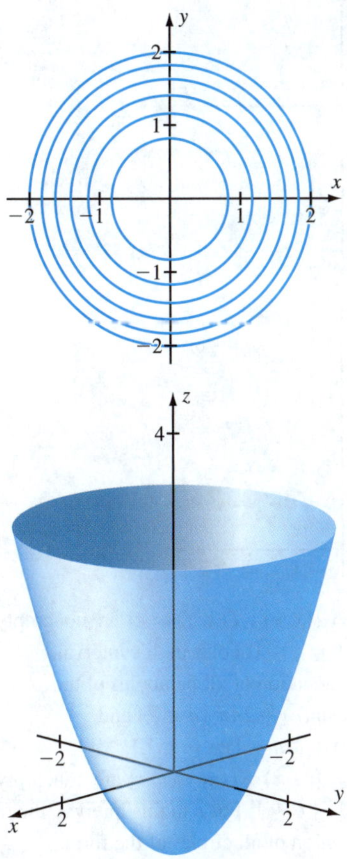
y
2
1
x
−2
−1
1
2
−1
−2
z
4
−2
−2
y
x
2
2

33.

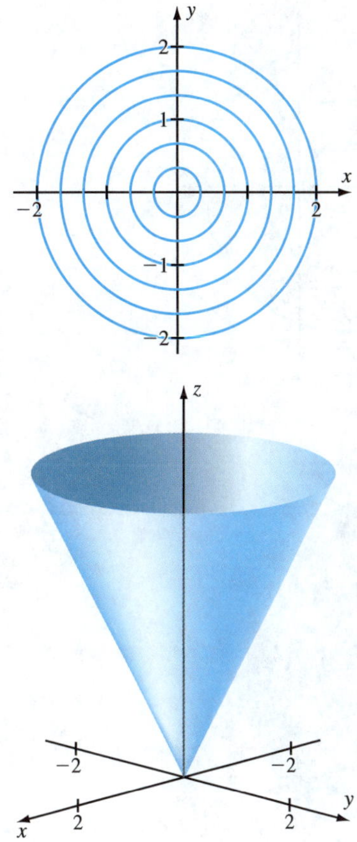
y
2
1
x
−2
2
−1
−2
z
−2
−2
y
x
2
2

35.

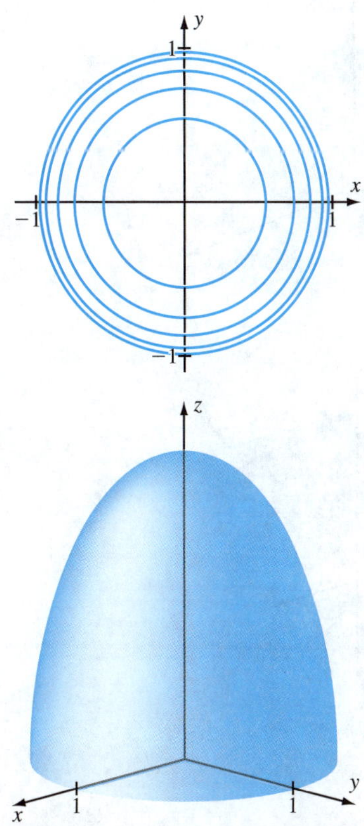
y
1
x
−1
1
−1
z
y
x
1
1

37.

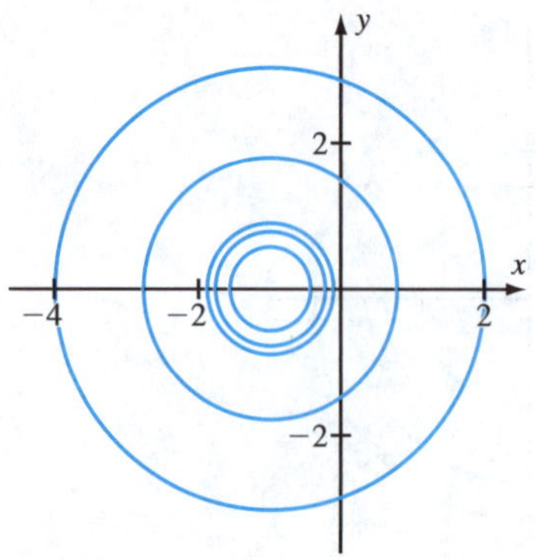

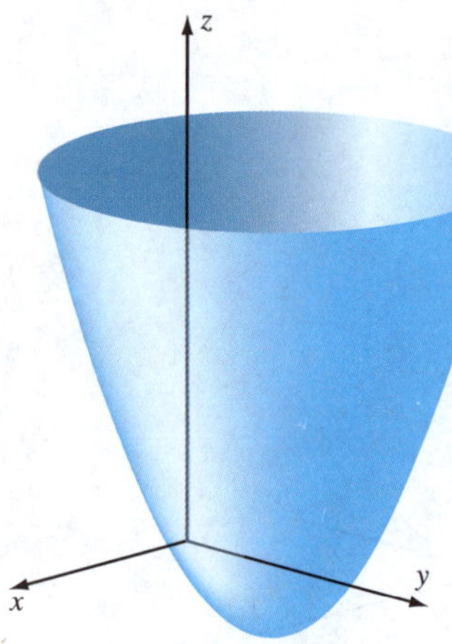

39.

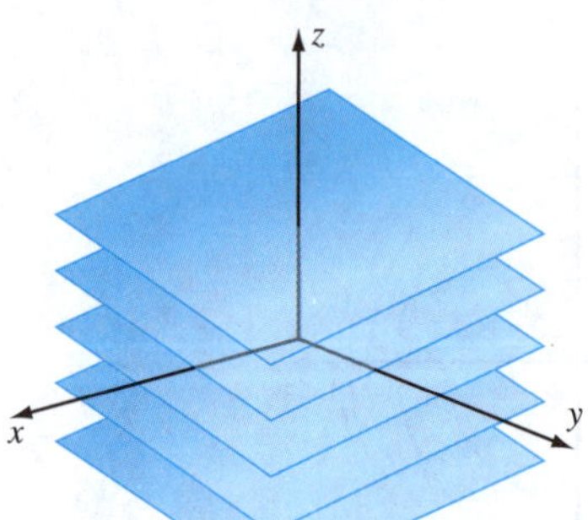

41.

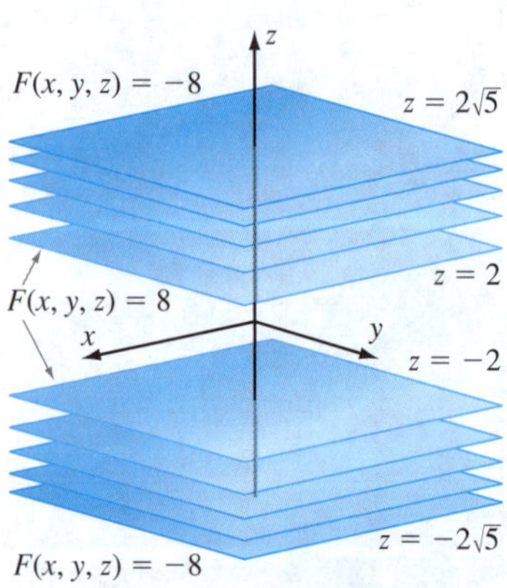

43.

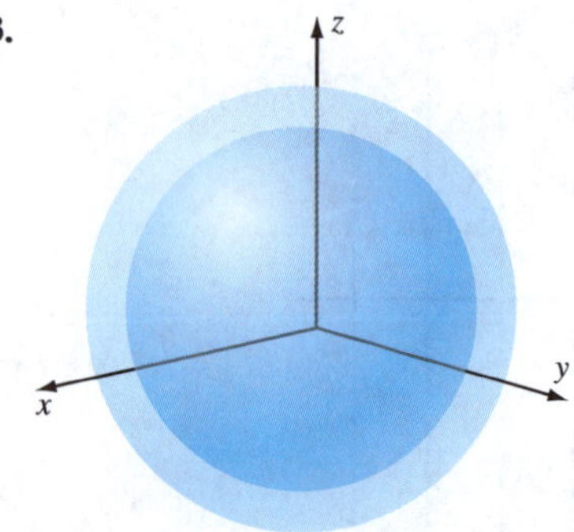

The level sets are concentric spheres. Two are shown.

45. $|xy|/\sqrt{2(x^2+y^2)}$

47. The level sets of f_1 are the lines $y = x + c$, $c \in \mathbb{R}$. The level sets of f_2 are the pairs of lines $y = x + c$ and $y = x - c$, $c \geq 0$.

49. $a(h, c) = c - 1 - h/\exp(c)$

51.

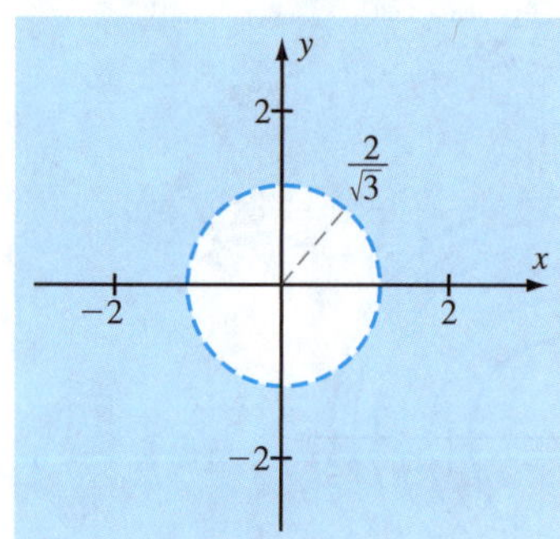

53.

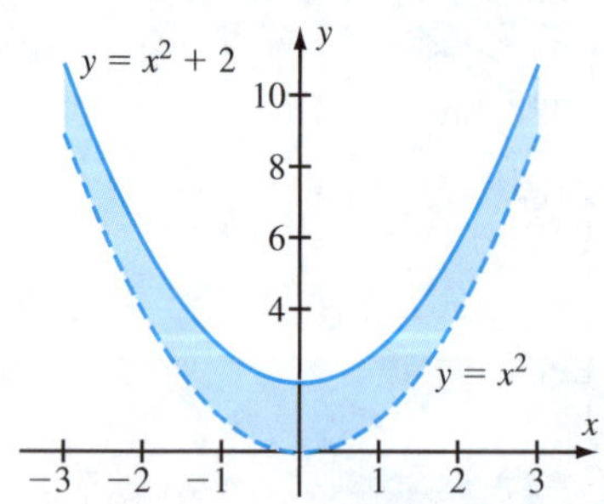

55. The level sets $L_c = \{(x, y) : f(x, y) = c\}$ are nonempty precisely when $-1 \leq c \leq 1$. For c in this interval, the level set L_c is the union of all parabolas of the form $y = x^2 + \arcsin(c) + 2n\pi$ $(n \in \mathbb{Z})$ and $y = x^2 - \arcsin(c) + (2n+1)\pi$ $(n \in \mathbb{Z})$.

57. The level sets $L_c = \{(x, y) : f(x, y) = c\}$ are nonempty precisely when $-1 \leq c \leq 1$. For c in this interval, the level set L_c is the union of all circles of the form $x^2 + y^2 = \arcsin(c) + 2n\pi$ $(n \in \mathbb{Z}, n \geq -\arcsin(c)/2\pi)$ and $x^2 + y^2 = -\arcsin(c) + (2n+1)\pi$ $(n \in \mathbb{Z}, n \geq \arcsin(c)/(2\pi) - 1/2)$.

59. The level sets $L_c = \{(x, y) : f(x, y) = c\}$ are nonempty precisely when $c \geq 0$. For c positive, the level set L_c is the hyperbola $y^2/c - x^2 = 1$. For $c = 0$, the level set L_c is the x-axis.

61. $f(x, y) = y/(y - x)$; domain: $\{(x, y) : |x| < |y|\}$

63. $f(x, y) = x^2 + y^3$

65. The level sets $L_c = \{(x, y) : f(x, y) = c\}$ are nonempty precisely when $c \geq 0$. For nonnegative c, we have $L_c = \{(\sqrt{c}(\cos(t) + \sin(t)), \sqrt{c}\sin(t)) : t \in [0, 2\pi)\}$.

67.

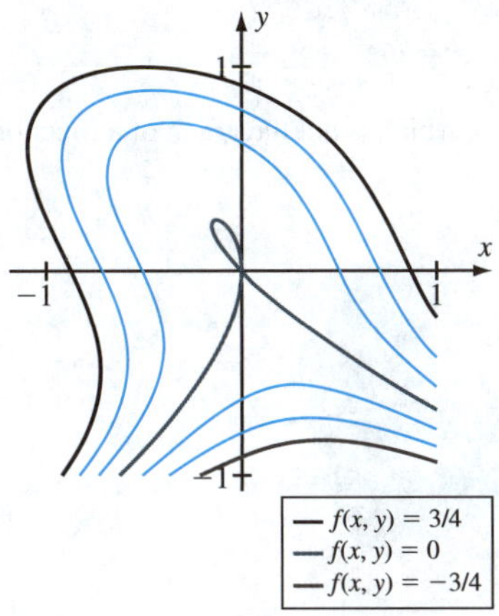

69.

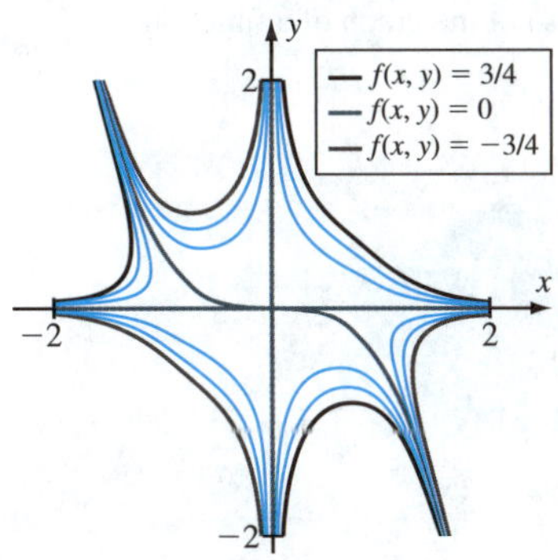

Section 13.2

1.

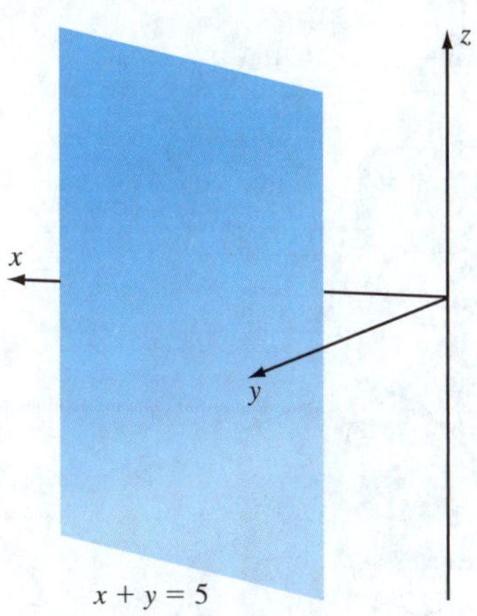

$x + y = 5$

3.

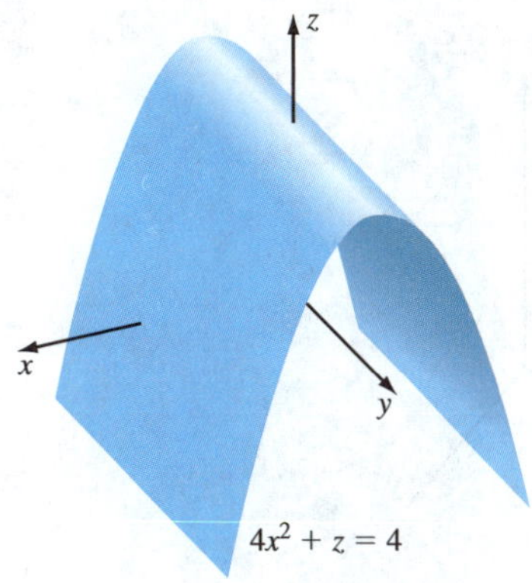

$4x^2 + z = 4$

5.

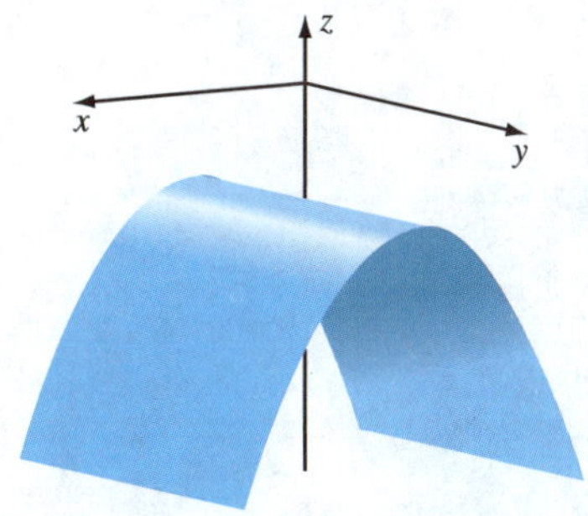

$x^2 + z = -4$

7.

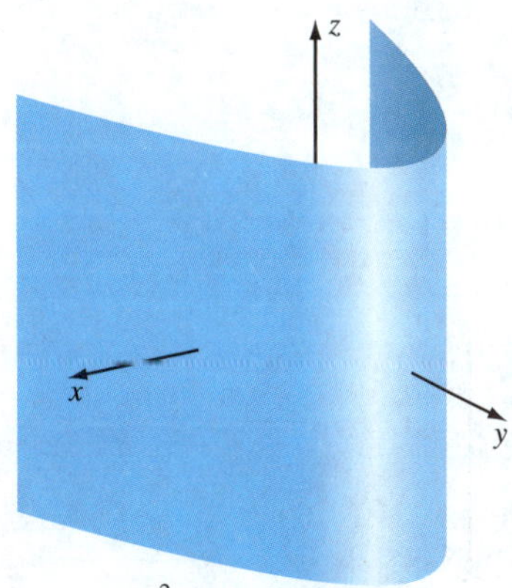

$x^2 + y = 4$

9.

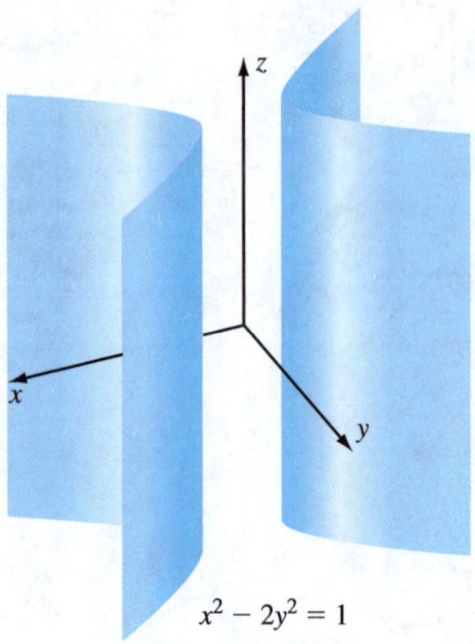

$x^2 - 2y^2 = 1$

11.

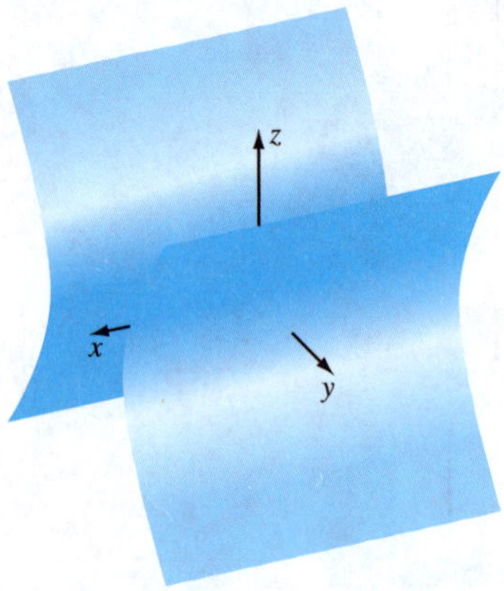

$y^2 - 4z^2 = 1$

13.

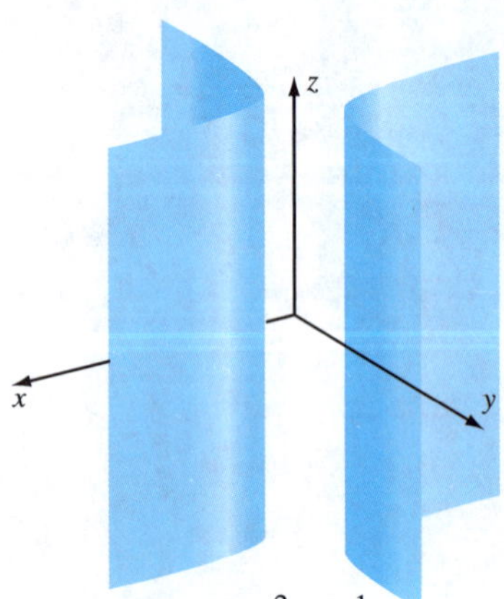

$x - 2xy = 1$

15. Hyperboloid of one sheet, which is not the graph of a function

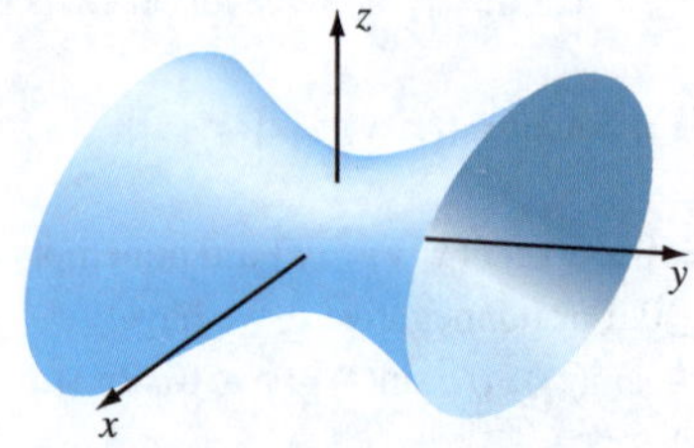

$x^2 - 4y^2 + 9z^2 = 36$

17. Circular paraboloid, which is not the graph of a function

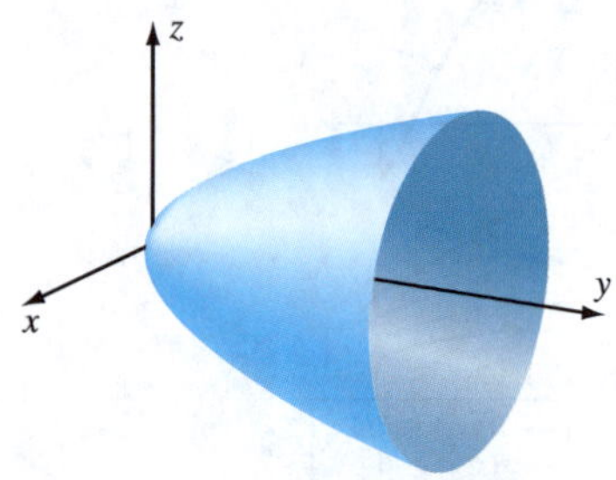

$y - x^2 = z^2$

19. Ellipsoid, which is not the graph of a function

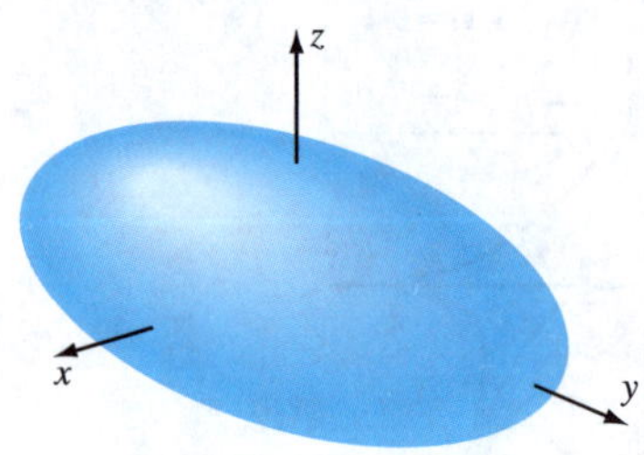

$4x^2 + y^2 + 9z^2 = 25$

21. Circular paraboloid, which is the graph of $f(x, y) = 4 - (x^2 + y^2)$

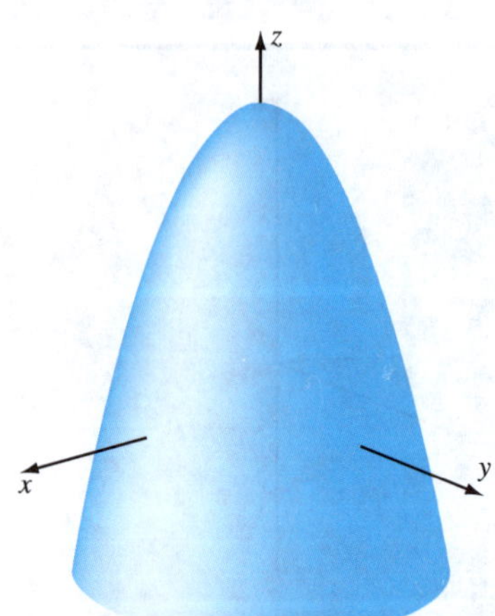

$x^2 + y^2 + z = 4$

23. Elliptic paraboloid, which is the graph of $f(x, y) = 4x^2 + y^2$

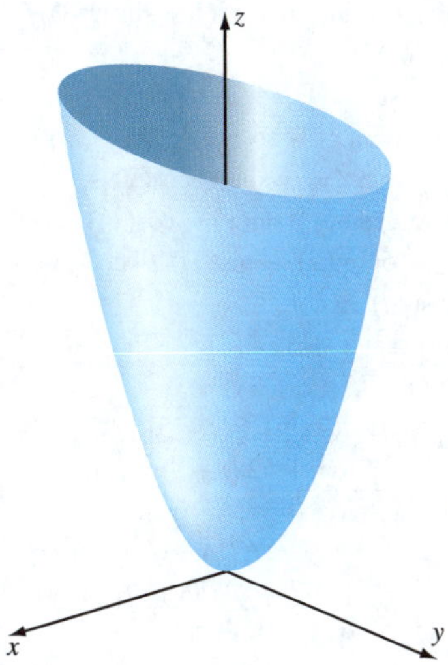

$4x^2 + y^2 = z$

25. Hyperboloid of two sheets, which is not the graph of a function

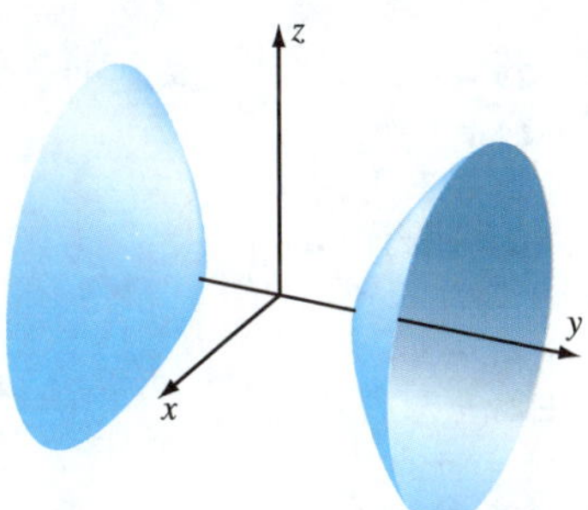

$y^2 - z^2 - x^2 = 1$

27. Elliptic cone, which is not the graph of a function

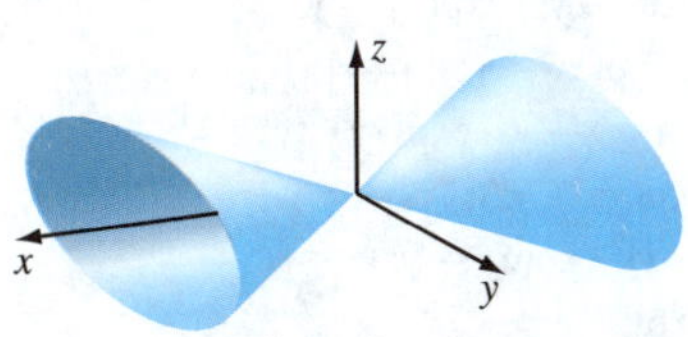

$y^2 + 8z^2 = x^2$

29. Hyperbolic paraboloid, which is not the graph of a function

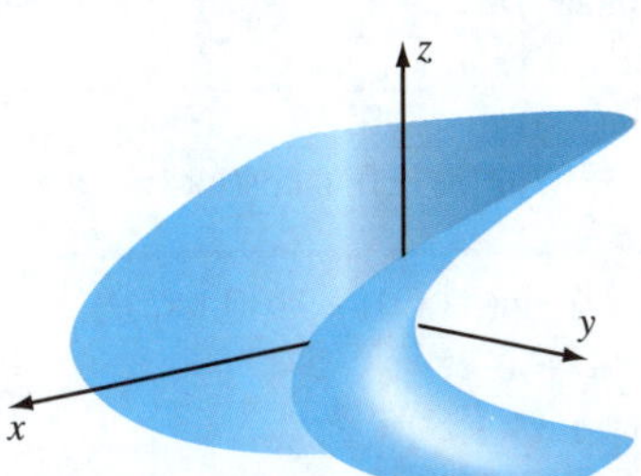

$y^2 - x - z^2 = 0$

31. False

35. The ellipse: $x^2 + y^2/4 = 1$

37.

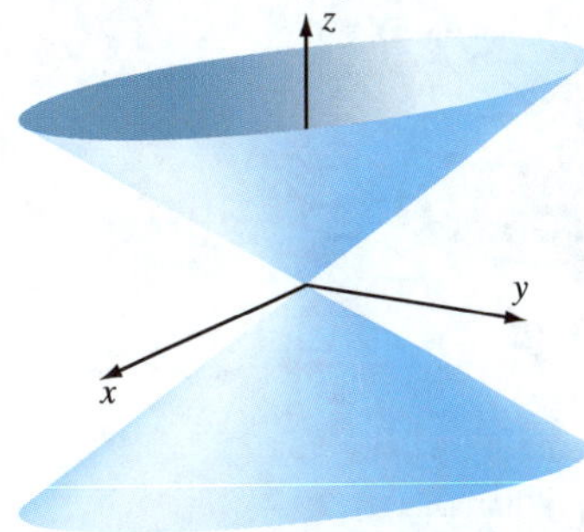

$x^2 + 2xy + 2y^2 - z^2 = 0$

39.

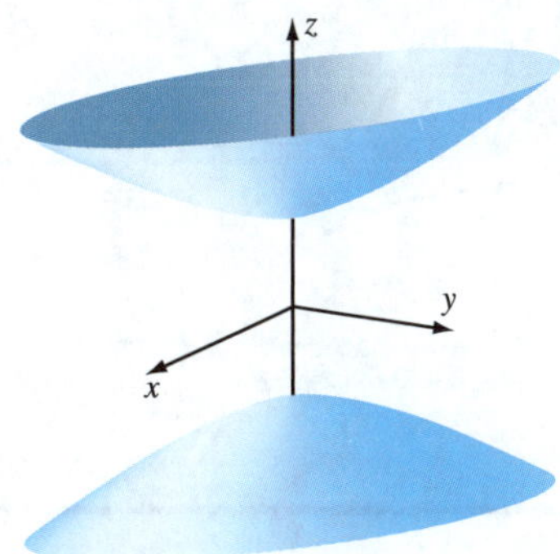

$x^2 + 2xy + 2y^2 - z^2 = -1$

41. $x = -1 + \sqrt{2}\cos(\theta)$, $y = \sin(\theta)$, $z = 2\sqrt{2} - 2\cos(\theta)$, $0 \leq \theta \leq 2\pi$

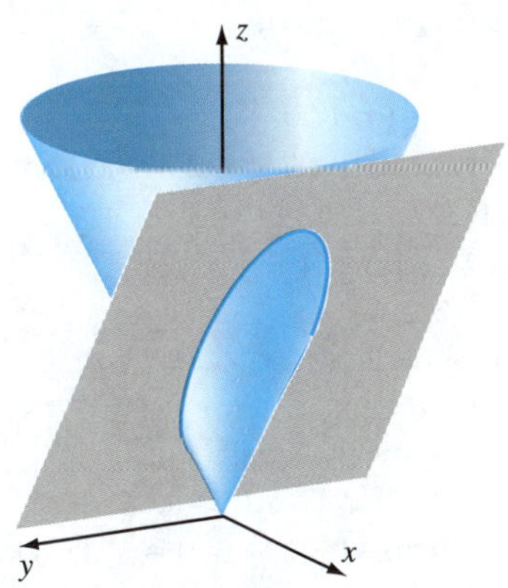

Section 13.3

1. 162
3. $2\sqrt{7}$
5. ln(12)
7. $\pi/2$
9. 2
11. 2
13. 9
21. Continuous
23. Not continuous
29. 0

31. $\lim_{y\to 0} f(y^2, y) = 1/2$

33.

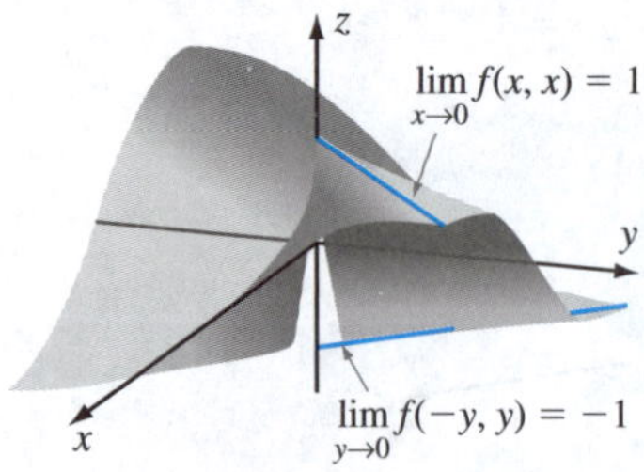

35.

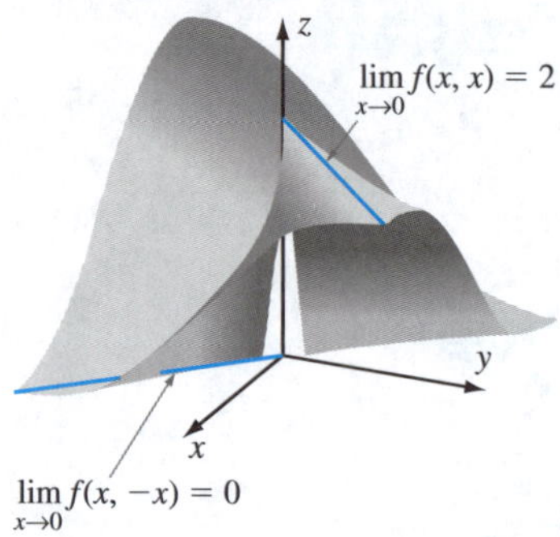

Section 13.4

1. $\varphi(x) = 2x + 1$; $\psi(y) = 10 + y^3$; $\varphi'(5) = 2 = \frac{\partial f}{\partial x}(5, 1)$; $\psi'(1) = 3 = \frac{\partial f}{\partial y}(5, 1)$

3. $\varphi(x) = 10$; $\psi(y) = 7 - 3y^4$; $\varphi'(2) = 0 = \frac{\partial f}{\partial x}(2, -1)$; $\psi'(-1) = 12 = \frac{\partial f}{\partial y}(2, -1)$

5. $\varphi(x) = \cos(x/4)$; $\psi(y) = \cos(\pi y^2)$; $\varphi'(\pi) = -\sqrt{2}/8 = \frac{\partial f}{\partial x}(\pi, 1/2)$; $\psi'(1/2) = -\pi\sqrt{2}/2 = \frac{\partial f}{\partial y}(\pi, 1/2)$

7. $\varphi(x) = x\ln(ex)$; $\psi(y) = \ln(y)$; $\varphi'(1) = 2 = \frac{\partial f}{\partial x}(1, e)$; $\psi'(e) = 1/e = \frac{\partial f}{\partial y}(1, e)$

9. $\varphi(x) = x^2$; $\psi(y) = 2^y$; $\varphi'(2) = 4 = \frac{\partial f}{\partial x}(2, 2)$; $\psi'(2) = 4\ln(2) = \frac{\partial f}{\partial y}(2, 2)$

11. $f_{xx}(x, y) = 2y$; $f_{yy}(x, y) = 20y^3$; $f_{xy}(x, y) = 2x - 1 = f_{yx}(x, y)$

13. $f_{xx}(x, y) = e^{xy}((y^2 - 1)\cos(x) - 2y\sin(x))$; $f_{yy}(x, y) = x^2e^{xy}\cos(x)$; $f_{xy}(x, y) = e^{xy}((1 + xy)\cos(x) - x\sin(x)) = f_{yx}(x, y)$

15. $f_{xx}(x, y) = -\sin(x)/\cos(y)$; $f_{yy}(x, y) = \sin(x)(1 + \sin^2(y))/\cos^3(y)$; $f_{xy}(x, y) = \cos(x)\sin(y)/\cos^2(y) = f_{yx}(x, y)$

17. $f_{xx}(x, y) = -1/(x - y)^2$; $f_{yy}(x, y) = -(x^2 - 2yx + 2y^2)/(xy - y^2)^2$; $f_{xy}(x, y) = 1/(x - y)^2 = f_{yx}(x, y)$

19. $f_{xx}(x, y) = -\cos(x)\sin(y)$; $f_{yy}(x, y) = -\cos(x)\sin(y)$; $f_{xy}(x, y) = -\sin(x)\cos(y) = f_{yx}(x, y)$

21. $D_y f(x, y) = x\sec(y)\tan(y)$; $D_{yy} f(x, y) = x\sec(y)(1 + 2\tan^2(y))$; $D_{yx} f(x, y) = \sec(y)\tan(y) = f_{12}(x, y)$; $f_{11}(x, y) = 0$

23. $D_y f(x, y) = x\cos(xy)/(2\sqrt{\sin(xy)})$; $D_{yy} f(x, y) = -x^2(1 + \sin^2(xy))/(4\sin^{3/2}(xy))$; $D_{yx} f(x, y) = (2\cos(xy)\sin(xy) - xy(1 + \sin^2(xy)))/(4\sin^{3/2}(xy)) = f_{12}(x, y)$; $f_{11}(x, y) = -y^2(1 + \sin^2(xy))/(4\sin^{3/2}(xy))$

25. $D_y f(x, y) = 5\sin(y)(\sin(x) - \cos(y))^4$; $D_{yy} f(x, y) = 5(\sin(x) - \cos(y))^3(4\sin^2(y) + \sin(x)\cos(y) - \cos^2(y))$; $D_{yx} f(x, y) = 20\cos(x)\sin(y)(\sin(x) - \cos(y))^3 = f_{12}(x, y)$; $f_{11}(x, y) = 5(\sin(x) - \cos(y))^3(4\cos^2(x) - \sin^2(x) + \sin(x)\cos(y))$

27. $D_y f(x, y) = 18(1 + x^2)(x + 3y)^5$; $D_{yy} f(x, y) = 3240(1 + x^2)(x + 3y)^3$; $D_{yx} f(x, y) = 18(x + 3y)^4(7x^2 + 6xy + 5) = f_{12}(x, y)$; $f_{11}(x, y) = 2(x + 3y)^4(28x^2 + 42xy + 9y^2 + 15)$

29. $D_y f(x, y) = 2y^3/\sqrt{1 + 3x^2 + y^4}$; $D_{yy} f(x, y) = 2y^2(y^4 + 3 + 9x^2)/(1 + 3x^2 + y^4)^{3/2}$; $D_{yx} f(x, y) = -6y^3x/(1 + 3x^2 + y^4)^{3/2} = f_{12}(x, y)$; $f_{11}(x, y) = 3(1 + y^4)/(1 + 3x^2 + y^4)^{3/2}$

31. $f_x(x, y, z) = 6x^2$; $f_y(x, y, z) = 5y^4z^8$; $f_z(x, y, z) = 8y^5z^7$; $f_{xy}(x, y, z) = 0$; $f_{xz}(x, y, z) = 0$; $f_{yz}(x, y, z) = 40y^4z^7$; $f_{xx}(x, y, z) = 12x$; $f_{yy}(x, y, z) = 20y^3z^8$; $f_{zz}(x, y, z) = 56y^5z^6$

33. $f_x(x, y, z) = 5(x + 2y + 3z)^4$; $f_y(x, y, z) = 10(x + 2y + 3z)^4$; $f_z(x, y, z) = 15(x + 2y + 3z)^4$; $f_{xy}(x, y, z) = 40(x + 2y + 3z)^3$; $f_{xz}(x, y, z) = 60(x + 2y + 3z)^3$; $f_{yz}(x, y, z) = 120(x + 2y + 3z)^3$; $f_{xx}(x, y, z) = 20(x + 2y + 3z)^3$; $f_{yy}(x, y, z) = 80(x + 2y + 3z)^3$; $f_{zz}(x, y, z) = 180(x + 2y + 3z)^3$

35. $f_x(x, y, z) = y\exp(xy - yz)$; $f_y(x, y, z) = (x - z)\exp(xy - yz)$; $f_z(x, y, z) = -y\exp(xy - yz)$; $f_{xy}(x, y, z) = (1 + xy - yz)\exp(xy - yz)$; $f_{xz}(x, y, z) = -y^2\exp(xy - yz)$; $f_{yz}(x, y, z) = -(1 + xy - yz)\exp(xy - yz)$; $f_{xx}(x, y, z) = y^2\exp(xy - yz)$; $f_{yy}(x, y, z) = (x - z)^2\exp(xy - yz)$; $f_{zz}(x, y, z) = y^2\exp(xy - yz)$

37. $f_x(x, y, z) = (y/x)\cos(\ln(xz))$; $f_y(x, y, z) = \sin(\ln(xz))$; $f_z(x, y, z) = (y/z)\cos(\ln(xz))$; $f_{xy}(x, y, z) = \cos(\ln(xz))/x$; $f_{xz}(x, y, z) = -y\sin(\ln(xz))/xz$; $f_{yz}(x, y, z) = \cos(\ln(xz))/z$; $f_{xx}(x, y, z) = -(y/x^2)(\sin(\ln(xz)) + \cos(\ln(xz)))$; $f_{yy}(x, y, z) = 0$; $f_{zz}(x, y, z) = -(y/z^2)(\sin(\ln(xz)) + \cos(\ln(xz)))$

39. $f_x(x, y, z) = 6x(x^2 + y^3 + z^4)^2$; $f_y(x, y, z) = 9y^2(x^2 + y^3 + z^4)^2$;

$f_z(x, y, z) = 12z^3(x^2 + y^3 + z^4)^2$;
$f_{xy}(x, y, z) = 36xy^2(x^2 + y^3 + z^4)$;
$f_{xz}(x, y, z) = 48xz^3(x^2 + y^3 + z^4)$;
$f_{yz}(x, y, z) = 72y^2z^3(x^2 + y^3 + z^4)$;
$f_{xx}(x, y, z) = 6(x^2 + y^3 + z^4)(5x^2 + y^3 + z^4)$;
$f_{yy}(x, y, z) = 18y(x^2 + y^3 + z^4)(x^2 + 4y^3 + z^4)$;
$f_{zz}(x, y, z) = 12z^2(x^2 + y^3 + z^4)(3x^2 + 3y^3 + 11z^4)$

41. $12xyz$ **43.** 0

51. **a.** Harmonic **b.** Harmonic **c.** Harmonic **d.** Not harmonic **e.** Not harmonic **f.** Harmonic

55. $B = -3E$; $C = -3A$

61. $f(x, y) = x^2y + y - 1$

63. $f(x, y) = x^2 + y - y^2 + 3$

69. $f_x(4, 3) = -0.02126$; $f_y(4, 3) = -0.09108$

71. $f_{xx}(4, 3) = 0.03337$; $f_{yy}(4, 3) = 0.07633$

Section 13.5

1. $5s^4 - 4s - s^{-2} + 3s^{-4}$

3. $-\sin(s)$ **5.** $3/2s$

7. $-(2\sin(s) + 3\cos(s))\exp(2\cos(s) - 3\sin(s))$

9. $4s/(\cos(s^2) - \sin(s^2))^2$

11. $z_s = 3t\sin(3st)\cos(3st)(2 + 7\cos^5(3st))$;
$z_t = 3s\sin(3st)\cos(3st)(2 + 7\cos^5(3st))$

13. $z_s = 0$; $z_t = 0$

15. $z_s = -2t(3s^4 + t^4)/(s^4 - t^4)^2$; $z_t = 2s(s^4 + 3t^4)/(s^4 - t^4)^2$

17. $z_s = 2(s - t)(s + 4t)/(2s + 3t)^2$;
$z_t = -(s - t)(7s + 3t)/(2s + 3t)^2$

19. $s^2(s - 2)^5 e^s(s^2 + 7s - 6)$

21. $(2s\ln(s) + 3s^{-2} + 2s^3\sec^2(s^2))/(s^2\sqrt{\ln^2(s) - s^{-3} + \tan(s^2)})$

23. $3s^2(s + 1)e^{3s}\cos(s^3e^{3s})$

25. $2^{4s+1}\ln(2)$

27. $2\pi(x - 1) + (y - \pi) - \pi(x - 1)^2/2 + (x - 1)(y - \pi)$

29. $1 - xy$

31. $1 + 4(x - \pi/4) - 2(y - \pi/4) + 8(x - \pi/4)^2 - 8(x - \pi/4)(y - \pi/4) + 2(y - \pi/4)^2$

35. Decreasing (at the rate of 0.31 cm^2/s)

37. $g_x(x, y) = 2x\ln(y)\cos(x^2\ln(y))$;
$g_y(x, y) = (x^2/y)\cos(x^2\ln(y))$

47. $D_1(f)(R) \cdot D_1(f)(P) + D_2(f)(R) \cdot D_1(f)(Q) + D_2(f)(R) \cdot D_2(f)(Q) \cdot D_1(f)(P)$ where $P = (x, y)$, $Q = (x, f(P))$, and $R = (f(P), f(Q))$

Section 13.6

1. $\langle \sin(y), x\cos(y)\rangle$

3. $\langle y^2\sec^2(xy^2), 2xy\sec^2(xy^2)\rangle$

5. $\langle -y\sin(xy)\sin(yz), -x\sin(xy)\sin(yz) + z\cos(xy)\cos(yz), y\cos(xy)\cos(yz)\rangle$

7. $\langle y^2/\cos(y), 2xy/\cos(y) + xy^2\sec(y)\tan(y)\rangle$

9. $\langle yz\cos(xy), xz\cos(xy), \sin(xy)\rangle$

11. -49 **13.** $-(3/2)\sqrt{\pi}$

15. $1/32$ **17.** $-1/16$

19. $1/\sqrt{2}$ **21.** $-3\sqrt{3}/8$

23. **a.** $\langle 5/13, 12/13\rangle$ **b.** $13/2$ **c.** $\langle -5/13, -12/13\rangle$ **d.** $-13/2$

25. **a.** $\langle 1/\sqrt{2}, 1/\sqrt{2}\rangle$ **b.** $\sqrt{2}(1 - \ln(2))/4$ **c.** $\langle -1/\sqrt{2}, -1/\sqrt{2}\rangle$ **d.** $-\sqrt{2}(1 - \ln(2))/4$

27. **a.** $\langle -1/\sqrt{5}, -2/\sqrt{5}\rangle$ **b.** $2\sqrt{5}/3$ **c.** $\langle 1/\sqrt{5}, 2/\sqrt{5}\rangle$ **d.** $-2\sqrt{5}/3$

29. **a.** $\langle 1, 0\rangle$ **b.** $1/e$ **c.** $\langle -1, 0\rangle$ **d.** $-1/e$

31. **a.** $\langle 2/\sqrt{5}, 1/\sqrt{5}\rangle$ **b.** $80\sqrt{5}$ **c.** $\langle -2/\sqrt{5}, -1/\sqrt{5}\rangle$ **d.** $-80\sqrt{5}$

33. 36

35. $-3/4$

37. $(1 + 3\sqrt{3})/8$

39. **a.** $\langle 3/\sqrt{11}, -1/\sqrt{11}, 1/\sqrt{11}\rangle$ **b.** $4\sqrt{11}$ **c.** $\langle -3/\sqrt{11}, 1/\sqrt{11}, -1/\sqrt{11}\rangle$ **d.** $-4\sqrt{11}$

41. **a.** $\langle 0, 1/\sqrt{2}, -1/\sqrt{2}\rangle$ **b.** $\sqrt{2}$ **c.** $\langle 0, -1/\sqrt{2}, 1/\sqrt{2}\rangle$ **d.** $-\sqrt{2}$

45. $47/15$

47. $-2/5 + 6\pi/5$

49. $f_x(P) = 3$; $f_y(P) = -1$

51. $f_x(P) = 3\sqrt{3} - 3$; $f_y(P) = -3\sqrt{3} + 3$

55. $1358^{-1/2}\langle -25, -27, -2\rangle$

61.

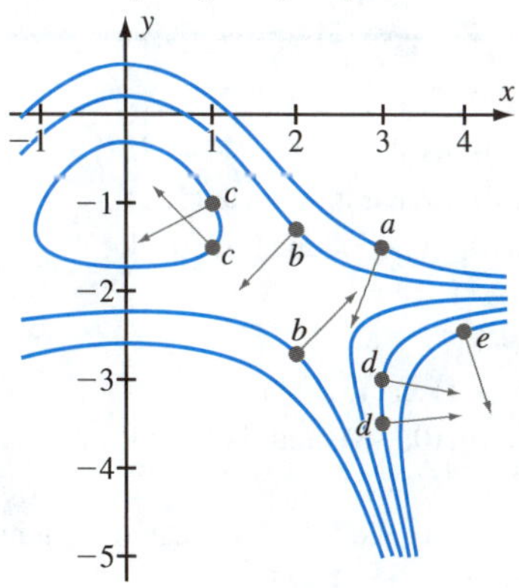

Section 13.7

1. Normal vector: $\langle 6, -47, -1\rangle$; tangent plane: $z = 6x - 47y + 123$

3. Normal vector: $\langle 2/7, 4/7, -1\rangle$; tangent plane: $z = 2x/7 + 4y/7 + \ln(7) - 10/7$

5. Normal vector: $\langle 1/9, -2/9, -1\rangle$; tangent plane: $z = x/9 - 2y/9 + 2/3$

7. Normal vector: $\langle 4e^3, 3e^3, -1\rangle$; tangent plane: $z = e^3(4x + 3y - 8)$

9. Normal vector: $\langle 1/2, 1/2, -1\rangle$; tangent plane: $z = (x + y)/2$

11. Normal vector: $\langle -1/48, 1/48, -1\rangle$; tangent plane: $z = (-x + y + 32)/48$

13. Normal vector: $\langle 2, -16, 2\rangle$; tangent plane: $2x - 16y + 2z = -28$

15. Normal vector: $\langle -1/2, 1/4, -1/2\rangle$; tangent plane: $2x - y + 2z = 0$

17. Normal vector: $\langle 1, 1, 1\rangle$; tangent plane: $x + y + z = e^3 - 1$

19. Normal vector: $\langle 6, 6, 12\rangle$; tangent plane: $x + y + 2z = 2$

21. $(x - 2)/12 = (y - 1)/-5 = (z - 7)/-1$

23. $x - \pi/4 = y = z - \pi/4$

25. $x = 0,\ y - \pi = 1 - z$

27. 5.26

29. 0.9.

31. 3750 ft^2

33. 1.08 cm^3

35. The summand $f(1, -1)$ has been omitted; correct equation: $z = 2 + 6(x - 1) + 6(y + 1)$.

37. The coefficients of x and y should be the evaluations $f_x(P_0)$ and $f_y(P_0)$, not $f_x(x, y)$ and $f_y(x, y)$; correct equation: $z = -3x + (y + 3)$.

41. $\pi/2$

43. $\pi/4$

45. $80p + 9r < 2112/25$ where $100p$ is the percentage of gold and r is the radius

Section 13.8

1. Local maximum at $(1, 2)$

3. Saddle point at $(0, 0)$; local maximum at $(-2, 2)$

5. Saddle point at $(0, 0)$; local minimum at $(1, 1)$

7. Saddle points at $(0, 0)$, $(0, -3)$, and $(-3, 0)$; local maximum at $(-1, -1)$

9. Local minimum at $(3, 3)$

11. Saddle points at $(0, -1)$ and $(2, -1)$

13. Saddle points at $(0, 0)$, $(0, -8)$, and $(-8, 0)$; local maximum at $(-8/3, -8/3)$

15. Saddle points at $(-2, -1)$ and $(2, -1)$; local maximum at $(0, -\sqrt{5})$; local minimum at $(0, \sqrt{5})$

17. Saddle point at $(0, 0)$; local minima at $(-\sqrt{2}, -\sqrt{2})$ and $(\sqrt{2}, \sqrt{2})$

19. Local minimum at $(0, 0)$

21. Each number is $100/3$.

23. Base: a square with side length $4/3^{1/3}$ in.; height: $4 \cdot 3^{2/3}$ in.

25. Equilateral triangle with side length $20/3^{1/4}$

27. Regression line: $y = 0.001956x + 0.2577$; $y(100) \approx 0.4533$ g/mi

29. $y = 6.12x + 21.2$

31. Least squares line: $\ln(y) = -0.14t + 0.0093$; $y(10) \approx 0.25$

39. Saddle point: $(-0.50608337, 0.47573876)$

41. Critical point: $(0, \pi/2)$, which is a global minimum

Section 13.9

1. Maximum value: $6 + \sqrt{34}$; minimum value: $6 - \sqrt{34}$

3. Maximum value: 25; minimum value: 11/3

5. Maximum value: $128/\sqrt{3} - 24$; minimum value: $-128/\sqrt{3} - 24$

7. Maximum value: 37/4; minimum value: -3

9. Maximum value: $16\sqrt{2}$; minimum value: 16

11. Maximum value: $8/(3\sqrt{3})$; minimum value: $-8/(3\sqrt{3})$

13. 223/22

15. -4

17. $4\sqrt{3}/3$

19. $3\sqrt[3]{2\pi V_0^2}$

21. 144

23. $\sqrt{2/3}$

25. $256\sqrt{2}/3$

27. 3

29. Highest: $150 + 180\sqrt{2}$; lowest: $150 - 180\sqrt{2}$

31. $3\sqrt[3]{2\pi k V_0^2}$ times the unit side cost

33. 156/7

35. 11/7

37. $x = pT/(A(p + q)),\ y = qT/(B(p + q))$

41. $(x(10000), y(10000)) = (8000/21, 2500/7)$; $M(10000) \approx 31296.961$; $\lambda(10000) = (2500/7)^{3/4}/15 \approx 5.477$; $(x(10001), y(10001)) = (40004/105, 10001/28)$; $M(10001) \approx 31302.438$; $M(10001) - M(10000) \approx 5.477$ (the same value as $\lambda(10000)$)

49. 3.15694

51. 0.0342767

Section 14.1

1. 56

3. 96

5. $\mathcal{I}(x) = 4x + 24$; $\mathcal{J}(y) = 2 + 4y$

7. $\mathcal{I}(x) = 12x^2$; $\mathcal{J}(y) = 24y^2 + 4$

9. $\mathcal{I}(x) = (e^x - 1)/(2x)$; $\mathcal{J}(y) = \left(e^{y^2} - 1\right)/y$

11. 161/2

13. 0

15. $e^3 - e^2 - e + 1$

17. 0

19. 0

21. 5/3

23. $\pi^2/8$

25. 4/3

27. 1/12

29. $(e + 2)\ln(e + 2) - e - 3\ln(3)$

31. $2e - 2$

33. $(e^4 - 1)/8$

35. $e - 2$

37. $424\ln(2) - 105$

39. $4(\sin(1) - \cos(1))$

41. $(e^4 - 1)/4$

43. $\pi^2/32$

45. 1/2

47. 2/3

49. 0.2248

51. 2.231

Section 14.2

1. $\mathcal{R}$ is x-simple only; boundary curves: $\alpha_1(y) = -2y - 1$ and $\alpha_2(y) = -y^2 + 2$ for $-1 \le y \le 3$

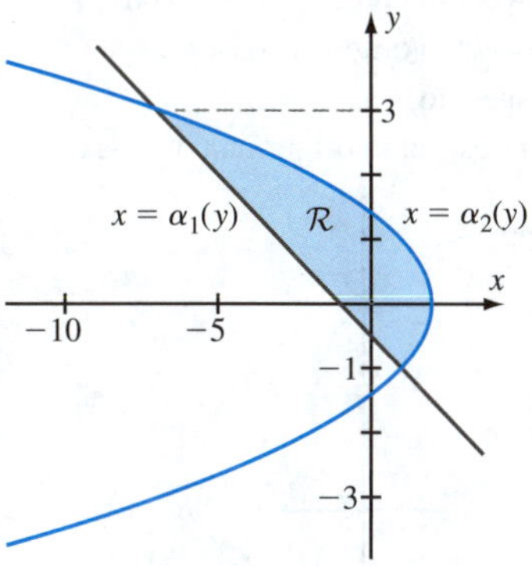

3. $\mathcal{R}$ is x-simple only; boundary curves: $\alpha_1(y) = y^2$ and $\alpha_2(y) = -2y^2 - 3y + 18$ for $-3 \le y \le 2$

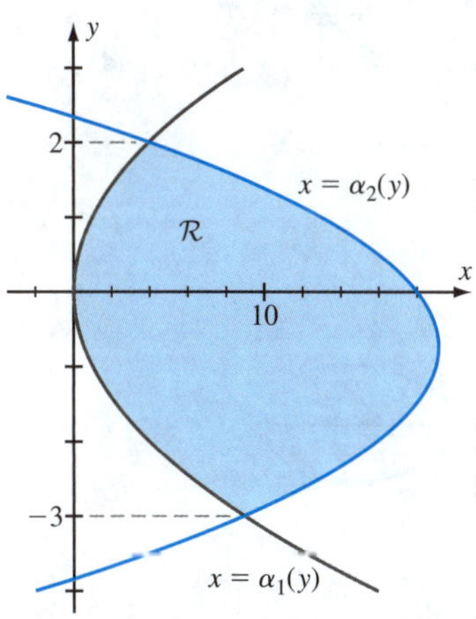

5. $\mathcal{R}$ is y-simple only; boundary curves: $\beta_1(x) = x^{3/2}$ and $\beta_2(x) = 6x - 1$ for $1 \le x \le 3$

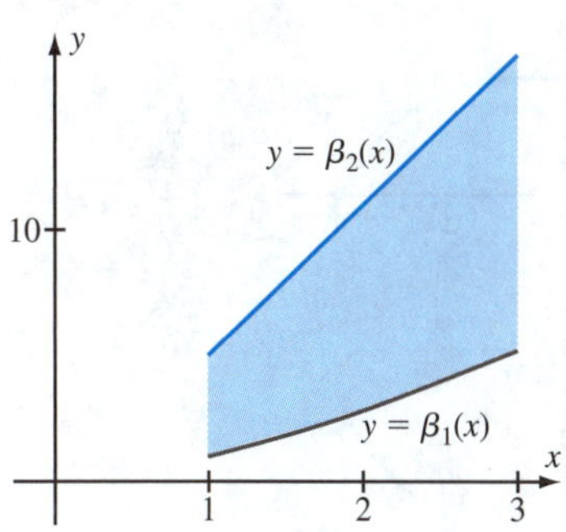

7. -8
9. $78\sqrt{2}/5$
11. $4\pi - 4$
13. $5/\sqrt{2}$
15. $112/15$
17. $e^2(13e^6 - 7)/4 - 33/2$
19. $640 - 145\sqrt{3}$
21. $5/6$
23. 1
25. $e^4 - 1$
27. 1
29. $1/12$
31. $(\sqrt{2}\pi - 5)/4$
33. $(1 - \sin(1))/2$
35. $11/15$
37. $-1/24$
39. $9/4$
41. $2/3$
43. $1/4$
45. $(143a + 91b)/3$
47. $68/15$
49. $2/3$
51. $1/6$
53. $1/3$
57. 6.6525
59. 0.27694
61. 0.62
63. 0.66
65. Exact: $8/15$; approximation: $17/32$; absolute error: 0.00208
67. Exact: $4/35$; approximation: $(3 + 4\sqrt{2} + 3\sqrt{3})/128$; absolute error: 0.0060591

Section 14.3

1. $1029/4$
3. $99/4$
5. $48/5$
7. $375/4$
9. 12
11. 728
13. $813/2$
15. $1960/3$
17. 108
19. $4e^2 + e^{-3}$
21. $1288/15$
23. $2\sin(1) - \sin(2)$
25. $136/15$
27. $2744/3$
29. $4576/675$
31. $64/3$
33. $2/3$
35. 81π
37. $1/8$
39. $\pi/2$
41. $\ln(54/25) - 2\arctan(2) + \pi/2$
43. $1/8$
45. 4.7918
47. 3.4931

Section 14.4

1.

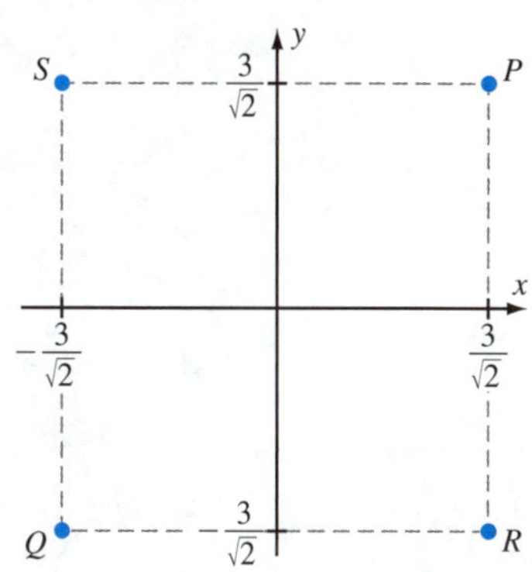

$P = \left(3, \frac{\pi}{4}\right)$ $Q = \left(-3, \frac{\pi}{4}\right)$

$R = \left(3, -\frac{\pi}{4}\right)$ $S = \left(-3, -\frac{\pi}{4}\right)$

3.

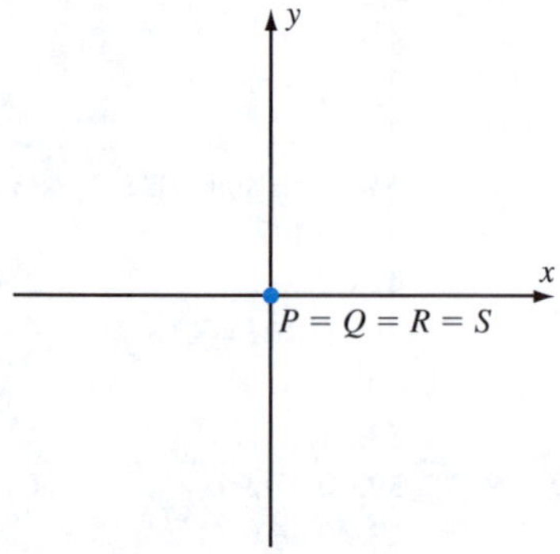

$P = \left(0, \frac{\pi}{2}\right) \quad Q = (0, \pi)$

$R = (0, -\pi) \quad S = \left(0, -\frac{\pi}{2}\right)$

5.

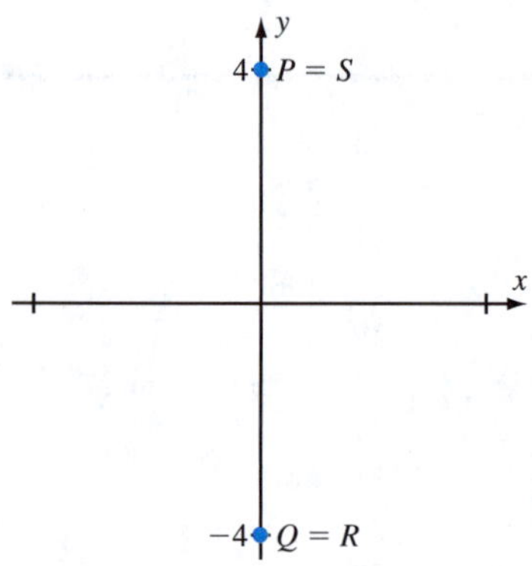

$P = \left(4, -\frac{7\pi}{2}\right) \quad Q = \left(-4, -\frac{7\pi}{2}\right)$

$R = \left(4, \frac{7\pi}{2}\right) \quad S = \left(-4, \frac{7\pi}{2}\right)$

7.

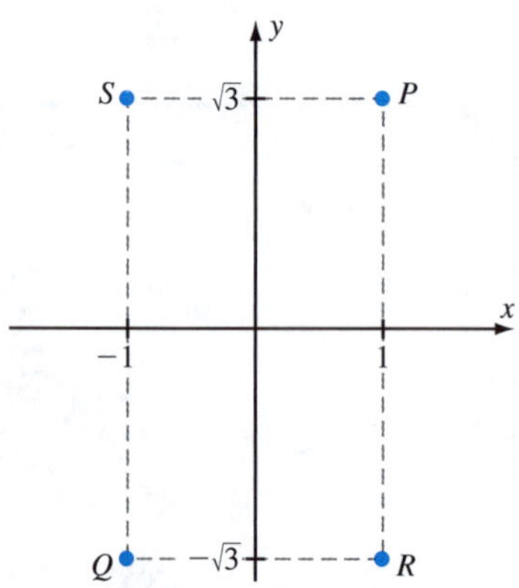

$P = \left(2, \frac{7\pi}{3}\right) \quad Q = \left(-2, \frac{7\pi}{3}\right)$

$R = \left(2, -\frac{7\pi}{3}\right) \quad S = \left(-2, -\frac{7\pi}{3}\right)$

9. $(2\sqrt{2}, 2\sqrt{2})$ **11.** $(0, 0)$

13. $(2\sqrt{3}, 2)$ **15.** $(-3, -3\sqrt{3})$

17. $(4\sqrt{2}, -\pi/4 + 2n\pi)$, $n \in \mathbb{Z}$; $(-4\sqrt{2}, 3\pi/4 + 2n\pi)$, $n \in \mathbb{Z}$

19. $(9, -\pi/2 + 2n\pi)$, $n \in \mathbb{Z}$; $(-9, \pi/2 + 2n\pi)$, $n \in \mathbb{Z}$

21. $(1, \pi + 2n\pi)$, $n \in \mathbb{Z}$; $(-1, 2n\pi)$, $n \in \mathbb{Z}$

23. Symmetric with respect to origin, x-axis, y-axis

25. Symmetric with respect to origin, x-axis, y-axis

27. Symmetric with respect to origin, x-axis, y-axis

29. Symmetric with respect to x-axis

31. Not symmetric with respect to origin, x-axis, y-axis

33.

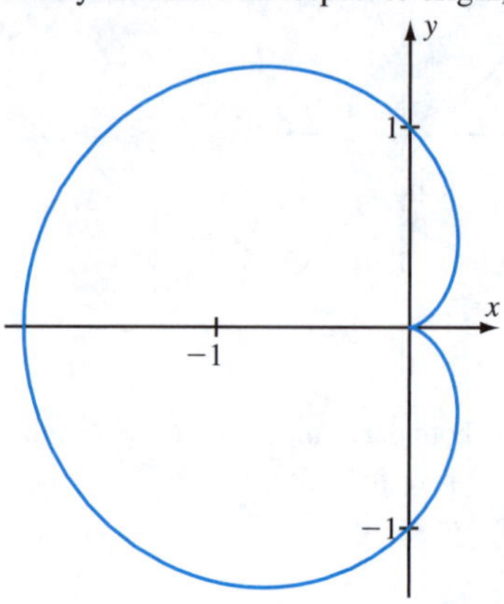

$r = 1 - \cos(\theta)$

35.

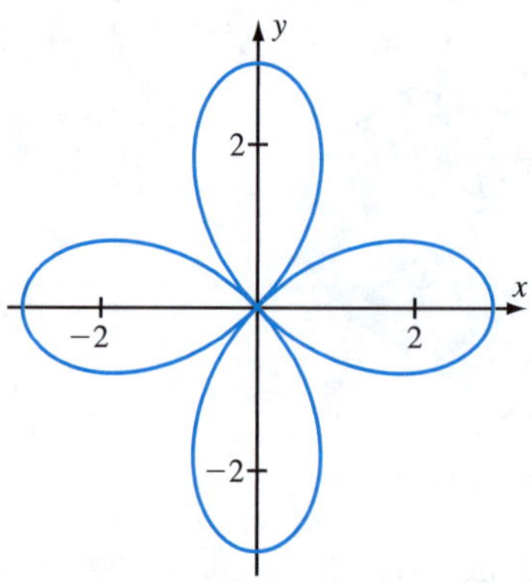

$r = 3\cos(2\theta)$

37.

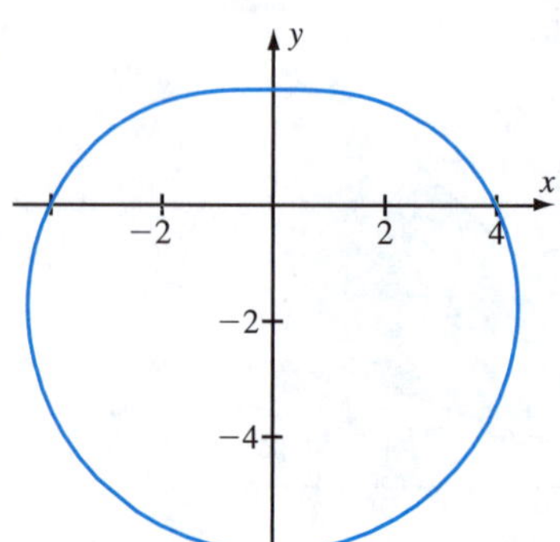

$r = 4 - 2\sin(\theta)$

39.

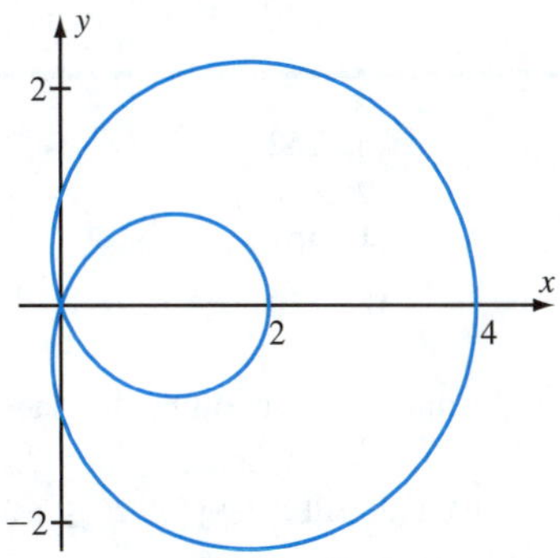

$r = -1 + 3\cos(\theta)$

41.

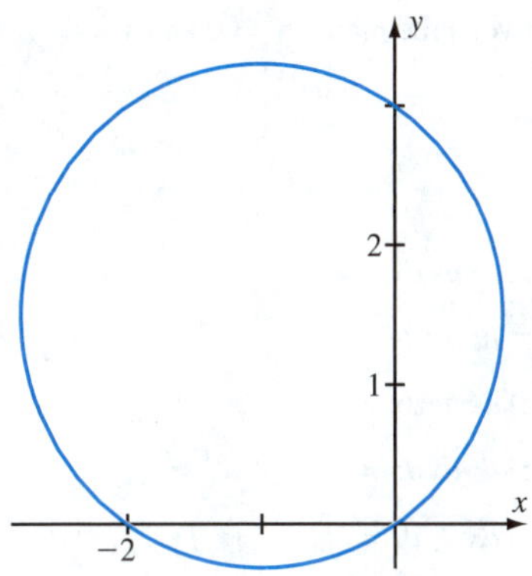

$r = 3\sin(\theta) - 2\cos(\theta)$

43.

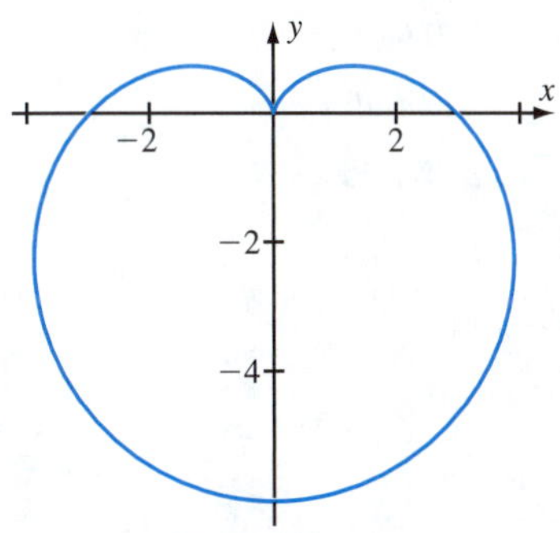

$r = 3 - 3\sin(\theta)$

45. **a.** $(\sqrt{9+\pi^2/4}, \arctan(\pi/12) + 2n\pi)$ and $(-\sqrt{9+\pi^2/4}, \arctan(\pi/12) + (2n+1)\pi)$ for $n \in \mathbb{Z}$
b. $(\sqrt{36+49\pi^2/4}, -\arctan(7\pi/12) + (2n+1)\pi)$ and $(-\sqrt{36+49\pi^2/4}, -\arctan(7\pi/12) + 2n\pi)$ for $n \in \mathbb{Z}$
c. $(\pi\sqrt{1033}/12, -\arctan(32/3) + 2n\pi)$ and $(-\pi\sqrt{1033}/12, -\arctan(32/3) + (2n+1)\pi)$ for $n \in \mathbb{Z}$

47. Distance: $\sqrt{r_1^2 + r_2^2 - 2r_1r_2\cos(\theta_1 + \theta_2)}$; polar equation of circle: $r^2 - 6r\cos(\theta + \pi/4) + 8 = 0$

49. In rectangular coordinates: $(0, 0)$, $(1/4, \sqrt{3}/4)$, and $(1/4, -\sqrt{3}/4)$

51. In rectangular coordinates: $(\pm 1/\sqrt{2}, \pm 1/\sqrt{2})$

53. In rectangular coordinates: $(0, 0)$ and $(-1/4, \pm\sqrt{3}/4)$

55. The points of the x-axis

57. $\sqrt{2}(\exp(\pi/2) - \exp(-\pi/2))$

59. $\pi/4 - 3\sqrt{3}/8$

61.

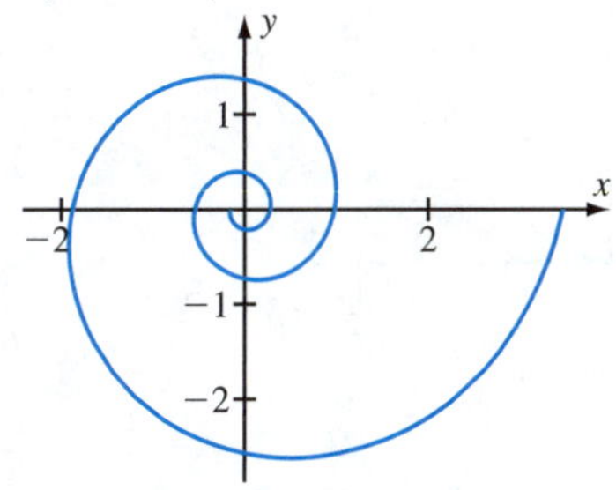

$r = \exp\left(\frac{\theta}{5}\right)$

63.

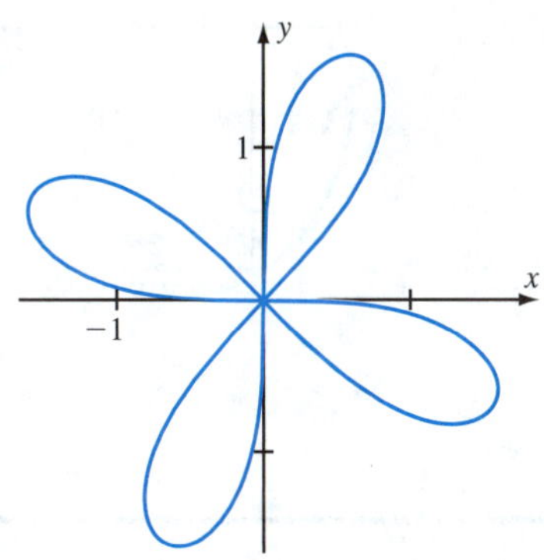

$r^2 = -3\sin(4\theta)$

65.

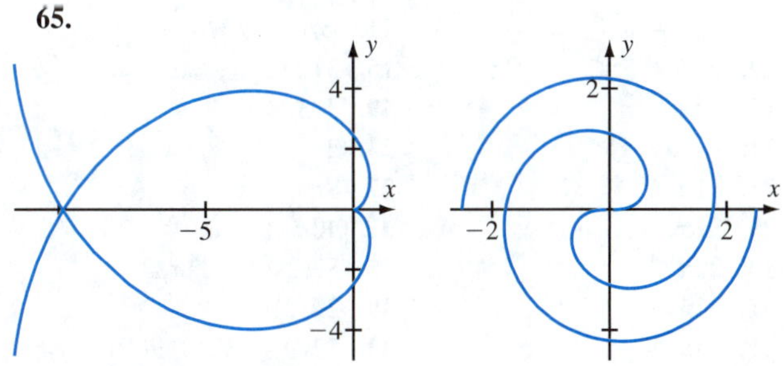

$r = \theta^2$; $-\frac{9\pi}{8} \le \theta \le \frac{9\pi}{8}$ $\qquad$ $\theta = r^2$; $0 \le \theta \le 2\pi$

67.

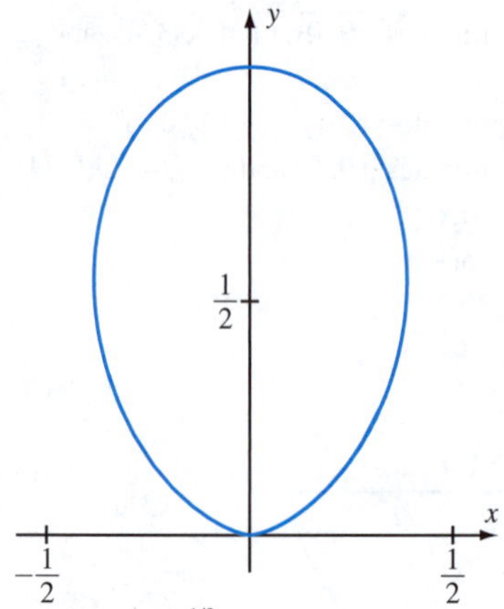

$r^{1/2} = \sin(\theta)$

69.

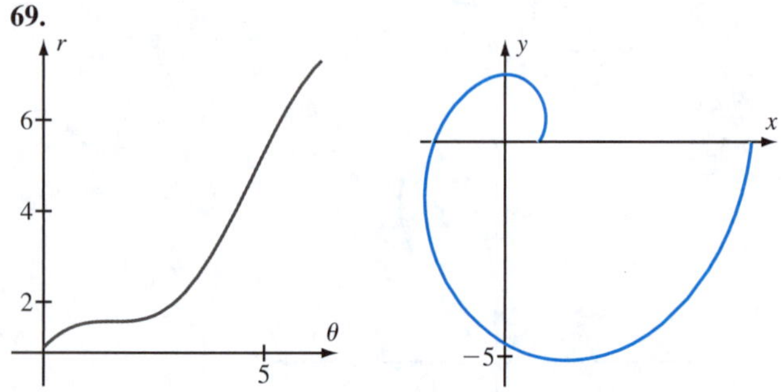

$r = \theta + \cos(\theta)$ $r = \theta + \cos(\theta)$

Section 14.5

1. $3\pi/2$
3. 4π
5. 2
7. 35π
9. $\pi/8 - 1/4$
11. $\pi/6 + \sqrt{3}/4$
13. $\pi + 4$
15. $31\pi/2$
17. 15π
19. $2/3$
21. $-4\sqrt{2}/15$
23. $2/3$
25. 11π
27. 8π
29. 32π
31. $40\pi/3$
33. $22\pi/3$
35. $32(2\sqrt{2} - 1)\pi/3$
37. $32\pi/3$
39. $29/210$
41. $248/27$
43. $52/9 + (32/3)\ln(2/3)$
45. 12π
47. $7\pi/12 - \sqrt{3}$
49. $\pi/3 - \sqrt{3}/4$
51. $\pi - 3\sqrt{3}/2$
53. $2(\sqrt{5} - 1)\pi$
55. $3^{11/3}\pi/16$
57. $\pi^2/32 - 1/8$
59. $\ln(2)$
63. $\theta_0 \approx 0.967025$; area: 2.4393
65. 0.24

Section 14.6

1. 27
3. 252
5. 3
7. $\pi - 4$
9. $\cos(\pi^2) - 1$
11. 30π
13. $16/15$
15. 24π
17. 128π
19. 3; $\mathcal{U}$ is the region in the first octant cut off by the plane $3x + 2y + z = 6$.
21. 1/15; $\mathcal{U}$ is that part of the unit ball $x^2 + y^2 + z^2 \leq 1$ that lies in the first octant.
23. 1/8; $\mathcal{U}$ is that part of the cylinder $x^2 + z^2 \leq 1$ that lies in the first octant between the planes $y = 0$ and $y = x$.
33. $\pi(a^2 + b^2 + 4c)^2/32$
35. $\pi m^6(3m^2 + 4)/60$
37. $(1 - \cos(1))/24$
39. 216
41. $\int_0^1 \int_x^1 \int_0^1 f(x, y, z)\, dz\, dy\, dx =$
$\int_0^1 \int_0^y \int_0^1 f(x, y, z)\, dz\, dx\, dy =$
$\int_0^1 \int_0^1 \int_x^1 f(x, y, z)\, dy\, dz\, dx =$
$\int_0^1 \int_0^1 \int_x^1 f(x, y, z)\, dy\, dx\, dz =$
$\int_0^1 \int_0^1 \int_0^y f(x, y, z)\, dx\, dy\, dz = \int_0^1 \int_0^1 \int_0^y f(x, y, z)\, dx\, dz\, dy$
43. $\int_{-1}^0 \int_2^{37} \int_{x+1}^{\sqrt{x+1}} y\sqrt{1-x}\, dy\, dz\, dx =$
$\int_2^{37} \int_{-1}^0 \int_{x+1}^{\sqrt{x+1}} y\sqrt{1-x}\, dy\, dx\, dz =$
$\int_{-1}^0 \int_{x+1}^{\sqrt{x+1}} \int_2^{37} y\sqrt{1-x}\, dz\, dy\, dx =$
$\int_0^1 \int_{y^2-1}^{y-1} \int_2^{37} y\sqrt{1-x}\, dz\, dx\, dy =$
$\int_0^1 \int_2^{37} \int_{y^2-1}^{y-1} y\sqrt{1-x}\, dx\, dz\, dy =$
$\int_2^{37} \int_0^1 \int_{y^2-1}^{y-1} y\sqrt{1-x}\, dx\, dy\, dz$
47. 2.0333

Section 14.7

1. 108
3. $1027/2$
5. $-(26/3)\ln(2)$
7. $(183/448, 65/128)$
9. $(156/55, 29/55)$
11. $5/3$
13. 13
15. $(-193/335, 25/201, 778/1005)$
17. $(31/75, 31/25, 124/75)$
19. $I_x = 650$; $I_y = 1160$; $I_z = 1530$
21. $I_x = 186$; $I_y = 194$; $I_z = 64$
23. $\mathcal{R} = \{(x, y) : 0 < x^2 + y^2 < 1\}$

25. $KE_{\mathcal{U}} = I_x\omega^2/2$
27. $2\pi^2 r^2 R$
29. (1.16866, 3.52217)
31. (0.52368, 0.48447)

Section 14.8

1. $(4, \pi/6, 5)$
3. $(2\sqrt{2}, 3\pi/4, -1)$
5. $(3, 3\pi/2, -2)$
7. $(2, 5\pi/3, -2)$
9. $(2, \pi/4, \pi/4)$
11. $(4\sqrt{2}, \pi/4, 0)$
13. $(2\sqrt{2}, \pi/6, \pi/4)$
15. $(8\sqrt{2}, 5\pi/6, 3\pi/4)$
17. Rectangular coordinates: $(-3\sqrt{3}/4, 9/4, 3/2)$; cylindrical coordinates: $(3\sqrt{3}/2, 2\pi/3, 3/2)$
19. Cylindrical coordinates: $(2\sqrt{2}, 7/4\pi, -2\sqrt{2})$; spherical coordinates: $(4, 3\pi/4, 7\pi/4)$
21. Rectangular coordinates: $(-\sqrt{2}, -\sqrt{2}, -2)$; spherical coordinates: $(2\sqrt{2}, 3\pi/4, 5\pi/4)$
23. Cylindrical coordinates: $(3, 0, -3)$; spherical coordinates: $(3\sqrt{2}, 3\pi/4, 0)$
25. $32\pi/3$
27. $512\pi/15$
29. $10\pi \ln(2)$
31. 180π
33. 65π
35. -2π
37. 2π
39. $32\pi/3$
41. Rectangular coordinates: $(3\cos(5)\sin(2), 3\sin(5)\sin(2), 3\cos(2))$; cylindrical coordinates: $(3\sin(2), 5, 3\cos(2))$
43. Points in the xz-plane
45.

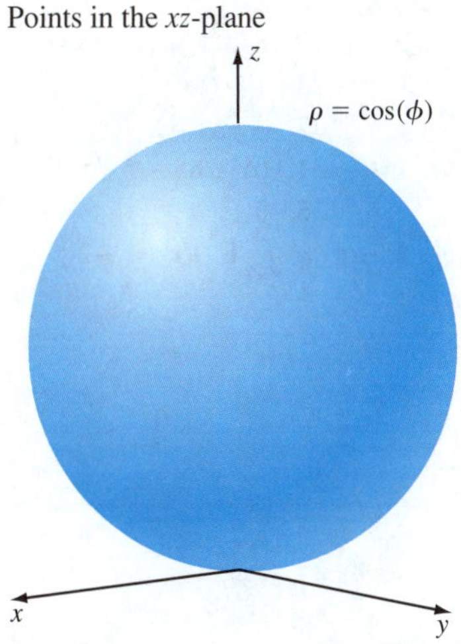

47.

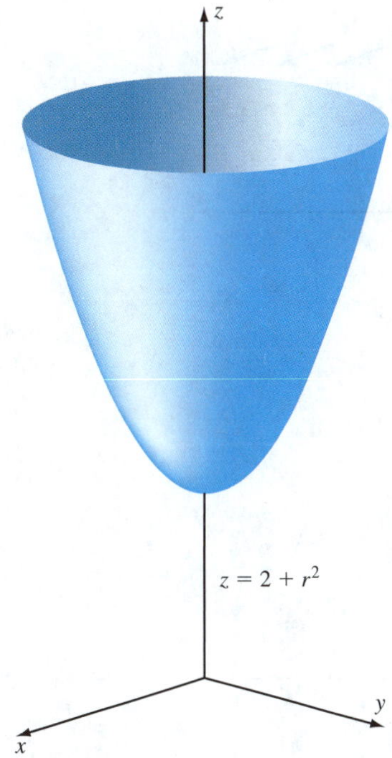

49. 4.797453714

Section 15.1

In the answers for this section, A, B, C, C_1, and C_2 represent arbitrary constants.

1.

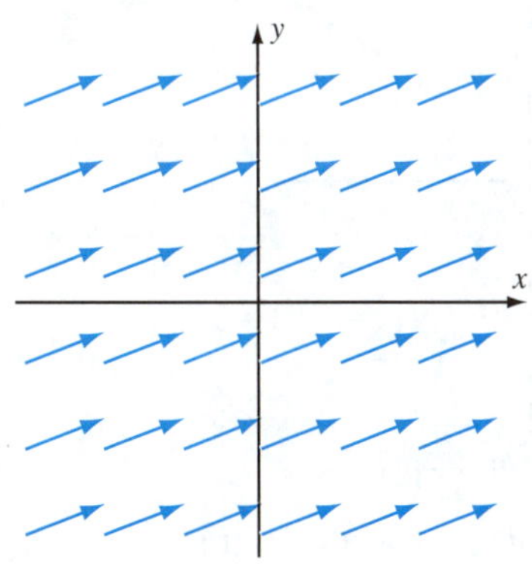

3.

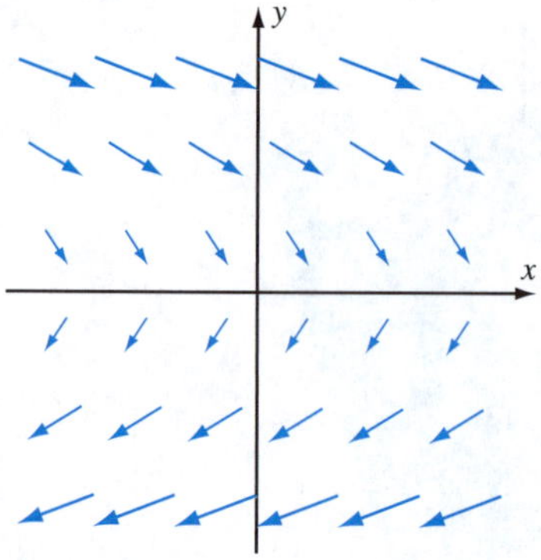

$\mathbf{F}(x, y) = 2y\mathbf{i} - 3\mathbf{j}$

5.

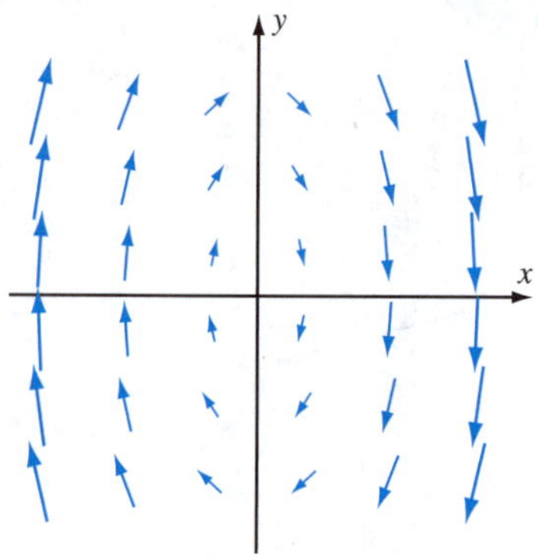

$\mathbf{F}(x, y) = y\mathbf{i} - 5x\mathbf{j}$

7.

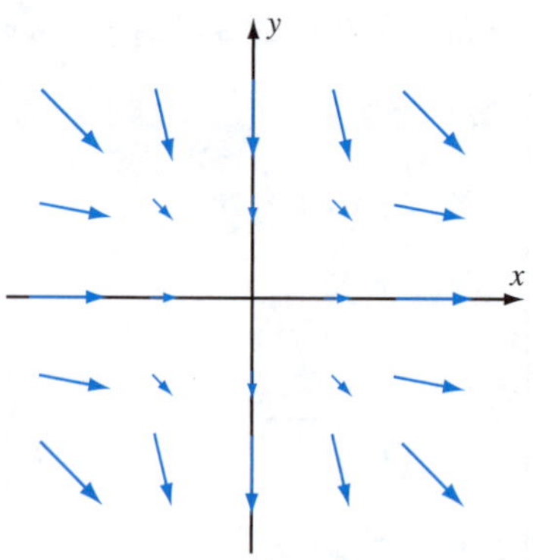

$\mathbf{F}(x, y) = x^2\mathbf{i} - y^2\mathbf{j}$

9. $\alpha\sqrt{x^2 + z^2}\mathbf{j}$ for $-4 < x < 4, 0 < z < 10$
11. $\alpha(y\mathbf{i} - x\mathbf{j})$ for $x^2 + y^2 < 16, 0 < z < 5$
13. $(y + 3)\mathbf{i} + (x - 2y)\mathbf{j}$
15. $(3x^2/(x^3 - y^2))\mathbf{i} + (-2y/(x^3 - y^2))\mathbf{j}$
17. $2xy^3z\mathbf{i} + 3x^2y^2z\mathbf{j} + x^2y^3\mathbf{k}$
19. $((x - y)\ln(z))^{-1}\mathbf{i} - ((x - y)\ln(z))^{-1}\mathbf{j} - (z\ln^2(z))^{-1}\ln(x - y)\mathbf{k}$
21. $x + y + C$
23. $xy + C$
25. $xy^3 + x^2y + C$
27. $x^2 + y + C$
29. $x^2yz^3 + 5y + C$
31. $xz^3 - x^2y^2 + yz^2 + C$
33. $(t + C_1)\mathbf{i} + (t + C_2)\mathbf{j}$
35. $C_1e^t\mathbf{i} + (2t + C_2)\mathbf{j}$
37. $(t^2 + 2C_1t + C_2)\mathbf{i} + (t + C_1)\mathbf{j}$
39. $x\exp(2x) + y + C$
41. $y\sin(xy) + C$
43. $xz(y + z)^{-1} + C$
45. $x^y + z\ln(z) - z + C$
49. $-r^{-3}(x\mathbf{i} + y\mathbf{j} + z\mathbf{k})$
51.

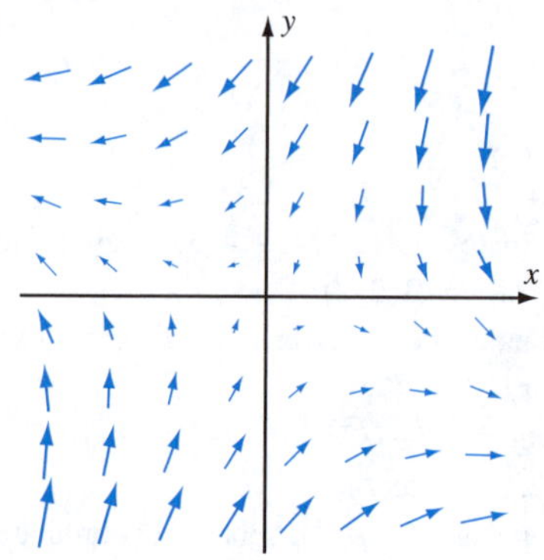

The vector field $-\nabla h(x, y)$

53. $\langle A\exp(2t) + B\exp(3t), -A\exp(2t)\rangle$
55. $\langle A\exp(4t) + B\exp(-t), (-A/4)\exp(4t) + B\exp(-t)\rangle$
57. $-y\mathbf{i} + x\mathbf{j}$

Section 15.2

1. 0
3. $2 - \sin(\pi^{3/2})$
5. 67/896
7. $255/4 + \ln(2/17)$
9. $-2/5$
11. 3/4
13. $\pi/2$
15. 2
17. 1069/35
19. 8/3
21. 2/11
23. 179/180
25. 3
27. -3
29. 2π
31. $(65\sqrt{65} - 5\sqrt{5})/6$
33. 4
35. 0
37. $\int_{C_1} \mathbf{F} \cdot d\mathbf{r}_1 = -4$; $\int_{C_2} \mathbf{F} \cdot d\mathbf{r}_2 = \int_{C_3} \mathbf{F} \cdot d\mathbf{r}_3 = -2$
39. 2π
41. 0
43. $-\pi/4$
45. 3.33904
47. 1.54122
49. 214.420

Section 15.3

In the answers for this section, C represents an arbitrary constant.

1. Potential function: $V(x, y) = C - (x^2/2 + \pi x + y)$
3. Potential function: $V(x, y) = C - xy^3 + y^7$
5. **F** is *not* closed on Q.
7. Potential function: $V(x, y) = (x - 2)^{-1} - (y - 1/2)^3/3 + C$
9. Potential function: $V(x, y) = \cos(x^2 - y) + C$
11. **F** is *not* closed on Q.
13. Potential function: $V(x, y, z) = C - (xyz + x + 2y + 3z)$
15. **F** is *not* closed on Q.
17. Potential function: $V(x, y, z) = C - z^2/(x + y - 4)$
19. **F** is *not* closed on Q.
21. Potential function: $V(x, y, z) = C - z\tan(x + y)$
23. $-1/36$
25. $\exp(-1) - 2$
27. $-\exp(4)$
29. $\int_{\overrightarrow{P_0P_1}} \mathbf{F} \cdot d\mathbf{r} = 0$; $\int_{\overrightarrow{P_1P_2}} \mathbf{F} \cdot d\mathbf{r} = 3$; $\int_{\overrightarrow{P_2P_0}} \mathbf{F} \cdot d\mathbf{r} = -3$
31. $\int_{\overrightarrow{P_0P_1}} \mathbf{F} \cdot d\mathbf{r} = 0$; $\int_{\overrightarrow{P_1P_2}} \mathbf{F} \cdot d\mathbf{r} = 1$; $\int_{\overrightarrow{P_2P_0}} \mathbf{F} \cdot d\mathbf{r} = -1$
37. $\{(c, d) : c \le -1, c < d < 1\} \cup \{(c, d) : -1 < c, \max\{c, 1\} < d\}$
39.

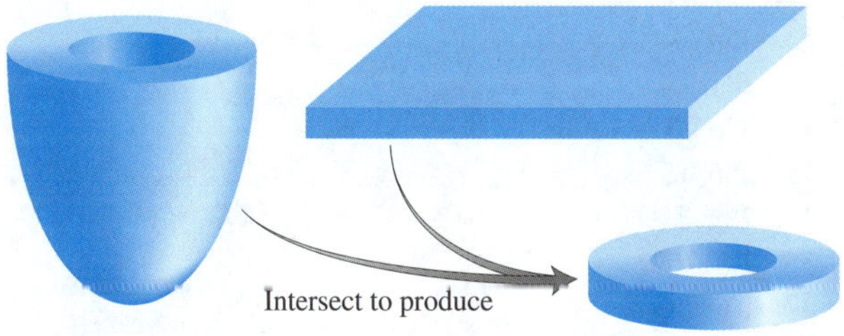

41. 0.6118

Section 15.4

1. $-\sin(x) + 2x\sin(xy)\cos(xy)$
3. $y/(x + y)^2 - 2x/(x - y)^2$
5. $-(x/y^2)\sec^2(x/y) + (y/x^2)\csc^2(y/x)$
7. $2(x + y + z)$
9. $e^{x-z} - e^{z-y}$
11. $-y\sin(xy) - z\cos(yz)$
13. $xz\mathbf{i} + (x - yz)\mathbf{j}$
15. $(x + ye^{yz})\mathbf{i} + (1/(x + z) - y)\mathbf{j}$
17. $-x\cos(xy)\mathbf{i} + y\cos(xy)\mathbf{j} - (y\sin(xy) + x\sec^2(xy))\mathbf{k}$
19. $0\mathbf{i} + 0\mathbf{j} + 0\mathbf{k}$; closed
21. $0\mathbf{i} + 0\mathbf{j} + 0\mathbf{k}$; closed
23. $(x\sin(z) + \sin(x))\mathbf{i} - y\sin(z)\mathbf{j} + (x\sin(y) - z\cos(x))\mathbf{k}$; *not* closed
25. 0
27. $2x(3y^2 + z + 1)/(z - y^2)^3$
29. 0
39. u *is* harmonic.
45. $(1/2\pi)\int_0^{2\pi} \exp(1 + \cos(t))\cos(3 + \sin(t))\,dt = -2.6910786\ldots = e^1\cos(3)$

Section 15.5

1. -3π
3. 2π
5. π
7. 0
9. $3\pi/4$
11. $-32/3$
13. $12\ln(3) - 20\ln(2)$
15. $648/5$
17. 9
19. $64/3$
21. $\pi/3 - \sqrt{3}/4$
23. $1/2$
25. $82/693$
27. 6π
29. $-1/30$
31. $4/3$

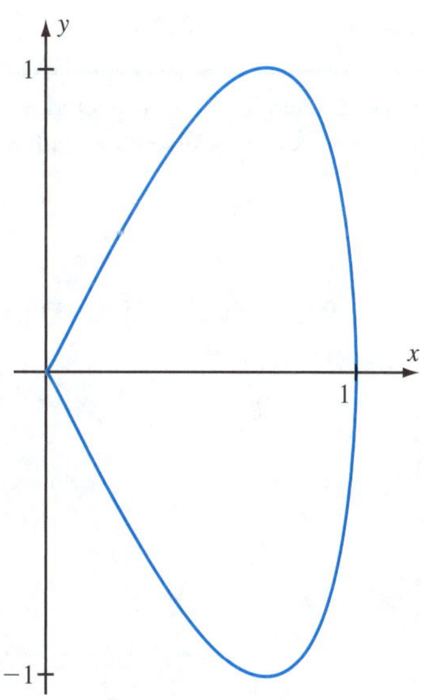

$\mathbf{r}(t) = \sin(t)\mathbf{i} + \sin(2t)\mathbf{j}$

33. $3/2$
37. -0.070393
39. 5.3755

Section 15.6

1. $3\sqrt{26}$
3. $8\sqrt{2}\pi$
5. $(17^{3/2} - 1)\pi/6$
7. $(37^{3/2} - 5^{3/2})/3$
9. $9\sqrt{38}$
11. $(145^{3/2} - 65^{3/2})\pi/24$
13. $5\pi\sqrt{74}$
15. $84\pi\sqrt{2}$
17. $4(275 - 6^{5/2} - 7^{5/2})/15$
19. $2\pi\sqrt{19}$
21. $3574\pi/5$
23. 525π
25. $52/3$
27. $135\sqrt{59}/32$
29. $\int_0^1 \int_0^4 \sqrt{1 + 4u^2 + (4uv + 1)^2}\, dv\, du$
31. $\int_0^1 \int_{-1}^1 \sqrt{4\exp(4u) + \exp(2u + 2v) + \exp(2u - 2v)}\, dv\, du$
33. $27\sqrt{26}$
35. $8\sqrt{3}$
37. $2\pi \int_a^b f(x)\sqrt{1 + f'(x)^2}\, dx$
41. $5\pi/2$
43. $(0, 0, 8/3)$
45. $\left(0, 0, 10(17\sqrt{17} - 1)/(289\sqrt{17} - 41)\right)$
47. $2\pi a^2$
49. $2\pi a^2 + 2\pi(ac/e)\arcsin(e)$ where $e = \sqrt{c^2 - a^2}/c$
51. 8.510466
53. 4.74121

Section 15.7

Each of Exercises 1–11, 19–29, and 35–37 is answered with the common value assumed by both sides of Stokes's equation.

1. 136
3. $2 - 2/e^2 + \ln(2)$
5. $-\pi$
7. -42
9. $-22/3$
11. -24
13. -9π
15. 63π
17. -4
19. $-3\pi/4$
21. 8π
23. $\pi/2$
25. $175\pi/4$
27. -6π
29. 20π
31. $-\pi$
35. -3.550999
37. -0.4626517

Section 15.8

1. 0
3. π
5. π
7. 1
9. -8
11. $4/15$
13. $45\pi/4$
15. 65
17. $32\pi/3 - 4\sqrt{3}\pi$
19. $256/15$
27. $28\pi/3$
35. 2.81762
37. 1.68773

Index

A

B

C

E

H

I

O

P

Q

R

S

T

U

V

W

X

Y

Z